CRC SERIES IN AGING

Editors-in-Chief: **Richard C. Adelman, Ph.D. and George S. Roth, Ph.D.**

VOLUMES AND VOLUME EDITORS

HANDBOOK OF BIOCHEMISTRY IN AGING
James Florini, Ph.D.
Syracuse University
Syracuse, New York

HANDBOOK OF IMMUNOLOGY IN AGING
Marguerite M. B. Kay, M.D. and Takashi
Makinodan, Ph.D.
Geriatric Research Education and Clinical Center
V.A. Wadsworth Medical Center
Los Angeles, California

SENESCENCE IN PLANTS
Kenneth V. Thimann, Ph.D.
The Thimann Laboratories
University of California
Santa Cruz, California

ALCOHOLISM AND AGING: ADVANCES IN
RESEARCH
W. Gibson Wood, Ph.D.
Geriatric Research Education and Clinical Center
V.A. Medical Center
St. Louis, Missouri
Merrill F. Elias, Ph.D.
University of Maine at Orono
Orono, Maine

TESTING THE THEORIES OF AGING
Richard C. Adelman, Ph.D.
University of Michigan
Ann Arbor, Michigan
George S. Roth, Ph.D.
Gerontology Research Center
National Institute on Aging
Baltimore City Hospitals
Baltimore, Maryland

HANDBOOK OF PHYSIOLOGY IN AGING
Edward J. Masoro, Ph.D.
University of Texas Health Science Center
San Antonio, Texas

IMMUNOLOGICAL TECHNIQUES APPLIED
TO AGING RESEARCH
William H. Adler, M.D. and
Albert A. Nordin, Ph.D.
Gerontology Research Center
National Institute on Aging
Baltimore City Hospitals
Baltimore, Maryland

CURRENT TRENDS IN MORPHOLOGICAL
TECHNIQUES
John E. Johnson, Jr., Ph.D.
Gerontology Research Center
National Institute on Aging
Baltimore City Hospitals
Baltimore, Maryland

NUTRITIONAL APPROACHES TO AGING
RESEARCH
Gairdner B. Moment, Ph.D.
Goucher College, and
Gerontology Research Center
National Institute on Aging
Baltimore, Maryland

ENDOCRINE AND NEUROENDOCRINE
MECHANISMS OF AGING
Richard C. Adelman, Ph.D.
University of Michigan
Ann Arbor, Michigan
George S. Roth, Ph.D.
Gerontology Research Center
National Institute on Aging
Baltimore City Hospitals
Baltimore, Maryland

HANDBOOK OF PHARMACOLOGY OF
AGING
Paula B. Goldberg, Ph.D. and
Jay Robert, Ph.D.
The Medical College of Pennsylvania
Philadelphia, Pennsylvania

ALTERED PROTEINS AND AGING
Richard C. Adelman, Ph.D.
University of Michigan
Ann Arbor, Michigan
George S. Roth, Ph.D.
Gerontology Research Center
National Institute on Aging
Baltimore City Hospitals
Baltimore, Maryland

INVERTEBRATE MODELS IN AGING
RESEARCH
Thomas E. Johnson
University of California
Irvine, California
David H. Mitchell
Sensor Diagnostics, Inc.
Pocasset, Massachusetts

Additional topics to be covered in this series include Microbiology of Aging, Evolution and Genetics, Animal
Models for Aging Research, and Insect Models.

CRC Handbook
of
Cell Biology of Aging

Editor

Vincent J. Cristofalo, Ph.D.

Professor, The Wistar Institute
Professor of Biochemistry, University of Pennsylvania
Philadelphia, Pennsylvania

CRC Series in Aging

Editors-in-Chief

Richard C. Adelman, Ph.D.,

Director
Institute of Gerontology
Professor of Biological Chemistry
University of Michigan
Ann Arbor, Michigan

George S. Roth, Ph.D.

Chief, Molecular Physiology and Genetics Section
Gerontology Research Center
National Institute of on Aging
Baltimore City Hospitals
Baltimore, Maryland

CRC Press, Inc.
Boca Raton, Florida

Library of Congress Cataloging in Publication Data

Main entry under title:

CRC handbook of cell biology of aging.

  (CRC series in aging)
  Includes bibliographies and index.
  1. Cells--Aging. 2. Aging.   I. Cristofalo,
Vincent J., 1933-   . II. Title: Cell biology of aging. III. Title: C.R.C. handbook of cell biology of aging. IV.
Series. [DNLM: 1. Aging. 2. Cytology.
WT 104 C9105]
QH608.C73 1985   591.87'6   84-14312
ISBN O-8493-3142-0

Direct all inquires to CRC Press, Inc., 2000 Corporate Blvd., N.W., Boca Raton, Florida, 33431.

© 1985 by CRC Press, Inc.

International Standard Book Number 0-8493-3142-2

Library of Congress Card Number 84-14312
Printed in the United States

FOREWORD

The purpose of this volume is to provide a summary of current information on aging in cells.

The view that aging in individuals is based on cellular changes is essentially as old as the recognition of cells as the basic unit of biological organization. However the modern era of research on aging at the cellular level was probably ushered in by the seminal work of Hayflick and Moorhead in the late 1950s and early 1960s. In their experiments on human cells in culture, they recognized that aging was a cellular phenomenon as well as a supra-cellular one and that, for the most part, aging changes in the individual represented the summation of aging changes in the multitude of cell types of which the individual is composed. The publication of this view and its gradual acceptance by the scientific community coincided with an explosion in research in the general area referred to as cell biology. Thus the field of cytogerontology developed in a fortuitous atmosphere. It can be said that gerontologic studies in the cell have lagged behind the field of cell biology but certainly not far behind.

This volume begins with an introductory section on general changes with age including changes in cell number, cell proliferation, and cell morphology. There follows a review of the age-associated changes in the various cell types in mammals. Finally, there is a section addressing the most studied of the invertebrates with respect to aging: the protozoa, the nematodes, and the insects.

In our view the selection of chapters and authors represents the most developed areas of the cell biology of aging. Clearly there are both omissions and overlaps. There are overlaps both within the chapters of this volume as well as with other volumes of this series. However, this overlap, which allows individual authors' points of view to emerge, is preferable to gaps either in knowledge or philosophy.

It is my sincere hope that this volume will be a valued source of information for workers in the cell biology of aging as well as those in other disciplines. Perhaps the information contained here will serve to focus our knowledge and stimulate new studies on the cellular basis of aging. I would like to gratefully acknowledge the editorial assistance of Ms. Marina Hoffman and the secretarial assistance of Ms. Joyce Robinson during the preparation of this volume. I would also like to acknowledge the partial support of this volume from the National Institute on Aging, Grant AG-00378.

EDITORS-IN-CHIEF

Richard C. Adelman, Ph.D., is currently Director of the Institute of Gerontology at the University of Michigan, Ann Arbor, as well as Professor of Biological Chemistry in the Medical School. An active gerontologist for more than 10 years, he has achieved international prominence as a researcher, educator, and administrator. These accomplishments span a broad spectrum of activities ranging from the traditional disciplinary interests of the research biologist to the advocacy, implementation, and administration of multidisciplinary issues of public policy of concern to elderly people. He is the author and/or editor of more than 95 publications, including original research papers in refereed journals, review chapters, and books. His research efforts have been supported by grants from the National Institutes of Health for the past 13 consecutive years, and he continues to serve as an invited speaker at seminar programs, symposiums, and workshops all over the world. He is the recipient of the IntraScience Research Foundation Medalist Award, an annual research prize awarded by peer evaluation for major advances in newly emerging areas of the life sciences; and the recipient of an Established Investigatorship of the American Heart Association.

Dr. Adelman serves on the editorial boards of the *Journal of Gerontology, Mechanisms of Ageing and Development,* and *Gerontological Abstracts.* He chaired a subcommittee of the National Academy of Sciences Committee on Animal Models for Aging Research. As an active Fellow of the Gerontological Society, he was Chairman of the Biological Sciences section; a past Chairman of the Society Public Policy Committee; and is currently Chairman of the Research, Education, and Practice Committee. He serves on National Advisory Committees which impact on diverse key issues dealing with the elderly, including a 4-year appointment as member of the NIH Study Section on Pathobiological Chemistry; the Executive Committee of the Health Resources Administration Project on publication of the recent edition of *Working with Older People — A Guide to Practice;* and a 4-year appointment on the Veterans Administration Advisory Council for Geriatrics and Gerontology.

George S. Roth, Ph.D., is chief of the Molecular Physiology and Genetics Section of the Gerontology Research Center of the National Institute on Aging in Baltimore, Md., where he has been affiliated since 1972. Dr. Roth received his B.S. in Biology from Villanova University in 1968 and his Ph.D. in microbiology from Temple University School of Medicine in 1971. He received postdoctoral training in Biochemistry at the Fels Research Institute in Philadelphia, Pa. Dr. Roth has also been associated with the graduate schools of Georgetown University and George Washington University where he has sponsored two Ph.D. students.

He has published more than 100 papers in the area of aging and hormone/neurotransmitter action, and has lectured, organized meetings, and chaired sessions throughout the world on this subject.

Dr. Roth's other activities include fellowship in the Gerontological Society of America, where he has served in numerous capacities, including chairmanship of the 1979 midyear conference of "Functional Status and Aging." He is a past Chairman of the Biological Sciences Section and a past Vice President of the Society. He has three times been selected as an exchange scientist by the National Academy of Sciences and in this capacity has established liaisons with gerontologists, endocrinologists, and biochemists in several Eastern European countries. Dr. Roth serves as an editor of *Neurobiology of Aging* and *The Journal of Gerontology and Experimental Aging Research* and is a frequent reviewer for many other journals including *Mechanisms in Aging and Development, Life Sciences, Science,* and *Endocrinology.* He also serves as a grant reviewer for several funding agencies including the National Science Foundation, and the research advisory boards for several medical schools. In 1981 Dr. Roth was awarded the Annual Research Award of the American Aging Association. He has also received the NIH Merit Award, and is currently Chairman of the Gordon Research Conference on the Biology of Aging.

THE EDITOR

Vincent J. Cristofalo, Ph.D. is a Professor of The Wistar Institute, Philadelphia, and Director of the Center for the Study of Aging, University of Pennsylvania, also in Philadelphia.

Dr. Cristofalo holds other positions at the University of Pennsylvania, as follows: he is Professor of Biochemistry in the Division of Animal Biology in the School of Veterinary Medicine and he is a member of the Graduate Faculties of Physiology, Genetics, and Molecular Biology.

Dr. Cristofalo is a member of the Gerontological Society of America, The American Society for Cell Biology, The Tissue Culture Association, and The Society for Experimental Biology and Medicine.

Dr. Cristofalo graduated in 1955 from St. Joseph's College, Philadelphia, with a B.S. degree in biology and obtained his M.S. degree in physiology from Temple University, Philadelphia; he received his Ph.D. in physiology/biochemistry from the University of Delaware in 1962. In 1982 he was awarded an honorary M.A. from the University of Pennsylvania.

Among many awards won by Dr. Cristofalo for his research have been the Brookdale Award for Scientific Research (1983) and the Kleemeier Award for Scientific Research (1982); both were awarded for his work on cellular senescence.

In 1977 Dr. Cristofalo was Chairman of the Gordon Conference on the Biology of Aging. He has served as Vice President of the Gerontological Society and has served on many committees for the scientific organizations to which he belongs. In 1972 he was a Delegate to the White House Conference on Aging. Dr. Cristofalo is an elected Fellow of the Gerontological Society.

Dr. Cristofalo is associated with a number of scientific publications. He is Editorial Consultant to *Geriatric Consultant, Experimental Aging Research* and *Biological Abstracts and Biological Abstracts/RRM*. He is on the editorial Board of *The Annual Review of Gerontology and Geriatrics, Cell Biology International Reports, In Vitro* and *Mechanisms of Aging and Development*. He has served as Biological Sciences Editor for *The Journal of Gerontology*. Dr. Cristofalo is an author of over 170 scientific publications. His current interests include the biology of cellular aging, the regulation of cell proliferation, and intermediary carbohydrate metabolism.

Vincent J. Cristofalo, Ph.D.
August 1983

ADVISORY BOARD

CONTRIBUTORS

William H. Adler, M.D.
Chief, Clinical Immunology Section
Gerontology Research Center
National Institute on Aging
Baltimore, Maryland

M. A. Alnaqeeb, Ph.D.
Muscle Research Unit
Department of Zoology
University of Hull
North Humberside, England

George T. Baker, III, Ph.D
Director, Center on Aging
University of Maryland
College Park, Maryland

Gretchen H. Bean, Ph.D.
Director, Tissue Culture Laboratory
Boston University School of Medicine
Boston, Massachusetts

Phillip D. Bowman, Ph.D
Department of Pediatrics
University of Michigan
Ann Arbor, Michigan

Jerome S. Brody, M.D.
Professor of Medicine
Boston University School of Medicine
Boston, Massachusetts

Dennis E. Buetow, Ph.D
Professor of Physiology, and Head of
  Department of Physiology and
  Biophysics
University of Illinois
Urbana, Illinois

Charles W. Daniel, Ph.D.
Professor of Biology
University of California
Santa Cruz, California

David Danon, M.D.
Professor of Biology
Department of Membrane Research
Weizmann Institute of Science
Rehovot, Israel

Kerry L. Dearfield, Ph.D.
Department of Radiology
George Washington University School of
  Medicine and Health Sciences
Washington, D.C.

Geoffrey Goldspink, Ph.D., Sc.D.
Professor of Anatomy and Cellular
  Biology
Tufts University School of Medicine
Boston, Massachusetts

Ronald H. Goldstein, M.D.
Assistant Professor of Medicine
Boston University School of Medicine
Boston, Massachusetts

Elizabeth Hamilton, Ph.D.
Research Associate
Department of Oncology
Middlesex Hospital Medical School
London, England

Mark Jacobson, M.S.,
Department of Biological Sciences
Drexel University
Philadelphia, Pennsylvania

Robert J. Johnson, M.D.
Professor of Anatomy
School of Medicine
University of Pennsylvania
Philadelphia, Pennsylvania

Thomas E. Johnson
Assistant Professor of Molecular Genetics
Department of Molecular Biology and
  Biochemistry
University of California
Irvine, California

Dick L. Knook, Ph.D.
Deputy Director
TNO Institute for Experimental
  Gerontology
Rijswijk, Netherlands

Elliot M. Levine
Wistar Institute
Philadelphia, Pennsylvania

Richard A. Lockshin, Ph.D.
Professor and Chairman
Department of Biological Sciences
St. John's University
Jamaica, New York

**Leah M. Lowenstein, M.D., D. Phil.
(Deceased)**
Director, Basic and Clinical Sciences
Gerontology Center, and
Professor of Medicine, both at
Boston University School of Medicine
Boston, Massachusetts

Merry C. McArthur, Ph.D.
Southern Methodist University
Dallas, Texas

Gergory Mokrynski, M.S.
Department of Biological Sciences
Drexel University
Philadelphia, Pennsylvania

Stephen N. Mueller
Wistar Institute
Philadelphia, Pennsylvania

James E. Nagel, M.D.
Clinical Immunology Section
Gerontology Research Center
National Institute on Aging
Baltimore, Maryland

Olivia M. Pereira-Smith
Assistant Professor
Department of Virology and
  Epidemiology
Baylor College of Medicine
Houston, Texas

E. L. Schneider
Deputy Director
National Institute on Aging
Bethesda, Maryland

Gary B. Silberstein
Biology Board of Studies
University of California
Santa Cruz, California

Victoria J. Simpson, Ph.D.
Postdoctoral Associate
Department of Molecular Biology and
  Biochemistry
University of California
Irvine, California

James R. Smith
Professor
Department of Virology and
  Epidemiology
Baylor College of Medicine
Houston, Texas

Joan Smith-Sonneborn, Ph.D.
Professor of Zoology and Physiology
University of Wyoming
Laramie, Wyoming

R. S. Sohal, Ph.D.
Professor of Biology
Southern Methodist University
Dallas, Texas

Phyllis Strickland
Biology Board of Studies
University of California
Santa Cruz, California

Edgar A. Tonna, Ph.D., F.R.M.S.
Professor, Chairman Dept. Histology and
  Cell Biology
Director, Institute for Dental Research
New York University Dental Center
New York, New York

Richard F. Walker, Ph.D.
Associate Professor
University of Kentucky Medical Center
Lexington, Kentucky

Jerry R. Williams, D.Sc.
Professor
Johns Hopkins Oncology Center
Baltimore, Maryland

R. H. Yanagihara, M.D.
Medical Oncology Section
University of Texas Health Center at
  Tyler
Tyler, Texas

TABLE OF CONTENTS

Cell Numbers Versus Age in Mammalian Tissues and Organs 1
D.E. Buetow

Aging and the Cell Cycle In Vivo and In Vitro 117
Phillip D. Bowman

Aging, Cell Death, and Lysosomes 137
Richard A. Lockshin

Anatomy of the Aging Nerve Cell... 149
Robert Johnson

Aging of Skeletal Muscle 179
Geoffrey Goldspink and M. S. Alnaqeeb

Aging of the Skeletal System and Supporting Tissues................................ 195
Edgar A. Tonna

Aging Liver Cells 229
Dick L. Knook

The Effect of Aging on the Kidney 245
Leah M. Lowenstein and Gretchen H. Bean

Intestinal Tissue and Age... 255
Elizabeth Hamilton

Aging of the Reproductive System 271
Richard F. Walker

Mammary Cells 289
Charles W. Daniel, Gary B. Silberstein, and Phyllis Strickland

Cultured Endothelial Cells as an In Vitro Model System for Cellular Senescence 303
Elliot M. Levine and Stephen N. Mueller

Aging of the Erythrocyte 317
David Danon

Cells of the Immune Response 341
J. E. Nagel, R. H. Yanagihara, and W. H. Adler

Biology of the Aging Lung.. 365
Ronald H. Goldstein and Jerome S. Brody

Lung-Derived Fibroblast-Like Human Cells in Culture............................... 375
James R. Smith and Olilvia M. Pereira-Smith

Human Skin-Derived Fibroblast-Like Cells in Culture............................... 425
E. L. Schneider

Non-Human Fibroblast-Like Cells in Culture..433
Jerry R. Williams and Kerry L. Dearfield

Genome Interactions in the Pathology of Aging.......................................453
Joan Smith-Sonneborn

Aging Studies in *Caenorhabditis Elegans* and Other Nematodes.......................481
Thomas E. Johnson and Victoria J. Simpson

Cellular Aspects of Aging in Insects 497
R. S. Sohal and M. C. McArthur

Aging in Drosophila..511
George T. Baker, III, Mark Jacobson, and Gregory Mokrynski

Index 579

CELL NUMBERS VS. AGE IN MAMMALIAN TISSUES AND ORGANS

Dennis E. Buetow

Senescence is associated with a loss of adaptability and a progressive decrease in the capacity of the individual to maintain homeostasis.[1-5] Indeed, longitudinal studies of age changes in the human show that on the average, a variety of physiological functions decline linearly with age, starting with age 30 to 40 years.[6,7] Accompanying these changes in physiological function are changes in the cellular content[8] and proliferative capacity[8,9] of many tissues as well as alterations in cellular and subcellular structure and function.[10,11] Clearly, the degeneration and loss of cells is a concomitant of aging. The present chapter summarizes the known data on cell number vs. age in mammalian tissues and organs and is an expanded and updated version (through 1982) of a similar summary published in 1971.[8] The data covered in this chapter were extracted from the gerontological literature on the following bases: (1) The data include normal (free of disease) subjects which are very old, or at least, beyond early adulthood; (2) the data from any one publication include enough subjects and/or cell counts, etc. so that any differences between different age groups are, at least, suggested; and (3) the ages of the subjects are known. Some of the studies on rats, mice, and other animals use weight, for example, as a criterion of age. Such studies are not referenced here.

HUMAN BODY MASS

Total body mass declines with age in the human adult, male or female, regardless whether this parameter is measured as fat-free body weight, lean body mass, fat-free mass, body cell mass, or total body nitrogen (Table 1*). The loss of body mass begins on the average in the decade 45 to 55 years. However, the relative loss is greater for adult males than for females. Body fat increases as total body mass declines in males and females. The decline in fat-free body mass from middle age onward in the human suggests the loss of some parenchymal tissue in the elderly, a loss documented in Tables 2 through 11.

NERVOUS SYSTEM

Hodge[20] was the first to suggest that the loss of neurons is a feature of senescence. In a careful review of the data on normal subjects, Hanley[21] concluded that neuronal "fall-out" does not occur throughout the brain with aging but may occur in particular regions of the brain in the human and in certain animal species. Dayan[22] studied the brains of 47 species of animals and concluded that the pathologies which occur with age differ in different animals. The data in Table 2 support the conclusions of Hanley[21] and Dayan.[22] In the human, neuronal loss occurs usually by age 60 to 70 in the spinal nerves[23,24] and in some[30-32] but not all[33-35] cranial nerves. One study with a limited number of subjects per age group indicated no change with age in the number of total cells in the human cerebral cortex.[39] Other studies show a loss of both total cells[40] and total neurons[40,41] in the cerebral cortex of aged humans. However, this loss of neurons occurs in many[45,49,50,53] but not all[53] areas of the cerebrum. Also, a loss of neurons occurs in aged humans in certain regions of the brain stem[71] and hypothalamus[61] and in some[62,63] but not all[64,65] regions of the cerebellum. Other areas of the human nervous system[68,72,73] so far measured do not lose neurons with age (Table 2).

---

* Tables 1 to 12 appear at end of this text.

In rats 2 years of age or older, neuron loss occurs in the olfactory bulb,[36] in the cerebellum,[51,66] in the spiral ganglion of the cochlea,[74] in some[47,51,54,55] but not all[43,52] areas of the cerebrum, in some[44,47,48,56] but not all[57,58] regions of the hippocampus, and in some[59] but not all[59,60] regions of the hypothalamus. No loss was observed in the spinal nerves,[25-27] the brain stem,[51,69,70] or motor nerves ennervating the soleus muscle[80] in aged rats. The mouse loses neurons in the spinal cord by 25 months of age.[29] One study shows a loss of total brain neurons in the mouse by 24 months of age[38] whereas another study shows no change in total brain nuclei (neurons plus glia) up to 36 months.[37] The guinea pig shows no loss of Purkinje cells in the cerebellar cortex up to an age of 7.5 years,[67] and the cat shows no change up to 18 years of age in the number of fibers found in two spinal nerves.[28] Fewer neurons are found in various regions of the cerebral cortex and hippocampus of rhesus monkeys at 18 to 28 years of age than at 4 to 7 years of age.[46]

Changes in the numbers of glial cells with age have been studied mainly in the rat nervous system. No changes in number occur in the cerebral hemisphere,[51] the visual cortex after 17 months of age,[47,48] the brain stem,[51] or the cerebellum.[51] The total number of glial cells shows no change[44] or in one case is reported to increase by 2 years of age in the cerebral cortex.[42] The latter increase in the cerebral cortex appears accounted for by increased numbers of astrocytes and oligodendrocytes[43,52] but not of microglia.[43] In contrast, the microglia but not the astrocytes or oligodendrocytes increase by 27 months of age in the auditory cerebral cortex.[77] Astrocytes increase in number in some but not all regions of the dentate gyrus.[78] By 25 months of age, total glial cells are increased and hypertrophied astrocytes are greatly increased in number in the mid-polar regions of the hippocampus.[79] In contrast, total glial cells in the CA1 zone of the hippocampus[47,48] and the motor cortex,[76] astrocytes, and oligodendrocytes in the occipital cortex[55] decrease in number in the aged rat. Glial cells in the motor cortex increase in number in the human aged 64 to 70 years.[76] Glial cells also increase in number in the human visual cortex to age 61 to 69 years but then decrease at 73 to 87 years.[50] Total glial cells overall do not change in number with age in mouse brain,[38] but do decrease in number in the motor cortex of the old mouse.[78] The various types of glial cells increase in number in certain areas of the cerebrum but decrease in other areas at various times during the lifespan of the mouse.[75] The monkey shows an increased number of glial cells in at least one cerebral cortex area with age.[46]

EYE

The number of cells in the trabecular meshwork of the human eye[81] decreases by an average 42% from birth to age 80 years (Table 3). In the rat, the number of lens epithelial cells on the average slowly increases from age 4 to 5 months to an age of over 3 years.[82] The number of mast cells in the exorbital lacrimal gland of the rat decrease by 1 month and then shows little change up to 16 months.[201]

SKIN AND BONE

Few studies have been done on changes in number of skin cells with age (Table 4). The aging human loses connective tissue cells from the abdominal corium layer[83] and melanocytes from the skin of the forehead,[88] thigh,[88] abdominal wall,[89] and buttocks.[90] The rat at 30 months of age shows an increased number of lymphocytes in the stratum germinativum of the abdomen[84] but a decreased number in the interscapular region[85] compared to 10-month-old animals. The number of mast cells in the dorsal dermis of the mouse increases, decreases, or does not change depending upon the sex and/or the strain of animal studied.[86] The dark keratinocytes represent a decreasing percent of the total number of basal layer cells in the dorsal epidermis of the mouse from 1 to 15 months of age but not older.[87]

Bone loss occurs in the aging human (Table 4). This subject has been reviewed recently by Mazess.[91] Although bone loss in old age is not evident in ash weight measurements of the total skeleton or the long bones,[92] it is evident in measurements of the dry, fat-free weight of the whole skeleton[93] and of the combined skull-cranium-mandible,[93] in the cortical thickness of the femur[94-96] and second metacarpal,[102] the cortical area of the ribs,[101] the cortical density of the third metacarpal,[99] the volumetric density of the pelvic iliac crest,[97] the mineral content of the ulna and humerus,[9] the bone density[99] and the mineral content[98,100] of the radius, and the bone density of the fifth mid-phalanx.[13] Though the mandible of the human female seems to increase in thickness with age,[103] its absolute bone mass decreases.[104] In a consideration of the effects of dietary and hormonal factors on human bone loss, Smith[111] concluded that bone loss occurs with age in both males and females beyond that caused by abnormal mineral nutrition or by various endocrine diseases. Further, in the adult female bone loss is unrelated to the age of menarche or onset of menopause.[111] Similarly, Goldsmith[100] concluded that age is the main determining factor in the loss of bone in the human male and female.

In the mouse, the cortical thickness of the tibia does not decrease even in the very old (40 months) animal but that of the femur and lumbar vertebrae decrease by 18 months of age and again by 30 months[105] (Table 4). Also, the number of osteocytes in the femur cortex is decreased in the 2-year-old mouse compared to younger mice.[107] In the rat, osteoblasts in the periosteum of the femur decrease in number by 6 months and again at 2 years of age.[106]

In the human, the thickness of the cartilage at the head of the femur does not decrease up to age 90 years but the number of cells in this cartilage is lower in 30 to 90-year-old compared to younger subjects.[108] Cell number in knee-joint cartilage decreases by age 13 to 26 years and then remains constant on the average up to age 89.[109] Cell number in the fourth costal cartilage also decreases by 13 to 26 years of age and then shows an additional small decrease after 50 years.[109] In the cow, the number of phalangeal cartilage cells decreases by age 1 to 7 years and again at 8 to 11 years.[110]

BLOOD AND BLOOD VESSELS

Shapleigh et al.[149] reviewed the literature on human blood cell counts to 1952 and concluded that the reported numbers of erythrocytes and leukocytes vs. age of the subjects varied somewhat in different reports. Therefore, it is possible that some of the studies prior to 1952 included subjects who had diseases other than those common to the aging process.[149] Pre-1952 studies included here are those which stated that normal subjects were used.[112,113]

In the human, the number of erythrocytes decreases at age 40 to 60 years[113,114,116] compared to younger individuals (Table 5). By about age 80 years, this number declines again at least in the male.[112,113,115] In the male rat, no change in erythrocyte number occurs between 11 and 25 months of age.[117] In contrast, this number decreases with age in the mouse,[118-123] particularly by 21 months of age or older.[119-123] In guinea pigs, bulls, and cows, erythrocyte number is lower in all cases in the older animals measured.[124,125] In sheep and goats, no change is observed after 2 years of age.[125] Adult golden tamarins show a small increase in erythrocyte number compared to juveniles.[126]

In the human, total leukocytes either do not change in number up to age 65 years[127] or show only a small decrease in number by 50 to 69 years of age compared to younger individuals.[114,116,127,128] However, the number of total leukocytes has been reported to increase in old age[112,115,129] though two studies show a decrease at the oldest ages examined.[130,131] The number of granulocytes in the human does not change up to 99 years when data from males and females are considered together,[133] but does appear to increase in the aging female[127] and decrease slightly in the aging male.[131] There is some change in number of the different types of granulocytes in old age. Neutrophils increase in number during ages 75 to 94 years,[112,115] whereas both eosinophils and basophils decrease after age 70 years.[112] Zacharski et al.,[150] in a study of 292 healthy men and 215 healthy women aged 15 to >70 years, concluded that there were no changes with age or sex in numbers of leukocytes, neutrophils, eosinophils, and basophils. In general, the data in Table 5 tend to support this conclusion in that any changes measured in these blood cells in the human are quite small.

In the human, total lymphocytes either do not change in number[112,115,127,129,138,140,141] or show only a small decrease in old age.[128,130,133-137,139] of the lymphocytes, total T cells and "null" cells generally are reported to decrease in number beyond about age 50 to 60 years.[128,130,134,136,142,143] Total B cells show a small decrease in number after age 60,[128,134,136] but an increase after age 90 years.[134] One study shows that total T cells and total B cells decrease in number in the aging male but not in the aging female.[141] Lymphocytes other than T cells or B cells show a small decrease in number beyond age 60.[136] As regards subsets of human T cells, Portaro et al.[151] reported no changes with age but noted that changes had been reported by others. Therefore, Portaro et al.[151] suggested that the response of the human immune system to age may be polymorphic. Data on subsets of T cells in Table 5 tend to support this suggestion. For example, different investigators report different results with E-rosette-forming T cells,[129,137,145] including both early[135,137,144] and late[135,144] E-rosette formers, and with HTLA-antigen-bearing T cells.[129,137] However, one study[145] indicates that subsets of E-rosette formers may behave differently from one another with age with the number of E_A-rosette formers increasing and the number of EP- and E_{ab}-rosette formers showing no change in old age. Other studies indicate possible age and sex effects on T_G lymphocytes.[138,141] Further, on the average, T_M, T3$^+$, and T8$^+$ lymphocytes seem to decrease a little in number in old age whereas T4$^+$ show no change.[138,141,146] Subsets of B cells may also behave differently from one another with age: the numbers of E_{AC}-rosette formers either do not change[129] or slightly decrease,[137] Fc-receptor-positive cells increase slightly on the average,[130] and single-spotted ANAE-positive cells do not change while granular-scattered ANAE-positive cells increase.[145] In one study,[112] monocytes increased in number in very old subjects whereas in two studies no changes with age were noted.[130,150] Platelets are higher in number from ages 67 to 94 years vs. ages 20 to 63 years,[147] but no further change occurs in the older group itself.[115]

The male rat shows no change in number of erythrocytes from 11 to 25 months of age.[117] The female mouse shows fewer erythrocytes by about 20 months compared to younger ages[118,121,122] and fewer again at about 30 months.[122] The male mouse shows fewer erythrocytes at 20 to 28 months compared to younger ages[120,122,123] and fewer again at 31 to 35 months.[122,123] One study shows a slow, but continuous decline in the number of erythrocytes in male mice as they age from 4 to 24 months.[119] The guinea pig, bull, cow, sheep, and goat all appear to have fewer erythrocytes at some point in their lifespan compared to younger ages.[125]

In the same strain of male mouse, two reports show no change in number of total leukocytes up to 28 to 34 months.[120,122] A third report shows that this number increases by 35 to 37 months.[123] Another study on the same strain of male mice, however, shows a decrease in total leukocytes starting by 24 to 25 months.[132] In the female, leukocytes decrease by 15 to 21 months[118,122] and then increase by 34 months.[122] In a hybrid strain of female mice, the number of leukocytes does not change up to 26 months but then declines by 30 months.[121] Granulocytes appear to be greatly increased in number after 16 to 21 months compared to 4 to 7 months.[118] Neutrophils are increased at 22 to 25 months[120,122] and are highest in number at 33 to 34 months.[122] Results with eosinophils vary. In the female, one study shows no change up to 21 months,[118] whereas another shows a decrease at 7 months and a further decrease at 16 to 34 months.[122] In the male, one study shows a decrease in eosinophils at ages 14 to 34 months compared to younger animals,[122] whereas another study shows an increase at ages 25 to 28 months vs. 8 to 10 months.[120] Since these latter studies were done on the same strain,[118,120,122] differences may be due to diet and/or environmental conditions. Monocytes are increased at 25 to 28 months compared to 8 to 10 months.[120] In the female, lymphocytes are decreased at ages 16 to 34 months.[118,122] In the male, lymphocytes either do not change in number over the age span of 4 to 34 months[122] or possibly decrease slightly at 25 to 28 months compared to younger animals.[120]

A few studies have determined the number of leukocytes vs. age in other mammals (Table 5). In the rat, total leukocytes and eosinophils decrease by 25 months of age.[117] Total leukocytes in the guinea pig increase up to 5 years.[124] In the bull, total leukocytes decrease after 2 years of age with lymphocytes decreasing after 6 years, monocytes after 10 years, and basophils after 12 years.[125] Neutrophils and eosinophils do not change up to 15 years.[125] In the cow, total leukocytes and monocytes, lymphocytes, and basophils are decreased in the period of 5 to 12 years of age compared to younger animals whereas neutrophils increase at 5 to 12 years and eosinophils show no change up to 12 years.[125] In sheep, total leukocytes and monocytes and lymphocytes are decreased at ages 4 to 9 years whereas neutrophils, eosinophils, and basophils all increase at ages 2 to 9 years.[125] Goats aged 2 or 3 to 6 years old have fewer total leukocytes and fewer lymphocytes, monocytes, and basophils compared to younger animals.[125] However, the number of neutrophils possibly increases at 1 year and then again at 2 to 6 years and the number of eosinophils does not change up to 6 years of age.[125]

The number of mast cells with age in the tunica adventitia of various human blood vessels has been measured (Table 5). This number does not change in the aorta abdominalis in the male to age 78 years or the female to age 88 years, whereas it decreases in both sexes after age 60 years in several veins.[148] In the female rat, the numbers of endothelial cells and pericytes in capillary walls in the cerebral cortex decrease slightly at 22 to 27 months of age compared to 3 months.[44]

BONE MARROW TOTAL AND HEMATOPOIETIC CELLS

The number of nucleated bone marrow cells in the whole mouse skeleton increases at about 5 months of age and again in old animals 30 months of age or older[152,153] (Table 6). The total number of femoral bone marrow nucleated cells is higher on the average in animals 12 to 26 months old[121] compared to younger animals, with increases occurring at about 15 to 18 months[155,156] and 20 to 26 months of age.[154,156] Further increases occur in old animals 33 to 35 months of age.[154,155] The number of stem cells in the total skeletal bone marrow increases at 5 months of age and possibly again at 36 months.[153] However, variable results[121,152,154-157] have been reported for the number of femoral bone marrow stem cells in the mouse. This number, therefore, may be strain dependent since different strains were used in different studies. On the average, the number of stem cells appears higher in the 12 to 26 month age group compared to younger animals in one study[121] but higher in those aged 3 weeks to 18 months compared to older animals in another study.[157] Other studies report no clear change in number from 3 to 33 months[154-156] with a possible increase at 35 months.[156] When expressed as number of femoral bone marrow stem cells per 10^5 bone marrow cells, the number is decreased in the age group 7 to 23 months and again at 35 months compared to younger animals.[152] The number of total skeletal bone marrow cells containing cytoplasmic immunoglobulin (C-Ig) increases with age up to 2 years.[158]

In the guinea pig, the total number of femoral bone marrow stem cells appears to be slightly lower in the age group 1.1 to 5 years compared to animals 3 days to 7.5 months of age[124] (Table 6). The total number of bone marrow erythrocytes and lymphocytes also decreases in the older age group. In contrast, the total femoral bone marrow neutrophils, eosinphils, and monocytes are increased in the older age group.[124]

SPLEEN, THYMUS, AND LYMPH NODES

The total number of spleen cells in the mouse vs. age (Table 7) varies with the report.[121,153,159] This number may be strain dependent since each study was done on a different strain. Silini and Andreozzi[121] report no clear change in number up to 26 months of age whereas Brennan and Jaroslow[159] show that on the average this number is higher in mice 13 to 26 months old compared to those 3 to 10 months old. In contrast again, Coggle et al.[153] report that this number increases up to 3 to 4 months and then decreases at 17 months.

The total number of spleen stromal cells in the mouse increases slightly at about 20 months compared to animals 2 to 6 months old.[160] The total number of spleen stem or antibody cells generally are reported to be decreased in mice 21 to 36 months old.[153,155,161,162] However, two studies indicate a possible increase in this number at about 20 to 25 months.[121,160] The numbers of mouse spleen plaque-forming and total C-Ig cells decrease at about 2 years of age[158,163] and the number of total T cells decreases progressively at 13, 20, and 29 months.[159] The number of IgG-bearing cells increases at 26 to 30 months.[164]

In rat spleen, the number of megakaryocytes shows no clear changes between 50 days and 38 months of age, whereas the number of macrophages on the average increases up to 10 to 24 months and then decreases in the period 26 to 38 months.[166] The numbers of various types of T cells in rat spleen either do not change or decrease when animals 27 months old or older are compared to animals 3 to 4 months old.[165]

Compared to the 4.5-month-old mouse, at 32 months the total number of thymus cortex cells as well as the numbers of lymphocyte and epithelial cells per unit area are decreased, the number of megakaryocytes is increased, and the numbers of fibroblasts and macrophages are greatly increased.[167] The number of IgG-bearing cells in the thymus also are increased in the 26- to 30-month-old mouse compared to young mice.[164]

The number of C-Ig cells in the mesenteric lymph nodes of the mouse decrease at 2 years[158] while IgG-bearing cells increase at 26 to 30 months in the brachial and submandibular lymph nodes.[164] In the old rat, the total number of nucleated cells recoverable from the cervical lymph nodes greatly decreases.[164] Of the various types of T cells present in these latter lymph nodes, only the Thy-1-positive cells show a relative decrease in the old animal.[165]

DIGESTIVE SYSTEM

Gingival epithelial cells are increased in number in the human at 50 to 78 years of age compared to 25 to 34 years[168] (Table 8). Basal cells of the epithelium of the rat's palate and tongue show a small decrease in number by 27 months.[169] Stem cells in the same tissue also decrease in the older animals.[169] In the rat's incisor, pulp tissue cells do not change in number up to 25 months[170] and ameloblasts and odontoblasts do not change up to 19 months.[171]

Connective tissue cells in the tunica muscularis layer of the small intestine of the human (Table 8) increase each decade from age 10 to age 90 years.[172] The number of lymphocytes in the epithelium of the small intestine of the mouse are decreased in number at 19 to 27 months compared to 4 to 8 months of age.[173] Also, C-Ig cells in Peyer's patches in the rat decrease at 1 year and again at 2 years of age.[158]

The number of nucleated cells in mouse liver (Table 8) declines with age.[174] The number of hepatocytes declines with age in human males, but the onset and pattern of the decline differs for individuals of different national origins. In Caucasians from the U.S., hepatocytes are progressively lost from age 49 years through 80 years and older,[175] in Japanese after age 70,[175] and in Costa Ricans at age 50 to 79 years compared to younger individuals.[176] The latter group then shows a further loss of hepatocytes after age 80.[176] The female rat shows increased numbers of hepatocytes up to 12 months and then a possible decreased number by 27 months,[178] whereas the male rat shows increased numbers throughout the lifespan[179] or at least up to 7 months with further increases at 20 and at 26 months.[177] "Littoral" cells in male rat liver increase in number at 24 months compared to younger ages.[179]

The number of mast cells in the mesentery of the lower small intestine of the mouse varies with the species (Table 8). Males and females of the BALB/c strain show a large increase in these cells at 18 months compared to 2 to 3 months, whereas only a small, if any, increase is seen at 18 months in the C57BL strains.[86]

MUSCLE

The number of myocardial muscle fibers in the mouse (Table 9) decreases from 3 to 7 months and then remains constant to 24 months of age.[180] In contrast, the number of mouse myocardial reticular fibers increases at 7 to 9 months and then decreases at 14 to 24 months.[180] In the rat, the number of fibers in the heart is the same at 26 to 27 months as it is at 4 months of age,[181] but the number of myocardial mast cells decreases at 18 months and again at 24 months.[184] In the human, there is a small decrease in these mast cells at ages 60 to 90 years compared to 20 to 35 years.[183]

The mass of various skeletal muscles and the numbers of fibers in them at different ages have been determined in several studies (Table 9). In the rat, the mass of the hind limb thigh muscle is less in the male at 24 to 27 months than at 12 to 14 months but is only slightly less, if at all, in the female at 24 to 27 months.[184] In rats aged 28 to 29 months vs. those aged 3 months, the number of fibers in the extensor digitorum longus[186] and soleus muscles[80,186] decreases while the number in the diaphragm increases.[186] Tauchi et al.[185] have determined the number of red and white fibers in two skeletal muscles in two species of rat. In the tibialis anterior muscle of both male and female Wistar rats, the number of red fibers is decreased at 24 months of age, whereas white fibers increase with age. Males show increased numbers of white fibers at 6 months and again at 24 months, whereas females show increases only at 12 months of age. In the same muscle in the male Donryu rat, red fibers are also decreased at 24 months but white fibers are unchanged in number from 3 to 24 months. The psoas major muscle in the male Wistar rat shows progressively fewer red fibers at 6, 12, and 24 months of age but more white fibers at 6 and 12 months. The number of white fibers then decreases at 24 months.

In the mouse, changes in numbers of skeletal muscle fibers (Table 9) vary with the muscle and the sex of the animals.[187,188] A decreased number of fibers at 16 to 25 months of age is seen in the male mouse for the extensor digitorum longus[187] and the biceps brachii[187,188] and possibly also for the soleus[187] and the anterior tibialis,[187,188] but apparently no change occurs in the number of fibers of the sternomastoid.[187,188] In the female mouse at 25 months, decreased fibers occur in the extensor digitorum longus and the soleus and possibly the sternomastoid, but no change or possibly a small increase in the number of fibers occurs in anterior tibialis and the biceps brachii.[187] In the human male, the density of fibers in the quadriceps is greater at 72 to 74 years of age than at 21 years regardless whether or not the subjects were athletically trained.[15]

As for smooth muscle (Table 9), the proportion of the ciliary body which is muscle decreases in the human male each decade up to 80 years.[172] A similar decrease also occurs in the human female but to a lesser extent than in the male.[172] As the proportion of muscle decreases the proportion of connective tissue in the ciliary body increases in both males and females.

REPRODUCTIVE SYSTEM AND KIDNEY

In the human (Table 10), the number of Leydig cells decreases at 40, 60 and again at 80 years of age.[189,190] In the rat, one study shows some increase in numbers of Leydig cells at 24 months compared to 3 months[191] while a second study shows fewer cells at 12 months compared to 5 months and fewer still at 29 months.[192] In the stallion, the number of Leydig cells increases with age.[193] In the mouse, the number of Sertoli cells does not change, but the number of spermatozoa in various stages of development decreases at 22 months of age compared to 4 to 5 months.[194] In the rabbit, the number of spermatids and spermatozoa increases to 24 months but decreases by 36 months.[195] In the stallion, the number of spermatozoa increases with age.[193] The number of epididymal fat cells in the rat increases at 30 month of age.[196,197]

In an early study, Arai[198] showed that the number of ova decreased through the life of the rat, at least up to 31 months of age. Similarly, Jones and Krohn[199] showed that the number of oocytes in several strains of mice decreased progressively with age during the 16 to 24 months measured.

In the rat, the number of fat cells in the perirenal depot of the kidney increases from 6 to 18 months of age.[197]

LUNG

In the alveoli of the lungs (Table 11) of male rats, the total, type II, interstitial, capillary endothelial, and macrophage cells show lower numbers at 14 to 26 months of age vs. 1 week to 5 months.[200] Type I cells increase in number to 14 months and then possibly decrease to a small extent at 26 months. In the alveoli of the lungs of female rats, the total, type II, interstitial, and capillary endothelial cells show no changes in number from 5 to 26 months of age.[200] However, the number of type I cells increases to 14 months and then does not change. Macrophages, on the average, are fewer at 14 to 26 months than at 5 months.

DISCUSSION AND CONCLUSIONS

The data in Tables 1 to 11 demonstrate that there is a high degree of variation in cell counts in a particular tissue among individuals of the same chronological age. This variation correlates with the high degree of variation also noted for multiple physiological and biochemical measurements as a function of age.[10] Therefore, average values at different chronological ages are used to assess biological changes with advancing age, e.g., Shock,[7] though the gerontologist should not forget the high degree of individual variation.

Degenerative changes in a tissue with age do not occur uniformly in all cells of that tissue, a finding repeatedly reported in the literature. Such changes are not uniform even in all cells of the same type in a tissue. Some cells degenerate but are still surrounded by cells which appear normal even in very old individuals. There is a decline in absolute weights of brain, kidney, liver, and spleen but not other organs in the aging human.[202] However, measurements of the weight of an organ or tissue may not always reflect a loss of functional elements with age since parenchymal cells which have degenerated are replaced often by interstitial substances (e.g., collagen) which, with time, increase both in amount and in density.[10,11] Some method of direct enumeration, as used in the references in Tables 1 to 11, appears necessary to determine any changes in quantities of cells, etc. with age.

Table 12 summarizes the cell number changes documented in Tables 1 to 11 for the three most commonly studied organisms, the human, the rat, and the mouse. First, all three lose some parenchymal cells especially in old age. Secondly, though the data available do not allow many comparisons to be made, the overall complement of parenchymal cells that is lost with age seems to differ in the three organisms. In the human, cells are lost on the average from the nervous system, eye, skin, bone, blood, liver, smooth muscle, and testis; in the rat from the nervous system, bone, blood, spleen, lymph nodes, palate, tongue, skeletal muscle, reproductive system, kidney, and lung; in the mouse from the nervous system, bone, blood, spleen, lymph nodes, small intestine, liver, cardiac muscle, and reproductive system. Clearly, the loss of bone, neurons, blood, and male reproductive cells is a common attribute of aging in all three organisms. Losses of cells in other tissues or of the types of cells in a given tissue or organ may differ with the organism, however. Data on the mouse (Table 12) show it to be the most variable of the three organisms. Numbers of blood cells, bone marrow stem cells, spleen cells, small intestine mesentery mast cells, and skeletal muscle fibers in the mouse at a given age vary with the sex and/or the strain of the animal and possibly also with the diet and/or environmental conditions. A good correlation is seen when cell loss in the brain of the human and the rat is compared. The loss of neurons in different areas of the brain is similar in both organisms (Tables 2, 12).

The loss of parenchymal cells with age alone is unlikely to explain the limitation on lifespan, as noted by Franks.[203] Cellular reserves can be large, e.g., if one kidney is removed, the organism still survives. However, cell loss with age is likely to be a contributing factor to lifespan limitation along with a lowered proliferation of cells with age[8,9] and a decline in physiological function of at least some of the cells remaining in certain aged tissues.[6,7] Further, the loss of a cell results not only in the loss of the intrinsic function(s) of that cell itself but also the loss of any function(s) associated with the interactions of that cell with other cells in a tissue. The brain is a case in point. When a neuron is lost so are its associated dendrites. Thus, with the loss of dendritic processes, a link between neurons forming a conducting pathway is broken. Similarly, functions dependent on the interactions of cells in other tissues can be affected as cells are lost with age. Therefore, the loss of cells from multiple tissues with age (Table 12) is likely to contribute to a destabilization of the whole organism.

Table 1
HUMAN BODY MASS WITH INCREASING AGE

	No.	Age (years)	Sex	Value[a]			Unit	Ref.
Body weight, fat-free	27	23—29	M	60.4 + 5.1			kg	12
	44	48—52		57.7 + 5.4				
	34	53—57		56.8 ± 5.3				
Body weight, fat-free	23	30—39	M	55.4	55.7	56.1	Calculated from data given: total weight (kg) minus fat (kg)	13
	35	40—49		55.1	56.2	59.6		
	30	50—59		52.7	53.9	57.6		
	26	60—69		54.0	54.4	59.8		
	21	70—79		50.3	50.1	55.1		
Lean body mass	585	25	M	59			kg	14
	881	45		56				
	835	55		52				
	234	65—70		47				
Lean body mass	267	25	F	40			kg	14
	391	45		39				
	373	55		39				
	144	65—70		35				
Lean body mass	7	20.7 (average)	M	67.3 ± 8.0			kg; (a) active, athletically trained, (b) relatively inactive	15
	10	73.9 (a)		54.7 ± 3.6				
	8	72.4 (b)		56.4 ± 5.5				
Fat-free mass	27	18—25	M	59.5 ± 6.69			kg	16
	58	25—35		60.6 ± 5.90				
	33	35—45		60.4 ± 8.32				
	37	45—55		55.4 ± 6.24				
	42	55—65		53.0 ± 6.36				
	18	65—85		46.5 ± 3.44				
Fat-free mass	89	18—25	F	37.5 ± 4.02			kg	16
	33	25—35		38.6 ± 5.17				
	44	35—45		39.0 ± 5.16				
	72	45—55		37.8 ± 4.23				
	54	55—65		35.3 ± 4.21				
	13	65—85		34.5 ± 2.94				
Body cell mass	27	18—25	M	33.7 ± 3.79			kg	16
	58	25—35		34.4 ± 3.35				
	33	35—45		34.2 ± 4.72				
	37	45—55		31.4 ± 3.54				
	32	55—65		30.0 ± 3.61				
	18	65—85		26.4 ± 1.95				
Body cell mass	89	18—25	F	21.3 ± 2.28			kg	16
	33	25—35		21.9 ± 2.94				
	44	35—45		22.1 ± 2.93				
	72	45—55		21.4 ± 2.42				
	54	55—65		20.0 ± 2.39				

Table 1 (continued)
HUMAN BODY MASS WITH INCREASING AGE

	No.	Age (years)	Sex	Value[a]			Unit	Ref.
				Subjects				
	13	65—85		19.6 ± 1.67				
Body cell mass	4	21.5 ± 2.4	M	34 ± 4.2			kg	17
	6	70.3 ± 1.5		25 ± 3.0				
Body cell mass	4	20.0 ± 2.4	F	23 ± 3.4			kg	17
	5	76.4 ± 10.8		16 ± 2.4				
Total body nitrogen	22	20—29	M	2023 ± 256			gm; subjects whose weight	18
	14	30—39		1813 ± 164			was within 20% of ex-	
	10	40—49		1825 ± 187			pected for age	
	10	50—59		1818 ± 232				
	10	60—69		1696 ± 203				
	10	70—79		1549 ± 171				
Total body nitrogen	10	20—29	F	1454 ± 179			gm; subjects whose weight	18
	10	30—39		1402 ± 185			was within 20% of ex-	
	10	40—49		1339 ± 192			pected for age	
	9	50—59		1243 ± 97				
	13	60—69		1145 ± 172				
	8	70—79		1089 ± 125				
Body fat	25	23—29	M	10.2 ± 5.1			kg	12
	44	48—52		18.2 ± 5.1				
	34	53—57		19.2 ± 5.6				
Body fat	23	30—39	M	21.9	21.6	21.2	Kg calculated from data	13
	35	40—49		23.9	22.8	19.4	given	
	30	50—59		24.5	23.3	19.6		
	26	60—69		25.0	24.6	19.2		
	21	70—79		21.1	21.3	16.3		
Body fat	585	25	M	14			kg	14
	881	45		20				
	835	55		22				
	234	65—70		26				
Body fat	267	25	F	19			kg	14
	391	45		28				
	373	55		31				
	144	65—70		34				
Body fat	7	20.7 (average)	M	11.2 ± 4.6			Percent of body weight;	15
	10	73.9 (a)		19.8 ± 3.6			(a) active, athletically	
	8	2.4 (b)		18.3 ± 3.3			trained, (b) relatively inactive	
Total body fat	27	18—25	M	13.4 ± 7.37			kg	16
	58	25—35		17.2 ± 6.72				
	33	35—45		17.9 ± 4.85				
	37	45—55		21.1 ± 5.75				
	42	55—65		23.1 ± 6.68				
	18	65—85		26.9 ± 6.01				

Table 1 (continued)
HUMAN BODY MASS WITH INCREASING AGE

	Subjects					
	No.	Age (years)	Sex	Value[a]	Unit	Ref.
Total body fat	89	18—25	F	18.8 ± 4.94	kg	16
	33	25—35		19.8 ± 10.36		
	44	35—45		22.5 ± 7.94		
	72	45—55		29.3 ± 9.16		
	54	55—65		27.9 ± 7.93		
	13	65—85		28.5 ± 6.57		
Total body fat	17	15—17	M	16.6 ± 8.2	Percent calculated as 100	19
	23	18—23		14.9 ± 7.0	× (body weight minus	
	24	24—38		21.5 ± 6.2	lean body mass/body	
	16	40—48		24.8 ± 6.3	weight); K40 method	
	12	50—58		31.3 ± 3.2		
	8	60—87		36.2 ± 9.6		

[a] Where three numbers are given, three different formulas were used to estimate body fat.

Table 2
NUMBER OF CELLS OR FIBERS VS. AGE IN THE NERVOUS SYSTEM

Tissue	Type	(no.)	Age	Sex	Total or mean	Range	Measure	Ref.
					Spinal Nerves			
VIII and IX Thoracic spinal ganglia	Human	(2)	30—36 years	M	66,947	61,559—72,335	Total cells	23
		(2)	41—49 years	M,F	76,648	68,890—82,424		
		(4)	51—58 years	M,F	66,694	58,086—74,197		
		(5)	63—68 years	M	52,930	42,818—65,780		
		(5)	72—78 years	M	52,635	46,171—59,349		
		(2)	80—85 years	M	52,715	45,415—60,015		
VIII and IX Thoracic, dorsal root	Human	(2)	26—27 years	M	11,181	11,058—11,304	Total no. myelinated fibers	23,24
		(2)	34—36 years	M	10,530	9,665—11,396		
		(8)	41—49 years	M,F	9,844	8,180—11,757		
		(14)	50—59 years	M,F	9,250	7,098—11,829		
		(13)	61—69 years	M,F	8,346	6,432—10,930		
		(16)	71—89 years	M	8,442	6,262—9,748		
		(5)	80—89 years	M	8,710	7,923—9,710		
VIII and IX Thoracic, ventral root	Human	(2)	26—27 years	M	6,204	6,204—6,205	Total no. myelinated fibers	23,24
		(2)	34—36 years	M	5,732	5,648—5,816		
		(9)	41—49 years	M,F	5,500	4,997—5,971		
		(14)	50—59 years	M,F	5,327	4,309—6,370		
		(13)	61—69 years	M,F	4,825	3,942—5,611		
		(16)	71—79 years	M	4,524	3,487—5,258		
		(5)	80—89 years	M	5,004	3,924—5,517		
VIII Thoracic, ventral root	Rat	(12)	18—50 days	M	1,107	1,005—1,351	Total no. fibers (axis cylinders)	25
		(4)	170—300 days	M	1,268	1,190—1,364		
		(4)	548—864 days	M	1,099	1,103—1,229		
VIII Thoracic, ventral root	Rat	(5)	14—50 days	F	944	908—992	Total no. fibers (axis cylinders)	25
		(4)	150 days	F	986	904—1,079		
		(4)	730—1,060 days	F	1,023	930—1,200		

Table 2 (continued)
NUMBER OF CELLS OR FIBERS VS. AGE IN THE NERVOUS SYSTEM

Tissue	Type	(no.)	Age	Sex	Total or mean	Range	Measure	Ref.
II Cervical, ventral root	Rat	(4)	7 days	M,F	367	360—372	Total no. myelinated fibers	26
		(4)	14 days	M,F	554	518—594		
		(4)	36 days	M,F	633	536—689		
		(4)	75 days	M,F	614	505—726		
		(4)	132 days	M,F	655	569—731		
		(4)	180 days	M,F	564	510—662		
		(4)	270 days	M,F	697	576—853		
		(3)	640 days	M	864	758—934		
IV or V Lumbar, largest, ventral root	Rat	(8)	18—40 days	M	1,900	1,753—2,011	Total no. fibers (axis cylinders)	25
		(6)	136—300 days	M	1,935	1,637—2,142		
		(4)	548—864 days	M	1,799	1,726—1,930		
IV or V Lumbar, largest, ventral root	Rat	(9)	14—50 days	F	1,620	1,439—1,891	Total no. fibers (axis cylinders)	25
		(2)	95—142 days	F	1,794	1,783—1,805		
		(2)	856—1,060 days	F	1,632	1,500—1,764		
Sciatic nerve	Rat	(9)	50—250 days	—	61%	50—69%	Proportion of nerve occupied by fibers	27
		(8)	250—650 days	—	64%	51—70%		
		(7)	650—850 days	—	60%	46—65%		
I Thoracic, ventral root	Cat	(32)	1 day—14 weeks	M,F	4,542	3,664—5,522	Total no. fibers (axis cylinders)	28
		(16)	1 year—5 years	M,F	4,981	4,403—5,926		
		(16)	10—18 years	M,F	5,194	4,181—5,746		
VIII Cervical, ventral root	Cat	(37)	1 day—14 weeks	M,F	5,404	4,889—6,358	Total no. fibers (axis cylinders)	28
		(16)	1 year—5 years	M,F	5,671	4,970—6,387		
		(16)	10—18 years	M,F	5,844	4,850—6,441		
Spinal cord, large anterior horn cells	Mouse	(5)	25 weeks	M	6,412	5,647—7,745	Total cells	29
		(3)	50 weeks	M	7,032	6,557—7,342		
		(4)	110 weeks	M	5,614	5,246—5,981		
Spinal cord, larger anterior horn cells	Mouse	(3)	25 weeks	F	6,267	5,688—6,634	Total cells	29
		(3)	50 weeks	F	6,024	5,505—6,571		
		(4)	110 weeks	F	4,905	4,225—5,589		

Cranial Nerves

	Species	(n)	Age	Sex	A	B		Description	
Olfactory nerve	Human	(27)	16—30 years	M,F	20%	0—100%		No. nerve fibers (mean % loss from value at birth)	30
		(20)	31—45 years	M,F	33%	0—100%			
		(45)	46—60 years	M,F	57%	0—100%			
		(55)	61—75 years	M,F	68%	0—100%			
		(30)	76—91 years	M,F	73%	0—100%			
Optic nerve	Human	(2)	47—51 years	M,F	1.11×10^6	1.02—1.20×10^6		Total no. nerve fibers	31
		(6)	61—68 years	M,F	1.02×10^6	0.92—1.14×10^6			
		(2)	70—71 years	F	0.89×10^6	0.87—0.90×10^6			
Facial nucleus	Human	(3)	22—27 years	—	—	5,712—6,100		Total neurons	32
		(4)	33—41 years	—	—	5,850—6,271			
		(4)	54—64 years	—	—	5,518—6,152			
		(2)	70—76 years	—	—	4,758—5,433			
		(2)	80—86 years	—	—	5,169—5,225			
Abducens nucleus	Human	(1)	Newborn	M	6,500	—		Total no.	33
		(1)	14 years	M	5,040	—			
		(1)	31—35 years	M	6,660	6,510—6,820			
		(2)	43—49 years	M	7,510	4,980—10,040			
		(3)	54—57 years	M	—	5,190—8,110			
		(5)	60—68 years	M	—	3,940—8,460			
		(3)	71—79 years	M	—	4,250—7,630			
		(3)	83—87 years	M	—	5,190—6,920			
Gasserian ganglion cells	Human	(1)	27 months	M	8,593	—		Total no. in 5 sections	34
		(1)	26 years	F	7,700	—			
		(2)	36—48 years	M,F	14,136	11,708—16,564			
		(4)	55—66 years	M,F	8,421	6,073—11,827			
		(2)	75—81 years	F	7,588	6,518—8,658			

	Species	(n)	Age	Sex	A	B	A	B	Description	
Vagus nerve, mid-cervical vagus	Human	(1)	2 years	—	146,738	81,261	—	—	Total fibers: (A) right nerve, (B) left nerve	35
		(1)	10 years	—	122,000	96,566	—	—		
		(3)	49—58 years	—	—	—	69,951—153,123	45,110—102,048		
		(4)	61—68 years	—	—	—	80,717—143,476	85,982—106,642		
		(7)	70—86 years	—	—	—	75,708—111,605	75,929—111,807		

Table 2 (continued)
NUMBER OF CELLS OR FIBERS VS. AGE IN THE NERVOUS SYSTEM

Tissue	Type	(no.)	Age	Sex	Total or mean	Range	Measure	Ref.
Vagus nerve, esophageal plexus	Human	(1)	10 years	—	85,936	—	Total fibers	35
		(1)	49 years	—	16,013	—		
		(1)	65 years	—	59,921	—		
		(5)	70—79 years	—	—	35,086—101,492		
		(2)	82—86 years	—	—	43,891—49,054		
					A **B**	**A** **B**		
Vagus nerve, recurrent laryngeal nerve	Human	(1)	2 years	—	7,285 8,627	— —	Total fibers: (A) right nerve, (B) left nerve	35
		(1)	10 years	—	16,806 7,669	— —		
		(1)	16 years	—	10,159 —	— —		
		(3)	49—58 years	—	—	5,398—10,850 5,444—7,142		
		(3)	65—68 years	—	—	6,079—11,306 6,899—9,522		
		(6)	70—79 years	—	—	4,704—12,990 5,689—15,697		
		(1)	82 years	—	3,651	— —		
		(1)	86 years	—	7,870	—		
Olfactory bulb, mitral cells	Rat	(12)	3 months	—	72,248	—	Total no.	36
		(12)	2.5 years	—	58,795	—		

Brain

Tissue	Type	(no.)	Age	Sex	Total or mean	Range	Measure	Ref.
Whole brain, total cells	Mouse	(4)	4 weeks	—	$0.85 \times 10^8 \pm 0.05 \times 10^8$	—	Total no. nuclei per brain	37
		(4)	16 weeks	—	$0.80 \times 10^8 \pm 0.03 \times 10^8$	—		
		(4)	28 weeks	—	$1.02 \times 10^8 \pm 0.05 \times 10^8$	—		
		(4)	32 weeks	—	$0.93 \times 10^8 \pm 0.08 \times 10^8$	—		
		(2)	72 weeks	—	$0.80 \times 10^8 \pm 0.02 \times 10^8$	—		
		(4)	96 weeks	—	$0.80 \times 10^8 \pm 0.05 \times 10^8$	—		
		(4)	144 weeks	—	$0.89 \times 10^8 \pm 0.03 \times 10^8$	—		
Whole brain, neurons	Mouse	(5)	1 month	—	5.5×10^6	—	Total no.: estimated from data given	38
		(5)	8 months	—	5.2×10^6	—		
		(4)	13 months	—	5.1×10^6	—		

Region	Species	(n)	Age	Sex	Value	Range	Description	Ref.
Cerebral cortex	Human	(8)	24 months	—	4.1×10^6	—	No. cells per cm³	39
		(4)	29 months	—	1.9×10^6	—		
		(1)	15 years	F	13.5×10^6	—		
		(1)	26 years	M	16.7×10^6	—		
		(1)	38 years	M	17.8×10^6	—		
		(3)	43—49 years	M	—	10.5×10^6—20.0×10^6		
		(1)	54 years	F	—	—		
		(2)	65—66 years	M,F	16.6×10^6	16.2×10^6—18.2×10^6		
		(4)	70—76 years	M,F	17.2×10^6	9.8×10^6—19.3×10^6		
		(1)	89 years	F	14.0×10^6	—		
Cerebral cortex	Human	(5)	40—50 years	M	$83{,}000 \pm 6{,}000$	—	No. cells per mm³	40
		(5)	70—85 years	M	$65{,}000 \pm$ —	—		
Cerebral cortex	Human	(5)	19—28 years	—	100%	—	Av. nerve cells in % of control (19—28 years) in 7 diff. regions of cortex	41
		(10)	77 years	—	79%	71—88%		
Cerebral cortex, neurons	Rat	(-)	10 days	—	145,000	—	Total no. per mm³; 4—6 animals at each age, fixed tissue	42
		(-)	20 days	—	110,000	—		
		(-)	50 days	—	110,000	—		
		(-)	100 days	—	90,000	—		
		(-)	2 years	—	90,000	—		
Cerebral cortex, neurons	Rat	(6)	109—113 days	M	97,000	—	Cells per mm³ per 20 depth levels	43
		(6)	763—972 days	M	98,000	—		
Cerebral cortex, neurons,	Human	(5)	40—50 years	M	$30{,}000 \pm 5{,}000$	—	Total per mm³	40
		(5)	70—85 years	M	$21{,}000 \pm$ —	—		
Cerebral cortex, neurons	Rat	(4)	3 months	F	$40.30 \times 10^3 \pm 2.21 \times 10^3$	—	Total per mm³	44
		(4)	22—27 months	F	$30.71 \times 10^3 \pm 1.22 \times 10^3$	—		
Cerebral cortex, superior temporal gyrus	Human	(4)	16—48 years	M,F	2,079	1,950—2,187	Total nerve cells	45
		(6)	70—78 years	M,F	1,140	1,055—1,271		
		(3)	80—95 years	M	949	887—991		

Table 2 (continued)
NUMBER OF CELLS OR FIBERS VS. AGE IN THE NERVOUS SYSTEM

Tissue		Subjects			Cells or fibers		Measure	Ref.
	Type	(no.)	Age	Sex	Total or mean	Range		
Cerebral cortex, right frontal lobe, lateral gyrus, neurons	Monkey (rhesus)	(10)	4—7 years	M,F	61	—	No. in total depth of cortex; estimated from data given	46
		(10)	18—28 years	M,F	51	—		
Cerebral cortex, area striata	Human	(7)	16—48 years	M,F	2,355	1,740—2,648	Total nerve cells	45
		(5)	70—78 years	M,F	1,704	1,243—1,840		
		(3)	80—95 years	M	1,691	1,489—1,810		
Cerebral cortex, postcentral gyrus, 3rd quarter	Human	(7)	16—48 years	M,F	1,363	1,126—1,512	Total nerve cells	45
		(6)	73—95 years	M,F	1,540	1,241—1,932		
Cerebral cortex, postcentral gyrus, 3rd quarter	Human	(7)	16—48 years	M,F	1,363	1,126—1,512	Total nerve cells	45
		(6)	73—95 years	M,F	1,540	1,241—1,932		
Visual cortex, area 17, neurons	Rat	(5)	11 months	M	101×10^3	—	No. per mm^3 per 20 μm; 11 month value est. from data given	47,48
		(5)	17 months	M	93.8×10^3	—		
		(5)	29 months	M	77.0×10^3	—		
Visual cortex, macular projection area	Human	(-)	3rd decade	—	4×10^7	—	No. per g visual cortex; 23 subjects total used	49
		(-)	8th decade	—	2×10^7	—		
Visual cortex, macular projection area, neurons	Human	(5)	20—29 years	M,F	4.352×10^7	3.984×10^7—4.745×10^7	Total no. per g tissue	50
		(3)	32—41 years	M,F	4.514×10^7	3.560×10^7—5.593×10^7		
		(4)	50—56 years	M,F	2.926×10^7	2.348×10^7—3.199×10^7		
		(5)	61—69 years	M,F	2.660×10^7	1.969×10^7—3.177×10^7		
		(3)	73—76 years	M,F	2.548×10^7	1.724×10^7—3.026×10^7		
		(3)	81—87 years	M,F	2.612×10^7	2.432×10^7—2.850×10^7		

				Sex	Value		Remarks	Ref.
Cerebral hemisphere, neurons	Rat	(16) 148—156 days	(12) 806—878 days	M,F / M,F	$24.4 \times 10^6 \pm 1.5 \times 10^6$ / $17.2 \times 10^6 \pm 0.9 \times 10^6$	— / —	Total no.	51
Lissencephalic cortex, pyramidal neurons	Rat	(4) 6 months	(9) 25 months	F / F	4,642 / 5,044	— / —	No. per 7000 counts	52
Lissencephalic cortex, nonpyramidal neurons	Rat	(4) 6 months	(9) 25 months	F / F	451 / 610	— / —	No. per 7000 counts	52
Amygdala, neurons	Human	(4) 20—50 years	(4) 67—76 years	M / M	A B C 115 97 80 111 95 81	— / —	Total no./mm³; (A) lateral, dorsolateral portion, (B) lateral, ventromedial portion, (C) lateral, basal portion	53
Amygdala, neurons	Human	(4) 20—50 years	(4) 67—76 years	M / M	A B C D 74 93 146 310 72 95 150 300	— / —	Total no./mm³; (A) accessory basal, magnocellular portion; (B) accessory basal, parvocellular portion, (C) medial, basal, deep portion, (D) medial basal, superficial portion	53

Table 2 (continued)
NUMBER OF CELLS OR FIBERS VS. AGE IN THE NERVOUS SYSTEM

Tissue	Type	Subjects (no.)	Age	Sex	Total or mean	Range	Measure	Ref.
Amygdala, neurons	Human	(4)	20—50 years	M	A 182, B 267, C 140, D 210	—	Total no/mm³; (A) cortical, medial portion, (B) cortical, lateral portion, (C) cortical, deep portion, (D) cortical transition area	53
		(4)	67—76 years	M	168, 248, 134, 214	—		
Amygdala, neurons	Human	(4)	20—50 years	M	A 391, B 39, C 206	—	Total no/mm³ (A) medial, (B) central, large-celled fusiform portion, (C) central, lateral portion	53
		(4)	67—76 years	M	316, 34, 206	—		
Lateral amygdala, neurons	Rat	(4)	90 days	M	119,404	—	Total no.	54
		(5)	360 days	M	139,512	—		
		(5)	520—580 days	M	135,513	—		
		(4)	820—850 days	M	92,521	—		
Corticoamygdaloid nucleus, neurons	Rat	(4)	90 days	M	213,388	—	Total no.	54
		(5)	360 days	M	166,404	—		
		(5)	520—580 days	M	160,540	—		
		(4)	820—850 days	M	112,596	—		

Region	Species	Age (n)	Sex	Value	Range	Measure	Ref.
Substantia nigra, neurons	Rat	(4) 90 days	M	224,383	—	Total no.	54
		(5) 360 days	M	155,876	—		
		(5) 520—580 days	M	190,930	—		
		(4) 850—880 days	M	100,992	—		
Dorsal lateral septum, neurons	Rat	(4) 90 days	M	616,684	—	Total no.	54
		(5) 360 days	M	460,492	—		
		(5) 520—580 days	M	332,002	—		
		(4) 850—880 days	M	197,579	—		
Occipital cortex, medial portion, neurons	Rat	(14) 26 days	M	667 ± 57	—	Total number, medial portion of section B	55
		(15) 41 days	M	730 ± 62	—		
		(14) 108 days	M	611 ± 37	—		
		(13) 650 days	M	582 ± 56	—		
Hippocampus, CA1 region to dentate gyrus, neurons	Rat	(11) 8 months	M	7.4	—	Total neuronal nucleoli per µm; est. from data given	56
		(14) 27 months	M	5.0	—		
Hippocampus, right dentate gyrus, granule cells	Rat	(3) 3 months	M	0.488	0.462—0.505	Total no. per 100 µm^2	57
		(3) 25 months	M	0.494	0.471—0.539		
Hippocampus, dentate gyrus, supragranular zone, granule cells	Rat	(5) 3 months	—	19.2 ± 0.7	—	No. per 100 µm	58
		(5) 25 months	—	18.3 ± 1.0	—		
Hippocampus, CA1 zone, neurons	Rat	(5) 11 months	M	28.63	—	No. in a 110-µm segment	47,48
		(5) 17 months	M	25.79	—		
		(5) 29 months	M	24.75	—		
Hippocampus, CA1 zone, neurons	Monkey (rhesus)	(10) 4—7 years	M,F	60	—	Total per transverse section; est. from data given	46
		(10) 18—28 years	M,F	34	—		
Hippocampus, CA1 pyramidal layer, neurons	Rat	(5) 3 months	F	62.52 ± 5.95	—	Total per mm	44
		(2) 12 months	F	69.59 ± 9.57	—		
		(5) 22—27 months	F	42.53 ± 1.67	—		

Table 2 (continued)
NUMBER OF CELLS OR FIBERS VS. AGE IN THE NERVOUS SYSTEM

Tissue	Subjects				Cells or fibers		Measure	Ref.
	Type	(no.)	Age	Sex	Total or mean	Range		
Hippocampus CA2 zone, neurons	Monkey (rhesus)	(10)	4—7 years	M,F	32	—	Total per transverse section; est. from data given	46
		(10)	18—28 years	M,F	26	—		
Hippocampus, CA3 zone, neurons	Monkey (rhesus)	(10)	4—7 years	M,F	49	—	Total per transverse section; est. from data given	46
		(10)	18—28 years	M,F	30	—		
Hypothalamus, anterior area	Rat	(8)	3.5—5 months	F	$33.5 \times 10^3 \pm 2.5 \times 10^3$	—	Total no.[a]	59
		(4)	24—25 mo.(A)	F	$(A)25.9 \times 10^3 \pm 2.0 \times 10^3$	—		
		(4)	24—30 mo.(B)	F	$(B)24.0 \times 10^3 \pm 2.3 \times 10^3$	—		
Hypothalamus, anterior area	Rat	(5)	3.5—5 months	M	$31.6 \times 10^3 \pm 1.9 \times 10^3$	—	Total number	60
		(4)	26—27 months	M	$36.1 \times 10^3 \pm 3.9 \times 10^3$	—		
Hypothalamus, ventromedial nucleus	Rat	(8)	3.5—5 months	F	$81.3 \times 10^3 \pm 5.7 \times 10^3$	—	Total no.[a]	59
		(3)	24—25 mo.(A)	F	$(A)87.5 \times 10^3 \pm 2.6 \times 10^3$	—		
		(5)	24—30 mo.(B)	F	$(B)69.9 \times 10^3 \pm 8.2 \times 10^3$	—		
Hypothalamus, ventromedial nucleus, neurons	Rat	(4)	90 days	M	58,635	—	Total no.	54
		(5)	360 days	M	57,638	—		
		(5)	520—580 days	M	50,601	—		
		(4)	850—880 days	M	23,086	—		
Hypothalamus, ventromedial nucleus	Rat	(5)	3.5—5 months	M	$80.0 \times 10^3 \pm 8.2 \times 10^3$	—	Total no.	60
		(4)	26—27 months	M	$82.4 \times 10^3 \pm 3.8 \times 10^3$	—		
Hypothalamus, arcuate nucleus	Rat	(6)	3.5—5 months	F	$54.4 \times 10^3 \pm 3.3 \times 10^3$	—	Total no.[a]	59
		(4)	24—25 mo.(A)	F	$(A)41.8 \times 10^3 \pm 3.7 \times 10^3$	—		
		(4)	24—30 mo.(B)	F	$(B)27.0 \times 10^3 \pm 3.4 \times 10^3$	—		
Hypothalamus, arcuate nucleus	Rat	(4)	3.5—5 months	M	$44.6 \times 10^3 \pm 4.0 \times 10^3$	—	Total no.	60
		(4)	26—27 months	M	$43.9 \times 10^3 \pm 4.3 \times 10^3$	—		

Region	Species	(n)	Age	Sex	Value	Range	Measure	Ref.
Hypothalamus, dorsomedial nucleus	Rat	(8)	3.5—5 months	F	52.3 × 10³ ± 4.6 × 10³	—	Total no.ᵃ	59
		(4)	24—25 mo.(A)	F	(A)42.2 × 10³ ± 4.1 × 10³	—		
		(4)	24—30 mo.(B)	F	(B)38.7 × 10³ ± 1.3 × 10³	—		
Hypothalamus, dorsomedial nucleus	Rat	(4)	3.5—5 months	M	48.4 × 10³ ± 3.1 × 10³	—	Total no.	60
		(4)	26—27 months	M	46.6 × 10³ ± 6.2 × 10³	—		
Hypothalamus, supraoptic nucleus	Rat	(10)	3.5—5 months	F	13.6 × 10³ ± 0.6 × 10³	—	Total no.ᵃ	59
		(5)	24—25 mo.(A)	F	(A)15.1 × 10³ ± 1.0 × 10³	—		
		(5)	24—30 mo.(B)	F	(B)13.5 × 10³ ± 0.6 × 10³	—		
Hypothalamus, supraoptic nucleus	Rat	(5)	3.5—5 months	M	15.1 × 10³ ± 1.9 × 10³	—	Total no.	60
		(5)	26—27 months	M	14.7 × 10³ ± 0.9 × 10³	—		
Hypothalamus, paraventricular nucleus	Rat	(10)	3.5—5 months	F	15.5 × 10³ ± 0.9 × 10³	—	Total no.	59
		(5)	24—25 mo.(A)	F	(A)15.0 × 10³ ± 1.1 × 10³	—		
		(5)	24—30 mo.(B)	F	(B)13.3 × 10³ ± 1.4 × 10³	—		
Hypothalamus, paraventricular nucleus	Rat	(5)	3.5—5 months	M	16.3 × 10³ ± 1.6 × 10³	—	Total no.	60
		(5)	26—27 months	M	16.5 × 10³ ± 1.1 × 10³	—		
Hypothalamus, medial preoptic area	Rat	(4)	3.5—5 months	F	42.5 × 10³ ± 1.0 × 10³	—	Total no.	59
		(4)	24—25 mo.(A)	F	(A)29.8 × 10³ ± 4.7 × 10³	—		
		(4)	24—30 mo.(B)	F	(B)25.8 × 10³ ± 2.9 × 10³	—		
Hypothalamus, medial preoptic area	Rat	(4)	3.5—5 months	M	37.7 × 10³ ± 2.7 × 10³	—	Total no.	60
		(4)	26—27 months	M	33.0 × 10³ ± 4.8 × 10³	—		
Hypothalamus, medial mammillary nucleus	Human	(5)	26—46 years	F	291,162	176,988—410,496	Total cells	61
		(5)	71—84 years	F	259,733	196,812—374,784		
Cerebellum, total neurons	Rat	(16)	148—156 days	M,F	41.8 × 10⁶ ± 2.1 × 10⁶	—	Total no.	51
		(12)	806—878 days	M,F	25.2 × 10⁶ ± 1.4 × 10⁶	—		
Cerebellum, Purkinje cells right and left hemisphere	Human	(4)	22—42 years	M	607	591—624	No. cells per equivalent unit area anterior to primary sulcus and anterior to great horizontal sulcus	62,63
		(5)	50—65 years	M	575	509—667		
		(5)	73—100 years	M	466	403—500		

Table 2 (continued)
NUMBER OF CELLS OR FIBERS VS. AGE IN THE NERVOUS SYSTEM

	Subjects				Cells or fibers		Measure	Ref.
Tissue	Type	(no.)	Age	Sex	Total or mean	Range		
Cerebellum, Purkinje cells, right and left hemisphere	Human	(6)	19—32 years	F	543	511—612	No. cells per equivalent unit area anterior to primary sulcus and anterior to great horizontal sulcus	62,63
		(3)	50—65 years	F	478	462—499		
		(5)	71—94 years	F	418	334—505		
Cerebellum, Purkinje cells,	Human	(2)	2 years	—	10.18	10.13—10.22	No. cells per linear mm	64,65
		(8)	15—88 years	—	8.05	7.15—8.83		
Cerebellar cortex, Purkinje cells	Rat	(1)	200 days	M	547,413	—	Total normal and "pathological" cells	66
		(1)	730 days	M	473,532	—		
		(1)	1,017 days	M	447,545	—		
Cerebellar cortex, Purkinje cells	Rat	(1)	200 days	F	565,027	—	Total normal and "pathological" cells	66
		(1)	730 days	F	477,628	—		
		(1)	1,085 days	F	452,343	—		
Cerebellar cortex Purkinje cells	Guinea pig	(16)	139—1,460 days	—	11.10 ± 0.5559	10.00—12.28	No. cells per 0.22-mm strip; mean of 80 strips per animal	67
		(10)	1,461—2,765 days	—	11.44 ± 0.5040	10.20—12.54		
Medulla, inferior olive main nucleus	Human	(1)	23 wks gestation	F	233,360	—	Total no.	68
		(1)	24 wks gestation	F	247,240	—		
		(1)	30 wks gestation	M	300,790	—		
		(1)	34 wks gestation	M	346,850	—		
		(1)	Birth	F	330,610	—		
		(1)	3 days	M	327,330	—		
		(1)	3 months	M	380,460	—		
		(3)	9—19 years	M,F	—	366,100—370,370		
		(2)	32—44 years	M	360,240	355,560—364,930		

Brain stem, total neurons — Rat

	(n)	Age	Sex	A	B	Ref.
	(2)	56—65 years	M	383,020	376,420—389,610	51
	(2)	73—77 years	M	382,250	358,920—405,580	
	(2)	82—89 years	M	349,260	317,600—380,920	

Brain stem, red nucleus — Rat
Mean total cell no. of every 6th section: 144 animals total used

	(n)	Age	Sex	A	B	Ref.
	(16)	148—156 days	M,F	$12.8 \times 10^6 \pm 0.4 \times 10^6$	—	69
	(12)	806—878 days	M,F	$11.6 \times 10^6 \pm 0.5 \times 10^6$	—	
	(—)	Newborn	M,F	384,666	—	
	(—)	7 days	M,F	384,500	—	
	(—)	15 days	M,F	384,500	—	
	(—)	1 month	M,F	384,333	—	
	(—)	3 months	M,F	384,333	—	
	(—)	6 months	M,F	384,166	—	
	(—)	1 year	M,F	384,166	—	
	(—)	2 years	M,F	384,166	—	

Brain stem, locus coeruleus, neurons — Rat
Total no. (A) right side, (B) left side

(n)	Age	Sex	A	B	A	B	Ref.
(2)	12 months	M	1,474	1,472	1,446—1,503	1,441—1,503	70
(3)	20 months	M	2,093	2,098	1,889—2,352	1,946—2,255	
(3)	27 months	M	1,859	1,812	1,703—2,074	1,544—2,114	
(3)	30 months	M	1,832	1,717	1,553—2,149	1,580—1,858	
(3)	32 months	M	1,847	1,806	1,581—2,291	1,405—2,207	

Brain stem, pars cerebellaris loci coerulei, neurons — Human
Total no. (A) right side, (B) left side

(n)	Age	Sex	A	B	A	B	Ref.
(2)	15—16 years	M	834	822	678—989	592—1,051	71
(5)	21—29 years	M	1,078	1,048	734—1,339	851—1,212	
(4)	31—36 years	M	1,074	1,139	624—1,480	656—1,550	
(3)	44—49 years	M	812	853	346—1,147	319—1,289	
(4)	54—59 years	M	812	813	579—1,128	569—1,043	
(6)	60—67 years	M	625	658	200—898	219—928	
(9)	71—79 years	M	634	656	348—958	326—1,051	
(2)	81—90 years	M	268	342	131—404	205—479	

Brain stem, pars cerebellaris loci coerulei, neurons — Human
Total no. (A) right side, (B) left side

(n)	Age	Sex	A	B	A	B	Ref.
(2)	11—18 years	F	927	987	889—965	969—1,005	71
(2)	25—29 years	F	1,006	1,053	861—1,150	865—1,241	
(3)	32—35 years	F	1,033	1,032	590—1,511	569—1,523	
(3)	42—43 years	F	1,064	1,135	855—1,306	902—1,401	
(3)	51—58 years	F	882	857	817—965	800—939	
(7)	60—68 years	F	864	892	569—1,344	620—1,314	
(7)	73—76 years	F	659	648	455—989	469—869	
(5)	84—88 years	F	541	629	252—963	316—1,250	

Table 2 (continued)
NUMBER OF CELLS OR FIBERS VS. AGE IN THE NERVOUS SYSTEM

Tissue	Type	(no.)	Age	Sex	Total or mean	Range	Measure	Ref.
					Auditory			
Ventral cochlear nucleus	Human	(3)	0.01—0.08 years	M,F	—	52,010—63,920	Total no.	72,73
		(3)	0.33—1.5 years	M,F	—	63,680—70,980		
		(2)	7—19 years	M	65,105	64,210—66,000		
		(3)	20—27 years	M,F	—	52,150—64,330		
		(2)	31—35 years	M,F	62,620	56,520—68,720		
		(3)	40—48 years	M	—	54,290—83,830		
		(3)	60—68 years	M	—	64,850—69,040		
		(3)	80—82 years	M,F	—	55,680—62,720		
		(1)	90 years	F	65,470	—		
Cochlea, spiral ganglion	Rat	(9)	1—2 months	—	15,800	13,800—16,900	Total cells	74
		(4)	6 months	—	14,500	13,500—15,500		
		(3)	12 months	—	16,000	14,100—16,300		
		(3)	23 months	—	13,600	13,500—13,800		
		(8)	27—29 months	—	12,700	11,100—13,900		
		(6)	33-34 months	—	13,100	11,100—15,000		
Cochlea, spiral ganglion, Type II cells	Rat	(9)	1—2 months	—	1,059	800—1,294	Total cells	74
		(4)	6 months	—	1,000	826—1,198		
		(3)	12 months	—	1,066	855—1,145		
		(3)	23 months	—	980	855—990		
		(8)	27—29 months	—	763	578—896		
		(6)	33-34 months	—	757	690—1,088		
					Glial Cells			
Whole brain, glial cells	Mouse	(5)	1 month	—	62×10^6	—	Total no. est. from data given	38
		(5)	8 months	—	65×10^6	—		
		(4)	13 months	—	46×10^6	—		
		(8)	24 months	—	56×10^6	—		
		(4)	29 months	—	65×10^6	—		

			Age	Sex	No. per 100 neurons		Notes	Ref.
Cerebrum, neostriatum, astrocytes	Mouse	(—)	6 months	M	22	—	No. per 100 neurons; est. from data given	75
		(—)	9 months	M	27	—		
		(—)	12 months	M	26	—		
		(—)	15 months	M	27	—		
		(—)	18 months	M	29	—		
		(—)	22 months	M	27	—		
Cerebrum, neostriatum, oligodendrocytes	Mouse	(—)	6 months	M	14	—	No. per 100 neurons; est. from data given	75
		(—)	9 months	M	15.5	—		
		(—)	12 months	M	20	—		
		(—)	15 months	M	18	—		
		(—)	18 months	M	18.3	—		
		(—)	22 months	M	18.5	—		
Cerebrum, neo-striatum, microglia	Mouse	(—)	6 months	M	12.5	—	No. per 100 neurons; est. from data given	75
		(—)	9 months	M	13.2	—		
		(—)	12 months	M	13.3	—		
		(—)	15 months	M	14.2	—		
		(—)	18 months	M	12	—		
		(—)	22 months	M	18	—		
Cerebrum, indusium griseum, astrocytes	Mouse	(—)	6 months	M	37	—	No. per 100 neurons; est. from data given	75
		(—)	9 months	M	42	—		
		(—)	12 months	M	41	—		
		(—)	15 months	M	42.5	—		
		(—)	18 months	M	45.2	—		
		(—)	22 months	M	53	—		
Cerebrum, indusium griseum, oligodendrocytes	Mouse	(—)	6 months	M	12.2	—	No. per 100 neurons; est. from data given	75
		(—)	9 months	M	15	—		
		(—)	12 months	M	14	—		
		(—)	15 months	M	13	—		
		(—)	18 months	M	11.7	—		
		(—)	22 months	M	11.7	—		
Cerebrum, indusium griseum, microglia	Mouse	(—)	6 months	M	11	—	No. per 100 neurons; est. from data given	75
		(—)	9 months	M	12	—		
		(—)	12 months	M	12.4	—		
		(—)	15 months	M	12.5	—		
		(—)	18 months	M	13.5	—		
		(—)	22 months	M	20	—		

Table 2 (continued)
NUMBER OF CELLS OR FIBERS VS. AGE IN THE NERVOUS SYSTEM

Subjects					Cells or fibers			
Tissue	Type	(no.)	Age	Sex	Total or mean	Range	Measure	Ref.
Cerebrum, anterior limb of anterior commissure, astroyctes	Mouse	(—)	6 months	M	2.9×10^3	—	Total no.; est. from data given	75
		(—)	9 months	M	2.5×10^3	—		
		(—)	12 months	M	2.0×10^3	—		
		(—)	15 months	M	1.6×10^3	—		
		(—)	18 months	M	1.2×10^3	—		
		(—)	22 months	M	1.4×10^3	—		
Cerebrum, anterior limb of anterior commissure, oligodendrocytes	Mouse	(—)	6 months	M	17.9×10^3	—	Total no.; est. from data given	75
		(—)	9 months	M	19.1×10^3	—		
		(—)	12 months	M	17.4×10^3	—		
		(—)	15 months	M	14.2×10^3	—		
		(—)	18 months	M	12.5×10^3	—		
		(—)	22 months	M	14.2×10^3	—		
Cerebrum, anterior limb of anterior commissure, microglia	Mouse	(—)	6 months	M	1.4×10^3	—	Total no.; est. from data given	75
		(—)	9 months	M	1.5×10^3	—		
		(—)	12 months	M	1.2×10^3	—		
		(—)	15 months	M	1.1×10^3	—		
		(—)	18 months	M	0.8×10^3	—		
		(—)	22 months	M	1.4×10^3	—		
Cerebrum, anterior limb of anterior commissure, glioblasts	Mouse	(—)	6 months	M	0.5×10^3	—	Total no.; est. from data given	75
		(—)	9 months	M	0.5×10^3	—		
		(—)	12 months	M	0.25×10^3	—		
		(—)	15 months	M	0.35×10^3	—		
		(—)	18 months	M	0.35×10^3	—		
		(—)	22 months	M	0.25×10^3	—		
Cerebrum, posterior limb of anterior commissure, glioblasts	Mouse	(—)	6 months	M	0.45×10^3	—	Total no.; est. from data given	75
		(—)	9 months	M	0.25×10^3	—		
		(—)	12 months	M	0.20×10^3	—		
		(—)	15 months	M	0.25×10^3	—		
		(—)	18 months	M	0.25×10^3	—		
		(—)	22 months	M	0.12×10^3	—		

Region, cells	Species	(n)	Age	Sex	Value		Notes	Ref.
Cerebrum, posterior limb of anterior commissure, astrocytes	Mouse	(—)	6 months	M	2.4×10^3	—	Total no.; est. from data given	75
		(—)	9 months	M	1.9×10^3	—		
		(—)	12 months	M	1.1×10^3	—		
		(—)	15 months	M	0.9×10^3	—		
		(—)	18 months	M	0.8×10^3	—		
		(—)	22 months	M	0.9×10^3	—		
Cerebrum, posterior limb of anterior commissure, oligodendrocytes	Mouse	(—)	6 months	M	10.9×10^3	—	Total no.; est. from data given	75
		(—)	9 months	M	13.6×10^3	—		
		(—)	12 months	M	10.5×10^3	—		
		(—)	15 months	M	9.8×10^3	—		
		(—)	18 months	M	7.2×10^3	—		
		(—)	22 months	M	8.6×10^3	—		
Cerebrum, posterior limb of anterior commissure, microglia	Mouse	(—)	6 months	M	1.05×10^3	—	Total no.; est. from data given	75
		(—)	9 months	M	1.12×10^3	—		
		(—)	12 months	M	0.5×10^3	—		
		(—)	15 months	M	0.5×10^3	—		
		(—)	18 months	M	0.5×10^3	—		
		(—)	22 months	M	0.8×10^3			
Cerebral hemisphere, glial cells	Rat	(16)	148—156 days	M,F	$106.2 \times 10^6 \pm 3.5 \times 10^6$	—	Total no.	51
		(12)	806—878 days	M,F	$105.2 \times 10^6 \pm 3.3 \times 10^6$	—		
Cerebral cortex, neuroglia	Rat	(—)	10 days	—	30,000	—	No. per mm³; 4—6 animals at each age; fixed tissue	42
		(—)	50 days	—	50,000			
		(—)	100 days	—	50,000			
		(—)	2 years	—	85,000			
Cerebral cortex, glia	Rat	(4)	3 months	F	35.23 ± 1.81	—	Total no. per core sample	44
		(4)	22—27 months	F	32.70 ± 1.76			
Cerebral cortex, astrocytes	Rat	(4)	3 months	F	13.38 ± 1.41	—	Total no. per core sample	44
		(4)	22—27 months	F	12.55 ± 0.67			
Cerebral cortex, microglia	Rat	(6)	109—113 days	M	6,700	—	Cells per mm³ per 20 depth levels	43
		(6)	763—972 days	M	6,700			
Cerebral cortex, microglia and oligodendrocytes	Rat	(4)	3 months	F	21.85 ± 1.61	—	Total no. per core sample	44
		(4)	22—27 months	F	20.15 ± 1.88	—		

Table 2 (continued)
NUMBER OF CELLS OR FIBERS VS. AGE IN THE NERVOUS SYSTEM

Subjects					Cells or fibers			Ref.
Tissue	Type	(no.)	Age	Sex	Total or mean	Range	Measure	
Cerebral cortex, oligodendroglia plus astrocytes	Rat	(6) (6)	109—113 days 763—972	M M	42,000 53,000	— —	Cells per mm^3 per 20 depth levels	43
Cerebral cortex, right frontal lobe, lateral gyrus, glia	Monkey (rhesus)	(10) (10)	4—7 years 18—28 years	M,F M,F	13 19	— —	No. in total depth of cortex; est. from data given	46
Lissencephalic cortex, glia	Rat	(4) (9)	6 months 25 months	F F	360 488	— —	No. per 7000 counts	52
Lissencephalic cortex, astrocytes	Rat	(4) (9)	6 months 25 months	F F	193 258	— —	No. per 7000 counts	52
Lissencephalic cortex, oligodendrocytes	Rat	(4) (9)	6 months 25 months	F F	167 231	— —	No. per 7000 counts	52
Precentral primary motor cortex glial cells	Human	(5) (5)	49—59 years 64—70 years	— —	0.49 0.64	0.38—0.58 0.52—0.76	Av. no. glial cells per large neuron	76
Precentral primary motor cortex, glial cells	Rat	(3) (2) (3)	17 days 210 days 720 days	— — —	1.06 0.72 0.50	1.02—1.12 0.68—0.76 0.34—0.62	Av. no. glial cells per large neuron	76
Precentral primary motor cortex, glial cells	Mouse	(2) (2) (6)	21 days 129 days 200—700 days	— — —	0.63 0.70 0.47	0.56—0.70 0.64—0.76 0.32—0.56	Av. no. glial cells per large neuron	76

Region / cell type	Species	Age	(n)	Sex	Value	Range	Description	Ref
Occipital cortex, medial portion, astrocytes plus oligodendrocytes	Rat	26 days	(15)	M	465 ± 100	—	Total no. medial portion of section B	55
		41 days	(13)	M	582 ± 138	—		
		108 days	(13)	M	382 ± 91	—		
		650 days	(13)	M	367 ± 52	—		
Auditory cerebral cortex, astrocytes	Rat	3 months	(2)	M	1.06 ± 0.27	—	No. per 55 μm × 270-μm field; 39—43 fields counted per animal	77
		12 months	(2)	M	0.99 ± 0.19	—		
		24 months	(2)	M	0.97 ± 0.17	—		
		27 months	(4)	M	0.98 ± 0.16	—		
Auditory cerebral cortex, oligodendrocytes	Rat	3 months	(2)	M	0.61 ± 0.12	—	No. per 55 μm × 270-μm field; 39—43 fields counted per animal	77
		12 months	(2)	M	0.67 ± 0.14	—		
		24 months	(2)	M	0.64 ± 0.13	—		
		27 months	(4)	M	0.68 ± 0.16	—		
Auditory cerebral cortex, microglia	Rat	3 months	(2)	M	0.38 ± 0.12	—	No. per 55 μm × 270-μm field; 39—43 fields counted per animal	77
		12 months	(2)	M	0.39 ± 0.14	—		
		24 months	(2)	M	0.53 ± 0.12	—		
		27 months	(4)	M	0.63 ± 0.11	—		
Visual cortex, area 17, glia	Rat	11 months	(5)	M	52.6×10^3	—	No. per mm^3 per 20 μm; est. from data given	47,48
		17 months	(5)	M	35.8×10^3	—		
		29 months	(5)	M	35.3×10^3	—		
Visual cortex, macular projection area, glia	Human	20—29 years	(5)	M,F	6.234×10^7	$5.247 \times 10^7 - 7.218 \times 10^7$	Total no. per g of tissue	50
		32—41 years	(3)	M,F	7.044×10^7	$6.355 \times 10^7 - 7.490 \times 10^7$		
		50—56 years	(4)	M,F	8.380×10^7	$7.361 \times 10^7 - 10.104 \times 10^7$		
		61—69 years	(5)	M,F	9.098×10^7	$7.053 \times 10^7 - 11.520 \times 10^7$		
		73—76 years	(3)	M,F	7.717×10^7	$6.510 \times 10^7 - 9.030 \times 10^7$		
		81—87 years	(3)	M,F	7.959×10^7	$7.684 \times 10^7 - 8.341 \times 10^7$		
Hippocampus CA1 zone, glia	Rat	11 months	(5)	M	11.5	—	No. in a 110-μm segment; est. from data given	47,48
		17 months	(5)	M	11.2	—		
		29 months	(5)	M	8.8	—		
Hippocampus, dentate gyrus, total molecular layer, astrocytes	Rat	3 months	(5)	M	0.182 ± 0.010	0..160—0.209	No. per 1000 μm^2	78
		25 months	(5)	M	0.183 ± 0.007	0.163—0.206		

Table 2 (continued)
NUMBER OF CELLS OR FIBERS VS. AGE IN THE NERVOUS SYSTEM

Tissue	Type	(no.)	Age	Sex	Total or mean	Range	Measure	Ref.
			Subjects			**Cells or fibers**		
Hippocampus, dentate gyrus, supragranular zone of molecular layer, astrocytes	Rat	(5)	3 months	M	0.120 ± 0.011	0.095—0.156	No. per 1000 μm^2	78
		(5)	25 months	M	0.132 ± 0.019	0.056—0.162		
Hippocampus, midpolar region, glia	Rat	(6)	4 months	M	20.7 ± 2.5	—	No. per grid square	79
		(9)	13 months	M	20.1 ± 1.3	—		
		(6)	25 months	M	23.7 ± 2.2	—		
Hippocampus, midpolar region, hypertrophied astrocytes	Rat	(6)	4 months	M	1.21 ± 0.25	—	No. per grid square	79
		(9)	13 months	M	3.19 ± 0.45	—		
		(6)	25 months	M	7.02 ± 1.1	—		
Cerebellum, glial cells	Rat	(16)	148—156 days	M,F	$147.8 \times 10^6 \pm 5.2 \times 10^6$	—	Total no.	51
		(12)	806—878 days	M,F	$143.1 \times 10^6 \pm 4.9 \times 10^6$	—		
Brain stem, glial cells	Rat	(16)	148—156 days	M,F	$71.2 \times 10^6 \pm 1.4 \times 10^6$	—	Total no.	51
		(12)	806—878 days	M,F	$73.2 \times 10^6 \pm 1.5 \times 10^6$	—		
					Muscle, Motor Nerves			
Soleus	Rat	(—)	4 months	—	120 ± 3.80	—	Total no. fibers innervating muscle	80
		(—)	24 months	—	115 ± 2.31	—		

Note: est., estimated.

[a] A, old with prolonged vaginal cornification; B, old with continuous diestrous with atrophic ovaries.

Table 3
NUMBER OF CELLS VS. AGE IN THE EYE

Cell type	Subjects				Count (total or mean)	Measure	Ref.
	Type	no.	Age	Sex			
Whole trabecular meshwork, cells	Human	(—)	Birth	M,F	166	No. per section; calculated from regression equation given; total of 34 subjects; presence of tumors did not affect	81
		(—)	20 years	M,F	148		
		(—)	30 years	M,F	140		
		(—)	40 years	M,F	131		
		(—)	50 years	M,F	122		
		(—)	60 years	M,F	113		
		(—)	70 years	M,F	104		
		(—)	80 years	M,F	96		
Lens epithelial cells	Rat	(6)	19 weeks	M	0.83×10^6	Total no.; estimated from data given	82
		(6)	25 weeks	M	0.94×10^6		
		(6)	59 weeks	M	1.06×10^6		
		(5)	73 weeks	M	1.10×10^6		
		(2)	172 weeks	M	1.25×10^6		
Exorbital lacrimal gland, mast cells	Rat	(8)	15 days	M	278	Av. density per mm^2 of surface area	201
		(7)	30 days	M	104		
		(7)	60 days	M	155		
		(5)	90 days	M	134		
		(7)	150 days	M	108		
		(7)	300 days	M	123		
		(7)	500	M	107		

Table 4
NUMBER OF CELLS VS. AGE IN SKIN AND BONE

Tissue	Type	(no.)	Age	Sex	Mean	Range	Measure	Ref.
					Cells			
				Skin				
Connective tissue, abdominal corium layer	Human	(11)	<1 year	M	90.90 ± 4.86	—	No. nuclei in 34,810 μm² area; Caucasian subjects	83
		(10)	1—10 years	M	83.50 ± 4.01	—		
		(10)	11—20 years	M	76.70 ± 2.42	—		
		(11)	21—30 years	M	69.18 ± 2.39	—		
		(12)	31—40 years	M	58.66 ± 4.39	—		
		(11)	41—50 years	M	50.90 ± 4.48	—		
		(10)	51—60 years	M	54.20 ± 4.08	—		
		(24)	61—70 years	M	45.50 ± 1.79	—		
		(25)	71—80 years	M	50.72 ± 1.86	—		
		(20)	81+ years	M	46.15 ± 1.60	—		
Connective tissue, abdominal corium layer	Human	(7)	<1 year	M	109.14 ± 7.09	—	No. nuclei in 34,810 μm² area; Japanese subjects	83
		(7)	1—10 years	M	85.85 ± 5.10	—		
		(15)	11—20 years	M	72.40 ± 3.33	—		
		(17)	21—30 years	M	64.23 ± 2.68	—		
		(23)	31—40 years	M	63.08 ± 2.57	—		
		(20)	41—50 years	M	64.30 ± 2.44	—		
		(35)	51—60 years	M	58.17 ± 1.69	—		
		(29)	61—70 years	M	55.48 ± 1.64	—		
		(16)	71—80 years	M	42.18 ± 1.20	—		
		(3)	81+ years	M	45.33 ± 1.35	—		
Connective tissue, abdominal corium layer	Human	(8)	<1 year	F	106.59 ± 4.95	—	No. nuclei in 34,810 μm² area; Japanese subjects	83
		(10)	1—10 years	F	73	—		
		(7)	11—20 years	F	66.42 ± 4.53	—		
		(10)	21—30 years	F	61.10 ± 2.44	—		
		(16)	31—40 years	F	62.93 ± 2.55	—		
		(14)	41—50 years	F	66.71 ± 3.12	—		
		(12)	51—60 years	F	61.00 ± 2.22	—		

Cell type, location	Animal	(n)	Age	Sex	Value	Range	Units	Ref.
Lymphocytes, stratum germinativum from midline of abdomen	Rat	(16)	61—70 years	F	55.68 ± 2.16	—	No. per 1000 cells	84
		(13)	71—80 years	F	50.07 ± 2.30	—		
		(6)	81+ years	F	39.33 ± 1.41	—		
		(9)	21 days	M,F	4.33 ± 0.73	1—12		
		(6)	300 days	M,F	18.16 ± 1.94	9—29		
		(11)	900 days	M,F	30.27 ± 6.45	4—111		
Lymphocytes, stratum germinitivum from interscapular region	Rat	(9)	21 days	M,F	23.7 ± 1.43	13—38	No. per 1000 cells	85
		(6)	300 days	M,F	33.67 ± 2.55	24—48		
		(2)	600—650 days	M	—	28—33		
		(11)	900 days	M,F	24.8 ± 2.72	13—65		
Mast cells, dorsal dermis	Mouse	(5)	2—3 months	M	42.4	22—65	Av. total no. in ten 0.25-mm wide strips (full depth of dermis); C57BL strain	86
		(4)	18 months	M	144.8	70—179		
Mast cells, dorsal dermis	Mouse	(3)	2—3 months	F	101.1	78—135	Av. total no. in ten 0.25-mm wide strips (full depth of dermis); C57BL strain	86
		(3)	18 months	F	77.3	63—89		
Mast cells, dorsal dermis	Mouse	(4)	2—3 months	M	39.5	15—54	Av. total no. in ten 0.25-mm wide strips (full depth of dermis); BALB/c strain	86
		(4)	18 months	M	42.3	37—46		

Table 4 (continued)
NUMBER OF CELLS VS. AGE IN SKIN AND BONE

Tissue	Type	(no.)	Age	Sex	Mean	Range	Measure	Ref.
			Subjects		**Cells**			
Mast cells, dorsal dermis	Mouse	(4)	2—3 months	F	65.5	57—75	Av. total no. in ten 0.25-mm wide strips (full depth of dermis); BALB/c strain	86
		(4)	18 months	F	62.3	58—71		
Dark keratinocytes, dorsal epidermis	Mouse	(3)	1 month	F	2.5 ± 1.3	—	Percent of total no. basal layer cells	87
		(3)	6 months	F	1.6 ± 1.0	—		
		(3)	15 months	F	0.2 ± 0.2	—		
		(3)	23 months	F	0.2 ± 0.2	—		
Melanocytes, forehead	Human	(8)	16—70 years	—	$2,010 \pm 20$	—	No. per mm^2 skin	88
		(4)	70+ years	—	$1,145 \pm 85$	—		
Melanocytes, thigh	Human	(34)	16—70 years	—	$1,000 \pm 70$	—	No. per mm^2 skin	88
		(4)	70+ years	—	560 ± 70	—		
Melanocytes (dopa-positive), anterior abdominal wall	Human	(28)	0—12 years	M	1,380	740—2,200	No. per mm^2 skin; Caucasian subjects	89
		(11)	20—59 years	M	813	618—1,274		
		(7)	60 and over	M	759	385—1,173		
Melanocytes (dopa-positive), anterior abdominal wall	Human	(9)	0—12 years	F	1,445	995—2,137	No. per mm^2 skin; non-pregnant Caucasian subjects	89
		(13)	20—59 years	F	741	452—1,068		
		(7)	60 and over	F	617	475—1,050		
Melanocytes (dopa positive), buttock	Human	(4)	27—39 years	M	1,289	950—1,630	No. per mm^2 skin; Caucasian subjects	90
		(4)	40—47 years	M	1,024	875—1,215		
		(4)	53—65 years	M	715	655—760		

Bone

Material	Source	(n)	Age	Sex	Value	Ash weight, % of fat-free dry weight	Notes	Ref
Total skeleton	Human	(3)	45—60 years	M	65.8	65.2—66.9	Ash weight, % of fat-free dry weight	92
		(3)	72—89 years	M	65.1	64.8—65.6		
Whole skeleton except patellae	Human	(1)	16 weeks, fetus	M	3.4	—	Grams; dry, fat free; white males	93
		(11)	22—28 weeks, fetus	M	20.1	—		
		(13)	29—36 weeks, fetus	M	44.6	—		
		(4)	37—44 weeks, fetus	M	108.3	—		
		(7)	Newborn-0.5 years	M	95.8	—		
		(5)	>0.5—3 years	M	315.8	—		
		(7)	5—12 years	M	932.6	—		
		(9)	17—22 years	M	4,004.4	—		
		(1)	30 years	M	5,415.2	—		
		(11)	45—64 years	M	3,538.4	—		
		(18)	65—86 years	M	3,184.7	—		
Whole skeleton except patellae	Human	(2)	19 weeks, fetus	F	7.7	—	Grams; dry, fat free; white females	93
		(6)	21—28 weeks, fetus	F	19.5	—		
		(8)	29—36 weeks, fetus	F	67.1	—		
		(15)	37—42 weeks, fetus	F	85.1	—		
		(3)	Newborn-0.5 years	f	71.0	—		
		(4)	>0.5—3 years	F	286.4	—		
		(10)	4—13 years	F	1,132.4	—		
		(3)	17—22 years	F	2,724.3	—		
		(2)	25—44 years	F	3,084.4	—		
		(13)	45—64 years	F	2,336.2	—		
		(15)	65—90 years	F	2,130.0	—		
Whole skeleton except patellae	Human	(2)	17—20 weeks, fetus	M	9.3	—	Grams; dry, fat free; Negro males	93
		(9)	23—27 weeks, fetus	M	20.4	—		
		(13)	29—36 weeks, fetus	M	56.6	—		
		(9)	37—44 weeks, fetus	M	82.5	—		
		(4)	Newborn-0.5 years	M	110.0	—		
		(13)	>0.5—3 years	M	260.2	—		
		(9)	6—11 years	M	1,456.6	—		
		(29)	14—21 years	M	4,228.5	—		
		(2)	25—44 years	M	3,910.1	—		
		(15)	45—64 years	M	3,973.2	—		
		(13)	65—80 years	M	3,852.4	—		

Table 4 (continued)
NUMBER OF CELLS VS. AGE IN SKIN AND BONE

Tissue	Subjects				Cells			
	Type	(no.)	Age	Sex	Mean	Range	Measure	Ref.
Whole skeleton except patellae	Human	(1)	16 weeks, fetus	F	7.6	—	Grams; dry, fat free; Negro females	93
		(4)	21—28 weeks, fetus	F	22.5	—		
		(8)	30—36 weeks, fetus	F	61.7	—		
		(11)	37—44 weeks, fetus	F	89.3	—		
		(4)	Newborn-0.5 months	F	102.3	—		
		(7)	>0.5—3 years	F	382.3	—		
		(11)	4—13 years	F	1,305.6	—		
		(19)	14—22 years	F	3,256.2	—		
		(9)	30—44 years	F	3,127.2	—		
		(7)	50—64 years	F	3,005.2	—		
		(14)	65—95 years	F	2,620.5	—		
Femur	Human	(14)	41—56 years	M	64.8	62.3—66.6	Ash weight, percent of fat-free dry weight	92
		(13)	62—89 years	M	64.0	58.4—67.2		
Femur	Human	(286)	45—49 years	F	18.67 ± 17	—	Cortical thickness, mm	94
		(303)	50—54 years	F	18.69 ± 0.14	—		
		(501)	55—59 years	F	18.17 ± 0.12	—		
		(424)	60—64 years	F	17.95 ± 0.14	—		
		(291)	65—69 years	F	18.08 ± 0.16	—		
		(162)	70—74 years	F	17.32 ± 0.22	—		
		(63)	75—90 years	F	17.68 ± 0.37	—		
Femur	Human	(11)	20—40 years	M	(a)0.706 ± 0.067	—	Mean cortical thickness (cm) left femur, at (a) 26% of total femoral	95
					(b)0.761 ± 0.076	—		
					(c)0.718 ± 0.054	—		
					(d)0.586 ± 0.044	—		
					(e)0.426 ± 0.037	—		
		(8)	41—55 + years	M	(a)0.643 ± 0.073	—		

Bone	Species	(N)	Age	Sex	Value	Measurement	Ref
Femur	Human	(11)	20—40 years	F	(b)0.738 ± 0.080 (c)0.660 ± 0.083 (d)0.513 ± 0.075 (e)0.329 ± 0.048 (a)0.597 ± 0.061 (b)0.641 ± 0.054 (c)0.587 ± 0.057 (d)0.463 ± 0.055 (e)0.327 ± 0.040	length, (b) 38%, (c)50%, (d)62% (e)74% Mean cortical thickness (cm) left femur, at (a)26% of total femoral length, (b)38%, (c)50%, (d)62%, (e)74%	95
		(10)	41—55 + years	F	(a)0.475 ± 0.074 (b)0.479 ± 0.085 (c)0.465 ± 0.077 (d)0.387 ± 0.067 (e)0.255 ± 0.065		
Femur	Human	(4) (11) (21) (14)	50—59 years 60—69 years 70—79 years 80—89 years	M M M M	5.53 ± 1.17 5.09 ± 1.15 5.01 ± 2.97 4.52 ± 0.97	Cortical thickness (mm) from bone core of anterior midshaft; right leg	96
Femur	Human	(7) (5) (8) (12)	50—59 years 60—69 years 70—79 years 80—90 years	F F F F	4.84 ± 1.55 3.35 ± 0.69 3.13 ± 0.77 3.39 ± 0.94	Cortical thickness (mm) from bone core of anterior midshaft; right leg	96
Femur	Human	(4) (11) (21) (14)	50—59 years 60—69 years 70—79 years 80—89 years	M M M M	1.868 ± 0.056 1.818 ± 0.048 1.844 ± 0.099 1.830 ± 0.056	Cortical bone density (g/cm^3) from bone core of anterior midshaft; right leg	96

Table 4 (continued)
NUMBER OF CELLS VS. AGE IN SKIN AND BONE

	Subjects				Cells			
Tissue	Type	(no.)	Age	Sex	Mean	Range	Measure	Ref.
Femur	Human	(7)	50—59 years	F	1.873 ± 0.078	—	Cortical bone density (g/cm³) from bone core of anterior midshaft; right leg	96
		(5)	60—69 years	F	1.842 ± 0.097	—		
		(8)	70—79 years	F	1.795 ± 0.093	—		
		(12)	80—89 years	F	1.803 ± 0.132	—		
Pelvis, iliac crest	Human	(16)	20—30 years	M,F	22.8 ± 4.29	—	Volumetric density, percent trabecular volume to total volume of cancellous bone	97
		(19)	30—40 years	M,F	22.1 ± 4.05	—		
		(17)	40—50 years	M,F	21.0 ± 4.01	—		
		(23)	50—60 years	M,F	19.6 ± 4.15	—		
		(20)	60—70 years	M,F	19.1 ± 4.9	—		
		(18)	70—80 years	M,F	17.1 ± 3.13	—		
Tibia	Human	(14)	41—56 years	M	65.1	62.9—66.5	Ash weight, % of fat-free dry weight	92
		(14)	62—89 years	M	64.3	60.8—67.4		
Fibula	Human	(13)	41—56 years	M	64.8	57.6—66.9	Ash weight, % of fat-free dry weight	92
		(13)	62—89 years	M	62.8	54.0—67.0		
Humerus	Human	(14)	41—56 years	M	65.0	62.4—68.1	Ash weight % of fat-free dry weight	92
		(14)	62—89	M	64.1	58.7—68.0		
Humerus, midshaft	Human	(14)	6 years	M	1,009 ± 128	—	Mineral content in mg/cm; U.S. white subjects	98
		(27)	7 years	M	1,112 ± 168	—		
		(39)	8 years	M	1,268 ± 137	—		
		(42)	9 years	M	1,311 ± 155	—		
		(47)	10 years	M	1,470 ± 178	—		
		(44)	11 years	M	1,531 ± 198	—		

				Mineral content in mg/cm; U.S. white subjects
Humerus, midshaft	Human	(37)	12 years	M	1,684 ± 191
		(34)	13 years	M	1,786 ± 264
		(28)	14 years	M	2,070 ± 312
		(43)	15 years	M	2,311 ± 299
		(30)	16 years	M	2,672 ± 400
		(35)	17 years	M	2,754 ± 372
		(30)	18 years	M	2,948 ± 363
		(16)	19 years	M	3,161 ± 258
		(100)	20—29 years	M	2,766 ± 333
		(118)	30—39 years	M	2,732 ± 383
		(118)	40—49 years	M	2,731 ± 340
		(49)	50—59 years	M	2,779 ± 335
		(28)	60—69 years	M	2,579 ± 408
		(16)	70—79 years	M	2,622 ± 415
		(18)	80—89 years	M	2,287 ± 456
		(11)	6 years	F	943 ± 107
		(20)	7 years	F	1,004 ± 121
		(15)	8 years	F	1,125 ± 165
		(22)	9 years	F	1,215 ± 137
		(24)	10 years	F	1,181 ± 203
		(20)	11 years	F	1,368 ± 247
		(13)	12 years	F	1,533 ± 238
		(15)	13 years	F	1,661 ± 280
		(14)	14 years	F	1,818 ± 154
		(24)	15 years	F	1,896 ± 174
		(10)	16 years	F	1,898 ± 164
		(18)	17 years	F	1,948 ± 265
		(6)	18 years	F	2,102 ± 292
		(3)	19 years	F	1,940 ± 236
		(31)	20—29 years	F	2,098 ± 243
		(11)	30—39 years	F	2,117 ± 260
		(19)	40—49 years	F	2,081 ± 240
		(12)	50—59 years	F	1,811 ± 307
		(22)	60—69 years	F	1,539 ± 464
		(47)	70—79 years	F	1,384 ± 300
		(39)	80—89 years	F	1,296 ± 211
		(10)	90—99 years	F	1,271 ± 227

98

Table 4 (continued)
NUMBER OF CELLS VS. AGE IN SKIN AND BONE

Subjects					Cells			
Tissue	Type	(no.)	Age	Sex	Range	Mean	Measure	Ref.
Radius	Human	(14)	41—56 years	M	63.2—68.0	65.6	Ash weight, % of fat-free dry weight	92
		(14)	62—89 years	M	60.3—69.5	64.4		
Radius, distal portion	Human	(10)	20—29 years	M	—	1.500 ± 0.014	Bone density, g/cm³	99
		(10)	30—49 years	M	—	1.530 ± 0.018		
		(17)	50—59 years	M	—	1.512 ± 0.022		
		(27)	60—69 years	M	—	1.509 ± 0.019		
		(34)	70—80 years	M	—	1.485 ± 0.018		
Radius, distal portion	Human	(12)	20—29 years	F	—	1.430 ± 0.016	Bone density, g/cm³	99
		(12)	30—49 years	F	—	1.440 ± 0.020		
		(39)	50—59 years	F	—	1.420 ± 0.016		
		(31)	60—69 years	F	—	1.395 ± 0.016		
		(34)	70—80 years	F	—	1.356 ± 0.012		

Mineral content rows (no. and Mean given for (a) white race, (b) black, (c) yellow):

Subjects							Cells				Measure	Ref.
Tissue	Type	(a)	(b)	(c)	Age	Sex	(a) Mean	(b) Mean	(c) Mean		Measure	Ref.
Radius, distal portion	Human	(31)	(4)	(—)	10—19 years	M	1.129 ± 0.241	1.182 ± 0.240	—		Mineral content in g/cm; (a) white race, (b) black, (c) yellow	100
		(165)	(39)	(5)	20—29 years	M	1.284 ± 0.195	1.442 ± 0.233	1.132 ± 0.170			
		(243)	(58)	(22)	30—39 years	M	1.329 ± 0.238	1.456 ± 0.172	1.267 ± 0.181			
		(254)	(64)	(25)	40—49 years	M	1.309 ± 0.207	1.370 ± 0.239	1.226 ± 0.195			
		(245)	(49)	(12)	50—59 years	M	1.290 ± 0.231	1.455 ± 0.233	1.194 ± 0.155			
		(111)	(8)	(3)	60—69 years	M	1.241 ± 0.237	1.335 ± 0.216	1.170 ± 0.148			
		(38)	(1)	(1)	70—79 years	M	1.120 ± 0.260	1.13	1.08			
Radius, distal portion	Human	(24)	(11)	(2)	10—19 years	F	1.008 ± 0.180	0.961 ± 0.186	0.970 ± 0.085		Mineral content in g/cm; (a) white race (b) black, (c) yellow	100
		(313)	(102)	(17)	20—29 years	F	0.911 ± 0.134	1.015 ± 0.156	0.927 ± 0.217			
		(281)	(99)	(23)	30—39 years	F	0.946 ± 0.173	1.036 ± 0.167	0.938 ± 0.119			
		(402)	(135)	(34)	40—49 years	F	0.948 ± 0.154	1.029 ± 0.192	0.880 ± 0.169			
		(344)	(73)	(11)	50—59 years	F	0.886 ± 0.186	0.955 ± 0.114	0.982 ± 0.135			
		(194)	(20)	(1)	60—69 years	F	0.781 ± 0.171	0.762 ± 0.179	0.74			
		(48)	(—)	(1)	70—79 years	F	0.720 ± 0.162	—	0.68			
		(2)	(—)	(—)	80—89 years	F	0.630 ± 0.170	—	—			

		(a)	(b)			(a)	(b)	Mineral content in mg/cm; (a) distal third of shaft, (b) distal radius; U.S. white subjects	98
Radius	Human	(16)	(—)	6 years	M	472 ± 59	—		
		(27)	(—)	7 years	M	509 ± 73	—		
		(38)	(—)	8 years	M	565 ± 71	—		
		(39)	(—)	9 years	M	592 ± 70	—		
		(52)	(—)	10 years	M	640 ± 78	—		
		(43)	(—)	11 years	M	702 ± 107	—		
		(39)	(—)	12 years	M	746 ± 95	—		
		(39)	(—)	13 years	M	813 ± 112	—		
		(35)	(—)	14 years	M	898 ± 126	—		
		(43)	(—)	15 years	M	1,048 ± 146	—		
		(36)	(—)	16 years	M	1,154 ± 143	—		
		(39)	(—)	17 years	M	1,196 ± 130	—		
		(31)	(—)	18 years	M	1,247 ± 116	—		
		(19)	(—)	19 years	M	1,296 ± 187	—		
		(105)	(27)	20—29 years	M	1,307 ± 173	1,386 ± 204		
		(72)	(19)	30—39 years	M	1,322 ± 138	1,317 ± 186		
		(64)	(12)	40—49 years	M	1,304 ± 146	1,298 ± 154		
		(31)	(10)	50—59 years	M	1,313 ± 159	1,330 ± 177		
		(46)	(30)	60—69 years	M	1,226 ± 229	1,191 ± 288		
		(22)	(14)	70—79 years	M	1,256 ± 188	1,187 ± 279		
		(16)	(14)	80—89 years	M	1,182 ± 219	1,134 ± 282		
Radius	Human	(14)	—	6 years	F	434 ± 79	—	Mineral content in mg/cm; (a) distal third of shaft, (b) distal radius; U.S. white subjects	98
		(20)	—	7 years	F	452 ± 63	—		
		(18)	—	8 years	F	485 ± 66	—		
		(20)	—	9 years	F	548 ± 70	—		
		(24)	—	10 years	F	564 ± 93	—		
		(22)	—	11 years	F	653 ± 125	—		
		(15)	—	12 years	F	745 ± 114	—		
		(18)	—	13 years	F	738 ± 93	—		
		(28)	—	14 years	F	844 ± 112	—		
		(25)	—	15 years	F	869 ± 76	—		
		(13)	—	16 years	F	882 ± 86	—		
		(22)	—	17 years	F	893 ± 99	—		
		(16)	—	18 years	F	911 ± 134	—		
		(32)	—	19 years	F	937 ± 107	—		
		(126)	(54)	20—29 years	F	952 ± 108	984 ± 126		

Table 4 (continued)
NUMBER OF CELLS VS. AGE IN SKIN AND BONE

Tissue	Type	(no.) (a)	(no.) (b)	Age	Sex	Mean (a)	Mean (b)	Range	Measure	Ref.
		(29)	(6)	30—39 years	F	1000 ± 121	955 ± 84	—		
		(42)	(11)	40—49 years	F	977 ± 95	933 ± 174	—		
		(63)	(37)	50—59 years	F	885 ± 136	886 ± 146	—		
		(93)	(71)	60—69 years	F	769 ± 142	742 ± 116	—		
		(134)	(96)	70—79 years	F	722 ± 126	718 ± 175	—		
		(121)	(87)	80—89 years	F	681 ± 143	645 ± 144	—		
		(29)	(18)	90—99 years	F	675 ± 129	668 ± 173	—		
Ulna	Human	(14)		41—56 years	M	64.7		62.1—66.7	Ash weight, % of fat-free dry weight	92
		(14)		62—89	M	63.9		52.2—68.5		
Ulna	Human	(22)	(20)	20—29 years	F	843 ± 121	511 ± 154	—	Mineral content in mg/cm; (a) mid-shaft (b) distal ulna; U.S. white subjects	98
		(6)	(6)	30—49 years	F	868 ± 95	510 ± 86	—		
		(18)	(17)	50—59 years	F	804 ± 126	452 ± 89	—		
		(24)	(22)	60—69 years	F	702 ± 106	364 ± 60	—		
		(38)	(38)	70—79 years	F	615 ± 121	330 ± 96	—		
		(30)	(29)	80—89 years	F	590 ± 116	303 ± 78	—		
		(7)	(7)	90—99 years	F	560 ± 86	287 ± 113	—		
Rib	Human	(8)		20—29 years	M	26.5		—	Cortical area, mm²	102
		(6)		30—39 years	M	23.3		—		
		(8)		40—49 years	M	22.1		—		
		(17)		50—59 years	M	21.2		—		
		(12)		60—69 years	M	20.3		—		
		(10)		70—84 years	M	21.0		—		
Rib	Human	(6)		20—29 years	F	24.0		—	Cortical area, mm²	102
		(14)		30—39 years	F	19.4		—		
		(12)		40—49 years	F	18.7		—		
		(12)		50—59 years	F	16.9		—		
		(7)		60—84 years	F	18.8		—		

		(a)	(b)	(c)		Sex	(a)	(b)	(c)		Description	Ref.
Second metacarpal	Human	(62)	(89)	(57) 30 years		M	5.9	5.45	5.34	—	Cortical thickness, mm; subjects from (a) U.S., (b) Guatemala, (c) El Salvador	102
		(92)	(92)	(46) 40 years		M	5.8	5.19	5.40	—		
		(60)	(62)	(22) 50 years		M	5.7	5.38	5.51	—		
		(35)	(42)	(29) 60 years		M	5.3	5.12	5.11	—		
		(23)	(24)	(14) 70 years		M	5.0	4.72	4.80	—		
		(12)	(11)	(11) 80 years		M	4.9	4.80	4.44	—		
Second metacarpal	Human	(153)	(159)	(101) 30 years		F	5.4	5.03	5.19	—	Cortical thickness, mm; subjects from (a) U.S., (b) Guatemala, (c) El Salvador	102
		(85)	(137)	(80) 40 years		F	5.5	4.77	5.05	—		
		(61)	(101)	(71) 50 years		F	5.2	4.73	4.87	—		
		(40)	(51)	(48) 60 years		F	4.6	3.91	4.02	—		
		(32)	(38)	(29) 70 years		F	3.9	3.73	3.74	—		
		(22)	(17)	(8) 80 years		F	3.3	3.22	3.11	—		
Third metacarpal	Human	(10)		20—29 years		M	0.799 ± 0.032			—	Cortical index: ratio of cortical area to bone area at center of metacarpal shaft	99
		(10)		30—49 years		M	0.725 ± 0.026			—		
		(17)		50—59 years		M	0.763 ± 0.022			—		
		(27)		60—69 years		M	0.714 ± 0.019			—		
		(34)		70—80 years		M	0.703 ± 0.017			—		
Third metacarpal	Human	(12)		20—29 years		F	0.719 ± 0.032			—	Cortical index: ratio of cortical area to bone area at center of metacarpal shaft	99
		(12)		30—49 years		F	0.825 ± 0.027			—		
		(39)		50—59 years		F	0.718 ± 0.013			—		
		(31)		60—69 years		F	0.677 ± 0.018			—		
		(34)		70—80 years		F	0.635 ± 0.015			—		
Skull, cranium and mandible	Human	(1)		16 weeks, fetus		M	1.2			—	Grams; dry fat-free; white males	93
		(11)		22—28 weeks, fetus		M	8.3			—		
		(13)		29—36 weeks, fetus		M	18.8			—		
		(4)		37—44 weeks, fetus		M	43.9			—		
		(7)		Newborn-0.5 years		M	48.2			—		
		(5)		>0.5—3 years		M	133.4			—		
		(7)		5—12 years		M	365.6			—		
		(9)		17—22 years		M	697.3			—		
		(1)		30 years		M	978.0			—		
		(11)		45—64 years		M	611.6			—		
		(18)		65—85 years		M	554.0			—		

Table 4 (continued)
NUMBER OF CELLS VS. AGE IN SKIN AND BONE

Tissue	Type	(no.)	Age	Sex	Mean	Range	Measure	Ref.
Skull, cranium and mandible	Human	(2)	19 weeks, fetus	F	2.8	—	Grams; dry, fat-free; white females	93
		(6)	21—28 weeks, fetus	F	8.7	—		
		(8)	29—36 weeks, fetus	F	29.4	—		
		(15)	37—42 weeks, fetus	F	35.7	—		
		(3)	Newborn-0.5 years	F	30.1	—		
		(4)	>0.5—3 years	F	129.3	—		
		(10)	4—13 years	F	312.0	—		
		(3)	17—22 years	F	524.0	—		
		(2)	25—44 years	F	514.7	—		
		(13)	45—64 years	F	505.2	—		
		(15)	65—90 years	F	519.4	—		
Skull, cranium and mandible	Human	(2)	17—20 weeks, fetus	M	3.7	—	Grams; dry, fat-free; Negro males	93
		(9)	23—27 weeks, fetus	M	8.4	—		
		(13)	29—36 weeks, fetus	M	22.8	—		
		(9)	37—44 weeks, fetus	M	35.2	—		
		(4)	Newborn-0.5 years	M	48.7	—		
		(13)	>0.5—3 years	M	125.9	—		
		(9)	6—11 years	M	438.9	—		
		(29)	14—21 years	M	690.1	—		
		(2)	25—44 years	M	575.4	—		
		(15)	45—64 years	M	685.4	—		
		(13)	65—80 years	M	701.9	—		
Skull, cranium and mandible	Human	(1)	16 weeks, fetus	F	3.1	—	Grams; dry fat-free; Negro females	93
		(4)	21—28 weeks, fetus	F	9.0	—		
		(8)	30—36 weeks, fetus	F	25.4	—		
		(11)	37—44 weeks, fetus	F	37.7	—		
		(4)	Newborn-0.5 yrs	F	45.4	—		
		(7)	>0.5—3 years	F	182.4	—		
		(11)	4—13 years	F	379.5	—		

47

Bone	Species	(n)	Age	Sex	Value	Measurement	Ref.
Mandible	Human	(19)	14—22 years	F	629.8		
		(9)	30—44 years	F	729.3		
		(7)	50—64 years	F	626.5		
		(14)	65—95 years	F	594.8		
Mandible	Human	(18)	6—8 years	M	3.1	Cortical thickness (mm) at inferior border of mandible	103
		(16)	9—11 years	M	3.3		
		(10)	12—14 years	M	4.5		
		(29)	15—24 years	M	5.6		
		(22)	25—34 years	M	5.6		
		(14)	35—44 years	M	5.6		
		(10)	45—54 years	M	5.9		
Mandible	Human	(20)	6—8 years	F	3.2	Cortical thickness (mm) at inferior border of mandible	103
		(10)	9—11 years	F	3.4		
		(14)	12—14 years	F	3.8		
		(13)	15—24 years	F	4.4		
		(11)	25—34 years	F	4.2		
		(9)	35—44 years	F	4.7		
		(8)	45—54 years	F	5.0		
		(8)	>55 years	F	4.9		
Mandible	Human	(7)	20—29 years	M	2.260 ± 0.144	Absolute bone mass in mm³ periosteal surface	104
		(6)	30—39 years	M	2.143 ± 0.265		
		(8)	40—49 years	M	2.034 ± 0.193		
		(9)	50—59 years	M	1.927 ± 0.321		
		(10)	60—69 years	M	1.631 ± 0.214		
		(5)	70—79 years	M	1.460 ± 0.216		
		(5)	80—90 years	M	1.482 ± 0.556		
Mandible	Human	(5)	20—29 years	F	2.192 ± 0.248	Absolute bone mass in mm³ periosteal surface	104
		(6)	30—39 years	F	2.015 ± 0.364		
		(9)	40—49 years	F	1.981 ± 0.352		
		(11)	50—59 years	F	1.704 ± 0.448		
		(5)	60—69 years	F	1.381 ± 0.504		
		(8)	70—79 years	F	1.071 ± 0.423		
		(6)	80—90 years	F	1.478 ± 0.498		
Left 5th mid-phalanx, midshaft	Human	(23)	30—39 years	M	1.588 ± 0.20	Bone density (X-ray mass coefficient)	13
		(35)	40—49 years	M	1.434 ± 0.23		
		(30)	50—59 years	M	1.242 ± 0.22		
		(26)	60—69 years	M	1.355 ± 0.16		

Table 4 (continued)
NUMBER OF CELLS VS. AGE IN SKIN AND BONE

Subjects					Cells			
Tissue	Type	(no.)	Age	Sex	Mean	Range	Measure	Ref.
Femur	Mouse	(21)	70—79 years	M	1.166 ± 0.18	—	Cortical thickness (mm) at midpoint	105
		(10)	6 months	F	0.288 ± 0.132	—		
		(10)	18 months	F	0.150 ± 0.030	—		
		(10)	30 months	F	0.108 ± 0.030	—		
		(10)	40 months	F	0.112 ± 0.038	—		
Tibia	Mouse	(10)	6 months	F	0.098 ± 0.018	—	Cortical thickness (mm) at a point 3 mm from proximal epiphyseal cartilage	105
		(10)	18 months	F	0.094 ± 0.016	—		
		(10)	30 months	F	0.078 ± 0.018	—		
		(10)	40 months	F	0.090 ± 0.017	—		
4th and 5th lumbar vertebrae	Mouse	(10)	6 months	F	0.086 ± 0.023	—	Cortical thickness (mm) at narrowest point	105
		(10)	18 months	F	0.035 ± 0.016	—		
		(10)	30 months	F	0.021 ± 0.007	—		
		(10)	40 months	F	0.027 ± 0.011	—		
Periosteum osteoblasts, mid-diaphyseal region of femur	Rat	(—)	1 week	M	10.0 ± 0.7	—	No. per unit area; 60 animals total	106
		(—)	5 weeks	M	10.7 ± 0.5	—		
		(—)	8 weeks	M	10.5 ± 0.3	—		
		(—)	26 weeks	M	2.9 ± 0.9	—		
		(—)	52 weeks	M	4.3 ± 0.7	—		
		(—)	104 weeks	M	0.8 ± 0.3	—		
Periosteum osteoblasts, mid-diaphyseal region of femur	Rat	(—)	1 week	F	10.4 ± 0.3	—	No. per unit area; 60 animals total	106
		(—)	5 weeks	F	10.4 ± 0.6	—		
		(—)	8 weeks	F	10.5 ± 0.3	—		
		(—)	26 weeks	F	2.8 ± 1.5	—		
		(—)	52 weeks	F	2.9 ± 0.7	—		
		(—)	104 weeks	F	1.2 ± 0.2	—		
Osteocytes, femur cortex	Mouse	(—)	7 weeks	F	12.7 ± 0.26	—	No. per 0.014 mm² area; 25	107
		(—)	28 weeks	F	12.7 ± 0.20	—		

Material	Species	Ref	Age	Sex	Value	Cells (range)	Parameter	Ref
Cartilage, femoral head	Human	(—)	54 weeks	F	11.1 ± 0.24	—	animals total	—
		(—)	106 weeks	F	10.2 ± 0.44	—	Thickness in mm	108
Cartilage, femoral head	Human	(17)	6—25 years	—	1.76 ± 0.36	—		
	Human	(48)	30—90 years	—	1.84 ± 0.43	—	Cells per 0.137 mm²	108
	Human	(17)	6—25 years	—	59	—		
	Human	(48)	30—90 years	—	40	—	Cells per 0.22 mm²	109
Articular cartilage, knee joint	Human	(2)	2 days—6 weeks	—	68	61—76		
		(4)	13—26 years	—	11	7—20		
		(6)	31—39 years	—	10	8—13		
		(3)	42—48 years	—	11	10—14		
		(6)	51—59 years	—	10	7—11		
		(10)	60—69 years	—	9	7—12		
		(4)	75—89 years	—	12	9—16		
Fourth costal cartilage	Human	(2)	2 days—6 weeks	—	54	52—56	Cells per 0.22 mm²	109
		(1)	1.5 years	—	39	—		
		(6)	13—26 years	—	8	5—10		
		(8)	31—39 years	—	7	5—9		
		(4)	42—48 years	—	6	4—10		
		(7)	51—59 years	—	6	4—8		
		(12)	60—69 years	—	6	4—9		
		(5)	75—89 years	—	5	4—7		
Cartilage, metatarso-phalangeal and metacarpal-phalangeal articulations	Bovine	(4)	up to 0.5 year	—	133×10^3 ($\pm 23 \times 10^3$)	—	Number cells per mm³	110
		(3)	1—7 years	—	47.2×10^3 ($\pm 3.7 \times 10^3$)	—		
		(11)	8—11 years	—	34.0×10^3 ($\pm 5.9 \times 10^3$)	—		

Table 5
NUMBER OF CELLS VS. AGE IN BLOOD AND BLOOD VESSELS

Cell	Subjects				Count		Measure	Ref.
	Type	(no.)	Age	Sex	Mean	Range		
					Blood			
Erythrocytes	Human	(13)	60—64 years	M	4.74×10^6	—	No. per mm^3	112
		(36)	65—69 years	M	4.49×10^6	—		
		(39)	70—74 years	M	4.32×10^6	—		
		(36)	75—79 years	M	4.44×10^6	—		
		(26)	80—84 years	M	4.51×10^6	—		
		(7)	85—89 years	M	4.64×10^6	—		
		(3)	94—104 years	M	4.23×10^6	—		
Erythrocytes	Human	(39)	20—29 years	M	0.475 ± 0.005	—	Proportion of cells in 1 cc total blood	113
		(10)	40—49 years	M	0.478 ± 0.022	—		
		(26)	50—59 years	M	0.441 ± 0.007	—		
		(54)	60—69 years	M	0.448 ± 0.004	—		
		(45)	70—79 years	M	0.434 ± 0.005	—		
		(17)	80—89 years	M	0.415 ± 0.007	—		
Erythrocytes	Human	(173)	20—29 years	M	4.94×10^6	—	No. per mm^3	114
		(52)	30—39 years	M	4.90×10^6	—		
		(42)	40—49 years	M	4.81×10^6	—		
		(24)	50—59 years	M	4.74×10^6	—		
		(14)	60—69 years	M	4.39×10^6	—		
Erythrocytes	Human	(51)	65—69 years	M	$5.05 \times 10^{12} \pm 0.07 \times 10^{12}$	—	No. per liter	115
		(43)	70—74 years	M	$4.81 \times 10^{12} \pm 0.10 \times 10^{12}$	—		
		(29)	75—79 years	M	$4.95 \times 10^{12} \pm 0.10 \times 10^{12}$	—		
		(15)	80—84 years	M	$4.64 \times 10^{12} \pm 0.11 \times 10^{12}$	—		
		(8)	85+ years	M	$4.47 \times 10^{12} \pm 0.17 \times 10^{12}$	—		
Erythrocytes	Human	(56)	65—69 years	F	$4.68 \times 10^{12} \pm 0.06 \times 10^{12}$	—	No. per liter	115
		(41)	70—74 years	F	$4.74 \times 10^{12} \pm 0.08 \times 10^{12}$	—		
		(29)	75—79 years	F	$4.57 \times 10^{12} \pm 0.08 \times 10^{12}$	—		
		(19)	80—84 years	F	$4.69 \times 10^{12} \pm 0.07 \times 10^{12}$	—		
		(10)	85+ years	F	$4.61 \times 10^{12} \pm 0.20 \times 10^{12}$	—		

	Species	(n)	Age	Sex	Mean	Range	Comments	Ref.
Erythrocytes	Human	( )	20 years	M	5.31×10^{12}	4.53×10^{12}—6.09×10^{12}	No./ℓ; data from parametric statistical analysis of 638 subjects 16—89 years old; range given equal 95% confidence intervals; nonparametric analysis gives similar results	116
		( )	60 years	M	5.05×10^{12}	4.27×10^{12}—5.83×10^{12}		
Erythrocytes	Human	( )	20 years	F	4.64×10^{12}	3.95×10^{12}—5.31×10^{12}	No./ℓ; data from parametric statistical analysis of 1,106 subjects 16—89 years old; range given equals 95% confidence intervals; nonparametric analysis gives similar results	116
		( )	60 years	F	4.60×10^{12}	3.92×10^{12}—5.28×10^{12}		
Erythrocytes	Rat	(11)	326 days	M	$7.9 \times 10^6 \pm 0.19 \times 10^6$	—	No. per mm^3	117
		(11)	560 days	M	$8.4 \times 10^6 \pm 0.13 \times 10^6$	—		
		(11)	762 days	M	$7.9 \times 10^6 \pm 0.33 \times 10^6$	—		
Erythrocytes	Mouse	(30)	4—7 months	F	789×10^4	—	No. per mm^3	118
		(29)	16.5—21 months	F	718×10^4	—		
Erythrocytes	Mouse	(10)	4 months	M	$9.97 \times 10^6 \pm 0.14 \times 10^6$	—	No. per mm^3	119
		(10)	6 months	M	$9.75 \times 10^6 \pm 0.14 \times 10^6$	—		
		(9)	9 months	M	$9.61 \times 10^6 \pm 0.14 \times 10^6$	—		
		(9)	12 months	M	$9.52 \times 10^6 \pm 0.14 \times 10^6$	—		
		(9)	16 months	M	$9.41 \times 10^6 \pm 0.14 \times 10^6$	—		
		(7)	20 months	M	$9.25 \times 10^6 \pm 0.16 \times 10^6$	—		
		(5)	24 months	M	$9.08 \times 10^6 \pm 0.19 \times 10^6$	—		
Erythrocytes	Mouse	(44)	8—10 months	M	51.7 ± 2.8	46—58	% of total volume of hematocrit tube occupied by RBCs	120
		(14)	25—28 months	M	45.8 ± 2.0	42—50		
Erythrocytes	Mouse	(10)	85 days	F	$8.41 \times 10^6 \pm 0.12 \times 10^6$	—	No. per mm^3	121
		(10)	134 days	F	$8.21 \times 10^6 \pm 0.27 \times 10^6$	—		
		(6)	192 days	F	$8.01 \times 10^6 \pm 0.20 \times 10^6$	—		

Table 5 (continued)
NUMBER OF CELLS VS. AGE IN BLOOD AND BLOOD VESSELS

Cell	Type	Subjects (no.)	Age	Sex	Count Mean	Count Range	Measure	Ref.
Erythrocytes	Mouse	(10)	308 days	F	$7.85 \times 10^6 \pm 0.41 \times 10^6$	—	No. per mm^3	122
		(10)	472 days	F	$8.66 \times 10^6 \pm 0.19 \times 10^6$	—		
		(10)	518 days	F	$8.23 \times 10^6 \pm 0.09 \times 10^6$	—		
		(6)	575 days	F	$7.99 \times 10^6 \pm 0.21 \times 10^6$	—		
		(9)	649 days	F	$7.00 \times 10^6 \pm 0.13 \times 10^6$	—		
		(9)	800 days	F	$7.64 \times 10^6 \pm 0.14 \times 10^6$	—		
		(9)	921 days	F	$7.32 \times 10^6 \pm 0.13 \times 10^6$	—		
		(30)	110 days	M	$9.91 \times 10^6 \pm 0.12 \times 10^6$	—		
		(30)	200 days	M	$9.04 \times 10^6 \pm 0.24 \times 10^6$	—		
		(30)	286 days	M	$8.91 \times 10^6 \pm 0.21 \times 10^6$	—		
		(36)	373 days	M	$9.34 \times 10^6 \pm 0.19 \times 10^6$	—		
		(57)	417 days	M	$9.32 \times 10^6 \pm 0.13 \times 10^6$	—		
		(59)	473 days	M	$9.26 \times 10^6 \pm 0.09 \times 10^6$	—		
		(56)	528 days	M	$9.26 \times 10^6 \pm 0.13 \times 10^6$	—		
		(49)	586 days	M	$9.90 \times 10^6 \pm 0.13 \times 10^6$	—		
		(55)	626 days	M	$8.67 \times 10^6 \pm 0.12 \times 10^6$	—		
		(53)	670 days	M	$8.29 \times 10^6 \pm 0.10 \times 10^6$	—		
		(51)	710 days	M	$8.55 \times 10^6 \pm 0.15 \times 10^6$	—		
		(52)	761 days	M	$8.38 \times 10^6 \pm 0.16 \times 10^6$	—		
		(53)	822 days	M	$8.41 \times 10^6 \pm 0.21 \times 10^6$	—		
		(54)	864 days	M	$8.24 \times 10^6 \pm 0.21 \times 10^6$	—		
		(51)	906 days	M	$8.01 \times 10^6 \pm 0.19 \times 10^6$	—		
		(52)	948 days	M	$7.58 \times 10^6 \pm 0.26 \times 10^6$	—		
		(50)	991 days	M	$7.25 \times 10^6 \pm 0.23 \times 10^6$	—		
		(53)	1044 days	M	$6.28 \times 10^6 \pm 0.22 \times 10^6$	—		
Erythrocytes	Mouse	(30)	118 days	F	$10.21 \times 10^6 \pm 0.09 \times 10^6$	—	No. per mm^3	122
		(30)	216 days	F	$9.85 \times 10^6 \pm 0.12 \times 10^6$	—		

		(n)	Age	Sex		Units	Ref
Erythrocytes	Mouse	(30)	302 days	F	$9.10 \times 10^6 \pm 0.18 \times 10^6$	No. per $\mu\ell$	123
		(30)	375 days	F	$9.47 \times 10^6 \pm 0.15 \times 10^6$		
		(57)	417 days	F	$9.55 \times 10^6 \pm 0.12 \times 10^6$		
		(60)	473 days	F	$9.63 \times 10^6 \pm 0.12 \times 10^6$		
		(58)	530 days	F	$9.15 \times 10^6 \pm 0.13 \times 10^6$		
		(58)	587 days	F	$8.90 \times 10^6 \pm 0.11 \times 10^6$		
		(58)	627 days	F	$8.43 \times 10^6 \pm 0.14 \times 10^6$		
		(53)	670 days	F	$8.56 \times 10^6 \pm 0.14 \times 10^6$		
		(54)	711 days	F	$8.79 \times 10^6 \pm 0.27 \times 10^6$		
		(52)	762 days	F	$8.00 \times 10^6 \pm 0.18 \times 10^6$		
		(52)	823 days	F	$8.13 \times 10^6 \pm 0.21 \times 10^6$		
		(50)	866 days	F	$8.23 \times 10^6 \pm 0.23 \times 10^6$		
		(51)	907 days	F	$7.78 \times 10^6 \pm 0.24 \times 10^6$		
		(52)	950 days	F	$7.50 \times 10^6 \pm 0.22 \times 10^6$		
		(50)	991 days	F	$6.52 \times 10^6 \pm 0.22 \times 10^6$		
		(34)	1045 days	F	$6.79 \times 10^6 \pm 0.30 \times 10^6$		
Erythrocytes	Guinea pig	(3)	8—11 months	M	7.7×10^6	Number per mm³; est. from data given	124
		(3)	16—19 months	M	9.4×10^6		
		(3)	27—30 months	M	8.0×10^6		
		(3)	35—37 months	M	6.9×10^6		
		(5)	1—2 days	M,F	6.0×10^6		
		(61)	3—38 days	M,F	4.8×10^6		
		(25)	46—229 days	M,F	5.0×10^6		
		(13)	406—1820 days	M,F	4.5×10^6		
Erythrocytes	Bull	(9)	1—2 years	M	$9.09 \times 10^6 \pm 0.47 \times 10^6$	No. per mm³	125
		(8)	2—4 years	M	$9.16 \times 10^6 \pm 0.46 \times 10^6$		
		(11)	4—6 years	M	$9.33 \times 10^6 \pm 0.36 \times 10^6$		
		(12)	6—8 years	M	$8.62 \times 10^6 \pm 0.21 \times 10^6$		
		(10)	8—10 years	M	$9.06 \times 10^6 \pm 0.58 \times 10^6$		
		(7)	10—12 years	M	$8.69 \times 10^6 \pm 0.54 \times 10^6$		
		(7)	12—15 years	M	$8.14 \times 10^6 \pm 0.39 \times 10^6$		
Erythrocytes	Cow	(25)	1—3 years	F	$8.18 \times 10^6 \pm 0.25 \times 10^6$	No. per mm³	125
		(10)	5—12 years	F	$5.71 \times 10^6 \pm 0.20 \times 10^6$		
Erythrocytes	Sheep	(9)	0.25—0.5 years	M,F	$11.74 \times 10^6 \pm 0.45 \times 10^6$	No. per mm³	125
		(10)	2 years	M,F	$10.27 \times 10^6 \pm 0.56 \times 10^6$		
		(10)	4—9 years	M,F	$10.68 \times 10^6 \pm 0.65 \times 10^6$		

Table 5 (continued)
NUMBER OF CELLS VS. AGE IN BLOOD AND BLOOD VESSELS

Cell	Type	(no.)	Age	Sex	Mean	Range	Measure	Ref.
					Count			
			Subjects					
Erythrocytes	Goat	(3)	0.5 year	F	$24.60 \times 10^6 \pm 0.40 \times 10^6$	—	No. per mm^3	125
		(4)	1—2 years	F	$22.40 \times 10^6 \pm 1.00 \times 10^6$	—		
		(7)	2—3 years	F	$20.00 \times 10^6 \pm 0.90 \times 10^6$	—		
		(2)	3—6 years	F	$20.20 \times 10^6 \pm 3.10 \times 10^6$	—		
Erythrocytes	Golden tamarin	(33)	up to 1 year (juveniles)	M,F	$5.4 \times 10^6 \pm 0.7 \times 10^6$	—	No. per mm^3	126
		(129)	>1 year (adults)	M,F	$5.8 \times 10^6 \pm 0.6 \times 10^6$	—		
Leukocytes	Human	(13)	60—64 years	M	8,842	—	No. per mm^3	112
		(36)	65—69 years	M	7,742	—		
		(39)	70—74 years	M	7,752	—		
		(36)	75—79 years	M	7,114	—		
		(26)	80—84 years	M	8,032	—		
		(7)	85—89 years	M	8,042	—		
		(3)	94—104 years	M	10,816	—		
Leukocytes	Human	(173)	20—29 years	M	6,555	—	No. per mm^3	114
		(52)	30—39 years	M	6,960	—		
		(42)	40—49 years	M	6,795	—		
		(24)	50—59 years	M	6,892	—		
		(14)	60—69 years	M	5,850	—		
Leukocytes	Human	(41)	18—29 years	M	5,200	—	No. per mm^3	127
		(52)	30—39 years	M	5,500	—		
		(52)	40—49 years	M	5,300	—		
		(33)	50—65 years	M	5,600	—		
Leukocytes	Human	(43)	18—29 years	F	5,700	—	No. per mm^3	127
		(22)	30—39 years	F	5,100	—		
		(48)	40—49 years	F	5,200	—		
		(34)	50—65 years	F	4,800	—		

Cell	Species	(N)	Age	Sex	Mean	Range	Comments	Ref.
Leukocytes	Human	(51)	65—69 years	M	$6.49 \times 10^9 \pm 0.18 \times 10^9$	—	No. per liter	115
		(43)	70—74 years	M	$6.87 \times 10^9 \pm 0.31 \times 10^9$	—		
		(29)	75—79 years	M	$6.94 \times 10^9 \pm 0.33 \times 10^9$	—		
		(15)	80—84 years	M	$7.48 \times 10^9 \pm 0.59 \times 10^9$	—		
		(8)	85 + years	M	$7.79 \times 10^9 \pm 0.63 \times 10^9$	—		
Leukocytes	Human	(56)	65—69 years	F	$6.46 \times 10^9 \pm 0.22 \times 10^9$	—	No. per liter	115
		(41)	70—74 years	F	$6.15 \times 10^9 \pm 0.20 \times 10^9$	—		
		(29)	75—79 years	F	$6.85 \times 10^9 \pm 0.32 \times 10^9$	—		
		(19)	80—84 years	F	$5.98 \times 10^9 \pm 0.37 \times 10^9$	—		
		(10)	85 + years	F	$7.02 \times 10^9 \pm 0.70 \times 10^9$	—		
Leukocytes	Human	(—)	20 years	M	6.59×10^9	2.71×10^9—10.47×10^9	No. per liter; data from parametric statistical analysis of 638 subjects 16—89 years old; range given equals 95% confidence intervals; nonparametric analysis gives similar results	116
		(—)	60 years	M	7.56×10^9	3.67×10^9—11.44×10^9		
Leukocytes	Human	(—)	20 years	F	7.28×10^9	3.39×10^9—11.16×10^9	No./ℓ; data from parametric statistical analysis of 1,106 subjects 16—89 years old; range given equals 95% confidence intervals; nonparametric analysis gives similar results	116
		(—)	60 years	F	7.00×10^9	3.12×10^9—10.89×10^9		
Leukocytes	Human	(12)	3—12 years	M,F	$6,340 \pm 1,900$	3,800—8,000	Total no. per $\mu\ell$	128
		(20)	19—25 years	M,F	$6,150 \pm 1,100$	4,900—8,100		
		(20)	33—50 years	M,F	$5,940 \pm 1,560$	4,500—9,200		
		(20)	65—83 years	M,F	$6,080 \pm 1,460$	4,100—7,500		
Leukocytes	Human	(56)	20—39 years	M,F	$6,195 \pm 1,825$	—	Total no. per mm³	129
		(71)	40—59 years	M,F	$6,616 \pm 2,028$	—		
		(77)	60—79 years	M,F	$6,488 \pm 1,737$	—		

Table 5 (continued)
NUMBER OF CELLS VS. AGE IN BLOOD AND BLOOD VESSELS

Cell	Type	Subjects (no.)	Age	Sex	Count Mean	Count Range	Measure	Ref.
Leukocytes	Human	(40)	80—99 years	M,F	$7{,}040 \pm 2{,}349$	—		
		(15)	20—39 years	M,F	$4{,}927 \pm 368$	—	No. per mm^3	130
		(15)	76—93 years	M,F	$3{,}993 \pm 222$	—		
Leukocytes	Human	(5)	20—29 years	M	$5{,}686 \pm 1{,}296$	—	No. per mm^3	131
		(19)	30—39 years	M	$7{,}013 \pm 3{,}586$	—		
		(50)	40—49 years	M	$6{,}964 \pm 2{,}058$	—		
		(67)	50—59 years	M	$6{,}815 \pm 3{,}144$	—		
		(43)	60—69 years	M	$7{,}059 \pm 3{,}298$	—		
		(47)	70—79 years	M	$7{,}267 \pm 2{,}448$	—		
		(8)	80—89 years	M	$6{,}376 \pm 1{,}326$	—		
Leukocytes	Rat	(11)	326 days	M	$14.4 \times 10^3 \pm 0.92 \times 10^3$	—	No. per mm^3	117
		(11)	560 days	M	$14.4 \times 10^3 \pm 0.78 \times 10^3$	—		
		(11)	762 days	M	$11.9 \times 10^3 \pm 0.89 \times 10^3$	—		
Leukocytes	Mouse	(30)	4—7 months	F	$12{,}900$	—	No. per mm^3	118
		(29)	16.5—21 months	F	$10{,}900$	—		
Leukocytes	Mouse	(31)	8—10 months	M	$9{,}700 \pm 3{,}300$	4,000—18,000	No. per mm^3	120
		(14)	25—28 months	M	$10{,}400 \pm 3{,}300$	4,000—14,000		
Leukocytes	Mouse	(10)	85 days	F	$4.21 \times 10^3 \pm 0.34 \times 10^3$	—	No. per mm^3	121
		(10)	134 days	F	$4.21 \times 10^3 \pm 0.12 \times 10^3$	—		
		(6)	192 days	F	$4.48 \times 10^3 \pm 0.54 \times 10^3$	—		
		(10)	308 days	F	$4.60 \times 10^3 \pm 0.33 \times 10^3$	—		
		(10)	472 days	F	$3.29 \times 10^3 \pm 0.43 \times 10^3$	—		
		(10)	518 days	F	$3.09 \times 10^3 \pm 0.21 \times 10^3$	—		
		(10)	575 days	F	$5.13 \times 10^3 \pm 0.43 \times 10^3$	—		
		(6)	649 days	F	$4.25 \times 10^3 \pm 0.43 \times 10^3$	—		
		(9)	800 days	F	$3.70 \times 10^3 \pm 0.58 \times 10^3$	—		
		(9)	921 days	F	$2.98 \times 10^3 \pm 0.31 \times 10^3$	—		

Leukocytes	Mouse				No. per mm³		122
	(30)	110 days	M	10,470 ± 700		—	
	(30)	200 days	M	10,280 ± 573		—	
	(30)	286 days	M	9,910 ± 662		—	
	(36)	373 days	M	9,460 ± 668		—	
	(57)	417 days	M	10,180 ± 407		—	
	(59)	473 days	M	10,510 ± 379		—	
	(56)	528 days	M	7,900 ± 376		—	
	(49)	586 days	M	7,300 ± 301		—	
	(55)	626 days	M	7,510 ± 359		—	
	(53)	670 days	M	8,980 ± 399		—	
	(51)	710 days	M	9,940 ± 429		—	
	(52)	761 days	M	8,440 ± 378		—	
	(53)	822 days	M	10,510 ± 590		—	
	(54)	864 days	M	10,540 ± 579		—	
	(51)	906 days	M	9,150 ± 608		—	
	(52)	948 days	M	8,080 ± 378		—	
	(50)	991 days	M	10,010 ± 461		—	
	(53)	1044 days	M	10,410 ± 557		—	

Leukocytes	Mouse				No. per mm³		122
	(30)	118 days	F	9,903 ± 611		—	
	(30)	216 days	F	9,197 ± 686		—	
	(30)	302 days	F	8,640 ± 523		—	
	(30)	375 days	F	9,800 ± 555		—	
	(57)	417 days	F	8,754 ± 322		—	
	(60)	473 days	F	7,760 ± 270		—	
	(58)	530 days	F	7,393 ± 210		—	
	(58)	587 days	F	6,350 ± 306		—	
	(58)	627 days	F	6,114 ± 232		—	
	(53)	670 days	F	8,008 ± 416		—	
	(54)	711 days	F	6,341 ± 400		—	
	(52)	762 days	F	7,498 ± 323		—	
	(52)	823 days	F	8,988 ± 453		—	
	(50)	866 days	F	7,042 ± 362		—	
	(51)	907 days	F	7,898 ± 456		—	
	(52)	950 days	F	6,576 ± 315		—	
	(50)	991 days	F	6,638 ± 354		—	
	(34)	1045 days	F	8,837 ± 555		—	

Table 5 (continued)
NUMBER OF CELLS VS. AGE IN BLOOD AND BLOOD VESSELS

Cell	Type	Subjects (no.)	Age	Sex	Count Mean	Range	Measure	Ref.
Leukocytes	Mouse	(—)	16.5 months	M	24,000	—	No. per mm^3; est. from data given; 10 animals total used	132
		(—)	19.5 months	M	27,000	—		
		(—)	20 months	M	23,400	—		
		(—)	20.5 months	M	22,600	—		
		(—)	23.5 months	M	24,000	—		
		(—)	24.5 months	M	20,300	—		
		(—)	27.5 months	M	19,900	—		
		(—)	28 months	M	18,200	—		
		(—)	31.5 months	M	16,700	—		
Leukocytes	Mouse	(3)	8—11 months	M	10.0×10^3	—	No. per $\mu\ell$	123
		(3)	16—19 months	M	7.8×10^3	—		
		(3)	27—30 months	M	8.5×10^3	—		
		(3)	35—37 months	M	13.4×10^3	—		
Leukocytes	Guinea pig	(5)	1—2 days	M,F	1.6×10^6	—	No. per mm^3; est. from data given	124
		(61)	3—38 days	M,F	3.0×10^6	—		
		(25)	46—229 days	M,F	8.9×10^6	—		
		(13)	406—1820 days	M,F	9.5×10^6	—		
Leukocytes	Bull	(9)	1—2 years	M	$12,350 \pm 1,318$	—	No. per mm^3	125
		(8)	2—4 years	M	$8,500 \pm 661$	—		
		(11)	4—6 years	M	$7,393 \pm 576$	—		
		(12)	6—8 years	M	$7,692 \pm 413$	—		
		(10)	8—10 years	M	$7,125 \pm 314$	—		
		(7)	10—12 years	M	$6,925 \pm 471$	—		
		(7)	12—15 years	M	$6,400 \pm 323$	—		
Leukocytes	Cow	(25)	1—3 years	F	$9,461 \pm 472$	—	No. per mm^3	125
		(10)	5—12 years	F	$7,400 \pm 509$	—		

Cell type	Species	(n)	Age	Sex	Value	Units	Ref.
Leukocytes	Sheep	(9)	0.25—0.5 years	M,F	6,272 ± 806	No. per mm³	125
		(10)	2 years	M,F	8,000 ± 236		
		(10)	4—9 years	M,F	6,755 ± 507		
Leukocytes	Goat	(3)	0.5 years	F	7,150 ± 1,280	No. per mm³	125
		(4)	1—2 years	F	9,690 ± 663		
		(7)	2—3 years	F	9,450 ± 1,176		
		(2)	3—6 years	F	7,725 ± 1,175		
Leukocytes	Golden tamarin	(35)	Up to 1 year (juveniles)	M,F	$6.14 \times 10^3 \pm 2.54 \times 10^3$	No. per mm³	126
		(135)	<1 year (adults)	M,F	$7.39 \times 10^3 \pm 2.86 \times 10^3$		
Polymorphonuclear leukocytes	Human	(41)	18—29 years	M	2,900	No. per mm³	127
		(52)	30—39 years	M	3,100		
		(52)	40—49 years	M	3,100		
		(33)	50—65 years	M	3,200		
Polymorphonuclear leukocytes	Human	(43)	18—29 years	F	3,300	No. per mm³	127
		(22)	30—39 years	F	2,900		
		(48)	40—49 years	F	3,100		
		(34)	50—65 years	F	4,800		
Granulocytes (polymorphonuclear leukocytes)	Human	(96)	0—9 years	M,F	4,839 ± 3,237	No. per µℓ	133
		(241)	10—19 years	M,F	4,990 ± 3,082		
		(353)	20—29 years	M,F	4,915 ± 2,540		
		(425)	30—39 years	M,F	5,450 ± 2,730		
		(262)	40—49 years	M,F	4,878 ± 2,402		
		(259)	50—59 years	M,F	4,900 ± 2,524		
		(373)	60—69 years	M,F	4,887 ± 2,576		
		(360)	70—79 years	M,F	5,078 ± 2,782		
		(157)	80—89 years	M,F	4,619 ± 1,977		
		(43)	90—99 years	M,F	5,673 ± 2,798		
Polymorphonuclear leukocytes	Human	(3)	20—29 years	M	2,913 ± 190	No. per mm³	131
		(14)	30—39 years	M	4,170 ± 904		
		(50)	40—49 years	M	4,020 ± 450		
		(67)	50—59 years	M	4,222 ± 414		
		(43)	60—69 years	M	4,071 ± 500		
		(47)	70—79 years	M	4,542 ± 416		
		(8)	80—89 years	M	3,697 ± 468		
Polymorphonuclear leukocytes	Mouse	(29)	4—7 months	F	811	No. per mm³	118
		(29)	16.5—21 months	F	2,224		

Table 5 (continued)
NUMBER OF CELLS VS. AGE IN BLOOD AND BLOOD VESSELS

Cell	Type	Subjects (no.)	Age	Sex	Count Mean	Range	Measure	Ref.
Neutrophils	Human	(51)	65—69 years	M	$3.68 \times 10^9 \pm 0.13 \times 10^9$	—	No. per liter	115
		(43)	70—74 years	M	$4.07 \times 10^9 \pm 0.29 \times 10^9$	—		
		(29)	75—79 years	M	$4.62 \times 10^9 \pm 0.40 \times 10^9$	—		
		(15)	80—84 years	M	$4.84 \times 10^9 \pm 0.40 \times 10^9$	—		
		(8)	85+ years	M	$4.44 \times 10^9 \pm 0.43 \times 10^9$	—		
Neutrophils	Human	(56)	65—69 years	F	$3.74 \times 10^9 \pm 0.17 \times 10^9$	—	No. per liter	115
		(41)	70—74 years	F	$3.62 \times 10^9 \pm 0.18 \times 10^9$	—		
		(29)	75—79 years	F	$4.15 \times 10^9 \pm 0.24 \times 10^9$	—		
		(19)	80—84 years	F	$3.46 \times 10^9 \pm 0.29 \times 10^9$	—		
		(10)	85+ years	F	$4.72 \times 10^9 \pm 0.62 \times 10^9$	—		
Neutrophils	Human	(13)	60—64 years	M	4,952	—	No. per mm³; calculated from data given	112
		(36)	65—69 years	M	4,490	—		
		(39)	70—74 years	M	4,496	—		
		(36)	75—79 years	M	4,126	—		
		(26)	80—84 years	M	4,418	—		
		(7)	85—89 years	M	4,423	—		
		(3)	94—104 years	M	6,598	—		
Neutrophils (juvenile forms)	Mouse	(31)	8—10 months	M	320 ± 275	0—1,000	No. per mm³	120
		(14)	25—28 months	M	540 ± 240	300—1,000		
Neutrophils (mature forms)	Mouse	(31)	8—10 months	M	1,020 ± 1,040	100—5,000	No. per mm³	120
		(14)	25—28 months	M	1,530 ± 980	500—3,000		
Neutrophils	Mouse	(30)	110 days	M	1,880 ± 125	—	No. per mm³	122
		(30)	200 days	M	1,490 ± 83	—		
		(30)	286 days	M	1,540 ± 102	—		
		(36)	373 days	M	1,390 ± 98	—		
		(57)	417 days	M	1,670 ± 67	—		
		(59)	473 days	M	1,380 ± 50	—		

				No. per mm³	
Neutrophils	Mouse	(56)	528 days	M	1,380 ± 66	122
		(49)	586 days	M	1,380 ± 57	
		(55)	626 days	M	1,550 ± 74	
		(53)	670 days	M	1,990 ± 89	
		(51)	710 days	M	1,930 ± 84	
		(52)	761 days	M	2,180 ± 98	
		(53)	822 days	M	2,100 ± 118	
		(54)	864 days	M	2,460 ± 135	
		(51)	906 days	M	2,100 ± 139	
		(52)	948 days	M	2,070 ± 74	
		(50)	991 days	M	2,880 ± 133	
		(53)	1044 days	M	3,190 ± 170	
		(30)	118 days	F	1,220 ± 75	
		(30)	216 days	F	1,340 ± 93	
		(30)	302 days	F	1,410 ± 85	
		(30)	375 days	F	1,730 ± 97	
		(57)	417 days	F	1,580 ± 58	
		(60)	473 days	F	1,260 ± 44	
		(58)	530 days	F	1,610 ± 46	
		(58)	587 days	F	1,460 ± 70	
		(58)	627 days	F	1,420 ± 54	
		(53)	670 days	F	1,860 ± 97	
		(54)	711 days	F	1,410 ± 89	
		(52)	762 days	F	1,640 ± 71	
		(52)	823 days	F	1,910 ± 96	
		(50)	866 days	F	1,830 ± 90	
		(51)	907 days	F	1,930 ± 111	
		(52)	950 days	F	1,590 ± 76	
		(50)	991 days	F	1,780 ± 95	
		(34)	1045 days	F	2,420 ± 152	

				No. per mm³	
Neutrophils	Bull	(9)	1—2 years	M	3,754	125
		(8)	2—4 years	M	3,009	
		(11)	4—6 years	M	3,735	
		(12)	6—8 years	M	3,300	
		(10)	8—10 years	M	3,334	
		(7)	10—12 years	M	3,061	
		(7)	12—15 years	M	2,944	

Table 5 (continued)
NUMBER OF CELLS VS. AGE IN BLOOD AND BLOOD VESSELS

Cell	Type	(no.)	Age	Sex	Mean	Range	Measure	Ref.
						Count		
		Subjects						
Neutrophils	Cow	(25)	1—3 years	F	1,987	—	No. per mm³	125
		(10)	5—12 years	F	3,101	—		
Neutrophils	Sheep	(9)	0.25—0.5 years	M,F	1,254	—	No. per mm³	125
		(10)	2 years	M,F	2.830	—		
		(10)	4—9 years	M,F	2,310	—		
Neutrophils	Goat	(3)	0.5 years	F	1,737	—	No. per mm³	125
		(4)	1—2 years	F	2,597	—		
		(7)	2—3 years	F	3,941	—		
		(2)	3—6 years	F	3,283	—		
Neutrophils	Golden tamarin	(34)	Up to 1 year (juveniles)	M,F	$3.26 \times 10^3 \pm 1.67 \times 10^3$	—	No. per mm³	126
		(135)	>1 year (adults)	M,F	$4.72 \times 10^3 \pm 2.61 \times 10^3$	—		
Eosinophils	Human	(13)	60—64 years	M	265	—	No. per mm³; calculated from data given	112
		(36)	65—69 yeas	M	232	—		
		(39)	70—74 years	M	155	—		
		(36)	75—79 years	M	178	—		
		(26)	80—84 years	M	241	—		
		(7)	85—89 years	M	161	—		
		(3)	94—104 years	M	108	—		
Eosinophils	Rat	(11)	326 days	M	501 ± 118	—	No. per mm³	117
		(11)	560 days	M	500 ± 80	—		
		(11)	762 days	M	427 ± 48	—		
Eosinophils	Mouse	(30)	4—7 months	F	101	—	No. per mm³	118
		(29)	16.5—21 months	F	112	—		
Eosinophils	Mouse	(31)	8—10 months	M	130 ± 145	0—600	No. per mm³	120
		(14)	25—28 months	M	225 ± 205	0—700		

Eosinophils

Mouse — No. per mm³ — 122

			No. per mm³
(30)	110 days	M	262
(30)	200 days	M	288
(30)	286 days	M	258
(36)	373 days	M	303
(57)	417 days	M	122
(59)	473 days	M	84
(56)	528 days	M	63
(49)	586 days	M	44
(55)	626 days	M	75
(53)	670 days	M	81
(51)	710 days	M	70
(52)	761 days	M	68
(53)	822 days	M	105
(54)	864 days	M	74
(51)	906 days	M	73
(52)	948 days	M	73
(50)	991 days	M	90
(53)	1044 days	M	83

Eosinophils

Mouse — No. per mm³ — 122

			No. per mm³
(30)	118 days	F	238
(30)	216 days	F	184
(30)	302 days	F	173
(30)	375 days	F	157
(57)	417 days	F	105
(60)	473 days	F	70
(58)	530 days	F	15
(58)	587 days	F	38
(58)	627 days	F	67
(53)	670 days	F	72
(54)	711 days	F	57
(52)	762 days	F	82
(52)	823 days	F	72
(50)	866 days	F	63
(51)	907 days	F	55
(52)	950 days	F	53
(50)	991 days	F	53
(34)	1045 days	F	44

Table 5 (continued)
NUMBER OF CELLS VS. AGE IN BLOOD AND BLOOD VESSELS

| | Subjects | | | | Count | | | |
Cell	Type	(no.)	Age	Sex	Mean	Range	Measure	Ref.
Eosinophils	Bull	(9)	1—2 years	M	630	—	No. per mm^3	125
		(8)	2—4 years	M	433	—		
		(11)	4—6 years	M	577	—		
		(12)	6—8 years	M	708	—		
		(10)	8—10 years	M	520	—		
		(7)	10—12 years	M	436	—		
		(7)	12—15 years	M	506	—		
Eosinophils	Cow	(25)	1—3 years	F	643	—	No. per mm^3	125
		(10)	5—12 years	F	636	—		
Eosinophils	Sheep	(9)	0.25—0.5 years	M,F	113	—	No. per mm^3	125
		(10)	2 years	M,F	427	—		
		(10)	4—9 years	M,F	608	—		
Eosinophils	Goat	(3)	0.5 years	F	72	—	No. per mm^3	125
		(4)	1—2 years	F	48	—		
		(7)	2—3 years	F	123	—		
		(2)	3—6 years	F	77	—		
Eosinophils	Golden tamarin	(34)	Up to 1 year (juveniles)	M,F	$0.19 \times 10^3 \pm 0.26 \times 10^3$	—	No. per mm^3	126
		(133)	>1 year (adults)	M,F	$0.33 \times 10^3 \pm 0.43 \times 10^3$	—		
Basophils	Human	(13)	60—64 years	M	26	—	No. per mm^3; calculated from data given	112
		(36)	65—69 years	M	39	—		
		(39)	70—74 years	M	78	—		
		(36)	75—79 years	M	36	—		
		(26)	80—84 years	M	80	—		
		(7)	85—89 years	M	80	—		
		(3)	94—104 years	M	0	—		

Cell type	Animal	(n)	Age	Sex	Value	Range	Units	Ref.
Basophils	Bull	(9)	1—2 years	M	31	—	No. per mm^3	125
		(8)	2—4 years	M	32	—		
		(11)	4—6 years	M	33	—		
		(12)	6—8 years	M	45	—		
		(10)	8—10 years	M	45	—		
		(7)	10—12 years	M	35	—		
		(7)	12—15 years	M	19	—		
Basophils	Cow	(25)	1—3 years	F	38	—	No. per mm^3	125
		(10)	5—12 years	F	15	—		
Basophils	Sheep	(9)	0.25—0.5 years	M,F	25	—	No. per mm^3	125
		(10)	2 years	M,F	89	—		
		(10)	4—9 years	M,F	94	—		
Basophils	Goat	(3)	0.5 years	F	72	—	No. per mm^3	125
		(4)	1—2 years	F	73	—		
		(7)	2—3 years	F	13	—		
		(2)	3—6 years	F	0	—		
Basophils	Golden tamarin	(34)	Up to 1 year (juveniles)	M,F	$0.04 \times 10^3 \pm 0.09 \times 10^3$	—	No. per mm^3	126
		(133)	>1 year (adults)	M,F	$0.09 \times 10^3 \pm 0.16 \times 10^3$	—		
Monocytes	Human	(13)	60—64 years	M	619	—	No. per mm^3; calculated from data given	112
		(36)	65—69 years	M	619	—		
		(39)	70—74 years	M	543	—		
		(36)	75—79 years	M	569	—		
		(26)	80—84 years	M	723	—		
		(7)	85—89 years	M	724	—		
		(3)	94—104 years	M	1,190	—		
Monocytes	Human	(15)	20—39 years	M,F	278 ± 25	—	No. per mm^3	130
		(15)	76—93 years	M,F	290 ± 39	—		
Monocytes	Mouse	(31)	8—10 months	M	565 ± 300	0—1,500	No. per mm^3	120
		(14)	25—28 months	M	895 ± 410	200—1,500		
Monocytes	Bull	(99)	1—2 years	M	321	—	No. per mm^3	125
		(8)	2—4 years	M	425	—		
		(11)	4—6 years	M	251	—		
		(12)	6—8 years	M	377	—		
		(10)	8—10 years	M	271	—		
		(7)	10—12 years	M	194	—		
		(7)	12—15 years	M	186	—		

Table 5 (continued)
NUMBER OF CELLS VS. AGE IN BLOOD AND BLOOD VESSELS

Cell	Type	Subjects			Count		Measure	Ref.
		(no.)	Age	Sex	Mean	Range		
Monocytes	Cow	(25)	1—3 years	F	303	—	No. per mm^3	125
		(10)	5—12 years	F	170	—		
Monocytes	Sheep	(9)	0.25—0.5 years	M,F	144	—	No. per mm^3	125
		(10)	2 years	M,F	169	—		
		(10)	4—9 years	M,F	94	—		
Monocytes	Goat	(3)	0.5 years	F	307	—	No. per mm^3	125
		(4)	1—2 years	F	223	—		
		(7)	2—3 years	F	217	—		
		(2)	3—6 years	F	116	—		
Total lymphocytes	Human	(13)	60—64 years	M	2,829	—	No. per mm^3; calculated from data given	112
		(36)	65—69 years	M	2,323	—		
		(39)	70—74 years	M	2,326	—		
		(36)	75—79 years	M	2,063	—		
		(26)	80—84 years	M	2,490	—		
		(7)	85—89 years	M	2,493	—		
		(3)	94—104 years	M	2,920	—		
Total lymphocytes	Human	(41)	18—29 years	M	2,000	—	No. per mm^3	127
		(52)	30—39 years	M	1,900	—		
		(52)	40—49 years	M	1,800	—		
		(33)	50—65 years	M	1,900	—		
Total lymphocytes	Human	(43)	18—29 years	F	2,100	—	No. per mm^3	127
		(22)	30—39 years	F	2,000	—		
		(48)	40—49 years	F	1,800	—		
		(34)	50—65 years	F	1,900	—		
Total lymphocytes	Human	(51)	65—69 years	M	$2.14 \times 10^9 \pm 0.09 \times 10^9$	—	No. per liter	115
		(43)	70—74 years	M	$2.03 \times 10^9 \pm 0.10 \times 10^9$	—		
		(29)	75—79 years	M	$1.66 \times 10^9 \pm 0.08 \times 10^9$	—		
		(15)	80—84 years	M	$2.24 \times 10^9 \pm 0.27 \times 10^9$	—		

		(n)	Age	Sex	Mean ± SD	Range	Units	Ref
Total lymphocytes	Human	(8)	85 + years	M	$2.42 \times 10^9 \pm 0.41 \times 10^9$	—	No. per liter	115
		(56)	65—69 years	F	$2.24 \times 10^9 \pm 0.10 \times 10^9$	—		
		(41)	70—74 years	F	$1.91 \times 10^9 \pm 0.10 \times 10^9$	—		
		(29)	75—79 years	F	$1.99 \times 10^9 \pm 0.14 \times 10^9$	—		
		(19)	80—84 years	F	$2.11 \times 10^9 \pm 0.19 \times 10^9$	—		
		(10)	85 + years	F	$1.78 \times 10^9 \pm 0.21 \times 10^9$	—		
Total lymphocytes	Human	(96)	0—9 years	M,F	4,980 ± 2,448	—	No. per μℓ	133
		(241)	10—19 years	M,F	2,654 ± 2,350	—		
		(353)	20—29 years	M,F	2,023 ± 902	—		
		(425)	30—39 years	M,F	2,056 ± 889	—		
		(262)	40—49 years	M,F	2,008 ± 925	—		
		(259)	50—59 years	M,F	1,948 ± 1,002	—		
		(373)	60—69 years	M,F	1,872 ± 916	—		
		(359)	70—79 years	M,F	1,634 ± 854	—		
		(157)	80—89 years	M,F	1,485 ± 783	—		
		(43)	90—99 years	M,F	1,580 ± 853	—		
Total lymphocytes	Human	(26)	20—40 years	—	2,548 ± 154	—	No. per mm³	134
		(18)	60—69 years	—	2,314 ± 245	—		
		(20)	70—79 years	—	2,399 ± 281	—		
		(16)	80—89 years	—	1,779 ± 116	—		
		(18)	90—96 years	—	2,246 ± 320	—		
Total lymphocytes	Human	(23)	18—40 years	M,F	2,761 ± 850	—	No. per mm³	135
		(30)	>60 years	M,F	2,013 ± 663	—		
Total lymphocytes	Human	(29)	Newborn—10 years	M,F	4,266 ± 1,874	—	No. per μℓ	136
		(50)	11—60 years	M,F	2,320 ± 769	—		
		(22)	61—98 years	M,F	2,032 ± 768	—		
Total lymphocytes	Human	(26)	3—21 days	M,F	5,200.3 ± 1,738.3	—	Total no. per mm³	137
		(12)	3 months	M,F	4,463.8 ± 1,552.6	—		
		(40)	20—50 years	M,F	1,767.5 ± 526.9	—		
		(40)	75—97 years	M,F	1,542.2 ± 300.5	—		
Total lymphocytes	Human	(56)	20—39 years	M,F	2,034 ± 787	—	Total no. per mm³	129
		(71)	40—59 years	M,F	2,037 ± 630	—		
		(77)	60—79 years	M,F	1,931 ± 667	—		
		(40)	80—99 years	M,F	2,092 ± 824	—		
Total lymphocytes	Human	(12)	3—12 years	M,F	2,490 ± 940	870—4,190	Total no. per μℓ	128
		(20)	19—25 years	M,F	1,870 ± 500	1,040—3,080		

Table 5 (continued)
NUMBER OF CELLS VS. AGE IN BLOOD AND BLOOD VESSELS

Cell	Type	Subjects (no.)	Age	Sex	Count Mean	Count Range	Measure	Ref.
Total lymphocytes	Human	(20)	33—50 years	M,F	1,890 ± 560	1,100—2,730		138
		(20)	65—83 years	M,F	1,560 ± 610	690—2,590		
		(59)	22—52 years	M,F	2,001 ± 898	—	Total no. per mm³	
		(61)	60—80 years	M,F	1,940 ± 473	—		
Total lymphocytes	Human	(26)	10—19 years	—	2,134 ± 222	—	Total no. per mm³	139
		(36)	20—29 years	—	2,417 ± 211	—		
		(36)	30—39 years	—	1,923 ± 155	—		
		(55)	40—49 years	—	1,974 ± 138	—		
		(53)	50—59 years	—	1,791 ± 94	—		
		(41)	60—69 years	—	1,890 ± 138	—		
		(32)	70—79 years	—	1,563 ± 197	—		
		(21)	80—89 years	—	1,621 ± 124	—		
Total lymphocytes	Human	(261)	23—44 years	M	$2.64 \times 10^3 \pm 0.04 \times 10^3$	—	No. per mm³	140
		(135)	45—54 years	M	$2.58 \times 10^3 \pm 0.06 \times 10^3$	—		
		(40)	55+ years	M	$2.58 \times 10^3 \pm 0.12 \times 10^3$	—		
Total lymphocytes	Human	(10)	3rd—4th decades	M	2,139 ± 707	—	No. per mm³	141
		(21)	5th—8th decades	M	1,836 ± 538	—		
Total lymphocytes	Human	(10)	3rd—4th decades	F	2,011 ± 829	—	No. per mm³	141
		(23)	5th—9th decades	F	2,292 ± 939	—		
Total lymphocytes	Human	(15)	20—39 years	M,F	1,679 ± 136	—	No. per mm³	130
		(15)	76—93 years	M,F	1,375 ± 116	—		
Lymphocytes, "thymic-derived"	Human	(20)	35 years	M,F	66% ± 4%	—	% of circulating lymphocytes	142
		(20)	72—96 years	M,F	59% ± 5%	—		
T lymphocytes	Human	(10)	16—19 years	M,F	61.2% ± 2.9%	—	% of total lymphocytes capable of forming rosettes	143
		(8)	20—29 years	M,F	54.5% ± 8.1%	—		
		(9)	30—39 years	M,F	54.0% ± 5.4%	—		
		(8)	40—49 years	M,F	54.7% ± 5.4%	—		

Cell type	Species		Age	Sex	Value	Range	Units	Ref.
T lymphocytes	Human	(9)	50—59 years	M,F	42.7% ± 4.5%	—		
		(7)	60—69 years	M,F	33.6% ± 4.4%	—		
		(29)	Newborn—10 years	M,F	1,948 ± 1,034	—	No. per μℓ	136
T lymphocytes	Human	(50)	11—60 years	M,F	1,402 ± 602	—		
		(22)	61—98 years	M,F	1,213 ± 396	—	No. per mm³	134
T lymphocytes	Human	(26)	20—40 years	—	1,564 ± 95	—		
		(18)	60—69 years	—	1,215 ± 157	—		
		(20)	70—79 years	—	1,393 ± 208	—		
		(16)	80—89 years	—	895 ± 87	—		
		(18)	90—96 years	—	1,198 ± 114	—		
T lymphocytes	Human	(12)	3—12 years	M,F	1,890 ± 710	720—3,400		
		(20)	19—25 years	M,F	1,310 ± 380	700—2,200		
		(20)	33—50 years	M,F	1,210 ± 390	600—1,750		
		(20)	65—83 years	M,F	1,030 ± 440	430—1,800	Total no. per μℓ	128
T lymphocytes	Human	(10)	3rd—4th decades	M	1,723 ± 538			
		(23)	5th—8th decades	M	1,437 ± 456		No. per mm³	141
T lymphocytes	Human	(10)	3rd—4th decades	F	1,603 ± 718			
		(23)	5th—9th decades	F	1,862 ± 826		No. per mm³	141
T lymphocytes	Human	(15)	20—39 years	M,F	1,019 ± 108			
		(15)	76—93 years	M,F	734 ± 70		No.per mm³	130
T lymphocytes, E-rosette formers	Human	(26)	3—21 days	M,F	2,875.7 ± 147.8			
		(12)	3 months	M,F	2,379.2 ± 163.0			
		(40)	20—50 years	M,F	1,253.3 ± 112.7			
		(40)	75—79 years	M,F	1,021.3 ± 28.0		Total no. per mm³	137
T lymphocytes, E-rosette formers	Human	(56)	20—39 years	M,F	1,184 ± 442			
		(71)	40—59 years	M,F	1,115 ± 376			
		(77)	60—79 years	M,F	1,117 ± 432			
		(40)	80—99 years	M,F	1,313 ± 526		Total no. per mm³	129
T lymphocytes (E-RFC)	Human	(52)	20—45 years	M,F	49.8 ± 1.44		% that are E-rosette-forming cells in serum-free medium with untreated sheep erythrocytes	145
		(56)	70—98 years	M,F	45.3 ± 2.16			
T lymphocytes, early E-rosette formers	Human	(23)	18—40 years	M,F	940			
		(30)	>60 years	M,F	800		Total no. per mm³; est. from data given	135
T lymphocytes, early	Human	(12)	19—27 years	M	458.6 ± 134.3	—	No. per standard no.	144

Table 5 (continued)
NUMBER OF CELLS VS. AGE IN BLOOD AND BLOOD VESSELS

Cell	Type	Subjects (no.)	Age	Sex	Count Mean	Range	Measure	Ref.
E-rosette formers	Human	(8)	60—89 years	M	852.3 ± 239.4	—	lymphocytes	144
T lymphocytes, early		(5)	19—27 years	F	717.8 ± 259.6	—	No. per standard no. lymphocytes	
E-rosette formers		(9)	60—89 years	F	1,263 ± 781.9	—	lymphocytes	
T lymphocytes, early	Human	(26)	3—31 days	M,F	670.8 ± 35.6	—	Total no. per mm³	137
E-rosette formers		(12)	3 months	M,F	611.5 ± 125.3	—		
		(40)	20—50 years	M,F	387.6 ± 52.7	—		
		(40)	75—97 years	M,F	450.2 ± 31.1	—		
T lymphocytes, late	Human	(23)	18—40 years	M,F	1,620	—	Total no. per mm³; est. from data given	135
E-rosette formers		(30)	>60 years	M,F	1,187	—		
T lymphocytes, late	Human	(12)	19—27 years	M	1,618.9 ± 334.4	—	No. per standard no. lymphocytes	144
E-rosette formers		(8)	60—89 years	M	1,211.9 ± 271.2	—		
T lymphocytes, late	Human	(5)	19—27 years	F	1,105.6 ± 536.4	—	No. per standard no. lymphocytes	144
E-rosette formers		(9)	60—89 years	F	1,831.5 ± 1,228.1	—		
T lymphocytes (Ep-RFC)	Human	(52)	20—45 years	M,F	58.4 ± 1.15	—	% that are E-rosette-forming cells with papain-pretreated sheep erythrocytes	145
		(56)	70—98 years	M,F	62.1 ± 2.14	—		
T lymphocytes, (E$_{AB}$-RFC)	Human	(52)	20—45 years	M,F	70.8 ± 1.73	—	% that are E-rosette-forming cells in medium with heat inactivated, E-absorbed, pooled human AB serum, and untreated sheep erythrocytes	145
		(56)	70—98 years	M,F	68.3 ± 2.29	—		
T lymphocytes, (EA-RFC)	Human	(22)	20—45 years	M	11.7 ± 1.11	—	% that are E-rosette-forming cells in medium with human D	145
		(20)	70—98 years	M	17.2 ± 1.73	—		

	Species	(n)	Age	Sex	Value			Method	Ref.
T lymphocytes, (EA-RFC)	Human	(40)	20—45 years	F	9.7 ± 0.81	—	—	serum and sheep erythrocytes sensitized by incomplete anti-D antibody	145
		(24)	70—98 years	F	13.5 ± 0.79	—	—	% that are E-rosette-forming cells in medium with human D serum and sheep erythrocytes sensitized by incomplete anti-D antibody	
T_G lymphocytes	Human	(10)	3rd—4th decades	M	238 ± 103	—	—	No. per mm^3 binding 3 or more IgG-sensitized ox erythrocytes	141
		(21)	5th—8th decades	M	249 ± 117	—	—		
T_G lymphocytes	Human	(10)	3rd—4th decades	F	225 ± 140	—	—	No. per mm^3 binding 3 or more IgG-sensitized ox erythrocytes	141
		(23)	5th—9th decades	F	319 ± 144	—	—		
T_γ (T_G) lymphocytes	Human	(31)	Young	M	230 ± 128	—	—	Total no. per mm^3; young from group of M and F aged 22—52 years; old from group of M and F aged 60—80 years	138
		(28)	Old	M	279 ± 130	—	—		
T_γ (T_G) lymphocytes	Human	(28)	Young	F	176 ± 116	—	—	Total no. per mm^3; young from group of M and F aged 22—52 years; old from group of M and F aged 60—80 years	138
		(33)	Old	F	380 ± 241	—	—		
T_M lymphocytes	Human	(10)	3rd—4th decades	M	894 ± 250	—	—	No. per mm^3 binding 3 or more IgM-sensitized ox erythrocytes	141
		(21)	4th—8th decades	M	538 ± 245	—	—		
T_M lymphocytes	Human	(10)	3rd—4th decades	F	795 ± 436	—	—	No. per mm^3 binding 3 or more IgM-sensitized ox erythrocytes	141
		(23)	5th—9th decades	F	645 ± 309	—	—		

Table 5 (continued)
NUMBER OF CELLS VS. AGE IN BLOOD AND BLOOD VESSELS

Cell	Subjects				Count		Measure	Ref.
	Type	(no.)	Age	Sex	Mean	Range		
T_μ (T_M) lymphocytes	Human	(31)	Young	M	878 ± 415	—	Total no. per mm³; young from group of M and F aged 22—52 years; old from group of M and F aged 60—80 years	138
		(28)	Old	M	668 ± 394	—		
T_μ (T_M) lymphocytes	Human	(28)	Young	F	827 ± 371	—	Total no. per mm³; young from group of M and F aged 22—52 years; old from group of M and F aged 60—80 years	138
		(33)	Old	F	592 ± 289	—		
T lymphocytes, HTLA-antigen-bearing cells	Human	(26)	3—21 days	M,F	3,297.0 ± 278.1	—	Total no. per mm³	137
		(12)	3 months	M,F	2,870.2 ± 218.9	—		
		(40)	20—50 years	M,F	1,341.5 ± 16.3	—		
		(40)	75—97 years	M,F	950.3 ± 39.2	—		
T lymphocytes, HTLA-antigen-bearing cells	Human	(56)	20—39 years	M,F	1,226 ± 507	—	Total no. per mm³	129
		(71)	40—59 years	M,F	1,218 ± 434	—		
		(77)	60—69 years	M,F	1,213 ± 524	—		
		(40)	80—99 years	M,F	1,420 ± 599	—		
Lymphocytes (T3+)	Human	(19)	<40 years	M,F	5,088 ± 239	—	No. per mm³	146
		(31)	>60 years	M,F	4,523 ± 197	—		
Lymphocytes, helper/inducer cells (T4+)	Human	(19)	<40 years	M,F	3,559 ± 222	—	No. per mm³	146
		(31)	>60 years	M,F	3,400 ± 178	—		
Lymphocytes, suppressor/cytotoxic cells (T8+)	Human	(19)	<40 years	M,F	1,670 ± 84	—	No. per mm³	146
		(31)	>60 years	M,F	1,368 ± 117	—		

Cell type	Species	(No. tested)	Age	Sex	Value	Range	Units	Ref.
B lymphocytes	Human	(29)	Newborn—10 years	M,F	339 ± 233	—	No. per μℓ	136
B lymphocytes	Human	(50)	11—60 years	M,F	325 ± 190	—	No. per mm³	134
		(22)	61—98 years	M,F	269 ± 209	—		
		(26)	20—40 years	—	512 ± 49	—		
		(18)	60—69 years	—	490 ± 58	—		
		(20)	70—79 years	—	552 ± 41	—		
		(16)	80—89 years	—	399 ± 42	—		
		(18)	90—96 years	—	653 ± 154	—		
B lymphocytes	Human	(12)	3—12 years	M,F	300 ± 140	110—660	Total no. per μℓ	128
		(20)	19—25 years	M,F	350 ± 70	250—500		
		(20)	33—50 years	M,F	330 ± 100	220—520		
		(20)	65—83 years	M,F	270 ± 140	170—650		
B lymphocytes	Human	(10)	3rd—4th decades	M	138 ± 72	—	No. per mm³	141
		(20)	5th—8th decades	M	74 ± 44	—		
B lymphocytes	Human	(10)	3rd—4th decades	F	92 ± 68	—	No. per mm³	141
		(23)	5th—9th decades	F	83 ± 60	—		
B lymphocytes, EA-rosette formers	Human	(26)	3—21 days	M,F	314.6 ± 50.4	—	Total no. per mm³	137
		(12)	3 months	M,F	486.6 ± 49.6	—		
		(40)	20—50 years	M,F	217.4 ± 18.4	—		
		(40)	75—97 years	M,F	161.7 ± 15.6	—		
B lymphocytes, EAC-rosette formers	Human	(26)	3—21 days	M,F	1,060.9 ± 109.5	—	Total no. per mm³	137
		(12)	3 months	M,F	821.3 ± 43.5	—		
		(40)	20—50 years	M,F	279.3 ± 17.5	—		
		(40)	75—97 years	M,F	168.6 ± 9.3	—		
B lymphocytes, EAC-rosette formers	Human	(56)	20—39 years	M,F	252 ± 105	—	Total no. per mm³	129
		(71)	40—59 years	M,F	289 ± 117	—		
		(77)	60—69 years	M,F	298 ± 148	—		
		(40)	80—99 years	M,F	274 ± 118	—		
B lymphocytes, with surface Ig	Human	(26)	3—31 days	M,F	796.6 ± 78.2	—	Total no. per mm³	137
		(12)	3 months	M,F	602.6 ± 65.1	—		
		(40)	20—50 years	M,F	203.3 ± 16.8	—		
		(40)	75—97 years	M,F	225.1 ± 10.5	—		
Lymphocytes, Fc-receptor positive	Human	(15)	20—39 years	M,F	121 ± 19	—	No. per mm³	130
		(15)	76—93 years	M,F	170 ± 23	—		
Lymphocytes, Fc-receptor positive	Human	(15)	20—39 years	M,F	50 ± 9	—	No. per mm³	130
		(15)	76—93 years	M,F	69 ± 11	—		

Table 5 (continued)
NUMBER OF CELLS VS. AGE IN BLOOD AND BLOOD VESSELS

Cell	Type	Subjects (no.)	Age	Sex	Count Mean	Count Range	Measure	Ref.
Lymphocytes, ANAE-positive	Human	(52)	20—45 years	M,F	62.3 ± 2.46	—	% stained by non-specific α-naphthyl-acetate esterase	145
		(56)	70—98 years	M,F	68.0 ± 2.26	—		
Lymphocytes, ANAE-positive	Human	(52)	20—45 years	M,F	(a) 51.8 ± 1.84 (b) 11.6 ± 1.21	—	% stained by non-specific α-naphthyl-acetate esterase; (a) single spotted. (b) granular scattered	145
		(56)	70—98 years	M,F	48.1 ± 2.01 18.2 ± 1.86	—		
"Null" lymphocytes	Human	(26)	20—40 years	—	448 ± 55	—	No. per $\mu\ell$	134
		(18)	60—69 years	—	516 ± 83	—		
		(20)	70—79 years	—	345 ± 83	—		
		(16)	80—89 years	—	331 ± 64	—		
		(18)	90—96 years	—	218 ± 62	—		
Lymphocytes, (other than T and B cells)	Human	(29)	Newborn—10 years	M,F	1,899 ± 1,274	—	No. per $\mu\ell$	136
		(50)	11—60 years	M,F	607 ± 310	—		
		(22)	61—98 years	M,F	542 ± 369	—		
Total lymphocytes	Mouse	(29)	4—7 months	F	12,179	—	No. per mm^3	118
		(28)	16.5—21 months	F	8,632			
Total lymphocytes	Mouse	(31)	8—10 months	M	7,630 ± 2,870	4,000—14,000	No. per mm^3	120
		(14)	25—28 months	M	7,100 ± 2,450	4,000—12,000		
Total lymphocytes	Mouse	(30)	110 days	M	8,330 ± 550	—	No. per mm^3	122
		(30)	200 days	M	8,500 ± 480	—		
		(30)	286 days	M	8,120 ± 540	—		
		(36)	373 days	M	7,770 ± 550	—		
		(57)	417 days	M	8,390 ± 340	—		

		(n)	Age	Sex	No. per mm³	
Total lymphocytes	Mouse	(59)	473 days	M	9,050 ± 330	122
		(56)	528 days	M	6,450 ± 310	
		(49)	586 days	M	5,880 ± 240	
		(55)	626 days	M	5,890 ± 280	
		(53)	670 days	M	6,910 ± 310	
		(51)	710 days	M	7,940 ± 340	
		(52)	761 days	M	6,200 ± 280	
		(53)	822 days	M	8,300 ± 470	
		(54)	864 days	M	8,010 ± 440	
		(51)	906 days	M	6,980 ± 460	
		(52)	948 days	M	5,940 ± 280	
		(50)	991 days	M	7,040 ± 280	
		(53)	1044 days	M	7,140 ± 380	
		(30)	118 days	F	8,440 ± 520	
		(30)	216 days	F	8,330 ± 580	
		(30)	302 days	F	7,060 ± 430	
		(30)	375 days	F	7,910 ± 450	
		(57)	417 days	F	7,060 ± 260	
		(60)	473 days	F	6,430 ± 220	
		(58)	530 days	F	5,770 ± 160	
		(58)	587 days	F	4,850 ± 230	
		(58)	627 days	F	4,630 ± 180	
		(53)	670 days	F	6,080 ± 320	
		(54)	711 days	F	4,870 ± 310	
		(52)	762 days	F	5,770 ± 250	
		(52)	823 days	F	7,010 ± 350	
		(50)	866 days	F	5,510 ± 270	
		(51)	907 days	F	5,920 ± 340	
		(52)	950 days	F	4,930 ± 240	
		(50)	991 days	F	4,810 ± 260	
		(34)	1045 days	F	6,370 ± 400	
Total lymphocytes	Bull	(9)	1—2 years	M	7,608	125
		(8)	2—4 years	M	4,598	
		(11)	4—6 years	M	3,800	
		(12)	6—8 years	M	3,261	
		(10)	8—10 years	M	2,978	
		(7)	10—12 years	M	3,199	
		(7)	12—15 years	M	2,733	

Table 5 (continued)
NUMBER OF CELLS VS. AGE IN BLOOD AND BLOOD VESSELS

Cell	Subjects				Count		Measure	Ref.
	Type	(no.)	Age	Sex	Mean	Range		
Total lymphocytes	Cow	(25)	1—3 years	F	6,471	—	No. per mm³	125
		(10)	5—12 years	F	3,478	—		
Total lymphocytes	Sheep	(9)	0.25—0.5 years	M,F	4,729	—	No. per mm³	125
		(10)	2 years	M,F	5,384	—		
		(10)	4—9 years	M,F	3,654	—		
Total lymphocytes	Goat	(3)	0.5 years	F	4,955	—	No. per mm³	125
		(4)	1—2 years	F	6,764	—		
		(7)	2—3 years	F	5,141	—		
		(2)	3—6 years	F	4,249	—		
Platelets	Human	(27)	20—63 years	M,F	190×10^3	—	No. per mm³	147
		(14)	67—94 years	M,F	253×10^3	—		
Platelets	Human	(56)	65—69 years	F	$2.68 \times 10^{11} \pm 0.09 \times 10^{11}$	—	No. per liter	115
		(41)	70—74 years	F	$2.47 \times 10^{11} \pm 0.09 \times 10^{11}$	—		
		(29)	75—79 years	F	$2.93 \times 10^{11} \pm 0.21 \times 10^{11}$	—		
		(19)	80—84 years	F	$2.29 \times 10^{11} \pm 0.16 \times 10^{11}$	—		
		(10)	85+ years	F	$2.56 \times 10^{11} \pm 0.26 \times 10^{11}$	—		
Platelets	Human	(51)	65—69 years	M	$2.44 \times 10^{11} \pm 0.11 \times 10^{11}$	—	No. per liter	115
		(43)	70—74 years	M	$2.49 \times 10^{11} \pm 0.10 \times 10^{11}$	—		
		(29)	75—79 years	M	$2.23 \times 10^{11} \pm 0.10 \times 10^{11}$	—		
		(15)	80—84 years	M	$2.29 \times 10^{11} \pm 0.21 \times 10^{11}$	—		
		(8)	85+ years	M	$2.70 \times 10^{11} \pm 0.28 \times 10^{11}$	—		
Blood Vessels								
Mast cells, tunica adventitia of vena ilica externa	Human	(8)	1 day—1 year, 9 months	M	830	321—1,436	No. per mm³; average of right and left vein	148
		(5)	3—14 years	M	986	461—2,314		

77

	(n)	Age	Sex	No.	Range	Notes	Ref.
Mast cells, tunica adventitia of vena ilica externa — Human	(10)	20—35 years	M	1,226	707—1,778		
	(13)	41—59 years	M	1,157	493—2,560		
	(13)	61—77 years	M	745	107—1,692		
	(1)	85 years	M	370			
	(7)	2 days—1 year, 5 months	F	803	257—1,392	No. per mm³; average of right and left vein	148
	(4)	2.5—15 years	F	846	428—1,189		
	(8)	21—39 years	F	1,580	1,136—2,239		
	(5)	44—58 years	F	1,151	578—1,424		
	(6)	65—77 years	F	745	311—1,382		
	(2)	82—85 years	F	863	638—1,211		
Mast cells, tunica adventitia of aorta abdominalis — Human	(19)	1 day—1 year, 9 months	M	953	278—1,864	No. per mm³	148
	(10)	2—14 years	M	1,168	353—2,849		
	(8)	20—35 years	M	1,484	332—2,560		
	(10)	41—57 years	M	1,189	707—1,821		
	(4)	61—78	M	1,023	546—1,553		
Mast cells, tunica adventitia of aorta abdominalis — Human	(13)	1.5 day—1 year, 11 months	F	851	471—1,328	No. per mm³	148
	(3)	4—15 years	F	1,146	600—1,436		
	(7)	21—38 years	F	1,660	1,049—3,235		
	(9)	44—59 years	F	969	396—1,842		
	(12)	60—77 years	F	1,157	471—2,613		
	(3)	81—88 years	F	1,173	771—1,735		
Mast cells, tunica adventitia of vena cava caudalis — Human	(16)	1 day—1 year, 9 months	M	1,018	289—1,856	No. per mm³	148
	(9)	2—14 years	M	1,039	300—2,517		
	(10)	20—35 years	M	1,178	621—2,303		
	(20)	41—59 years	M	1,055	332—2,078		
	(19)	61—78 years	M	728	278—1,232		
	(1)	85 years	M	536	—		
Mast cells, tunica adventitia of vena cava caudalis — Human	(14)	1.5—1 year, 5 months	F	894	375—1,660	No. per mm³	148
	(4)	2.5—15 years	F	1,055	482—1,810		
	(8)	21—39 years	F	1,751	1,264—2,024		
	(11)	44—59 years	F	1,141	718—2,282		
	(19)	60—79 years	F	771	107—1,895		
	(4)	82—88 years	F	595	353—932		

Table 5 (continued)
NUMBER OF CELLS VS. AGE IN BLOOD AND BLOOD VESSELS

Cell	Type	(no.)	Age	Sex	Count		Measure	Ref.
					Mean	Range		
Mast cells, tunica adventitia of vena subclavia	Human	(8)	3 days—1 year, 9 months	M	884	246—1,553	No. per mm³; average of right and left vein	148
		(5)	3—14 years	M	1,082	514—1,649		
		(10)	20—35 years	M	1,039	257—2,078		
		(13)	41—59 years	M	1,109	75—5,366		
		(12)	61—77 years	M	959	193—1,778		
Mast cells, tunica adventitia of vena subclavia	Human	(5)	2 days—1 year, 9 months	M	1,028	525—1,436	No. per mm³; average of right and left vein	148
		(4)	2.5—15 years	M	938	246—1,296		
		(7)	21—38 years	M	1,436	921—2,549		
		(3)	46—58 years	M	1,216	557—1,885		
		(7)	64—77 years	M	771	321—1,328		
		(2)	82—85 years	M	691	428—986		
Endothelial cells, walls of capillaries in cerebral cortex	Rat	(4)	3 months	F	20.40 ± 0.43	—	No. cells per "core" sample of cerebral cortex	44
		(4)	22—27 months	F	18.55 ± 1.34	—		
Pericytes, walls of capillaries in cerebral cortex	Rat	(4)	3 months	F	5.73 ± 0.72	—	No. cells per "core" sample of cerebral cortex	44
		(4)	22—27 months	F	4.00 ± 0.38	—		

Note: est., estimated.

Table 6
NUMBER OF BONE MARROW TOTAL AND HEMATOPOIETIC CELLS VS AGE

Tissue and cell type	Type	Subjects (no.)	Age	Sex	Mean cell count	Measure	Ref.
Total skeleton marrow cells	Mouse	(11)	5 weeks	—	$3.08 \times 10^9 \pm 0.26 \times 10^9$	No. nucleated cells per 100 g body weight	152
		(12)	6 weeks	—	$2.10 \times 10^9 \pm 0.10 \times 10^9$		
		(12)	20 weeks	—	$2.62 \times 10^9 \pm 0.19 \times 10^9$		
		(11)	30 weeks	—	$2.48 \times 10^9 \pm 0.17 \times 10^9$		
		(10)	52 weeks	—	$2.19 \times 10^9 \pm 0.17 \times 10^9$		
		(9)	94 weeks	—	$2.35 \times 10^9 \pm 0.16 \times 10^9$		
		(12)	130 weeks	—	$3.69 \times 10^9 \pm 0.39 \times 10^9$		
Total bone marrow, nucleated cells	Mouse	(—)	1.4 months	M,F	5.0×10^8	Total bone marrow nucleated cells per mouse; est. from data given; 12—24 animals per age group	153
		(—)	1.5 months	M,F	4.6×10^8		
		(—)	5.0 months	M,F	7.9×10^8		
		(—)	7.6 months	M,F	8.6×10^8		
		(—)	13.1 months	M,F	8.6×10^8		
		(—)	23.2 months	M,F	8.0×10^8		
		(—)	36.0 months	M,F	10.8×10^8		
Femur marrow, nucleated cells	Mouse	(85)	73 days	F	1.20×10^7	Total no. per femur	121
		(27)	89 days	F	1.28×10^7		
		(3)	100 days	F	1.82×10^7		
		(3)	100 days	F	2.00×10^7		
		(3)	105 days	F	1.78×10^7		
		(50)	109 days	F	1.45×10^7		
		(3)	121 days	F	2.00×10^7		
		(3)	127 days	F	2.17×10^7		
		(3)	147 days	F	1.93×10^7		
		(3)	160 days	F	2.03×10^7		
		(3)	169 days	F	2.50×10^7		
		(3)	185 days	F	1.88×10^7		
		(4)	271 days	F	2.48×10^7		
		(3)	293 days	F	1.85×10^7		
		(3)	308 days	F	2.23×10^7		

Table 6 (continued)
NUMBER OF BONE MARROW TOTAL AND HEMATOPOIETIC CELLS VS AGE

Tissue and cell type	Type	Subjects (no.)	Age	Sex	Mean cell count	Measure	Ref.
		(3)	315 days	F	1.68×10^7		
		(3)	329 days	F	2.37×10^7		
		(3)	365 days	F	2.90×10^7		
		(3)	404 days	F	2.31×10^7		
		(3)	418 days	F	2.35×10^7		
		(3)	420 days	F	3.12×10^7		
		(3)	462 days	F	1.96×10^7		
		(3)	604 days	F	2.53×10^7		
		(3)	638 days	F	3.23×10^7		
		(3)	641 days	F	3.00×10^7		
		(3)	652 days	F	2.73×10^7		
		(3)	666 days	F	2.96×10^7		
		(5)	763 days	F	2.84×10^7		
		(3)	777 days	F	3.56×10^7		
Femur marrow, total cells	Mouse	( )	3 months	F	28.026×10^6	Total no. per femur	154
		( )	12 months	F	24.062×10^6		
		( )	18 months	F	26.333×10^6		
		( )	20 months	F	38.101×10^6		
		( )	24 months	F	35.310×10^6		
Femur marrow, total cells	Mouse	( )	3 months	M,F	23.2×10^6	Total no. per femur; est. from data given; 12—15 animals per age group	155
		( )	7 months	M,F	22.5×10^6		
		( )	18 months	M,F	32.7×10^6		
		( )	27 months	M,F	34.5×10^6		
		( )	33 months	M,F	38.8×10^6		
Femur marrow, nucleated cells	Mouse	( )	3.5 months	M	$16.8 \times 10^6 \pm 0.1 \times 10^6$	Total no. per femur; 5—9 animals per age group	156
		( )	6.5 months	M	$18.9 \times 10^6 \pm 0.6 \times 10^6$		
		( )	12 months	M	$18.5 \times 10^6 \pm 0.2 \times 10^6$		
		( )	15 months	M	$23.4 \times 10^6 \pm 0.3 \times 10^6$		
		( )	20 months	M	$25.9 \times 10^6 \pm 0.6 \times 10^6$		

		(No.)	Age	Sex	Value	Comments	Ref.
Total bone marrow, stem cells	Mouse	(—)	26 months	M	$29.8 \times 10^6 \pm 0.2 \times 10^6$	Total colony-forming units in bone marrow per mouse; est. from data given; 12—24 animals per age group	153
		(—)	32 months	M	$29.2 \times 10^6 \pm 0.5 \times 10^6$		
		(—)	33 months	M	$27.5 \times 10^6 \pm 0.1 \times 10^6$		
		(—)	35 months	M	$32.9 \times 10^6 \pm 1.0 \times 10^6$		
		(—)	1.4 months	M,F	6.9×10^4		
		(—)	1.5 months	M,F	6.0×10^4		
		(—)	5.0 months	M,F	11.2×10^4		
		(—)	7.6 months	M,F	10.8×10^4		
		(—)	13.1 months	M,F	10.2×10^4		
		(—)	23.2 months	M,F	10.0×10^4		
		(—)	36.0 months	M,F	12.0×10^4		
Femur marrow, stem cells	Mouse	(—)	3 months	F	5.12	Total no. per femur	154
		(—)	12 months	F	6.38		
		(—)	18 months	F	10.26		
		(—)	20 months	F	6.48		
		(—)	24 months	F	5.90		
Femur marrow, stem cells	Mouse	(18)	4 weeks	—	13.99 ± 2.01	Total no. "colony forming units" per 10^5 bone marrow cells	152
		(20)	6 weeks	—	13.08 ± 1.18		
		(8)	20 weeks	—	14.23 ± 1.14		
		(22)	30 weeks	—	12.44 ± 1.09		
		(10)	52 weeks	—	11.72 ± 1.10		
		(15)	100 weeks	—	12.22 ± 1.58		
		(12)	150 weeks	—	10.88 ± 0.86		
Femur marrow, stem cells	Mouse	(15—20)	1 week	M,F	9.76 ± 0.92	Total no. colony units formed in a 5-μm thick longitudinal spleen section per inoculum of 10^5 bone marrow cells	157
		(15—20)	2 weeks	M,F	12.40 ± 1.64		
		(15—20)	3 weeks	M,F	20.00 ± 2.56		
		(15—20)	11 weeks	M,F	19.80 ± 2.30		
		(15—20)	52 weeks	M,F	22.10 ± 2.60		
		(15—20)	58 weeks	M,F	19.40 ± 2.50		
		(15—20)	78 weeks	M,F	6.62 ± 1.51		
		(15—20)	100 weeks	M,F	5.25 ± 1.67		
		(15—20)	113 weeks	M,F	4.20 ± 1.80		
Femur marrow, stem cells	Mouse	(—)	3 months	M,F	6,010	Total spleen colony-forming units per femur; est. from data given; 12—15 animals per age group	155
		(—)	7 months	M,F	5,380		
		(—)	18 months	M,F	4,690		
		(—)	27 months	M,F	5,250		
		(—)	33 months	M,F	5,400		

Table 6 (continued)
NUMBER OF BONE MARROW TOTAL AND HEMATOPOIETIC CELLS VS AGE

Tissue and cell type	Type	Subjects (no.)	Age	Sex	Mean cell count	Measure	Ref.
Femur marrow, stem cells	Mouse	(—)	3 months	M,F	27,000	Total agar colony-forming units per femur; at least 5 animals per age group	155
		(—)	7 months	M,F	12,600		
		(—)	27 months	M,F	24,800		
Femur marrow, stem cells	Mouse	(—)	3.5 months	M	4,284	Total no. per femur from histological method: 5—9 animals per age group	156
		(—)	6.5 months	M	5,084		
		(—)	12 months	M	3,959		
		(—)	15 months	M	5,382		
		(—)	20 months	M	4,817		
		(—)	26 months	M	5,304		
		(—)	32 months	M	5,402		
		(—)	33 months	M	5,665		
		(—)	35 months	M	7,370		
Femur marrow, stem cells	Mouse	(85)	73 days	F	$3.71 \times 10^3 \pm 0.47 \times 10^3$	Total colony-forming units per femur	121
		(27)	89 days	F	$4.25 \times 10^3 \pm 0.27 \times 10^3$		
		(3)	100 days	F	$6.62 \times 10^3 \pm 0.47 \times 10^3$		
		(3)	100 days	F	$4.94 \times 10^3 \pm 0.52 \times 10^3$		
		(3)	105 days	F	$3.74 \times 10^3 \pm 0.67 \times 10^3$		
		(50)	109 days	F	$4.53 \times 10^3 \pm 0.45 \times 10^3$		
		(3)	121 days	F	$5.02 \times 10^3 \pm 0.28 \times 10^3$		
		(3)	127 days	F	$5.19 \times 10^3 \pm 0.40 \times 10^3$		
		(3)	147 days	F	$4.25 \times 10^3 \pm 0.33 \times 10^3$		
		(3)	160 days	F	$5.68 \times 10^3 \pm 0.67 \times 10^3$		
		(3)	169 days	F	$4.00 \times 10^3 \pm 0.57 \times 10^3$		
		(3)	185 days	F	$4.49 \times 10^3 \pm 0.25 \times 10^3$		
		(4)	271 days	F	$7.56 \times 10^3 \pm 0.64 \times 10^3$		
		(3)	293 days	F	$3.75 \times 10^3 \pm 0.23 \times 10^3$		
		(3)	308 days	F	$5.42 \times 10^3 \pm 0.36 \times 10^3$		
		(3)	315 days	F	$4.99 \times 10^3 \pm 0.48 \times 10^3$		
		(3)	329 days	F	$6.36 \times 10^3 \pm 0.56 \times 10^3$		

83

Tissue, cell type	Species	(No.)	Age	Sex	Value	Notes	Ref.
		(3)	365 days	F	$8.70 \times 10^3 \pm 0.34 \times 10^3$		
		(3)	404 days	F	$4.87 \times 10^3 \pm 0.49 \times 10^3$		
		(3)	418 days	F	$5.43 \times 10^3 \pm 1.49 \times 10^3$		
		(3)	420 days	F	$7.49 \times 10^3 \pm 0.96 \times 10^3$		
		(3)	462 days	F	$4.03 \times 10^3 \pm 0.29 \times 10^3$		
		(3)	604 days	F	$4.41 \times 10^3 \pm 0.69 \times 10^3$		
		(3)	638 days	F	$6.07 \times 10^3 \pm 0.75 \times 10^3$		
		(3)	641 days	F	$4.83 \times 10^3 \pm 0.64 \times 10^3$		
		(3)	652 days	F	$6.01 \times 10^3 \pm 1.05 \times 10^3$		
		(3)	666 days	F	$6.59 \times 10^3 \pm 0.50 \times 10^3$		
		(5)	763 days	F	$6.56 \times 10^3 \pm 0.53 \times 10^3$		
		(3)	777 days	F	$6.56 \times 10^3 \pm 0.63 \times 10^3$		
Femur marrow, "blast" (precursor) cells	Guinea pig	(5)	1—2 days	M,F	1.25×10^6	Total no.; est. from data given	124
		(61)	3—38 days	M,F	3.4×10^6		
		(25)	46—229 days	M,F	3.4×10^6		
		(13)	406—1820 days	M,F	2.8×10^6		
Total bone marrow, C-Ig cells	Mouse	(—)	6 weeks	M	62×10^3	Total containing cytoplasmic immunoglobulin (C-Ig) per total body bone marrow; est. from data given; at least 3 animals per age group	158
		(—)	26 weeks	M	269×10^3		
		(—)	1 year	M	344×10^3		
		(—)	2 years	M	362×10^3		
Femur marrow, erythrocytes	Guinea pig	(5)	1—2 days	M,F	7.5×10^6	Total no.; est. from data given	124
		(61)	3—38 days	M,F	66×10^6		
		(25)	46—229 days	M,F	70×10^6		
		(13)	406—1820 days	M,F	41×10^6		
Femur marrow, neutrophils	Guinea pig	(5)	1—2 days	M,F	19×10^6	Total no.; est. from data given	124
		(61)	3—38 days	M,F	49×10^6		
		(25)	46—229 days	M,F	94×10^6		
		(13)	406—1820 days	M,F	120×10^6		
Femur marrow, eosinophils	Guinea pig	(5)	1—2 days	M,F	0.4×10^6	Total no.	124
		(61)	3—38 days	M,F	2.2×10^6		
		(25)	46—229 days	M,F	10×10^6		
		(13)	406—1820 days	M,F	15×10^6		
Femur marrow, lymphocytes	Guinea pig	(5)	1—2 days	M,F	39×10^6	Total no.; est. from data given	124
		(61)	3—38 days	M,F	50×10^6		
		(25)	46—229 days	M,F	61×10^6		
		(13)	406—1820 days	M,F	36×10^6		

Table 6 (continued)
NUMBER OF BONE MARROW TOTAL AND HEMATOPOIETIC CELLS VS AGE

Tissue and cell type	Subjects		Age	Sex	Mean cell count	Measure	Ref.
	Type	(No.)					
Femur marrow, monocytes	Guinea pig	(5)	1—2 days	M,F	5.3×10^6	Total no.; est. from data given	124
		(61)	3—38 days	M,F	16×10^6		
		(25)	46—229 days	M,F	20×10^6		
		(13)	406—1820 days	M,F	26×10^6		

Note: est., estimated.

Table 7
NUMBER OF CELLS VS. AGE IN THE SPLEEN, THYMUS AND LYMPH NODES

Tissue and cell type	Subjects		Age	Sex	Cell counts		Measure	Ref.
	Type	(no.)			Mean	Range		
Spleen								
Total cells	Mouse	(—)	100—300 days	M,F	3.0×10^7	2.3×10^7—4.7×10^7	Total no. per spleen. range = 95% confidence limits	159
		(—)	400—800 days	M,F	4.0×10^7	2.4×10^7—5.6×10^7		
Total nucleated cells	Mouse	(85)	73 days	F	1.77×10^8	—	Total no. per spleen	121
		(27)	89 days	F	3.65×10^8	—		
		(3)	100 days	F	2.09×10^8	—		
		(3)	100 days	F	2.57×10^8	—		
		(3)	105 days	F	2.77×10^8	—		
		(50)	109 days	F	2.60×10^8	—		
		(3)	121 days	F	2.34×10^8	—		
		(3)	127 days	F	2.31×10^8	—		
		(3)	135 days	F	2.06×10^8	—		

Total nucleated cells	Mouse	(3)	147 days	F	2.01 × 10⁸	—		
		(3)	160 days	F	2.23 × 10⁸	—		
		(3)	169 days	F	2.35 × 10⁸	—		
		(3)	185 days	F	1.99 × 10⁸	—		
		(4)	271 days	F	2.12 × 10⁸	—		
		(3)	293 days	F	1.78 × 10⁸	—		
		(3)	308 days	F	2.16 × 10⁸	—		
		(3)	315 days	F	2.75 × 10⁸	—		
		(3)	329 days	F	2.13 × 10⁸	—		
		(3)	365 days	F	3.80 × 10⁸	—		
		(3)	404 days	F	2.08 × 10⁸	—		
		(3)	418 days	F	1.98 × 10⁸	—		
		(3)	420 days	F	2.55 × 10⁸	—		
		(3)	462 days	F	2.40 × 10⁸	—		
		(3)	604 days	F	3.22 × 10⁸	—		
		(3)	638 days	F	2.81 × 10⁸	—		
		(3)	641 days	F	2.72 × 10⁸	—		
		(3)	652 days	F	2.72 × 10⁸	—		
		(3)	666 days	F	2.72 × 10⁸	—		
		(5)	763 days	F	3.64 × 10⁸	—		
		(3)	777 days	F	3.54 × 10⁸	—		
Total nucleated cells	Mouse	(1)	2.0 weeks	M,F	44.4 × 10⁹	—	Total nucleated cells per spleen; est. from data given; 10—16 animals per age group	153
		(1)	4.9 weeks	M,F	171 × 10⁶	—		
		(1)	7.2 weeks	M,F	180 × 10⁶	—		
		(1)	15.3 weeks	M,F	335 × 10⁶	—		
		(1)	33.3 weeks	M,F	302 × 10⁶	—		
		(1)	74.8 weeks	M,F	160 × 10⁶	—		
		(1)	92.3 weeks	M,F	240 × 10⁶	—		
		(1)	109 weeks	M,F	158 × 10⁶	—		
		(1)	122 weeks	M,F	200 × 10⁶	—		
Total stromal cells	Mouse	(1)	9 weeks	M	2.25 × 10⁸	—	Total no. per spleen	160
		(1)	15 weeks	M	2.05 × 10⁸	—		
		(1)	24 weeks	M	2.33 × 10⁸	—		
		(1)	85 weeks	M	2.62 × 10⁸	—		
Total stem cells	Mouse	(—)	2.0 weeks	M,F	6.8 × 10⁵	—	Total colony-forming units per spleen; est. from data given; 10—	153
		(—)	4.9 weeks	M,F	27.4 × 10⁵	—		
		(—)	7.2 weeks	M,F	21.2 × 10⁵	—		

Table 7 (continued)
NUMBER OF CELLS VS. AGE IN THE SPLEEN, THYMUS AND LYMPH NODES

Tissue and cell type	Subjects			Sex	Cell counts		Measure	Ref.
	Type	(no.)	Age		Mean	Range		
		(—)	15.3 weeks	M,F	23.6×10^5	—	16 animals per age group	
		(—)	33.3 weeks	M,F	16.0×10^5	—		
		(—)	74.8 weeks	M,F	4.1×10^5	—		
		(—)	92.3 weeks	M,F	6.2×10^5	—		
		(—)	109 weeks	M,F	5.4×10^5	—		
		(—)	122 weeks	M,F	9.4×10^5	—		
Total stem cells	Mouse	(85)	73 days	F	$2.19 \times 10^3 \pm 0.25 \times 10^3$	—	Total colony-forming units per spleen	121
		(27)	89 days	F	$5.10 \times 10^3 \pm 0.37 \times 10^3$	—		
		(3)	100 days	F	$5.21 \times 10^3 \pm 0.50 \times 10^3$	—		
		(3)	100 days	F	$2.68 \times 10^3 \pm 0.23 \times 10^3$	—		
		(3)	105 days	F	$1.73 \times 10^3 \pm 0.23 \times 10^3$	—		
		(50)	109 days	F	$3.51 \times 10^3 \pm 0.57 \times 10^3$	—		
		(3)	121 days	F	$1.25 \times 10^3 \pm 0.23 \times 10^3$	—		
		(3)	127 days	F	$2.48 \times 10^3 \pm 0.23 \times 10^3$	—		
		(3)	135 days	F	$2.18 \times 10^3 \pm 0.35 \times 10^3$	—		
		(3)	147 days	F	$1.63 \times 10^3 \pm 0.24 \times 10^3$	—		
		(3)	160 days	F	$1.47 \times 10^3 \pm 0.11 \times 10^3$	—		
		(3)	169 days	F	$1.31 \times 10^3 \pm 0.14 \times 10^3$	—		
		(3)	185 days	F	$2.55 \times 10^3 \pm 0.26 \times 10^3$	—		
		(4)	271 days	F	$4.53 \times 10^3 \pm 0.43 \times 10^3$	—		
		(3)	293 days	F	$1.91 \times 10^3 \pm 0.25 \times 10^3$	—		
		(3)	308 days	F	$2.47 \times 10^3 \pm 0.33 \times 10^3$	—		
		(3)	315 days	F	$0.96 \times 10^3 \pm 0.28 \times 10^3$	—		
		(3)	329 days	F	$2.20 \times 10^3 \pm 0.10 \times 10^3$	—		
		(3)	365 days	F	$8.06 \times 10^3 \pm 1.13 \times 10^3$	—		
		(3)	404 days	F	$2.39 \times 10^3 \pm 0.22 \times 10^3$	—		
		(3)	418 days	F	$1.29 \times 10^3 \pm 0.24 \times 10^3$	—		
		(3)	420 days	F	$1.66 \times 10^3 \pm 0.20 \times 10^3$	—		
		(3)	462 days	F	$2.19 \times 10^3 \pm 0.09 \times 10^3$	—		
		(3)	638 days	F	$1.63 \times 10^3 \pm 0.19 \times 10^3$	—		

Parameter	Species	(n)	Age	Sex	Value		Ref.	Comments
Total stem cells	Mouse	(3)	641 days	F	$1.97 \times 10^3 \pm 0.45 \times 10^3$	—	160	Total colony-forming units per spleen
		(3)	652 days	F	$2.60 \times 10^3 \pm 0.38 \times 10^3$	—		
		(3)	666 days	F	$2.12 \times 10^3 \pm 0.16 \times 10^3$	—		
		(5)	763 days	F	$5.58 \times 10^3 \pm 0.41 \times 10^3$	—		
		(3)	777 days	F	$3.05 \times 10^3 \pm 0.44 \times 10^3$	—		
Total antibody-forming cells	Mouse	(1)	9 weeks	M	$4,200 \pm 378$	—	161	Total no. per spleen; est. from data given; 10—20 animals per age group
		(1)	15 weeks	M	$4,500 \pm 366$	—		
		(1)	24 weeks	M	$4,250 \pm 326$	—		
		(1)	85 weeks	M	$5,833 \pm 560$	—		
Total antibody-forming cells	Mouse	(—)	1 week	—	2.2×10^3	—		
		(—)	5 weeks	—	7.1×10^3	—		
		(—)	8 weeks	—	4.9×10^3	—		
		(—)	12 weeks	—	3.1×10^3	—		
		(—)	15 weeks	—	3.0×10^3	—		
		(—)	100 weeks	—	2.3×10^3	—		
		(—)	120 weeks	—	2.0×10^3	—		
Total antibody-forming cells	Mouse	(8)	Birth	—	0	—	162	Total no. per spleen; calculated from data given
		(7)	1 week	—	96	—		
		(7)	2 weeks	—	316	—		
		(7)	3 weeks	—	5,620	—		
		(9)	1 month	—	20,400	—		
		(12)	3 months	—	77,600	—		
		(9)	7 months	—	91,200	—		
		(10)	12 months	—	22,400	—		
		(6)	22 months	—	17,400	—		
		(5)	36 months	—	1,510	—		
Total antibody-forming cells	Mouse	(—)	3 months	M,F	2,670	—	155	Total agar colony-forming units per spleen; at least 5 animals per age group
		(—)	7 months	M,F	1,060	—		
		(—)	27 months	M,F	470	—		
Total direct plaque-forming cells	Mouse	(—)	18—20 weeks	F	2.2×10^5	—	163	Total no. per spleen, nonrestricted diet; est. from data given; mean of 5—8 animals per age group
		(—)	52—55 weeks	F	1.9×10^5	—		
		(—)	108—118 weeks	F	0.1×10^5	—		

Table 7 (continued)
NUMBER OF CELLS VS. AGE IN THE SPLEEN, THYMUS AND LYMPH NODES

Tissue and cell type	Subjects				Cell counts		Measure	Ref.
	Type	(no.)	Age	Sex	Mean	Range		
Total indirect plaque-forming	Mouse	(—)	18—20 weeks	F	0.6×10^5	—	Total no. per spleen, nonrestricted diet; est. from data given; mean of 5—8 animals per age group	163
		(—)	52—55 weeks	F	1.4×10^5	—		
		(—)	108—118 weeks	F	0.1×10^5	—		
Total C-Ig cells	Mouse	(—)	6 weeks	M	209×10^3	—	Total no. cells containing cytoplasmic immunoglobulin (C-Ig) per spleen; est. from data given; at least 3 animals per age group	158
		(—)	26 weeks	M	100×10^3	—		
		(—)	1 year	M	99×10^3	—		
		(—)	2 years	M	83×10^3	—		
IgG-bearing cells	Mouse	(12)	3—4 months	F	26 ± 6	—	% of total spleen cells	164
		(12)	26—30 months	F	32 ± 10	—		
Total T cells	Mouse	(—)	1—7 days	M,F	14%	—	% of total spleen cells, est. from data given; 5—15 animals each age group	159
		(—)	14 days	M,F	22%	—		
		(—)	25 days	M,F	32%	—		
		(—)	183 days	M,F	32%	—		
		(—)	400 days	M,F	23%	—		
		(—)	600 days	M,F	16%	—		
		(—)	893 days	M,F	11%	—		
Total T cells, W3/13-positive	Rat	(—)	3—4 months	M,F	38 ± 2	3—19%	% of total spleen cells	165
		(—)	≥27 months	M,F	39 ± 2	—		
Total T cells, W3/25-positive	Rat	(—)	3—4 months	M,F	32 ± 1	—	% of total spleen cells	165
		(—)	≥27 months	M,F	33 ± 1	—		
Total T cells, Thy-1-positive	Rat	(—)	3—4 months	M,F	17 ± 2	—	% of total spleen cells	165
		(—)	≥27 months	M,F	10 ± 1	—		
Macrophages	Rat	(8)	21 days	M,F	0	0	Av. no. in 0.30 mm² section of 8 μm thickness	166
		(19)	50—150 days	M,F	90	0—431		
		(13)	200 days	M,F	326	170—325		

89

		(n)	Age	Sex	Value		Units	Ref
Megakaryocytes	Rat	(25)	300—726 days	M,F	429	121—684	Av. no. in 0.0676 mm² area	166
		(35)	800—1170 days	M,F	343	4—692		
		(8)	21 days	M,F	59	44—105		
		(19)	50—150 days	M,F	19	4—38		
		(13)	200 days	M,F	16	4—26		
		(25)	300—726 days	M,F	22	2—90		
		(35)	800—1170 days	M,F	16	4—90		

Thymus, Cortex

		(n)	Age	Sex	Value		Units	Ref
Total cells	Mouse	(22)	4.5 months	M	319.36 ± 13.41	—	No. per 80 μm² area	167
		(22)	32 months	M	186.04 ± 11.33	—		
Lymphocytes	Mouse	(22)	4.5 months	M	295.86 ± 12.68	—	No. per 80 μm² area	167
		(22)	32 months	M	137.17 ± 11.27	—		
IgG-bearing cells	Mouse	(12)	3—4 months	F	0 ± 1	—	% of total thymus cells	164
		(12)	26—30 months	F	10 ± 5	—		
Macrophages	Mouse	(22)	4.5 months	M	1.95 ± 0.30	—	No. per 80 μm² area	167
		(22)	32 months	M	17.54 ± 2.11	—		
Plasma cells	Mouse	(22)	4.5 months	M	0.53 ± 0.17	—	No. per 80 μm² area	167
		(22)	32 months	M	1.41 ± 0.42	—		
Epithelial cells	Mouse	(22)	4.5 months	M	19.00 ± 2.41	—	No. per 80 μm² area	167
		(22)	32 months	M	13.08 ± 2.59	—		
Fibroblasts	Mouse	(22)	4.5 months	M	0.41 ± 0.14	—	No. per 80 μm² area	167
		(22)	32 months	M	6.08 ± 1.19	—		

Lymph Nodes

		(n)	Age	Sex	Value		Units	Ref
Mesenteric lymph nodes, C-Ig cells	Mouse	(—)	6 weeks	M	55×10^3	—	Total no. containing cytoplasmic immunoglobulin (C-Ig) per total mesenteric lymph nodes; est. from data given; at least 3 animals per age group	158
		(—)	26 weeks	M	14×10^3	—		
		(—)	1 year	M	13×10^3			
		(—)	2 years	M	3×10^3			
Brachial and submandibular lymph nodes, IgG-bearing cells	Mouse	(12)	3—4 months	F	10 ± 4	—	% of total cells	164
		(12)	26—30 months	F	20 ± 6	—		

Table 7 (continued)
NUMBER OF CELLS VS. AGE IN THE SPLEEN, THYMUS AND LYMPH NODES

Tissue and cell type	Type	(no.)	Age	Sex	Cell counts		Measure	Ref.
					Mean	Range		
Cervical lymph nodes, total cells	Rat	(—)	3—4 months	F	$1.2 \times 10^8 \pm 0.2 \times 10^8$	—	Total no. of recoverable nucleated cells	165
		(—)	≥27 months	F	$0.28 \times 10^8 \pm 0.2 \times 10^8$	—		
Cervical lymph nodes, total T cells, W3/13-positive	Rat	(—)	3—4 months	F	64 ± 2	—	% of total cervical lymph node cells	165
		(—)	≥27 months	F	65 ± 2	—		
Cervical lymph nodes, total T cells, W3/25-positive	Rat	(—)	3—4 months	F	60 ± 2	—	% total cervical lymph node cells	165
		(—)	≥27 months	F	59 ± 2	—		
Cervical lymph nodes, total T cells, Thy-1-positive	Rat	(—)	3—4 months	F	12 ± 1	—	% total cervical lymph node cells	165
		(—)	≥27 months	F	5 ± 1	—		

Note: est., estimated.

Table 8
NUMBER OF CELLS VS. AGE IN THE DIGESTIVE SYSTEM

Tissue and cell type	Subjects				Cell count		Measure	Ref.
	Type	(no.)	Age	Sex	Mean	Range		
Oral								
Gingival epithelium	Human	(30)	25—34 years	M	55	42—73	No. per $10^4 \ \mu m^2$	168
		(30)	50—78 years	M	73	50—98		
Palate, epithelium, basal cells	Rat	(3)	2 months	M	158	—	No. per mm length of epithelium	169
		(3)	9 months	M	168	—		
		(3)	19 months	M	162	—		
		(3)	27 months	M	131	—		

Tissue/cells	Species	(n)	Age	Sex	Value		Measurement	Ref.
Tongue, epithelium, basal cells	Rat	(3)	2 months	M	212	—	No. per mm length of epithelium	169
		(3)	9 months	M	226	—		
		(3)	19 months	M	233	—		
		(3)	27 months	M	172	—		
Palate, epithelium, stem cells	Rat	(3)	2 months	M	231	—	No. per mm length of epithelium	169
		(3)	9 months	M	231	—		
		(3)	19 months	M	186	—		
		(3)	27 months	M	149	—		
Tongue, epithelium, stem cells	Rat	(3)	2 months	M	255	—	No. per mm length of epithelium	169
		(3)	9 months	M	270	—		
		(3)	19 months	M	253	—		
		(3)	27 months	M	185	—		
Incisor, pulp tissue cells	Rat	(4)	10 days	M	$27,723 \pm 317$	—	Total no. cells	170
		(4)	28 days	M	$29,815 \pm 815$	—		
		(4)	30 days	M	$31,455 \pm 552$	—		
		(4)	90 days	M	$29,908 \pm 808$	—		
		(4)	99 days	M	$26,028 \pm 495$	—		
		(4)	135 days	M	$31,227 \pm 989$	—		
		(4)	203 days	M	$30,207 \pm 449$	—		
		(4)	495 days	M	$33,411 \pm 362$	—		
		(4)	751 days	M	$28,046 \pm 532$	—		
Incisor, right mandibular, ameloblasts	Rat	(14)	9 days	M	63.08 ± 0.16	—	No. in 5 serial 5-μm thick mid-sagittal sections	171
		(14)	27 days	M	60.62 ± 0.31	—		
		(14)	54 days	M	64.46 ± 0.21	—		
		(14)	89 days	M	60.77 ± 0.13	—		
		(14)	495 days	M	63.53 ± 0.52	—		
		(14)	574 days	M	61.51 ± 0.31	—		
Incisor, right mandibular, labial odontoblasts	Rat	(14)	9 days	M	57.38 ± 0.78	—	No. in 5 serial 5-μm thick mid-sagittal sections	171
		(14)	27 days	M	55.55 ± 0.12	—		
		(14)	54 days	M	56.05 ± 0.56	—		
		(14)	89 days	M	58.05 ± 0.25	—		
		(14)	495 days	M	54.70 ± 0.42	—		
		(14)	574 days	M	56.89 ± 0.39	—		
Incisor, right mandibular, lingual odontoblasts	Rat	(14)	9 days	M	59.44 ± 0.68	—	No. in 5 serial 5-μm thick mid-sagittal sections	171
		(14)	27 days	M	64.16 ± 0.54	—		
		(14)	54 days	M	60.38 ± 0.63	—		
		(14)	89 days	M	57.65 ± 0.42	—		

Table 8 (continued)
NUMBER OF CELLS VS. AGE IN THE DIGESTIVE SYSTEM

Tissue and cell type	Type	Subjects (no.)	Age	Sex	Cell count Mean	Cell count Range	Measure	Ref.
		(14)	495 days	M	58.65 ± 0.58	—		
		(14)	574 days	M	59.51 ± 0.69	—		
Small Intestine								
Connective tissue, tunica muscularis longitudinal layer	Human	(—)	1 year	M,F	0.84	—	Proportion of t.m. which is connective tissue relative to proportion found in 30-year-olds; calculated from regression equation given; 60 subjects total	172
		(—)	10 years	M,F	0.89	—		
		(—)	20 years	M,F	0.94	—		
		(—)	30 years	M,F	1.00	—		
		(—)	50 years	M,F	1.10	—		
		(—)	70 years	M,F	1.22	—		
		(—)	90 years	M,F	1.32	—		
Epithelial layer, lymphocytes	Mouse	(8)	128—248 days	—	227.75 ± 12.59	—	No. per 20 oil immersion fields per animal	173
		(10)	573—818 days	—	109.00 ± 4.64	—		
Peyer's patches, C-Ig cells	Mouse	(—)	6 weeks	M	67×10^3	—	No. containing cytoplasmic immunoglobulin (C-Ig) per total Peyer's patches; est. from data given; at least 3 animals per age group	158
		(—)	26 weeks	M	60×10^3	—		
		(—)	1 year	M	20×10^3	—		
		(—)	2 years	M	2×10^3	—		
Liver								
Nucleated cells	Mouse	(5)	2 months	M	116×10^6	—	No. per unit volume (cm³) of liver; est. from data given	174
		(5)	12 months	M	98×10^6	—		
		(5)	24 months	M	88×10^6	—		

Cell type	Species	(n)	Age	Sex	Value		Description	Ref
Hepatocytes	Human	(—)	<49 years	M	549 ± 11	—	No. per 176,000 μm² area; 109 Caucasian subjects	175
		(—)	50—59 years	M	516 ± 15	—		
		(—)	60—69 years	M	490 ± 12	—		
		(—)	70—79 years	M	474 ± 15	—		
		(—)	Above 80 years	M	449 ± 13	—		
Hepatocytes	Human	(—)	<49 years	M	526 ± 12	—	No. per 176,000 μm² area; 185 Japanese subjects	175
		(—)	50—59 years	M	537 ± 21	—		
		(—)	60—69 years	M	527 ± 17	—		
		(—)	70—79 years	M	501 ± 20	—		
		(—)	Above 80 years	M	424 ± 14	—		
Hepatocytes	Human	(10)	<49 years	M	571	—	No. per 176,000 μm² area; 47 Costa Rican subjects	176
		(10)	50—59 years	M	525	—		
		(10)	60—69 years	M	517	—		
		(10)	70—79 years	M	528	—		
		(7)	80 and over	M	477	—		
Hepatocytes	Rat	(—)	21 days	M	$585 \times 10^6 \pm 49 \times 10^6$	—	Total cells per liver; at least 10 animals per age	177
		(—)	121 days	M	$1,772 \times 10^6 \pm 81 \times 10^6$	—		
		(—)	221 days	M	$2,252 \times 10^6 \pm 67 \times 10^6$	—		
		(—)	385 days	M	$1,988 \times 10^6 \pm 120 \times 10^6$	—		
		(—)	621 days	M	$2,563 \times 10^6 \pm 149 \times 10^6$	—		
		(—)	795 days	M	$2,891 \times 10^6 \pm 175 \times 10^6$	—		
Hepatocytes	Rat	(5)	1 month	F	239×10^6	—	Total mononuclear cells per liver; calculated from data given	178
		(5)	12 months	F	1094×10^6	—		
		(5)	27 months	F	812×10^6	—		
Hepatocytes	Rat	(12)	3 months	M	144	—	Total no. per 10⁶ μm³; data pooled from 6 diets	179
		(12)	6 months	M	158	—		
		(12)	12 months	M	167	—		
		(12)	18 months	M	175	—		
		(12)	24 months	M	192	—		
"Littoral" cells (endothelial, Kupffer, and Ito)	Rat	(12)	3 months	M	80	—	Total no. per 10⁶ μm³; data pooled from 6 diets	179
		(12)	6 months	M	87	—		
		(12)	12 months	M	73	—		
		(12)	18 months	M	70	—		
		(12)	24 months	M	108	—		

Table 8 (continued)
NUMBER OF CELLS VS. AGE IN THE DIGESTIVE SYSTEM

Tissue and cell type	Subjects				Cell count			Measure	Ref.
	Type	(no.)	Age	Sex	Mean	Range			
Mesentery, Lower Small Intestine									
Mast cells	Mouse	(3)	2—3 months	M	8.8 ± 0.9	—		No. per 1 mm^2 fat-free, avascular area; C57BL strain	86
		(4)	18 months	M	10.5 ± 1.2	—			
Mast cells	Mouse	(4)	2—3 months	F	11.8 ± 1.4	—		No. per 1 mm^2 fat-free, avascular area; C57BL strain	86
		(5)	18 months	F	12.9 ± 0.9	—			
Mast cells	Mouse	(4)	2—3 months	M	8.9 ± 0.8	—		No. per 1 mm^2 fat-free, avascular area; BALB/c strain	86
		(5)	18 months	M	22.1 ± 1.8	—			
Mast cells	Mouse	(5)	2—3 months	F	8.3 ± 0.7	—		No. per 1 mm^2 fat-free, avascular area; BALB/c strain	86
		(5)	18 months	F	18.2 ± 1.7	—			

Note: est., estimated.

Table 9
NUMBER OF MUSCLE CELLS OR FIBERS VS. AGE

Tissue	Type	Subjects (no.)	Age	Sex	Mean count	Measure	Ref.
				Cardiac			
Myocardium, muscle fibers	Mouse	(4)	1 hour	—	37	Av. no. in 10 oil immersion fields (970 ×)	180
		(5)	1 day	—	35		
		(5)	1 week	—	26		
		(4)	3 weeks	—	18		
		(6)	5 weeks	—	19		
		(6)	13 weeks	—	14		
		(6)	7 months	—	10		
		(4)	9 months	—	12		
		(5)	14 months	—	13		
		(4)	20 months	—	14		
		(2)	24 months	—	14		
Myocardium, reticular fibers	Mouse	(4)	1 hour	—	15	Av. no. in 10 oil immersion fields (970 ×)	180
		(5)	1 day	—	23		
		(5)	1 week	—	23		
		(4)	3 weeks	—	29		
		(6)	5 weeks	—	37		
		(6)	13 weeks	—	37		
		(6)	7 months	—	40		
		(4)	9 months	—	38		
		(5)	14 months	—	33		
		(4)	20 months	—	31		
		(2)	24 months	—	30		
Heart	Rat	(16)	4 months	—	2,465 ± 32	No. fibers per mm^3	181
		(7)	26—27 months	—	2,504 ± 75		
Myocardium, mast cells	Rat	(—)	1 month	M	44.2 ± 3.6	No. per 7 mm^2; 7—9 animals per age group	182
		(—)	18 months	M	20.2 ± 1.9		
		(—)	24 months	M	12.2 ± 0.5		

Table 9 (continued)
NUMBER OF MUSCLE CELLS OR FIBERS VS. AGE

Tissue	Type	Subjects (no.)	Age	Sex	Mean count	Measure	Ref.
Myocardium, mast cells	Rat	(—)	1 month	F	35.7 ± 2.6	No. per 7 mm^2; 7—9 animals per age group	182
		(—)	18 months	F	17.4 ± 2.6		
		(—)	24 months	F	12.3 ± 1.1		
Myocardium, mast cells	Human	(46)	20—35 years	M,F	297 ± 18.2	No. per cm^2	183
		(48)	60—90 years	M,F	273 ± 23.9		
Skeletal							
Quadriceps	Human	(7)	20.7 years av.	M	329.2 ± 45.4	No. fibers per mm^2: (a) active, athletically trained, (b) relatively inactive	15
		(10)	73.9 years (a)	M	556.2 ± 108.3		
		(8)	72.4 years (b)	M	532.8 ± 154.1		
Hind limb thigh, muscle mass	Rat	(10)	12—14 months	M	28.05 ± 2.01	Muscle mass (g), freed of fat and connective tissue	184
		(10)	24—27 months	M	18.45 ± 1.83		
Hind limb thigh, muscle mass	Rat	(10)	12—14 months	F	14.56 ± 0.67	Muscle mass (g), freed of fat and connective tissue	184
		(10)	24—27 months	F	13.90 ± 0.50		
Tibialis anterior, red fibers	Rat	(5)	3 months	M	$15,754 \pm 1,472$	Total per muscle, Wistar rat	185
		(5)	6 months	M	$15,625 \pm 758$		
		(5)	12 months	M	$15,020 \pm 835$		
		(8)	24 months	M	$7,616 \pm 554$		
Tibialis anterior, white fibers	Rat	(5)	3 months	M	$8,881 \pm 906$	Total per muscle, Wistar rat	185
		(5)	6 months	M	$10,425 \pm 1,565$		
		(5)	12 months	M	$10,932 \pm 626$		
		(8)	24 months	M	$13,673 \pm 1,302$		
Tibialis anterior, red fibers	Rat	(5)	3 months	F	$15,984 \pm 1,272$	Total per muscle, Wistar rat	185
		(5)	6 months	F	$15,850 \pm 360$		
		(5)	12 months	F	$15,663 \pm 687$		
		(5)	24 months	F	$6,993 \pm 672$		

Muscle	Species	(n)	Age	Sex	Value	Description	Ref.
Tibialis anterior, white fibers	Rat	(5)	3 months	F	9,436 ± 540	Total per muscle, Wistar rat	185
		(5)	6 months	F	8,900 ± 482		
		(5)	12 months	F	11,147 ± 853		
		(5)	24 months	F	10,350 ± 530		
Tibialis anterior, red fibers	Rat	(5)	3 months	M	21,930 ± 2,693	Total per muscle, Donryu rat	185
		(5)	6 months	M	23,883 ± 1,578		
		(5)	24 months	M	7,191 ± 997		
Tibialis anterior, white fibers	Rat	(5)	3 months	M	11,962 ± 1,047	Total per muscle, Donryu rat	185
		(5)	6 months	M	11,262 ± 600		
		(5)	24 months	M	10,574 ± 1,672		
Psoas major, red fibers	Rat	(5)	3 months	M	18,960 ± 2,150	Total per muscle, Wistar rat	185
		(35)	6 months	M	13,763 ± 1,252		
		(7)	12 months	M	11,686 ± 1,333		
		(34)	24 months	M	6,591 ± 561		
Psoas major, white fibers	Rat	(5)	3 months	M	5,781 ± 594	Total per muscle, Wistar rat	185
		(35)	6 months	M	9,309 ± 585		
		(7)	12 months	M	12,394 ± 915		
		(34)	24 months	M	8,603 ± 704		
Extensor digitorum longus	Rat	(—)	3 months	M	3,042 ± 350	No. fibers per whole cross-sectional area	186
		(—)	28—29 months	M	1,710 ± 256		
Extensor digitorum longus	Mouse	(—)	137 ± 13 days	M	1,020 ± 6	Total fibers	187
		(—)	750 ± 0 days	M	902 ± 64		
Extensor digitorum longus	Mouse	(—)	126 ± 15 days	F	1,010 ± 22	Total fibers	187
		(—)	750 ± 0 days	F	847 ± 28		
Soleus	Rat	(—)	3 months	M	2,416 ± 110	No. fibers per whole cross-sectional area	186
		(—)	28—29 months	M	1,486 ± 325		
Soleus	Rat	(—)	4 months	—	2,357 ± 78	Total fibers	80
		(—)	24 months	—	1,758 ± 39		
Soleus	Mouse	(—)	137 ± 13 days	M	753 ± 10	Total fibers	187
		(—)	750 ± 0 days	M	708 ± 28		
Soleus	Mouse	(—)	126 ± 15 days	F	737 ± 11	Total fibers	187
		(—)	750 ± 0 days	F	582 ± 16		
Diaphragm	Rat	(—)	3 months	M	785 ± 59	Total no. fibers per 1 mm^2 cross-sectional area	186
		(—)	28—29 months	M	894 ± 44		
Anterior tibialis	Mouse	(—)	137 ± 13 days	M	3,022 ± 22	Total fibers	187
		(—)	750 ± 0 days	M	2,967 ± 102		

Table 9 (continued)
NUMBER OF MUSCLE CELLS OR FIBERS VS. AGE

Tissue	Type	Subjects (no.)	Age	Sex	Mean count	Measure	Ref.
Anterior tibialis	Mouse	(15)	26 weeks	M	2,730	Total fibers	188
		(12)	70 weeks	M	2,220		
Anterior tibialis	Mouse	(—)	126 ± 15 days	F	2,835 ± 34	Total fibers	187
		(—)	750 ± 0 days	F	2,944 ± 118		
Biceps brachii	Mouse	(—)	137 ± 13 days	M	2,061 ± 23	Total fibers	187
		(—)	750 ± 0 days	M	1,784 ± 56		
Biceps brachii	Mouse	(—)	126 ± 15 days	F	2,140 ± 106	Total fibers	187
		(—)	750 ± 0 days	F	2,221 ± 106		
Biceps brachii	Mouse	(15)	26 weeks	M	2,218	Total fibers	188
		(12)	70 weeks	M	1,887		
Sternomastoideus	Mouse	(—)	137 ± 13 days	M	1,277 ± 17	Total fibers	187
		(—)	750 ± 0 days	M	1,211 ± 35		
Sternomastoideus	Mouse	(—)	125 ± 15 days	F	1,202 ± 10	Total fibers	187
		(—)	750 ± 0 days	F	1,118 ± 60		
Sternomastoid	Mouse	(15)	26 weeks	M	1,505	Total fibers	188
		(12)	70 weeks	M	1,396		
			Smooth				
Ciliary body, muscle	Human	(—)	1 year	M	1.16	Proportion of c.b. which is muscle tissue relative to proportion found in 30-year-olds; calc. from linear equation given; 55 subjects total	172
		(—)	10 years	M	1.11		
		(—)	20 years	M	1.06		
		(—)	30 years	M	1.00		
		(—)	50 years	M	0.89		
		(—)	70 years	M	0.77		
		(—)	80 years	M	0.72		
Ciliary body, muscle	Human	(—)	1 year	F	1.025	Proportion of c.b. which is muscle tissue relative to proportion found in 30-year-olds;	172
		(—)	10 years	F	1.02		
		(—)	20 years	F	1.01		

Cell	Type	(no.)		Age	Mean	Measure	Ref.
Ciliary body, connective tissue	Human	(—)	F	30 years	1.00	calc. from linear equation given; 55 subjects total	
		(—)	F	50 years	0.98		
		(—)	F	70 years	0.96		
		(—)	F	80 years	0.955		
		(—)	M.F	1 year	0.65	Proportion of c.b. which is connective tissue relative to proportion found in 30-year-olds; calc. from regression equation given; 55 subjects total	172
		(—)	M.F	10 years	0.76		
		(—)	M.F	20 years	0.88		
		(—)	M.F	30 years	1.00		
		(—)	M.F	50 years	1.24		
		(—)	M.F	70 years	1.48		
		(—)	M.F	80 years	1.61		

Table 10
NUMBER OF CELLS VS. AGE IN THE REPRODUCTIVE SYSTEM AND KIDNEY

Cell	Subjects		Age	Cell count		Measure	Ref.
	Type	(no.)		Mean	Range		

Male Reproductive System

Cell	Type	(no.)	Age	Mean	Range	Measure	Ref.
Leydig cells	Human	(—)	0—9 years	0.99	—	Total no. relative to age 20—29 years, both testes; est. from data given; 504 total subjects	189
		(—)	10—19 years	1.04	—		
		(—)	20—29 years	1.00	—		
		(—)	30—39 years	0.90	—		
		(—)	40—49 years	0.78	—		
		(—)	50—59 years	0.73	—		
		(—)	60—69 years	0.70	—		
		(—)	70—79 years	0.70	—		
		(—)	80—89 years	0.54	—		
Leydig cells	Human	(—)	20 years	72.2×10^7	—	Total per individual calc. from linear regression equation given; 25 subjects total	190
		(—)	40 years	56.2×10^7	—		
		(—)	60 years	40.2×10^7	—		
		(—)	80 years	24.2×10^7	—		

Table 10 (continued)
NUMBER OF CELLS VS. AGE IN THE REPRODUCTIVE SYSTEM AND KIDNEY

Cell	Subjects Type	(no.)	Age	Cell count Mean	Range	Measure	Ref.
Leydig cells	Rat	(16)	3 months	4.39×10^7	—	Total no.	191
	Rat	(16)	24 months	$5.18 \times 10^7 \pm 0.24 \times 10^7$	—		
Leydig cells	Rat	(8)	5 months	19.6×10^7	—	Total per testis; est. from data given	192
		(7)	12 months	13.0×10^7	—		
		(7)	29 months	7.2×10^7	—		
Leydig cells	Stallion	(10)	2—3 years	$1.41 \times 10^9 \pm 0.11 \times 10^9$	—	Total no. per testis	193
		(10)	4—5 years	$3.06 \times 10^9 \pm 0.38 \times 10^9$	—		
		(4)	13—20 years	$4.66 \times 10^9 \pm 0.42 \times 10^9$	—		
Sertoli cells	Mouse	(8)	4—5 months	9	7—11	No. in a 6-μm thick cross-sectional profile of seminiferous tubule	194
		(7)	22 months	9	7—13		
Spermatozoa	Mouse	(8)	4—5 months	4.3×10^6	3.4×10^6—7.8×10^6	Total no. per left testis	194
		(8)	22 months	1.2×10^6	0.4×10^6—8.8×10^6		
Spermatozoa	Mouse	(8)	4—5 months	17.7×10^6	12.4×10^6—22.8×10^6	Total no. per left epididymis	194
		(8)	22 months	4.7×10^6	0.6×10^6—12.4×10^6		
Spermatozoa	Stallion	(10)	2—3 years	$1.27 \times 10^9 \pm 0.19 \times 10^9$	—	Total number produced daily	193
		(10)	4—5 years	$2.67 \times 10^9 \pm 0.26 \times 10^9$	—		
		(4)	13—20 years	$3.18 \times 10^9 \pm 0.50 \times 10^9$	—		
Spermatids and spermatozoa	Rabbit	(10)	6 months	$355 \times 10^6 \pm 11 \times 10^6$	—	Total number, both testes	195
		(10)	12 months	$582 \times 10^6 \pm 76 \times 10^6$	—		
		(10)	24 months	$743 \times 10^6 \pm 77 \times 10^6$	—		
		(7)	36 months	$397 \times 10^6 \pm 41 \times 10^6$	—		
Developing spermatozoa	Mouse	(8)	4—5 months	(a) 21.6 (b) 38.7 (c) 92.7 (d) 95.4	—	No. in 6 μm-thick cross-sectional	194
		(7)	22 months	9.9 13.5 19.8 9.0	—		

	Species	(n)	Age		profile of seminiferous tubule, calc. from data given: (a) spermatogonia (b) spermatocytes, (c) early spermatids, (d) late spermatids Total no.	196
Epididymal depot, fat cells	Rat	(12)	9 weeks	12.0×10^6	—	
		(13)	13 weeks	10.6×10^6	—	
		(12)	26 weeks	10.3×10^6	—	
		(13)	52 weeks	12.3×10^6	—	
		(14)	104 weeks	12.9×10^6	—	
		(9)	130 weeks	16.9×10^6	—	

	Species	(n)	Age		Total no. in left depot; est. from data given	197
Epididymal depot, adipocytes	Rat	(10)	6 months	7.3×10^6	—	
		(10)	12 months	7.5×10^6	—	
		(10)	18 months	7.8×10^6	—	

Female Reproductive System

	Species	(n)	Age		Total no., both ovaries	198
Ova	Rat	(4)	1—7 days	27,483	20,985—35,105	
		(2)	10—15 days	15,641	15,406—15,976	
		(2)	20—26 days	10,746	10,422—11,076	
		(3)	30—36 days	10,196	8,998—12,541	
		(3)	41—46 days	10,836	10,266—11,971	
		(2)	50 days	10,553	10,073—11,033	
		(3)	60—64 days	10,185	10,030—10,452	
		(1)	70 days	6,606	—	
		(3)	80—84 days	7,850	5,268—8,566	
		(4)	95—100 days	7,548	5,744—10,764	
		(2)	110 days	6,568	5,392—7,744	
		(3)	140—198 days	6,950	2,750—9,077	
		(2)	206—262 days	5,909	3,747—8,066	
		(2)	318—385 days	5,102	4,502—5,702	
		(2)	454—559 days	4,826	4,759—4,893	
		(1)	947 days	1,919	—	

Table 10 (continued)
NUMBER OF CELLS VS. AGE IN THE REPRODUCTIVE SYSTEM AND KIDNEY

Cell	Subjects			Cell count		Measure	Ref.
	Type	(no.)	Age	Mean	Range		
Oocytes	Mouse	(1)	155 days	950	—	Total no.; multiparous CBA strain	199
		(3)	238—253 days	—	300—410		
		(2)	278—283 days	160	135—186		
		(3)	350—373 days	—	26—138		
		(7)	423—439 days	—	2—54		
		(2)	464—465 days	2	0—4		
		(3)	612—735 days	0	—		
Oocytes	Mouse	(1)	125 days	3,260	—	Total no. multiparous A strain	199
		(3)	210—285 days	—	1,210—1,580		
		(3)	303—343 days	—	1,000—1,830		
		(3)	351—371 days	—	800—1,180		
		(5)	403—422 days	—	660—940		
		(1)	477 days	500	—		
		(1)	492 days	179	—		
Oocytes	Mouse	(2)	93—98 days	3,940	3,420—4,460	Total no.; multiparous RIII strain	199
		(5)	111—159 days	—	2,680—5,010		
		(3)	184—264 days	—	2,700—3,030		
		(5)	302—371 days	—	1,190—1,980		
		(4)	407—476 days	—	240—880		
		(1)	635 days	50	—		
Oocytes	Mouse	(6)	320 days	—	1,770—3,530	Total no.; multiparous CBA × A hybrid strain	199
		(2)	370 days	1,520	1,340—1,700		
		(3)	431—491 days	—	700—1,700		
		(2)	517—602 days	610	590—630		

Kidney

Cell	Subjects			Cell count		Measure	Ref.
	Type	(no.)	Age	Mean	Range		
Perirenal depot, adipocytes	Rat (male)	(10)	6 months	6.0×10^6	—	Total no. in left depot; est. from data given	197
		(10)	12 months	8.4×10^6	—		
		(10)	18 months	11.4×10^6	—		

Table 11
NUMBER OF CELLS VS. AGE IN THE ALVEOLUS OF THE LUNG

Cell type	Subjects				Mean count	Measure	Ref.
	Type	(no.)	Age	Sex			
Total cells	Rat	(4)	1 week	M	$224 \times 10^6 \pm 19 \times 10^6$	Total no., both lungs	200
		(4)	6 weeks	M	$460 \times 10^6 \pm 33 \times 10^6$		
		(4)	5 months	M	$668 \times 10^6 \pm 16 \times 10^6$		
		(4)	14 months	M	$554 \times 10^6 \pm 27 \times 10^6$		
		(4)	26 months	M	$527 \times 10^6 \pm 56 \times 10^6$		
Total cells	Rat	(4)	5 months	F	$466 \times 10^6 \pm 34 \times 10^6$	Total no., both lungs	200
		(4)	14 months	F	$437 \times 10^6 \pm 12 \times 10^6$		
		(4)	26 months	F	$476 \times 10^6 \pm 12 \times 10^6$		
Type I cells	Rat	(4)	1 week	M	$15 \times 10^6 \pm 1 \times 10^6$	Total no., both lungs	200
		(4)	6 weeks	M	$30 \times 10^6 \pm 4 \times 10^6$		
		(4)	5 months	M	$54 \times 10^6 \pm 1 \times 10^6$		
		(4)	14 months	M	$63 \times 10^6 \pm 6 \times 10^6$		
		(4)	26 months	M	$59 \times 10^6 \pm 7 \times 10^6$		
Type I cells	Rat	(4)	5 months	F	$39 \times 10^6 \pm 5 \times 10^6$	Total no., both lungs	200
		(4)	14 months	F	$48 \times 10^6 \pm 4 \times 10^6$		
		(4)	26 months	F	$49 \times 10^6 \pm 5 \times 10^6$		
Type II cells	Rat	(4)	1 week	M	$18 \times 10^6 \pm 1 \times 10^6$	Total no., both lungs	200
		(4)	6 weeks	M	$54 \times 10^6 \pm 3 \times 10^6$		
		(4)	5 months	M	$81 \times 10^6 \pm 7 \times 10^6$		
		(4)	14 months	M	$59 \times 10^6 \pm 7 \times 10^6$		
		(4)	26 months	M	$57 \times 10^6 \pm 8 \times 10^6$		
Type II cells	Rat	(4)	5 months	F	$50 \times 10^6 \pm 7 \times 10^6$	Total no., both lungs	200
		(4)	14 months	F	$57 \times 10^6 \pm 5 \times 10^6$		
		(4)	26 months	F	$51 \times 10^6 \pm 4 \times 10^6$		
Interstitial cells	Rat	(4)	1 week	M	$118 \times 10^6 \pm 9 \times 10^6$	Total no., both lungs	200
		(4)	6 weeks	M	$114 \times 10^6 \pm 7 \times 10^6$		
		(4)	5 months	M	$163 \times 10^6 \pm 10 \times 10^6$		
		(4)	14 months	M	$134 \times 10^6 \pm 10 \times 10^6$		
		(4)	26 months	M	$130 \times 10^6 \pm 15 \times 10^6$		

Table 11 (continued)
NUMBER OF CELLS VS. AGE IN THE ALVEOLUS OF THE LUNG

Cell type	Type	Subjects (no.)	Age	Sex	Mean count	Measure	Ref.
Interstitial cells	Rat	(4)	5 months	F	$133 \times 10^6 \pm 16 \times 10^6$	Total no., both lungs	200
		(4)	14 months	F	$120 \times 10^6 \pm 4 \times 10^6$		
		(4)	26 months	F	$137 \times 10^6 \pm 3 \times 10^6$		
Capillary endothelial cells	Rat	(4)	1 week	M	$71 \times 10^6 \pm 8 \times 10^6$	Total no., both lungs	200
		(4)	6 weeks	M	$253 \times 10^6 \pm 22 \times 10^6$		
		(4)	5 months	M	$341 \times 10^6 \pm 9 \times 10^6$		
		(4)	14 months	M	$284 \times 10^6 \pm 16 \times 10^6$		
		(4)	26 months	M	$263 \times 10^6 \pm 26 \times 10^6$		
Capillary endothelial cells	Rat	(4)	5 months	F	$223 \times 10^6 \pm 11 \times 10^6$	Total no., both lungs	200
		(4)	14 months	F	$205 \times 10^6 \pm 8 \times 10^6$		
		(4)	26 months	F	$223 \times 10^6 \pm 8 \times 10^6$		
Macrophage	Rat	(4)	1 week	M	$2 \times 10^6 \pm 0.5 \times 10^6$	Total no., both lungs	200
		(4)	6 weeks	M	$9 \times 10^6 \pm 1 \times 10^6$		
		(4)	5 months	M	$29 \times 10^6 \pm 7 \times 10^6$		
		(4)	14 months	M	$14 \times 10^6 \pm 3 \times 10^6$		
		(4)	26 months	M	$17 \times 10^6 \pm 6 \times 10^6$		
Macrophage	Rat	(4)	5 months	F	$20 \times 10^6 \pm 6 \times 10^6$	Total no., both lungs	200
		(4)	14 months	F	$8 \times 10^6 \pm 1 \times 10^6$		
		(4)	26 months	F	$17 \times 10^6 \pm 3 \times 10^6$		

Table 12

**SUMMARY OF CHANGES IN AVERAGE CELL NUMBER VS. AGE IN
VARIOUS TISSUES (TABLES 1—11) IN THE HUMAN, THE RAT,
AND THE MOUSE**

Tissue	Human	Rat	Mouse
Nervous system			
Spinal nerves	Decrease after 40—60 years[23,24]	—[a]	Decrease in large anterior horn cells of spinal cord by 25 months[29]
Cranial nerves	Decrease after 60—70 years in facial nucleus[32] and optic nerves;[31] decrease through life in olfactory nerve[30]	Decrease in olfactory bulb at 30 months[36]	—
Brain			
Total cells	—	—	One study shows loss of neurons but not glial cells by 24 months;[38] second study shows no change in total brain nuclei up to 33 months[37]
Cerebrum	Decrease in total neurons[40,41] and in some[45,49,50,53] but not all[53] subareas of cortex by 70—80 years	Decrease in total neurons of hemisphere[51] and in some[42,47,51,54,55] but not all[43,52] subareas of cortex including some[44,47,48,56] but not all[58,59] areas of the hippocampus and some[59] but not all[59,60] areas of the hypothalamus by 24—29 months	—
Cerebellum	Decrease in Pukinje cells in some[62,63] but not all[64,65] areas by 50—70 years	Decrease in total neurons by 26—28 months[51] and in Purkinje cells of cortex by 25 months[66]	—
Brain stem	No change in medulla up to 89 years[68] but some loss by age 50 in the loci coerulei[71]	—	—
Auditory nerves	No change in ventral cochlear nucleus up to 60—90 years[72,73]	Decrease in cochlear spiral ganglion Type II cells by 27 months[74]	—
Eye	Loss in trabecular meshwork cells throughout lifespan[81]	Increase in lens epithelial cell from 4 to 40 months[82]	—
Skin	Decrease in abdominal connective tissue cells through 81 years[83] and in melanocytes from various regions of the skin after 50—70 years[88-90]		

Table 12 (continued)
SUMMARY OF CHANGES IN AVERAGE CELL NUMBER VS. AGE IN
VARIOUS TISSUES (TABLES 1—11) IN THE HUMAN, THE RAT,
AND THE MOUSE

Tissue	Human	Rat	Mouse
Bone	Loss in whole skeleton[93] and in femur, ribs, fingers, radius, ulna, humerus, illiac crest, and mandible starting at 30—60 years and continuing to 80 years;[13,94-102,104] cartilage decreases in femoral head and in fourth costal region by 30—50 years[108,109]	Loss of osteoblasts and osteocytes of femur at 0.5, 1, and 2 years[106,107]	Loss in lumbar vertebrae and femur but not tibia by 18 and 30 months[105]
Blood			
Erythrocytes	No. decreases at 40—60 years[113,114,116] and again, at least in males, at 80 years[112-115]	—	No decreases by 21 months or older[118-123]
Leukocytes	Total shows little or no change up to 69 years[114,116,127,128] but may increase after 85 years;[112,115,129] neutrophils may increase but eosinophils and basophils may decrease after 70 years;[112,115] T cells and B cells decrease at 60 years but B cells increase at 90 years;[128,130,134,136,142,143] platelets increase by 67 years[115]	Total as well as eosinophils decrease by 25 months[117]	Total no. vs. age may be strain dependent;[118,121-123] no. of eosinophils may be dependent on diet and/or environmental conditions[118,120,122]
Bone marrow			
Total nucleated cells	—	—	Increase in total skeleton at 5 months and 30 months or older;[152,153] increase in femoral bone marrow at 15—18, 20—26, and again at 33—35 months[154-156]
Stem cells	—	—	Increase in total skeleton at 5 months and possibly again at 36 months;[153] increase in total bone marrow C-Ig cells up to 2 years;[158] however, variable in femur and may be strain dependent[121,152,154-157]

Table 12 (continued)
SUMMARY OF CHANGES IN AVERAGE CELL NUMBER VS. AGE IN VARIOUS TISSUES (TABLES 1—11) IN THE HUMAN, THE RAT, AND THE MOUSE

Tissue	Human	Rat	Mouse
Spleen			
Total cells	—	—	Variable, may be strain-dependent[121,153,159]
Stem cells	—	—	Total decreases at 21 to 36 months;[153,155,161,162] plaque-forming, total C-Ig and total T-cells decrease by 2 years[158,159,163]
Other cells	—	Macrophages decrease by 26—29 months[166]	—
Thymus cortex	—	—	At 32 months, total cells, lymphocytes and epithelial cells per unit area decrease;[167] megakaryocytes, macrophages, fibroblasts, and IgG-bearing cells increase by 26—32 months[164,167]
Lymph nodes			
Mesenteric	—	—	C-Ig cells decrease at 2 years[158]
Brachial and submandibular	—	—	IgG-bearing cells increase at 26—30 months[164]
Cervical	—	Total cells and Thy-1-positive T cells decrease at 27 months and older[165]	—
Digestive System			
Gingival epithelium	Increase by 50—78 years[168]	—	—
Palate and tongue	—	Basal and stem cells decrease by 27 months[169]	—
Small intestine	Connective tissue increases each decade 10 to 90 years[172]	—	Lymphocytes in epithelial layer decrease by 19—27 months;[173] C-Ig cells in Peyer's patches decrease by 1 year and again at 2 years[158]
Liver	Hepatocytes decline by age 50—70 years and again after 80 years[175,176]	Hepatocytes show small decrease in female by 27 months[178] and increase in male up to 26 months;[177,179] "littoral" cells increase in male at 24 months[179]	Total nucleated cells decrease at 12 months and again at 24 months[174]
Mesentery, small intestine	—	—	Increase in mast cells is strain-dependent[86]

Table 12 (continued)
SUMMARY OF CHANGES IN AVERAGE CELL NUMBER VS. AGE IN
VARIOUS TISSUES (TABLES 1—11) IN THE HUMAN, THE RAT,
AND THE MOUSE

Tissue	Human	Rat	Mouse
Muscle			
Cardiac	Small decrease in my-ocardial mast cells at 60—90 years[183]	Myocardial mast cells decrease at 18 months and at 24 months[182]	Myocardial muscle fibers decrease at 3—7 months and then constant to 24 months;[180] reticular fibers decrease at 14—24 months[180]
Skeletal	Quadriceps muscle fibers increase at 72—74 years vs. 21 years[15]	Fibers decrease in several muscles by 24—29 months[80,184-186] but increase in diaphragm;[186] red fibers decrease with age but white fibers variable and dependent on sex and strain[185]	Increase or decrease in fibers of several muscles is sex-dependent[187,188]
Smooth	Muscle portion of ciliary body decreases and connective tissue portion increases each decade to 80 years[172]	—	—
Reproductive system and kidney			
Reproductive system	Leydig cells decrease progressively from 40—60 years[189,190]	One study shows small increase in Leydig cells by 24 months;[191] second study shows decrease at 12 and again at 29 months;[192] epididymal fat cells increase at 30 months;[196,197] ova decrease through life[198]	Oocytes decrease at least to 24 months;[199] developing spermatozoa decrease at 22 months[194]
Kidney	—	Fat cells in kidney perirenal depot increase up to 18 months[197]	—
Lung	—	Increase or decrease in various alveolar cells is sex dependent:[200] alveolar macrophages decrease at 14—26 months vs. 5 months[200]	—

[a]   —, No change detected in studies done so far or no studies done.

REFERENCES

1. **Cannon, W. B.,** Aging of homeostatic mechanisms, in *Problems of Ageing,* 2nd. ed., Cowdry, E. V., Ed., Williams & Wilkins, Baltimore, 1942, 567.
2. **Shock, N. W.,** Ageing of homeostatic mechanisms, in *Cowdry's Problems of Ageing,* 3rd. ed., Lansing, A. I., Ed., Williams & Wilkins, Baltimore, 1952, 415.
3. **Verzár, F.,** Studies on adaptation as a method of gerontological research, *Ciba Colloq. Ageing,* 3, 60, 1957.
4. **Comfort, A.,** Physiology, homeostasis and ageing, *Gerontologia,* 14, 224, 1968.
5. **Timiras, P. S.,** *Developmental Physiology and Aging,* MacMillan, New York, 1972, 542.
6. **Shock, N. W.,** Age changes in physiological functions in the total animal: the role of tissue loss, in *The Biology of Aging,* Strehler, B. L., Ed., American Institute of Biological Sciences Publication No. 6, Washington, D.C., 1960, 258.
7. **Shock, N. W.,** Systems integration, in *Handbook of the Biology of Aging,* Finch, C. E. and Hayflick, L., Eds., van Nostrand Reinhold, New York, 1977, 639.
8. **Buetow, D. E.,** Cellular content and cellular proliferation changes in the tissues and organs of the aging mammal, in *Cellular and Molecular Renewal in the Mammalian Body,* Cameron, I. L. and Thrasher, J. D., Eds., Academic Press, New York, 1971, 87.
9. **Cameron, I. L. and Thrasher, J. D.,** Cell renewal and cell loss in the tissues of mammals, *Interdisciplinary Top. Gerontol.,* 10, 108, 1976.
10. **Finch, C. E. and Hayflick, L., Eds.,** *Handbook of the Biology of Aging,* van Nostrand Reinhold, New York, 1977.
11. **Strehler, B. L.,** *Time, Cells and Aging,* 2nd ed., Academic Press, New York, 1977.
12. **Brožek, J.,** Changes of body composition in man during maturity and their nutritional implications, *Fed. Proc.,* 11, 784, 1952.
13. **Norris, A. H., Lundy, T., and Shock, N. W.,** Trends in selected indices of body composition in men between the ages of 30 and 80 years, *Ann. N.Y. Acad. Sci.,* 110, 623, 1963.
14. **Forbes, G. B. and Reina, J. C.,** Adult lean body mass declines with age: some longitudinal observations, *Metabolism,* 19, 653, 1970.
15. **Parízková, J., Eiselt, E., Šprynarová, Š., and Wachtlová, M.,** Body composition, aerobic capacity, and density of muscle capillaries in young and old men, *J. Appl. Physiol.,* 31, 323, 1971.
16. **Novak, L. P.,** Aging, total body potassium, fat-free mass, and cell mass in males and females between ages 17 and 85 years, *J. Gerontol.,* 27, 438, 1972.
17. **Uauy, R., Winterer, J. C., Bilmazes, C., Haverberg, L. N., Scrimshaw, N. S., Munro, H. N., and Young, V. R.,** The changing pattern of whole body protein metabolism in aging humans, *J. Gerontol.,* 33, 663, 1978.
18. **Ellis, K. J., Yasumura, S., Vartsky, D., Vaswani, A. N., and Cohn, S. H.,** Total body nitrogen in health and disease: effects of age, weight, height and sex, *J. Lab. Clin. Med.,* 99, 917, 1982.
19. **Myhre, L. G. and Kessler, W. V.,** Body density and potassium 40 measurements of body composition as related to age, *J. Appl. Physiol.,* 21, 1351, 1966.
20. **Hodge, C. D.,** Changes in ganglion cells from birth to senile death. Observations on man and honey-bee, *J. Physiol.,* 17, 129, 1894/95.
21. **Hanley, T.,** "Neuronal fall-out" in the aging brain: a critical review of the quantitative data, *Age Ageing,* 3, 133, 1974.
22. **Dayan, A. D.,** Comparative neuropathology of aging. Studies on the brains of 47 species of vertebrates, *Brain,* 94, 31, 1971.
23. **Gardner, E.,** Decrease in human neurones with age, *Anta. Rec.,* 77, 529, 1940.
24. **Corbin, K. B. and Gardner, E.,** Decreases in number of myelinated fibers in human spinal roots with age, *Anat. Rec.,* 68, 63, 1937.
25. **Duncan, D.,** A determination of the number of nerve fibers in the eighth thoracic and the largest lumbar ventral roots of the albino rat, *J. Comp. Neurol.,* 59, 47, 1934.
26. **Dunn, E. H.,** The influence of age, sex, weight and relationship upon the number of medullated nerve fibers and on the size of the largest fibers in the ventral root of the second cervical nerve of the albino rat, *J. Comp. Neurol.,* 22, 131, 1912.
27. **Birren, J. E. and Wall, P. D.,** Age changes in conduction velocity, refractory period, number of fibers, connective tissue space and blood vessels in the sciatic nerve of rats, *J. Comp. Neurol.,* 104, 1, 1956.
28. **Moyer, E. K. and Kaliszewski, B. F.,** The number of nerve fibers in motor spinal nerve roots of young, mature and aged cats, *Anat. Rec.,* 131, 681, 1958.
29. **Wright, E. A. and Spink, J. M.,** A study of the loss of nerve cells in the central nervous system in relation to age, *Gerontologia,* 3, 277, 1959.
30. **Smith, C. G.,** Age incidence of olfactory nerves in man, *J. Comp. Neurol.,* 77, 589, 1942.

31. **Breuch, S. R. and Arey, L. B.,** The number of myelinated and unmyelinated fibers in the optic nerve of vertebrates, *J. Comp. Neurol.,* 77, 631, 1942.

32. **Maleci, O.,** Contributo alla conoscenze della variazioini quantitative delle cellule nervose nella senescenza, *Arch. Ital. Anat. Embriol.,* 33, 883, 1934.

33. **Vijayashankar, N. and Brody, H.,** A study of aging in the human abducens nucleus, *J. Comp. Neurol.,* 173, 433, 1977.

34. **Truex, R. C.,** Morphological alterations in the Gasserian ganglion cells and their association with senescence in man, *Am. J. Pathol.,* 16, 255, 1940.

35. **Hoffman, H. H. and Schnitzlein, H. M.,** The numbers of nerve fibers in the vagus nerve of man, *Anat. Rec.,* 139, 4329, 1961.

36. **Meisami, E. and Shafa, F.,** Changes in total counts of mitral cells and glomeruli in the rat olfactory bulb with aging, *Fed. Proc.,* 36, 292, 1977, abstract.

37. **Franks, L. M., Wilson, P. D., and Whelan, R. D.,** The effects of age on total DNA and cell number in the mouse brain, *Gerontologia,* 20, 21, 1974.

38. **Johnson, H. A. and Erner, S.,** Neuron survival in the aging mouse, *Exp. Gerontol.,* 7, 111, 1972.

39. **Cragg, B. G.,** The density of synapses and neurons in normal, mentally defective and ageing human brain, *Brain,* 98, 81, 1975.

40. **DeKosky, S. T. and Bass, N. H.,** Effects of aging and senile dementia on the microchemical pathology of human cerebral cortex, in *Aging,* Vol. 13, Amaducci, L., Davison, A. N., and Antuono, P., Eds., Raven Press, New York, 1980, 33.

41. **Shefer, V. F.,** Absolute number of neurons and thickness of the cerebral cortex during aging, senile and vascular dementia, and Pick's and Alzheimer's diseases, (transl.), *Neurosci. Behav. Physiol.,* 6, 319, 1973.

42. **Brizzee, K. R., Vogt, J., and Kharetchko, X.,** Postnatal changes in glia/neuron index with a comparison of methods of cell enumeration in the white rat, *Prog. Brain Res.,* 4, 136, 1964.

43. **Brizzee, K. R., Sherwood, N., and Timiras, P. S.,** A comparison of cell populations at various depth levels in cerebral cortex of young adult and aged Long-Evans rats, *J. Gerontol.,* 23, 289, 1968.

44. **Knox, C. A.,** Effect of aging and chronic arterial hypertension on the cell populations in the neocortex and archicortex of the rat, *Acta Neuropathol. (Berlin),* 56, 139, 1982.

45. **Brody, H.,** Organization of the cerebral cortex. III. A study of aging in the human cerebral cortex, *J. Comp. Neurol.,* 102, 511, 1955.

46. **Brizzee, K. R., Odry, J. M., and Bartus, R. T.,** Localization of cellular changes within multimodal sensory regions in aged monkey brain: possible implications for age-related cognitive loss, *Neurobiol. Aging,* 1, 45, 1980.

47. **Ordy, J. M., Brizzee, K. R., Kaack, B., and Hansche, J.,** Age differences in short-term memory and cell loss in the cortex of the rat, *Gerontology,* 24, 276, 1978.

48. **Brizzee, K. R. and Ordy, J. M.,** Age pigments, cell loss and hippocampal function, *Mech. Ageing Dev.,* 9, 143, 1979.

50. **Devaney, K. O. and Johnson, H. A.,** Neuron loss in the aging visual cortex of man, *J. Gerontol.,* 35, 836, 1980.

51. **Peng, M. T. and Lee, L. R.,** Regional differences of neuron loss of rat brain in old age, *Gerontology,* 25, 205, 1979.

52. **Klein, A. W. and Michel, M. E.,** A morphometric study of the neocortex of young and old-maze-differentiated rats, *Mech. Ageing Dev.,* 6, 441, 1977.

53. **Herzog, A. G. and Kemper, T. L.,** Amygdaloid changes in aging and dementia, *Arch. Neruol.,* 37, 625, 1980.

54. **Sabel, B. A. and Stein, D. G.,** Extensive loss of subcortical neurons in the aging rat brain, *Exp. Neurol.,* 73, 507, 1981.

55. **Diamond, M. C., Johnson, R. E., and Gold, M. W.,** Changes in neuron number and size and glia number in the young, adult and aging rat medial occipital cortex, *Behav. Biol.,* 20, 409, 1977.

56. **Landfield, P. W., Baskin, R. K., and Pitler, T. A.,** Brain aging correlates: retardation by hormonal-pharmacological treatment, *Science,* 214, 581, 1981.

57. **Bondareff, W. and Geinesman, Y.,** Loss of synapses in the dentate gyrus of the senescent rat, *Am. J. Anat.,* 145, 129, 1976.

58. **Bondareff, W.,** Synaptic atrophy in the senescent hippocampus, *Mech. Ageing Dev.,* 9, 163, 1979.

59. **Hsü, H. K. and Peng, M. T.,** Hypothalamic neuron number of old female rats, *Gerontology,* 24, 434, 1978.

60. **Peng, M. J. and Hsü, H. K.,** No neuron loss from hypothalamic nuclei of old male rats, *Gerontology,* 28, 19, 1982.

61. **Wilkinson, A. and Davies, I.,** The influence of age and dementia of the neurone population of the mammillary bodies, *Age Ageing,* 7, 151, 1978.

62. **Ellis, R. S.,** A preliminary quantitative study of the Purkinje cells in normal, subnormal and senescent human cerebella, with some notes on functional localization, *J. Comp. Neurol., 30,* 229, 1919.

63. **Ellis, R. S.,** Norms for some structural changes in the human cerebellum from birth to old age, *J. Comp. Neurol., 32,* 1, 1920.

64. **Delorenzi, E.,** Constanza numerica della cellule del Purkinje in individui di varia età, *Boll. Soc. Ital. Biol. Sper., 6,* 80, 1931.

65. **Delorenzi, E.,** Costanzae numerica della cellule di Purkinje del cervelletto del 'uomo in individui di varia età, *Z. Zellforsch. Mikroskop. Anat., 14,* 310, 1931.

66. **Inukai, T.,** On the loss of Purkinje cells, with advancing age, from the cerebellar cortex of the albino rat, *J. Comp. Neurol., 45,* 1, 1928.

67. **Wilcox, H. H.,** A quantitative study of Purkinje cells in guinea pigs, *J. Gerontol., 11,* 442, 1956.

68. **Monagle, R. D. and Brody, H.,** The effects of age upon the main nucleus of the inferior olive in the human, *J. Comp. Neurol., 155,* 61, 1974.

69. **Boseila, A.-W. A., Hashem, S. M., and Badawy, Y. H.,** Volumetric studies on the red nucleus of the rat at different ages, *Acta Anat., 91,* 175, 1975.

70. **Goldman, G. and Coleman, P. D.,** Neuron numbers in locus coeruleus do not change with age in Fisher 344 rat, *Neurobiol. Aging, 2,* 33, 1981.

71. **Wree, A., Braak, H., Schleicher, A., and Zilles, K.,** Biomathematical analysis of the neuronal loss in the aging human brain of both sexes, demonstrated in pigment preparations of the pars cerebellaris loci coerulei, *Anat. Embryol., 160,* 105, 1980.

72. **Konigsmark, B. W. and Murphy, E. A.,** Neuronal populations in the human brain, *Nature (London), 228,* 1335, 1970.

73. **Konigsmark, B. W. and Murphy, E. A.,** Volume of the ventral cochlear nucleus in man: its relationship to neuronal population with age, *J. Neuropathol. Exp. Nerol., 31,* 304, 1972.

74. **Keithly, E. M. and Feldman, M. L.,** Spiral ganglion cell counts in an age-graded series of rat cochleas, *J. Comp. Neurol., 188,* 429, 1979.

75. **Sturrock, R. R.,** A comparative quantitative and morphological study of ageing in the mouse neostriatum, indusium griseum and anterior commissure, *Neuropathol. Appl. Neurobiol., 6,* 51, 1980.

76. **Brownson, R. H.,** Perineuronal satellite cells in the motor cortex of aging brains, *J. Neuropathol. Exp. Neurol., 14,* 424, 1955.

77. **Vaughn, D. W. and Peters, A.,** Neuroglial cells in the cerebral cortex of rats from young adulthood to old age: an electron microscope study, *J. Neurocytol., 3,* 405, 1974.

78. **Geinesman, Y., Bondareff, W., and Dodge, J. T.,** Hypertrophy of astroglial processes in the dentate gyrus of the senescent rat, *Am. J. Anat., 153,* 537, 1978.

79. **Lindsey, J. D., Landfield, P. W., and Lynch, G.,** Early onset and topographical distribution of hypertrophied astrocytes in hippocampus of aging rats: a quantitative study, *J. Gerontol., 34,* 661, 1979.

80. **Gutmann, E. and Hanzlíková, V.,** Motor unit in old age, *Nature (London), 209,* 921, 1966.

81. **Alvarado, J., Murphy, C., Polansky, J., and Juster, R.,** Age-related changes in trabecular meshwork cellularity, *Invest. Opthalmol. Vis. Sci., 21,* 714, 1981.

82. **Treton, J. A. and Courtois, Y.,** Evolution of the distribution, proliferation and ultraviolet repair capacity of rat lens epithelial cells as a function of maturation and aging, *Mech. Ageing Dev., 15,* 251, 1981.

83. **Andrew, W., Behnke, R. H., and Sato, T.,** Changes with advancing age in the cell population of human dermis, *Gerontologia, 10,* 1, 1964/65.

84. **Andrew, W. and Andrew, N. V.,** An age difference in proportions of cell types in the epidermis of abdominal skin of the rat, *J. Gerontol., 11,* 18, 1956.

85. **Andrew, W. and Andrew, N. V.,** Lymphocytes in normal epidermis of young, older middle-aged, and senile rats, *J. Gerontol., 9,* 412, 1954.

86. **Simpson, W. L. and Hayashi, Y.,** Distribution of mast cells in the skin and mesentery of BALB/C and C57BL mice, *Anat. Rec., 138,* 193, 1960.

87. **Klein-Szanto, A. J. P. and Slaga, T. J.,** Numerical variation of dark cells in normal and chemically induced hyperplastic epidermis with age of animal and efficiency of tumor promoter, *Cancer Res., 41,* 4437, 1981.

88. **Fitzpatrick, T. B., Szabó, G., and Mitchell, R. E.,** Age changes in the human melanocyte system, *Adv. Biol. Skin, 6,* 35, 1965.

89. **Snell, R. S. and Bischitz, P. G.,** The melanocytes and melanin in human abdominal wall skin: a survey made at different ages in both sexes and during pregnancy, *J. Anat., 97,* 361, 1963.

90. **Quevedo, W. C., Szabó, G., and Virks, J.,** Influence of age and UV on the population of DOPA-positive melanocytes in human skin, *J. Invest. Dermatol., 52,* 287, 1969.

91. **Mazess, R. B.,** On aging bone loss, *Clin. Opthopaed. Relat. Res., 165,* 239, 1982.

92. **Trotter, M. and Peterson, R. R.,** Ash weight of human skeletons in per cent of their dry, fat-free weight, *Anat. Rec., 123,* 341, 1955.

93. **Trotter, M. and Hixon, B. B.,** Sequential changes in weight, density and percentage ash weight of human skeletons from an early fetal period through old age, *Anat. Rec.,* 179, 1, 1974.
94. **Smith, R. W., Jr. and Walker, R. W.,** Femoral expansion in aging women: implications for osteoporosis and fractures, *Science,* 145, 156, 1964.
95. **Carlson, D. S., Armelagos, G. J., and Van Gerven, D. P.,** Patterns of age-related cortical bone loss (osteoporosis) within the femoral diaphysis, *Hum. Biol.,* 48, 295, 1976.
96. **Thompson, D. D.,** Age changes in bone mineralization, cortical thickness, and Haversian canal area, *Calcif. Tissue Int.,* 31, 5, 1980.
97. **Merz, W. A. and Schenk, R. K.,** Quantitative structural analysis of human cancellous bone, *Acta Anat.,* 75, 54, 1970.
98. **Mazess, R. B. and Cameron, J. R.,** Bone mineral content in normal U.S. whites, in *Int. Conf. Bone Mineral Measurement,* Publ. No. (NIH) 75-683, Mazess, R. B., Ed., U.S. Department of Health, Education and Welfare, Washington, D.C., 1973, 228.
99. **Leichter, I., Weinreb, A., Hazan, G., Loewinger, E., Robin, G. C., Steinberg, R., Menczel, J., and Makin, M.,** The effect of age and sex on bone density, bone mineral content and cortical index, *Clin. Orthopaed. Relat. Res.,* 156, 232, 1981.
100. **Goldsmith, N. F.,** Normative data from the osteoporosis prevalence survey, Oakland, California, 1969—1970. Bone mineral at the distal radius: variation with age, sex, skin color, and exposure to oral contraceptives and exogenous hormones; relation to aortic calcification, osteoporosis, and hearing loss, in *Int. Conf. Bone Mineral Measurement,* Publ. No. (NIH) 75-683, Mazess, R. B., Ed., U.S. Department of Health, Education and Welfare, Washington, D.C., 1973, 228.
101. **Sedlin, E. D., Frost, H. M., and Villanueva, A. R.,** Variations in cross-section area of rib cortex with age, *J. Gerontol.,* 18, 9, 1963.
102. **Garn, S. M., Rohmann, C. G., and Wagner, B.,** Bone loss as a general phenomenon in man, *Fed. Proc.,* 26, 1729, 1967.
103. **Israel, H.,** Loss of bone and remodeling-redistribution in the craniofacial skeleton with age, *Fed. Proc.,* 26, 1723, 1967.
104. **Wowern, N. von and Stoltze, K.,** The pattern of age related bone loss in mandibles, *Scand. J. Dent. Res.,* 88, 134, 1980.
105. **Krishna Rao, G. V. G. and Draper, H. H.,** Age related changes in the bones of adult mice, *J. Gerontol.,* 24, 149, 1969.
106. **Tonna, E. A. and Pillsbury, N.,** Mitochondrial changes associated with aging of periosteal osteoblasts, *Anat. Rec.,* 134, 739, 1959.
107. **Tonna, E. A.,** A study of osteocyte formation and distribution in aging mice complemented with H³-proline autoradiography, *J. Gerontol.,* 21, 124, 1966.
108. **Lothe, K., Spycher, M. A., and Ruttner, J. R.,** Human articular cartilage in relation to age. A morphometric study, *Exp. Cell Biol.,* 47, 22, 1979.
109. **Stockwell, R. A.,** The cell density of human articular and costal cartilage, *J. Anat.,* 101, 753, 1967.
110. **Rosenthal, O., Bowie, M. A., and Wagoner, G.,** Studies in the metabolism of articular cartilage. I. Respiration and glycolysis of cartilage in relation to its age, *J. Cell. Comp. Physiol.,* 17, 221, 1941.
111. **Smith, R. W., Jr.,** Dietary and hormonal factors in bone loss, *Fed. Proc.,* 26, 1737, 1967.
112. **Miller, I.,** Normal hematologic standards in the aged, *J. Lab. Clin. Med.,* 24, 1172, 1939.
113. **Shock, N. W. and Yiengst, M. J.,** Age changes in the acid-base equilibrium of the blood of males, *J. Gerontol.,* 5, 1, 1950.
114. **Das, B. C.,** Indices of blood biochemistry in relation to age, height and weight, *Gerontologia,* 9, 179, 1964.
115. **Attwood, E. C., Robey, E., Kramer, J. J., Ovenden, N., Snape, S., Ross, J., and Bradley, F.,** A survey of the haematological, nutritional and biochemical state of the rural elderly with particular reference to vitamin C, *Age Ageing,* 7, 46, 1978.
116. **Giorno, R., Clifford, J. H., Beverly, S., and Rossing, R. G.,** Hematology reference values. Analysis by different statistical technics and variations with age and sex, *Am. J. Clin. Pathol.,* 74, 765, 1980.
117. **Everitt, A. V. and Webb, C.,** The blood picture of the aging male rat, *J. Gerontol.,* 13, 255, 1958.
118. **Grad, B. and Kral, V. A.,** The effect of senescence on resistance to stress. I. Response of young and old mice to cold, *J. Gerontol.,* 12, 172, 1957.
119. **Ewing, K. L. and Tauber, O. D.,** Hematological changes in aging male C57BL/6 Jax mice, *J. Gerontol.,* 19, 165, 1964.
120. **Finch, C. E. and Foster, J. R.,** Hematologic and serum electrolyte values of the C57BL/6J mouse in maturity and senescence, *Lab. Anim. Sci.,* 23, 339, 1973.
121. **Silini, G. and Andreozzi, U.,** Haematological changes in the ageing mouse, *Exp. Gerontol.,* 9, 99, 1974.
122. **Leuenberger, H.-G. W. and Kunstýř, I.,** Gerontological data of C57BL/6J mice. II. Changes in blood cell counts in the course of natural aging, *J. Gerontol.,* 31, 648, 1976.

123. **Abraham, E. C., Taylor, J. F., and Lang, C. A.,** Influence of mouse age and erythrocyte age on gluthathione metabolism, *Biochem. J., 174,* 819, 1978.

124. **Fand, I. and Gordon, A. S.,** A quantitative study of bone marrow in the guinea pig throughout life, *J. Morphol.,* 100, 473, 1957.

125. **Reigle, G. D. and Nellor, J. E.,** Changes in blood cellular and protein components during aging, *J. Gerontol.,* 21, 435, 1966.

126. **Bush, M., Custer, R. S., Whitla, J. C., and Smith, E. E.,** Hematologic values of captive golden lion tamarins *(Leontopithecus rosalia)* : variation with sex, age, and health status, *Lab. Anim. Sci.,* 32, 294, 1982.

127. **Allan, R. N. and Alexander, M. K.,** A sex difference in the leucocyte count, *J. Clin. Pathol.,* 21, 691, 1968.

128. **Cohnen, G., Augener, W., Reuter, A., and Brittinger, G.,** Peripheral blood T and B lymphocytes in men in different age groups, *Z. Immunol. Forsch.,* 149, 463, 1975.

129. **Hallgren, H. M., Kersey, J. H., Dubey, D. P., and Yunis, E. J.,** Lymphocyte subsets and integrated immune function in aging humans, *Clin. Immunol. Immunopathol.,* 10, 65, 1978.

130. **Onsrud, M.,** Age dependent changes in some human lymphocyte sub-populations. Changes in natural killer cell activity, *Acta Pathol. Microbiol. Scand. Sect. C,* 89, 55, 1981.

131. **Nagel, J. E., Pyle, R. S., Chrest, F. J., and Adler, W. H.,** Oxidative metabolism and bactericidal capacity of polymorphonuclear leukocytes from normal young and aged adults, *J. Gerontol.,* 37, 529, 1982.

132. **Rolsten, C., Claghorn, J., and Samorajski, T.,** Long-term treatment with clozapine on aging mice, *Life Sci.,* 25, 865, 1979.

133. **MacKinney, A. A., Jr.,** Effect of aging on the peripheral blood lymphocyte count, *J. Gerontol.,* 33, 213, 1978.

134. **Reddy, M. M. and Goh, K.,** B and T lymphocytes in man. IV. Circulating B, T and ''null'' lymphocytes in aging population, *J. Gerontol.,* 34, 5, 1979.

135. **Czlonkowska, A. and Korlak, J.,** The immune response during aging, *J. Gerontol.,* 34, 9, 1979.

136. **Davey, F. R. and Huntington, S.,** Age-related variation in lymphocyte subpopulations, *Gerontology,* 23, 381, 1977.

137. **Clot, J., Charmasson, E., and Brochier, J.,** Age-dependent changes of human blood lymphocyte sub-populations, *Clin. Exp. Immunol.,* 32, 346, 1972.

138. **Gupta, S. and Good, R. A.,** Subpopulations of human T lymphocytes. X. Alterations in T, B, third population cells, and T cells with receptors for immunoglobulin M (T_μ) or G(T_γ) in aging humans, *J. Immunol.,* 122, 1214, 1979.

139. **Kishimoto, S., Tomino, S., Inomata, K., Kotegawa, S., Saito, T., Kuroki, M., Mitsuya, H., and Hisamitsu, S.,** Age-related changes in the subsets and functions of human T lymphocytes, *J. Immunol.,* 121, 1773, 1978.

140. **Sparrow, D., Gilbert, J. E., and Rowe, J. W.,** The influence of age on peripheral lymphocyte count in men: a cross-sectional and longitudinal study, *J. Gerontol.,* 35, 163, 1980.

141. **Cobleigh, M. A., Braun, D. P., and Harris, J. E.,** Age-dependent changes in human peripheral blood B cells and T-cell subsets: correlation with mitogen responsiveness, *Clin. Immunol. Immunopathol.,* 15, 162, 1980.

142. **Henschke, P. J., Bell, D. A., and Cape, R. D. T.,** Immunologic indices in Alzheimer dementia, *J. Clin. Exp. Gerontol.,* 1, 23, 1979.

143. **Carosella, E. D., Mochanko, K., and Braun, M.,** Rosette-forming T cells in human blood at various ages, *Cell. Immunol.,* 12, 323, 1974.

144. **Mysliwska, J., Witkowski, J., and Mýsliwski, A.,** Proportion of early E rosettes formed by peripheral blood lymphocytes from young and aged subjects, *Gerontology,* 27, 140, 1981.

145. **Batory, G., Benczur, M., Varga, M., Garam, T., Önody, C., and Petranyi, G. Gy.,** Increased killer activity in aged humans, *Immunobiology,* 158, 393, 1981.

146. **Nagel, J. E., Chrest, F. J., and Adler, W. H.,** Enumeration of T lymphocyte subsets by monoclonal antibodies in young and aged humans, *J. Immunol.,* 127, 2086, 1981.

147. **Banerjee, A. K. and Etherington, M.,** Platelet function in old age, *Age Ageing,* 3, 29, 1974.

148. **Sundberg, M.,** On the mast cells in the human vascular wall, *Acta Pathol. Microbiol. Scand.,* Suppl. 7, 1, 1955.

149. **Shapleigh, J. B., Mayer, S. B., and Moore, C. V.,** Hematologic values in the aged, *J. Gerontol.,* 7, 207, 1952.

150. **Zacharski, L. R., Elveback, L. R., and Linman, J. W.,** Leukocyte counts in healthy adults, *Am. J. Clin. Pathol.,* 56, 148, 1971.

151. **Portaro, J. K., Glick, G. I., and Zighelboim, J.,** Population immunology: age and immune cell parameters, *Clin. Immunol. Immunopathol.,* 11, 339, 1978.

152. **Coggle, J. E. and Proukakis, C.,** The effect of age on the bone marrow cellularity of the mouse, *Gerontologia,* 16, 25, 1970.

153. **Coggle, J. E., Gordon, M. Y., Proukakis, C., and Bogg, C. E.,** Age-related changes in the bone marrow and spleen of SAS/4 mice, *Gerontology,* 21, 1, 1975.

154. **Yuhan, J. M. and Storer, J. B.,** The effect of age on two modes of radiation death and on hematopoietic cell survival in the mouse, *Radiat. Res.,* 32, 596, 1967.

155. **Chen, M. G.,** Age-related changes in hematopoietic stem cell populations of a long-lived hybrid mouse, *J. Cell. Physiol.,* 78, 225, 1971.

156. **Toya, R. E. and Davis, M. L.,** Age-related changes in bone marrow hemopoiesis potential in mice, *Biomédicine,* 19, 244, 1973.

157. **Davis, M. L., Upton, A. C., and Satterfield, L. C.,** Growth and senescence of the bone marrow stem cell pool in RFM/Um mice, *Proc. Soc. Exp. Biol. Med.,* 137, 1452, 1971.

158. **Haaijman, J. J., Schuit, H. R. E., and Hijmans, W.,** Immunoglobulin-containing cells in different lymphoid organs of the CBA mouse during its life span, *Immunology,* 32, 427, 1977.

159. **Brennan, P. C. and Jaroslow, B. N.,** Age-associated decline in the theta antigen on spleen thymus-derived lymphocytes of B6CF₁ mice, *Cell. Immunol.,* 15, 51, 1975.

160. **Ploemacher, R. E.,** Haemopoietic stroma in aged mice: splenic stroma, *IRCS Med. Sci: Biochem.,* 9, 929, 1981.

161. **Albright, J. F. and Makinodan, T.,** Growth and senescence of antibody-forming cells, *J. Cell. Physiol.,* Suppl. 67, 185, 1966.

162. **Wigzell, H. and Stjernswärd, J.,** Age-dependent rise and fall of immunological reactivity in the CBA mouse, *J. Natl. Cancer Inst.,* 37, 513, 1966.

163. **Gerbase-DeLima, M., Liu, R. K., Cheney, K. E., Mickey, R., and Walford, R. L.,** Immune function and survival in a long-lived mouse strain subjected to undernutrition, *Gerontologia,* 21, 184, 1975.

164. **Joncourt, F., Bettens, F., Kristensen, F., and deWeck, A. L.,** Age-related changes of mitogen responsiveness in different lymphoid organs from outbred NMRI mice, *Immunobiology,* 158, 439, 1981.

165. **Gilman, S. C., Woda, B. A., and Feldman, J. D.,** T lymphocytes of young and aged rats. I. Distribution, density, and capping of T antigens, *J. Immunol.,* 127, 149, 1981.

166. **Andrew, W.,** Age changes in the vascular architecture and cell content in the spleens of 100 Wistar Institute rats, including comparisons with human material, *Am. J. Anat.,* 79, 1, 1946.

167. **Bellamy, D. and Alkufaishi, H.,** The cellular composition of thymus: a comparison between cortisol-treated and aged C57/BL mice, *Age Ageing,* 1, 88, 1972.

168. **Meyer, J., Marwah, A. S., and Weinmann, J. P.,** Mitotic rate of gingival epithelium in two age groups, *J. Invest. Derm.,* 27, 237, 1956.

169. **Sharav, Y. and Massler, M.,** Age changes in oral epithelia. Progenitor population, synthesis index and tissue turnover, *Exp. Cell Res.,* 47, 132, 1967.

170. **Lavelle, C. L. B.,** The effect of age on the pulp tissue of rat incisors, *J. Gerontol.,* 24, 155, 1969.

171. **Lavelle, C.,** some age changes in rat incisor teeth, *J. Gerontol.,* 23, 393, 1968.

172. **Rother, P. and Leutert, G.,** Morphometrical and mathematical analysis of the ageing changes of the muscle-connective tissue relation in smooth muscle, in *Cell Impairment in Aging and Development,* Cristofalo, V. J. and Holěcková, E., Eds., Plenum Press, New York, 1974, 441.

173. **Andrew, W.,** Ageing changes in the alimentary tract, in *Structural Aspects of Aging,* Bourne, G. H., Ed., Hofner, New York, 1961, 61.

174. **Pieri, C., Giuli, C., del Moro, M., and Piantanelli, L.,** Electron-microscopic morphometric analysis of mouse liver. II. Effect of ageing and thymus transplantation in old animals, *Mech. Ageing Dev.,* 13, 275, 1980.

175. **Sato, T., Miwa, T., and Tauchi, H.,** Age changes in the human liver of the different races, *Gerontologia,* 16, 368, 1970.

176. **Sato, T., Cespedes, R. F., Goyenaga, P. H., and Tauchi, H.,** Age changes in the livers of Costa Ricans, *Mech. Ageing Dev.,* 11, 171, 1979.

177. **Ross, M. H.,** Aging, nutrition and hepatic enzyme activity patterns in the rat, *J. Nutr.,* 97 (Suppl. 1, Part 2), 565, 1969.

178. **Pieri, C., Zs.-Nagy, I., Mazzufferi, G., and Giuli, C.,** The aging of rat liver as revealed by electron microscopic morphometry. I. Basic parameters, *Exp. Gerontol.,* 10, 291, 1975.

179. **Porta, E. A., Keopuhiwa, L., Joun, N. S., and Nitta, R. T.,** Effects of the type of dietary fat at two levels of vitamin E in Wistar male rats during development and aging. III. Biochemical and morphometric parameters of the liver, *Mech. Ageing Dev.,* 15, 297, 1981.

180. **Bacon, R. L.,** Changes with age in the reticular fibers of the myocardium of the mouse, *Am. J, Anat.,* 82, 469, 1948.

181. **Rakŭsan, K. and Poupa, O.,** Capillaries and muscle fibers in the heart of old rats, *Gerontologia,* 9, 107, 1964.

182. **Constantinides, P. and Rutherdale, J.,** Effects of age and endocrines on the mast cell counts of the rat myocardium, *J. Gerontol.,* 12, 264, 1957.
183. **Cairns, A. and Constantinides, P.,** Mast cells in human atherosclerosis, *Science,* 120, 31, 1954.
184. **Yiengst, M. J., Barrows, C. H., and Shock, N. W.,** Age changes in the chemical composition of muscle and liver in the rat, *J. Gerontol.,* 14, 400, 1959.
185. **Tauchi, H., Yoshioka, T., and Kobayashi, H.,** Age change of skeletal muscles of rats, *Gerontologia,* 17, 219, 1971.
186. **Tůcek, S. and Gutmann, E.,** Choline acetyltransferase activity in muscle of old rats, *Exp. Neurol.,* 38, 349, 1973.
187. **Rowe, R. W. D.,** The effect of senility on skeletal muscles in the mouse, *Exp. Gerontol.,* 4, 119, 1969.
188. **Hooper, A. C. B.,** Length, diameter and number of ageing skeletal muscle fibres, *Gerontology,* 27, 121, 1981.
189. **Teem, M. van B.,** The relation of the interstitial cells of the testis to prostatic hypertrophy, *J. Urol.,* 34, 692, 1935.
190. **Kaler, L. W. and Neaves, W. B.,** Attrition of the human Leydig cell population with advancing age, *Anat. Rec.,* 192, 513, 1978.
191. **Kaler, L. W. and Neaves, W. B.,** Androgen status of aging rats (Abstr.), in 32nd Annu. Meet. Gerontological Society, Washington, D.C., 1979, 98.
192. **Bethea, C. L. and Walker, R. F.,** Age-related changes in reproductive hormones and in Leydig cell responsivity in the male Fischer 344 rat, *J. Gerontol.,* 34, 21, 1979.
193. **Johnson, L. and Neaves, W. B.,** Age-related changes in the Leydig cell population, seminiferous tubules, and sperm production in stallions, *Biol. Reprod.,* 24, 703, 1981.
194. **Gosden, R. G., Richardson, D. W., Brown, N., and Davidson, D. W.,** Structure and gametogenic potential of seminiferous tubules in aging mice, *J. Reprod. Fertil.,* 64, 127, 1982.
195. **Ewing, L. L., Johnson, B. H., Desjardins, C., and Clegg, R. F.,** Effect of age upon the spermatogenic and steroidogenic elements of rabbit testes, *Proc. Soc. Exp. Biol. Med.,* 140, 907, 1972.
196. **Stiles, J. W., Francendese, A. A., and Masoro, E. J.,** Influence of age on size and number of fat cells in the epididymal depot, *Am. J. Physiol.,* 229, 1561, 1975.
197. **Bertrand, H. A., Masoro, E. J., and Yu, B. P.,** Increasing adipocyte number as the basis of perirenal depot growth in adult rats, *Science,* 201, 1234, 1978.
198. **Arai, H.,** On the postnatal development of the ovary (albino rat) with especial reference to the number of ova, *Am. J. Anat.,* 27, 405, 1920.
199. **Jones, E. C. and Krohn, P. L.,** The relationship between age, numbers of oocytes and fertility in virgin and multiparous mice, *J. Endocrinol.,* 21, 469, 1961.
200. **Pinkerton, K. E., Barry, B. E., O'Neil, J. J., Raub, J. A., Pratt, P. C., and Crapo, J. D.,** Morphologic changes in the lung during the lifespan of Fischer 344 rats, *Am. J. Anat.,* 164, 155, 1982.
201. **Bolden, T. E.,** Age, a factor in the growth of the exorbital lacrimal gland, in *Control of Cellular Growth in Adult Organisms,* Teir, H. and Rytomaa, T., Eds., Academic Press, New York, 1967, 275.
202. **Rossman, I.,** Anatomic and body composition changes with aging, in *Handbook of the Biology of Aging,* Finch, C. E. and Hayflick, L., Eds., Reinhold, New York, 1977, 189.
203. **Franks, L. M.,** Ageing in differentiated cells, *Gerontologia,* 20, 51, 1974.

AGING AND THE CELL CYCLE IN VIVO AND IN VITRO

Phillip D. Bowman

INTRODUCTION

Prominent among the constellation of time-dependent deteriorative changes which are collectively termed aging and which lead to organismic death are alterations in cellular reproductive capacity. Diminished cellular reproductive capacity, or clonal senescence, is the hallmark of aging cells in vitro and ultimately leads to demise of the culture. In vivo, diminished cellular reproductive capacity is evident in many renewing tissues, although it is increased in a few and, in contrast to this background of generalized diminished cellular reproductive capacity, focal increases occur more frequently with age leading to neoplasia. The period of time between cellular reproductions is termed the cell cycle. As a fundamental unit of time at the cellular level, it defines the life cycle of cells, and within rather broad limits, its average length is characteristic of differentiated cell types. Since cellular reproduction is a major defining property of life, and given its central role in the dynamics of tissue homeostasis, it is important to know to what extent cellular reproduction is involved in the aging process. Here will be considered the role that loss of cellular reproductive capacity plays in aging, principally in mammalian tissues in vivo and in vitro, by examining the available information on changes in parameters of the cell cycle.

The development of an adult organism from a single fertilized ovum is the result of numerous cellular divisions. During early development increases in cell numbers exceed losses, and net growth occurs. The process of cellular reproduction does not cease, however, with the attainment of the adult condition, as maintenance requires continual cellular reproduction in a balanced way so that new cells are reproduced as old ones die off or are lost through attrition. The magnitude of this process of maintenance is not always appreciated. Prescott[1] has estimated that cellular renewal requires something on the order of 20×10^6 cell divisions per second in an adult human composed of about 10^{13} cells, most of these cellular reproductions occurring in intestine, skin, hemopoietic, and immune systems. Chalones, hormones, and growth factors are clearly involved, singly and in combination, to achieve an exquisite balance between cell loss and gain. However, the means by which such large numbers of cell divisions are regulated and integrated into the functional organism are largely unknown.

The concept that diminished cellular renewal capacity is involved in the aging process is not new. As early as 1907 Minot[2] noted that the decline of growth rate throughout life in some or all tissues appeared to be a new universal feature of metazoa. It has only been during the past 2 decades, however, that attention has been focused on a role of loss of cellular reproductive capacity in aging. Early work utilizing cell culture suggested that cells freed from the constraints of multicellular organization and placed in cell culture, or in vitro, were immortal. That this is not the case was unequivocally demonstrated by Hayflick and Moorhead[3] and Hayflick,[4] who showed that normal cells with a diploid chromosome complement exhibit a finite lifespan and further suggested that the limited capacity for cellular reproduction might be the expression at the cellular level of aging in vivo.

The hypothesis that diminished cellular reproductive capacity plays a significant role in senescence does not require that massive cell loss be the cause of aging and death. Instead, it predicts that important "functional decrements" occur with age that are attributable to an intrinsic cellular mechanism that limits a cell's reproductive capacity.[5] Diminished cellular reproductive capacity need not affect all tissues equally, as all cell populations do not renew themselves at the same rate. Since organismic death usually results from mechanical failure of cardiac or cerebral vasculature, it would predict that an important site for the functional decrement due to reduced reproductive capacity lies in the cardiovascular system.

It is useful to consider the types of cells whose diminished reproductive capacity would effect such a functional decrement. Table 1 classifies populations of cells from rat, mouse, and man according to their reproductive behavior. These are (1) static, or nonrenewing cell populations that are terminally differentiated (neurons, cardiac muscle cells, etc.), (2) renewing (rapid and slow) cell populations, (3) cell populations that proliferate but at a rate slower than the life of the animal, and (4) neoplastic.

A comparison of turnover times (defined as the time necessary to replace the number of cells present in the entire cell population) is also included in Table 1. As noted by Cameron and Thrasher, the turnover times of rapidly renewing cell populations are about the same regardless of lifespan of mice (2.5 years), rats (3.5 years), and man (70 years). One concludes that rapidly renewing cell populations are replaced more times during the lifespan of man than mice or rats. For example, colonic epithelium would renew itself roughly 365 times during the lifespan of mice and rats and 5110 times in man.[6] In addition, consideration of the number of cellular reproductions required to achieve the adult condition in man vs. rat and mouse indicates than an important difference among species may lie in developmental mechanisms that dictate cellular reproductive capacity. Some evidence for such a mechanism has been indicated by the positive correlation of species lifespan with capacity to repair damage to genetic material produced by exposure to UV light in cultured fibroblasts.[7]

Static cell populations will not be considered further, as the functional decrements which occur with age in these tissues do not express themselves at the level of cellular reproduction. Neither will neoplastic cells be considered to any extent, although they are of considerable interest with regard to cellular senescence because they seem to have escaped from the mechanism that limits cellular lifespan.

THE CELL CYCLE

To the early histologist observing cells in the light microscope, the cell cycle began with one mitosis and ended with the next. Between mitoses an approximate doubling of cell size could be observed, but other than this, no clear events were discernible, and the rest of the cell cycle was simply termed interphase. Fleming, Strasburger, and van Beneden described the separation of chromosomes and their distribution into daughter cells in the 1880s, and Roux shortly thereafter suggested that the chromosomes might be the bearers of heredity. This was confirmed in 1900 after rediscovery of Mendel's laws, laying the basis for the chromosome theory of inheritance.

Excellent reviews on the cell cycle are available.[8-11] The present concept of the cell cycle was largely developed by Howard and Pelc.[12] Utilizing ^{32}P incorporation into nucleic acid and autoradiography, they found that discrete intervals existed between mitosis (M) and DNA synthesis (S) and between the completion of S and the subsequent mitosis. These gaps in time were termed Gap 1 (G1) and Gap 2 (G2), respectively, and to a large extent, they represent gaps in our knowledge about the biochemical events occurring here.

Methods for determining the lengths of the phases of the cell cycle generally make use of cells that are in M and/or S. For a number of reasons cultured cells are much simpler to work with than tissues in vivo with respect to cell cycle analysis. For instance, problems peculiar to in vivo studies such as diurnal fluctuations in metabolism, stress, nutritional status, and rapid depletion of isotopes are generally not important considerations with cell cultures. Also, quantitation of population dynamics is often simpler with cultured cells; increases in cell number can be readily followed by cell counting. Furthermore, length of mitosis and interdivision time of daughter cells can be followed directly by time-lapse microcinematography. A recently developed technique, flow system microspectrofluorometry, can rapidly supply cell cycle information from cultured cells.

Table 1
CLASSIFICATION OF CELL POPULATIONS OF ADULT MAMMALS BASED ON THEIR PROLIFERATIVE BEHAVIOR

Population type	Examples	Turnover times in days (range)		
		Mouse	Man	Rat
Static	Neurons of all types, cardiac muscle cells, molar odontoblasts, sertoli cells of the testis			
Renewing				
Rapid (renewal in less than 30 days)	Malpighian layer of epidermis (varies	7—28	13—100	13—34
	with area sampled)	5—7	7	7
	Cornea	4—8		3—6
	Oral epithelium, tongue	2—3		2
	Duodenum epithelium	2—3	4—6	3
	Colon epithelium	34	74	48
	Seminiferous tubules (spermatogenesis)			
	Female genital system	4		4
	Vagina			
	Cervix		6	6
	Hemapoietic — bone marrow	1—2		1—3
	Lymphopoietic — small lymphocytes			
	Thymus	3—4		~7
	Spleen	14—21		~15
Slow (renewal in more than 30 days but less than the mean lifespan of the animal)	Respiratory tract epithelium			
	Trachea	20—90		48—111
	Bronchioles	50—100		200
	Lung wall cells	~125		21—29
	Kidney — cortical tubules	170—190		
	Liver			
	Hepatocytes	480—620		400—450
	Littoral cells	160		
	Pancreatic acinar cells	520		
	Islet cells	150		
	Submandibular			
	Alveolar cells	180		
	Granular cells	210		
	Adrenal cortex			
	Glomerulosa	30—125		
	Fasciculata	373—380		
	Reticularis	560—1040		
	Fibroblasts in the dermal connective tissue	60		
	Lamina propria cells of the alimentary tract	125		
	Parietal cells of the stomach	60		60
Cell populations which demonstrate some renewal but at such a slow rate that not all of the cells renew during the lifespan of the animal	Harderian gland cells, smooth muscle cells, glial cells in most areas of the brain, brown fat cells, osteocytes, interstitial cells of Leydig, zymogenic (chief) cells of the stomach, kidney medullary tubule cells, transitional epithelial cells			
Neoplastic	Solid tumor cells, metastatic cancer cells			

From Cameron I. L. and Thrasher, J. D., *Interdiscipl. Top. Gerontol.*, 10, 108, 1976. With permission.

Such methods as those above have shown that S, G2, and M tend to be relatively fixed periods among those cell types that have been analyzed, whereas G1 is quite variable. In vivo, G1 is fairly uniform for a given differentiated cell type; however, it is 33 hr in basal esophageal cells and 8000 hr in basal and intermediate cells of the ureter.[13] In cultured cell strains the interdivision time is generally 10 to 30 hr, with G1 lasting 5 to 15 hr.

The G_0 state as a subsection of G1 was introduced by Lajtha[14] to describe the state in which a cell may be withdrawn from the cell cycle for extended periods and exist in a nongrowing, quiescent state. A G_0 cell can return to traverse the cell cycle as circumstances require. In vivo examples of G_0 tissues include liver and kidney which, under normal circumstances, are slow renewing tissues but when damaged, may undergo sudden, rapid growth by activating a population of cells in G_0 to divide. In vitro, human diploid cells can be placed in the G_0 state by decreasing the serum concentration. Readdition of fresh medium with normal amounts of serum causes them to reenter the cell cycle. Perhaps the best evidence that the number of cellular reproductions determines the rate of cellular aging comes from the ability to maintain human diploid fibroblasts in G_0 for extended periods (with reduced serum) and, following restimulation, their completion of the same number of population doublings had they not entered G_0.[15,16] As is true for G1, however, the biochemical nature of G_0 is largely unknown except that such cells possess a G1 amount of DNA.

The crucial events for regulation of cellular reproduction seem to reside in G1. Pardee[17] has presented evidence for the existence of a restriction point in mid to late G1 at which a normal cell will not enter S unless conditions such as availability of nutrients or cell density are met. Transformed cells, on the other hand, are generally less sensitive to this restriction point and may lose it entirely. Rossow et al.[18] and Das[19] have recently presented evidence that the ability to proceed past the restriction point requires the synthesis of a labile protein which must accumulate to some critical level before S can begin. The transformed cell either makes large amounts of it or the restriction point is no longer operative.

Additionally, G1 is of special interest because expression of differentiated function occurs most prominently during this time, and old cells, those incapable of further reproductions, tend to end up in G1. These observations are consistent with the hypothesis that the loss of ability to divide in culture may be the result of a terminal differentiation such as occurs with hemopoietic cells or epidermal keratinocytes in vivo.[20-22] Furthermore, transformed cells, which do not lose their ability to divide, seem impaired in their ability to express differentiated functions as well. Unfortunately, there is too little information about the biochemistry of G1 to confirm such a hypothesis at this time.

AGING AND THE CELL CYCLE IN VIVO

Mitoses in a section of normal tissue or in culture is indicative of cell turnover. Because mitosis in vivo occupies only a short period of the cell cycle and tends to occur in waves, diurnal variation, nutritional status, and sex must be carefully controlled to obtain meaningful data. Some data exist for different cell types at different ages utilizing mitotic index as an indication of cellular reproductive capacity. This information, compiled by Buetow,[23] appears in Table 2. With the exception of abdominal skin and gingival epithelia, there is generally an age-related decline in mitotic index with age. Bullough[31] first noted that mitotic activity in mouse epidermis is higher in middle-aged than in young animals, although it becomes low in very old animals. It does not decrease significantly in human sebaceous gland either, although other aspects of sebaceous gland function change with age.[32] Some other accessory structures of skin do exhibit decreased reproductive capacity with age although this is not well worked out. For example, the graying of hair that occurs with age appears to result from not only the loss of melanocytes but also from the decreased production of melanin by those that remain.[33-35]

Table 2
MITOSIS VS. AGE IN VARIOUS TISSUES AND ORGANS

Tissue or organ	Subjects		Age	Mitoses			Measure	Ref.
	Type	No.		Sex	No. or %	Range		
Liver	Rat		1 day		3.8		% mitoses; 2000 or more random nuclei	24
			3 weeks		1.1			
			8 weeks		0.6			
			24—156 weeks		0.1			
Kidney								
Proximal convoluted tubules	Rat	6	120—125 days		2.00 (±0.37)		No. mitoses per 2000 cells	25
		4	1160—1170 days		0.50 (±0.28)			
Skin								
Abdominal skin epidermis	Human	8	2 days—19 years	M,F	24.5	8—64	No. mitoses per 10^5 cells	26
		21	23—39 years	M,F	36.8	14—106		
		13	41—60 years	M,F	49.7	17—88		
		12	61—77 years	M,F	48.9	19—84		
Ear epithelium	Mouse	10	1—3 months	M	16.9 (±4.3)		No. mitoses per cm^2	27
		9	3—6 months	M	17.3 (±4.3)			
		9	6—9 months	M	15.8 (±3.4)			
		10	9—12 months	M	14.7 (±4.5)			
		10	15—18 months	M	13.7 (±3.1)			
		9	18—24 months	M	12.2 (±9.6)			
		10	27—30 months	M	11.0 (±7.9)			
		4	30—33 months	M	2.4 (±2.1)			
		9	1—3 months	F	8.9 (±2.8)			
		10	3—6 months	F	18.9 (±2.5)			
		10	6—9 months	F	16.6 (±3.7)			
		10	9—12 months	F	12.8 (±4.0)			
		10	15—18 months	F	13.5 (±3.1)			
		10	18—24 months	F	10.0 (±2.9)			
		10	27—30 months	F	15.4 (±2.9)			
		3	30—33 months	F	5.6 (±4.1)			
		3	33—36 months	F	1.9 (±0.6)			

Table 2 (continued)
MITOSIS VS. AGE IN VARIOUS TISSUES AND ORGANS

Tissue or organ	Subjects				Mitoses			
	Type	No.	Age	Sex	No. or %	Range	Measure	Ref.
Eye								
Exorbital lacrimal gland	Rat	6	21 days	M	8.50		No. mitoses per mm²	28
		2	50 days	M	1.40			
		2	100 days	M	0.23			
		6	300 days	M	0.29			
		12	700 days	M	0.07			28
		9	900 days	M	0.05			
		6	1000 days	M	0.01			
		6	21 days	F	12.20		No. mitoses per mm²	
		2	50 days	F	6.30			
		2	100 days	F	0.01			
		6	300 days	F	0.04			
		13	700 days	F	0.06			
		16	900 days	F	0.03			
		6	1000 days	F	0.02			
Epithelium								
Gingival epithelium	Human	30	25—34 years	M	0.098		No. mitoses per 1000 cells	29
		30	50—78 years	M	0.156			
Thyroid	Guinea pig	23	4 days—1 month	M	217	80—960		30
		26	1—4 months	M	134	40—420		
		24	4—8 months	M	96	40—180		
		16	8—12 months	M	48	0—140		
		13	12—18 months	M	23	0—100	No. mitoses per gland, both lobes	
		9	1.5—3 years	M	20	0—60		
		23	4 days—1 month	F	560	120—1600		
		23	1—4 months	F	380	100—1480		

			Age	Sex				
Parathyroid	Guinea pig	31	4—8 months	F	225	0—1080	30	
		21	8—12 months	F	135	20—330		
		19	12—18 months	F	76	0—280		
		20	1.5—3 years	F	58	0—170		
		26	4 days—1 month	M	3.0	0—8.0		
		26	1—4 months	M	1.6	0—3.7		
		21	4—8 months	M	0.8	0—2.0		
		19	8—12 months	M	0.7	0—4.0		
		12	12—18 months	M	0.6	0—2.3		No. mitoses
		4	1.5—3 years	M	0.3	0—1.0		per 10^5 cells
		26	4 days—1 month	F	4.1	0.9—16.8		
		22	1—4 months	F	2.8	0.4—8.8		
		25	4—8 months	F	1.3	0—3.3		
		27	8—12 months	F	1.2	0—8.5		
		21	12—18 months	F	0.7	0—3.6		
		19	1.5—3 years	F	0.6	0—2.6		
Adrenal cortex	Guinea pig	26	4 days—1 month	M	3.3	1.2—9.6	30	
		26	1—4 months	M	2.8	0.8—10.0		
		31	4—8 months	M	2.3	0—7.4		
		24	8—12 months	M	1.5	0—8.7		
		6	12—18 months	M	1.2	0.2—1.8		Av. no. mi-
		9	1.5—3 years	M	0.9	0—2.5		toses per lon-
		28	4 days—1 month	F	5.9	1.0—16.6		gitudinal
		20	1—4 months	F	4.7	1.1—7.6		section
		20	4—8 months	F	2.8	0—10.0		
		17	8—12 months	F	2.1	0—8.4		
		19	12—18 months	F	2.0	0—6.8		
		25	1.5—3 years	F	1.1	0—3.8		

From Buetow, D. E. in *Cellular and Molecular Renewal in the Mammalian Body*, Cameron, I. L. and Thrasher, J. D., Eds., Academic Press, New York, 1971, chap. 4. With permission.

Because DNA synthesis takes approximately ten times longer than does mitosis, autoradiographs made from an animal following a single injection of [³H]thymidine show about ten times more labeled cells than mitotic cells. Since the duration of DNA synthesis is rather constant regardless of species or age of the animal DNA synthetic index tends to average out diurnal variation associated with mitotic indices and is more representative of the proliferative pool of cells. Table 3 summarizes the available data on labeling index as a function of age. Here again there is a generalized, age-related decline in cellular reproductive capacity in most tissues. Tongue, like skin, is a rapidly renewing tissue derived embryologically from ectoderm, and it also experiences an increase in cellular reproductive capacity with age. Appended to Table 3 are data from an experiment described by Martin et al.[40] They removed the aorta from mice of five ages, maintained them in organ culture for 3 to 4 days, and then labeled them for 24 hr with [³H]thymidine followed by autoradiography. Presumably, during the 3 to 4 days in culture, variables such as diurnal fluctuation in labeling indices would disappear, but during this short culture period, more or less normal turnover might be expected. Their data demonstrated a clear decreased reproductive capacity for six cell types in the aorta. The small sample size notwithstanding, the approach is novel in aging research and obviates many of the problems associated with in vivo work, such as loss of isotope. More importantly, it represents evidence for declining reproductive capacity with age of a portion of the vascular system which may be an important target for functional decrement leading to death in old organisms.

Table 4 presents data on changing cell population kinetics of mouse intestine with age. From the data available, it is shown that cellular reproductive capacity decreases with time, while the length of the cell cycle increases. M, S, and G2 remain relatively constant, while G1 gradually lengthens. Similar results have been documented with aging cells in culture, suggesting a similarity in the mechanism which halts further proliferation.

A possible basis for the failure to find a decrease in cellular reproductive capacity with age in some cell populations may result from the presence of a stem cell population which nonrandomly segregates DNA at mitosis so that one daughter cell retains all the new DNA.[47] Though the data are not conclusive the suggestion would explain the ability of certain tissues such as skin and the hemopoietic system to exhibit lifespans greater than the lifespan of the organism.

SERIAL TRANSPLANTATION STUDIES

A reasonable criticism of in vivo studies of aging and the cell cycle, especially those demonstrating decreased reproductive capacity, is that the observed decrement might merely be the result of the presence of essentially young cells in an old organism, that is, decreased cellular reproductive capacity is observed only because mechanisms exclusive of cell growth are limiting cellular reproduction. For example, hormone-dependent tissues might fail to reproduce because of insufficient hormone production by appropriate tissues.

One approach to circumvent this difficulty is the examination of the reproductive capacity of tissue serially passaged into young, histocompatible animals. Daniel[48] has recently reviewed the results available using this technique on several different tissues. Although subject to other interpretation, the data generally suggest that normal tissues cannot be serially passaged indefinitely. This result has been demonstrated most convincingly in mouse mammary gland where explants of mammary epithelium can be carried for about 2 years, the normal lifespan of the animal, before further transplants cease growing. This is in contrast to neoplastic mammary tissue which can be serially passaged indefinitely, suggesting a good analogy to the indefinite lifespan of transformed cells in vitro.[48]

Both skin and hemopoietic systems, however, can be serially passaged for periods considerably longer than the normal lifespan of their rodent hosts. Krohn[49] has shown that skin

Table 3
DNA SYNTHETIC INDEX VS. AGE IN VARIOUS TISSUES AND ORGANS

Species	Sex	Tissue	Age	DNA synthetic index (%)	Labeled cell nuclei/10^5 cells		Number animals	Ref.
					n	SD		
Mouse	F	Esophageal epithe-	10 days	12.6			6	
		lium basal layer	30—70 days	8.3			4	36
			380—399 days	7.0			6	
			579—638	6.6			7	
Mouse	M	Kidney cells of ne-	2 months	0.3			6	37
		phron tubules	4 months	0.2			6	
			13 months	0.1			12	
Rat	M	Liver	8 weeks	1.7			1.6	24
			24 weeks	0.2			1.6	
			104 weeks	0.9			1.6	
			156 weeks	0.5			1.6	
Rat	M	Palate basal cell layer	2 months	20.5			3	38
			9 months	11.4			3	
			19 months	14.1			3	
			27 months	16.5			3	
		Tongue basal cell	2 months	16.8			3	
		layer	9 months	12.3			3	
			19 months	13.5			3	
			27 months	19.0			3	
Mouse Strain A/Grb	M	Alveolar wall cells	3 months		305	± 104	10	
			12 months		171	± 68	8	
	F		3 months		380	± 29	5	
			12 months		245	± 179	5	
Strain A	M		3 months		490	± 444	5	39
			12 months		296	± 118	5	
			24 months		297	± 179	5	
Strain C57BL	M		3 months		464	± 188	13	
			12 months		200	± 60	10	
			24 months		211	± 132	9	
	F		3 months		565	± 235	10	39
			12 months		353	± 98	10	
			24 months		255	± 191	10	
Strain ICR[a]	M	Endothelial lining of	6 months	3.9[b]			4	40
		intima	12 months	7				
			18 months	1.2				
			24 months	<0.1				
			25 months	0.3				
		Endothelium of ad-	6 months	7				
		ventitial blood	12 months	<2.1				
		vessels	18 months	<0.25				
			24 months	<0.6				
			25 months	<0.23				
		Adventitial	6 months	8.3				
		fibroblasts	12 months	7.8				
			18 months	4.1				
			24 months	2.8				
			25 months	1.3				
		Adventitial areolar	6 months	10.9				
		connective tissue	12 months	9.7				
			18 months	4.5				

Table 3 (continued)
DNA SYNTHETIC INDEX VS. AGE IN VARIOUS TISSUES AND ORGANS

Species	Sex	Tissue	Age	DNA synthetic index (%)	Labeled cell nuclei/10⁵ cells		Number animals	Ref.
					n	SD		
			24 months	3				
			25 months	5				
		Smooth muscle of intima	6 months	0.5				
			12 months	0.5				
			18 months	0.15				
			24 months	0.07				
			25 months	<0.03				
		Adventitial mesothelium	6 months	46.9				
			12 months	29.8				
			18 months	21				
			24 months	14.8				
			25 months	13.8				

[a]   Appended to Cameron and Thrasher.[6]
[b]   Labeled for 24 hr in explant culture; average percent labeled cells labeled at 3—4 days and 4—5 days.

From Cameron, I. L. and Thrasher, J. D., *Interdiscipl. Top. Gerontol.*, 10, 108, 1976. With permission.

can be serially transplanted and survive for up to 7 years, about three times longer than the lifespan of the mouse. Similar results have been obtained with erythropoietic stem cells by Harrison,[50] and he has recently attributed the observed decline to an artifact of the transplantation technique.[51] It may be that skin and hemopoietic tissue require a large reserve of cellular reproductive capacity, given their exposure to the environment or in the face of hemorrhage, and are not important targets for age-related decline in reproductive capacity.

Williamson and Askonas[52] have investigated the age-related decline of immune function by estimating the lifespan of a clone of antibody-secreting cells followed by serial passage into irradiated syngeneic mice. Antibody production by the cells, quantitated by appropriate techniques, remained relatively constant during the first four passages but then declined exponentially until passage seven. They calculated that about 90 cell doublings occurred before senescence, although simple loss of ability to produce antibody cannot be ruled out.

AGING AND THE CELL CYCLE IN VITRO

Early in vitro work suggested that cultured cells possessed an indeterminate or indefinite lifespan. By 1911 Alexis Carrel and associates[53] were able to keep explants of chick heart alive for extended periods, one such culture purported to have been maintained for 34 years. These findings led to the generalization that all cells in culture are immortal and that the limited lifespan of the whole organism is the price paid for multicellularity. However, it is now thought probable that Carrel's ability to maintain chick cells in culture was due to inadvertent inoculation of the culture with fresh chick cells rather than to their inherent immortality.[5]

The work of Earle and colleagues[54] in the 1940s with rodent cell cultures and, later, the derivation by Gey et al.[55] of the well-known HeLa cell line appeared to confirm Carrel's early findings and reinforced the concept that cells in culture are immortal. Rodent cells, however, present a special case among mammalian cell cultures. Whereas these cells initially exhibit an in vitro aging phenomenon, a few cells spontaneously transform to give rise to

Table 4

AGE-RELATED CHANGES IN MUCOSAL CELLS OF THE MAMMALIAN INTESTINE AS DETERMINED BY [³H] THYMIDINE INCORPORATION

Area	Subjects Type	Age (days)	Sex	Generation time (hr)	Transit time (hr) Crypt	Villus	Total	DNA synthetic index (%)	Mitotic index (%)	S phase (hr)	G$_2$ and prophase (hr)	Ref.
Duodenum												
Epithelial cells	Mouse	89	M,F		5.5	35.5	41					
		362	M,F		6.5	41.5	48					
		945	M,F		>10.0	<43	53					41
Crypt cells (stem or progenitor cells)	Mouse	93	M,F	11—11.5								
		372	M,F	11—11.5								
		940	M,F	13—15								41
		10		11.4				63.3	5.5	7.2	0.75—2.0	43,44
		30—70		12.4				57.9	5.9	7.4	0.75—2.0	
		380—399		14.0				53.4	5.8	7.5	0.75—2.0	
		579—638		15.0				50.2	4.9	7.4	0.75—2.0	
Jejunum												
Epithelial cells	Mouse	93	M,F		3.2	38.8	42					
		372	M,F		5.1	46.9	52					
		940	M,F		6.8	47.2	54					45
Crypt cells (stem or progenitor cells)	Mouse	93	M,F	11								
		372	M,F	11								
		940	M,F	13								42
Ileum												
Epithelial cells	Mouse	93	M,F		4.4	26.6	31					
		372	M,F		5.3	27.7	33					
		940	M,F		8.0	23	31					45

Table 4 (continued)
AGE-RELATED CHANGES IN MUCOSAL CELLS OF THE MAMMALIAN INTESTINE AS DETERMINED BY [³H]
THYMIDINE INCORPORATION

| Area | Subjects | | | Generation time (hr) | Transit time (hr) | | | DNA synthetic index (%) | Mitotic index (%) | S phase (hr) | G₂ and prophase (hr) | Ref. |
	Type	Age (days)	Sex		Crypt	Villus	Total					
Crypt cells (stem or progenitor cells)	Mouse	93	M,F	11								42
		372	M,F	11								
		940	M,F	13								
Colon												
Crypt cells (stem or progenitor cells)	Mouse	10		15				23.9 (±3.3)		7.3	1—2	46
		30—70		19				22.1 (±4.1)		8.0	1—2	
		380—399		19				20.0 (±2.8)		8.0	1—2	
		579—638		21				15.8 (±3.8)		7.7	1—2	

From Buetow, D. E., in *Cellular and Molecular Renewal in the Mammalian Body*, Cameron, I. L. and Thrasher, J. D., Eds., Academic Press, New York, 1971, chap. 4. With permission.

cultures with an indefinite lifespan. Likewise, the HeLa cell line is not representative of either tumor or normal human cells because of its unusual ability to establish itself in culture.[56]

In 1957 Swim and Parker[57] published results from cultures derived from a variety of human tissues grown on different media and reported that 49 of 51 cultures could be propagated for only 3 to 34 passages. Hayflick and Moorhead[3] confirmed this work, extending it to show that cultures with a limited lifespan possess a normal complement of chromosomes. These results have been reproduced by many laboratories throughout the world and have led to the hypothesis that the limited reproductive capacity of normal cells might be the expression at the cellular level of aging in vitro.

Hayflick[4] subdivided the in vitro aging phenomenon into three parts: Phase I, the period of outgrowth of cells, Phase II, the period of rapid proliferation, and Phase III, the period when the time for the cells to reach confluence is increased. During this last phase the final cell density is decreased, and morphological changes such as an increase in cell size and changes in cell shape are observed.

The usual method for monitoring culture age is determination of the population doubling level (PDL) which is affixed at each subcultivation.[58] This number is useful when relatively rigid passaging schedules are followed, but it makes several assumptions which are not valid. For instance, plating efficiency is never 100% so that at a 1:2 split ratio some cells must undergo more than one doubling before confluency is reached. As the cultures age, fewer cells are capable of reproducing. Therefore the number of divisions by the proliferating pool is again greater than the PDL. Furthermore, its application to cell types in which a certain number of daughter cells terminally differentiate and do no divide again is questionable. Recently, it has become possible to more quantitatively assess the progress of the in vitro aging process by determination of the percentage of cells capable of incorporating [^{3}H]thymidine[59] or by a colony size distribution assay.[60-62]

The strongest evidence for the relevance of in vitro aging to in vivo aging comes from the inverse correlation between donor age and in vitro cellular reproductive capacity. Hayflick[4] first provided some evidence for this. Martin et al.[63] found a significant correlation between in vitro lifespan and age of the donor in an extensive study of skin fibroblasts. More recently, Schneider and Mitsui,[64] in similar experiments, examined several parameters that change with age in vitro and found them to occur earlier in cultures derived from older individuals. Further evidence for the relevance of in vitro to in vivo aging comes from the studies of cultures derived from individuals with diseases that are characterized by premature aging, such as diabetes, Down's Syndrome, and progeroid diseases such as Hutchinson-Gilford Syndrome and Werner's Syndrome.[65-69] Fibroblast-like cells from individuals with these diseases exhibit decreased lifespans compared to those from normal individuals. There is also a general correlation between species lifespan and lifespan of cells in culture, e.g., 10 to 20 population doublings for mouse (2.5 years), 30 population doublings for chicken (10 years), and 50 doublings for human (70).

Table 5 summarizes the available information on cultured cells for which reasonably good markers for cell types exist and which have been studied with regard to a possible in vitro aging effect. It is noteworthy that, with the exception of the fibroblast-like cells, most other reports have appeared since 1975. Many of the cells listed require specific growth factors such as epidermal growth factor (EGF) or fibroblast growth factor (FGF) or both, or an irradiated feeder layer in the case of human epidermal keratinocytes. These requirements for specific growth factors for cultivation in vitro indicate that other cell types could perhaps be cultured too if appropriate conditions were determined. Actually, very few of the renewing tissues listed in Table 1 have been grown in vitro.

Undoubtedly, the ability to culture differentiated cells will play an important role in future aging research. A serious drawback exists with respect to fibroblast-like cells in culture because neither the exact cell of origin or presence of specific markers has been determined.

Table 5
CELLS EXHIBITING IN VITRO AGING

Classification of cell types	Tissue of origin	Species	Probable cell type	PDL[a] attained	Differentiated function markers	Remarks	Ref.
Adrenal cortical cells	Adrenal cortex	Bovine		55—65	PGE-responsive steroids	Requires FGF[b]	74—77
Endothelial cells	Aorta	Bovine	Aortic endothelial	80	Factor VIII antigen angiotensin converting enzyme	Epithelial-like cells	78—80
Fibroblast-like	Various	Human	Fibroblast	3—34	?		57—73
Fibroblast-like	Embryonic lung	Human	Pericyte or endothelial	50—80	?		3,4,81
Fibroblast-like	Skin	Human	Fibroblast	50	?		63,73,82
Fibroblast-like	Embryo	Chick	Fibroblast	30	?		83—85
Fibroblast-like	Newborn skin	Mouse *Mus musculus*	Fibroblast	10—20	?	Ultimately establish into cell line	86—89,90
		Peromyscus leucopus	Fibroblast	25	?	Do not establish into cell line	90
Fibroblast-like	Embryo	Rat	Fibroblast	10—20	?		91—93
Granulosa cells	Ovary	Bovine		60	Progesterone	Requires FGF,[b] EGF[c]	94
Glial cells	Brain	Human	Glial	50	S100 Protein		62,95
Hepatocytes	Liver	Human	Hepatocyte	70	Albumin	Fibroblast-like	96,97
Keratinocytes	Skin	Human	Epidermal keratinocyte	50—150	Keratin	3T3 Feeder layer required; EGF responsive	21,98—100
Lens epithelium	Eye	Human	Lens epithelium	2—6		Only from old individuals	101
Lens epithelium	Eye	Bovine	Lens epithelium	15—20			102
Lens epithelium	Eye	Rat	Lens epithelium	2—35	α-Crystallin		103,104

[a]   Population doubling level.
[b]   Fibroblast growth factor.
[c]   Epidermal growth factor.

Neither collagen synthesis nor morphology as determined by light microscopy appears to be a specific marker for fibroblasts. Many cell types in culture make collagen, and Franks and Cooper,[70] by studying the outgrowth of explants of fetal lung at the ultrastructural level, have concluded that the fibroblast-like cells from this tissue are actually endothelial cells, pericytes, or both. They produce type I, III, and IV collagens,[71] and exhibit a small amount of angiotensin converting enzyme activity,[72] although no reports have yet appeared on the presence of endothelial cell-specific factor VIII antigen. Differences between human embryonic lung fibroblasts and dermal fibroblasts may result from the fact that they are different cell types.[73] Aortic endothelial cells, which grow readily and exhibit limited lifespan in vitro, express two well-accepted markers of endothelial cells, factor VIII antigen and angiotensin converting enzyme, up until the end of their life span.

No clear relationship can be drawn at this time from in vitro lifespans and the origin of tissue with respect to in vivo renewal rates. Glial cells, which exhibit almost no turnover in vivo, have an in vitro life span similar to embryonic fibroblast-like cells, possibly because little reproductive capacity has been utilized in vivo, and this is consistent with about 60 population doublings for normal tissues. However, one would expect greater cellular reproductive capacity in vitro from epidermal keratinocytes based on their reproductive capacity in vivo, but only 50, or at best 150, are attained. Their reported low plating efficiency and undetermined loss because of terminal differentiation may explain this rather low reproductive potential.

The report by Sacher and Hart[90] on skin fibroblast-like cells from the house mouse, *Mus musculus,* and the white-footed mosue, *Peromyscus leucopus,* is of interest as the white-footed mouse exhibits twice the lifespan of the house mouse, and its fibroblasts can be cultivated for twice as long. Skin fibroblasts from the white-footed mouse do not seem to spontaneously transform as do those of the house mouse.

Most information on alterations in parameters of the cell cycle as a function of in vitro age has come from studies on human diploid fibroblasts. Macieira-Coelho et al.[105] first investigated this in human embryonic lung-derived WI38 cells by examination of labeled mitoses curves. They demonstrated that: (1) fewer cells synthesize DNA as they approach Phase III, (2) fewer cells are capable of division after subcultivation, and (3) interdivision times become heterogeneous and lengthened with age, concluding that this occurred as a consequence of a prolonged G2. Macieira-Coelho and Pontén[82] later showed that human diploid fibroblast strains derived from adult tissue exhibited the same phenomena but that the changes occurred earlier in their in vitro passage levels.

More recent studies utilizing cytophotometric analysis of nuclear DNA content of WI38 cells have demonstrated that the majority of nondividing cells in old cultures are arrested in G1, not G2.[106-108] S, G2, and M remained constant, but G1 was lengthened from 2.5 to 14.3 hr.[106] In addition, this technique demonstrated an increase in the ploidy level of a fraction of cells with age. Polyploidization also occurs in vivo with age in some tissues such as liver and salivary gland. Schneider and Fowlkes[109] have applied flow microfluorometry to cell cycle analysis in WI38 cells and were able to show an increase in the proportion of G1 cells. They also found increased heterogeneity in DNA content, usually slight decreases, which might explain the aneuploidy observed in senescent WI38 cultures, especially hypodiploidy.[110-112] Increased aneuploidy also occurs in vivo as a function of age, but the significance of this to the aging process is presently unknown.[113]

Time-lapse microcinematographic studies of cell division patterns of WI38 cells have confirmed that the length and variability of interdivision time increases with age.[114,115] Presumably, this increase in interdivision time is due to a lengthening of G1 and may indicate disturbances in the mechanism of cellular entry into S phase.

Kapp and Klevecz[116] have examined the cell cycle in young and old cultures synchronized by mitotic selection. Seventy-five percent of mitotic cells derived from both young and old cultures entered S phase following selection, and large differences were not observed in any parameters of the cell cycle during the first cell cycle. This observation seems to contradict results obtained by other methods. However, this technique tends to select for the most rapidly dividing cells in a population. Analysis of clones by the distribution of colony size[60,61] demonstrated that some cells in old cultures are present which are still capable of division patterns similar to those of young cultures, albeit with decreasing frequency. Mitotic selection may prove valuable in understanding the mechanism of aging because it would seem that the important cells to study are those that are capable of dividing but that are being altered in some subtle, unknown way so that they are less likely to divide again.

SUMMARY

A considerable body of evidence exists to suggest that diminished cellular reproductive capacity occurs with age in vivo and in vitro. Except for relatively rare transformation events which lead to neoplasia and cells that are often immortal, somatic cells do not seem to be able to multiply indefinitely. Diminished cellular reproductive capacity produces a functional decrement that is part of the aging process, but it is not possible at the present time to estimate the extent of this involvement.

Little information is available on changes in cell cycle parameters with age. In general, S, G2, and M seem relatively unaffected both in vivo and in vitro, while G1 becomes lengthened or infinite. This increase in G1 accounts for the increase in interdivision time observed in aging cells.

The mechanism by which this intrinsic limitation of cellular reproductive capacity is mediated is unknown. A genetic basis is indicated by the characteristic lifespans of species, and results from cultured cell studies suggest that the timing mechanism is based on number of cellular reproductions rather than on metabolic or chronologic time. Recent reports by Strehler et al.[122,123] have demonstrated a loss of ribosomal DNA from post mitotic tissues during aging. These and the report of Shmookler Reis and Goldstein[124] on diminution of a specific family of tandemly repeated DNA sequences in serially passaged human diploid fibroblasts with age suggest that loss of genetic information with time in somatic cells could provide the basis for the mechanism of aging.

REFERENCES

1. **Prescott, D. M.,** *Reproduction of Eukaryotic Cells,* Academic Press, New York, 1976, chap. 1.
2. **Minot, C. S.,** The problem of age, growth and death, *Pop. Sci. Mon.,* 193, 1907.
3. **Hayflick, L. and Moorhead, P. S.,** The serial cultivation of human diploid cell strains, *Exp. Cell Res.,* 25, 585, 1961.
4. **Hayflick, L.,** The limited in vitro lifetime of human diploid cell strains, *Exp. Cell Res.,* 37, 614, 1965.
5. **Hayflick, L.,** The cellular basis for biological aging, in *Handbook of the Biology of Aging,* Finch, C. E. and Hayflick, L., Eds., Van Nostrand Reinhold, Cincinnati, 1977, chap. 7.
6. **Cameron, I. L. and Thrasher, J. D.,** Cell renewal and cell loss in tissues of aging mammals, *Interdiscipl. Top. Gerontol.,* 10, 108, 1976.
7. **Hart, R. W. and Setlow, R. B.,** Correlation between DNA excision repair and lifespan in a number of mammalian species, *Proc. Natl. Acad. Sci. U.S.A.,* 71, 2169, 1974.
8. **Mitchison, J. M.,** *The Biology of the Cell Cycle,* Cambridge University Press, Cambridge, 1971.
9. **Baserga, R.,** *Multiplication and Division in Mammalian Cells,* Marcel Dekker, New York, 1976.
10. **Baserga, R. L.,** Cell division and the cell cycle, in *Handbook of the Biology of Aging,* Finch, C. E. and Hayflick, L., Eds., Van Nostrand Reinhold, Cincinnati, 1977, chap. 5.

11. **Pardee, A. B., Dubrow, R., Hamlin, J. L., and Kletzien, R. F.,** Animal cell cycle, *Ann. Rev. Biochem.,* 47, 715, 1978.
12. **Howard, A. and Pelc, S. R.,** Synthesis of deoxyribonucleic acid in normal and irradiated cells and its relation to chromosome breakage, *Heredity (London),* 6(Suppl.), 261, 1953.
13. **Blenkinsopp, W. K.,** Cell proliferation in the epithelium of the oesophagus, trachea and ureter in mice, *J. Cell Sci.,* 5, 393, 1969.
14. **Lajtha, L. G.,** On the concept of the cell cycle, *J. Cell. Comp. Physiol.,* 62, 143, 1963.
15. **Dell 'Orco, R. T., Mertens, J. G., and Kruse, P. F.,** Doubling potential, calendar time, and donor age of human diploid cells in culture, *Exp. Cell Res.,* 84, 363, 1974.
16. **Kaji, K. and Matsuo, M.,** Doubling potential and calendar time of human diploid cells in vitro, *Exp. Gerontol.,* 14, 329, 1979.
17. **Pardee, A. B.,** A restriction point for control of normal animal cell proliferation, *Proc. Natl. Acad. Sci. U.S.A.,* 71, 1286, 1974.
18. **Rossow, P. W., Riddle, V. G. H., and Pardee, A. B.,** Synthesis of a labile, serum-dependent protein in early G_1 controls animal cell growth, *Proc. Natl. Acad. Sci. U.S.A.,* 76, 4446, 1979.
19. **Das, M.,** Mitogenic hormone-induced intracellular message: assay and partial characterization of an activator of DNA replication induced by epidermal growth factor, *Proc. Natl. Acad. Sci. U.S.A.,* 77, 112, 1980.
20. **Cristofalo, V. J.,** Animal cell cultures as a model system for the study of aging, in *Advances in Gerontological Research* Vol. 4, Strehler, B., Ed., Academic Press, New York, 1972, 45.
21. **Rheinwald, J. G. and Green, H.,** Epidermal growth factor and the multiplication of cultured human epidermal keratinocytes, *Nature (London),* 265, 421, 1977.
22. **Martin, G. M., Sprague, C. A., Norwood, T. H., Pendergrass, W. R., Bornstein, P., Hoehn, H., and Arend, W. P.,** Do hyperplastoid cell lines "differentiate themselves to death"?, *Adv. Exp. Med. Biol.,* 53, 67, 1975.
23. **Buetow, D. E.,** Cellular content and cellular proliferation changes in the tissues and organs of the aging mammal, in *Cellular and Molecular Renewal in the Mammalian Body,* Cameron, I. L. and Thrasher, J. D., Eds., Academic Press, New York, 1971, chap. 4.
24. **Post, J. and Hoffman, J.,** Changes in the replication times and patterns of the liver cell during the life of the rat, *Exp. Cell Res.,* 36, 111, 1964.
25. **McCreight, C. E. and Sulkin, N. M.,** Cellular proliferation in the kidneys of young and senile rats following unilateral nephrectomy, *J. Gerontol.,* 14, 440, 1959.
26. **Thuringer, J. M. and Katzberg, A. A.,** The effect of age on mitosis in the human epidermis, *J. Invest. Dermatol.,* 33, 35, 1959.
27. **Whitely, H. J. and Horton, D. L.,** The effect of age on the mitotic activity of the ear epithelium in the CBA mouse, *J. Gerontol.,* 18, 335, 1963.
28. **Walker, R.,** Age changes in the rat's exorbital lacrimal gland, *Anat. Rec.,* 132, 49, 1958.
29. **Meyer, J., Marwah, A. S., and Weinmann, J. P.,** Mitotic rate of gingival epithelium in two age groups, *J. Invest. Dermatol.,* 27, 237, 1956.
30. **Blumenthal, H. T.,** Aging processes in the endocrine glands of the guinea pig. I. The influence of age, sex, and pregnancy on the mitotic activity and the histologic structure of the thyroid, parathyroid and adrenal glands, *Arch. Pathol.,* 40, 264, 1945.
31. **Bullough, W. S.,** Age and mitotic activity in the male mouse, *Mus muscularus* L., *J. Exp. Biol.,* 26, 261, 1949.
32. **Plewig, G. and Kligman, A. M.,** Proliferative activity of the sebaceous glands of the aged, *J. Invest. Dermatol.,* 70, 314, 1978.
33. **Solomon, L. M. and Virtue, C.,** The biology of cutaneous aging, *Int. J. Dermatol.,* 14, 172, 1975.
34. **Montagna, W.,** Morphology of aging skin, in *Advances in Biology of Skin: Aging,* Vol. 6, Montagna, W., Ed., Pergamon Press, New York, 1965, chap. 1.
35. **Fitzpatrick, T. B., Szabo, G., and Mitchell, R. E.,** Age changes in the human melanocyte system, in *Advances in Biology of Skin: Aging,* Vol. 6, Montagna, W., Ed., Pergamon Press, New York, 1965, chap. 3.
36. **Thrasher, J.,** Age and the cell cycle of the mouse esophageal epithelium, *Exp. Gerontol.,* 6, 19, 1971.
37. **Litvak, R. and Baserga, R.,** An autoradiographic study of the uptake of ^{3}H-thymidine by kidney cells at different ages, *Exp. Cell Res.,* 33, 540, 1964.
38. **Sharav, Y. and Massler, M.,** Age changes in oral epithelia, *Exp. Cell Res.,* 47, 132, 1967.
39. **Simmett, J. and Heppleston, A.,** Cell renewal in the mouse lung, *Lab. Invest.,* 15, 1793, 1966.
40. **Martin, G., Ogburn, C., and Sprague, C.,** Senescence and vascular disease, in *Explorations in Aging,* Cristofalo, V. J., Roberts, J., and Adelman, R. C., Eds., Plenum Press, New York, series in *Adv. Exp. Med. Biol.,* 61, 163, 1975.
41. **Lesher, S., Fry, R., and Kohn, H.,** Aging and the generation cycle of intestinal epithelial cells of the mouse, *Gerontologia,* 5, 176, 1961.

42. **Lesher, S., Fry, R., and Kohn, H.,** Age and the generation time of the mouse duodenal epithelial cell, *Exp. Cell Res.,* 24, 334, 1961.

43. **Thrasher, J. and Greulich, R.,** The duodenal progenitor population. I. Age related increase in duration of the cryptal progenitor cycle, *J. Exp. Zool.,* 159, 39, 1965.

44. **Thrasher, J. and Greulich, R.,** The duodenal progenitor population. II. Age related changes in size and distribution, *J. Exp. Zool.,* 159, 385, 1965.

45. **Fry, R., Lesher, S., and Kohn, H.,** Age effect on cell transit time in mouse jejunal epithelium, *Am. J. Physiol.,* 201, 213, 1961.

46. **Thrahser, J.,** Age and the cell cycle of the mouse colonic epithelium, *Anat. Rec.,* 157, 621, 1967.

47. **Potten, C. S., Hume, W. J., Reid, P., and Cairns, J.,** The segregation of DNA in epithelial stem cells, *Cell,* 15, 899, 1978.

48. **Daniel, C. W.,** Cell longevity; in vivo, in *Handbook of the Biology of Aging,* Finch, C. E. and Hayflick, L., Eds., Van Nostrand Reinhold, Cincinnati, 1977, chap. 6.

49. **Krohn, P. L.,** Review lectures on senescence. II. Heterochronic transplantation in the study of aging, *Proc. R. Soc. (London) Ser. B,* 157, 128, 1962.

50. **Harrison, D. E.,** Normal production of erythrocytes by mouse marrow continuous for 73 months, *Proc. Natl. Acad. Sci. U.S.A.,* 70, 3184, 1973.

51. **Harrison, D. E., Astle, C. M., and Delaittre, J. A.,** Loss of proliferative capacity in immunohemopoietic stem cells caused by serial transplantation rather than aging, *J. Exp. Med.,* 147, 1526, 1978.

52. **Williamson, A. R. and Askonas, B. A.,** Senescence of an antibody-forming cell clone, *Nature (London),* 238, 337, 1972.

53. **Carrel, A.,** On the permanent life of tissues outside of the organism, *J. Exp. Med.,* 15, 516, 1912.

54. **Earle, W. R.,** Production of malignancy in vitro. IV. The mouse fibroblast cultures and changes seen in the living cells, *J. Natl. Cancer Inst.,* 4, 165, 1943.

55. **Gey, G. O., Coffman, W. D., and Kubicek, M. T.,** Tissue culture studies of the proliferative capacity of cervical carcinoma and normal epithelium, *Cancer Res.,* 12, 264, 1952.

56. **Rafferty, K. A.,** Epithelial cells: growth in culture of normal and neoplastic forms, *Adv. Cancer Res.,* 21, 261, 1975.

57. **Swim, H. E. and Parker, R. F.,** Culture characteristics of human fibroblasts propagated serially, *Am. J. Hyg.,* 66, 235, 1957.

58. **Hayflick, L.,** Subculturing human diploid fibroblast cultures, in *Tissue Culture: Methods and Application,* Kruse, P. and Patterson, M. K., Eds., Academic Press, New York, 1973, 220.

59. **Cristofalo, V. J. and Sharf, B. B.,** Cellular senescence and DNA synthesis, *Exp. Cell Res.,* 76, 419, 1973.

60. **Smith, J. R., Pereira-Smith, O. M., and Schneider, E. L.,** Colony size distributions as a measure of in vivo and in vitro aging, *Proc. Natl. Acad. Sci. U.S.A.,* 75, 1353, 1978.

61. **Smith, J. R., Pereira-Smith, O. M., Braunschweiger, K. I., Roberts, T. W., and Whitney, R. G.,** A general method for determining the replicative age of normal animal cell cultures, *Mech. Ageing Dev.,* 12, 355, 1980.

62. **Blomquist, E., Westermark, B., and Pontén, J.,** Ageing of human glial cells in culture; increase in the fraction of non-dividers as demonstrated by minicloning technique, *Mech. Ageing Dev.,* 12, 173, 1980.

63. **Martin, G. M., Sprague, C. A., and Epstein, C. J.,** Replicative life-span of cultivated human cells, *Lab. Invest.,* 23, 86, 1970.

64. **Schneider, E. L. and Mitsui, Y.,** The relationship between in vitro cellular aging and in vivo human age, *Proc. Natl. Acad. Sci. U.S.A.,* 73, 3584, 1976.

65. **Goldstein, S., Littlefield, S., and Soeldner, J. S.,** Diabetes mellitus and aging: diminished plating efficiency of cultured human fibroblasts, *Proc. Natl. Acad. Sci. U.S.A.,* 64, 155, 1969.

66. **Vracko, R. and Benditt, E. P.,** Restricted replicative lifespan of diabetic fibroblasts in vitro: its relation to microangiopathy, *Fed. Proc.,* 34, 68, 1975.

67. **Vracko, R. and McFarland, B. H.,** Lifespans of diabetic and non-diabetic fibroblasts in vitro, *Exp. Cell Res.,* 129, 345, 1980.

68. **Schneider, E. L. and Epstein, C. J.,** Replication rate and lifespan of cultured fibroblasts in Down's syndrome, *Proc. Soc. Exp. Biol. Med.,* 141, 1092, 1972.

69. **Tice, R. R. and Schneider, E. L.,** In vitro aspects of human genetic disorders which feature accelerated aging, *Interdisc. Top. Gerontol.,* 9, 60, 1976.

70. **Franks, L. M. and Cooper, T. W.,** The origin of human embryo lung cells in culture: a comment on cell differentiation, in vitro growth and neoplasia, *Int. J. Cancer,* 9, 19, 1972.

71. **Alitalo, K.,** Production of both interstitial and basement membrane procollagens by fibroblastic WI-38 cells from human emybronic lung, *Biochem. Biophys. Res. Commun.,* 93, 873, 1980.

72. **Rubin, D. B. and Dobbs, L. G.,** Angiotensin converting enzyme activity in fibroblasts and endothelial cells, *J. Cell Biol.,* 83, 98a, 1979.

73. **Schneider, E. L., Mitsui, Y., Au, K. S., and Shorr, S. S.,** Tissue-specific differences in cultured human diploid fibroblasts, *Exp. Cell Res.,* 108, 1, 1977.

74. **Gospodarowicz, D., Ill, C. R., Hornsby, P. J., and Gill, G. N.,** Control of bovine adrenal cortical cell proliferation by fibroblast growth factor. Lack of effect of epidermal growth factor, *Endocrinology,* 100, 1080, 1977.

75. **Hornsby, P. J. and Gill, G. N.,** Characterization of adult bovine adrenocortical cells throughout their life span in tissue culture, *Endocrinology,* 102, 926, 1978.

76. **Simonian, M. H., Hornsby, P. J., Ill, C. R., O'Hare, M. J., and Gill, G. N.,** Characterization of cultured bovine adrenocortical cells and derived clonal lines: regulation of steroidogenesis and culture life span, *Endocrinology,* 105, 99, 1979.

77. **Hornsby, P. J., Simonian, M. H., and Gill, G. N.,** Aging of adrenocortical cells in culture, *Int. Rev. Cytol.,* 10, (suppl.), 131, 1979.

78. **Levine, E. M. and Mueller, S. N.,** Cultured vascular endothelial cells as a model system for the study of cellular senescence, *Int. Rev. Cytol.,* 10, (Suppl.), 67, 1979.

79. **Mueller, S. N., Rosen, E. M., and Levine, E. M.,** Cellular senescence in a cloned strain of bovine fetal aortic endothelial cells, *Science,* 207, 889, 1980.

80. **Duthu, G. S. and Smith, J. R.,** In vitro proliferation and lifespan of bovine aortic endothelial cells: effect of culture conditions and fibroblast growth factor, *J. Cell. Physiol.,* 103, 385, 1980.

81. **Nichols, W. W., Murphy, D. G., Cristofalo, V. J., Toji, L. H., Greene, A. E., and Dwight, S. A.,** Characterization of a new human diploid cell strain, IMR-90, *Science,* 196, 60, 1976.

82. **Macieira-Coelho, A. and Pontén, J.,** Analogy in growth between late passage human embryonic and early passage human adult fibroblasts, *J. Cell Biol.,* 43, 374, 1969.

83. **Hay, R. J. and Strehler, B. L.,** The limited growth span of cell strains isolated from the chick embryo, *Exp. Gerontol.,* 2, 123, 1967.

84. **Lima, L. and Macieira-Coelho, A.,** Parameters of aging in chicken embryo fibroblasts cultivated in vitro, *Exp. Cell Res.,* 70, 279, 1972.

85. **Ryan, J. M.,** The kinetics of chick cell population aging in vitro, *J. Cell. Physiol.,* 99, 67, 1979.

86. **Todaro, G. J. and Green, H.,** Quantitative studies on the growth of mouse embryo cells in culture and their development into established lines, *J. Cell Biol.,* 17, 299, 1963.

87. **Tuffery, A. A. and Baker, R. S. V.,** Alterations of mouse embryo cells during in vitro aging, *Exp. Cell Res.,* 76, 186, 1973.

88. **Meek, R. L., Bowman, P. D., and Daniel, C. W.,** Establishment of mouse embryo cells in vitro, *Exp. Cell Res.,* 107, 277, 1977.

89. **Weisman-Shomer, P., Kaftory, A., and Fry, M.,** Replicative activity of isolated chromatin from proliferating and quiescent early passage and aging cultured mouse cells, *J. Cell. Physiol.,* 101, 219, 1979.

90. **Sacher, G. A. and Hart, R. W.,** Longevity, aging, and comparative cellular and molecular biology of the house mouse, Mus musculus, and the white-footed mouse, Peromyscus leucopus, in *Genetic Effects on Aging,* Bergsma, D. and Harrison, D. E., Eds., Alan R. Liss, New York, 1978.

91. **Peterson, G., Coughlin, J. I., and Meyhan, C.,** Long-term cultivation of diploid rat cells, *Exp. Cell Res.,* 33, 60, 1964.

92. **Kontermann, K. and Bayreuther, K.,** The cellular aging of rat fibroblasts in vitro is a differentiation process, *Gerontology,* 25, 261, 1979.

93. **Meek, R. L., Bowman, P. D., and Daniel, C. W.,** Establishment of rat embryonic cells in vitro. Relationship of DNA synthesis, senescence, and acquisition of unlimited growth potential, *Exp. Cell Res.,* 127, 127, 1980.

94. **Gospodarowicz, D. and Bialecki, H.,** The effects of the epidermal and fibroblast growth factors on the replicative lifespan of cultured bovine granulosa cells, *Endocrinology,* 103, 854, 1978.

95. **Pontén, J.,** Human glia cells, in *Tissue Culture: Methods and Application,* Kruse, P. and Patterson, M. K., Eds., Academic Press, New York, 1973, 50.

96. **Kaighn, M. E. and Prince, A. M.,** Production of albumin and other serum proteins by clonal cultures of normal liver, *Proc. Natl. Acad. Sci. U.S.A.,* 68, 2396, 1971.

97. **Le Guilly, Y., Simon, M., Lenoir, P., and Bourel, M.,** Long-term culture of human adult liver: morphological changes related to in vitro senescence and effect of donor's age on growth potential, *Gerontologia,* 19, 303, 1973.

98. **Rheinwald, J. G. and Green, H.,** Serial cultivation of strains of human epidermal keratinocytes: the formation of keratinizing clones from single cells, *Cell,* 6, 331, 1975.

99. **Rheinwald, J. G.,** The role of terminal differentiation in the finite culture lifetime of the human epidermal keratinocyte, *Int. Rev. Cytol.,* 10, (Suppl.), 25, 1979.

100. **Milo, G. E., Ackerman, A., and Noyes, I.,** Growth and ultrastructural characterization of proliferating human keratinocytes in vitro without added extrinsic factors, *In Vitro,* 16, 20, 1980.

101. **Tassin, J., Malaise, E., and Courtois, Y.,** Human lens cells have an in vitro proliferative capacity inversely proportional to the donor age, *Exp. Cell Res.,* 123, 388, 1979.

102. **Taylor-Papadimitriou, J., Shearer, M., and Watling, D.,** Growth requirements of calf lens epithelium in culture, *J. Cell. Physiol.,* 95, 95, 1978.

103. **Rink, H., Vornhagen, R., and Koch, H. R.,** Rat lens epithelial cells in vitro, I. Observations on aging, differentiation and culture alterations, *In Vitro,* 16, 15, 1980.

104. **Rink, H. and Vornhagen, R.,** Rat lens epithelial cells in vitro. II. Changes of protein patterns during aging and transformation, *In Vitro,* 16, 277, 1980.

105. **Macieira-Coelho, A., Pontén, J., and Philipson, L.,** The division cycle and RNA-synthesis in diploid human cells at different passage levels in vitro, *Exp. Cell Res.,* 42, 673, 1966.

106. **Yanishevsky, R., Mendelsohn, M. L., Mayall, B. H., and Cristofalo, V. J.,** Proliferative capacity and DNA content of aging human diploid cells in culture: A cytophotometric and autoradiographic analysis, *J. Cell. Physiol.,* 84, 165, 1974.

107. **Grove, G. L., Kress, E. D., and Cristofalo, V. J.,** The cell cycle and thymidine incorporation during aging in vitro, *J. Cell Biol.,* 70, 133a, 1976.

108. **Grove, G. L. and Cristofalo, V. J.,** Characterization of the cell cycle of cultured human diploid cells: effects of aging and hydrocortisone, *J. Cell. Physiol.,* 90, 415, 1977.

109. **Schneider, E. L. and Fowlkes, B. J.,** Measurement of DNA content and cell volume in senescent human fibroblasts utilizing flow multiparameter single cell analysis, *Exp. Cell Res.,* 98, 298, 1976.

110. **Saksela, E. and Moorhead, P. S.,** Aneuploidy in the degenerative phase of serial cultivation of human cell strains, *Proc. Natl. Acad. Sci. U.S.A.,* 50, 390, 1963.

111. **Benn, P. A.,** Specific chromosome aberrations in senescent fibroblast cell lines derived from human embryos, *Am. J. Hum. Genet.,* 28, 465, 1976.

112. **Miller, R. C., Nichols, W. W., Pottash, J., and Aronson, M. M.,** In vitro aging, cytogenetic comparison of diploid human fibroblast and epithelioid cell lines, *Exp. Cell Res.,* 110, 63, 1977.

113. **Mattevi, M. S. and Salzano, F. M.,** Senescence and human chromosome changes, *Humangenetik,* 27, 1, 1975.

114. **Absher, P. M., Absher, R. G., and Barnes, W. D.,** Genealogies of clones of diploid fibroblasts, *Exp. Cell Res.,* 88, 95, 1974.

115. **Absher, P. M. and Absher, R. G.,** Clonal variation and aging of diploid fibroblasts, *Exp. Cell Res.,* 103, 247, 1976.

116. **Kapp, L. N. and Klevecz, R. R.,** The cell cycle of low passage and high passage human diploid fibroblasts, *Exp. Cell Res.,* 101, 154, 1976.

117. **Kay, H. E. M.,** How many cell generations?, *Lancet,* 2, 418, 1965.

118. **Kay, M. M. B.,** An overview of immune aging, *Mech. Ageing Dev.,* 9, 39, 1979.

119. **Fernandez, L. A. and MacSween, J. M.,** Decreased autologous mixed lymphocyte reaction with aging, *Mech. Ageing Dev.,* 12, 245, 1980.

120. **Hori, Y., Perkins, E. H., and Halsall, M. K.,** Decline in phytohemagglutinin responsiveness of spleen cells from aging mice, *Proc. Soc. Exp. Biol. Med.,* 144, 48, 1973.

121. **Tice, R. R., Schneider, E. L., Kram, D., and Thorne, P.,** Cytokinetic analysis of the impaired proliferative response of peripheral lymphocytes from aged humans to phytohemagglutinin, *J. Exp. Med.,* 149, 1029, 1979.

122. **Strehler, B. L., Chang, M. P., and Johnson, L. K.,** Loss of hybridizable ribosomal DNA from human post-mitotic tissues during aging. I. Age-dependent loss in human myocardium, *Mech. Ageing Dev.,* 11, 371, 1979.

123. **Strehler, B. L. and Chang, M. P.,** Loss of hybridizable ribosomal DNA from human post-mitotic tissues during aging. II. Age-dependent loss in human cerebral cortex-hippocampal and somatosensory cortex comparison, *Mech. Ageing Dev.,* 11, 379, 1979.

124. **Shmookler-Reis, R. J. and Goldstein, S.,** Loss of reiterated DNA sequences during serial passage of human diploid fibroblasts, *Cell,* 21, 739, 1980.

AGING, CELL DEATH, AND LYSOSOMES

Richard A. Lockshin

Cells carry the genetic capability of destroying themselves, and they do so in large numbers during periods of rapid change such as organogenesis and metamorphosis. One may ask to what extent scheduled or unscheduled cell death plays a part in senescence, to what extent cell loss is a physiological process rather than the result of trauma, and to what extent lysosomes or lysosomal enzymes control the lysis of the cell.

There is little evidence that programmed cell death, that is, the destruction of cells according to an inexorable clock, is a primary process of aging. Except in the unusual circumstance of organisms that have a fixed number of postmitotic cells from the late embryonic period, it is unlikely that aging and death result from the clocked march of individual cells to their own destruction.[1-20] Nevertheless, there is a steady loss of nonreplicative cells throughout life (Figure 1)[8,21-32] and loss of neurons or other cells may pose limits to survival, though the loss may be secondary to pathology. It is therefore of interest to understand how a cell dies. This report considers several issues related to cell death. The orientation is comparative since the following points are often confused: the relation of cell death to the limited potential of cells in culture; the significance of cell death among replicative and postreplicative cells; death as a physiological or as a traumatic event; and the role of lysosomes and lipofuscin in cell death (Table 1).

CELL DEATH AND THE HAYFLICK LIMIT

Diploid chordate cells possess limited viability in culture.[11,33-36] It is unclear to what extend this limit is related to cell loss in aging organisms or even to the question of aging. Several arguments can be raised against such a relationship: first, the number of divisions that fibroblasts can undergo is too massive to be related to aging; second, a cell line differentiates to a nonreplicative state rather than dying;[37-39] and third, failure of the culture is an artifact deriving from accidental loss of a small number of stem cells at transfer bottlenecks.[5,39] However, a small asynchrony in the division of stem cells can easily accommodate the natural history of fibroblasts, lymphocytes, or hematopoietic cells; dying cells are seen in phase III (terminal) cultures; and, to date, in no laboratory has the putative stem cell been seen.[40] Furthermore, although incessant pressure for reproduction in culture is highly artificial and selective, it is argued that the ability to reproduce is a complex and therefore sensitive measure of function. The terminal differentiation of the phase III cell, it is argued, is a functional property related to aging.

The argument seems defensible with two limitations. First, it is not at all clear that the oft-cited correlation between number of doublings in various species and the recorded maximum lifespan of the species is meaningful, since the relationship of doublings in vitro to either metabolic time or doublings in vivo must be vastly different for mammals or birds and for the slowly metabolizing Galapagos tortoise. Second, there is no evidence that anything analogous to the Hayflick limit applies outside of the phylum Chordata. Meristematic cells of plants can persist through uncounted numbers of divisions and, even if the living tissue of an ancient tree consists only of a thin outer shell, the descendants of one zygote can survive, apparently unaltered, for 2000 or more years. Insect cells can also be transplanted indefinitely, although they undergo a peculiar transformation: imaginal disc cells of known destiny (such as antenna) ultimately convert via an orderly sequence to new destinies.[41] Perhaps the loss of the ability to metamorphose into the original commitment represents a loss in function akin to the loss of the ability of vertebrate cells to continue dividing. Among

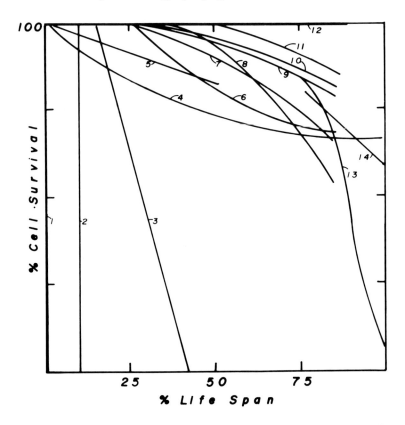

FIGURE 1.   Cell death during aging in vivo. The various lines indicate loss of specific types of cells as reported in the literature. (1) Embryonic cells such as posterior necrotic zone, Wolffian or mullerian ducts, various neurons;[21,22] (2) all tissues of tadpole tail at metamorphosis (or equivalent, but at differing points in total life, for insects); (3) primordial follicles, human female;[8] (4) neurons in brain of adult honeybee;[23] (5) neurons, dorsal and ventral roots, spinal ganglia;[25] (7) neurons, depth 8, rat cerebral cortex;[26] (8) superior frontal gyrus and other locales, human brain;[27-29] (9) muscle fibers, rat soleus;[30] (10) Purkinje cells, human cerebellum;[8] (11) small neurons, nucleus dentatus, human;[26] (12) hypothalamic cells, female hamster;[31] (13) total neurons, mouse brain;[32] (14) Total cerebrum, human.[27]

mammals, the Hayflick limit appears to be obeyed under the most natural conditions currently obtainable, growth within a syngenic host.[11,42] In this situation, the transplanted cells are capable of surviving for two or three species lifetimes, but ultimately they die.

The fate of phase III cells is poorly understood, in part because of increased heterogeneity in the culture during its final stages. Numerous metabolic features change (Table 2),[11] but the changes cannot be characterized as either causal or descriptive of the changed life style of the cells.[7] It is noteworthy that the levels of most lysosomal enzymes rise, presumably suggesting increased degradative activity.[43] Whether the cells undergo any process akin to apoptosis (see below) appears to be unreported.

Although death in mammals does not occur because of loss of fibroblasts, and indeed in many tissues age brings increased proliferation,[42] decrement in function of tissues such as lymphatic cells is significant. The correlative association of age and species of donor with potential survival in culture is such that the study remains one of presumed importance.

Table 1
OCCURRENCE OF LYSOSOMES, LIPOFUSCIN, AND CELL DEATH[a]

Tissue	Lysosomes	Lipofuscin	Cell death in aging
Mammals			
Liver (parenchyma)	+ + +	+	0.1%/day (does not change)
Fibroblasts	+ +	+	+ / + + +
Skeletal muscle	+ +	+ + +	+ +
Heart	+ +	+ + +	+
Retinal pigment epithelium	+ +	+ +	+
Intestinal epithelium	0	0	0
Nervous System			
	+ +	+ + +	0
	+ +	+ + +	0
	+ +	+ + +	0
	+ +	0- +	+ + +
Invertebrates			
Insect			
Nervous system	+ +	+ + +	+ + +
Fat body	+	+	?
Oenocytes	+ +	+ + +	?
Mid gut	+	+	+ + +
Muscle	+ + +	0	
Paramecium	+ + +	+ + +	+ + +
Campanularia	+ + +	0	+ + +

[a]   Increase of organelles or event with aging. The values represent subjective evaluations of the literature cited.

MITOSIS AND CELL DEATH

Many researchers feel that postmitotic cells rather than actively dividing cells are subject to cell death resulting from senescence,[34,44-49] although the argument from the culture of WI-38 cells appears to contradict this hypothesis. Certainly loss of cells from a nonrenewing population such as muscle or nerve is more striking, more readily quantifiable, and potentially more serious than is loss from a tissue that can regenerate. Furthermore, lipofuscin, a possible sign of pathology or cellular infirmity, is seen only in cell populations that undergo few or no mitoses.[50] Rather than an aged cell's having suffered intrinisic damage, it may be that these nonmitotic cells are trapped and smothered in an ever-stiffening web of connective tissue. The necessity for mitosis may well be correlated with the necessity for sexual recombination in *Paramecium*[51-53] and in bacteria.[54] In both of these instances, clones that are prevented from undergoing sexual recombination ultimately die out. (*Tetrahymena,* however, can survive indefinitely without sexual recombination, and the loss of a clone appears to derive from random, perhaps mutational, events.[55]) Sexual recombination may provide the cells with an opportunity to repair or replace defective genes; indeed, C. and H. Bernstein[56] have postulated that the function of synapsis in meiosis is the repair of allelic genes, one of which may have been damaged. According to those authors, even the sperm of a non-agenarian fertilizing the ovum of a quinquagenarian will produce an appropriately young infant. The alternative explanation, that DNA has nothing to do with the aging process, is dubious. The necessity for mitosis among somatic cells might be analogous or related in

Table 2
CELL DEATH IN AGING AND IN VITRO

Cell death in aging	Cell death in vitro
Most important in postmitotic cells: muscles, some nerves, oocytes; some cells proliferate with aging	Death is actually cessation of mitosis; cells live much longer; population becomes or is heterogeneous with respect to divisions
Perhaps results from trauma: nucleus shrinks, Golgi fragments, number of histones and lysosomes ↑, amount of lipofuscin ↑, mitochondria become abnormal	Cells repeatedly trypsinized, nuclei become lobulated, cytoplasm enlarges, vacuoles form, increased aneuploidy, ↑ lysosomes; some lipofuscin; ↑ protein degradation with ↓ Selectivity
Membrane folding, T-system expands (abortive regeneration?)	Cells become flat rather than fusiform
Paramecium: fission rate ↓, RNA ↓ (programed: RNA synth ↓ → 0?)	RNA ↓; DNA synth ↓ but still present

one of several ways: mitosis might allow the "stripping" from DNA of otherwise unremovable blocking proteins, or allow the repair of single-chain damage; mitosis might simply allow the death of the cell carrying the damaged allele and the survival of the now-unburdened other daughter cell; or mitosis, by freeing the cell temporarily from the constraints of its surroundings, might give it the opportunity to eliminate damaged and potentially toxic matter; or the transient shift in ionic milieu of the chromosomes might permit various allosteric rearrangements. Relevant to the possibility of elimination of cells at mitosis are the findings for intestinal crypt cells that cell loss is greatest at G_2.[34] This result suggests that packaging of the chromosome for mitosis, rather than synthesis of cell products, is a critical step.

CLASSIFICATION OF CELL DEATH

Cells die in one of two basic ways: either they are struck by a sudden and violent challenge to their integrity, resulting in prompt lysis (traumatic cell death) or, because of milder trauma, they gradually condense, spontaneously dismantle a good part of their structure, and in an apparently controlled manner destroy themselves or facilitate their own phagocytosis (physiological or programmed cell death; apoptosis).[19,20] Concerning the first way, the mechanism is reasonably unequivocal; following an injury such as osmotic shock, physical disruption of the cell membrane, or interference with energy supply, membrane ion pumps fail to correct for ion leaks, and usually the cell lyses. Lysosomal enzymes play little or no role. Their escape from their confining membranes is dependent on osmotic rupture of the lysosome secondary to osmotic rupture of the cell (Figure 2).

The situation is much less clear in physiological cell death. In all instances, the alterations observed cannot be simply and easily attributed to gross physical forces, and the cell is involved in its own destruction, requiring energy, isolating parts of its cytoplasm within enveloping membranes, and in an unknown manner, extruding water and budding off parts of its own cytoplasm.[20] Particularly in cells of ectodermal origin, much of the cytoplasm is destroyed within autophagic vacuoles which arise by unknown means in response to unknown signals.

In many other cells, and especially in muscle, the role of the lysosome in destroying the cytoplasm is less evident. Organelles such as mitochondria are often seen to be destroyed within autophagic vacuoles, but myofilaments are almost never seen within the confines of a conventionally described autophagic vacuole or isolation membrane.[57-60] In tadpole metamorphosis, large fragments of tail muscle break free, apparently confined by membranes of the sarcoplasmic reticulum, and these "sarcolytes" are phagocytosed.[19,61] In many other situations of muscle involution, the presence of any form of lysosome identifiable in thin

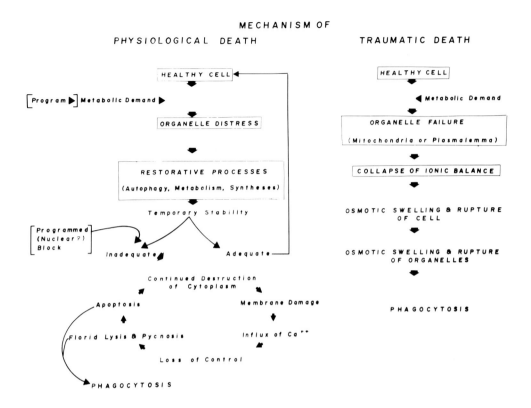

FIGURE 2. Comparison of generalized modes of traumatic and physiological or nontraumatic death. In the latter situation, it is suggested that prolonged or excessive demands on a cell can lead to its ultimate collapse. Programing, where it occurs, may act by initiating these demands or blocking the response of the cell to them. References for the arguments are found in the last section of the text.

sections is flatly denied.[62] A comparison of traumatic and physiological cell death is presented in Table 3 and Figure 2.

In physiological cell death, one cannot with any assurance claim an unprovoked attack by a lysosome on an organelle. Although the lysosomal system is certainly deeply involved in the destruction of the cytoplasm, and the control of its activity is therefore certainly worthy of study, activation of the lysosomal system is most likely secondary to prior changes. Lysosomal activity appears to be indicative of cell death rather than causal.

Various factors have been postulated as the initiating event in cell death, including small but irreparable membrane lesions[63] that may allow the influx of Ca^+. In immunological killing, the initial attack is at the lymphocyte-target cell interface but is complete within 5 min, allowing the target cell to undergo a series of unknown metabolic changes that ultimately cause it to lyse.[67] Such ideas are consistent with the concept of metabolic insult. In embryological or developmental cell death, one of the earliest events detected is a decrease or cessation in the uptake of tritiated thymidine.[68-70] (Such a claim cannot be made for most of the insect cellular systems that have been studied, since these cells are postmitotic long before the program for death is evoked.) If blockage of nuclear mechanisms is an initiating event (whether the blockage involves complete shut-down of the chromosome or blockage of response mechanisms to stress) we could better understand the difference between atrophy and programed death. A programed cell could respond to the challenge deliberately provoked by the body with only the repertoire not dependent on new synthesis of RNA, and it would inevitably succumb to the overwhelming pressures on it. Such a model would also be fully

Table 3
A COMPARISON OF APOPTOSIS AND TRAUMATIC DEATH

Apoptosis	Traumatic death
Perhaps early membrane lesion	Drastic membrane lesion (osmotic or energy)
Shrinkage, condensation	Osmotic swelling
Nuclear condensation, pycnosis	Nuclear swelling
Mitochondria digested in autophagic vacuoles; also ribosomes, glycogen	Ca^{++} precipitation in mitochondria; mitochondrial swelling
Lysosomal derivatives may digest many organelles	Lysosomes rupture secondarily
Cell divides into small pieces, which are phagocytosed	Remnants of lysed cell phagocytosed
Programmed (?) shutdown of synthesis of nucleic acids, this perhaps through a synthetic event	No control of synthetic or lytic reactions
Requires energy	Does not require energy
Perhaps a result of excess demands on tissues during aging	Result of infarct or other drastic event in aging

consistent with the idea of metabolic challenge and could also be rationalized to accommodate the apparently contradictory concept that RNA synthesis is an early event in programed cell death. The putative new RNA would engineer the shut-down of nuclear response mechanisms. The question remains unresolved.[19,71,72]

LYSOSOMES AND CELL DEATH

An increase in lysosomal activity, whether in phagocytes or dying cells, is usually considered presumptive evidence of increased cell death. The activity of lysosomes is under complex genetic control.[73] Although a large portion of the literature on the increase in lysosomal activity in aging cites the apparent increase in "soluble" or "free" vs. sedimentable or latent activity of one or more lysosomal enzymes such as acid phosphatase,[74-97] it is difficult to relate these findings to documented physiological processes. In most tissues, the differentiation of cell types or the amount of connective tissue with age is frequently not considered in biochemical analyses,[95] and there is little evidence that enzymes leak from their containing membranes except as a late event which is connected with the actual lysis of the cell.[96,97] In general, when a precipitate such as lead phosphate is found scattered throughout the cell, the investigator must prove that the precipitation is not artifactual. However, there is evidence that some of the lysosomal enzymes are located on the external surface of the membrane,[98] and supposedly lysosomal enzymes can be recognized in other organelles, presumably in transit from the site of their synthesis. The site of synthesis is a matter of some dispute,[99] but at least one group claims that acid phosphatase may reside at least temporarily in the cytosol.[100] Otherwise, a conservative assumption would be that rather than the in vivo existence of free enzymes, an increase with age in soluble enzymes reflects different physical characteristics of young and old tissues. It may also reflect differing sizes of classes of intact lysosomes so that, during homogenization, lysosomes of old animals tend to rupture more easily.

LIPOFUSCIN AND LYSOSOMES

Lipofuscin is a complex lipoidal material, assumed generally to be the remnants of membraneous materials digested within lysosomes. Although the pigment is usually considered to be inert, there is no direct evidence that it is,[101,103] but its insoluble nature renders study difficult. Although there is poor correlation between the presence of lipofuscin and the function of a cell (for instance, in cardiac conduction),[105,106] the potential for disruption of function by simply physical problems is very clear. The pigment in some cells, for instance, can occupy 90% of the cytoplasmic space;[107-110] in such a situation, the mass could disrupt diffusional gradients, absorb and bind lipophilic materials such as hormones, and distort and stress the cell membrane. In excitable cells, the presence of a large body of dielectric material would slow conduction time and, depending on the physical location, could alter the characteristics of synaptic excitation. Since lipofuscin accumulates in the axon hillock,[109] it would probably interfere with the spread of an action potential from soma to axon, or enhance propagation in the opposite direction.

Lipofuscin is frequently correlated with physiological demands on a cell[103,111-115] and is probably generated by interaction of lipids with oxygen in one of several possible excited states.[116] Thus, either as a pathological result of potentially injurious oxidative reactions, or as the symbol of the past occurrence of these reactions, the pigment can be considered to be a reasonably good marker of physiological age in postmitotic cells.[104,111,112] However, it is interesting that in aged neural tissue, lipofuscin granules are most common in those regions of the brain that show little or no loss of cells.[27-29] It is possible that those cells able to form lipofuscin, and thus disarm singlet electrons from more dangerous interactions, can survive. The cells that cannot protect themselves suffer disruption of nucleic acids or other less reparable macromolecules by the electrons. It would therefore appear that lipofuscin is undoubtedly a metabolic waste product quantitatively related to the cumulative metabolic demands on postmitotic cells. It remains to be determined whether its formation or presence has any biological significance other than marking the passage of an event.

UNIFYING CONCEPTS OF CELL DEATH

There has been a pervasive philosophy that the processes of physiological cell death would ultimately funnel through a common pathway, such as the influx of Ca^{++} or depletion of intracellular energy reserves.[61-64,117] That assumption has recently been challenged in two systems that are perhaps the most thoroughly and analytically studied as models of cell death: glucocorticoid-induced destruction of thymocytes and death of neurons resulting from lack of either nerve growth factor or synaptic contact.[118-126] Munck especially considers that death is the final response to accumulated defects generated by glucocorticoids in a variety of ways.[127] He emphasizes that the metabolic changes induced by glucocorticoids do not irreversibly lead to death and that there is little or no latent period between the final commitment to death and the actual lysis of the cell. Similarly, the death of neurons may derive from a variety of pathways, not all of which can be unmistakably associated with the physiological crisis (inability to export material accumulating in the endoplasmic reticulum to a postsynaptic cell, for example).[19,126] In a much more tentative and less articulate manner, the same concept has been broached for the programed death of the intersegmental muscles. In our laboratory we have been unable to distinguish by study of intermediary metabolism the process of atrophy from cell death,[19,128] and recently we have hypothesized that cell death results from the failure to resolve a metabolic crisis.[116,129-136] This idea relates to the others in that the authors assume the erection of physiological defenses and the ultimate inability of the cell to stave off an assault on its metabolic integrity. The significance of this concept is the derivative idea that all cells have homeostatic, probably catabolic, mechanisms to cope with physiological stress. These mechanisms capitalize on whatever reserve the cell contains and in a sense return that reserve to the body. Death ensues whenever that reserve is drained to a finite limit, whether through pathology or prearrangement (programing). A cell in homeostatic equilibrium would thus be capable of indefinite or at least prolonged survival, and the challenge to gerontologists is to ensure that the physiological demands at the cell level are kept within bounds.

Recently, several articles and reviews have appeared which are related to this subject. Sohal[137] has published an excellent book on age pigments, and Beaulaton and Lockshin[138] have reviewed programed cell death as a developmental event. Oppenheim[139] has reviewed neuronal cell death, emphasizing the variability from system to system and noting that many neurons differentiate before dying. Also, in some experimental situations, pharmacological studies indicate that the trophic effect of establishing peripheral connections appears to be related to target cell activity and thus a peripheral rather than central effect. Cunningham[140] emphasizes the importance for the neuron of establishing adequate peripheral connections. In the nematode Caenorhabditis elegans, however, the selection of cells for programed death appears not to involve competition for peripheral targets.[141] Also, dying insect muscle cells clearly contain higher levels of calcium than previously.[142]

ACKNOWLEDGMENTS

This research was supported in part by The National Science Foundation (PCM 77-15687). Many of the ideas presented here are derived and condensed from thoughts recently expressed by several others. These ideas are developed by the authors in a recent book.[136]

REFERENCES

1. **Miguel, J.,** Aging of male *Drosophila melanogaster:* histological, histochemical, and ultrastructural observations, *Adv. Gerontol. Res.,* 3, 39, 1971.
2. **Bullough, W. S.,** Ageing of mammals, *Z. Alternsforsch,* 27, 247, 1973.
3. **Cristofalo, V. J. and Holeckova, E., Eds.,** Cell impairment on aging and development, *Advances in Experimental Medical Biology,* Vol. 53, Plenum Press, New York, 1975.
4. **Gevers, W.,** Biochemical aspects of cell death, *Forensic Sci.,* 6, 25, 1975.
5. **Good, P. I.,** Aging in mammalian cell populations — a review, *Mech. Ageing Dev.,* 4, 239, 1975.
6. **Hinsull, S. M., Bellamy, D., and Franklin, A.,** A quantitative histological assessment of cellular death in relation to mitosis and rat thymus during growth and age involution, *Age Ageing,* 6, 77, 1977.
7. **Martin, G. M.,** Cellular aging — clonal senescence. A review. I, *Am. J. Pathol.,* 89, 984, 1977.
8. **Martin, G. M.,** Cellular aging — postreplicative cells. A review. II, *Am. J. Pathol.,* 89, 513, 1977.
9. **Wallace, P. J.,** The biology of aging in 1976 — a review, *J. Am. Geriatr. Soc.,* 25, 104, 1977.
10. **Griffin, W. S. T., Woodward, D. J., and Chanda, R.,** Quantification of cell death in developing cerebellum by a ^{14}C tracer method, *Brain Res. Bull.,* 3, 369, 1978.
11. **Hayflick, L.,** Progress in cytogerontology, *Mech. Ageing Dev.,* 9, 353, 1979.
12. **Abernethy, J. D.,** The exponential increase in mortality rate with age attributed to wearing-out of biological components, *J. Theor. Biol.,* 80, 333, 1979.
13. **Hocman, G.,** Biochemistry of ageing, *Int. J. Biochem.,* 10, 867, 1979.
14. **Markofosky, J. and Milstoc, M.,** Ageing changes in the liver of the male annual cyprinodont fish, *Nothobranchus guentheri, Exp. Gerontol.,* 14, 11, 1979.
15. **Miguel, J., Economos, A. C., Bensch, K. G., Atlan, H., and Johnson, J. E., Jr.,** Review of cell aging in *Drosophila* and mouse, *Age,* 2, 78, 1979.
16. **Nichols, W. W. and Murphy, D. G.,** Differentiated cells in aging research, *Int. Rev. Cytol., Suppl.,* 10, 1979.
17. **Skurnick, I. D. and Kemeny, G.,** Stochastic studies of aging and mortality in multicellular organisms. II. The Finite Theory, *Mech. Ageing Dev.,* 10, 157, 1979.
18. **Bowen, I. D.,** Techniques for demonstrating cell death, in *Cell Death,* Bowen, I. D. and Lockshin, R. A., Eds., Chapman and Hall, London, 1981, 381.
19. **Lockshin, R. A.,** Cell death in metamorphosis, in *Cell Death,* Bowen, I. D. and Lockshin, R. A., Eds., Chapman and Hall, London, 1981, 79.
20. **Kerr, J. F. R.,** Shrinkage necrosis: a distinct mode of cellular death, *J. Pathol.,* 105, 13, 1971.
21. **Saunders, J. W., Jr.,** Death in embryonic systems, *Science,* 154, 604, 1966.
22. **Lockshin, R. A. and Beaulaton, J.,** Programmed cell death, *Life Sci.,* 15, 1549, 1974.

23. **Rockstein, M.,** The relation of cholinesterase activity to decrease in cell number with age in the brain of the adult worker honeybee, *J. Cell. Comp. Physiol.,* 35, 110, 1950.

24. **Uleklova, L.,** Aging and loss of the auditory neuroepithelium in the guinea pig, *Adv. Exp. Med. Biol.,* 53, 257, 1974.

25. **Magladery, J. W.,** Neurophysiology of aging, in *Handbook of Aging and the Individual,* Birren, J. E., Ed., University of Chicago Press, Chicago, 1979, 173.

26. **Strehler, B. L.,** *Time Cells and Aging,* 2nd ed., Academic Press, New York, 1977, 220.

27. **Brody, H.,** Organization of the cerebral cortex. III. A study of aging in the human cerebral cortex, *J. Comp. Neurol.,* 102, 511, 1955.

28. **Brody, H. and Vijayashankar, N.,** Anatomical changes in the nervous system, in *Handbook of the Biology of Aging,* Finch, C. E. and Hayflick, L., Eds., Van Nostrand Reinhold, New York, 1977, 241.

29. **Vijayashankar, N. and Brody, H.,** A study of aging in the human abducens nucleus, *J. Comp. Neurol.,* 173, 433, 1977.

30. **Gutman, E. and Hanzlikova, V.,** Basic mechanisms of aging in the neuromuscular system, *Mech. Ageing Dev.,* 1, 327, 1972.

31. **Lamperti, A. and Blaha, G.,** The numbers of neurons in the hypothalamic nuclei of young and reproductively senescent female golden hamsters, *J. Gerontol.,* 35, 335, 1980.

32. **Johnson, H. A. and Erner, S.,** Neuron survival in the aging mouse, *Exp. Gerontol.,* 7, 111, 1972.

33. **Bradley, M. O., Hayflick, L., and Schimke, R. T.,** Protein degradation in human fibroblasts (WI-38). Effects of aging, viral transformation, and amino acid analogues, *J. Biol. Chem.,* 251, 3521, 1976.

34. **Rowlatt, C.,** Cell aging in the intestinal tract, *Adv. Exp. Med. Biol.,* 53, 215, 1975.

35. **Blumenthal, H. and Kasbekar, D. K.,** Nonreplicating cultures of frog gastric-tubular cells, *Int. Rev. Cytol.,* Suppl., 10, 191, 1979.

36. **Harrison, D. E.,** Proliferative capacity of erythropoietic stem cell liver and aging: a overview, *Mech. Ageing Dev.,* 9, 409, 1979.

37. **Martin, G. M., Sprague, C. A., Norwood, T. H., Pendergast, W. R., Bornstein, P., Hoehn, H., and Arend, W. P.,** Do hyperplastoid cell lines "differentiate themselves to death?", *Adv. Exp. Med. Biol.,* 53, 67, 1975.

38. **Bell, E., Marek, L. F., Levinstone, D. S., Merrill, C., Short, S., Young, I. T., and Edan, M.,** Loss of division potential in vitro: ageing or differentiation, *Science,* 202, 1158, 1978.

39. **Holliday, R., Huschtscha, L. I., Tarrant, G. M., and Kirkwood, T. B. L.,** Testing the commitment theory of cellular aging, *Science,* 198, 366, 1977.

40. **Hayflick, L.,** The cellular basis for biological aging, in *Handbook of the Biology of Aging,* Finch, C. and Hayflick, L., Eds., Van Nostrand Reinhold, New York, 1977, 159.

41. **Hadorn, E.,** Dynamics of determination, *Symp. Soc. Dev. Biol.,* 21, 85, 1966.

42. **Martin, G. M.,** Proliferative homeostasis and its age-related aberrations, *Mech. Ageing Dev.,* 9, 385, 1979.

43. **Cristofalo, V. J. and Kabakjian, J.,** Lysosomal enzymes and aging in vitro: subcellular enzyme distribution and effect of hydrocortisone on cell life span, *Mech. Ageing Dev.,* 4, 19, 1975.

44. **Balazs, A.,** Organismal differentiation, ageing, and rejuvenation, *Exp. Gerontol.,* 5, 305, 1970.

45. **Franks, L. M.,** Ageing in differentiatied cells, *Gerontologia,* 20, 51, 1974.

46. **Krook, D. L.,** Model systems for studies on cellular basis of organ ageing, *Aktuel Gerontol.,* 7, 1977.

47. **Rosen, R.,** Cells and senescence, *Int. Rev. Cytol.,* 54, 161, 1978.

48. **Houck, J. C.,** The relevance of growth control (chalones) to the aging process, *Mech. Ageing Dev.,* 9, 463, 1979.

49. **Rytomaa, T. and Toivonen, H.,** Chalones: concept and results, *Mech. Ageing Dev.,* 9, 471, 1979.

50. **Toth, S. E.,** The origin of lipofuscin age pigments, *Exp. Gerontol.,* 3, 19, 1968.

51. **Smith-Sonneborn, J. and Reed, J. C.,** Calendar life span vs. fission life span of *Paramecium aurelia.* I. Alterations in the cytoplasm, *Mech. Ageing Dev.,* 5, 139, 1976.

52. **Sundaraman, V. and Cummings, D. J.,** Morphological changes in aging cell lines of *Paramecium aurelia.* I. Alterations in the cytoplasm, *Mech. Ageing Dev.,* 5, 139, 1976.

53. **Takagi, Y. and Yoshida, M.,** Clonal death associated with the number of fissions in *Paramecium caudatum,* *J. Cell Sci.,* 41, 177, 1980.

54. **Gunn, J. M.,** Does the regulation of intracellular protein degradation require protein synthesis?, *Exp. Cell Res.,* 117, 448, 1978.

55. **Simon, E. M. and Nanney, D. L.,** Germinal aging in *Tetrahymena thermophilia, Mech. Ageing Dev.,* 11, 253, 1979.

56. **Bernstein, C. and Bernstein, H.,** Why are babies young? Meiosis may prevent aging of the germ line, *Perspect. Biol. Med.,* 22, 539, 1979.

57. **Hanzlikova, V. and Gutmann, E.,** Ultrastructural changes in senile muscle, *Adv. Exp. Med. Biol.,* 53, 421, 1975.

58. **Kasten, F. H.,** Functional capacity of neonatal mammalian myocardial cells during aging in tissue culture, *Adv. Exp. Med. Biol.,* 53, 420, 1975.
59. **Bird, J. W. C., Carter, J. H., Triemer, R. E., Brooks, R. M., and Spanier, A. M.,** Proteinases in cardiac and skeletal muscle, *Fed. Proc.,* 39, 20, 1980.
60. **Lockshin, R. A., Colon, A. D., and Dorsey, A. M.,** Control of muscle proteolysis in insects, *Fed. Proc.,* 39, 48, 1980.
61. **Weber, R.,** Tissue involution and lysosomal enzymes during anuran metamorphosis, in *Lysosomes in Biology and Pathology,* Vol. 2, Dingle, J. T. and Fell, H. B., Eds., North-Holland, Amsterdam, 1969, 437.
62. **Auber-Thomay, M.,** Modifications ultrastructurales au cours de la degenerescence et de la croissance de fibres muscuaires chez un insecte, *J. Microsc.,* 6, 622, 1967.
63. **Bank, H. and Mazur, P.,** Relation between ultrastructure and viability of frozen-thawed Chinese hamster tissue-culture cells, *Exp. Cell. Res.,* 71, 441, 1972.
64. **Farber, J. L. and El-Mofty, S. K.,** The biochemical pathology of liver cell necrosis, *Am. J. Pathol.,* 81, 237, 1975.
65. **Fozzard, H. A.,** How do cardiac cells die?, *Circ. Res.,* 53 (Suppl. 1), 40, 1976.
66. **Schanne, F. A. X., Kane, A. B., Young, E. E., and Farber, J. L.,** Calcium dependence of toxic cell death: a final common pathway, *Science,* 206, 700, 1979.
67. **Ross, M. W., Yamamoto, R. S., and Granger, G. A.,** The role of LT system in lymphocyte-induced cell destruction *in vitro,* in *Cell Death,* Bowen, I. D. and Lockshin, R. A., Eds., Chapman and Hall, London, 1981, 363.
68. **Pollack, R. D. and Fallon, J. F.,** Autoradiographic analysis of macromolecular synthesis in prospectively necrotic cells of the chick limb bud. II. Nucleic acids, *Exp. Cell Res.,* 100, 15, 1976.
69. **Gahan, P. B.,** Increased levels of euploidy as a strategy against rapid ageing in diploid mammalian systems: a hypothesis, *Exp. Gerontol.,* 12, 133, 1977.
70. **Gahan, P. B.,** Reduced incorporation of ³H thymidine into cytoplasmic DNA in aging cell populations *in vivo, Exp. Gerontol.,* 12, 13, 1977.
71. **Miguel, J. and Johnson, J. E., Jr.,** Senescent changes in the ribosomes of animal cells *in vivo* and *in vitro, Mech. Ageing Dev.,* 9, 247, 1979.
72. **Munck, A. and Leung, K.,** Glucocorticoid receptors and mechanism of action, in *Receptors and Mechanism of Action of Steroid Hormones,* Part II, Pasqualini, J. R., Ed., Marcel Dekker, New York, 1977, 311.
73. **Paigen, K.,** Acid hydrolases as models of genetic control, *Ann. Rev. Genet.,* 13, 417, 1979.
74. **Youhotsky-Gore, I. and Pathmanathan, K.,** Some comparative observations in the lysosomal status of muscle from young and old mice, *Exp. Gerontol.,* 3, 281, 1968.
75. **Goto, S., Takano, T., Mizuno, D., Nakano, T., and Imazumi, K.,** Aging and location of acid ribonuclease in liver of various animals, *J. Gerontol.,* 24, 305, 1969.
76. **Rahman, Y. E. and Cerney, E.,** Studies in rat liver ribonucleases. III. Further studies in heterogeneity of liver lysosomes — intracellular localization of acid ribonuclease and acid phosphatase in rats of various ages, *Biochim. Biophys. Acta,* 178, 61, 1969.
77. **Brock, M. A.,** Ultrastructural studies in the life cycle of a short-lived metazoan, *Campanularia flexuosa.* II. Structure of the old adult, *J. Ultrastruct. Res.,* 32, 118, 1970.
78. **Takahashi, A., Philpott, D. E., and Miguel, J.,** Electron microscope studies in aging *Drosophila melangaster.* I. Dense bodies, *J. Gerontol.,* 25, 210, 1970.
79. **Sohal, R. S. and Allison, V. F.,** Senescent changes in the cardiac myofiber of the housefly, *Musca domestica,* an electron microscopic study, *J. Gerontol.,* 26, 490, 1971.
80. **Brunk, U. and Brun, A.,** The effect of aging on lysosomal permeability in nerve cells of the central nervous system, an enzyme histochemical study in rat, *Histochemie,* 30, 315, 1972.
81. **Comolli, R., Ferioli, M. E., and Azzola, S.,** Protein turnover of the lysosomal and mitochondrial fractions of rat liver during aging, *Exp. Gerontol.,* 7, 369, 1972.
82. **Travis, D. F. and Travis, R.,** Ultrastructural changes in left ventricular rat myocardial cells with age, *J. Ultrastruct. Res.,* 39, 124, 1972.
83. **Carroll, F. E. and Carroll, G. C.,** Senescence and death of the conidiogenous cell in *Stemphylium botryosum* Wallbroth, *Arch. Mikrobiol,* 94, 109, 1973.
84. **Sohal, R. S.,** Fine structural alterations with age in the fat body of the adult male housefly, *Musca domestica, Z. Zellforsch. Mikrosk. Anat.,* 140, 169, 1973.
85. **Sohal, R. S. and McCarthy, J. L.,** Age-related changes in acid phosphatase activity in adult housefly *Musca domestica.* A histochemical and biochemical study, *Exp. Gerontol.,* 8, 223, 1973.
86. **Knook, D. L., Sleyster, E. C., and Van Noord, M. J.,** Changes in lysosomes during ageing of parenchymal and non-parenchymal liver cells, *Adv. Exp. Med. Biol.,* 53, 155, 1975.
87. **Leuenberger, P. M.,** The phagolysosomal system of the retinal pigment epithelium in aging rats, *Adv. Exp. Med. Biol.,* 53, 265, 1975.

88. **Traurig, H. H.,** Lysosomal and hydrolase activities in the lungs of fetal, neonatal, adult and senile mice, *Gerontology,* 22, 419, 1976.

89. **Webster, G. C. and Webster, S. L.,** Lysosomal enzyme activity during aging in *Drosophila melanogaster, Exp. Gerontol.,* 13, 343, 1978.

90. **Grinna, L. S.,** Changes in cell membranes during aging, *Gerontology,* 23, 452, 1977.

91. **Wideranders, B. and Romer, N.,** Aging changes in intracellular protein breakdown, *Acta Biol. Med. Germ.,* 36, 1837, 1977.

92. **Wildenthal, K., Decker, R. S., Poole, A. R., and Dingle, J. T.,** Age-related alterations in cardiac lysosomes, *J. Mol. Cell Cardiol.,* 9, 859, 1977.

93. **Schmucher, D. L. and Wang, R. K.,** Rat liver lysosomal enzymes: effect of animal age and phenobarbital, *Age,* 2, 93, 1979.

94. **Van Pelt-Verkuil, E.,** The induction of lysosomal enzyme activity in the fat body of *Calliphora erythrocephala:* changes in the internal environment, *J. Insect. Physiol.,* 26, 91, 1980.

95. **Elens, A. and Wattiaux, R.,** Age-correlated changes in lysosomal enzyme activities: an index of ageing?, *Exp. Gerontol.,* 4, 131, 1969.

96. **Hochschild, R.,** Effect of membrane stabilizing drugs on mortality in *Drosophila melanogaster, Exp. Gerontol.,* 6, 133, 1971.

97. **Schneider, D. L., Burnside, J., Gorga, F. R., and Nettleton, C. J.,** Properties of the membrane proteins of rat liver lsysosomes. The majority of lysosomal membrane proteins are exposed to the cytoplasm, *Biochem. J.,* 196, 75, 1977.

98. **Hochschild, R.,** Lysosomes, membranes and aging, *Exp. Gerontol.,* 6, 153, 1971.

99. **DeDuve, C.,** The lysosome in retrospect, in *Lysosomes in Biology and Pathology,* Vol. 1, Dingle, J. T. and Fell, H. B., Eds., Elsevier/North-Holland, Amsterdam, 1969, 3.

100. **Bowen, I. D. and Lewis, G. H.,** Acid phosphatase activity and cell death in mouse thymus, *Histochemistry,* 65, 173, 1980.

101. **Hasan, M. and Glees, P.,** Genesis and possible dissolution of neuronal lipofuscin, *Gerontologia,* 18, 217, 1972.

102. **Kent, S.,** Solving the riddle of lipofuscin's origin may uncover clues to the aging process, *Geriatrics,* 3, 128, 1976.

103. **Kocher, E.,** The function of chloragosomes, the specific age-pigment granules of annelids — a review, *Exp. Gerontol.,* 12, 69, 1977.

104. **Sohal, R. S. and Donato, H.,** Effects of experimentally altered life spans in the accumulation of fluorescent age pigment in the housefly, *Musca domestica, Exp. Gerontol.,* 13, 335, 1978.

105. **Strehler, L., Mark, D., and Mildvan, A. S.,** Rate and magnitude of age pigment accumulation in the human myocardium, *J. Gerontol.,* 14, 430, 1959.

106. **Biscardi, H. and Webster, G. C.,** Accumulation of fluorescent age pigments in different genetic strains of *Drosophila melanogaster, Exp. Gerontol.,* 12, 201, 1977.

107. **Miguel, J., Tappel, A. L., Dillard, C. J., Herman, M. M., and Bensch, K. G.,** Fluorescent products and lysosomal components in aging *Drosophila melanogaster, J. Gerontol.,* 29, 622, 1974.

108. **Saladino, C. F. and Getty, R.,** Some histologic and electron microscopic observations in the aged beagle pancreas, *Exp. Gerontol.,* 7, 409, 1972.

109. **Mann, D. M. and Yates, P. O.,** Lipoprotein pigments in their relationship to ageing in the human nervous system. I. The lipofuscin content of nerve cells, *Brain,* 97, 481, 1974.

110. **Nanda, B. S. and Getty, R.,** Lipofuscin pigment in the nervous system of aging pig, *Exp. Gerontol.,* 6, 447, 1971.

111. **Nandy, K., Baste, C., and Schneider, F. H.,** Further studies on the effects of centrophenoxine on lipofuscin pigment in neuroplastoma cells in culture: an electron microscopic study, *Exp. Gerontol.,* 13, 311, 1978.

112. **Donato, H. and Sohal, R. S.,** Age-related changes in lipofuscin-associated fluorescent substances in the adult housefly, *Musca domestica, Exp. Gerontol.,* 13, 171, 1978.

113. **Sohal, R. S. and Donato, H.,** Effects of experimentally altered life span on the accumulation of fluorescent age pigment in the housefly, *Musca domestica, Exp. Gerontol.,* 13, 335, 1978.

114. **Silberberg, D. H. and Kim, S. U.,** Studies of aging in cultured nervous system tissue, *Int. Rev. Cytol., Suppl.* 10, 117, 1979.

115. **Leibowitz, B. E. and Siegel, B. V.,** Aspects of free radical reactions in biological systems: aging, *J. Gerontol.,* 35, 45, 1980.

116. **Finn, A. F., Jr. and Lockshin, R. A.,** Activation by anoxia of latent lactic acid dehydrogenase isozymes in insect intersegmental muscles (*Manduca sexta,* Sphingidae) *Comp. Biochem. Physiol.,* 68c, 1, 1980.

117. **Herman, M. M., Miguel, J., and Johnson, M.,** Insect brain as a model for the study of ageing. Age-related changes in *Drosophila melanogaster, Acta Neuropathol. (Berlin),* 19, 167, 1971.

118. **Nandy, K.,** Neuronal degradation in aging and after experimental injury, *Exp. Gerontol.,* 7, 303, 1972.

119. **Decker, R. S.,** Influence of thyroid hormones on neuronal death and differentiation in larval *Rana pipiens,* *Dev. Biol.,* 49, 101, 1976.
120. **Gershon, S. and Terry, R., Eds.,** *Neurobiology of Aging,* Raven Press, New York, 1976.
121. **Geinisman, Y., Bondareff, W., and Dodge, J. T.,** Dendritic atrophy in the dentate gyrus of the senescent rat, *Am. J. Anat.,* 152, 321, 1975.
122. **Landmesser, L. and Pilar, G.,** Interactions between neurons and their targets during in vivo synaptogenesis, *Fed. Proc.,* 37, 2016, 1978.
123. **Oppenheim, R. W. and Majors-Willard, C.,** Neuronal cell death in the brachial spinal cord of the chick is unrelated to the loss of polyneuronal innervation in wing muscle, *Brain Res.,* 154, 148, 1978.
124. **Hughes, W. F. and McLoon, S. C.,** Ganglion cell death during normal retinal development in the chick: comparisons with cell death induced by early target field destruction, *Exp. Neurol.,* 66, 587, 1979.
125. **Aloe, L. and Levi-Montalcini, R.,** Nerve growth factor induced overgrowth of axotomized superior cervical ganglia in neonatal rats: similarities and differences with N.G.F. effects in chemically axotomized synapthetic ganglia., *Arch. Ital. Biol.,* 117, 287, 1979.
126. **Munck, A., Crabtree, G. R., and Smith, K. A.,** Glucorticoid receptors and actions in rat thymocytes and immunologically stimulated human peripheral lymphocytes, in *Glucorticoid Hormone Action,* Baker, J. B. and Rousseau, G. G., Eds., Springer-Verlag, Berlin, 1979, 341.
127. **Lockshin, R. A. and Beaulaton, J.,** Programmed cell death. Electrophysiological and ultrastructural correlations in metamorphosing muscles of lepidopteran insects, *Tissue Cell,* 11, 803, 1979.
128. **Sohal, R. S. and Allison, V. F.,** Age-related changes in the fine structure of flight muscle in the housefly, *Exp. Gerontol.,* 6, 167, 1971.
129. **Davies, I. and King, P. E.,** A morphometric and cytochemical analysis of aging changes in the flight muscle of *Nasonia vitripennis* (Walk.) (Hymenoptera, Pteromalidae), *Mech. Ageing Dev.,* 4, 459, 1975.
130. **Sacktor, B. and Shimada, Y.,** Degenerative changes in the mitochondria of flight muscle from aging blow flies, *J. Cell Biol.,* 52, 465, 1972.
131. **Miguel, J., Oro, J., Bensch, K. G., and Johnson, J. E., Jr.,** Lipofuscin: fine structural and biochemical studies, in *Free Radicals in Biology,* Vol. 3, Pryor, W., Ed., Academic Press, New York, 1977, 133.
132. **Sachs, M. G., Colgan, J. A., and Lazarus, M. D.,** Ultrastructure of the aging myocardium: a morphometric approach. *Am. J. Anat.,* 150, 63, 1977.
133. **Vann, A. C. and Webster, G. C.,** Age-related changes in mitochondrial function in *Drosophila melangaster, Exp. Gerontol.,* 12, 1, 1977.
134. **Tribe, M. and Webb, S.,** How far does radiation mimic ageing in insects? II. Ultrastructural changes, *Exp. Gerontol.,* 14, 255, 1979.
135. **Martinez, A. O. and McDaniel, R. G.,** Mitochondrial heterosis in aging *Drosophila* hybrids, *Exp. Gerontol.,* 14, 231, 1979.
136. **Bowen, I. D. and Lockshin, R. A., Eds.,** *Cell Death,* Chapman and Hall, London, 1981, 493 pp.
137. **Sohal, R. S., Ed.,** *Age Pigments,* Elsevier/North Holland, Amsterdam, 1981.
138. **Beaulaton, J. and Lockshin, R. A.,** The relation of programmed cell death to development and reproduction: comparative studies and an attempt at classification, *Int. Rev. Cytol.,* 79, 215, 1982.
139. **Oppenheim, R. W.,** Neuronal cell death and some related retrogressive phenomena during neurogenesis: a selective historical review and progress report, in *Studies in Developmental Neurobiology: Essays in Honor of Viktor Hamburger,* Cowan, W. M., Ed., Oxford University Press, New York, 1981, 74.
140. **Cunningham, T. J.,** Naturally occurring neuron death and its regulation by developing neural pathways, *Int. Rev. Cytol.,* 74, 163, 1982.
141. **Robertson, A. M. G. and Thomson, J. N.,** Morphology of programmed cell death in the ventral nerve cord of *Caenorhabditis elegans* larvae, *J. Embryol. Exp. Morphol.,* 67, 89, 1982.
142. **Jones, R. G., Davis, W. L., and Vinson, S. B.,** A histochemical and x-ray microanalysis study of calcium changes in insect flight muscle degeneration in *Solenopsis,* the queen five ant, *J. Histochem. Cytochem.,* 30, 293, 1982.

ANATOMY OF THE AGING NERVE CELL

Robert J. Johnson

In analyzing the physical expression of aging changes in the nerve cell, one faces the constant problem of differentiating those structural changes that are truly part of the aging process from those that are a consequence of diseases or events associated with aging. For example, there are nerve cell changes that are a consequence of degeneration and aging in the cardiovascular system. Such nerve cell changes may be age-related yet in no way a direct result of aging in the nerve cell itself. This complex interdependence of one body system upon another in the higher organisms, such as humans and other mammals, makes it extremely difficult to draw conclusions as to whether age itself or the accumulation of pathologic processes over many years is responsible for nerve cell and nervous system changes.

The very diversity of cells in the highly complex nervous system of humans and other mammals adds another problem, as some types of cells show changes associated with aging that are not seen to the same degree, if at all, in other types of cells within the same nervous system. Some structural changes that are quite evident in the total nervous system and that are certainly strongly age-related, such as generalized atrophy and loss of brain weight or size, are very difficult to quantify at the histological or cellular level. Cellular changes may be seen in such instances, but are they the ones responsible for the generalized loss of substance?

There are widely recognized functional changes in the human nervous system that are accepted as age-related, if not truly a direct consequence of aging, yet little is known of cellular expression of these functional losses. To some degree, this difficulty of associating cellular anatomic change with functional deficit rests in the complex organization of the nervous system and in the inherent diffusing of structural responsibility for all functional results.

Since nerve cells always exist as a system of cells and as part of a more general organism, they cannot be studied as isolated entities except in tissue cultures. Nerve cells in general, at least in higher organisms, are postmitotic and these organisms have lifespans ranging from several months to many years. There are, obviously, many difficulties in the comparison of aging changes from one species to another and even in sampling the progressive changes in long-lived species. Certainly, in humans, the data come from many diverse groups who have led different lives, lived in different places, been subjected to different diseases, and have different hereditary patterns for longevity.

At the present time there is much disagreement in the accumulated data, especially when experimental data derived from animal systems are compared with observations in humans. Most workers have accepted the idea of a certain loss of nerve cells and fibers with aging in humans. Whereas some animal studies support that notion, others indicate no loss of nerve cells or fibers associated with aging, as in certain rodents. Is this merely a species difference, experimental error, or has the selection of strains from which to gather data eliminated variables which were in fact irrelevant to aging?

Despite the difficulties inherent in investigating aging of human nerve cells, much biologic interest centers here and humans do offer one of the longest lifespans for these postmitotic cells. The best that can be done is to offer an assemblage of data from a variety of species, with a variety of lifespans, and from as many of the specialized areas of nervous systems as are available. Attention will focus on recognizable structural or anatomic changes that are age-related, or at least currently believed to be so, regardless of whether these changes have any known functional consequences.

Aging in the human brain poses three states or conditions, two of which should be considered pathological, and one as normal aging. Virtually all of the physical changes to which the aging brain is subjected can be found in each of these conditions. These changes are differentiated based on the extent to which they occur and by their time of onset. When the changes become sufficiently extensive only late in life, they cause senile dementia, while changes that become extensive in the middle years of life cause presenile dementia (Alzheimer's disease). There is no physical or pathological distinction between the brains in the two types of dementia. Only the time of onset of symptoms determines the classification. The brain of an aged person with normal mental function will show some of the changes seen so extensively in the brain of the demented patient, but the amount of damage done by these far fewer lesions will not cause a significant or easily recognized loss of mental facility.

The diffuse cortical lesions which cause the two types of dementia must be contrasted with the focal lesions of cerebral vascular disease. When abundant, the focal vascular lesions may cause the commonly seen cerebral deficit that is often erroneously labeled as senile brain disease.

Physical changes characteristic of the aging brain which must be considered are as follows:

1. Shrinkage of the brain and loss of total brain weight
2. Nerve cell loss and disorganization of cortical and nuclear cellular patterns
3. Shrinkage and ''sclerosis'' of the individual nerve cell
4. Irregularities of axons such as: torpedoes, spheroids, and changes in the myelin sheaths
5. Accumulation of intraneuronal (cytoplasmic), lipofuscin (lipochrome), melanin, and iron pigment
6. Presence of neurofibrillary tangles or changes
7. Presence of neuritic (senile) plaques and granulovacuolar degeneration
8. Other aging changes in intracellular organelles and cytoplasm of neurons
9. Age changes in neuroglia and the neuronal microenvironment
10. Changes in brain blood vessels which are more specifically correlated with aging, as opposed to ordinary arteriosclerosis
11. Changes in afferent and efferent nerve terminals outside of the central nervous system

SHRINKAGE OF THE BRAIN AND LOSS OF TOTAL BRAIN WEIGHT

This becomes evident in humans usually by the fourth or fifth decade, with a slowly progressive decline in brain weight thereafter and an increasing widening of the sulci and narrowing of the gyri on the surface of the brain. An associated enlargement of the lateral ventricles and an increase in the capacity of the pericerebral space also occur. This weight and volume loss is variable and greatly increased in senile dementia and Alzheimer's disease. Brain weight loss may be due to loss of neurons, myelin, and neuroglia and also to water loss from the intercellular space, or from the ground substance of the brain. The relative value for each is not known (see Table 1).

NERVE CELL LOSS WITH AGING

Nerve cell loss and consequent disorganization or change in the cortical or nuclear cellular pattern is a common finding in most areas of the gray substance. Although very difficult to quantify, and variable from region to region, this loss is believed to be a real decline in nerve cells, at least in the human brain. Certain areas and nuclei do not show a loss of cells with advancing years in the human brain. Some animal studies suggest there is no neuronal loss anywhere in those specific aging animals. This may be a species difference or it may

Table 1

SHRINKAGE AND WEIGHT LOSS OF THE TOTAL BRAIN WITH AGING

Physical change	Part of nervous system	Species	Ref.
Decline in total brain weight	Total brain	Man	1—13
No contrast in brain weight in young and old animals	Total brain	Mice	14
Decline in brain protein and lipid	Total brain	Man	2,3
Slight rise in water content of brain with aging	Total brain	Man	3,15
Increase in pericerebral space	Total brain	Man	3,16
Increase in size of ventricular system	Total brain	Man	3,17,18
No increase in size of ventricular system	Total brain	Dog	19,20
Early post-mortem increase in brain weight due to absorption of cerebrospinal fluid; effects of time and temperature after death	Total brain	Man	10,21
Narrowing of gyri and widening of sulci, particularly in frontal lobe	Cerebrum	Man	4
Reichardt's method of determining brain weight to skull volume ratio	Total brain	Man, animals	22,23
Areas of brain atrophy in Alzheimer's dementia	Total brain	Man	4—6,24
Loss of brain weight is not due to loss of neurons	Cerebrum	Man	25
Comparative shrinkage of gray and white matter with aging	Cerebrum	Man	1,4,26
Comparative shrinkage with age of male vs. female brains	Cerebrum	Man	11,12,26
Mean brain weight in young and old rhesus monkeys	Total brain	Monkey	12,27,28

represent errors in technique of counting. All such counts are, of course, actually only an estimate. In all experimental studies, it should be remembered that aged animals of 2 or 3 years are being compared with aged humans of 70 to 100 years (see Table 2).

SHRINKAGE AND SCLEROSIS OF THE INDIVIDUAL NERVE CELL

Neuron cell bodies and dendritic arborizations have been observed to shrink in the cerebral cortex. Both the number of dendritic branches and the number of spines per branch may be reduced. Such changes have been noted in both human and animal brains. The cytoplasm of shrunken neurons tend to show a decrease in the amount of Nissl substance and perhaps noticeable pallor on staining. Sclerosis of nerve cells indicates an increase in the basophilia of the Nissl substance and a decrease in the differentiation between the nucleus and the cytoplasm with increased basophilia of the nucleus and pallor of the nucleolus. The influence of the terminal illness, postmortem changes, and fixation artifact must never be overlooked in evaluating the cellular changes alleged to be due to age (see Table 3).

IRREGULARITIES OF AXONS SUCH AS TORPEDOES, SPHEROIDS, AND CHANGES IN THE MYELIN SHEATHS

Rarefaction and paleness of myelin sheaths of axons in cerebral and cerebellar cortex may be seen. Localized axonal changes may appear in aged brains. Such changes may take the form of torpedo-like swellings and round or spheroidal swellings. The term neuroaxonal dystrophy designates these changes. Such findings tend to occur in limited distribution at particular locations only (see Table 4).

Table 2
NERVE CELL AND FIBER LOSS WITH AGING

Physical change	Part of nervous system	Species	Ref.
Significant loss of Purkinje cells with aging	Cerebellum	Man, rat	26, 29—33
Loss of neurons in cerebral cortex	Superior temporal gyrus, superior frontal gyrus, precentral gyrus, postcentral gyrus, inferior temporal gyrus, cingulate gyrus, area striata, hippocampus	Man	24, 27, 34—49
Loss of neurons in thalamus and basal ganglia	Cerebrum	Man	36, 50, 51
Loss of neurons in locus coeruleus	Brain stem	Man, rat	52—56
Loss of neurons in brain stem	Brain stem	Man	36, 51
No loss of neurons in various brain stem nuclei	Inferior olivary nucleus, nucleus ambiguus, trochlear nucleus, facial nucleus, cochlear nuclei, abducens nucleus, trigeminal nuclei	Man	25, 52, 53, 58—65
Loss of neurons in total brain homogenate	Total brain	Mice	66
No loss of Purkinje cells	Cerebellum	Man	41, 67, 68
No loss of nerve cells with age	Cerebrum, cerebellum, brain stem	Man, guinea pig, rat, frog	25, 41, 69—78
No loss of cells of dentate nucleus	Cerebellum	Man	79
Marked loss of Purkinje cells with age in animals	Cerebellum	Rat, guinea pig	33, 80
Increased neuronphagia noted with nerve cell loss	Cerebral cortex	Man	81
Types and layers of cortical	Cerebral cortex	Man	35, 37, 41, 46, 53, 82, 83
Reduction of nerve fibers in spinal nerve roots	Ventral roots, dorsal roots	Man, rat	84—91
No reduction of nerve fibers in spinal nerve roots	Ventral roots, dorsal roots	Rats, cats	92, 93
Reduction of cells in dorsal root ganglia	Dorsal root ganglia	Man	86, 87, 94—96
No reduction of cells in dorsal root ganglia	Dorsal root	Man	97, 98
Reduction of fibers in peripheral nerves	Peroneal nerve, sciatic nerve, median nerve, femoral nerve, vagus nerve, optic nerve, vestibular nerve	Man, rat	94, 99—107
Reduction of cells in sympathetic ganglia	Autonomic nervous system	Man	108—111
No reduction of fibers in peripheral nerves	Sciatic nerve, vagus nerve	Man, rat	106, 112—114
Reduction of cells in intermediolateral cell column	Spinal cord	Man	115
Reduction of cells in anterior gray horn	Spinal cord	Man, mice	84, 89, 94, 116
Reduction of cells in spiral ganglion (8th nerve)	Cranial nerve sensory ganglion	Man	117—125
Changes in trigeminal ganglion cells	Cranial nerve sensory ganglion	Man	126, 127
Disorganization and loss of ependymal cells with increasing occlusion of central canal	Spinal cord	Man	128
Changes in intramural autonomic ganglia	Autonomic nervous system	Man	129
Compensatory changes in striatal DA receptors	Cerebrum	Rat	130

Table 2 (continued)
NERVE CELL AND FIBER LOSS WITH AGING

Physical change	Part of nervous system	Species	Ref.
Reduction of synapses	Cerebrum, brain, stem, spinal cord	Man, dog, rabbit, rat	44, 131—142
Selective loss of cholinergic neurons of hippocampus in Alzheimer's disease	Cerebrum	Man	43, 73, 143
Compensatory dendritic and synaptic growth in remaining neurons of aged animals after neuronal death	Cerebrum	Various animals	130, 144—148

Table 3
SHRINKAGE AND SCLEROSIS OF THE INDIVIDUAL NERVE CELL

Physical change	Part of nervous system	Species	Ref.
Shrunken Purkinje cells	Cerebellum	Man, dog	30, 32, 149, 150
Shrunken pyramidal cells and dendrite systems	Cerebral cortex	Man, rat	82, 151—160
No correlation of shrunken neurons with age	Cerebral cortex, dentate nucleus of cerebellum	Man, rat	79, 162
Decline of dendritic spine density with age	Cerebral cortex, hypothalamus, spinal cord	Man, rat, mouse, dog	73, 78, 82, 149—154, 157, 159, 160, 162—165
Comparison of declining spine density with age	Cerebral cortex	Rat, cat, mouse, monkey	151, 154, 165—175
Decline in diameters of apical dendrites	Cerebral cortex	Rat	153, 154, 156
Decline in spine density and/or changes in spine morphology associated with pathological conditions other than age	Cerebral cortex	Man, monkey	72, 73, 82, 139, 149, 152—154, 156, 176—180
Decline of spine density in presenile and senile dementia	Cerebral cortex	Man	82, 149, 152, 153, 156, 159
Intradendritic membranous inclusion bodies and other dendritic degenerations associated with aging	Cerebral cortex	Man, rat, monkey	151, 154, 181—186
Neuronal and dendritic specificity in reference to morphologic changes with aging	Cerebral cortex	Man, rat, mouse	151, 154, 170, 187
Shrinkage and chromatolysis of anterior horn cells and cells of Clark's column	Spinal cord	Man, mouse	128, 188
Reduction in number of synapses involving spines of dendrites	Cerebrum	Rat, man, dog, monkey	140—142, 150, 154, 174, 175, 182
Increase in arborization of dendrites of some neurons when others in region show reduction of dendrites and spines	Cerebral cortex	Man, rat, monkey	71, 73, 130, 143—148, 180, 189, 191—193
Degenerative changes in still present synapses	Cerebrum	Rat	142, 182

Table 4
**IRREGULARITIES OF AXONS (NEURAXONAL DYSTROPHY) SUCH AS
"TORPEDOES" AND "SPHEROIDS", AND CHANGES IN THE MYELIN SHEATHS**

Physical change	Part of nervous system	Species	Ref.
Segmental demyelination	Peripheral nerves	Man, rabbit, rat, monkey	184, 194—202
Neuronaxonal dystrophy in preterminal axons of nucleus gracilis and cuneatus	Medulla oblongata	Man, rat	202—209
Neuronaxonal dystrophy in preterminal axons near motor end plates and in neuromuscular spindles	Peripheral nerves	Man	211—213
Changes in myelin sheaths described by decade of life	Spinal cord	Man	128
Ultrastructural changes in growth of myelin sheaths	Spinal cord	Mouse	213
Reduction of maximum and average diameters of peripheral nerve fibers	Peripheral nerves	Man	214—216
Reduction in internodal length of fibers in peripheral nerves	Peripheral nerves	Man, rabbit	197, 217—221
Degeneration of unmyelinated axons	Peripheral nerves	Man	222—224
Presence of Renaut bodies	Peripheral nerves	Man	225
Decline in conduction velocity in peripheral nerve fibers	Peripheral nerves	Man	219, 226—230
No decline in conduction velocity in peripheral nerve fibers	Peripheral nerves	Rat	113

ACCUMULATION OF LIPOFUSCIN

This intracytoplasmic granular material of yellowish to brown color is found in increasing amounts in many types of aging neurons. The coarse granules are composed chiefly of oxidized lipids, yet are largely insoluble in the usual lipid solvents. It is probable that lipofuscin is formed from undegradable waste products derived from partially broken down membranes and other such cell components. Its quantity is variable among cell types or within a given area of an aged brain. For unknown reasons neurons of certain areas do not accumulate lipofuscin. The neurons of the inferior olivary nucleus are particularly interesting in their high degree of lipofuscin accumulation despite maintenance of cell number with advanced age. This finding indicates that lipofuscin is not necessarily a hazard to the cell that accumulated it. The substance is found widely in the nerve cells of many mammals, although the type of neurons and the extent to which they accumulate lipofuscin differs among species. A comparable, if not identical, intracellular material is also found in myocardial cells (see Table 5).

PRESENCE OF NEUROFIBRILLARY TANGLES

Neither the neurofibrillary tangles nor the neuritic plaques are solely age-related, although they do, in general, increase in numbers with advancing years. They are both very closely related to age as their frequency increases with the occurrence of senile mental deterioration and, most specifically, with the presenile dementia of the Alzheimer types. Neurofibrillary tangles are found in small numbers in the brains of many normal people in the seventh decade of life. In the ninth and tenth decades nearly all "normal" people will have such lesions in their brains, but obviously not in sufficient amount to impair mental function recognizably. This contrasts with the large numbers of tangles which are present in the even less aged brains of persons with Alzheimer's dementia. So far, this neurofibrillary change

Table 5
ACCUMULATION OF LIPOFUSCIN WITH AGE

Physical change	Part of nervous system	Species	Ref.
Cells involved, genesis, and staining reaction for lipofuscin	Total brain	Man, animals	4, 27, 28, 45, 231—244
Lipofuscin accumulates early in motor nuclei of cranial nerves	Brain stem	Guinea pig	69
Earliest appearance of lipofuscin in brain is in mesencephalic nucleus of trigeminal	Total brain	Guinea pig	69
Lipofuscin does not accumulate at all in cochlear nucleus, exteroceptive sensory nuclei, and motor nucleus of vagus	Brain stem	Guinea pig	69
Lipofuscin does not accumulate in human Purkinje cells	Cerebellum	Man	245
Lipofuscin does accumulate in nonhuman primate Purkinje cells	Cerebellum	Monkey	246
Lipofuscin accumulates in cells of dentate nucleus	Cerebellum	Man	79, 245, 247
Contrast in lipofuscin in cells of dentate nucleus and Purkinje cells in guinea pig	Cerebellum	Guinea pig	69
Lipofuscin accumulation in pyramidal cells of cortex of rat contrasted with man	Cerebrum	Rat	248—250
Thalamus as site of early and regular accumulation of lipofuscin	Thalamus	Man	4, 251
Lipofuscin accumulation in autonomic ganglion cells	Peripheral nerves	Man, rat	108, 109, 252—255
Lipofuscin accumulation in sensory ganglia	Peripheral nerves	Man	126, 256, 257
Onset, course, and degree of lipofuscin deposition is specific for different nuclear groups on cell clusters	Total brain	Man	53, 237, 244, 248, 258
Time of appearance and distribution of lipofuscin in neurons of globus pallidus and hypothalamus	Cerebrum	Man	259
Intracellular distribution of lipofuscin changes with advancing years	Cerebral cortex	Man, rat, mice	79, 244, 260—262
Patterns of deposition of lipofuscin in amygdaloid complex at various ages	Cerebrum	Man	263
Appearance of lipofuscin in specific layers, cells, and areas of cortex	Cerebral cortex	Man, monkey, dog, pig, rat	27, 35, 37, 49, 237, 238, 245, 248, 250, 264, 265
Five formative stages of lipofuscin accumulation in cells of dentate nucleus	Cerebellum	Man	79
Theory of aging and lipofuscin formation on colloidal solution principle	Cerebrum	Man	22, 266, 267
Role of lysosomes in genesis of lipofuscin	Cerebral cortex, spinal cord	Man, monkey, mice	238, 242, 248, 260, 268—276
Mitochondrial role in genesis of lipofuscin	Various sites	Rodents	22, 44, 237, 239—242, 253, 257, 277—279

Table 5 (continued)
ACCUMULATION OF LIPOFUSCIN WITH AGE

Physical change	Part of nervous system	Species	Ref.
Golgi complex role in genesis in lipofuscin	Various sites	Rodents	22, 237, 239—242, 253, 280—286
Lipofuscin as inert nonharmful material or as harmful product or as valuable to cell physiology or health	Various sites	Man	22, 79, 231, 236, 237, 239, 243, 266, 290—297
Chemistry, differential staining, and fine structure of lipofuscin	Various sites	Man, animals	237, 242, 243, 247, 273, 275
Appearance of lipofuscin in inferior olivary nucleus	Brain stem	Man	53, 64, 65, 247
Extraneural (intraglial) lipofuscin	Cerebral cortex	Monkey	27, 28, 239
Lipofuscin accumulation in neurons in tissue culture	Cerebral cortex, dorsal root, ganglion cells, spinal motoneurons	Man, rat, cat, rabbit	273, 274, 276, 278, 307
Increase of lipofuscin accumulation in cultured neurons when treated by agents which inhibit cell division	Cerebral cortex	Mouse	307
Experimental production of lipofuscin in nerve cells of intact young animal	Cerebrum	Rat	299, 316, 317
Experimental reduction of lipofuscin in nerve cells of intact old animal	Cerebrum	Guinea pig, rat	242, 243, 297
Appearance of lipofuscin in anterior horn cells and Clark's column cells by decade	Spinal cord	Man	128, 322
Ultrastructural (EM) studies on lipofuscin	Various sites	Man	44, 323—330, 238, 239, 241, 248, 253, 257, 260, 268, 272, 273, 278, 288, 290

has been found in the neurons of humans only, although the application of alumina gel to the cortex of animals can produce a similar toxic fibrillar change in the nerve cells.

In electron microscopy, the neurofibrillary tangle is found to comprise abnormal neurofibers or filaments arranged in excessive numbers within the neuron. This abnormal neurofiber is a paired helical filament without side arms. Each filament is 10 nm in diameter. This abnormal neurofilament contrasts with the two types of normal neurofilaments, both of which are single, have side arms and are 24 and 10 nm in diameter, respectively. It is not yet known whether the abnormal paired helical filaments are chemically different from the smaller of the two normal filaments, or whether they are primarily a morphological malformation.

Neurofibrillary tangles composed of paired helical filaments are found in at least three conditions not associated with age. They are noted in adult patients with Down's Syndrome, in "punch-drunk" boxers, and in the neurons of patients with postencephalitic Parkinson's disease (see Table 6).

Table 6
PRESENCE OF NEUROFIBRILLARY TANGLES

Physical change	Part of nervous system	Species	Ref.
Paired helical filaments in neurofibrillary tangles in senile and presenile dementia	Cerebral cortex	Man	1, 329, 331 — 342
Neurofibrillary tangles in Alzheimer's disease, senile dementia of Alzheimer's type, and in normal aged brain	Cerebral cortex	Man	1, 5, 6, 19, 46, 327, 332 — 339, 342 — 355
Neurofibrillary tangles and/or paired helical filaments found in a variety of nervous diseases	Cerebral cortex	Man	1, 46, 356 — 359, 342, 347, 349, 360 — 364
Neurofibrillary tangles cannot be correlated with any recognized metabolic disorder	Cerebral cortex	Man	18
Neurofibrillary tangles found in small pyramidal cells of superficial layers of cortex	Cerebral cortex	Man	4, 19, 233, 343, 348, 350
Neurofibrillary tangles occur early and regularly in Sommer's sector of hippocampus gyrus and in amygdala	Hippocampus and amygdala	Man	4, 5, 18, 19, 343 — 346, 348, 349, 365
Neurofibrillary tangles in neurons of cardiac sympathetic ganglia	Peripheral nerves	Man	252
Experimental induction of neurofibrillary tangles and of paired helical filaments	Cerebrum and spinal ganglia	Animals, cultured human neurons	342, 349, 366 — 369, 370 — 378
Neurofibrillary tangles do not occur in spinal cord or in cerebellum in normal aged persons nor in Alzheimer's disease	Central nervous system	Man	342, 347, 349
Neurofibrillary tangles do not occur in senescent rat brain	Cerebrum	Rat	379

PRESENCE OF NEURITIC PLAQUES

The senile or neuritic plaque is a conglomerate lesion with a central core of an abnormal, extracellular, fibrillar protein known as amyloid. Surrounding this core of amyloid are numerous abnormal synaptic elements and preterminal axons. The latter are distended in various places with masses of paired helical filaments comparable to those described in the neurofibrillary tangles. These preterminal axons also contain lamellated residual bodies and degenerating mitochondria. Additional findings common to the neuritic plaque are coarse lipofuscin granules in some of the neurons as well as astrocytic fibers interwoven among the other elements.

The neuritic plaque differs from the neurofibrillary tangle in that the former is found not only in humans, but in the neural tissue of aging dogs and monkeys. However the axons in dogs and monkeys show the abnormal paired helical filaments so characteristic of the plaque and the tangle in the human brain.

When only the central, fibrillar, amyloid material is present (without the surrounding halo of preterminal axons, abnormal synaptic elements, and astrocytic fibers, the lesion is called an amyloid body or amyloid plaque (see Table 7).

Table 7
PRESENCE OF NEURITIC PLAQUES AND GRANULOVACUOLAR
DEGENERATION

Physical change	Part of nervous system	Species	Ref.
Nature and frequency of neuritic (senile) plaques in normal aged brain and in senile and presenile dementia	Cerebral cortex	Man	1, 4, 46, 201, 327, 332, 334, 335, 342, 344— 347, 349, 350, 351, 372, 380—384
Relation of neuritic plaque to blood vessels	Cerebral cortex	Man, monkey	1, 327, 335, 342, 344, 347, 385
Extracellular amyloid formation in neuritic plaque			327, 335, 342, 347, 355, 373, 380, 381, 386—391
Neuritic plaques in scrapie-infected animals	Cerebral cortex	Mice, monkey	1, 342, 347, 392—397
Neuritic plaques not absolutely associated with neurofibrillary tangles	Cerebrum	Man	1, 4, 350, 365
Neuritic plaques most frequent in hippocampal gyrus, followed by frontal cortex	Cerebrum	Man	1, 4, 23, 232, 329, 344, 347
Neuritic plaques in basal ganglia	Cerebrum	Man	365
Neuritic plaques do not occur in cerebellum and spinal cord	Central nervous	Man	1, 232, 347, 349
Neuritic plaques do not occur in rat brain	Cerebrum	Rat	379
Spontaneously arising neuritic plaques in aging dogs and monkeys	Central nervous system	Dog, monkey	157, 201, 380, 398
Granulovacuolar degeneration chiefly in pyramidal cells of hippocampus and in Alzheimer's disease	Hippocampus	Man	4, 46, 343, 351, 399, 400, 401
Spongiform alterations in cortex of presenile dementia	Cerebrum	Man	402

OTHER PHYSICAL CHANGES OF AGING

Table 8 summarizes other changes in intracellular organelles and neuron cytoplasm, and Table 9 summarizes age changes in neuralgia and the neuronal microenvironment.

CHANGES IN BLOOD VESSELS OF THE BRAIN THAT ARE MORE SPECIFICALLY CORRELATED WITH AGING (AS OPPOSED TO ARTERIOSCLEROSIS)

An age-related change seen in the small arteries of the brain has been variously named "plaque-like degeneration" and "dyshoric angiopathy". There is a reduction in the caliber of the lumen and the degenerated vascular wall is thickened and composed of pale metachromatic material showing argentophilic, radially arranged fibrils. The elastic fibers of the

Table 8

OTHER AGING CHANGES IN INTRACELLULAR ORGANELLES AND CYTOPLASM OF NEURONS

Physical change	Part of nervous system	Species	Ref.
Fragmentation, displacement, and modification of Golgi apparatus	Various sites	Rat, mouse, rabbit, frog, toad	44,280—285,403—408
Modification and disappearance of mitochondria	Various sites	Guinea pig, fowl, rat, mice	213,257,409—416
Modification and aggregation of neurofibrils	Various sites	Man, chick	1,333,417—421
Modification of nucleus and nucleclus	Various sites	Man, animals	22,44,69,249,258,261, 263,266,411,422— 442
Histochemical changes indicative of enzymatic and metabolic changes in cytoplasm	Various sites	Man, dog, rat	13,306,406,407,409,416, 443—448
Modification in endoplasmic reticuhm (including amount, distribution, and staining reaction of Nissl substance)	Various sites	Man, animals	22,30,32,44,69,79,150, 184—186,233,249,258, 263,300,411,422,424, 426,435,449—458
Argyrophilic inclusions characteristic of Pick's disease	Hippocampus	Man	344,459
Topography of neuro-transmitters and of zinc in hippocampus	Hippocampus	Man	43,73,143,344,460—462

Table 9
AGE CHANGES IN NEUROGLIA AND THE NEURONAL MICROENVIRONMENT

Physical change	Part of nervous system	Species	Ref.
Decrement in volume of intercellular space	Cerebral cortex	Rat	22,231,463—473
No decrement in volume of intercellular space	Cerebrum	Mouse	474
Increase in neuroglial cell density associated with decrease in neuron density	Cerebral cortex	Man, monkey	27,45,72, 475—479
No correlation between glial cell counts and age	Cerebral cortex	Man, animals	46,78,131—133, 480—483
Lipofuscin accumulation in glial cells of cortex and in culture	Cerebral cortex	Man, monkey	238,250,330,484
Occurrence of binucleate oligodendrocytes	Hippocampus	Rat	44
Astrocytes show gliofibril changes and modified cytoplasmic form	Cerebral cortex	Man, rat	185,484,485
Glial changes in gray matter by decade of life	Spinal cord	Man	128
Increase in thickness of perineurium and endoneurium	Peripheral nerves	Man	94,341,486
Intranuclear inclusions in neuroglia cells	Dorsal root ganglia	Mouse	440
Intraglial location of corpora amylacea	Cerebral cortex	Monkey	73,184

Table 10A
CHANGES IN BLOOD VESSELS OF THE CENTRAL NERVOUS SYSTEM AND PERIPHERAL NERVES CORRELATED WITH AGING RATHER THAN ARTERIOSCLEROSIS

Physical change	Part of nervous system	Species	Ref.
Irregularities of course and caliber of small vessels	Cerebral cortex and subjacent white matter	Man	487—489
Opacification and loss of tubular resiliency	Circle of Willis	Man	488,489
Amyloid angiopathy of arterioles and terminal capillaries	Cerebrum	Man	344,385
Changes in capillary density and thickness of wall at different ages	Brain stem, cerebellum, cerebral cortex	Rat	147,247, 490—492
Attenuated lumen of small vessels	Peripheral nerves	Man	107,215,216, 218,486,493
Cerebral blood flow and cerebral metabolic rate for oxygen do not change with age in normal subjects	Cerebrum	Man, rats	147,494—497
Regional cerebral blood flow does decline with age in normal brains	Cerebrum	Man, dogs	147,498—500

Table 10B
CHANGES IN MOTOR NERVE ENDINGS

Physical change	Species	Ref.
Preterminal axon shows neuronaxonal dystrophy (spheroid and ovoid swellings)	Man	210,211,213,501
Elaborate and multiple motor end plates	Man	210
Changes in neuromuscular spindles	Man	211
No change in terminal innervation ratio	Man	501
Decrease in motor unit size plus muscle atrophy	Rat	502—504

Table 10C
CHANGES IN CUTANEOUS AND MUCOSAL NEURAL RECEPTORS

Physical change	Species	Ref.
Pacinian corpuscles modified, enlarged, lost, and replaced	Man	505—508
Meissner's corpuscles	Man	505,509,510
Decline in touch-pressure sensitivity	Man	511
Decline in number of taste buds	Man	512—517
No decline in ability to differentiate between basic tastes	Man	518, 519
Decline in number of olfactory cells and nerve fibers	Man	144,520,521
Decline in vibration sensitivity	Man	522

wall are destroyed, but the tunica intima appears normal. The degenerated wall is fluorescent and stains with Congo red.

Another characteristic senile change in smaller vessels of the brain is classified as glomerular loop formation in which there may be peculiar bundles of parallel vessels and meshwork patterns with twining and braiding of vessels. A capillary fibrosis is often found in aging brains and a hyaline degeneration of the tunica media is occasionally observed in vessels of the aged brain (see Tables 10A to 10C).

REFERENCES

1. **Wisniewski, H. M. and Terry, R. D.**, Neuropathology of the aging brain, in *Aging, Neurobiology of Aging*, Vol. 3, Terry, R. D. and Gershon, S., Eds., Raven Press, New York, 1976.
2. **Bürger, M.**, Die chemishe biomorphase des menschlichen zentralnerven system, *Medizinische*, 15, 561, 1956.
3. **Hunwick, H. E.**, Biochemistry of the nervous system in relation to the process of aging, in *The Process of Aging in the Nervous System*, Birren, J. E., Imsa, H. A., and Windle, W. F., Eds., Charles C Thomas, Springfield, Ill., 1959.
4. **Minchalen, F.**, *Pathology of the Nervous System*, Vols. 1 to 3, McGraw-Hill, New York, 1968.
5. **Corsellis, J. A. N.**, *Mental Illness and the Aging Brain*, Oxford University Press, London, 1962.
6. **Tomlinson, B. E., Blessed, G., and Roth, M.**, Observations of the brains of demented old people, *J. Neurol. Sci.*, 11, 205, 1970.
7. **Pearl, R.**, *The Biology of Death*, Lippincott, New York, 1982.
8. **Pakkenberg, H. and Voight, J.**, Brain weights of the Danes, *Acta Anat.*, 56, 297, 1964.
9. **Brody, H. and Vijayashankan, N.**, Anatomical changes in the nervous system, in *Handbook of the Biology of Aging*, Finch, C. E. and Hayflick, L., Eds., Van Nostrand Reinhold, New York, 1977, 241.
10. **Appel, F. W. and Appel, E. M.**, Intracranial variation in the weight of the human brain, *Hum. Biol.*, 14, 48, 1942.
11. **Peress, N. S., Kane, W. C., and Aronson, S. M.**, Central nervous system findings in a tenth decade autopsy population, in *Progress in Brain Research*, Vol. 40, Ford, D. H., Ed., Elsevier, New York, 1973.
12. **Ordy, J. M., Kaack, B., and Briggee, K. R.**, Lifespan neurochemical changes in the human and nonhuman primate brain, in *Aging*, Brody, H., Harman, D., and Ordy, J. M., Eds., Raven Press, New York, 1975, 133.
13. **Samorajski, T.**, Normal and pathologic aging of the brain, in *Brain Neurotransmitters and Receptors in Aging and Age-Related Disorders*, Vol. 17, Enna, S. J., Samorajski, T., and Beer, B., Eds., Raven Press, New York, 1981.
14. **Finch, C. E.**, Catecholamine metabolism on the brains of aging male mice, *Brain Res.*, 52, 261, 1973.
15. **Branté, G.**, Studies on lipids in the nervous system with special reference to quantitative chemical determination and topical distribution, *Acta Physiol. Scand.*, 18, 1, 1949.
16. **Borins, H.**, Zur kerntuis des spietrauma zwischen gehian und schädel, *Z. Neur.*, 94, 72, 1925.
17. **Heinrich, A.**, Alternsvorgänge in Röntgenfild, *Ergangungshnad*, 62, 1941.
18. **Morel, F. and Wildi, E.**, General and cellular pathochemistry of senile and presenile alterations of the brain, in *Proc. 1st Int. Congr. Neuropathol., Rome*, Rosenberg, I., Sellier, P., Torino, L., Eds., Casa Editoriae Libraria, Rome, 1952.
19. **Tomlinson, B.**, Morphological brain changes in non-demented old people, in *Ageing on the Central Nervous System: Biological and Psychological Aspects*, Van Praag, H. M. and Kalverbone, A. F., Eds., De Erven F. Bohn, N. V. Haarlen, 1972.
20. **Cammermeyer, J.**, Third round-table discussion, in *The Process of Aging in the Nervous System*, Birren, J. E., Imers, H. A., and Windle, W. F., Eds., Charles C Thomas, Springfield, Ill., 1959, 131.
21. **Baile, P. and von Bonin, G.**, The isacortex of man, in *Illinois Monographs in the Medical Science*, Vol. 6, University of Illinois Press, Urbana, 1951.
22. **Bondareff, W.**, Morphology of the aging nervous system, in *Handbook of Aging and the Individual. Psychological and Biological Aspects*, Birren, J. E., Ed., University of Chicago Press, 1959, 136.
23. **Hoff, H. and Sietelberger, F.**, Altersveranderrgen des menschlichem gelions, *Z. Alternforsch*, 10, 307, 1957.
24. **Sjorgren, T., Sjorgren, H., and Lindgren, A.**, Morbus alzheimer and morbus pick. A genetic and pathoanatomical study, *Acta Psychiatr. Neurol. Scand.*, Suppl. 82, 1952.
25. **Konigsmark, B. W. and Murphy, E. A.**, Neuronal populations in the human brain, *Nature (London)*, 299, 1335, 1970.
26. **Corsellis, J. A. N.**, Some observations on the Purkinje cell population and on brain volume in human aging, in *Aging*, Vol. 3, Terry, R. D. and Gershon, S., Eds., Raven Press, New York, 1976.
27. **Boiggee, K. R., Ordy, J. M., Hanshe, J., and Kaack, B.**, Quantitative assessment of changes in neuron and glia cell packing density of lipofusion accumulation with age in the cerebral cortex of a nonhuman primate *(Macaca mulatta)*, in *Aging*, Vol. 3, Terry, R. D. and Gershon, J., Eds., Raven Press, New York, 1976.
28. **Briggee, K. R., Ordy, J. M., and Kaack, B.**, Early appearance and regional differences in intraneuronal and extraneuronal lipofuscin accumulation with age in the brain of a nonhuman primate *(Macaca mulatta)*, *J. Gerontol.*, 29, 366, 1974.
29. **Hall, T. C., Miller, A. K. H., and Corsellis, J. A. N.**, Variations in the human Purkinje cell population according to age and sex, *Neuropathol. Appl. Neurobiol.*, 1, 267, 1975.

30. **Ellis, R. S.,** Norms for some structural changes in the human cerebellum from birth to old age, *J. Comp. Neurol.,* 32, 1, 1920.

31. **Harms, J. W.,** Altern und somatod der zellverbandstiere, *Z. Alternsforsch,* 5, 73, 1944.

32. **Andrew, W.,** The Purkinje cell in man from birth to senility, *Z. Zellforsch. Mikrosc. Anat.,* 28, 292, 1938.

33. **Inukai, T.,** On the loss of Purkinje cells with advancing age from the cerebella cortex of the albino rat, *J. Comp. Neurol.,* 45, 1, 1928.

34. **Brody, H.,** Aging of the vertebrate brain, in *Development and Aging in the Nervous System,* Rockstein, M. and Sussman, M. L., Eds., Academic Press, New York, 1973, 121.

35. **Brody, H.,** Organization of the cerebral cortex. III. A study of aging in the human cerebral cortex, *J. Comp. Neurol.,* 102, 511, 1955.

36. **Critchley, M.,** Ageing of the nervous system, in *Problems of Ageing,* Cowdroy, E. V., Ed., Williams & Wilkins, Baltimore, 1942, 518.

37. **Brody, H.,** Structural changes in the aging nervous system, in *The Regulatory Role of the Nervous System,* Blumenthal, H. T., Ed., S. Karger, Basel, 1970, 9.

38. **Cohon, E. J.,** The elderly brain. A quantitative analysis of the cerebral cortex in two cases, *Psychiatr. Neurol. Neurochir.,* 75, 261, 1972.

39. **Terry, R. D., Peck, A., De Teresa, R., Schechter, R., and Horoupian, D. S.,** *Ann. Neurol.,* 10, 184, 1981.

40. **Henderson, G., Tomlinson, B. E., and Gibson, P. H.,** Cell counts in human cerebral cortex in normal adults throughout life using an image analyzing computer, *J. Neurol Sci.,* 46, 113, 1980.

41. **Hanley, T.,** "Neuronal fall-out" in the ageing brain, *Age Ageing,* 3, 133, 1974.

42. **Hempel, K. J.,** Quantitative und topishe probleme der altersvorgäne in gehion (Quantitative and topical problems of cerebral aging), *Verk. Dtsch. Ges. Pathol.,* 52, 179, 1968.

43. **Davies, P. and Maloney, A. J. F.,** Selective loss of central cholinergic neurons in Alzheimer's disease, *Lancet,* 2, 1403, 1976.

44. **Hasan, M. and Glees, P.,** Ultrastructural age changes in hippocampal neurons, synapses and neuroglia, *Exp. Gerontol.,* 8, 75, 1973.

45. **Brizzee, K. R.,** Gross morphometric analyses and quantitative histology of the aging brain, in *Neurobiology of Aging,* Ordy, J. M. and Brizzee, Eds., Plenum Press, New York, 1975, 401.

46. **Tomlinson, B. E. and Henderson, G.,** Some quantitative cerebral findings in normal and demented old people, in *Aging,* Vol. 3, Terry, R. D. and Gershon, S., Eds., Raven Press, New York, 1976.

47. **Perry, E. K., Gibson, P. H., Blessed, G., Perry, R. H., and Tomlinson, B. E.,** Neurotransmitter enzyme abnormalities in senile dementia, *J. Neurol Sci.,* 34, 247, 1977.

48. **Bowen, D. M.,** Biochemical evidence for nerve cell changes in senile dementia, in *Aging,* Vol. 13, Annaducci, L., Davison, A. N., and Antvono, P., Eds., Raven Press, New York, 1980.

49. **Brizzee, K. R. and Ordy, J. M.,** Age pigments, cell loss and hippocampal function, *Mech. Ageing Dev.,* 9, 143, 1979.

50. **Vogt, C. and Vogt, O.,** *J. Psychiatr. Neurol.,* 50, 1, 1942.

51. **Bugiani, O., Salvarani, S., Perdelli, F., Mancardi, G., and Leonardi, A.,** Nerve cell loss with aging in the putamen, *Eur. Neurol.,* 17, 286, 1978.

52. **Vijayashankar, N. and Brody, H.,** The neuronal population of the nuclei of the trochlear nerve and the locus coeruleus in the human, *Anat. Rec.,* 172, 421, 1973.

53. **Brody, H.,** An examination of cerebral cortex and brainstem aging, in *Neurobiology of Aging,* Vol. 3, Terry, R. D. and Gershon, S., Eds., Raven Press, New York, 1976.

54. **Brody, H. and Vijayashankar, N.,** Neuronal loss in the human brainstem and its relation to sleep in the elderly, in *10th Int. Congr. Neurol., Abstracts,* 1975, Vol. 2.

55. **Vijayashankar, N. and Brody, H.,** A quantitative study of the pigmented neurons in the nuclei locus coeruleus and subcoeruleus in man as related to aging, *J. Neuropathol. Exp. Neurol.,* 38, 490, 1979.

56. **Coleman, P. D. and Goldman, G.,** Neuron counts in locus coeruleus of aging rat, in *Brain Neurotransmitters and Receptors in Aging and Age-Related Disorders,* Vol. 17, Enna, S. J., Samorajski, T., and Beer, B., Eds., Raven Press, New York, 1981.

57. **Maleci, O.,** Contributo all conoscenza delle variazioni quantitative delle cellule nervose nelle senescenza, *Arch. Ital. Anat. Estilos. Pathol.,* 33, 883, 1934.

58. **Vijayashankar, N. and Brody, H.,** A study of aging in the human abducens nucleus, *J. Comp. Neurol.,* 173, 433, 1977.

59. **van Buskirk, C.,** The seventh nerve complex, *J. Comp. Neurol.,* 82, 303, 1945.

60. **Konigsmark, B. W. and Murphy, E. A.,** Volume of ventral cochlear nucleus in man: Its relationship to neuronal population and age, *J. Neuropathol. Exp. Neurol.,* 31, 304, 1972.

61. **Vijayashankar, N. and Brody, H.,** Neuronal population of the human abducens nucleus, *Anat. Rec.,* 169, 447, 1971.

62. **Tomasch, J. and Ebnessajjade, D.,** The human nucleus ambiguus, *Anat. Rec.,* 141, 247, 1961.
53. **Tomasch, J. and Malpass, A. J.,** The human motor trigeminal nucleus. A quantitative study, *Anat. Rec.,* 130, 91, 1958.
64. **Moatamed, F.,** Cell frequencies in human inferior olivary complex, *J. Comp. Neurol.,* 128, 109, 1966.
65. **Monagle, R. D. and Brody, H.,** The effects of age upon the main nucleus of the inferior olivary in the human, *J. Comp. Neurol.,* 155, 51, 1974.
66. **Johnson, H. A. and Erner, S.,** Neuron survival in the aging mouse, *Exp. Gerontol.,* 7, 111, 1972.
67. **Delorenzi, E.,** Costanza numerica delle cellule del Purkinje in individui di varia età, *Boll. Soc. Ital. Biol. Sper.,* 6, 80, 1931.
68. **Delorenzi, E.,** Costanza numerica delle cellule di Purkinje del cervelletto dell' uomo in individui di varia età, *Z. Zellforsch.,* 14, 310, 1931.
69. **Wilcox, H. H.,** Structural changes in the nervous system relating to the process of aging, in *The Process of Aging in the Nervous System,* Birren, J. E., Innes, H. A., and Windle, W. F., Eds., Charles C Thomas, Springfield, Ill., 1959, chap. 2.
70. **Comfort, A.,** Neuromythology, *Nature (London),* 229, 1971.
71. **Klein, A. W. and Michel, M. E.,** A morphometric study of the neocortex of young adult and old image-differentiated rats, *Mech. Ageing Dev.,* 6, 441, 1977.
72. **Brizzee, K. R., Sherwood, N., and Timiras, P. S.,** A comparison of cell populations at various depth levels in cerebral cortex of young adult and aged Long-Evans rats, *J. Gerontol.,* 23, 289, 1968.
73. **Mervis, R.,** Cytomorphological alterations in the aging animal brain with emphasis on Golgi studies, in *Aging and Cell Structure,* Vol. 1, Johnson, J. E., Jr., Ed., Plenum Press, New York, 1981.
74. **Terry, R. D.,** Morphological changes in Alzheimer's disease-senile dementia: ultrastructural changes and quantitative studies, in *Congenital and Acquired Cognitive Disorders,* Raven Press, New York, 1979, 99.
75. **Cragg, B. G.,** The density of synapses and neurons in normal, mentally defective and aging human brains, *Brain,* 98, 81, 1975.
76. **Terry, R. D., Fitzgerald, C., Peck, R., Milner, J., and Farmer, P.,** Cortical cell counts in senile dementia, in Abstr. 53rd Annu. Meet. Am. Assoc. Neuropathologists, Chicago, 1977.
77. **Terry, R. D. and Davies, P.,** Dementia of the Alzheimer type, *Ann. Rev. Neurosci.,* 3, 77, 1980.
78. **Diamond, M. C. and Cannon, J. R., Jr.,** A search for the potential of the aging brain, in *Brain Neurotransmitters and Receptors in Aging and Age-Related Disorders,* Vol. 17, Enna, S. J. and Samorajski, T., Eds., Raven Press, New York, 1981.
79. **von Höpken, W.,** Das alteron des nucleus dentatus, *Z. Altersforsch,* 5, 256, 1951.
80. **Speigel, A.,** Ueber die degenerativen veranderungen in der kleinhionreinden im verlauf des individual-zyklus vom cavia cobaya marcgr, *Zool. Huz.,* 79, 173, 1928.
81. **Andrew, W. and Cardwell, C. S.,** Neuronophagia in the human cerebral cortex in senility and in pathologic conditions, *Arch. Pathol. (Lab. Med.),* 29, 400, 1940.
82. **Scheibel, M. and Scheibel, A.,** Structural changes in the aging brain, in *Aging,* Vol. 1, Brody, H., Harmon, D., and Ordy, J. M., Eds., Raven Press, New York, 1975.
83. **Conel, J. L.,** The postnatal development of the human cerebral cortex, in *The Cortex of the One-Month Infant,* Vol. 3, Harvard University Press, Cambridge, 1947.
84. **Buetow, D. E.,** Cellular content and cellular proliferation changes in the tissues and organs of the aging mammal, in *Cellular and Molecular Renewal in the Mammalian Body,* Cameron, I. J. and Thrasher, J. D., Eds., Academic Press, New York, 1971, 87.
85. **Corbin, K. B.,** Decrease in number of myelinated fibers in human spinal roots with age, *Anat. Rec.,* 68, 63, 1937.
86. **Gardner, E.,** Decrease in human neurons with age, *Anat. Rec.,* 77, 529, 1940.
87. **Duncan, D.,** A determination of the number of nerve fibers in the eighth thoracic and the largest lumbar and ventral roots of albino rat, *J. Comp. Neurol.,* 59, 47, 1934.
88. **Spencer, P. S. and Ochoa, J.,** The mammalian peripheral nervous system in old age, in *Aging and Cell Structure,* Vol. 1, Johnson, J. E., Ed., Plenum Press, New York, 1981.
89. **Kawamura, Y., Okazaki, H., O'Brien, P. C., and Dyck, P. J.,** Lumbar motoneurons of man. I. Number and diameter histogram of alpha and gamma axons of ventral root, *J. Neuropathol. Exp. Neurol.,* 36, 853, 1977.
90. **Takakashi, K.,** A clinicopathologic study on the peripheral nervous system of the aged. III. With special reference to the spinal nerve roots, *Clin. Neurol.,* 4, 151, 1964.
91. **Gilimore, S. A.,** Spinal nerve root degeneration in aging laboratory rats: A light microscopic study, *Anat. Rec.,* 174, 251, 1972.
92. **Moyer, E. K. and Kaliszewski, B. F.,** The number of nerve fibers in motor spinal nerve roots of young, mature, and aged cats, *Anat. Rec.,* 131, 681, 1958.
93. **Dunn, E. H.,** The influence of age, sex, weight, and the relationship upon the number of medullated fibers in the ventral root of two cervical nerve of albino rat, *J. Comp. Neurol.,* 22, 131, 1912.

94. **Rexel, B.,** Contribution to the knowledge of the postnatal development of the peripheral nervous system of man, *Acta Psychiatr. Neurol. Suppl.*, 33, 164, 1944.

95. **Scharf, J. H. and Blumenthal, H. J.,** Neuere aspekte zur altersabhangigen involution des sensiblen peripheren nervensystems, *Z. Zellforsch. Mikrosk. Anat.*, 78, 280, 1967.

96. **Nagashima, K. and Oata, K.,** A histopathological study of the human spinal ganglia. I. Normal variations in aging, *Acta Pathol. Jpn.*, 24, 333, 1974.

97. **Okita, M., Offord, K., and Dyck, P. J.,** Morphometric evaluation of first sacral ganglia of man, *J. Neurol. Sci.*, 22, 72, 1974.

98. **Emery, J. L. and Singhal, R.,** Changes associated with growth in the cells of the dorsal root ganglion in children, *Dev. Med. Child Neurol.*, 15, 460, 1973.

99. **Schnitzlein, H. N., Rowe, L. N., and Hoffmann, H.,** The myelinated component of the vagus nerve of man, *Anat. Rec.*, 131, 649, 1958.

100. **Cottrell, L.,** Histologic variations with age in apparently normal peripheral nerve trunks, *Arch. Neurol. Psychiatr.*, 43, 1138, 1940.

101. **Greenman, M. J.,** The number, size, and axis sheath relation of the large myelinated fibers in the peroneal nerve of inbred albino rat under normal conditions, in disease, and after stimulation, *J. Comp. Neurol.*, 27, 403, 1917.

102. **Duncan, D.,** The incidence of secondary (Wallerian) degeneration in normal mammals compared to that in certain experimental and disease conditions, *J. Comp. Neurol.*, 51, 197, 1930.

103. **Bruesh, S. R. and Arey, L. B.,** The number of myelinated and unmyelinated fibers in the optic nerve of vertebrates, *J. Comp. Neurol.*, 77, 631, 1942.

104. **Bergström, B.,** Morphology of the vestibular nerve. II. The number of myelinated vestibular nerve fibers in man at various ages, *Acta Otalargngol (Stockholm)*, 76, 173, 1973.

105. **Brown, W. F.,** A method for estimating the number of motor units in muscle and the changes in motor unit count with aging, *J. Neurol. Neurosurg. Psychiatr.*, 35, 845, 1972.

106. **Haynaard, M.,** Automatic analysis of the electromyogram in healthy subjects of different ages, *J. Neurol. Sci.*, 33, 397, 1977.

107. **Cottrell, L.,** Histologic variations with age in apparently normal nerve trunks of limbs, *Arch. Neurol. Psychiatr.*, 43, 1138, 1940.

108. **Kuntz, A.,** Histological variations in autonomic ganglia and ganglion cells associated with age, *Am. J. Pathol.*, 14, 783, 1938.

109. **Amprino, R.,** Modifications de la structure des neurones sympathetiques pendant l'accroisement et la senescence. Recherches sur le ganglion cervical supeoieur, *C. R. Assoc. Anat.*, 33, 3, 1938.

110. **Herman, H.,** Zusammenfassende ergebnisse über alteraveranderung am peripheren nervesystem, *Z. Alternsforsch*, 6, 197, 1952.

111. **Vandervael, F.,** Recherches sur l'evolution des neurones sympathetiques du ganglion cervical superieur chez l'homme, *Arch. Biol.*, 54, 53, 1943.

112. **Hoffman, H. H.,** The number of nerve fibers in the vagus nerve of man, *Anat. Rec.*, 139, 429, 1961.

113. **Birren, J. E. and Wall, P. D.,** Age changes in conduction velocity, refractory period, number of fibers, connective tissue space and blood vessels in sciatic nerve of rats, *J. Comp. Neurol.*, 104, 1, 1956.

114. **Sharma, A. K. and Thomas, P. K.,** Quantitative studies on age changes in unmyelinated nerve fibers in the vagus nerve in man, in *Studies on Neuromuscular Disease*, Kunze, K. and Desmedt, J. E., Eds., S. Karger, Basel, 1975.

115. **Dyck, P. J. et al.,** Reconstruction of motor, sensory, and autonomic neurons based on morphometric study of sampled levels, *Muscle Nerve*, 399, 1979.

116. **Wright, E. A. and Spink, J. M.,** A study of the loss of nerve cells in the central nervous system in relation to age, *Gerontologia*, 3, 277, 1959.

117. **Keithley, E. M. and Feldman, M. L.,** Spiral ganglion cell counts in age-graded series of rat cochleas, *J. Comp. Neurol.*, 188, 429, 1979.

118. **Otte, J., Schuknecht, H. F., and Kerr, A. G.,** Ganglion cell populations in normal and pathological human cochleae. Implications for cochlear implantation, *Laryngoscope*, 88, 1231, 1978.

119. **Suga, F. and Lindsay, J.,** Histopathological observations of presbycusis, *Ann. Otol. Rhinol. Laryngol.*, 85, 169, 1976.

120. **Schuknecht, H. F.,** Presbycusis, *Laryngoscope*, 65, 402, 1955.

121. **Johnsson, L. G. and Hawkins, J. E.,** Sensory and neural degeneration with age as seen in microdissections of the human inner ear, *Ann. Otol. Rhinol. Laryngol.*, 81, 179, 1972.

122. **Gredberg, G.,** Cellular pattern and nerve supply of the human organ of corti, *Acta Otol. Laryngol.* Suppl., 236, 1, 1968.

123. **Schuknecht, H. F.,** Further observations on the pathology of presbycusis, *Arch. Otolaryngol.*, 80, 369, 1964.

124. **Hansen, C. C. and Reske, E.,** Pathological studies in presbycusis, *Arch. Otolaryngol.*, 82, 115, 1965.

125. **Jorgensen, M. B.,** Changes of aging in the inner ear, *Arch. Otolaryngol.,* 74, 164, 1961.
126. **Malitskaya, G. A.,** Age changes in the trigeminal nerve system, *Vrach. Delo.,* 8, 123, 1971.
127. **Selby, G.,** Diseases of the fifth cranial nerve, in *Peripheral Neuropathy,* Dyke, P. J., Thomas, P. K., and Lambert, E. H., Eds., W. B. Saunders, Philadelphia, 1975, 533.
128. **Morrison, L. R.,** *The Effect of Advancing Age upon the Human Spinal Cord,* Harvard University Press, Cambridge, 1959.
129. **Lorenz, J.,** Observations compartives sur l'inneruation intramurale du cardia, du pylove et de la volvule ileo-coecale chez l'homme normal au cours de l'age, *Z. Mikrosk. Anat. Forsch.,* 68, 540, 1962.
130. **Cubells, J. F., Filburn, C. R., Roth, G., Engel, B. T., and Joseph, J. A.,** Specificity of age-related changes in striatal DA receptors: plasticity in the face of a deficit, *Abstr. Soc. Neurosci.,* 6, 739, 1980.
131. **Brownson, R. H.,** Perineuronal satellite cells in the motor cortex of the aging brains, *J. Neuropathol.,* 14, 424, 1955.
132. **Brownson, R. H.,** Perineuronal satellite cells in the motor cortex of the aging brains, *J. Neuropathol.,* 15, 190, 1956.
133. **Nurnberger, J. J. and Gordon, M. W.,** The cell density of neural tissues: direct counting method and possible applications as a biological reference, in *Progress in Neurobiology, II, Cytochemistry and Ultrastructure of Neural Tissue,* Korey, S. R. and Nurnberger, J., Eds., Paul B. Hoeber — Agathon Press, New York, 1957.
134. **Rasmussen, G. L.,** Discussion, in *The Process of Aging in the Nervous System,* Birren, J. E., Imus, H. A., and Windle, W. F., Eds., Charles C Thomas, Springfield, Ill., 1959, 136.
135. **Johnson, J. E. and Miquel, J.,** Fine structural changes in the lateral vestibular nucleus of aging rats, *Ageing Dev.,* 3, 203, 1974.
136. **Bourne, G.,** Recent discoveries concerning mitochondria and golgi apparatus and their significance in cellular physiology, *J. R. Microsc. Soc.,* 70, 368, 1950.
137. **Globus, A. and Scheibel, A. B.,** Pattern and field in cortical structure: The rabbit, *J. Comp. Neurol.,* 131, 155, 1967.
138. **Globus, A. and Scheibel, A. B.,** Loss of dendrite spines as an index of presynaptic terminal patterns, *Nature (London),* 212, 463, 1966.
139. **Globus, A. and Scheibel, A. B.,** Synaptic loci on parietal cortical neurons: terminations of corpus callosum fibers, *Science,* 156, 1127, 1967.
140. **Glick, R. and Bondareff, W.,** Loss of synapses in the cerebellar cortex of the senescent rat, *J. Gerontol.,* 36, 818, 1979.
141. **Artukhina, N. I.,** Electron microscopic study of aged changes of synapses in brain cortex in rats, *Tsitologiia,* 10, 1505, 1968.
142. **Geinisman, Y.,** Loss of axosomatic synapses in the dentaligyous of aged rats, *Brain Res.,* 168, 485, 1979.
143. **Davies, P.,** Neurotransmitter-related enzymes in senile dementia of the Alzheimer type, *Brain Res.,* 171, 319, 1979.
144. **Hinds, J. W. and McNelly, N. A.,** Correlation of aging changes in the olfactory epithelium and olfactory bulb in the rat, *Abstr. Soc. Neurosci.,* 6, 739, 1980.
145. **Cotman, C. W. and Scheff, S. W.,** Compensatory synapse growth in aged animals after neuronal death, *Mech. Ageing Dev.,* 9, 103, 1979.
146. **Connor, J. R., Beban, S. E., Hansen, B., Hopper, P., and Diamond, M. C.,** Dendritic increase in the aged rat somatosensory cortex, *Abstr. Soc. Neurosci.,* 6, 739, 1980.
147. **Rappoport, S. I. and London, E. D.,** Brain metabolism during aging of the rat and dog: implications for brain function in man during aging and dementia, in *Neural Aging and its Implications in Human Neurological Pathology,* Vol. 18, Terry, R. D., Bolis, C. L., and Toffano, G., Eds., Raven Press, New York, 1982.
148. **Scheibel, A. B. and Tomiyasu, U.,** Dendritic sprouting in Alzheimer's presenile dementia, *Exp. Neurol.,* 60, 1, 1978.
149. **Mehraein, P., Yamada, M., and Tarnowska-Dziduszko, E.,** Quantitative study on dendrites and dendritic spines in Alzheimer's disease and senile dementia, in *Advances in Neurobiology,* Kreutzberg, G. W., Ed., Raven Press, New York, 1975, 453.
150. **Mervis, R. F.,** Purkinje cell alterations in the cerebella of aged dogs, *Gerontol. Soc. Abstr.,* 19, 119, 1979.
151. **Feldman, M. L. and Dowd, C.,** Loss of dendritic spines in aging cerebral cortex, *Z. Anat. Entwichl. Gesch.,* 148, 279, 1975.
152. **Feldman, M. L. and Dowd, C.,** Aging in rat visual cortex: light microscopic observations on layer IV pyramidal apical dendrites, *Anat. Rec.,* 178, 355, 1974.
153. **Scheibel, M. E., Lindsay, R. D., Tormiyasu, Y., and Scheibel, A. B.,** Progressive dendritic changes in aging human cortex, *Exp. Neurol.,* 47, 392, 1975.

154. **Felman, M. L.,** Aging changes in the morphology of cortical dendrites, in *Aging,* Terry, R. D. and Gershon, S., Eds., Raven Press, New York, 1976, 211.

155. **Schulz, U. and Hunjiker, O.,** Comparative studies of neuronal perikaryon size and shape in the aging cerebral cortex, *J. Gerontol.,* 35, 483, 1980.

156. **Scheibel, M. E., Lindsay, R. D., Tomiyasu, U., and Scheibel, A. B.,** Progressive dendritic changes in the aging human limbic system, *Exp. Neurol.,* 53, 420, 1976.

157. **Mervis, R.,** Structural alterations in neurons of aged canine neocortex: A golgi study, *Exp. Neurol.,* 62, 417, 1978.

158. **Vaughan, D. W.,** Age-related deterioration of pyramidal cell basal dendrites in rat auditory cortex, *J. Comp. Neurol.,* 171, 501, 1977.

159. **Scheibel, A. B.,** The gerohistology of the aging human forebrain: some structuro-functional considerations, in *Brain Neurotransmitters and Receptors in Aging and Age-Related Disorders,* Enna, E. J., Samorajsk, T., and Beer, B., Eds., Raven Press, New York, 1981, 31.

160. **Scheibel, M. E. and Scheibel, A. B.,** Structural alterations in the aging brain, in *Aging: A Challenge to Science and Society,* Danon, D., Shock, N. W., and Morris, M., Eds., Oxford University Press, 1981, 4.

161. **Reise, W.,** The cerebral cortex in the very old human brain, *J. Neuropathol. Exp. Neurol.,* 5, 160, 1946.

162. **Scheibel, A. B.,** The hippocampus: organizational patterns in health and senescence, *Mech. Ageing Dev.,* 9, 89, 1979.

163. **Scheibel, A. B.,** Aging in human control systems, in *Sensory Systems and Communications in the Elderly,* Vol. 10, Raven Press, New York, 1979.

164. **Machado-Salas, J. P., Scheibel, M. E., and Scheibel, A. B.,** Morphological changes in the hyopthalamus of the old mouse, *Exp. Neurol.,* 57, 102, 1977.

165. **Scheibel, M. E., Lindsay, R. D., Tomigasu, U., and Scheibel, A. B.,** Progressive dendritic changes in aging human cortex, *Exp. Neurol.,* 47, 392, 1975.

166. **Valverde, F. and Estrella-Esteban, M.,** Peristriate cortex of mouse: Location and the effects of enucleation on the number of dendritic spines, *Brain Res.,* 9, 145, 1968.

167. **Ryugo, R., Ryugo, D. K., and Killackey, H.,** Differential effect of enucleation on two populations of layer V pyramidal cell, *Brain Res.,* 88, 554, 1975.

168. **Fifkova, E.,** The effect of unilateral deprivation on visual centers in rats, *J. Comp. Neurol.,* 140, 431, 1970.

169. **Rutledge, L. T., Duncan, J., and Cant, N.,** Long-term status of pyramidal cell axon collaterals and apical dendritic spines in denervated cortex, *Brain Res.,* 41, 249, 1972.

170. **Valverde, F.,** Rate and extent of recovery from dark rearing in the visual cortex of the mouse, *Brain Res.,* 33, 1, 1971.

171. **Hamida, C. B., de Pereda, G. R., and Hirsch, J. C.,** Les epines dendritiques du cortex de gyrus isole de chat, *Brain Res.,* 21, 313, 1970.

172. **Garey, L. J. and Powell, T. P. S.,** An experimental study of the termination of the lateral geniculocortical pathway in the cat and monkey, *Proc. R. Soc. London,* 179, 41, 1971.

173. **Globus, A. and Scheibel, A. B.,** The effect of visual deprivation on cortical neurons: A golgi study, *Exp. Neurol.,* 19, 331, 1967.

174. **Naranjo, N. and Greene, E.,** Use of reduced silver staining to show loss of connections in aged rat brain, *Brain Res.,* 2, 71, 1977.

175. **Bondareff, W.,** Synaptic atrophy in the senescent hippocampus, *Mech. Ageing Dev.,* 9, 163, 1979.

176. **Marin-Padilla, M.,** Structural abnormalities of the cerebral cortex in human chromosomal aberrations: A golgi study, *Brain Res.,* 44, 625, 1972.

177. **Marin-Padilla, M.,** Structural organization of the cerebral cortex (motor area) in human chromosomal aberrations: A golgi study, *Brain Res.,* 44, 375, 1974.

178. **Purpura, D. P.,** Dendritic spine "dysgenesis" and mental retardation, *Science,* 186, 1126, 1974.

179. **Gruner, J. E., Hirsch, J. C., and Sotelo, C.,** Ultrastructural features of the isolated suprasylvian gyrus in the cat, *J. Comp. Neurol.,* 154, 1, 1974.

180. **Uemura, E.,** Age-related changes in prefrontal cortex of *Macaca mulatta:* Synaptic density, *Exp. Neurol.,* 69, 164, 1980.

181. **Paula-Barbosa, M. M., Cardosa, R. M., Guimares, M. L., and Cruz, C.,** Dendritic degeneration and regrowth in the cerebral cortex of patients with Alzheimer's disease, *J. Neurol. Sci.,* 45, 129, 1980.

182. **Orlovskaya, D.,** The process of aging in light of clinical neuropathology, in *Neural Aging and its Implications in Human Neurological Pathology,* Terry, R. D., Bolis, C. L., and Toffano, G., Eds., Raven Press, New York, 1982, 33.

183. **Geinsiman, Y., Bondareff, W., and Dodge, J. T.,** Dendritic atrophy in the dentate gyrus of the senescent rat, *Am. J. Anat.,* 152, 321, 1978.

184. **Mervis, R. F., Terry, R. D., and Bowden, D.,** Morphological correlates of aging in the monkey brain — a light and electron microscope study, *Soc. Neurosci. Abstr.,* 5, 8, 1979.

185. **Rees, S.,** A quantitative electron microscopic study of the aging human cerebral cortex, *Acta Neuropathol.,* 36, 347, 1976.

186. **Rees, S.,** A quantitative electron microscopic study of atypical structures in normal human cerebral cortex, *Anat. Embryol.,* 148, 303, 1975.

187. **Sturrock, R. R.,** Quantitative and morphological changes in neurons and neuroglia in the indusium griseum of aging mice, *J. Gerontol.,* 32, 647, 1977.

188. **Mackado-Salas, J. P., Scheibel, M. E., and Scheibel, A. B.,** Neuronal changes in the aging mouse: Spinal cord and lower brainstems, *Exp. Neurol.,* 54, 504, 1977.

189. **Buell, S. J. and Coleman, P. D.,** Dendritic growth in the aged human brain and failure of growth in senile dementia, *Science,* 206, 854, 1979.

190. **Cupp, C. J. and Uemura, E.,** Age-related changes in prefrontal cortex of Macaca mulatta: Quantitative analysis of dendritic branching patterns, *Exp. Neurol.,* 69, 143, 1980.

191. **Connor, J. R., Jr., Diamond, M. C., and Johnson, R. E.,** Occipital cortical morphology of the rat: Alterations with age and environment, *Exp. Neurol.,* 68, 158, 1980.

192. **Connor, J. R., Jr., Diamond, M. C., and Johnson, R. E.,** Aging and environmental influences on two types of dendritic spines in the rat occipital cortex, *Exp. Neurol.,* 70, 371, 1980.

193. **Buell, S. J. and Coleman, P. D.,** Dendritic growth in the aged human brain and failure of growth in senile dementia, *Science,* 206, 854, 1979.

194. **Tascelles, R. G. and Thomas, P. K.,** Changes due to age in the internodal length on the sural nerve in man, *J. Neurol. Neurosurg. Psychiatr.,* 29, 40, 1966.

195. **Kemble, F.,** Conduction in the normal adult median nerve. The different effects of aging in men and woman, *Electromyography,* 7, 275, 1967.

196. **Asbury, A. and Johnson, P. C.,** *Pathology of Peripheral Nerve,* W. B. Saunders, Philadelphia, 1978.

197. **Arnold, N., Harriman, D. G., and Harriman, G. F.,** The incidence of abnormality in control human peripheral nerves studied by single axon dissection, *J. Neurol. Neurosurg. Psychiatr.,* 33, 55, 1970.

198. **Spritz, N., Singh, H., and Mariman, B.,** Decrease in myelin content of rabbit sciatic nerve with aging and diabetes, *Diabetes,* 24, 680, 1975.

199. **Griffiths, J. R. and Duncan, J. D.,** Age changes in the dorsal and ventral lumbar nerve roots of dogs, *Acta Neuropathol.,* 32, 75, 1975.

200. **Grover-Johnson, N. and Spencer, P. S.,** Peripheral nerve abnormalities in aging rats, *J. Neuropathol. Exp. Neurol.,* 40, 155, 1981.

201. **Wisniewski, H. M., Ghetti, B., and Terry, R. D.,** Neuritic (senile) plaques and filamentous changes in aged Rhesus monkeys,

202. **Johnson, J. E., Jr. and Miquel, J.,** Fine structure changes in the lateral vestibular nucleus of aging rats, *Mech. Ageing Dev.,* 3, 203, 1974.

203. **Sung, J. H.,** in Proc. 5th Int. Congr. Neuropathol., Zurich, 1966, 478.

204. **Jervis, G. A.,** Senile dementia, in *Pathology of the Nervous System,* Minckler, J., Ed., McGraw-Hill, New York, 1971, 1379.

205. **Sroka, C. H., Bornstein, Strulovici, N., and Sandbank, J.,** Neuroaxonal dystrophy: Its relation to age and central nervous system lesions, *Isr. J. Med. Sci.,* 5, 373, 1969.

206. **Fujisawa, K.,** A unique type of axonal alteration (so-called axonal dystrophy) as seen in Goll's nucleus in 277 cases of controls, *Acta Neuropathol.,* 8, 255, 1967.

207. **Brannon, W., McCormick, W., and Lampert, P.,** Axonal dystrophy in the gracile nucleus of man, *Acta Neuropathol.,* 9, 1, 1967.

208. **Fujisawa, K. and Shiraki, H.,** Study of axonal dystrophy. I. Pathology of the neurophil of the gracile and the cuneate nucleus in aging and old rats: A sterological study, *Neuropathol. Appl. Neurobiol.,* 4, 1, 1978.

209. **Farmer, P. M., Wisniewski, H. W., and Terry, R. D.,** Origin of dystrophic axons in the gracile nucleus, *J. Neuropathol. Exp. Neurol.,* 35, 366, 1976.

210. **Harriman, D. G., Taverner, D., and Woolf, A. L.,** Ekbom's syndrome and brunign parasthesiae. A biopsy study of vital staining and electron microscopy of the intramuscular innervation with a note on age changes in motor nerve endings in distal muscles, *Brain,* 93, 393, 1970.

211. **Swash, M. and Fox, K. P.,** The effect of age on human skeletal muscle. Studies of the morphology and innervation of muscle spindles, *J. Neurol. Sci.,* 16, 417, 1972.

212. **Woolf, A. L. and Coers, C.,** Pathological anatomy of the intramuscular nerve endings, in *Disorders of Voluntary Muscle,* Walton, J. N., Ed., Churchill Livingstone, Edinburgh, 1974, 274.

213. **Samorajski, T., Friede, R. L., and Ordy, J. M.,** Age differences in the ultrastructure of axons in the pyramidal tract of mouse, *J. Gerontol.,* 26, 542, 1971.

214. **Sharma, A. K., Bajada, S., and Thomas, P. K.,** Age changes in the tibial and plantar nerves of the rat, *J. Anat.,* 130, 417, 1980.

215. **Tokgi, H., Tsukagoski, H., and Toyokura, Y.,** Quantitative changes with age in normal sural nerves, *Acta Neuropathol.,* 38, 213, 1977.

216. **Tohgi, H.,** Quantitative variation with age in normal sural nerves, *Clin. Neurol.,* 12, 484, 1972.
217. **Lascelles, R. G. and Thomas, P. K.,** Changes due to age in internodal length in the sural nerve in man, *J. Neuropathol. Neurosurg. Psychiatr.,* 29, 40, 1966.
218. **Cragg, B. and Thomas, P. K.,** The conduction velocity of regenerated peripheral nerve, *J. Physiol.,* 171, 164, 1964.
219. **Ritchie, J. M.,** Some pathophysiological aspects of neuronal aging, in *Neural Aging and its Implications in Human Neurological Pathology,* Terry, R. D., Bolis, C. L., and Toffano, G., Eds., Raven Press, New York, 1982, 89.
220. **Vizosa, A. D.,** The relationship between internodal length and growth in human nerves, *J. Anat.,* 82, 110, 1950.
221. **Stevens, J. C., Lofgren, E. P., and Dyck, P. J.,** Histometric evaluation of branches of peroneal nerve: Technique for combined biopsy of muscle nerve and cutaneous nerve, *Brain Res.,* 52, 37, 1973.
222. **Schroder, J. M. and Gibbels, E.,** Marklose nervenfasen in serium and in spatstadium der thalidomid-polyneuropathue: quantitatives-elektromenmikro-skopische untersuchungen, *Acta Neuropathol.,* 39, 271, 1977.
223. **Ochao, J. and Mair, W. G. P.,** The normal sural nerve in man. II. Changes in the axons and Schwann cells due to aging, *Acta Neuropathol.,* 13, 217, 1969.
224. **Ochoa, J.,** Recognition of unmyelinated fiber disease: Morphologic criteria, *Muscle Nerve,* 1, 375, 1978.
225. **Dyck, P. J., Thomas, P. K., and Lamber, E.,** *Peripheral Neuropathy,* W. B. Saunders, Philadelphia, 1975.
226. **Downie, A. W. and Newell, D. J.,** Sensory nerve conduction in patients with diabetes mellitus and controls, *Neurology,* 11, 876, 1961.
227. **Buchthal, F. and Rosenfalck, A.,** Evoked action potentials and conduction nerves, *Brain Res.,* 3, 1, 1966.
228. **Norris, A. J., Shock, N. W., and Wagman, J. H.,** Age changes in the maximum conduction velocity of motor fibers of human ulnar nerves, *J. Appl. Physiol.,* 5, 589, 1953.
229. **Dorfman, L. J. and Bosley, T. M.,** Age-related changes in peripheral and central nerve conduction in man, *Neurology,* 29, 38, 1978.
230. **Chiu, S. Y. and Ritchie, J. M.,** Potassium channels in nodal and internodal axonal membrane of mammalian myelinated fibers, *Nature (London),* 284, 170, 1980.
231. **Mann, D. M. A. and Yates, P. O.,** Lipoprotein pigments — their relationship to aging in the human nervous system. I. The lipofuscin content of nerve cells, *Brain,* 97, 481, 1974.
232. **Dubbin, W. B.,** *Fundamentals of Neuropathology,* Charles C Thomas, Springfield, Ill., 1954.
233. **Andrew, W.,** *Cellular Changes with Age,* Charles C Thomas, Springfield, Ill., 1952.
234. **Dayan, A. D.,** Comparative neuropathology of aging studies on the brains of 47 species of vertebrate, *Brain,* 94, 31, 1971.
235. **Strehler, B. L. and Barrows, C. H.,** Senescence: Cell biological aspects of aging, in *Cell Differentiation,* Schjeide, O. A. and de Vallis, J., Eds., Van Nostrand Reinhold, New York, 1970.
236. **Timiras, P. S.,** Degenerative changes in cells, in *Developmental Physiology and Aging,* Macmillan, New York, 1972, 429.
237. **Siakotus, A. N. and Armstrong, D.,** Age pigment: A biochemical indicator of intracellular aging, in *Neurobiology of Aging,* Ordy, J. M. and Brizzee, K. R., Eds., Plenum Press, New York, 1975, 369.
238. **Brizzee, K. R., Cancilla, P. A., Sherwood, N., and Timiras, P. S.,** The amount and distribution of pigments in neurons and glia of the cerebral cortex. Autofluorescent and ultrastructural studies, *J. Gerontol.,* 24, 127, 1969.
239. **Porta, E. A. and Hartroft, W. S.,** Lipid pigments in relation to aging and dietary factors (lipofuscin), in *Pigments in Pathology,* Wolman, M., Ed., Academic Press, New York, 1969, 191.
240. **Toth, S. E.,** The origin of lipofuscin age pigments, *Exp. Gerontol.,* 3, 19, 1968.
241. **Bourne, G. H.,** Lipoprotein, in *Neurobiological Aspects of Maturation and Aging, Progress in Brain Research,* Ford, D. H., Ed., Elsevier, Amsterdam, 1973, 187.
242. **Hasan, M. and Glees, P.,** Genesis and possible dissolution of neuronal lipofuscin, *Gerontology,* 18, 217, 1972.
243. **Nandy, K.,** Further studies on the effect of centrophenoxine on the lipofuscin in neurons of senile guinea pigs, *J. Gerontol.,* 23, 82, 1968.
244. **Whiteford, R. and Getty, R.,** Distribution of lipofuscin in the canine and porcine brain as related to aging, *J. Gerontol.,* 21, 31, 1966.
245. **Altshul, R.,** Uber das sogenannte "Alterspigment" der nervenzellen, *Virchows Arch.,* 301. 273, 1938.
246. **Hunziker, O. and Samir, A.,** The aging human cerebral cortex: a sterological characterization of changes in the capillary net, *J. Gerontol.,* 34, 345, 1979.
247. **Braak, H.,** Uber das neurolipofuscin in der unteren olive und dem nucleus dentatus cerebelli in gehirn des menschen, *Z. Zellforsch. Mikrosk. Anat.,* 121, 573, 1971.

248. **Brizzee, K. R. and Johnson, F. A.,** Depth distribution of lipofuscin pigment in cerebral cortex of albino rat, *Acta Neuropathol.,* 16, 205, 1970.

249. **Kuhlenbeck, H.,** Some histologic age changes in the rat's brain and their relationship to comparable changes in the human brain, *Confinia Neurol.,* 14, 329, 1954.

250. **Reichel, W. S., Hollander, J., Clark, J. H., and Strehler, B. L.,** Lipofuscin pigment accumulation as a function of age and distribution in rodent brain, *J. Gerontol.,* 23, 71, 1968.

251. **Wolf, A.,** Clinical neuropathology in relation to the process of aging, in *The Process of Aging in the Nervous System,* Charles C Thomas, Springfield, Ill., 1959, 175.

252. **Hermann, H.,** Zusammenfassende ergebuisse uber altersreeranderungen an peripheren nervenasystems, *Z. Alternsforsch,* 6, 197, 1952.

253. **Bondareff, W.,** Genesis of intracellular pigment in the spinal ganglia of senile rats. An electron microscope study, *J. Gerontol.,* 12, 364, 1957.

254. **Sulkin, N.,** Histochemical studies of the pigments in human autonomic ganglion cells, *J. Gerontol.,* 8, 435, 1953.

255. **Kott, J., Schaumburg, H., Spencer, P., Ellenberg, M., and Jacobson, J.,** Lumbar sympathectomy in diabetic and arteriosclerotic subjects. Clinical and histopathologic studies, *Diabetes* (Suppl.), 1, 289, 1973.

256. **Truex, R. C.,** Morphological alterations in the Gasserian ganglion cells and their association with senescence in man, *Am. J. Pathol.,* 16, 255, 1940.

257. **Hess, A.,** The fine structure of young and old spinal ganglia, *Anat. Rec.,* 123, 399, 1955.

258. **Vogt, C. and Vogt, O.,** Aging of nerve cells, *Nature (London),* 158, 304, 1946.

259. **Wahren, W.,** Neurohistologischer beitrag zu fragen des alterns, *Z. Alternsforsch,* 10, 343, 1957.

260. **Samorajski, T., Keefe, J. R., and Ordy, J. M.,** Intracellular localization of lipofuscin age pigments in the nervous system of aged mice, *J. Gerontol.,* 19, 262, 1964.

261. **von Buttlar-Brentano, K.,** Zur lebensgeschichte des nucleus basalis, tuberomammillaris, supraopticus und paraventricularis unternormalen und pathogenen bedingungen, *J. Hiruforsch.,* 1, 337, 1954.

262. **Treff, W. M.,** Das involutionsmuster des nucleus dentatus cerebelli, in *Altern,* Schattauer, Stuttgart, 1974, 37.

263. **Sanides, F.,** Untersuchungen uber die histologische stouktus des mandelkerngebietes. I. Mittellung cytologie und involution des asuggdaleum profundum, *J. Hiruforsch.,* 3, 56, 1957.

264. **Brody, H.,** The deposition of aging pigment in the human cerebral cortex, *J. Gerontol.,* 15, 258, 1960.

265. **Balthasar, K.,** Lebensgeschichite der vier grossten pyramidenzellarten in der V Schicht der menschlicehn area giganto-pyramidalis, *J. Hiruforsch.,* 1, 281, 1954.

266. **Matzdorff, P.,** *Grundlagen zur Erforschung des Alterns,* Steinkopff, Frankfurt, 1948.

267. **Wiinscher, W.,** Die anatomie des alten gehirons, *Z. Alternsforsch,* 11, 60, 1957.

268. **Samorajski, T., Ordy, J. M., and Rady-Reimer, P.,** Lipofuscin pigment accumulation in the nervous system of aging mice, *Anat. Rec.,* 160, 555, 1968.

269. **Barden, H.,** Relationship of golgi chiamine-pyrophosphatase and lysosomal acid phosphatase to neuro-melanin and lipofuscin in cerebral neurons of the aging rhesus monkey, *J. Neuropathol.,* 29, 225, 1970.

270. **Pallis, C. A., Duckett, S., and Pearse, A. G. E.,** Diffuse lipofuscinosis of the central nervous system, *Neurology,* 17, 381, 1967.

271. **Bounk, U. and Ericsson, J. L. E.,** Electron microscopical studies on the rat brain neurons. Localization of acid phosphatase and mode of formation of lipofuscin bodies, *J. Ultrastruct. Res.,* 38, 1, 1972.

272. **Samorajski, T., Ordy, J. M., and Keefe, J. R.,** The fine structure of lipofuscin age pigment in the nervous system of aged mice, *J. Cell Biol.,* 26, 779, 1965.

273. **Chang, H-T. and Pao, X.,** Lipofuscin pigment formation in cultured neurons, in *Neural Aging and its Implications in Human Neurological Pathology,* Terry, R. D., Bolis, C. L., and Toffano, G., Eds., Raven Press, New York, 1972, 23.

274. **Chu, L. W.,** A cytological study of anterior horn cells isolated from human spinal cord, *J. Comp. Neurol.,* 100, 58, 1954.

275. **Koenig, H.,** Histochemical study of lysosomes and lipofuscin granules in the nervous system, *Anat. Rec.,* 148, 303, 1964.

276. **Schneider, H., Rehnberg, S. G., and Baer, M. P.,** Aging of neurons in culture, in *The Aging Brain and Senile Dementa,* Nandy, K. and Sherwin, J., Eds., Plenum Press, New York, 1976, 157.

277. **Miquel, J., Economos, A. C., Fleming, J., and Johnson, J. E., Jr.,** Mitochondrial role in cell aging, *Exp. Gerontol.,* 15, 575, 1980.

278. **Spoerri, P. E. and Glees, P.,** Neuronal aging in cultures: An electronmicroscopic study, *Exp. Gerontol.,* 8, 259, 1973.

279. **Miquel, J.,** Aging of male Drosophila melanogaster: Histological, histochemical and ultrastructural observations, *Adv. Gerontol. Res.,* 3, 39, 1971.

280. **Dalton, A. and Felix, M.,** A comparative study of the golgi complex, *J. Biophys. Biochem. Cytol.,* 2, 79, 1956.

281. **Gatenby, J. B.,** The golgi apparatus of the living sympathetic ganglion cells of the mouse, photographed by phase contrast microscopy, *J. R. Microsc. Soc.*, 73, 61, 1953.

282. **Gatenby, J. B. and Moussa, T.,** The sympathetic ganglion cell, with Sudan black and the Zernike microscope, *J. R. Microsc. Soc.*, 70, 342, 1950.

283. **Moussa, T.,** Senility changes in the spinal ganglion neurons of the toad, Bufo regularis, *Nature (London)*, 170, 206, 1952.

284. **Malhotra, S.,** The cytoplasmic inclusions of the aging neurons, of the frog, Rana tigrina, *La Cellule*, 58, 363, 1957.

285. **Gatenby, J. and Moussa, T.,** The neuron of the human autonomic system and the so-called senility pigment, *J. Physiol.*, 114, 252, 1951.

286. **Thomas, O.,** A study of the spheroid system of sympathetic neurons with special reference to the problem of neurosecretion, *Quant. J. Microsc. Sci.*, 89, 33, 1948.

287. **Murray, M. and Stout, A.,** Adult human sympathetic ganglion cells cultivated in vitro, *Am. J. Anat.*, 80, 225, 1947.

288. **Zeman, W.,** The neuronal ceroid lipofuscinosis — Batten Vogt syndrome. A model for human aging?, in *Advances in Gerontological Research*, Strehler, B. L., Ed., Academic Press, New York, 1971, 147.

289. **Burger, M.,** *Altern und Krankheit*, 2nd ed., Georg Thieme, Verlag, Stuttgart, 1952.

290. **Beregi, E.,** The significance of lipofuscin in the aging process, especially in the neurons, in *Neural Aging and its Implications in Human Neurological Pathology*, Terry, R. D., Bolis, C. L., and Toffano, G., Eds., Raven Press, New York, 1982, 15.

291. **Tappel, A. L.,** Biological antioxidant protection against lipid peroxidation damage, *Am. J. Clin. Nutr.*, 23, 1137, 1970.

292. **Hartroft, W. S. and Porta, E. A.,** Observation and interpretation of lipid pigments (lipofuscins) in the pathology of laboratory animals, *CRC Crit. Rev. Toxicol.*, 1, 379, 1972.

293. **Solovyova, Z. V. and Orbovskaya, D. D.,** Lipofuscin in the embryo brain cells in the early stages of the development (problems) of brain pathology in schizopherenia), *Zh. Neuropatol. Psikhiatr.*, 77, 1040, 1977.

294. **Nosal, G.,** Neuronal involution during aging, ultrastructural study in the rat cerebellum, *Mech. Ageing Dev.*, 10, 295, 1979.

295. **Hollander, C. F.,** Model systems in experimental gerontology, *Med. T. Gerontol.*, 1, 144, 1970.

296. **Ferrendelli, J. A., Sedgwick, W. G., and Suntzeff, V.,** Regional energy metabolism and lipofuscin accumulation in the mouse brain during aging, *J. Neuropathol. Exp. Neurol.*, 30, 638, 1971.

297. **Kormendy, C. G. and Bender, A. D.,** Chemical interference with aging, *Gerontologia*, 17, 52, 1971.

298. **Siakatos, A. N. and Koppang, N.,** Procedures for the isolation of lipopigments from brain, heart, and liver, and their properties, *Mech. Age Ageing*, 2, 177, 1973.

299. **Nishioka, N., Takakata, N., and Iizuka, R.,** Histochemical studies on the lipopigments in the nerve cells. A comparison with lipofuscin and ceroid pigment, *Adv. Neuropathol.*, 11, 174, 1968.

300. **Strehler, B. L.,** *Time, Cells and Aging*, Academic Press, New York, 1962.

301. **Siebert, G., Heidenreich, O., Bohring, R., and Lang, K.,** Isolierung und chemische untersuchung von lipofuscin, *Naturasenschaften*, 42, 156, 1955.

302. **Bjorkerud, S.,** Studies of lipofuscin granules of human cardiac muscle. II. Chemical analysis of isolated granules, *Exp. Mol. Pathol.*, 3, 377, 1964.

303. **Hendley, D. D., Mildvan, A. S., Reporter, M. C., and Strehler, B. L.,** The properties of isolated human cardiac age pigment. II. Chemical and enzymatic properties. *J. Gerontol.*, 18, 250, 1963.

304. **Dixon, K. C. and Herbertson, B. M.,** Cytoplasmic constituent of brain, *J. Physiol. (London)*, 111, 244, 1950.

305. **Woolf, A. and Pappenheimer, A. M.,** Occurrence and distribution of acid fast pigment in the central nervous system, *J. Neuropathol. Exp. Neurol.*, 4, 402, 1945.

306. **Sulkin, N.,** Properties and distribution of PAS-positive substances in the nervous system of the senile dog, *J. Gerontol.*, 10, 135, 1955.

307. **Nandy, K. and Schneider, H.,** Lipofuscin pigment formation in neuroblastoma cells in culture, in *Aging*, Terry, R. D. and Gershon, S., Eds., Raven Press, New York, 1976, 245.

308. **Nandy, K.,** Properties of neuronal lipofuscin pigment in mice, *Acta Neuropathol. (Berlin)*, 19, 25, 1971.

309. **Gedigk, P. and Bontke, E.,** Uber den Nachweis von hydrolytichen enzymen in lipopigmenten, *Z. Zellforsch.*, 55, 495, 1956.

310. **Hyden, H. and Lindstrom, B.,** Microspectragraphic studies on the yellow pigment in nerve cells, *Discuss. Faraday Soc.*, 9, 436, 1950.

311. **D'Angelo, C., Issidorides, M., and Shanklin, W.,** A comparative study of the staining reactions of granules in the human neuron, *J. Comp. Neurol.*, 106, 487, 1956.

312. **Lillie, R.,** A Nile blue staining technique for the differentation of melanin and lipofuscins, *Stain Technol.*, 31, 151, 1956.

313. **Heidenreich, O. and Siebert, G.,** Untersuchungen an isolierterm, unverandertem lipofuscin aus herzmuskulatur, *Virchows Arch. Pathol. Anat.,* 327, 112, 1955.

314. **Pearse, A. G. E.,** *Histochemistry, Theoretical and Applied,* Churchill, London, 1960.

315. **Bethe, A. and Fluck, M.,** Uber das gelbe Pigment der Ganglienzellen, seine Kolloidchenischen und Topographischen Beziehungen zu andern Zellstrukturen und eine elective Methode zu einer Darstellung, *Z. Zellforsch. Mikrosk. Anat.,* 27, 211, 1937.

316. **Sulkin, N. M. and Srivani, P.,** The experimental production of senile pigment in nerve cells of young rats, *J. Gerontol.,* 15, 2, 1960.

317. **Sulkin, N.,** The occurrence and duration of "senile" pigments experimentally induced in the nerve cells of the young rat, *Anat. Rec.,* 130, 377, 1958.

318. **Nandy, K. and Bourne, G. H.,** Effects of centrophenoxine on the lipofuscin pigment in the neurons of senile guinea pigs, *Nature (London),* 210, 313, 1966.

319. **Meier, C. and Glees, P.,** Effect of centrophenoxine on the old age pigment in satellite cells and neurons of senile rat spinal ganglia, *Acta Neuropathol.,* 17, 310, 1971.

320. **Tappel, A. L.,** Will antioxidant nutrients slow aging process?, *Geriatrics,* 23, 97, 1968.

321. **Tappel, A. L.,** Biological antioxidant protection against lipid peroxidation damage, *Am. J. Clin. Nutr.,* 23, 1137, 1970.

322. **Nandy, K.,** Histological and histochemical study of motor neurons with special reference to experimental degeneration, aging and drug actions, in *Motor Neuron Disease,* Norris, F. H. and Kurland, L. T., Eds., Grune & Stratton, New York, 1969, 319.

323. **Gonatas, S. H., Evangetista, I., and Baird, H. W.,** Cytoplasmic inclusions in juvenile amaurotic idiocy, *J. Pediatr.,* 75, 796, 1969.

324. **Strehler, B. L.,** On the histochemistry and ultrastructure of age pigment, in *Advances in Gerontological Research,* Strehler, B. L., Ed., Academic Press, New York, 1964, 343.

325. **Miyigashi, T., Takashata, M., and Iizuka, R.,** Electronmicroscopy studies on the lipopigments in the cerebral cortex nerve cells of senile and vitamin E-deficient rats, *Acta Neuropathol.,* 9, 7, 1971.

326. **Essner, E. and Novikoff, A. B.,** Human hepatocellular pigments and lysosomes, *J. Ultrastruct. Res.,* 3, 374, 1960.

327. **Wisniewski, H. M. and Terry, R. D.,** Morphology of the aging brain, human and animal, in *Progress in Brain Research,* Ford, D. H., Ed., Elsevier, Amsterdam, 1973, 167.

328. **Sekhon, S. S. and Maxwell, D. S.,** Ultrastructural changes in neurons of the spinal anterior horn of aging mice with particular reference to the accumulation of lipofuscin pigment, *J. Neurocytol.,* 3, 59, 1974.

329. **Duncan, D., Nall, D., and Morales, R.,** Observations of the fine structure of old age pigment, *J. Gerontol.,* 15, 366, 1960.

330. **Brunk, U., Ericsson, J., Pouter, J., and Westermark, B.,** Residual bodies and "aging" in cultured human glial cells, *Exp. Cell Res.,* 79, 1, 1973.

331. **Iqbal, K., Gundlke-Iqbal, I., Wisniewski, H. M., and Terry, R. D.,** Chemical relationship of the paired helical filaments of Alzheimer's dementia to normal human neurofilaments and neurotubules, *Brain Res.,* 142, 321, 1978.

332. **Terry, R. D. and Wisniewski, H. M.,** The ultrastructure of the neurofibrillary tangle and the senile plaque, in *Alzheimer's Disease and Related Conditions,* Wolstenholme, G. E. W. and O'Connor, M., Eds., Churchill, London, 1970.

333. **Wisniewski, H. M., Narang, H. K., and Terry, R. D.,** Neurofibrillary tangles of paired helical filaments, *J. Neurol. Sci.,* 27, 173, 1976.

334. **Kidd, M.,** Alzheimer's disease — an electron microscopical study, *Brain,* 87, 307, 1064.

335. **Terry, R. D., Gonatas, N. K., and Weiss, M.,** Ultrastructural studies in Alzheimer's presenile dementia, *Am. J. Pathol.,* 44, 269, 1964.

336. **Kidd, M.,** Paired helical filaments in electron microscopy in Alzheimer's disease, *Nature (London),* 197, 192, 1963.

337. **Terry, R. D.,** The fine structure of neurofibrillary tangles in Alzheimer's disease, *J. Neuropathol. Exp. Neurol.,* 22, 629, 1963.

338. **Wisniewski, H. M., Terry, R. D., and Hirano, A.,** Neurofibrillary pathology, *J. Neuropathol. Exp. Neurol.,* 29, 163, 1970.

339. **Iqbal, K., Wisniewski, H. M., Shelanski, M. L., Brostoff, S., Liwniez, B. H., and Terry, R. D.,** Protein changes in senile dementia, *Brain Res.,* 77, 337, 1974.

340. **Wisniewski, H. M., Korthals, J. K., and Terry, R. D.,** A new protein in neurons with neurofibrillary degeneration (Abstr. 7), in 51st Annu. Meet., *Am. Assoc. Neuropathol.,* 1975.

341. **DeBoni, U. and Crapper, D. R.,** Paired helical filaments of the Alzheimer type in cultured neurons, *Nature (London),* 271, 566, 1978.

342. **Wisniewski, H. M., Sinatra, R. S., Iqbal, K., and Gundke-Iqbal, I.,** in *Aging and Cell Structure,* Johnson, J. E., Jr., Ed., Plenum Press, New York, 1981, 105.

343. **Ball, M. J.,** Topographic distribution of neurofibrillary tangles and granulovacuolar degeneration in hippocampal cortex of aging and demented patients, *Acta Neuropathol. (Berlin),* 42, 73, 1978.

344. **Constantinidis, J. and Tissot, R.,** Degenerative encephalopathies in old age: neurotransmitters and zinc metabolism, in *Neural Aging and its Implications in Human Neurological Pathology,* Terry, R. D., Bolis, C. L., and Toffano, G., Eds., Raven Press, New York, 1982, 53.

345. **Constantinidis, J.,** Is Alzheimer's disease a major form of senile dementia? Clinical, anatomical and genetic data, in *Aging, Vol. 7,* Katzman, R., Terry, R. D., and Bick, K. L., Eds., Raven Press, New York, 1978, 15.

346. **Constantinidis, J., Richard, J., and de Hjuriaguerra, J.,** Dementias with senile plaques and neurofibrillary tangles, in *Studies in Geriatric Psychiatry,* Isaaks, A. D. and Post, F., Eds., John Wiley & Sons, London, 1978, 119.

347. **Terry, R. D.,** Brain disease in aging, especially senile dementia, in *Neural Aging and its Implications in Human Neurological Pathology,* Terry, R. D., Bolis, C. L., and Toffano, G., Eds., Raven Press, New York, 1982, 43.

348. **Tomlinson, B. E., Blessed, G., and Roth, M.,** Observations on the brains of non-demented old people, *J. Neuropathol. Sci.,* 7, 331, 1968.

349. **Hirano, A. and Zimmermann, H. M.,** Alzheimer's neurofibrillary changes. A topographic study, *Arch. Neurol.,* 7, 227, 1962.

350. **Blessed, G., Tomlinson, B. E., and Roth, M.,** The association between quantitative measures of dementia and of senile changes in the cerebral grey matter of elderly subjects, *Br. J. Psychiatr.,* 114, 797, 1968.

351. **Woodard, J. S.,** Alzheimer's disease in late adult life, *Am. J. Pathol.,* 49, 1157, 1966.

352. **Iqbal, K., Wisniewski, H. M., Grundke-Iqbal, I., and Terry, R. D.,** Neurofibrillary pathology, in *The Aging Brain and Senile Dementia; Advances in Behavioral Biology,* Nandy, K. and Sherwin, I., Eds., Plenum Press, New York, 1977, 218.

353. **von Braunmuhl, A.,** Alterserkrankungen des zentralnervensystems. Senile Involution. Senile demenz. Alsheimersche Krankheit, in *Handbuch der Speziellen Pathologischen Anatomie und Histologie XIII: Erster Teil, Bandteil A., Erkrankungen des zentralen Nervensystems,* Lubarsch, O., Henke, F., and Rossle, R., Eds., Springer-Verlag, Berlin, 1957, 337.

354. **Divry, P.,** De la nature de'alteration fibrillaire d' Alzheimer, *J. Belge Neurol. Psychiatr.,* 34, 197, 1934.

355. **Alexander, L. and Looney, J. M.,** Physiochemical properties of brain, especially in senile dementia and cerebral edema. Differential ratio of skull capacity to volume, specific weight, water content, water-binding capacity and pH of the brain, *Arch. Neurol. Psychiatr.,* 40, 877, 1938.

356. **Yase, Y.,** The basic process of amyotrophic lateral sclerosis as reflected in Kii Peninsula and Guam, in *Excerpta Medica Int. Congr. Series, Neurology, Proc. 11th World Congr. Neurobiology,* Excerpta Medica, Amsterdam, 1977, 413.

357. **Mandybur, T. J., Nagpaul, A. S., Pappas, Z., and Nikolwitz, W. J.,** Alzheimer neurofibrillary changes in subacute sclerosing panencephalitis, *Ann. Neurol.,* 1, 103, 1977.

358. **Hadfield, M. G., Martinez, A. H., and Gilmartin, R. C.,** Progressive multifocal leukoencephalopathy with paramyxovirus-like structures. Hiranobodies and neurofibrillary tangles, *Acta Neuropathol. (Berlin),* 27, 227, 1974.

359. **Hallervorden, J.,** Zur pathogeneses des post-encephalitischen Parkinsonismus, *Klin. Wochenschr.,* 12, 692, 1933.

360. **Horoupian, D. S. and Yang, S. S.,** Paired helical filaments in neurovisceral lipidosis (juvenile dystonic lipidosis), *Ann. Neurol.,* 4, 404, 1978.

361. **Mandybur, T. I., Nagpaul, A. S., Pappas, Z., and Niklowitz, W. J.,** Alzheimer's neurofibrillary change in subacute sclerosing panencephalitis, *Ann. Neurol.,* 1, 103, 1977.

362. **Kurland, L. T. and Zimmerman, H. M.,** The fine structure of some intraganglionic alterations, *J. Neuropathol. Exp. Neurol.,* 27, 167, 1968.

363. **Burger, P. C. and Vogel, F. S.,** The development of the pathologic changes of Alzheimer's disease and senile dementia in patients with Down's syndrome, *Am. J. Pathol.,* 73, 457, 1973.

364. **Alexander, L. and Looney, J. M.,** Histologic changes in senile dementia and related conditions; studied by silver impregnation and microincineration, *Arch. Neurol. Psychiatr.,* 40, 1075, 1938.

365. **Rothschild, D.,** Pathologic changes in senile psychoses and their psychobiologic significance, *Am. J. Psychiatr.,* 93, 757, 1937.

366. **Crapper, D. R. and Tomko, G. J.,** Neuronal correlates of an encephalopathy induced by aluminum neurofibrillary degeneration, *Brain Res.,* 97, 253, 1975.

367. **Crapper, D. R., Krishnan, S. S., and Dalton, A. J.,** Brain aluminum distribution in Alzheimer's disease and experimental neurofibrillary degeneration, *Science,* 180, 511, 1973.

368. **Wisniewski, H., Narkiewicz, O., and Wisniewska, K.,** Topography and dynamics of neurofibrillary degeneration in aluminium encephalopathy, *Acta Neuropathol.,* 9, 127, 1967.

369. **Crapper, D. R., Krishnan, S. S., De Boni, U., and Tomko, G. J.,** Aluminum: A possibly neurotoxic agent in Alzheimer's disease, *Arch. Neurol. (Abstr.),* 32, 356, 1975.

370. **Crapper, D. R. and Dalton, A. J.,** Aluminum induced neurofibrillary degeneration, brain electrical action and alterations in aquisition and retention, *Physiol. Behav.,* 10, 935, 1973.

371. **Volk, B.,** Paired helical filaments in rat spinal ganglia following chronic alcohol administration: an electronmicroscopic investigation, *Neuropathol. Appl. Neurobiol.,* 6, 143, 1980.

372. **Crapper, D. R., Lachlan, M. C., and De Boni, U.,** Models for the study of pathological neural aging, in *Neural Aging and its Implications in Human Neurological Pathology,* Terry, R. D., Bolis, C. L., and Toffano, G., Eds., Raven Press, New York, 1982, 61.

373. **Wisniewski, H. and Terry, R. D.,** An experimental approach to the morphogenesis of neurofibrillary degeneration and the argyrophilic plaque, in *Alzheimer's Disease and Related Conditions,* Ciba Foundation Symp., Wolstenholme, G. and O'Connor, M., Eds., Churchill, London, 1970, 223.

374. **Farnell, B. J., De Boni, U., and Crapper, D. R.,** Aluminum neurotoxicity in the absence of neurofibrillary degeneration in CA-1 hippocampal pyramidal neurons in vivo, *Neurosci. Abstr.,* 6, 248, 1980.

375. **Wisniewski, H. M., Sturman, J. A., and Shek, J. W.,** Aluminum chloride induced neurofibrillary changes in the developing rabbit: A chronic animal modal, *Ann. Neurol.,* 8, 479, 1980.

376. **Volk, B.,** Paired helical filaments in rat spinal ganglia following chronic alcohol administration: an electron microscopic investigation, *Neuropathol. Appl. Neurobiol.,* 6, 143, 1980.

377. **Wisniewski, H. M. and Soifer, D.,** Neurofibrillary pathology: current status and research perspectives, *Mech. Ageing Dev.,* 9, 119, 1980.

378. **Ghetti, B.,** Induction of neurofibrillary degeneration following treatment with marytansine in vivo, *Brain Res.,* 163, 9, 1979.

379. **Coleman, G. I., Barthold, S. W., Osbaldiston, G. W., Foste, S. J., and Jonas, A. M.,** Pathological changes during aging in barrier-reared Fischer 344 male rats, *J. Gerontol.,* 32, 258, 1977.

380. **Wisniewski, H. M., Johnson, A. B., Raine, C. S., Kay, W. J., and Terry, R. D.,** Senile plaques and cerebral amyloidosis in aged dogs. A histochemical and ultrastructural study, *Lab. Invest.,* 23, 287, 1970.

381. **Wisniewski, H. M. and Terry, R. D.,** Re-examination of the pathogenesis of the senile plaque, in *Progress in Neuropathology,* Zimmerman, H., Ed., Grune & Stratton, New York, 1973, 1.

382. **Tomlinson, B. E.,** Morphological changes and dementia in old age, in *Aging and Dementia,* Lynn-Smith, W. and Kinsbourne, M., Eds., Spectrum, New York, 1977, 31.

383. **Matsuyama, H., Namiski, H., and Watanabe, I.,** Senile changes in the brain in the Japanese. Incidence of Alzheimer's neurofibrillary change and senile plaques, in *Proc. 5th Int. Congr. Neuropathology,* Luthy, F. and Bischoff, A., Eds., Excerpta Medica, Amsterdam, 1966, 979.

384. **Wisniewski, H. M. and Iqbal, K.,** Aging of the brain and dementia, *Trends Neurosci.,* 3, 226, 1980.

385. **Constantinidis, J. and Tissot, R.,** Plaques seniles degenerescences neurofibrillaires et autres lesions cerebrales associees, *Schweiz. Arch. Neurol. Neurochir. Psychiatr.,* 124, 317, 1979.

386. **Glenner, G. G., Ein, D., and Terry, W. D.,** The immunoglobulin origin of amyloid, *Am. J. Med.,* 52, 141, 1972.

387. **Glenner, G. G., Terry, W. D., and Isersky, C.,** Amyloidosis: its nature and pathogenesis, *Sem. Hematol.,* 10, 65, 1973.

388. **Gonatas, N. K., Anderson, A., and Evangelista, I.,** The contribution of altered synapses in the senile plaques: an electronmicroscopic study in Alzheimer's dementia, *J. Neuropathol. Exp. Neurol.,* 26, 25, 1967.

389. **Schwartz, P.,** *Amyloidosis: Causes and Manifestations of Senile Deterioration,* Charles C Thomas, Springfield, Ill., 1970.

390. **Powers, J. M. and Spicer, S. S.,** Histochemical similarity of senile plaque of amyloid to apudamyloid, *Virchows Arch. (Pathol. Anat.),* 376, 107, 1977.

391. **Pauli, B. and Luginbuhi, H.,** Fluorescenzmikroskopische untersuchungen zer cerebralen amyloidase bei alten hunden und senilen menschen, *Acta Neuropathol.,* 17, 121, 1971.

392. **Wisniewski, H. M., Bruce, M., and Fraser, H.** Infectious etiology of neuritic (senile) plaques in mice, *Science,* 190, 1108, 1975.

393. **Fraser, H. and Bruce, M. E.,** Argyrophilic plaques in mice inoculated with scrapie from particular sources, *Lancet,* 1, 617, 1973.

394. **Fraser, H. and Dickenson, A. G.,** Scrapie in mice. Differences in the distribution and intensity of gray matter vacuolation, *J. Comp. Pathol.,* 83, 29, 1973.

395. **Gajdusek, D. C. and Gibbs, C. J., Jr.,** Slow virus infections and aging, in *Neural Aging and its Implications in Neurological Pathology,* Terry, R. D., Bolis, C. L., and Toffano, G., Eds., Raven Press, New York, 1982, 1.

396. **Bruce, M. A. and Fraser, H.,** Amyloid plaques in the brains of mice infected with scrapie: Morphological variation and staining properties, *Neuropathol. Appl. Neurobiol.,* 1, 189, 1975.

397. **Wisniewski, H. M.,** Possible viral etiology of neurofibrillary changes and neuritic plaques, in *Alzheimer's Disease: Senile Dementia and Related Disorders,* Katzman, R., Terry, R. D., and Bick, K. L., Eds., Raven Press, New York, 1978, 555.

398. **Brizzee, K. R., Ordy, J. M., Hofer, H., and Kaack, B.,** in *Alzheimer's Disease: Senile Dementia and Related Disorders,* Katzman, R., Terry, R. D., and Bick, K. L., Eds., Raven Press, New York, 1978, 515.

399. **Tomlinson, B. E. and Kitchener, D.,** Granulovacuolar degeneration of the hippocampal pyramidal cells, *J. Pathol.,* 106, 165, 1972.

400. **Woodward, J. S.,** Clinico-pathologic significance of granulovacuolar degeneration in Alzheimer's disease, *J. Neuropathol. Exp. Neurol.,* 21, 85, 1962.

401. **Tomlinson, B. E.,** in Proc. 5th Int. Congr. Neuropathol., Zurich, 1966.

402. **Flamant-Durand, J. and Couck, A. M.,** Spongiform alterations in brain biopsies of presenile dementia, *Acta Neuropathol. (Berlin),* 46, 159, 1979.

403. **Andrew, W.,** The golgi apparatus in the nerve cells of the mouse from youth to senility, *Am. J. Anat.,* 64, 351, 1939.

404. **Moussa, T. and Bankawy, M.,** Morphological and chemical changes in aging of the golgi apparatus of amphibian neurons, *J. R. Microsc. Soc.,* 74, 162, 1954.

405. **Baker, J.,** The structure and chemical compositions of the golgi element, *Q. J. Microsc. Sci.,* 85, 1, 1944.

406. **Bourne, G.,** Changes in dephosphorylating enzymes in young and old tissues of the rat, *Gerontologia,* 1, 50, 1957.

407. **Sulkin, N. and Kuntz, A.,** Histochemical alterations in autonomic ganglion cells associated with aging, *J. Gerontol.,* 7, 533, 1952.

408. **Sosa, J. M. and de Zorillas, N. B.,** Morphological variations of the golgi apparatus in spinal ganglion nerve cells, related to aging, *Acta Anat.,* 64, 475, 1966.

409. **Weinbach, E. and Garbus, J.,** Oxidative phosphorylation in mitochondria from aged rats, *J. Biol. Chem.,* 234, 412, 1959.

410. **Bondareff, W.,** Histophysiology of the aging nervous system, in *Advances in Gerontological Research,* Strehler, B. L., Ed., Academic Press, New York, 1964, 1.

411. **Andrew, W.,** Structural alterations with aging in the nervous system, in *Proc. Assoc. Res. Nervous and Mental Disease,* Vol. 35, Williams & Wilkins, Baltimore, 1956, 129.

412. **Huemer, R. P., Bicker, C., Lee, K. D., and Reeves, A. E.,** Mitochondrial studies in senescent mice. I. Turnover of brain mitochondrial lipids, *Exp. Gerontol.,* 6, 259, 1971.

413. **Ma, W.,** The relation of mitochondria and other cytoplasmic constituents to the formation of secretion granules, *Am. J. Anat.,* 41, 51, 1928.

414. **Payne, F.,** Cytological changes in the cells of the pituitary, thyroids, adrenals and sex glands of ageing fowl, in *Cowdry's Problems of Aging,* Lansing, A., Ed., Williams & Wilkins, Baltimore, 1952, 381.

415. **Desupsey, E.,** Variations in the structure of mitochondria, *J. Biophysiol. Biochem. Cytol.,* 2, 305, 1956.

416. **Weinbach, E. and Garbus, J.,** Age and oxidation phosphorylation in rat liver and brain, *Nature (London),* 178, 1225, 1956.

417. **Weiss, P. and Wang, H.,** Neurofibrils in living ganglion cells of the chick, cultivated in vitro, *Anat. Rec.,* 67, 105, 1936.

418. **Hughes, A.,** The effect of fixation on neurons of the chick, *J. Anat.,* 88, 192, 1954.

419. **Palay, S. and Palade, G.,** The fine structure of neurons, *J. Biophys. Biochem. Cytol.,* 1, 69, 1955.

420. **Sosa, J. M.,** Aging of neurofibrils, *J. Gerontol.,* 7, 191, 1952.

421. **Palay, S.,** Structure and function in the neuron, in *Neurochemistry,* Kovey, S. R. and Nurnberger, J. L., Eds., Paul B. Hoeber — Agathon Press, 1956, 64.

422. **Andrew, W.,** Structural alterations with aging in the nervous system, *J. Chronic Dis.,* 3, 575, 1956.

423. **Bailey, A.,** Changes with age in the spinal cord, *Am. Med. Assoc. Arch Neurol. Psychiatr.,* 70, 299, 1953.

424. **Bondareff, W.,** Morphology of aging in neurons, in *4th Congr. Int. Assoc. Gerontology,* Merano, Italy, 1957, 85.

425. **Caspersson, T. O.,** *Cell Growth and Cell Functions,* W. W. Norton, New York, 1950.

426. **Hyden, H.,** Nucleic acids and proteins, in *Neurochemistry: The Chemical Dynamics of Brain and Nerve,* Elliot, K. A. C., Page, I. H., and Quastel, J. H., Eds., Charles C Thomas, Springfield, Ill., 1955, 204.

427. **Klatzo, I.,** Uber das verhalten des nukleolarapparates und der menschlichen pallidumzellen, *J. Hirnforsch.,* 1, 47, 1954.

428. **Hassler, R.,** Zur pathologie des paralysis agitaus und des postencephalitischen Parkinsonisumus, *J. Psychol. Neurol.,* 49, 387, 1938.

429. **Beheim-Schwarzbach, D.,** Morphologische beobachtungen an nervenzellkernen, *J. Hirnforsch.,* 2, 1, 1955.

430. **Beheim-Schwarzbach, D.,** Weitere beobachtungen an nervenzellkernen, *J. Hirnforsch.,* 3, 105, 1957.

431. **Olszewski, J.,** Zur morphologie und entwicklung des arbeitskerns unter besonderer beruchsichtigung des nervenzellkerns, *Biol. Zentralbl.,* 66, 265, 1947.

432. **Schiffer, D.,** Mucopolisaccaridi nel sistema neruaso di individui senili, *Acta Neurovegetativa,* 15, 25, 1957.
433. **Andrews, W.,** Changes in the nucleus with advancing age of the organism, in *Advances in Gerontological Research,* Strehler, B. L., Ed., Academic Press, New York, 1964, 87.
434. **Cox, A.,** Ganglienzellschrunpfung im tierischen gehrin, *Beitr. Pathol. Anat.,* 98, 399, 1936.
435. **Cammermeyer, J.,** Cytological manifestations of aging in rabbit and chinchilla brains, *J. Gerontol.,* 18, 41, 1963.
436. **Hempel, K. J. and Numba, M.,** Die involution des supranucleus medialis dorsalis. Sowie der lamella medialis und der lamella interna thalami, *J. Hirnforsch.,* 4, 43, 1958.
437. **Schiffler, D.,** Sur l' action reparatrice du moyan des cellules nerveuses, *J. Hirnforsch.,* 1, 326, 1954.
438. **Sayk, J.,** Uber die kernhomogenisierung in nervenzellen der menschlichen hirurinde bei verschiedenen erkrankungen, *Arch. Psychiatr. Nervenkrank.,* 200, 197, 1960.
439. **Solcher, H.,** Die involutionsveranderungen des corpus subthalamicus und des niger reliculatus, *J. Hirnforsch.,* 2, 148, 1956.
440. **Field, E. J. and Peat, A.,** Intranuclear inclusions in neurons and glia: A study in the aging mouse, *Gerontology,* 17, 129, 1971.
441. **Nosal, G.,** Neuronal involution during aging. Ultrastructural study in the rat cerebellum, *Mech. Ageing Dev.,* 10, 295, 1979.
442. **Johnson, J. E., Jr.,** Fine structural alterations in the aging rat pineal gland, *Exp. Aging Res.,* 6, 189, 1980.
443. **Bourne, G. H.,** Histochemical evidence of increased activity of hydrolytic enzymes in the cells of old animals, *Nature (London),* 179, 472, 1957.
444. **McGeer, E. G.,** Aging and neurotransmitter metabolism in the human brain, in *Alzheimer's Disease: Senile Dementia and Related Disorders,* Katzman, R., Terry, R. D., and Bick, K. L., Eds., Raven Press, New York, 1978, 427.
445. **Davies, P.,** Studies of the neurochemistry of central cholinergic systems in Alzheimer's disease in aging, in *Alzheimer's Disease: Senile Dementia and Related Disorders,* Katzman, R., Terry, R. D., and Bick, K. L., Eds., Raven Press, New York, 1978, 453.
446. **Sladek, J. R., Jr. and Blanchard, B. C.,** Age-related declines in perikaryal monoamine histofluorescence in the Fischer 344 rat, in *Brain Neurotransmitters and Receptors in Aging and Age-Related Disorders,* Enna, E. J., Sarmorajaki, T., and Beer, B., Eds., Raven Press, New York, 1981, 13.
447. **Maggi, A., Schmidt, M. J., Ghetti, B., and Enna, S. J.,** Effect of aging on neurotransmitter receptor binding in rat and human brain, *Life Sci.,* 24, 367, 1979.
448. **Misra, C. H., Shelat, H. S., and Smith, R. C.,** Effect of age on adrenergic and dopaminergic receptor binding in rat brain, *Life Sci.,* 27, 521, 1980.
449. **Andrew, M.,** The effects of fatigue due to muscular exercise on the Purkinje cells of the mouse with special reference to the factor of age, *Z. Zellforsch. Mikrosk. Anat.,* 27, 534, 1937.
450. **Palionis, T.,** Die nissl-substanz in den ganglienzellen des riechkolbens, gyrus olfactorius, lobus piriforimis, und ammonshorn des hundes, *Z. Alternsforsch,* 6, 293, 1952.
451. **Kuntz, A.,** Effects of lesions of the autonomic ganglia, associated with age and disease, on the vascular system, *Biol. Symp.,* 11, 101, 1945.
452. **Gersh, J. and Bodian, D.,** Some chemical mechanisms in chomatolysis, *J. Cell. Comp. Physiol.,* 21, 253, 1943.
453. **Van Steenis, G. and Kroes, R.,** Changes in the nervous system and musculature of old rats, *Vet. Pathol.,* 8, 320, 1971.
454. **Hyden, H.,** *Chemische Komponente der Nervengelle und ihre Veranderungen in Alter und Während der Funktion. 3. Collogquium der Gesellschaft fur Physiologische Chemie,* Springer-Verlag, Berlin, 1952.
455. **Palade, C. and Porter, K. R.,** Studies on the endoplasmic reticulum, *J. Biophys. Biochem. Cytol.,* 1, 59, 1955.
456. **Bondareff, W.,** Distribution of Niss substance in the neurons of rat spinal ganglia as a function of age and fatigue, in *Biological Aspects of Aging,* Shock, N. W., Ed., Columbia University Press, New York, 1962, 147.
457. **Johnson, J. E., Jr., Mehler, N. R., and Miquel, J.,** A fine structural study of degenerative changes in the dorsal column nuclei of aging mice. Lack of protection by vitamin E, *J. Gerontol.,* 30, 395, 1975.
458. **Gonatas, N. K. and Moss, A.,** Pathologic axons and synapses in human neuropsychiatric disorders, *Hum. Pathol.,* 6, 571, 1975.
459. **Constantinidis, J., Richard, J., and Tissot, R.,** Pick's disease, histological and clinical correlations, *Eur. Neurol.,* 11, 208, 1974.
460. **Haug, F.,** Light microscopical mapping of the hippocampal region, the pyriform cortex and the corticomedial amygdaloid nuclei of the rat with Timm sulphide silver method, *Z. Anat. Entwickl. Gesch.,* 145, 1, 1974.

177

461. **Constantinidis, J., Richard, J., and Tissot, R.,** Maladie de Pick et metabolisine du zinc, *Rev. Neurol. (Paris),* 133, 685, 1977.
462. **Ibata, Y. and Otsuka, N.,** Electron microscopic demonstration of zinc in the hippocampal formation using Timm's sulfide-silver technique, *J. Histochem. Cytochem.,* 17, 171, 1969.
463. **Bondareff, W.,** An intercellular substance in rat cerebral cortex: Submicroscopic distribution of ruthenium red, *Anat. Rec.,* 157, 527, 1967.
464. **Bondareff, W.,** Submicroscopic morphology of connective tissue ground substance with particular regard to fibrillogenesis and aging, *Gerontologia,* 1, 222, 1957.
465. **Bondareff, W.,** Age changes in the neuronal microenvironment, in *Development and Aging in the Nervous System,* Rockstein, M. and Sussman, M., Eds., Academic Press, New York, 1973, 1.
466. **Bondareff, W. and Narotzky, R.,** Age changes in the neuronal microenvironment, *Science,* 176, 1135, 1972.
467. **Bondareff, W., Narotzky, R., and Routtenberg, A.,** *J. Gerontol.,* 26, 163, 1971.
468. **Cotman, C. W. and Taylor, D.,** Localization and characterization of Concanavalin A receptors in synaptic cleft, *J. Cell. Biol.,* 62, 236, 1974.
469. **Margolis, R. V.,** Acid mucopolysaccharides and proteins of bovine whole brain, white matter, and myelin, *Biochim. Biophys. Acta,* 141, 91, 1967.
470. **Margolis, R. V. and Margolis, R. V.,** Sulfated glycopeptides from rat brain glycoprotein, *Biochemistry,* 9, 4389, 1970.
471. **Margolis, R. V. and Margolis, R. V.,** Distribution and metabolism of mucopolysaccharides and glycoproteins in neuronal perikaryastrocytes, and oligodendroglia, *Biochemistry,* 13, 2849, 1974.
472. **Nicholls, J. G. and Kuffler, S. W.,** Extracellular space as a pathway for exchange between blood and neurons in the central nervous system of the leech: Ionic composition of the glial cells and neurons, *J. Neurophysiol.,* 27, 645, 1964.
473. **Schmitt, F. O. and Samson, F. E., Jr.,** Brain cell microenvironment, *Neurosci. Res. Prog. Bull.,* 7, 323, 1969.
474. **Finch, C. E., Jonec, V., Hody, G., Walker, J. P., Smith, M., Alper, A., and Dougher, G. J.,** Aging and passage of L-tyrosine, L-dopa and insulin into mouse brain slices in vitro, *J. Gerontol.,* 30, 33, 1975.
475. **Sturrock, R. R.,** Development of the indresium griseum. I. A quantitative light microscopic study of neurons and glia, *J. Anat.,* 125, 293, 1977.
476. **Vaugh, J. E. and Peters, A.,** The morphology and development of neuroglial cells, in *Cellular Aspects of Neural Growth and Differentiation,* Pease, D. C., Ed., University California Press, Berkeley, 1971, 103.
477. **Feldman, M. H. and Peters, A.,** Morphological changes in the aging brain, in Survey Report on the Aging Nervous System, Publ. No. NIH 74-296, Maletta, G. J., Ed., U.S. Government Printing Office, Washington, D.C., 1975, 5.
478. **Geinisman, Y., Bondareff, W., and Dodge, J. T.,** Hypertrophy of astroglial processes in the dentate gyrus of the senescent rat, *Am. J. Anat.,* 153, 537, 1978.
479. **Schechter, R., Yen, S., and Terry, R. D.,** Fibrous astrocytes in senile dementia of the Alzheimer type, *J. Neuropathol. Exp. Neurol.,* 40, 95, 1981.
480. **Vaughn, D. W. and Peters, A.,** Neuroglial cells in the cerebral cortex of rats from young adulthood to old age: An EM study, *J. Neurocytol.,* 3, 405, 1974.
481. **Sturrock, R. R.,** Changes in the total number of neuroglia, mitotic cells and necrotic cells in the anterior limb of the mouse anterior commissure following hypoxic stress, *J. Anat.,* 122, 447, 1976.
482. **Diamond, M. D., Johnson, R. E., and Gold, M. W.,** Changes in neuron number and size and glia number of young adult and aging medial occipital cortex, *Behav. Biol.,* 20, 409, 1977.
483. **Landfield, P. W., Rose, G., Sandles, L., Wholstadler, T. C., and Lynch, G.,** Patterns of astroglial hypertrophy and neuronal degeneration in the hippocampus of aged, memory-deficient rats, *J. Gerontol.,* 32, 3, 1977.
484. **Vaughan, D. and Peters, A.,** Neuroglial cells in the cerebral cortex of rats from young adulthood to old age: An electron microscopic study, *J. Neurocytol.,* 3, 405, 1974.
485. **Ravens, J. R. and Calvo, W.,** Neuroglia changes in the senile brain, in *Proc. 5th Int. Congr. Neuropathol.,* Excerpta Medica, Amsterdam, 1965, 506.
486. **Takahashi, K.,** A clinico-pathologic study on the peripheral nervous system of the aged. I. With special reference to the sciatic nerves, *Clin. Neurol.,* 3, 137, 1963.
487. **Courville, C. B.,** Vascular patterns of the encephalic grey matter in man, *Bull. Los Angeles Neurol. Soc.,* 23, 30, 1958.
488. **Fang, H. C. H.,** Observations on aging characteristics of cerebral blood vessels, macroscopic and microscopic features, in *Neurobiology of Aging,* Vol. 3, Raven Press, New York, 1976, 155.
489. **Fang, H. C. H.,** The studies of cerebral arterioles and capillaries in normal and pathologic status in man, in *Pathology of Cerebral Microcirculation,* Cervas-Navarro, J., Ed., De Guyter, New York, 1974, 431.

490. **Craigie, E. H.,** Changes in the vascularity in the brain stem and cerebellum of the albino rat between birth and maturity, *J. Comp. Neurol.,* 38, 27, 1924.

491. **Zeman, W. and Innes, J. R. M.,** in *Craigie's Neuroanatomy of the Rat,* Academic Press, New York, 1963.

492. **Burns, E. M., Kruckeberg, T. W., and Comerford, L. E.,** Cerebral microcirculation, in *Aging in Nonhuman Primates,* Bowden, D. M., Ed., Van Nostrand Reinhold, New York, 1979, 123.

493. **Takahashi, K.,** A clinicopathologic study on the peripheral nervous system of the aged, *Geriatrics,* 31, 123, 1966.

494. **Dastur, D. K., Lane, M. H., Hansen, D. B., Kety, S. S., Butler, R. N., Perlin, S., and Sololoff, L.,** Effects of aging on cerebral circulation and metabolism in man, in Human Aging, a Biological and Behavioral Study, USPHS Publ. No. 986, U.S. Government Printing Office, Washington, D.C., 1963, 59.

495. **Gottstein, U. and Held, K.,** Effects of aging on cerebral circulation and metabolism in man, *Acta Neurol. Scand.* (Suppl.), 72, 54, 1979.

496. **London, E. D., Nespor, S. M., Ohata, M., and Rapoport, S. I.,** Local cerebral glucose utilization (LCGU) during development, maturity and aging of the Fischer 344 rat, *J. Neurochem.,* 37, 217, 1981.

497. **Ohata, M., Sundaram, U., Fredericks, W. R., London, E. D., and Rapoport, S. I.,** Regional cerebral blood flow during development and aging of the rat brain, *Brain,* 104, 319, 1981.

498. **Obrist, W. D.,** Noninvasive studies of cerebral blood flow in aging and dementia, in *Alzheimer's Disease: Senile Dementia and Related Disorders,* Katzman, R., Terry, R. D., and Bick, K. L., Eds., Raven Press, New York, 1978, 213.

499. **Wang, H. S. and Busse, E. W.,** Correlates of regional cerebral blood flow in elderly community residents, in *Blood Flow and Metabolism in the Brain,* Harper, M., Jennett, B., Miller, D., and Rowan, J., Eds., Churchill Livingstone, Edinburgh, 1975, 17.

500. **Garden, A. S., Ohata, M., Rapoport, S. I., and London, E. D.,** Age-associated decrease in local cerebral glucose utilization (LCGU) in the beagle, *Abstr. Soc. Neurosci.,* 6, 768, 1980.

501. **Coers, C., Telerman-Toppet, N., and Gerard, J. M.,** Terminal innervation ratio in neuromuscular diseases. II. Disorders of lower motor neuron, peripheral nerve, and muscle, *Arch. Neurol.,* 29, 215, 1973.

502. **Gutmann, E. and Hanzlikova, V.,** Age changes in motor end-plates in muscle fibers of the rat, *Gerontologia,* 11, 12, 1965.

503. **Gutmann, E. and Hanzlikova, V.,** Motor unit in old age, *Nature (London),* 209, 921, 1966.

504. **Gutmann, E., Hanzlikova, V., and Jaboubek, B.,** Changes in the neuromuscular system during old age, *Exp. Gerontol.,* 3, 141, 1968.

505. **Canna, N.,** The effects of aging on the receptor organs of human dermis, in *Advances in Biology of Skin,* Vol. 6, Montagna, W., Ed., Pergamon Press, Oxford, 1965, 63.

506. **Canna, N. and Mannan, G.,** The structure of human digital Pacinian corpuscles (corpuscula lamellosa) and its functional significance, *J. Anat.,* 92, 1, 1958.

507. **Zelená, J., Sobotková, M., and Zelená, H.,** Age-modulated dependence of Pacinian corpuscles upon their sensory innervation, *Physiol. Bohemoslov.,* 27, 437, 1978.

508. **Zelená, J.,** Development, degeneration, and regeneration of receptor organs, *Prog. Brain Res.,* 13, 175, 1964.

509. **Bolton, C. F., Winkelmann, R. K., and Dyck, P. J.,** A quantitative study of Meissner's corpuscles in man, *Neurology,* 16, 1, 1966.

510. **Witkin, J.,** Peripheral tactile innervation, in *Aging in Nonhuman Primates,* Bowden, D. M., Ed., Van Nostrand Reinhold, New York, 1979, 158.

511. **Dyck, P. H., Schultz, P. W., and O'Brien, P. C.,** Quantitative of touch-pressure sensation, *Arch Neurol.,* 26, 465, 1972.

512. **Arey, L. B., Termaine, M. J., and Monzingo, F. L.,** The numerical and topological relations of taste buds to human circumvallate papillas throughout the lifespan, *Anat. Rec.,* 64, 9, 1935.

513. **Harris, W.,** Fifth and seventh cranial nerves in relation to the nervous mechanism of taste sensation: A new approach, *Br. Med. J.,* 1, 831, 1952.

514. **Smith, B. H. and Setki, P. K.,** Aging of the nervous system, *Geriatrics,* 30, 109, 1975.

515. **Allara, E.,** Richerche sull' organo del gusto dell' nomo. I. La strutura della papille gustative nelle varie eta della vita, *Arch. Ital. Anat. Embriol.,* 42, 406, 1939.

516. **Mochizuki, Y.,** Papilla foliata of Japanese, *Folia Anat. Jpn.,* 18, 337, 1939.

517. **El-Baradi, A. and Bourne, G.,** Theory of tastes and odors, *Science,* 113, 660, 1951.

518. **Byrd, E. and Gertman, S.,** Taste sensitivity in aging persons, *Geriatrics,* 14, 381, 1959.

519. **Cohen, T. and Gitman, L.,** Oral complaints and taste perception in the aged, *J. Gerontol.,* 14, 294, 1959.

520. **Smith, C. G.,** Age incidence of atrophy of olfactory nerves in man. A contribution to the study of the process of ageing, *J. Comp. Neurol.,* 77, 589, 1942.

521. **Smith, C. G.,** Incidence of atrophy of the olfactory nerves in man, *Arch. Otolaryngol.,* 34, 533, 1941.

522. **Goldberg, M. and Lindblom, U.,** Standardized method of determining vibratory perception thresholds for diagnosis and screening in neurological investigation, *J. Neurol.,* 42, 793, 1979.

AGING OF SKELETAL MUSCLE

Geoffrey Goldspink and M. A. Alnaqeeb

INTRODUCTION

In all groups of animals with the exception of the protozoans, muscle is the tissue that provides the force for locomotion. In the animal kingdom, the ability to move quickly and efficiently is one of the most important requisites for survival. Indeed, in wild populations this must be considered one of the important factors in determining lifespan. The animal which cannot move quickly enough to catch its prey or to escape its predator will soon come to die. Human beings are not quite so dependent on the locomotor systems for prolongation of life, but malfunction of the locomotor system usually implies dependence on others and often considerably reduces their ability to enjoy life.

During this century the age structure of society has changed so that the percentage of old people alive today is much higher than ever before. As life expectancy increases, so does the number of persons with locomotory handicaps, unless commensurate strides are made in rehabilitation and other branches of medicine. This is one of the reasons why it is important to understand aging in skeletal muscle and the whole locomotor system.

Muscle is a very adaptable tissue, and in younger individuals it responds well to exercise training. The decrease in muscular performance with advancing age has been documented by a number of workers.[1-4] To what extent this decline in performance is a consequence of the change in activity of people with age and vice versa, is difficult to quantify. Certainly, motivation to perform physical tasks must be an important factor.

At the turn of the century most of the population in Western countries made their living by carrying out manual tasks. Now only a small percentage do so, and with the advent of the "computer revolution" this percentage will drop even further. The relationship between physical exercise and the aging process is therefore an important area which needs further study. The increasing cost of health care emphasizes the need for a preventative approach to medicine; otherwise a disproportionate amount of the wealth of a country will be used in caring for its increasing proportion of infirm, elderly citizens. It can be argued that the prime goal of aging research should be to prolong the active period of life rather than life expectancy per se. In this context, studies of the changes in the locomotor system with age are of prime importance.

Muscle is an interesting tissue in which to study the aging process as it is a post-mitotic tissue; i.e., no further cell division takes place within the cellular units (muscle fibers) once embryonic differentiation is complete. Indeed, the aging processes in post-mitotic tissues such as muscle and nervous tissue may be of a different nature than in tissues such as liver and intestine where cell replacement is continued. The aging process can be considered to be continuous with development and growth, and once initiated it probably proceeds at different rates within different tissues. Though aging, like death, is inherent in life, we know much less about it than any of the other basic characteristics of life. For instance, we know almost nothing about how the aging processes are programmed, if indeed they are programmed. As muscle is a mechanical tissue, it would seem to be an obvious choice for testing the "wear and tear" theories of aging. It is also a tissue in which the metabolic rate can fluctuate widely between the resting and fully active states. Therefore any theory that is based on metabolic rate and biochemical "wear and tear" cannot afford to overlook what is happening in the musculature.

The control and supply systems must also be considered. Hence the scope must be broadened to include the study of changes in the peripheral nervous system and the vascular system that occur with age.

PHYSIOLOGICAL CHANGES

It has long been established that muscles in senile animals are less able to develop force as compared to muscles in younger animals.[1,2,4,5] This is true for both isometric[3] as well as dynamic activities.[4] Muscle mass also declines as estimated by creatinine excretion[6] or from the routine use of ultrasound or CT scanner measurements (R.T.H. Edwards, personal communication).

Investigations into the decline of muscle strength and coordination have implicated the nervous and vascular systems as well as myogenic factors. Oxygen diffusion through the plasma was reported to decline with age due to the increase in protein and cholesterol concentration.[7] This could induce hypoxia and be partly responsible for a change in the enzyme profile that is seen in some muscles with age. An increase in oxygen uptake with age was demonstrated in resting and active thoracic flight muscle of some insects.[8,9] It is suggested that this may be due to uncoupling of oxidation phosphorylation.[9] This has been noted in humans by Norris and Shock[10] who reported a reduced efficiency with which elderly people "burn" oxygen at low work loads. This conflicts to some extent with the findings that elderly people have relatively good powers of endurance, providing the exercise is of low intensity. When subjected to 40% maximum isometric tension, Larsson and Karlsson[11] found that endurance was actually higher in older people when expressed on this relative basis. However, the maximum isometric tension is reduced in older subjects,[11] and therefore in absolute terms the younger subjects would probably fatigue less rapidly, if exercised at similar levels. Nevertheless it has been noted that the successful long distance runners (50 to 100 miles) are often older men 40 to 60 years of age. Psychological factors are obviously important in this event, but this may be a reflection of the fact that it is only the fast-contracting motor units that are affected appreciably by the aging process. All contracting muscles are uniformly slow at birth.[12-19] During postnatal development, some muscles become fast-contracting because they acquire a large proportion of fast-contracting motor units.[14] The decline in strength of predominantly fast-contracting and slow-contracting muscles has been previously reviewed.[20]

A motor unit consists of those muscles fibers that are supplied by the same motorneuron. There is good evidence that postnatal differentiation of the different motor units depends on the type of motorneuron that establishes the innervation during the differentiation process. As to contractile properties, the general response of the senile muscle is that of a more prolonged contraction and more prolonged relaxation. The extensor digitorum longus, which is made up chiefly of fast motor units, shows an increased contraction time, relaxation time, and latent period. The levator ani and the diaphragm, both very specialized muscles, show similar trends in spite of the fact that the latter muscle is continuously used throughout the animal's lifetime.[21]

The changes in the contractile properties of the slow soleus muscle appear to be less clear. Syrovy and Gutmann[22] reported an unchanged contraction time for the older muscles as compared with young ones. This finding was supported in a report by Vyskocil and Gutmann.[23] However, a later study showed a decrease in contraction time (time to peak) for the aged muscle but a definite prolongation in latent relaxation and latent period times.[24] Alnaqeeb and Goldspink[25] found that there was a slowing down of contractile and relaxation processes in senile extensor digitorum longus of the rat, accompanied with a reduction in maximum tetanic tension. The intrinsic speed of contraction increased after reaching a minimum around adulthood (Table 1). Studies on motor units within the soleus indicate a progressive shift towards slow motor units with age,[26] suggesting that more than one factor determines the overall performance of the muscle.

The number of functional motor units in a senile muscle is thought to decrease with age. Indeed a considerable decrease in functioning motor units in thenar hypothenas, soleus, and

Table 1
CONTRACTION AND RELAXATION WITH AGE IN RAT EXTENSOR
DIGITORUM LONGUS

Age (days)	Latent period (msec)	Contract. time (msec)	Relax. time (msec)	Twitch tension (g/mm²)	Max tetanic tension (g/mm²)	Intrinsic contract. speed
380	2.10 ± 0.06	13.0 ± 0.3	13.9 ± 2.0	4.5 ± 0.4	21.2 ± 2.2	0.216 ± 0.037
714	2.27 ± 0.10	15.7 ± 1.2	16.6 ± 3.5	4.3 ± 1.1	14.3 ± 3.0	0.297 ± 0.031

extensor digitorum brevis was found in patients older than 60 years.[27] This means that the surviving units must bear the extra work, which may lead to hypertrophy of the remaining functional units. If hypertrophy of the other units does not occur, the result is a reduced force output. Because it may not necessarily be the same type of motor unit that undergoes compensatory hypertrophy, the muscle properties may change by this as well as other mechanisms.

Nerve conduction velocities also appear to be impaired in the elderly.[27] This is probably due to the loss or diminution of motor neurons with low threshold and rapid conduction times.[28]

Miniature end plate potentials (m.e.p.p.) have also been shown to change with age. They increase in early life[29] and then decrease to a comparatively low level with extreme age.[23,29,30] This is an indication of the reduced resting activity of the muscle end plate, which may be a reflection of either a decrease in the amount of acetycholine per quantum (vesicle) or change in the interaction of acetylcholine with the receptors or both. The situation regarding the membrane potential is unclear. Schwarz and Wichan[31] reported an increase in the muscle resting membrane potential at middle age and a decrease in later life in albino rats. However, the data of others[29,32] were not in accord with these results, and one group also found no significant age-associated change in either the amplitude or maximum rate of rise of the action potential.[32]

In summary, it appears that the most apparent physiological changes associated with aging in muscle are the decline in strength and the slowing down of the contractile process. It is therefore of interest to see how these changes might be explained in structural and biochemical terms.

HISTOCHEMICAL CHANGES

The muscles of most vertebrate muscles are heterogeneous, containing some fast-contracting muscle fibers and some slow-contracting. There are several classifications of mammalian fiber types in current use, but two of the most widely used are those of Brooke and Kaiser[33] and Peter et al.[34] Basically, both classifications distinguish between the slow- and fast-contracting twitch fibers. In addition, they divide fast-contracting fibers into two main categories based on their oxidative capacity. Brooke's fiber type divisions are Type I, Type IIB, and Type IIA. In the same order, and according to the second classification, they would be: slow twitch oxidative, fast twitch glycolytic, and fast twitch oxidative glycolytic. Some people also distinguish a Type IIC fiber, but this is believed to be an embryonic fiber type. These fiber types differ in their histochemical characteristics, their ultrastructure, and their physiological properties.

There is good evidence that these fiber types have specific physiological functions. Though motor units are made up of one type of fiber only, the fibers belonging to one motor unit are usually interspersed with those of other motor units.[26,35] Kugelburg[36] related the rate of contraction of a unit to the histochemical properties of its fiber. Type II fibers in the soleus were shown to have a contraction time (twitch, time to peak) of 15 to 26 msec, whereas Type I fibers were shown to be slow with a contraction time of 27 to 40 msec. Burke et al.[35] have demonstrated the relationship between fiber types and their fatigue resistance. The slow-contracting fibers were very resistant to fatigue, whereas the fast fibers fatigued much more rapidly, although the fast oxidative glycolytic (FOG IIA, FR) type, which have appreciable numbers of mitochondria, are more fatigue-resistant than the fast glycolytic type (FG, IIB, FF).

The relation between histochemical fiber types and contraction times of muscles was further confirmed with cross-innervation studies. Muscles with slow-contracting rates were shown to alter both their histochemical profiles and contraction rates when cross-innervated with a nerve originating from a fast muscle and vice versa.[37] By studying the depletion of glycogen in the different fiber types[38,39] and from data of the energetics of the fibers contracting under different conditions,[40] it is concluded that the slow fibers are used for low-intensity isometric contractions such as the maintenance of posture and also for slow isotonic contractions such as walking. The fast fibers, on the other hand, are recruited when very forceful or very rapid contractions are required, such as in lifting very heavy weights or in running.

With histochemical fiber typing, it is possible to relate histochemical changes in aging muscle with the physiological changes. It is now appreciated that the size and to some extent the number of the different fiber types changes with age. In some muscles the alteration in the fiber type profile starts a few weeks after birth,[41-43] although in most muscles studied the rate of change is slow.

In the slow soleus muscle in the guinea pig, the ratio of slow to fast oxidative fibers changes in a continuous way throughout life.[41] The rat soleus behaves in a similar way in that there is a continuous transition between Type II and Type I fibers; this change is consistent with the increased contraction time of aging muscle.[36] Caccia et al.[44] reported an intermediate fiber type with a lower ATPase activity. This they claim represented fibers that were undergoing the transition from Type II to Type I.

According to Bass et al.[45] there is an overall loss of glycolytic and aerobic enzyme activity in the fibers. However, the loss of glycolytic enzymes is greater than the loss of aerobic enzymes, and the result is an apparent shift toward aerobic metabolism. This trend was more pronounced in the extensor digitorum longus muscle which is predominantly composed of Type II fibers. The Type II fibers also showed a greater loss of ATPase activity with age. The work involving human subjects shows similar results. The Type II fibers in the vastus lateralis were found to decrease, although the ratio of Type IIA to Type IIB fibers remained unchanged.[46] The quadriceps was found to exhibit similar changes.[11]

Different fiber types have been noted to grow at different rates.[47] In the extensor digitorum longus (EDL) of the rat, the diameters of all three fiber types were closely matched in young animals with overlapping distributions. As the animal aged, FG fibers grew at a higher rate than either FOG or SO fibers (Figure 1). This gave rise to a bimodal distribution in adult and aging animals. In senile animals the bimodality was eroded because of changes in the distributions of individual fiber types. The three fiber populations showed fibers with abnormally small and large diameters. This probably corresponded to splitting and hypertrophied fibers observed in aging muscles (Figure 2).

The general conclusion from studies of the histochemistry of aging muscle is that the fast fibers are affected much more than the slow fibers. However, with the sampling techniques used, it is not usually possible to say whether there is an absolute decrease in the number of fast type fibers. The total cross-sectional area of fast (FG) fibers is reduced in senile muscles. This finding seems to be in accordance with observations mentioned above that although older people are less capable of producing powerful movements, they have relatively good powers of endurance.

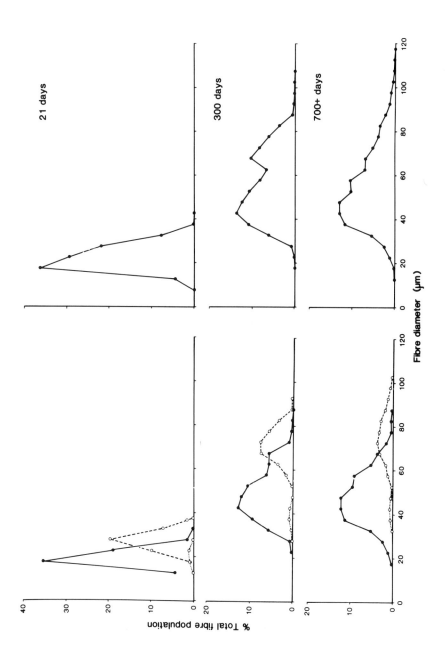

FIGURE 1. Left: The distribution of the three main fiber populations in the EDL of the rat, FOG (●———●), SO (○·····○) and FG (○----○). The frequency is expressed as a percentage of the total fiber population at ages of 21 days (weaning), 300 days (adult) and 700+ days (senile). Right: The combined reconstructed polygons of the three fiber types reveal a bimodal distribution at 300 days. This bimodality is generated by the higher growth rate of FG fiber diameters. Note the tails of the distribution in individual populations correspond to hypertrophied and splitting fibers.

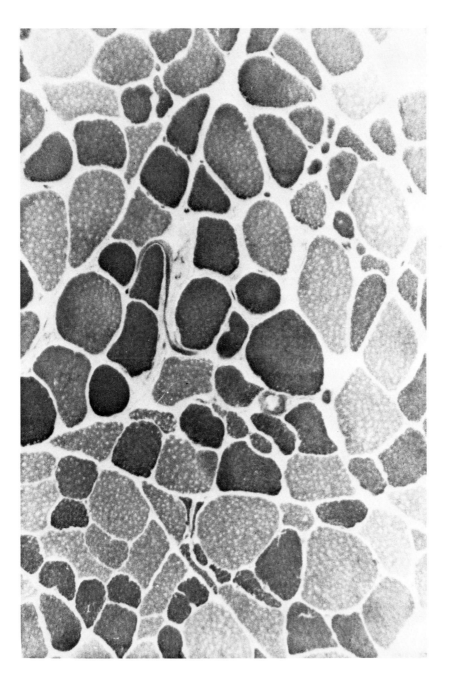

FIGURE 2.   Degenerating and hypertrophied fibers are a feature of senile extensor digitorum longus muscles in the rat. Both FOG and FG fibers undergo such changes.

MORPHOLOGICAL CHANGES

Morphometric Changes

Morphometric changes within a muscle correspond well with its reduction in weight. A reduction in weight can usually be explained by either a decrease in fiber number or a decrease in fiber cross-sectional area, or both. Loss of muscle fibers during development and aging has been reported for different muscles including the dog pectinius,[48] the rat soleus,[49] and the rat extensor digitorum longus and soleus muscles.[43]

Moore et al.[50] using several human muscles, demonstrated a reduction in fiber diameter in most muscles in subjects older than 40 years. Although he found that some muscles such as the superior rectus were resistant to fiber diameter changes, the exact relation between fiber number and fiber diameter is not clear. Despite fiber loss, some muscles tend to have an increased fiber diameter. There are several examples of these sorts of muscles in the mouse (e.g. soleus, extensor digitorum longus, and biceps brachii) in which fiber loss is accomplished by compensatory hypertrophy.[51] Both FOG and SO fibers in the senile soleus of the rat exhibited splitting (Figure 3). This phenomena has been reported to occur in both senile[47] and overloaded muscle. [52,53] The senile extensor digitorum longus showed hypertrophied and degenerating fibers (see Figure 2). The hypertrophy was probably in response to the disorganization and degeneration associated with age.

There appears to be a selective loss of certain fiber types with age in some muscles.[46,54] The decrease in cross-sectional area is of a similar nature,[11] with Type II fibers being the most susceptible to the aging process.

Satellite cells, which some workers believe to be residual myoblasts, show a significant decrease in number with age. It was reported that the percentage of nuclei that belonged to satellite cells decreased from 4 to 6% in young rat muscles to 2 to 4% in the muscles of senile animals.[55,56] This change may have important implications regarding the muscle's ability to repair itself or undergo compensatory hypertrophy.[57]

The blood supply of muscles is of great functional importance. However, capillary density does not appear to change with age[58,59] so the supply of oxygen and other substances to the muscle is presumably reasonably unimpaired. This is again in accord with the finding that the oxidative capacity and resistance to fatigue is relatively unaffected with age.

Ultrastructural Changes

When muscle fibers of senile animals are examined with the electron microscope, ultrastructural disorganization is evident. The regular cross-striations of the healthy muscle are sometimes replaced with longitudinal ones at irregular intervals, reflecting the disarray of the sarcomeres. In transverse sections, the outermost myofibrils are often seen to form a ring around the central core of myofibrils. This is sometimes referred to as "Ringbinden". In longitudinal sections, the myofibrils tend to exhibit a patchy appearance with frequent irregularities. Endomysial fibrosis manifests itself in senile muscle, although usually to a moderate extent.[60] The thickness of the old degenerating fibers is inconsistent along the longitudinal axis, and this is presumably a consequence of the myofibrillysis and endomysial fibrosis. Ultrastructurally, myofibrils begin to show degenerative change well before the disruptions are observable at light microscope level.[61]

Changes in the membrane systems have also been described, including a thickening of the sarcolemma (in this case taken to include plasma membrane and endomysium) which appears to be associated with an increase in collagenous material.[30] Those authors also reported that the general disorganization and disintegration of the myofibrils is accompanied by a proliferation of the sarcoplasmic reticulum. The myofibrillar degeneration appears to be limited to fibers rich in subsarcolemmal mitochondria. We now know these fibers to be FOG (IIA) type. Fibers with no appreciable number of subsarcolemmal mitochondria ex-

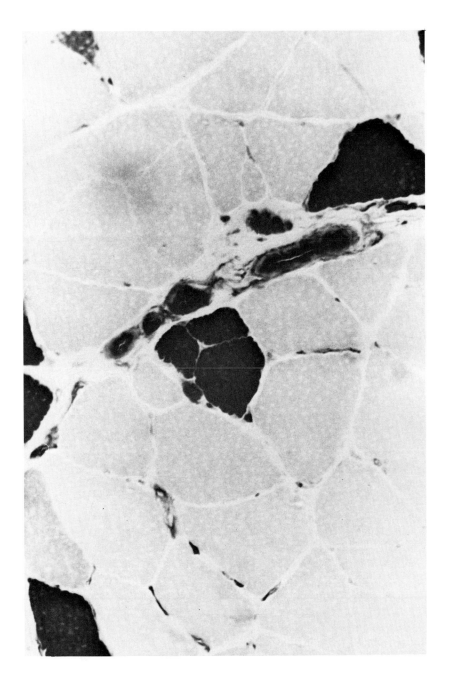

FIGURE 3.   The senile soleus of the rat demonstrates splitting of FOG (dark fibers) as well as SO fibers.

hibited streaming of Z-bands,[61] and probably belong to the FG (IIB) fiber type. Interestingly, the T-system in these fibers showed increased proliferation with age.[30] This may be a compensatory mechanism to ensure the spread of the excitation potential in the fibers where the sarcomeres are very much out of register with the sarcomeres of adjacent myofibrils due to the Z-band streaming.

In addition to the general disintegration reported earlier at the myofilament level, Steenis and Kroes[62] reported aggregation of muscle nuclei associated with a swelling of the cytoplasm which was more granular and more hyaline in nature. Lipofucsin, "the age pigment", showed an increase with age, and was almost exclusively located in the subsarcolemmal region.[63] A reduced mitochondrial fraction with age was also reported.[61]

The decrease in the number of large nerve fibers and the shift to smaller diameter axons was demonstrated as early as the 1940s.[64,65] The axons of aging nerves are more prone to demyelination and vaculation, a process which is usually preceded by swelling and dilation of myelin sheaths and the invasion of macrophages.[62]

The increase in the frequency at which lesions of the nervous system occur is directly related to age.[66] The dedifferentiation of end plates of both fast and slow muscle appears to be consistent with old age. Such dedifferentiation is accompanied with erratic cholinesterase activity and some actual degeneration of end plates.[67]

Ultrastructurally, Fujisawa[68] observed an increased number of vesicles that differ from the synaptic vesicles. Neurofilaments and mesaxons with degenerative changes and swollen terminal axons were found to lose their rounded shape.

Gutmann et al.[30] reported an increase in synaptic vesicles with an increased number of junctional folds. Synaptic clefts were larger than those in young animals, but the basement membrane was thicker. Neurotubules and neurofilaments were found to occur in peripheral axons. These have not been observed in younger animals. Collagen fibrils were more abundant in older animals. Cardasis[69] found that in the rat soleus muscle, the synaptic contact area of the neuromuscular junction decreased in aging. This was associated with insertion of Schwann cell processes between the axon terminals and the muscle fibers and numerous lysosomal-like structures, suggesting endocytic activity.

Structural changes in both the muscle fiber and the neuromuscular junction correspond well with the reduced function of the locomotor system. Some of these changes suggest that repair processes are taking place, but it appears that these processes are not completely successful in restoring full function in senile muscle.

Collagen in Muscle

The replacement of degenerating aging muscle tissue with connective tissue was reported by Bick.[70] Several muscle diseases in which wasting occur are known to undergo similar changes.[71-74]

Various muscles apparently accumulate collagen at different rates throughout life. Schaub[75] reported an increase of collagen in the hind leg and the abdominal musculature with age. Cardiac muscle shows a similar trend.[76,77] Indeed, according to Mohan and Radha,[78] cardiac muscle shows the greatest increase of collagen with age, followed by the fast extensor digitorum longus, and the slow soleus shows the least. Whether this has functional implications is hard to know, especially in light of the results published by Sasaki et al.,[79] who found no change in total collagen, but a shift toward an increased residual and trichloroacetic acid-soluble fraction.

Cross-linking of aging collagen molecules[80] is thought to reduce collagen breakdown rate, thus leading to the building up of residual collagen. "Old" collagen has been shown to be more resistant to degrading enzymes.[78] In addition, the activity of these enzymes is apparently reduced with age in all types of muscles. The hydrolysis time of collagen in vitro is also known to increase with age of the molecule.[81]

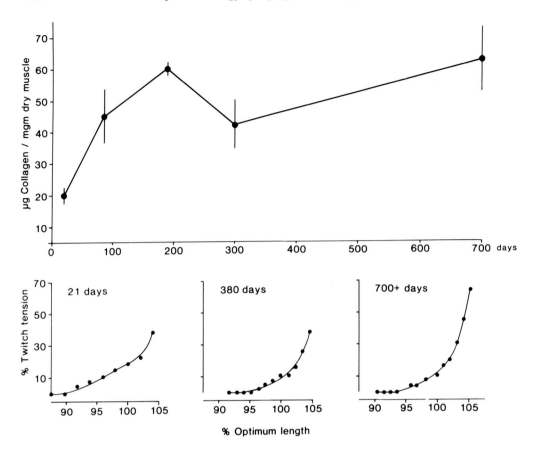

FIGURE 4.   Collagen concentrations of developing and aging muscles (top). The changes parallel the increase in passive tension (expressed as a percentage of active tension). The youngest muscle with the lowest collagen concentration possess the shallowest curve (21 days). As the collagen concentration increases in the senile extensor digitorum longus (380 days and 700+ days), so does the passive tension.

Alnaqeeb et al.[82] have shown that there is an increase in passive tension (stiffness) in the muscle which parallels the increase in collagen in the muscle (Figure 4). By selectively staining the collagen and using a video system image analyzer, Alnaqeeb et al. were able to show that in the rat, both the endomysium and perimysium thicken in relation to muscle fiber area, particularly after 500 days (Figure 5). Certainly the changes in the amount and nature of collagenous material in skeletal muscle are probably the main factors in the loss of suppleness in the musculature with age.

Biochemical Changes

The water and ion content of whole muscles have been measured by several workers, and it was found that those ions which are predominantly extracellular (e.g., K^+) tend to increase, whereas those that are predominantly intracellular (e.g., Na^+, Cl^-) tend to decrease with age.[71,83,84] This is what would be expected in view of the atrophy of fibers mentioned above.

The inability of the senile muscle to keep pace with the demands placed upon it may be the result of reduced enzyme activity. RNA concentrations show a rapid decrease during the first few weeks after birth in the mouse and thereafter continue to decrease, but at a slower rate. This is reflected in the microsomal activity of cell-free preparations[85] and a reduction in the rate of amino acid incorporation with age.[86] If RNA production continues to decline, it may well be that the enzymes, which tend to be turned over rapidly, cannot

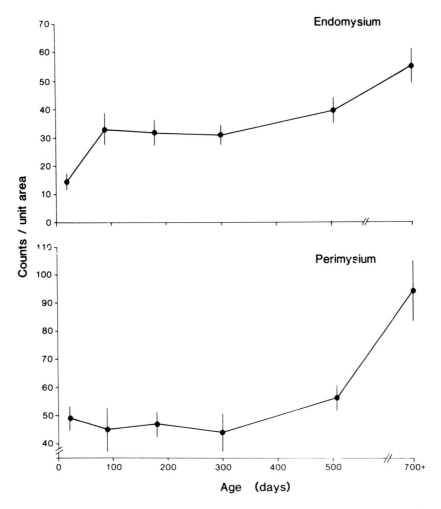

FIGURE 5.   Endomysial and perimysial connective tissue increases with age. The increase is greatest in senile muscles.

be synthesized sufficiently rapidly in old muscle. However, this may be only true of certain types of enzymes. Muscle mitochondria of old animals are known to retain their activity and to continue producing ATP at high rates if measured under optimum conditions.[87-89]

Froklis and Bogatskaya[90,91] noted that heart muscle tends to shift to a more anaerobic metabolism. Bass et al.[45] reported a decrease in the glycolytic activity of the fast extensor digitorum longus, in the form of reduced triosephosphate dehydrogenase (TPDH), lactate dehydrogenase (LDH), and glycerol-3-phosphate dehydrogenase (GPDH) activity with age. The slow soleus showed a decrease in TPDH and a decrease in the aerobic enzymes; malate dehydrogenase (MDH) and citrate synthase (CS). On the other hand Ermini[92] reported a reduced oxygen consumption associated with a reduced aerobic activity. The reduced aerobic activity reduces ATP production by 25 to 35%.[92] Both white and red muscles show a decreased creatine phosphate production but an unchanged creatine phosphokinase activity.[93] However, Orlander et al.[63] concluded that in humans the muscle cells are still capable of maintaining as great a metabolic flow up to 70 years of age as they are in young people. The decrease in the maximal O_2 uptake of the whole body which is associated with aging appears to be related to the decreased muscle mass rather than to biochemical changes. Hence it is clear that although enzyme activities change with age, there appears to be selectivity, as not all systems are affected to the same extent. Also, these enzyme changes may be the result of lack of activity rather than an aging change per se.

The enzyme which is central to the force production of muscle is myosin ATPase. It has been shown that there is a good correlation between the specific activity of this enzyme and the rate of contraction.[94] Syrovy and Gutmann[22] claimed that parallel changes in these two parameters do occur with age. However, Ermini[95] reported that neither Ca^{++}- or Mg^{++}–activated myofibrillar ATPase was affected by age. This is in contrast with work of Rockstein and Brandt[96] and Gutmann and Syrovy.[24] The latter workers reported an increase in Ca^{++}-activated myosin ATPase in the slow soleus muscle and a decrease in the fast extensor digitorum longus with age. The situation is therefore rather confusing, and the results seem to vary with the technique, the muscle used, and the laboratory.

The conflicting evidence concerning some enzyme activities is probably due to the complex makeup of mixed muscles and dependence of enzyme activities on the previous activity of the animal.

ADAPTABILITY OF MUSCLE WITH AGE

It is often difficult to decide which are primary aging changes and which are merely the result of decreased activity. The motivation to indulge in physical exercise undoubtedly declines with age. The reason for this is not known, but it must be mainly psychological in nature as it cannot be explained simply in terms of physical distress caused by the exercise or indeed inability to perform at a reasonable level. One important question that must be asked is to what extent can full muscle capability be restored in older individuals if the psychological barrier (the aversion to strenuous exercise) is overcome.

The results of a few studies on humans have been encouraging as they have indicated that old people were as trainable, with respect to their strength, as young people.[97] A reduction in heart rate at submaximal exercise in old men with training has also been noted by these and other workers.[98-101] As for the muscle itself, the changes are not as marked as the improvement in the strength measurements. One reason for this is that technique improves with training as well as muscle strength. However, the results of Aniansson and Gustafsson[97] suggest that the fast-contracting fibers FOG (Type IIA) and FG (Type IIB) respond well to exercise in the elderly. This is presumably because this fiber type is usually recruited so infrequently in older people because they do not use rapid or forceful movements. Instead, they use mainly their SO (I) fibers. Indeed, in Aniansson and Gustafsson's study, the Type I fiber actually decreased slightly in size, reflecting the change in the recruitment pattern of the fibers. Certainly, the impression one obtains from reviewing the available data on aging of muscle is that only in extreme old age are the changes degenerative and irreversible.

ACKNOWLEDGMENT

This work was supported by a grant from the National Institute on Aging NIH IROIAG04627-01 A1 to Professor Goldspink, and Dr. Alnaqueeb was in receipt of a scholarship from the University of Kuwait.

REFERENCES

1. **Ufland, J. M.,** Ginfluss des Lebensalters. Geschlechts der Konstitution und de's Berufs auf die Kraft verschiedener Muskelgruppen, *Arbeitsphysiologie,* 6, 653, 1933.
2. **Fisher, M. B. and Birren, J. E.,** Age and Strength, *J. Appl. Physiol.,* 3, 490, 1947.
3. **Amussen, E. and Heeboll-Nielsen, K.,** Isometric muscle strength of adult men and women, in *Communications from the Testing and Observation Institute of the Danish Association of Infantile Paralysis,* 11, 1961.
4. **Larsson, L., Grimby, G., and Karlsson, J.,** Muscle strength and speed of movement in relation to age and muscle morphology, *J. Appl. Physiol.,* 46, 451, 1979.
5. **Quetelet, A.,** Sur L'homme et le developpement de ses facultes, L. Hauman and Cie, Brussels, 1836.
6. **Tzankoff, S. P. and Norris, A. H.,** Effect of muscle mass decrease on age-related BMR change, *J. Appl. Physiol.,* 43, 1001, 1977.
7. **Chisholm, G. M., Terrado, E. M., and Gainer, J. L.,** Physiological transport in relation to ageing, *Nature (London),* 230, 390, 1971.
8. **Tribe, M. A.,** Some physiological studies in relation to age in the blowfly *Calliphora erythrocephala Meig, J. Insect Physiol.,* 12, 1577, 1966.
9. **Tribe, M. A.,** Age-related changes in the respiratory physiology of flight muscle from blowfly *Calliphora erythrocephala, Exp. Gerontol.,* 2, 113, 1967.
10. **Norris, A. H. and Shock, N. W.,** Age changes in ventilatory and metabolic response to submaximal exercise, *Proc. 4th Congr. Int. Assoc. Gerontol.,* Vol. II, 1957, 512.
11. **Larsson, L. and Karlsson, J.,** Isometric and dynamic endurance as a function of age and skeletal muscle characteristics, *Acta Physiol. Scand.,* 104, 129, 1978.
12. **Banu, G.,** Recherches Physiologiques sur le developpement neuromusculair, Paris, 1922.
13. **Denny-Brown, D.,** The historical features of striped muscle in relation to its functional activity, *Proc. Roy. Soc. Br.,* 104, 371, 1929.
14. **Buller, A. J., Eccles, J. C., and Eccles, R. M.,** Differentiation of fast and slow muscles in the cat hind limb, *J. Physiol.,* 150, 399, 1960.
15. **Buller, A. J., Eccles, J. C., and Eccles, R. M.,** Interactions between motoneurons and muscles in respect of the characteristic speeds of their responses, *J. Physiol.,* 150, 417, 1960.
16. **Close, R.,** Dynamic properties of fast and slow muscles of the rat during development, *J. Physiol.,* 173, 74, 1964.
17. **Buller, A. J. and Lewis, D. M.,** Further observations on the differentiation of skeletal muscles in the kitten hind limb, *J. Physiol.,* 176, 355, 1965.
18. **Mann, W. S. and Salafsky, B.,** Enzymic and physiological studies on normal and disused developing fast and slow cat muscles, *J. Physiol.,* 208, 33, 1970.
19. **Kelly, A. M. and Rubenstein, N. A.,** Why are fetal muscles slow? *Nature (London),* 288, 266, 1980.
20. **Gutmann, E. and Melichna, J.,** Changes in neuromuscular relationships in ageing, in *Neurobiology of Ageing,* Ordy, J. M. and Brizzee, K. R., Eds., Plenum Press, N.Y., 1975.
21. **Gutmann, E. and Melichna, J.,** Contractile properties of different skeletal muscles of the rat during development, *Physiologia Bohemoslovaca,* 21, 1, 1972.
22. **Syrovy, I. and Gutmann, E.,** Changes in speed of contraction and ATPase activity in striated muscles during old age, *Exp. Gerontol.,* 5, 31, 1970.
23. **Vyskocil, F. and Gutmann, E.,** Spontaneous transmitter release from nerve endings and contractile properties in the soleus and diaphragm muscles of senile rats, *Experientia,* 28, 280, 1972.
24. **Gutmann, E. and Syrovy, I.,** Contraction properties and myosin. ATPase activity of fast and slow senile muscles of the rat, *Gerontologia,* 20, 239, 1974.
25. **Alnaqeeb, M. A.,** Physiological changes associated with developing and ageing skeletal muscle, Ph.D. Thesis, University of Hull, England, 1981.
26. **Kugelberg, E. and Edstrom, L.,** Differential histochemical effects of muscle contractions on phosphorylase and glycogen in various types of fibres in relation to fatigue, *J. Neurol. Neurosurg. Psychiatr.,* 31, 415, 1968.
27. **Campbell, M. J., McComas, A. J., and Petito, F.,** Physiological changes in ageing muscles, *J. Neurol. Neurosurg. Psychiatr.,* 36, 174, 1973.
28. **Peterson, I. and Kugelburg, E.,** Duration and form of action potential in the normal human muscle, *J. Neurol. Neurosurg. Psychiatr.,* 19, 148, 1949.
29. **Kelly, S. S.,** The effect of age on neuromuscular transmission, *J. Physiol.,* 274, 51, 1978.
30. **Gutmann, E., Hanzlikova, V., and Vyskocil, F.,** Age changes in cross striated muscle of the rat, *J. Physiol.,* 219, 331, 1971.
31. **Schwarz, F. and Wichan, I.,** Membrane potentials of ageing muscle cells, *Acta Biol. Med.,* 17, 96, 1966.
32. **Gutmann, E. and Hanzlikova, V.,** *Age Changes in the Neuromuscular System,* Scientechnica, Bristol, 1972.

33. **Brooke, M. H. and Kaiser, K. K.,** The use and abuse of muscle histochemistry, in *The Trophic Functions of the Neuron,* Ann. N.Y. Acad. Sci., 228, 121, 1974.

34. **Peter, J. B., Barnard, R. J., Edgerton, V. R., Gillespie, C. A., and Stempel, K. E.,** Metabolic profiles of three fibre types of skeletal muscle in guinea pigs and rabbits, *Biochemistry,* 11, 26, 1972.

35. **Burke, R. G., Levine, D. N., Tsairis, P., and Zajac, F. E.,** Physiological types and histochemical profiles in motor units of the cat gastrocnemius, *J. Physiol.,* 234, 723, 1973.

36. **Kugleburg, F.,** Adaptive transformation of rat soleus motor units during growth, *J. Neurol. Sci.,* 27, 269, 1976.

37. **Dubowitz, V.,** Cross-innervated mammalian skeletal muscle; histochemical, physiological and biochemical observations, *J. Physiol.,* 193, 481, 1967.

38. **Gollnick, P., Armstrong, R. B., Saubert, C. W., IV, Sembrowich, W. L., and Shepherd, R. E.,** Glycogen depletion patterns in human skeletal muscle fibres during prolonged work, *Pflugers Arch.,* 244, 1, 1973.

39. **Armstrong, R. B., Saubert, C. W., IV, Sembrowich, W. L., Shepherd, R. F., and Gollnick, P. D.,** Glycogen depletion in rat skeletal muscle fibres at different intensities and durations of exercise, *Pflugers Arch.,* 52, 243, 1974.

40. **Goldspink, G.,** *Mechanics and Energetics of Animal Locomotion,* Alexander, M. and Goldspink, G., Eds., Chapman and Hall, London, Chap. 3, 1977.

41. **Maxwell, L. C., Faulkner, J. A., and Leiberman, D. A.,** Histochemical manifestations of age and endurance training in skeletal muscle fibres, *Am. J. Physiol.,* 224, 356, 1973.

42. **Goldspink, G. and Ward, P. S.,** Changes in rodent muscle fibre types during post-natal growth, under nutrition and exercise, *J. Physiol.,* 296, 453, 1979.

43. **Alnaqceb, M. A. and Goldspink, G.,** Interrelation of muscle fibre types, diameter and number in ageing white rats, *J. Physiol. Proc.,* 310, 56P, 1980.

44. **Caccia, M. R., Harris, J. B., and Johnson, M. A.,** Morphology and physiology of skeletal muscle in ageing rodents, *Muscle Nerve,* 2, 202, 1979.

45. **Bass, A., Gutmann, E., and Hanzlikova, V.,** Biochemical and histochemical changes in energy supply-enzyme pattern of muscle of the rat during old age, *Gerontologia,* 21, 31, 1975.

46. **Larsson, L., Sjodin, B., and Karlsson, J.,** Histochemical and biochemical changes in human skeletal muscle with age in sedentary males age 22-65 years, *Acta Physiol. Scand.,* 103, 31, 1978.

47. **Alnaqceb, M. A. and Goldspink, G.,** Morphometric changes associated with developing and ageing skeletal muscle (in preparation).

48. **Ihemelandu, E. C.,** Decrease in fibre number of dog pectineus muscle with age, *J. Anat.,* 130, 69, 1980.

49. **Gutmann, E., Hanzlikova, V., and Jackoubek, B.,** Changes in the neuromuscular system during old age, *Exp. Gerontol.,* 3, 141, 1968.

50. **Moore, M. J., Rebiez, J. J., Holden, M., and Adams, R. D.,** Biometric analysis of normal skeletal muscle atrophy, *Acta Neuropathol. Berl.,* 19, 51, 1971.

51. **Rowe, R. W. D.,** The effect of senility on the skeletal muscles in the mouse, *Exp. Gerontol.,* 4, 119, 1969.

52. **Van Linge, B.,** The response of muscle to strenuous exercise. An experimental study in the rat, *J. Bone Joint Surg.,* 44B, 7711, 1962.

53. **Vaughan, H. S. and Goldspink, G.,** Fibre number and fibre size in a surgically overloaded muscle, *J. Anat.,* 129, 293, 1979.

54. **Tauchi, H., Yoshioka, T., and Kobayashi, H.,** Age change of skeletal muscles of rats, *Gerontologia,* 17, 219, 1971.

55. **Snow, M. H.,** The effect of ageing on satellite cells in skeletal muscles of mice and rats, *Cell Tissue Res.,* 185, 399, 1979.

56. **Allbrook, D. B., Han, M. F., and Hellmuth, A. G.,** Population of muscle satellite cells in relation to age and mitotic activity, *Pathology,* 3, 233, 1971.

57. **Schiaffino, S., Bormioli, P., and Aloisi, M.,** The fate of newly formed satellite cells during compensatory muscle hypertrophy, *Virch. Arch. Abstr.,* B21, 113, 1976.

58. **Aniansson, A., Grimby, G., Hedberg, M., and Krotkiewski, L.,** Muscle morphology, enzyme activity and muscle strength in elderly men and women, *Clin. Physiol.* 1, 73, 1981.

59. **Parizkova, J., Eiselt, E., Spynarova, S., and Wachtlova, M.,** Body composition aerobic capacity and density of muscle capillaries in young and old men, *J. Appl. Physiol.,* 31, 323, 1971.

60. **Fujisawa, K.,** Some observations on the skeletal musculature of aged rats. I. Histochemical aspects, *J. Neurol. Sci.,* 22, 353, 1974.

61. **Fujisawa, K.,** Some observations on the skeletal musculature of aged rats. II. Fine morphology of diseased muscle fibres, *J. Neurol. Sci.,* 24, 447, 1975.

62. **Steenis, G. V. and Kroes, R.,** Changes in the nervous system and musculature of old rats, *Vet. Pathol.,* 8, 320, 1971.

63. **Orlander, J., Keissling, K. H., Larsson, L., Karlsson, J., and Aniansson, A.,** Skeletal muscle metabolism and ultrastructure in relation to age in sedentary men, *Acta Physiol. Scand.,* 104, 249, 1978.

64. **Cottrell, L.,** Histologic variations with age in apparently normal peripheral nerve trunks, *Arch. Neurol. Psychiatr.,* 43, 1138, 1940.

65. **Semenowa-Tjan-Schanskaja, W.,** Die morphologischen Veranderungen der peripheren Nerven beim Menschen im Greisenalter, *Z. Ges. Neurol. Psychiatr.,* 172, 587, 1941.

66. **Berg, B. N., Wolf, A., and Simms, H.,** Degenerative lesion of spinal roots and peripheral nerves in ageing rats, *Gerontologia,* 6, 72, 1962.

67. **Gutmann, E. and Hanzlikova, V.,** Age changes of motor end plates, *Gerontologia,* 11, 12, 1965.

68. **Fujisawa, K.,** Some observations on the skeletal musculature of aged rats. III. Abnormalities of terminal axons found in motor end-plates, *Exp. Gerontol.,* 11, 43, 1976.

69. **Cardasis, C. A.,** Ultrastructural evidence of continued reorganization of the aging rat soleus neuromuscular junction (Abstr.). American Association of Anatomists Meeting, 1981.

70. **Bick, E. M.,** Ageing in the connective tissues of the human musculoskeletal system, *Geriatrics,* 16, 448, 1961.

71. **Lowry, O. H., Hastings, A. B., Hull, T. Z., and Brown, A. N.,** Histochemical changes associated with ageing. II. Skeletal and cardiac muscle in the rat, *J. Biol. Chem.,* 143, 271, 1942.

72. **Dreyfus, J. C., Schapira, G., and Bourliene, F.,** Modifications chimiques du muscle au cours de la senescence chez le rat, *C. R. Soc. Biol. (Paris),* 148, 1065, 1954.

73. **Weinstock, J. M., Epstein, S., and Milhorat, A. T.,** Enzyme muscular dystrophy. III. In hereditary muscular dystrophy in mice, *Proc. Soc. Exp. Biol. N.Y.,* 99, 271, 1958.

74. **Dam, H., Prange, I., and Sundergaard, E.,** Muscular degeneration (white striation of muscles) in chicks reared on vitamin E-deficient, low fat diet, *Acta Pathol. Microbiol. Scand.,* 31, 172, 1952.

75. **Schaub, M. C.,** The ageing of collagen in the striated muscle, *Gerontologia,* 8, 16, 1963.

76. **Schaub, M. C.,** Degradation of young and old collagen by extracts of various organs, *Gerontologia,* 9, 52, 1964.

77. **Knorring, J. V.,** Effect of age on the collagen content of the normal rat myocardium, *Acta Physiol. Scand.,* 79, 216, 1970.

78. **Mohan, S. and Radha, E.,** Age-related changes in rat muscle collagen, *Gerontology,* 26, 61, 1980.

79. **Sasaki, R., Ichikawa, S., Yamagiwa, H., Ito, A., and Yamagata, S.,** Aging and hydroxyproline content in human heart muscle, *Tohoku J. Exp. Med.,* 118, 11, 1976.

80. **Verzar, F.,** Das Altern des Collagens, *Helv. Physiol. Pharmacol. Acta,* 14, 207, 1956.

81. **Harrison, D. E. and Archer, J. R.,** Measurement of changes in mouse tail collagen with age. Temperature dependence and procedural detail, *Exp. Gerontol.,* 13, 75, 1978.

82. **Alnaqeeb, M. A., Al Zaid, N., and Goldspink, G.,** Muscle stiffness and connective tissue in developing and ageing muscle (in preparation).

83. **Friedman, S. M., Sreter, F. A., and Friedman, C. L.,** The distribution of water sodium and potassium in the aged rat: a pattern of adrenal preponderance, *Gerontologia,* 7, 44, 1963.

84. **Mitolo, M.,** Biochemistry of muscle ageing, *J. Gerontol.,* 33 (suppl.), 63, 1964.

85. **Srivastava, U.,** Polyribosome concentration of mouse skeletal muscle as a function of age, *Arch. Biochem. Biophys.,* 130, 129, 1969.

86. **Narayanan, N. and Eapen, J.,** Age-related changes in the incorporation of ^{14}C-leucine into myofibrillar and sarcoplasmic proteins of red and white muscles of chicks, *Ajebak,* 53, 59, 1975.

87. **Gold, P. H., Gee, M., Nordgreen, R., and Strehler, B. L.,** A re-examination of the efficiency of oxidative phosphorylation versus age in rat liver, kidney and heart, *Proc. 7th Int. Congr. Gerontol. Wien,* Vol. 8, 1966, 103.

88. **Kment, S. A., Leibetseder, J., and Burger, H.,** Gerontologisch Untersuchungen an Rattenherzmitochrondrien, *Gerontologia,* 12, 193, 1966.

89. **Weinbach, E. C.,** Biochemical changes in mitochondria associated with age, in *The Biology of Ageing,* Strehler, B. L., Ed., 1960, 328.

90. **Froklis, V. V. and Bogatskaya, L. N.,** The energy metabolism of myocardium and its regulation in animals of various ages, *Exp. Gerontol.,* 3, 199, 1968.

91. **Froklis, V. V.,** The role of self-regulation and adaptation in the mechanism of ageing, *Proc. 8th Int. Congr. Gerontol.,* Washington, Vol. 1, 1969, 529.

92. **Ermini, M.,** Ageing changes in mammalian skeletal muscle, *Gerontology,* 22, 301, 1076.

93. **Ermini, M.,** The energy metabolism of the ageing skeletal muscle, *Aktuel Gerontol.,* 6, 151, 1976.

94. **Barany, M.,** ATPase activity of myosin correlated with speed of muscle shortening, *J. Gen. Physiol.,* 50, 197, 1967.

95. **Ermini, M.,** Das altern der skelettmuskulater. III. Untersuchungen an der Myofibrillen — Adenosin-5'-Triphosphatase bei verschieden alten Ratten; phil. — nat Diss. 1969 (Basel), *Gerontologia,* 16, 72, 1970.

96. **Rockstein, M. and Brandt, K. F.,** Changes in phosphorus metabolism of the gastrocnemius muscle in ageing white rats, *Proc. Soc. Exp. Biol. Med.,* 107, 377, 1961.

97. **Aniansson, A. and Gustafsson E.,** Physical training in elderly men with special reference to quadriceps muscle strength and morphology, *Clin. Physiol.* (in press).

98. **Adams, G. M. and deVries, H. A.,** Physiological effects of an exercise training regimen upon women aged 52 to 79, *J. Gerontol.,* 28, 50, 1973.

99. **Benestad, A. M.,** Trainability of old men, *Acta Med. Scand.,* 178, 321, 1965.

100. **Liesen, H., Heikkinen, E., Suominen, H., and Michel, D.,** Der Effekt eines Zwolfwochligen Ausdauertraining auf die Leistungsfoigkeit und den Muskelstoffwechsel bei untrainierten Mannern des 6 und 7 Lebensjahrzehuts, *Sportartzt Sportmed.,* 26, 26, 1975.

101. **Suominen, H., Heikkinen, E., Liesen, H., Michel, D., and Hollman, W.,** Effects of 8 weeks endurance training on skeletal muscle metabolism in 56-70 year-old sedentary men, *Eur. J. Appl. Physiol.,* 37, 173, 1977.

AGING OF THE SKELETAL SYSTEM AND SUPPORTING TISSUE

Edgar A. Tonna

Senile atrophy of bone, senile osteoporosis, degenerative joint disease (osteoarthritis), periodontal disease, and other connective tissue disorders are of wide occurrence in the human population and many studies have been focused on these medical problems. The parameter of aging may well be one of a number of essential components in the complex equation of each of these diseases. Surprisingly, however, the significance of age changes to the predisposition of the aging individual to connective tissue disorders is as obscure as the etiology. The specific study of the aging of skeletal tissues is presently quite limited.

The aim of this section is to outline the basic features of age-associated changes as have been observed to date in skeletal tissues.

GENERAL AGING OF BONE

Clinical studies[1] of the older segment of the human population reveal increased incidence of senile osteoporosis, long bone fractures, vertebral atrophy, and compression fractures (Figure 1). The rates of atrophy for cortical and trabecular bone differ in the axial and appendicular skeleton.[2] Between 40 and 80 years of age in the female, a bone loss of 40% occurs, while in the male it begins between 55 and 65 years of age.[3] The rates of bone loss are claimed by these authors to remain constant. In 30- to 40-year-olds, the rates of linear decline are about 10% per decade in the female and about 1 to 5% per decade in the male. Increased resorption of trabecular bone is more obvious than cortical bone resorption with age. The rates of atrophy of cortical thickness is greater in the female approaching the rate determined for trabecular bone. Cortical and trabecular bone mass at the iliac crest decrease significantly with age in women, but not in men.[4] Trotter et al.,[5] report that bone densities decrease at a uniform and parallel rate with increasing age in both sexes and race groups. A study of the fourth lumbar vertebra from 95 cases of sudden death shows that the percentage volume of the vertebral body occupied by bone decreases with age (Figure 2).[6] The difference between the sexes are small, of the order of 1%. Another study of vertebral bone reveals that the percentage of mineralized matrix within the medullary canal decreases more rapidly in females as a function of age.[7] The traverse trabeculae are more labile, while the vertical struts remain.[8] The ash content of vertebral and rib bone is markedly decreased with age (females somewhat more so than males) from a maximum at 21 to 30, to less than 50% (vertebrae) and 65% (rib) in 81 to 90 years of age.[2] An earlier study of 78 cases of sudden death notes that the ash content of vertebral cancellous bone shows little sex difference.[9] Sissons et al.,[10] studying the iliac crest and cortical femoral bone, also found little sex difference and arrived at a similar conclusion as Dunnill,[6] i. e., that much, if not all of the decrease in bone mass is a physiological manifestation of aging rather than a disease process peculiar to old age. Atkinson[11] goes so far as to claim that "senile osteoporosis, rather than being a degenerative disease of old age, seems to be an extension of a developmental process." Others attribute the etiology of senile osteoporosis to absence of exercise, lactase deficiency, diminished estrogen, an inability to adapt to low calcium diets, low calcium intake, or a small increase in parathyroid hormone secretion in response to a small decrement in the level of serum calcium.

Differences in laboratory data may well be due in part to the study of different bone structures, methods of analyses, and study of differing population samples. Clinical experiences, however, reveal the susceptibility of female bone to marked atrophy with age, accounting for the higher incidence of normal fractures in aged females. Although the rate

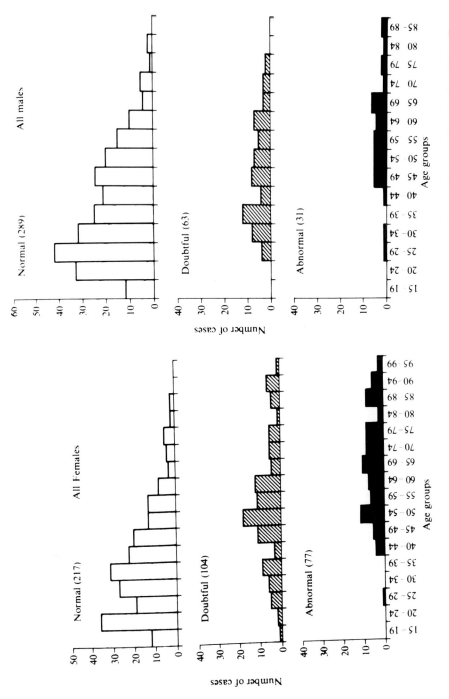

FIGURE 1.   Apparent international incidence of osteoporosis vs. age based on analyses of spinal X-ray films in females (left) and males (right). (From Nordin, B. E. C., *Clin. Orthoped.*, 45, 17, 1966. With permission.)

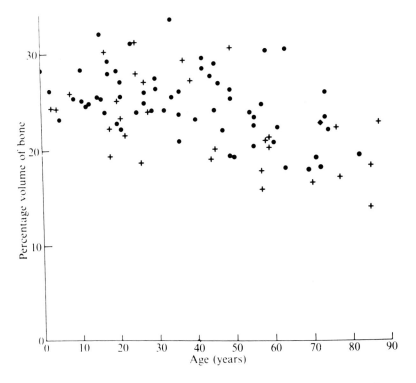

FIGURE 2. Percentage volume of human vertebral body bone plotted vs. age. Dots represent males; crosses represent females. (From Dunnill, M. S., Anderson, J. A., and Whitehead, R., *J. Pathol. Bacteriol.*, 94, 275, 1967. With permission.)

of vertebral atrophy appears to be similar in both sexes, in the female, trabecular bone is less mineralized. A higher incidence of vertebral collapse or compression fracture occurs in aging females.[2] Although, normally an increased level of bone resorption occurs with advancing age, in senile osteoporosis, the level is often increased above nonosteoporotic individuals of the same age. New bone formation, however, appears to remain normal.[12] Bartley et al.[13] report a high correlation between crushing strength and reduced ash content of vertebral bone with increasing age. Microscopically, bone resorption is seen to occur in discrete foci in adult bone during normal remodeling. The average size of the foci is altered only slightly throughout life;[14] however, the number of foci per unit of bone increases significantly with age.[15-17]

GENERAL AGING OF CARTILAGE

Progressive age changes associated with cartilage are known from numerous studies of animal and human material.[18] Cartilage changes start as early as 20 or 30 years,[19] converting the typical, translucent, bluish character of articular cartilage to an opaque, yellowish color. Cracking, fraying, and shredding of the surface are noted by age 30, eventually producing tears and fibrillation of the surface leading to deep vertical fissures. Chondrocytes become arranged into large clusters. Cell death is progressive and results in the formation of masses of irregular bodies. The thickness of articular cartilage diminishes as the removal of cartilage layers results in eventually removing the deeper calcified zone. Exposed subchondral bone is rubbed smooth and becomes highly polished (eburnation). The subchondral bone marrow becomes hyperemic, red blood cells extravasate, and fibrosis results. Osteophytes form at the margins of the condyles. The reparative process is limited, and the success of the process

Table 1

**INCIDENCE OF CARTILAGE CHANGES AND OSTEOARTHRITIS
IN JOINTS AT DIFFERENT AGES**[a]

Age groups (years)	Knee A (%)	Knee B (%)	Shoulder A (%)	Shoulder B (%)	Hip A (%)	Hip B (%)	Elbow A (%)	Elbow B (%)
15—19	3.1	0.0	0.0	0.0	0.0	0.0	0.0	0.0
20—29	9.2	0.0	0.8	0.0	0.8	0.0	1.7	0.0
30—39	48.1	1.0	2.0	0.0	7.8	0.0	18.0	0.0
40—49	74.0	0.8	4.7	0.0	16.7	0.8	26.9	0.0
50—59	87.1	2.6	14.3	0.7	44.8	0.7	61.7	1.3
60—69	92.6	12.0	22.1	2.6	60.0	2.7	73.7	5.2
70—79	97.5	33.3	44.1	9.7	76.1	12.2	87.4	10.5
80—95	100.0	39.4	60.0	15.7	89.4	16.7	95.8	15.5

[a]   Columns A: joints show naked eye evidence of cartilage destruction, i.e., osteoarthritis of all grades. Columns B: joints show "moderate or severe arthritis deformans", i.e., Grade III or Grade IV osteoarthritis.

Data from Collins, D. H., *The Pathology of Articular and Spinal Diseases,* Edward Arnold, Ltd., England, 1949, 74.

is dependent on depth of wound, proximity to blood vessels, protection from abrasion, and age of the organism.[20] With increasing age, the synovial membrane reveals formation of a villous structure with focal accumulation of mononuclear cells. Fibrosis is also observed. The viscosity of the synovial fluid is increased as the hyaluronate levels decrease beyond the fourth decade.[21]

The epiphyseal plates of long bones reveal age changes as proliferative and hypertrophic activities decline with increasing age leading to mineralization of the residual structure and closure of the plate; a process which signifies normal cessation of longitudinal bone growth. The percentage of water decreases with age at the intervertebral disk and progressive degenerative changes occur in the nucleus pulposus of man beyond 30 years.[22] The tissue loses its turgor and becomes friable in older individuals. Together with a decrease in mineral content and structural strength of bone, cartilage age changes lead to an increase in the incidence of vertebral atrophy and compression fractures.

The histologic description of osteoarthritic changes (degenerative joint disease) are similar to those recognized as age changes. Consequently, it is not known whether osteoarthritic changes will always occur with time should the organism live long enough or if etiologic factors in addition to aging cartilage are necessary. Osteoarthritis exhibiting the characteristics of degenerative joint disease in man have long been reported in different animal species.[23] In a study of human sternoclavicular joints, age alterations and osteoarthritis were noted to appear during the third decade.[24] The incidence and severity of osteoarthritis increase to the eighth decade, with the male exhibiting greater lesions than the female, and Negroes appear to be more susceptible than Caucasians. A significant decrease occurs after the ninth decade. Another human study involving the incidence of cartilage changes and osteoarthritis in knee, shoulder, hip, and elbow joints at different ages (Table 1) shows similar results.[25] Although degenerative changes are reported in regions that bear the greatest weight and shock, osteoarthritis occurs both in weight- and nonweight-bearing bones.[23]

Scattered foci of cell proliferation and hypertrophy of chondrocytes appear early in man and mouse.[20] This is associated with or followed by cellular and matrix degeneration.[26] Aging articular cartilage of mouse,[27] rabbit,[28] and man[29] have been studied at the ultrastructural level. Chemical studies[30] of human material show significant reduction in all glycosaminoglycan fractions in osteoarthritic cartilage, but no change in carbohydrate distribution.

Table 2
MAJOR EFFECTS OF AGING ON CELLS AND MATRIX[a,b]

Domains	Observed changes
Topographical/morphological	Cell distribution; cell-to-cell contact; cell to matrix ratios
	Cell size; shape; ultrastructural (e.g., organelles, membranes, granules, cytosol)
Replicative system/population	Cell proliferation; cell cycle kinetics; circadian rhythms; proliferative (reactivation) potential to perturbations (e.g., trauma), repair
	Cell numbers for growth, repair, remodeling
Metabolic/energy	Cell and tissue maintenance requirements, matrix precursor synthesis, replicative function
	Number, size and activity of mitochondria, associated respiratory enzymes, ATPase, etc.
Matrix/nutrition and waste	Fiber diameters, compactness, fiber fractions, turnover, cross-linking, type, carbohydrate fraction ratios, ground substance changes, mineralization, resorption
	Availability of cell nutrition, adequate provision for waste elimination

[a] Intrinsic changes occur at the biochemical level and have an effect on higher levels of organization (ultrastructural, cell, tissue, organ, system, organism, population). In turn, significant changes at higher levels through positive feedback enhance and affect additional biochemical changes.

[b] The listing implies complex interrelationships and multiple effects between the various domains.

Numerous factors have been attributed to the causation of osteoarthritis including matrical alterations leading to the release of the dormant growth potential of cartilage cells,[27] mechanical stress,[31] hormonal influence,[32] lysosomal enzyme access to cartilage matrix with age,[33] nutritional and hereditary factors,[34] etc. The relationships between age changes and osteoarthritis are obscure as are the inductive factors. Further studies are very much in order.

AGING OF CELLS AND MATRIX

The effects of aging on connective tissues, including skeletal tissues, are observed on various properties of cells and matrices. These properties are interdependent and essential for the viability of cartilage cells and maintenance of the matrix. Table 2 organizes the observed changes into four major domains, each of which are affected by age changes and, consequently, affect each other. Furthermore, changes at lower levels of organization affect higher orders and through positive feedback are seen to augment further changes at the lower levels, especially the biochemical level where negative feedback is operational.

Osteogenic Tissues

Figure 3 diagramatically reveals the general cellular complement of the periosteum and endosteum. The periosteum covers the external bone surfaces, while the endosteum covers the internal surfaces consisting of the medullary canal, trabecular surfaces, as well as the Volkmann and Haversian canals in higher mammalian forms. Unlike the periosteum, the endosteum is free of a fibrous layer which is functionally protective, supplies vascular elements to the osteogenic layer, and exhibits a significant role during bone repair.[35] The progenitive cellular elements of the periosteum and endosteum generate the bone cells which manufacture the collagen matrix of bone and the associated glycosaminoglycans, as well as maintain the matrical environment for mineralization and resorption of bone surfaces. Surface bone cells, for a time, become osteocytes as they are completely surrounded by matrix. This matrix subsequently becomes mineralized into bone. Canaliculi serve as passageways for numerous cytoplasmic projections extending from periosteal and endosteal osteoblasts, as

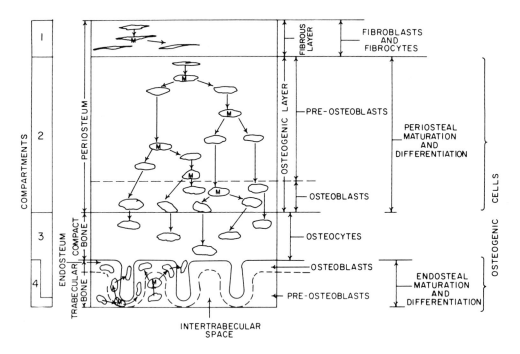

FIGURE 3.   The diagram illustrates a cross-section of the femoral cortex and the arrangement of its cellular components into compartments based on data from [³H-] thymidine studies. (From Tonna, E. A. and Cronkite, E. P., *Clin. Orthoped.,* 30, 218, 1963. With permission.)

well as osteocytes themselves. Consequently, cellular communication exists among cells maintaining bone viability. With increasing age, subtle alterations occur at all levels of organization. Subsequently these alterations become prominent, deleterious features, resulting in reduced rates of functional activity, repair, response to stress, and undoubtedly expose the bone surface to further skeletal disorders. As cellular communications break down, osteocyte death becomes a commonly observed feature (Figure 4). Aging studies of bone cells are scanty. Information is, however, available on the short-lived Brookhaven National Laboratory (BNL) Swiss albino mouse, extensively used by the author. The animal has a mean lifespan of less than 1 year. Only a few animals survive to over 2 years.

Cell Proliferative Activity

In the BNL aging mammalian model, cellular proliferative activity of the femoral periosteum is high at birth,[36-40] followed by a drastic reduction by 8 weeks of age to an insignificant level. In the rat, the number of functional osteoblasts is significantly reduced by 26 weeks of age and continues to diminish with increasing age (Table 3).[40] Cell number changes are preceded by reduced mitochondrial numbers (Table 4), mitochondrial surface, and volume (Table 5).[40] Histochemical analysis of mitochondrial associated enzymes reveals an increase in succinic dehydrogenase and cytochrome oxidase to 5 weeks of age, which is concomitant with the peak of femoral bone growth. This is followed by a significant decline to 8 weeks and further decline with increasing age.[41] Adenosine triphosphatase activity follows a similar pattern in the aging bone cells of mice.[42] The effects of time and aging are best appreciated in response to stress, such as midfemoral fractures. ³H-thymidine autoradiographic studies[43] show reactivation of cell proliferation at about 10 hr after fracture. A peak labeling index is reached at about 24 hr in the young mouse. Away from the fracture site, the labeling indices return to normal by about 4 to 5 days. At the fracture site, labeling indices decrease significantly until an elevated plateau is reached, also by about 5 days.

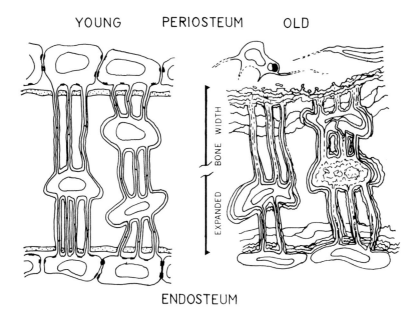

FIGURE 4.   The diagram illustrates the loss of bone cell communication and cellular biofeedback of periosteal bone surface with increasing age.

Table 3
MEAN NUMBER OSTEOBLASTS/UNIT AREA ± SE OF MEAN ESTIMATED FROM MIDFEMORAL REGIONS OF RATS AT DIFFERENT AGES

Group	I	II	III	IV	V	VI
Males	10.0 ± 0.7	10.7 ± 0.5	10.5 ± 0.3	2.9 ± 0.9	4.3 ± 0.7	0.8 ± 0.3
	(16)[a]	(11)	(11)	(9)	(10)	(10)
Females	10.4 ± 0.3	10.4 ± 0.6	10.5 ± 0.3	2.8 ± 1.5	2.9 ± 0.7	1.2 ± 0.2
	(9)	(14)	(14)	(16)	(15)	(24)
Total males and	10.2 ± 0.5	10.6 ± 0.4	10.5 ± 0.2	2.8 ± 1.9	3.5 ± 0.5	1.1 ± 0.2
females	(25)	(25)	(25)	(25)	(25)	(34)
Age (weeks)	1	5	8	26	52	104

[a]   Values in the parentheses indicate the number of areas examined.

From Tonna, E. A. and Pillsbury, N., *Anat. Rec.*, 134, 739, 1959. With permission.

Table 4
MEAN NUMBER MITOCHONDRIA ± SE OF PERIOSTEAL OSTEOBLASTS OF FEMORA OF RATS AT DIFFERENT AGES

Groups (age in weeks)	I (1)	II (5)	III (8)	IV (26)	V (52)	VI (104)
Males	29.0 ± 1.4	37.2 ± 2.5	16.6 ± 1.0	18.8 ± 1.4	9.3 ± 0.6	3.6 ± 0.6
Females	27.8 ± 1.1	35.6 ± 1.7	17.0 ± 0.9	17.1 ± 1.3	9.6 ± 0.6	3.0 ± 0.3
Total males and females	28.4 ± 1.3	37.1 ± 2.4	16.8 ± 0.9	17.9 ± 1.4	9.5 ± 0.6	3.3 ± 0.5

From Tonna, E. A. and Pillsbury, N., *Anat. Rec.*, 134, 739, 1959. With permission.

Table 5
MORPHOLOGICAL ANALYSIS OF THE MITOCHONDRIAL COMPLEMENT OF PERIOSTEAL OSTEOBLASTS OF THE FEMORA OF MALE RATS AT DIFFERENT AGES[a]

Groups (age in weeks)	Av. length ± SE (μm)	Av. width ± SE (μm)	Av. volume ± SE (μm³)	Av. mitochondrial vol/cell ± SE (μm³)	Av. surface ± SE (μm²)	Av. mitochondrial surface/cell ± SE (μm²)
Group I (1)	1.13 ± 0.06	0.54 ± 0.01	0.27 ± 0.02	7.99 ± 0.03	3.47 ± 0.14	100.77 ± 0.20
Group II (5)	1.00 ± 0.02	0.65 ± 0.01	0.35 ± 0.01	12.82 ± 0.03	3.90 ± 0.15	144.96 ± 0.26
Group III (8)	0.82 ± 0.02	0.52 ± 0.01	0.20 ± 0.01	3.23 ± 0.01	2.61 ± 0.08	43.20 ± 0.08
Group IV (26)	0.62 ± 0.01	0.42 ± 0.01	0.09 ± 0.00	1.90 ± 0.06	1.09 ± 0.05	24.31 ± 0.06
Group V (52)	0.51 ± 0.02	0.40 ± 0.01	0.07 ± 0.00	0.69 ± 0.00	0.92 ± 0.06	8.58 ± 0.03
Group VI (104)	0.49 ± 0.03	0.35 ± 0.02	0.06 ± 0.01	0.21 ± 0.00	0.74 ± 0.06	2.67 ± 0.03

[a]   60 Mitochondria were counted in each case.

From Tonna, E. A. and Pillsbury, N., *Anat. Rec.*, 134, 739, 1959. With permission.

Table 6
INCIDENCE OF PERIOSTEAL CELLS LABELED WITH TRITIATED THYMIDINE[a]

	Hours after fracture							
	1	2	4	8	16	24	32	48
Control femora								
Labeling index	0.018	0.011	0.025	0.019	0.015	0.008	0.023	0.020
No. cells counted	2000	2435	2000	2287	1939	1962	2169	2075
Fractured femora								
Labeling index	0.012	0.014	0.025	0.011	0.115	0.173	0.254	0.212
No. cells counted	2107	2055	2030	1940	2436	2499	2713	3814

[a]   In mouse femora with and without fracture (5-week-old).

From Tonna, E. A. and Cronkite, E. P., *J. Bone Jt. Surg.*, 43A, 352, 1961. With permission.

This level is maintained until repair is completed[39,44] (Tables 6,7). In old animals (Tables 8,9), cell proliferation is also initiated at about 10 hr post-trauma, but from a somewhat lower normal base line. A peak is reached by about day 2, after which time the labeling indices of the areas away from the fracture site return to normal by day 4. At the fracture site, activation of cell proliferation continues to a new peak at 4 days. Decreased cell numbers and increased bone surface size, together with osteocyte formation deplete periosteal cells with time. Consequently, a suitable population size adequate for repair is not achieved until later in older animals. This peak is, however, followed by a decrease in labeling indices to an elevated plateau which is gradually reduced as repair is completed, in a similar manner as occurs in young animals. Apparently, aging does not change the biological qualitative response of bone cells; only the quantitative response is effected. The organism is, therefore, assured of effective skeletal repair throughout its lifespan via an emergency mechanism wherein the cell proliferative potential can be reactivated. The proliferative potential is not lost with aging; only the proliferative activity is considerably diminished.[45-47] Respiratory enzyme studies in rats, however, show progressively diminished levels of enzyme activity when examined 1 to 2 weeks post-trauma.[47] This is not the case with acid and alkaline

Table 7
INCIDENCE OF PERIOSTEAL CELLS LABELED WITH TRITIATED THYMIDINE[a]

	Days after fracture					
	3	4	5	7	10	14
Control femora						
Labeling index	0.029	0.012	0.021	0.022	0.019	0.022
No. cells counted	2000	2400	2400	2350	2550	2500
Fractured femora						
Periosteum over fracture areas						
Labeling index	0.214	0.098	0.150	0.090	0.084	0.065
No. cells counted	2700	2000	1121	1800	2240	1650
Periosteum away from fracture areas						
Labeling index	0.069	0.033	0.018	0.019	0.011	0.014
No. cells counted	900	2000	2220	1569	2350	2300
Periosteum over and away from fracture areas						
Labeling index	0.178	0.065	0.061	0.059	0.047	0.036
No. cells counted	3600	4000	3341	3369	4590	3950

[a] In mouse femora with and without fracture (5-week-old).

From Tonna, E. A. and Cronkite, E. P., *J. Bone Jt. Surg.*, 43A, 352, 1961. With permission.

Table 8
INCIDENCE OF PERIOSTEAL CELL LABELING WITH TRITIATED THYMIDINE[a]

	Hours after fracture			
	8	16	24	32
Nonfractured femora				
Labeling index	0.003	0.005	0.003	0.002
No. cells counted	2400	3200	2400	2400
Fractured femora				
Labeling index	0.005	0.019	0.051	0.091
No. cells counted	1500	1649	1030	1400

[a] In 18-month-old mouse femora with and without fracture.

From Tonna, E. A. and Cronkite, E. P., *J. Bone Jt. Surg.*, 44A, 1557, 1962. With permission.

phosphatases in that enzyme activity is increased following trauma at all ages.[48] Furthermore, the depleted cell population is replenished and is responsible for fracture repair, not the original old population. ^{3}H-proline autoradiographic studies reveal that in old mice, the metabolic activity of newly formed cells is delayed 1 week. However, in time they reveal the capability of contributing to repair in measure equal to or exceeding that of younger animal cells (Tables 10, 11).[47]

Biochemical Activity

The utilization of a variety of ^{3}H-amino acids (histidine, glycine, proline) by skeletal cells has been studied extensively in the aging BNL mouse.[49-58] With increasing age, periosteal incorporation of these radiotracers into protein decreases while incorporation peaks are shifted to later time periods in older mice (Figure 5). The activity of the endosteum is

Table 9
INCIDENCE OF PERIOSTEAL CELL LABELING WITH TRITIATED THYMIDINE[a]

	Days after fracture							
	2	4	5	6	7	8	10	14
Nonfractured femora								
Labeling index	0.002	0.005	0.004	0.002	0.002	0.001	0.003	0.061
No. cells counted	2400	1188	1700	2800	2500	2000	2600	2200
Fractured femora								
Periosteum at fractured areas								
Labeling index	0.110	0.180	0.134	0.094	0.069	0.063	0.044	0.045
No. cells counted	1845	1866	1900	1800	1500	2600	2100	1500
Periosteum away from fractured areas								
Labeling index	0.105	0.005	0.011	0.012	0.007	0.002	0.007	0.005
No. cells counted	400	500	527	600	1370	500	1120	1670

[a]   In 18-month-old mouse femora with and without fracture.

From Tonna, E. A. and Cronkite, E. P., *J. Bone Jt. Surg.*, 44A, 1557, 1962. With permission.

Table 10
AVERAGE AUTORADIOGRAPHIC GRAIN COUNTS OVER MOUSE FEMORAL PERIOSTEAL OSTEOBLASTS 30 MIN AFTER ADMINISTRATION OF H^3-PROLINE

Age at start of experiment (weeks)	Grain counts over nonfractured femurs[a]	Grain counts over fractured femurs[a] Time after fracture				
		8 hr	16 hr	1 day	7 days	14 days
5	6.23	5.69	8.27	10.06	10.88	12.10
52	2.09	2.17	3.34	3.48	10.23	13.01

[a]   Each value represents an analysis of 300 osteoblasts.

From Tonna, E. A., in *The Healing of Osseous Tissues*, Natl. Acad. Sci. Symp., Warrington, Va., 1965, 100. With permission of the National Academy of Sciences, Washington, D.C.

Table 11
AVERAGE AUTORADIOGRAPHIC GRAIN COUNTS OVER MOUSE FEMORAL ENDOSTEAL OSTEOBLASTS 30 MIN AFTER ADMINISTRATION OF H^3-PROLINE

Age at start of experiment (weeks)	Grain counts over nonfractured femurs[a]	Grain counts over fractured femurs[a] Time after fracture				
		8 hr	16 hr	1 day	7 days	14 days
5	14.06	17.55	19.32	24.63	10.08	13.65
52	4.95	9.25	10.14	8.06	11.56	16.32

[a]   Each value represents an analysis of 300 osteoblasts.

From Tonna, E. A., in *The Healing of Osseous Tissues*, Natl. Acad. Sci. Symp., Warrington, Va., 1965, 100. With permission of the National Academy of Sciences, Washington, D.C.

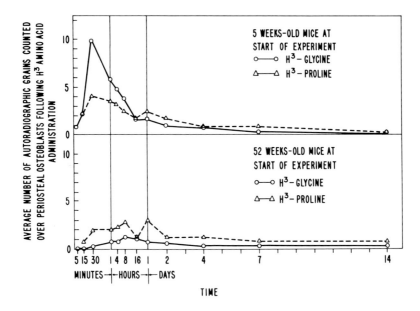

FIGURE 5. The average autoradiographic grain counts observed over femoral periosteal osteoblasts following the administration of [³H-]glycine and [³H-]proline in aging BNL mice. (From Tonna, E. A., *Biology and Clinical Medicine, Proc. 7th Int. Congr. Gerontol.*, 1, 225, 1966.[55] With permission.)

retained for some time beyond that of the periosteum because of the persistent involvement of the endosteum in remodeling (Figure 6). Precursor protein matrix synthesis by bone cells has also been investigated involving cytoplasmic staining for RNA[59] or its synthesis using tritiated precursors as radiomarkers. Brachet's method for RNA shows the staining intensity to increase during cell differentiation into functional osteoblasts and diminish when the cells are converted to osteocytes. With increasing age RNA staining diminishes.[45,46] Autoradiographic studies also reveal a decrease in the maximum incorporation of ³H-uridine, and a shifting of peaks (Figures 7,8).[60,63] A study of protein-bound sulfhydryl (PBSH) and disulfide groups (PBSS) exhibits gradual decreases with increasing age.[64] These substances are abundant in the cytoplasm of osteoblasts and osteoclasts involved in bone formation and resorption.

The glycosaminoglycan (mucopolysaccharide) composition of bone has been investigated extensively by Meyer.[65] This complex chemical group includes chondroitin-4-sulfate, chondroitin-6-sulfate, keratosulfate, hyaluronic acid, and neutral mucopolysaccharides, which are complexed with proteins forming the proteoglycans (glycoproteins). Vajlens[66] reports that the glycosaminoglycan concentration in association with collagens on a per weight basis is about 0.8 to 0.6% in older persons. Following maturity, the chondroitin sulfates isolated from different age groups show no difference in sulfation and in the distribution of molecular size. Aging does not reveal significantly altered concentrations in compact bone. A decrease in total hexosamine is, however, reported with advancing age in the cancellous bone of vertebral bodies.[67] These data may reflect changes in proteoglycans rather than sulfated glycosaminoglycans. ³⁵S-sulfate studies show diminished uptake with increasing age, implying reduced sulfated glycosaminoglycan synthesis.[68-70] Actually, little is known of the synthesis and degradation of glycosaminoglycans in aging bone. Current studies using ³H-galactose[71] reveal diminished cellular uptake and turnover rates of the carbohydrate with increasing age, but an autoradiographic analysis of the matrix surrounding old cells surprisingly shows an impressive continued utilization.

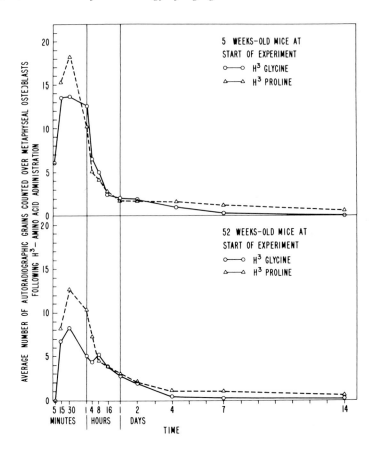

FIGURE 6. The average autoradiographic grain counts observed over femoral metaphyseal osteoblasts following the administration of [³H-]glycine and [³H-]proline in aging BNL mice. (From Tonna, E. A., *Biology and Clinical Medicine, Proc. 7th Int. Congr. Gerontol.*, 1, 225, 1966.[55] With permission.)

Connective Tissue Matrix

The physical properties and chemical composition of connective tissues change early in life from birth to about 5 years of age. Between 10 and 30 years, little alteration is noted. After about 40 years of age, a slow but continuous and progressive change occurs.[33] Changes that occur in tissue collagen also occur in the fibrous layer of the periosteum. These include alterations in fiber diameter, degree of intrafibrillar order of crystallinity, thermal shrinkage behavior, chemically induced swelling, solubility, and metabolism.[72,73] Observed changes in collagen with age have been correlated with increased intermolecular aggregation and cross-linking of collagen macromolecules and resulting age-associated loss of physiological function, e.g., loss of extensibility.[73,74] Furthermore, in many tissues where collagen turnover is very low, repair does not appear to occur. Collagen molecules are, therefore, exposed to their environment for long periods of time.[75] The turnover rate of collagen is more rapid in the uterus, cervix, and skin, but the low rate of turnover is least in bone as a consequence of continued skeletal remodeling.[76,77] Cannon and Davison[75] find that the content of reducible cross-links in selected human, bovine, and canine tissues decrease with age and that the ratios of individual components also change. Levels of specific radioactivity of ³H-borohydride-reduced human tendon are given in Table 12.[75] The cross-linking properties of compounds derived via reduction from borohydride treatment, however, have been placed in question.[78]

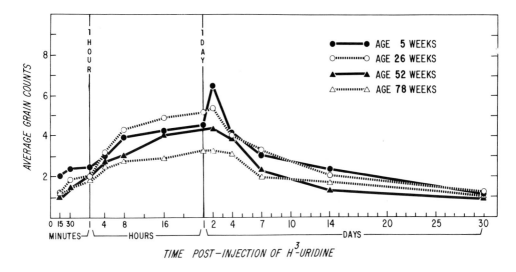

FIGURE 7. The average number of autoradiographic grains observed over the cytoplasm of femoral diaphyseal periosteal osteoblasts following [³H-]uridine administration in aging BNL mice. (From Singh, I. J. and Tonna, E. A., *J. Gerontol.*, 29, 1, 1974. With permission.)

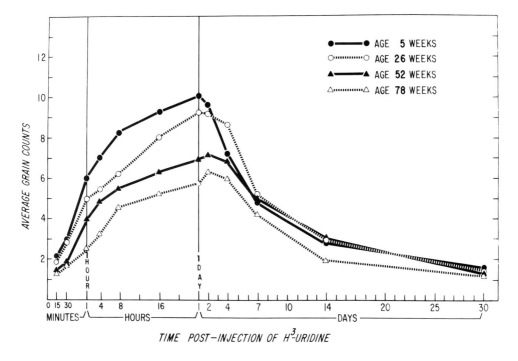

FIGURE 8. The average number of autoradiographic grains observed over the cytoplasm of femoral metaphyseal endosteal osteoblasts following [³H-]uridine administration in aging BNL mice. (From Singh, I. J. and Tonna, E. A., *J. Gerontol.*, 29, 1, 1974. With permission.)

Some of the compounds, namely, hexosyllysine and hexosylhydroxylysine increase in aging tissues and are derived from condensation of lysine and hydroxylysine with glucose and mannose. The compounds are involved with cross-linking of collagen to glycoproteins rather than collagen to collagen and are age dependent. Concerning collagen aging, some investigators regard the time associated changes to represent normal growth and develop-

Table 12
HUMAN TENDON SPECIFIC ACTIVITY[a]

Foot Tendon

Age	6 days	2.5 yrs	10 yrs	19 yrs	31 yrs	48 yrs
^{3}Hcpm/mg $\times$ 10^{-1}	842.1	323.0	377.3	158.4	80.0	106.5
Percent	100	38.4	44.8	18.8	9.5	12.7

Biceps Tendon

Age	2.5 mo	2.5 yrs	7 yrs	21 yrs	32 yrs	66 yrs
^{3}Hcpm/mg $\times$ 10^{-1}	346.7	373.5	347.1	158.9	154.5	150.3
Percent	100	108	100	45.8	44.6	43.4

[a]   Values are expressed as cpm/mg of hydrolyzate. Percent values are specific radioactivity percent compared to youngest sample.

From Cannon, D. J. and Davison, P. F., *Exper. Aging Res.*, 3, 1977. Copyright Buck Hill Enterprises. With permission.

mental requirements necessary to achieve the optimal functional state of supporting tissues.[79] Lysine-derived cross-links and cross-link precursor pattern of collagen change with time. As these changes appear to be anatomically and species specific, it becomes difficult to consider collagen cross-linking as a primary event in the aging process.[75]

Elastic tissue also reveals age changes, but much less information is available on this subject. Cross-linking components including desmosine, isodesmosine, and lysinonorlucine increase with tissue maturation[80] and possibly during aging.[81] Morphologically, fraying, fragmentation, and changes in staining properties are known for elastic tissue from blood vessels and skin.[82] The yellow color and fluorescence of elastin increase during aging in contrast to the desmosine content, which remains constant from maturity to old age.[83,84] Changes in the elastic tissue of bone is yet to be explored.[85] Elastic fibers consist of two morphologically distinct components.[86,87] One is the protein elastin and appears to be amorphous under the electron microscope, staining with anionic metal stains, e.g., phosphotungstic acid. The other component is a structural glycoprotein forming microfibrils of 110- to 120-nm diameter. These fibrils stain with uranyl and lead acetate and other cationic electron microscope stains. During embryogenesis only microfibrils are observed. In older organisms the amorphous component appears to permeate the microfibrils, finally maturing to an arrangement where the microfibrils occupy a peripheral position. In an ultrastructural study[88] of the periosteum, fragmentation was not observed, as has been reported at the histological level; however, this may be due to the limited field of observation at the electron microscope level. With increasing age, microfibrils appear to permeate the entire thickness of the elastin fiber bundle. Very little information exists on elastin synthesis and degradation, and renewed synthesis in damaged tissue is yet to be reported.[72] Elastic tissue is known to undergo calcification with increasing age.[89,90] In bone,[88] periosteal elastic fibers become mineralized together with collagen matrix during bone calcification and thus form part of the organic framework of the skeleton.

Age changes in the ground substances have been reported, but research has been aimed more to its carbohydrate moieties than to noncollagenous proteins and lipids.[91-93]

Morphological Changes

All skeletal cell types undergo both cellular and distributional (tissue) morphological changes. Hard data are yet unavailable for the cellular aging of osteoclasts; nevertheless, population numbers and distribution changes are known to occur with increasing age. Unfortunately, existing information has been derived largely from cytological, histological, and electron microscopic studies of laboratory animal material. Human bone cell morphological age changes are yet to be described adequately. Such descriptions must include changes occurring in the fibrous and osteogenic layers of the periosteum, endosteum, and the osteocytic and osteoclastic cell compartments.

In man, it has been reported[33] that the composition and physical properties of connective tissue change rapidly from birth to 5 years of age, are altered little from 10 to 30 years, and after 40 years, only a slow progressive change is noted. This includes the typical chronologic changes described for collagen and elastin.[72] Fibroblast ultrastructure has been reported.[94-96] In young mice and in other animals including man, the cellular complement of the fibrous periosteum consists of functionally active fibroblasts separated by large bundles of collagen which are permeated by smaller elastic bundles. With increasing age, maturation of collagen and elastin occurs, and fibroblastic cell proliferative activity is diminished significantly. Functional fibrogenic cell synthesis and release of matrical precursors is also diminished together with extensive ultrastructural cell changes. The fibroblasts are converted to very elongated fibrocytes. By 2 years of age, mouse residual fibrogenic cells become smaller in size, exhibiting small, hyperchromatic, pyknotic nuclei. The rough endoplasmic reticulum disappears, leaving a scattering of few free ribosomes. The once large Golgi complex is reduced to a small, simple structure. One to several lipofuscin granules (age pigment) appear (from 1 year of age on) within a thin rim of residual cytoplasm containing a few small mitochondria.[88,97] Numerous reports on the ultrastructural changes of skeletal cells exist, though few articles are devoted to morphological changes in skeletal cells during aging.[88,97-101] In both the short-lived BNL mouse and rat, osteogenic cell morphological changes occur rapidly. At the periosteum, the large polyhedral functional osteoblasts which are in intimate contact with one another via small desmosomes form a membrane-like structure covering the bone surface. Thus, a secondary compartment is formed between the cells and bone surfaces. The cells actively monitor organismic needs by cellular biofeedback and provide the necessary enzymatic and other ingredients for bone surface apposition and resorption. Osteoblasts actively synthesize and secrete the collagen matrix precursors and complex carbohydrates. Consequently, active collagen microfibril and fiber assembly is in progress. The bone surface is covered by a relatively thick osteoid layer where continued mineralization leads to bone formation.

The cellular complement of the young osteogenic layer consists of active osteoprogenitive elements (preosteoblasts), differentiating progeny (differentiating osteoblasts), and mature functional osteoblasts. Progenitive cells exhibit a different morphology from their progeny. The cells are elongated, with oval nuclei which are more hyperchromatic than those of the progeny. Cytoplasmic contents reveal a paucity of rough endoplasmic reticulum, few mitochondria, and a simple Golgi structure. The ground substance appears to be more electron dense, and pinocytosis is evident along the plasma membrane. By comparison, differentiating osteoblasts, which are oval to angular as maturation progresses, exhibit extensive development of endoplasmic reticulum and a more complex Golgi system and larger mitochondria. The ground substance is less electron dense, while the nuclei are oval and less hyperchromatic. Functional osteoblasts are large polyhedral cells containing a large nucleus, a large clearly defined nucleolus and numerous chromatin packets (karyosomes) arranged peripherally with the nuclear membrane. This structure exhibits numerous nuclear pores. Cytoplasmic microfilaments are in abundance within the ground substance of the cell and appear to be more concentrated near the plasma membrane, extending to the small desmosomes and hemidesmosomes which allow cell-to-cell and cell-to-matrix anchorage. The cytoplasm is very expansive and packed with an extensive arrangement of large rough endoplasmic reticulum, a large complex Golgi system, and few large to oval mitochondria. Lysosomes are few in number.

With skeletal maturation and increasing age, dramatic morphological changes occur which result in almost complete cessation of cell proliferation and significantly diminished precursor matrix synthesis. Periosteal thickness is reduced considerably to a one- to two-cell layer along bone shafts. This results from the utilization of bone cells in expansion of the bone surface during growth and their conversion to osteocytes during cortical bone thickening. Osteoblasts, as well as osteogenic progenitive cells, become morphologically indistinguishable. Cessation of cellular division eliminates the presence of differentiating osteoblasts. All the cells of the osteogenic cell line become fibrocytic in appearance. Such cells are elongated and spindle shaped, with elongated hyperchromatic nuclei, a small amount of rough endoplasmic reticulum, some scattered free ribosomes, a simplified Golgi system, a more electron dense cytosol, and fewer, but elongated mitochondria. The secondary compartment space is very much reduced and evidence of precursor matrix formation eventually disappears as does the thickness and presence of the osteoid layer. Finally, cell surfaces and mineralized bone surfaces come in close proximity and contact. An electron dense osmophilic lamina makes its appearance signifying cessation of bone surface activity.

Along the bone surfaces, where in younger animals functional osteoblasts were being converted to osteocytes, one now finds cells in a stage of arrested and incomplete osteocyte formation. These cells have aged and appear only partially embedded into the bone surface. Cell-to-cell contact exists along some stretches of the periosteum and appears to break up at some foci. With continued aging (104 to 130 weeks of age), cell-to-cell contact is entirely lost, and extensive cellular membrane destruction occurs along the plasma membrane. Residual endoplasmic reticulum and mitochrondria show swelling and disruption. The space between nuclear and cytoplasmic membranes also enlarge. Eventually, the cells are observed as small, effete, irregular-shaped structures with little cytoplasm undergoing extensive disintegration. The nuclei are small crenated, pyknotic, and hyperchromatic. One to several large lypofuscin granules make their appearance (from about 1 year on). Lysosomes are never in abundance. Cytoplasmic extensions, which once projected through bone canaliculi to serve as cell-to-cell contacts, degenerate during their "fibrocytic stage". The cells become far removed from the bone surfaces, exposing bone to physiochemical changes in the absence of control via cellular biofeedback. Such surfaces often exhibit regions of bone resorption or irregular apposition and micro-ossicle formation typical of the absence of cellular regulation. Occasionally, periosteal cells exhibiting a characteristically functional nucleus, centrioles, and cytoplasmic organelles with some level of residual integrity, are encountered. Such cells may retain the propensity to divide in response to trauma. The periosteum in the oldest animals studied retains the ability to undergo reactivated cell proliferation and replenish the functional cell compartment with active progeny in response to emergencies throughout the lifespan of the organism. It becomes the function of the differentiated progeny to provide skeletal repair.[46,53] Morphological changes with age in endosteal osteogenic cells are similar to periosteal cell changes; however, they occur later in time, since functional endosteal osteoblasts remain active in subcortical and trabecular remodeling. Furthermore, functional osteoblasts are encountered in the endosteum whose cytoplasm is more electron dense than adjacent functional osteoblasts.

In young animals, osteocytes have been reported to undergo degenerative changes,[102,103] similar to those observed in old animals.[100] Initially, young osteocytes resemble their functional osteoblastic precursors. When first formed near periosteal and endosteal surfaces, lacunae do not exist. Subsequently, the ultrastructural complement of rough endoplasmic reticulum and Golgi complex, together with the mitochondrial numbers, are diminished. These concomitant time changes result in the formation of lacunae and smaller cells with significantly reduced cytoplasm. During this process, osteocytes participate in the maintenance of mineral homeostasis through perilacunar osteoclasis and osteoplasis. Cessation of either process leads to the formation of osmiophilic laminae. With increasing age multiple laminae are often observed even in cross-sections of bone canaliculi.[98] Subsequently, few lysosomes and one to several lipofuscin granules are found together with cellular degeneration of the organelles and membranes. Cytoplasmic vacuoles are encountered. The cells undergo necrosis, swell, and finally lyse, forming an area cellular debris within the lacunae.

The large specialized multinucleated syncytial cells called osteoclasts are responsible for the removal of bone, both mineral and matrix. Cartilage is also removed by large, multinucleated cells often called ''chondroclasts''. These may, in fact, be osteoclasts or variants. Such cells are necessary to skeletal remodeling, and are normally observed at the perichondrial regions of long bones, along metaphyseal trabeculae immediately below the epiphyseal plate, at secondary centers of ossification, and wherever remodeling occurs. They are abundant in fracture repair, and where mesial drift of the dentition occurs. Their activity is marked by the presence of large numbers of mitochondria and abundance in adenosine triphosphatase[41,104] and acid phosphatase.[105,106] Active cells reside in bone resorptive depressions called Howship's lacunae and exhibit ''ruffled borders''. At the electron microscopic level, these borders are noted to be cytoplasmic modifications occurring at the resorptive surface and consisting of folds and numerous vacuoles of varying size, many opening to the bone surface.[107] Administration of parathyroid hormone (PTH) induces the formation of large numbers of active osteoclasts.[108] The property of active osteoclastogenesis in response to PTH remains throughout life.[109] An osteoclast activating factor (OAF) has also been demonstrated.[110] Calcitonin appears to reverse osteoclastic resorptive activity. Such cells appear to lose contact with bone surfaces and the ruffled border disappears.[111] Large numbers of clasts appear in response to fractures at all ages.[112] Osteoclasts are present throughout the lifespan of the organism, but their population decreases with age (Figures 9,10).[112-114] Osteoclast formation normally does not occur in adult animals. Existing cells in older animals appear to be ''holdovers'' from a younger age and are not as active metabolically as newly generated cells in older animals in response to skeletal trauma. In an autoradiographic study using [^{3}H-]histidine, older cells turned over the radiotracer more slowly than osteoclasts formed following fracture. Furthermore, the property of osteoclastogenesis is not diminished by aging, since a suitable mechanism is available for the production of large numbers in response to skeletal trauma. Morphological variations have been observed among osteoclasts. However, such variations are believed to represent functional variations rather than age changes.[109,114] The size of the populations of the different morphological types may be attributed to aging. Two morphologically different types were also observed at the electron microscopic level.[115]

Osteoclasts do not divide. Their origin resides in the fusion of cells of the osteogenic layer.[112-114,116-118] Fishman and Hay[119] demonstrated that mononuclear leucocytes also give rise to osteoclasts in newt limbs. Increasing evidence indicates that osteoclasts are also derived from a hematopoietic precursor cell.[120-123] Monocytes are considered excellent candidates for this function based on in vivo[124,125] and in vitro[126,127] experimental findings. Kahn et al.,[127] however, report that these cells assume some, but not all of the morphological characteristics of osteoclasts, e.g., well-developed ruffled membranes. They are, nevertheless, capable of resorbing mature bone.

Chondrogenic Tissues

The cellular composition of cartilage tissue proper consists of chondroblasts which serve as the progenitive precursors of chondrocytes yielding chondrogenic clones. The chondrocytes differentiate into functional cells which synthesize both chondroitin sulfate and type [$\alpha 1$ (II)]$_3$ collagen necessary for extracellular matrix formation.[128-130] The synthetic processes are under independent control[131] and could be inhibited independently.[132] However, the concentrations of glycosaminoglycans and collagen in extracellular cartilage matrices show close correlation,[133] and certain interactions between collagen and chondroitin sulfate are necessary before either can be deposited into the insoluble matrical complex.[134]

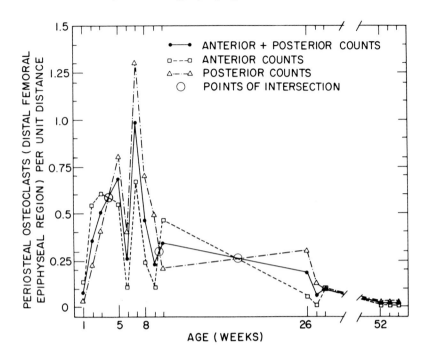

FIGURE 9.   Cell counts of the number of osteoclasts encountered per unit distance (80 μm) of femoral periosteum at the distal femoral region of aging BNL mice. (From Tonna, E. A., *Anat. Rec.*, 137, 251, 1960. With permission.)

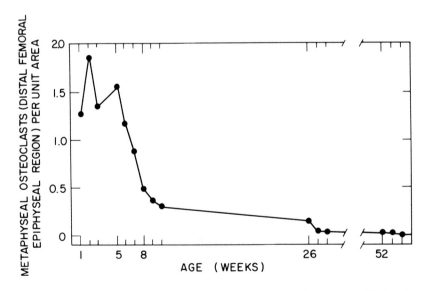

FIGURE 10.   Cell counts of the number of osteoclasts encountered per unit area (6400 μm²) of the distal femoral and metaphyseal region of aging BNL mice. (From Tonna, E. A., *Anat. Rec.*, 137, 251, 1960. With permission.)

Maintenance of the differentiated state requires continued synthesis and deposition of the cartilaginous matrix. The integrity of cartilage is dependent upon cellular viability and maintenance of the matrix.[56,57] This in turn results from the responsiveness of the cells to their extracellular environment via biofeedback mechanisms. The microenvironment plays a significant role in the synthetic response and phenotypic expression of the chondrocyte.[135]

Cell Proliferative Activity

The cell population size of human femoral articular cartilage, nonarticular hyaline cartilage from fourth costal cartilage,[136] and the epiphyseal plate decrease with increasing age. The DNA content also decreases with age.[137] The rate of this decrease is limited to the first and second decades of life, after which only minimal changes occur. These chemical findings agree with the histometric data.[136-139] A study of the articular and costal cartilage of 45 subjects 2 days to 89 years of age shows the cell density of the whole tissue to diminish significantly during the first 30 years[136] (Table 13). During the period of 31 to 89 years, no significant change in cell density is recorded for articular cartilage, but a significant decrease continues to occur in costal cartilage to about 25% (Table 14). In articular cartilage, however, growth continues at an extremely slow pace. The proliferative rates are noticeably slower in older organisms, but the initial proliferative activity of the epiphyseal plate is twice the value of the articular cartilage.[140] In the aging mouse, cell turnover rates for the plate are actually much higher, since peak labeling is 17.5% for articular cartilage and 58% for the plate. Within 14 days, the percent labeling diminishes to 11.3 for articular cartilage, and to 0.33 for the plate. The labeling in the knee joint of the immature rabbit is 0.3% following intraarticular injection of [³H-]thymidine. In the mature animal it is zero.[141,142] Articular cartilage exhibits the lower proliferative activity of all skeletal tissues studied.[45] In human adult articular cartilage, mitotic figures are not seen and synthesis of DNA cannot be demonstrated.[143] Total DNA and RNA, and their synthesis decrease with age in normal articular cartilage.[144] This is also true in mouse cartilage.[60] A [³H-]thymidine study of the labeling index of aging epiphyseal plate cartilage shows a progressive decrease with increasing age (Table 15).[36] As a consequence of the observed low proliferative activity of adult cartilage, many investigators maintain that mammalian articular chondrocytes lose their ability to undergo mitosis when the organism reaches adulthood.[145] Studies involving trauma in rabbits show some labeling of cartilage cells, but the number is small and is not significant to repair.[141-143] It should not be assumed, however, that because terminally differentiated chondrocytes do not divide, that cartilage cells, under proper circumstances, are incapable of responding and thus leading to renewed cell division. It may also be possible that mutual exclusion of division and differentiation is not universal. Impressive evidence is accumulating that indicates that articular chondrocytes of osteoarthritic patients or cells taken from experimentally induced degenerative states in animal joints resume DNA synthesis and divide. Telhag[146] reports a DNA increase in osteoarthritic cartilage. In rabbits, if prolonged degeneration of the knee joints is induced by division of the joint ligaments or surgical excision, [³H-]thymidine-labeled chondrocytes are observed by 5 days postoperatively.[144,147] Even slight movements of an immobilized joint may induce some chondrocytic division, although the numbers are negligible.

Mammalian articular chondrocytes, and perhaps all mammalian chondrocytes, normally become reversible postmitotics. They remain in this state, but retain the potential to respond to degenerative changes by initiation of DNA synthesis and proliferation during aging. The best evidence for retaining the proliferative potential comes from in vitro studies. Isolated, cultured articular chondrocytes from rabbits and individuals up to 84 years of age following collagenase digestion and mechanical disruption continue to produce extracellular matrix and proliferate when they are released from their matrices.[149] Cultured articular chondrocytes can also modify their synthetic activity in response to environmental perturbations.[150,151]

To date, too few studies have been devoted to the repairability of cartilage during aging and the significance of the latent growth potential. The limited repair process is dependent upon a number of factors including wound depth, proximity to blood vascular supply, protection from abrasion, and the age of the organism.[20]

Table 13
CELL DENSITY OF ARTICULAR AND COSTAL CARTILAGE DURING MATURATION

Age (years)	Articular cartilage/0.22mm²			Costal cartilage	
	Total cells	Superficial cells	Deep cells	Cells/ 0.22 mm²	Cells/section ($\times 10^{-3}$)
2 days	76	157	64	56	10.1
6 weeks	61	137	49	52	14.7
1.5	—	—	—	39	13.7
13	20	74	11	10	10.1
16	—	—	—	10	11.1
16	9	28	7	8	12.9
17	—	—	—	10	11.5
18	9	24	7	8	15.8
26	7	17	7	5	14.7
53	—	—	—	5	9.2
56	11	23	9	6	12.5
58	11	23	10	4	8.3
59	10	21	9	5	13.8
60	8	18	6	6	14.4
60	9	16	8	4	3.8
60	—	—	—	8	8.2
63	8	16	7	8	13.7
63	9	27	7	6	8.4
64	9	22	7	6	8.9
64	7	14	7	4	8.7
65	—	—	—	9	12.3
66	9	23	8	6	11.4
68	8	25	6	5	6.2
69	10	23	9	7	11.1
69	12	29	10	4	13.0
75	16	31	13	4	6.4
77	9	19	8	5	5.2
78	9	15	8	6	9.8
89	9	21	7	7	7.2
89	—	—	—	5	12.1

From Stockwell, R. A., *J. Anat.*, 101, 753, 1967. Cambridge University Press. With permission.

Biochemical Activity

Significant changes are known to occur during aging in cartilage cells and their secretory products, as well as within the matrix and its ground substance. In human costal cartilage the glycosaminoglycan content decreases with age.[152] The chondroitin sulfate content also decreases continuously from birth to old age, but the keratosulfate content increases to a plateau which is maintained even in senescence.[153,154] In costal cartilage of young individuals, the keratosulfate content is negligible, while in samples of older individuals it is greater than 50% of the total glycosaminoglycans. A marked decrease in hexuronic acid occurs by the third decade and is due primarily to the loss of chondroitin-4-sulfate, since chondroitin-6-sulfate decreases somewhat. Mathews[154] reports a continued loss of chondroitin-4-sulfate into the fourth decade, and an inverse relationship with keratosulfate. Greiling and Baumann[155] show a continous decrease in chondroitin-4-sulfate beyond the eighth decade and a continuous rise in chondroitin-6-sulfate into the seventh decade in knee joint cartilage. However, the total chondroitin sulfate and nonsulfated chondroitin continue to decrease with age. Gal-

Table 14
CELL DENSITY OF ARTICULAR AND COSTAL
CARTILAGE DURING AGING

Age (years)	Articular cartilage/0.22 mm²			Costal cartilage	
	Total cells	Superficial cells	Deep cells	Cells/ 0.22 mm²	Cells/section ($\times\ 10^{-3}$)
31	8	19	7	7	11.3
32	—	—	—	7	11.5
32	9	21	7	7	10.1
33	12	32	10	7	16.3
33	9	22	8	8	12.2
34	13	47	8	8	8.9
35	—	—	—	9	11.1
39	9	24	7	5	7.4
42	14	39	11	4	4.9
46	—	—	—	10	11.0
46	10	37	8	7	8.5
48	10	29	8	5	7.1
51	10	18	9	6	12.1
52	7	19	6	5	7.2
53	10	27	8	8	14.4
53	—	—	—	5	9.2
56	11	23	9	6	12.5
58	11	23	10	4	8.3
59	10	21	9	5	13.8
60	8	18	6	6	14.4
60	9	16	8	4	3.8
60	—	—	—	8	8.2
63	8	16	7	8	13.7
63	9	27	7	6	8.4
64	9	22	7	6	8.9
64	7	14	7	4	8.7
65	—	—	—	9	12.3
66	9	23	8	6	11.4
68	8	25	6	5	6.2
69	10	23	9	7	11.1
69	12	29	10	4	13.0
75	16	31	13	4	6.4
77	9	19	8	5	5.2
78	9	15	8	6	9.8
89	9	21	7	7	7.2
89	—	—	—	5	12.1

From Stockwell, R. A., *J. Anat.*, 101, 753, 1967. Cambridge University Press. With permission.

actosamine or uronic acid measurements show that the chondroitin sulfate at 85 years of age is about one sixth the value of the newborn, while the glucosamine which reflects the keratosulfate content increases about four times to age 30 and remains constant beyond this age.[156] A high ratio (2:2) of light to heavy molecular fractions of proteoglycans is maintained from birth to 16 years of age. At 24 years, the ratio is 1:2, and diminishes to 0.63 by 63 years of age.[157] Earlier reports of articular cartilage from the knee joint show that the hexosamine content is lower at 5 years of age than at age 40.[158] Anderson et al.[159] find no change between 10 and 80 years of age, leading to some confusion. Furthermore, the epiphyseal plate of rabbits exhibits a higher content of chondroitin-4-sulfate, lower chondroitin-6-sulfate, and a constant keratin sulfate in a comparison between young and old samples.[160] This makes extrapolation of laboratory animal data to human more difficult.

Table 15
H³TDR LABELED CELLULAR POPULATION OF FEMORAL DISTAL EPIPHYSEAL DISK OF MICE[a]

Age (weeks)	1	5	8	26	52
Labeling index	0.054	0.048	0.018	0.002	0
% Labeling	5.4	4.8	1.8	0.2	0
No. cells counted	1493	1784	1889	1507	1432

[a] Only cells making up the proliferative and hypertrophic zones were counted.

From Tonna, E. A., *J. Biophys. Biochem Cytol.*, 9(4), 813, 1961. With permission.

[³⁵S-]Sulfate uptake, which largely reflects chondroitin sulfate, decreases in the aging epiphyseal plate[68,161] and the articular and epiphyseal cartilage of mice.[70] However, Lindner[161] points out that the decrease is greatest during maturation. PBSS and PBSH groups have been studied in aging mouse articular and epiphyseal cartilage.[64] A similar distribution occurs between PBSH/PBSS in articular and epiphyseal plate cartilage. PBSS can also be detected in cartilage matrix. Both groups reveal a decrease as the demands for growth diminish. The levels are further decreased with increasing age.

A series of autoradiographic studies in which the utilization by aging mouse cartilage of ³H-amino acids (histidine, glycine, and proline) were traced show that the total uptake diminishes, the rates of uptake and turnover are reduced, and the half-times are prolonged with increasing age.[49-53,55-58] Peak uptake shifts occur to later time periods. These shifts are attributed to cellular aging. [³H-]Proline uptake time by articular cartilage exhibits no shifting, indicative of the significance of this amino acid in retaining the integrity of aging cartilage, i.e., a reflection of the physiological requirement for normal joint function despite accumulating cellular and matrical age changes (Figures 11, 12). [³H-]Uridine autoradiographic studies[60,62] also show a significant decrease in peak values and flattening of the curves with increasing age. These changes reveal a decline in the rates of radiotracer uptake.

Little work has been reported on enzyme changes in aging chondrocytes. The studies of Silberberg et al.[162-164] using the guinea pig articular cartilage reveal the following: (1) The glycolytic enzymes (hexokinase, glucose-6-phosphate dehydrogenase, phosphofructokinase, phosphoglucomutase, aldolase, α-glycerphosphate dehydrogenase, pyruvate kinase, and lactate dehydrogenase), calculated on a per dry weight basis, decline with the exception of pyruvate kinase, which increases significantly beyond 1 year of age. Calculated on the basis of DNA per gram, the activity of most glycolytic enzymes diminishes slightly at 3 months of age, but increases significantly (especially pyruvate kinase) beyond 1 year of age; (2) phosphorylase also decreases slightly to 3 months, increasing to a peak at 1 year, followed by a continuous decrease; (3) myokinase and creatine phosphokinase both decrease with increasing age on a per gram dry weight basis. However, on a DNA per gram basis, the enzyme activity increases consistently after cessation of growth; (4) the activity of lysosomal enzymes (β-glucosidase, β-xylosidase, β-galactosidase, β-glucuronidase, β-acetylglucosaminidase, β-acetylgalactosaminidase, β-fucosidase, cathepsins B, C, D, and sulfatase), on a dry weight basis, is generally lower in adult animals. Sulfatase activity is the exception. This enzyme increases steadily from 3 months of age on. During guinea pig midlife (about 2.5 years) enzyme activities increase especially β-galactosidase and β-xylosidase. On a DNA per gram basis, sulfatase and β-galactosidase increase steadily, whereas β-glucuronidase, β-xylosidase, and cathepsin D peak at 2.5 years, then decrease. Silberberg et al.[162-164]

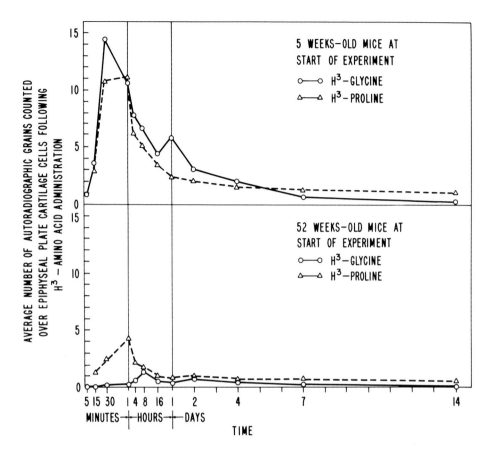

FIGURE 11. The average autoradiographic grain counts observed over femoral epiphyseal plate cartilage cells following the administration of [³H-]glycine and [³H-]proline in aging BNL mice. (From Tonna, E. A., *Biology and Clinical Medicine, Proc. 7th Int. Congr. Gerontol.*, 1, 225, 1966.[55] With permission.)

conclude that with decreasing number of chondrocytes, following cessation of growth, the cells generate higher levels of glycolysis and oxidative phosphorylation when DNA becomes stabilized. The individual chondrocytes may under excessive work load become prematurely exhausted and die. The compensatory hyperactivity of surviving chondrocytes apparently does not normally suffice to maintain tissue enzyme levels. However, uridine diphosphate glucose dehydrogenase (UDPGDH), pyruvate kinase, and sulfatase reach peak levels in old age. Silberberg et al.[162-164] believe that UDPGDH activity changes relate to a high rate of glycosaminoglycan synthesis while pyruvate kinase and sulfatase relate to degenerative changes in both cells and matrix. In the articular cartilage of mice, an increase of ATPase activity is noted while at the epiphyseal plate, a decrease in activity is recorded with increasing age.[42,104] Increased ATPase activity in residual aging chondrocytes is believed to be related to compensatory hyperactivity necessary for maintenance of aging cartilage integrity.

Intercellular Matrix and Synovial Fluid

The integrity of the viability of cartilaginous structures, throughout life, depends upon both the cellular and matrical compartments. The biological role of the intercellular matrix is varied[92,165] and includes mechanical stabilization and protection against mechanical stress (shock absorber, biolubricant, maintains cellular shape and organization, generates mechanical energy via the viscoelastic polyanionic chains of the proteoglycans). Furthermore it regulates transport of cells, particles, macromolecules, ions and water, supplying cellular

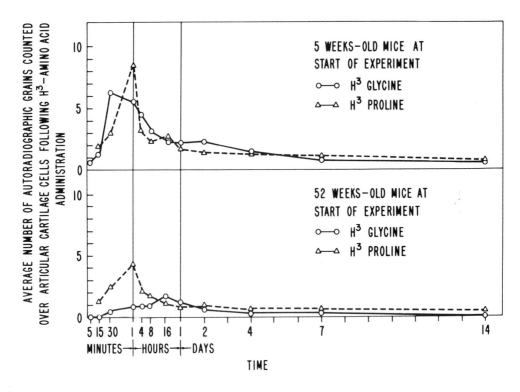

FIGURE 12.   The average autoradiographic grain counts observed over femoral articular cartilage cells following the administration of [³H-]glycine and [³H-]proline in aging BNL mice. (From Tonna, E. A., *Biology and Clinical Medicine, Proc. 7th Int. Congr. Gerontol.*, 1, 225, 1966.[55] With permission.)

nutrition, possibly biofeedback through regulatory factors, and elimination of cellular waste, as well as other functions not clearly understood.

In articular cartilage the concentration of glycosaminoglycans of the matrix varies with the depth of the structure. Balazs et al.[166] report that bovine carpometacarpal and tarso-metatarsal joints are covered by a layer of hyaluronic acid forming a surface cushion,[167] and which consists of 40 to 50% of the total glycosaminoglycan content of the surface layer (1 to 10 μm) of the joint. Morphologically this layer reveals a dense felt-like filamentous arrangement that is significantly different from the underlying collagen fibrils. The layer is absent in the newborn calf and rat, but present in adult animals, and is associated with weight bearing. With increasing age the layer can become 10 to 20 μm thick. This structure contains a considerable amount of protein and is believed to account for the low frictional coefficient properties of movable joints.[165] Balazs[92] reports that the surface cushion conceivably represents a connective tissue matrix surface which maintains the cell-free status of normal articular cartilage surfaces, since it can prevent fibroblasts, mononuclear phagocytes, granulocytes, and lymphocytes from neighboring tissues and synovial fluid from adhering to and proliferating on the cartilage surface. Experiments show that fibroblasts, leukocytes, and peritoneal exudate cells do not adhere to or move on the surface of high polymer jellies of hyaluronic acid of high concentration greater than 1%. In human femoral condyles of individuals 31 to 78 years of age,[168] chondroitin sulfate is in greater proportion within the surface layer, while keratin sulfate is in larger proportion in deeper layers. With increasing age, the glycosaminoglycan content of the total organic dry mass diminishes markedly, with the chondroitin sulfate content decreasing and the keratin sulfate content increasing.[169]

Table 16
EFFECT OF AGING ON MACROMOLECULAR COMPOSITION
AND VISCOELASTIC PROPERTIES OF HUMAN SYNOVIAL FLUID
FROM KNEE JOINT

	Young (av age 36 years)	Old (av age 64 years)	Osteoarthritic (av age 67 years)
Hyaluronic acid[a]			
mg/mℓ[b]	2.49 ± 0.01	2.51 ± 0.17	1.44 ± 0.12
[η] mℓ/g	5500 ± 200	5400 ± 230	3740 ± 300
Huggins' constant	0.29	0.27 ± 0.01	0.29 ± 0.02
Proteins[c]			
mg/mℓ	21.0	18.5	35.7 ± 2.5
Sialic acid (%)[d]	0.62	0.57 ± 0.04	0.64 ± 0.03
Dynamic shear moduli			
At 2.5 cps |G*|[e]	233 ± 4	203 ± 33	35 ± 11
G'' dynes cm^{-2}	45 ± 4	99 ± 13	21 ± 7
Energy stored (%)	77	52	42

[a] Hyaluronic acid concentration was calculated from hexuronic acid concentration.

[b] The limiting viscosity number was determined with a Cannon-Ubbelohde® semimicro dilution viscometer in solutions that were dialyzed against 0.1 N NaCℓ.

[c] Protein concentration was determined by using Folin's phenol reaction.

[d] Sialic acid determination was made by using the thiobarbituric acid test.

[e] Dynamic shear moduli were determined in a Couette® oscillating rheometer.

Hyaluronic acid in human synovial fluid decreases significantly between the ages of 22 and 27.[92,165] A statistical difference is found in hyaluronic concentration between 18- to 29-year-old individuals (2.17 ± 0.3 mg/mℓ) and 27- to 39-year-old individuals (2.45 ± 0.04 mg/mℓ). After the age of 30, the change is minimal except in knee joints which exhibit osteoarthritic involvement. Here, the hyaluronic acid concentration is markedly lower than in normal knee joints of individuals of the same age (Table 16). Synovial fluids[165] from young (average age 36 years) and older individuals (average age 64 years) do not show significant differences in concentration, limiting viscosity number, and Huggins' constant of the hyaluronic acid. Furthermore, no significant differences are observed in protein and sialic acid concentration. Age changes occur, however, in the viscoelastic properties of the fluid. The data is taken to indicate that the hyaluronic acid molecular chain is less flexible in the synovial fluid of young subjects than from older subjects. In studies of synovial fluid from bovine carpal and hock joints from animals 1.5 to 2.5, and 10 to 15 years of age,[165] considerable differences are noted between the two joints investigated (Table 17). No significant difference occurs in the synovial fluids from hock joints between young and old groups in the hyaluronic acid and protein concentration, and in the sialic acid and hexosamine content of proteins, limiting viscosity number, and Huggins' constant of hyaluronic acid. In synovial fluids from carpal joints, protein concentration is significantly higher in the old group. Other parameters show no significant differences, except for a slight decrease in the sialic acid content of the proteins of old carpal joints. This is indicative of a qualitative change with increasing age. In all young and old animals studied, the hyaluronic acid concentration, its limiting viscosity number, and protein concentration are higher in the carpal than the hock joint. Such differences are ascribed to the greater weight load carried by the carpal joint of cattle.[92,165]

Although the rheological age changes noted in the fluid of the bovine joint differ from that of the human joint, suggesting differences in the aging process between the two organisms, osteoarthritic changes in the fluids of both species produce similar changes. Balazs[92] points to the possibility that the recorded changes in normal human synovial fluid may, in

Table 17
EFFECT OF AGING ON THE HYALURONIC ACID AND PROTEIN
CONTENT OF THE BOVINE CARPAL AND HOCK SYNOVIAL FLUIDS

	1.5- to 2.5-year-old cows (12 animals)		10- to 15-year old cows (12 animals)	
	Carpal joint	Hock joint	Carpal joint	Hock joint
Synovial fluid per joint				
$\quad$ mℓ	$4.7 \pm 0.3^{a,1}$	13.1 ± 1.2^{1}	$3.3 \pm 0.2^{a,2}$	12.0 ± 2.4^{2}
Hyaluronic acid				
$\quad$ mg/mℓ	0.8 ± 0.05^{3}	0.33 ± 0.03^{3}	0.71 ± 0.08^{4}	0.35 ± 0.04^{4}
$\quad$ [η] mℓ/g	9200 ± 1100^{5}	6400 ± 400^{5}	8300 ± 600	6800 ± 500
$\quad$ Huggins' constant	0.11 ± 0.01^{6}	0.20 ± 0.01^{6}	0.17 ± 0.01	0.19 ± 0.01
Proteins				
$\quad$ mg/mℓ	$14.2 \pm 1.7^{b,7}$	7.7 ± 0.8^{7}	$23.2 \pm 3.0^{b,8}$	9.4 ± 0.9^{8}
$\quad$ Hexosamine %	1.34 ± 0.18	1.26 ± 0.20	1.43 ± 0.25	0.97 ± 0.19
$\quad$ Sialic acid %	$0.66 \pm 0.06^{c,9}$	0.83 ± 0.05^{9}	$0.47 \pm 0.05^{c,10}$	0.74 ± 0.05^{10}

Note: Comparing the data between the old and young animals and the carpal and hock joints, the following differences were significant:

[a]   Probability < 0.02
[b]   Probability < 0.02
[c]   Probability < 0.05
[1]   Probability < 0.001
[2]   Probability < 0.02
[3]   Probability < 0.001
[4]   Probability < 0.005
[5]   Probability < 0.001
[6]   Probability < 0.02
[7]   Probability < 0.001
[8]   Probability < 0.001
[9]   Probability < 0.01

fact, represent the accumulation of chronic, subclinical traumatic experiences of the joint concomitant with aging rather than a causal relationship.

Cell Morphological Changes

Morphological changes which occur during the life cycle of the cartilage cell have been reported at the ultrastructural level for man,[29] rabbit,[170] guinea pig,[164] and mouse.[27] In the latter, dying chondrocytes are an uncommon finding during growth, but are observed frequently after cessation of growth and during aging. Necrotic and lysed chondrocytes are replaced by microscars. The production of ground substance is altered with time, while the capacity to produce matrical precursors persists. Superficial chondrocytes of old mice elaborate a dense matrix consisting of atypical, short fibrils. The surface matrix shows vacuolization and blister formation with increasing age, while the cells remain largely intact. Aging influences the pattern of lipids which accumulate within cells of progressively older cartilage zones.[171] Intracellular lipid accumulation is not considered evidence of a degenerative change.

The upper midzonal chondrocytes of femoral heads of male guinea pigs, aged 2 weeks to 5.75 years have been studied.[164] Chondrocytes from young 2-year-old animals exhibit large, smooth nuclei surrounded by a narrow rim of cytoplasm having scattered organelles. The rough endoplasmic reticulum is scanty, few mitochondria are noted, while the Golgi complex is inconspicuous. These cells are involved with replication and may best be called chondroblasts. Some chondrocytes at 3 months of age exhibit undeveloped, vesiculated rough endoplastic reticulum, a delicate Golgi complex, and few mitochondria, while other chondrocytes show more advanced development. These cells contain numerous rough endoplasmic reticulum, densely lined by ribosomes and stacked in parallel arrangement. The Golgi apparatus is well developed, and few mitochondria are observed. Synthesis of matrical precursors is evident. By 1 year of age, the chondrocytes reveal irregular-shaped nuclei. The cytoplasm is abundant, and the nuclei appear smaller in proportion. Endoplasmic reticulum is much less abundant, but the Golgi complex is large and conspicuous. Mitochondria are numerous at this age. Lipid inclusions and glycogen deposits make their appearance. Microtubules are also scattered within the cytoplasm. At 2.5 years of age, guinea pig chondrocytes exhibit irregular-shaped or elongated nuclei. Little change is observed in the endoplasmic reticulum and Golgi complex. However, lipid inclusions are more numerous as are mitochondria. In old guinea pigs (5.75 years), the nuclei of cells are typically elongated. Endoplasmic reticulum present in moderate amounts consists of short-length organelles. The Golgi remains conspicuous and mitochondria remain abundant in number. Large deposits of glycogen are in evidence. Cellular degeneration and disintegration, as well as cellular debris, present from earlier ages, are often observed in all zones. In the matrical environment surrounding the cartilage cells, the collagen fibrils undergo time changes leading to a clear increase in average cross-sectional thickness with maturation. A proportional increase in the number of thicker collagen fibrils appear, as well as increased collagen packing during the aging of cartilage.[172] Readers interested in the aging of collagen as reflected in its physical properties are referred to Viidik.[173]

CONCLUSIONS

The sum knowledge of the aging of the skeletal system is at best exiguous, except in those areas, e.g., osteoporosis and degenerative joint disease (osteoarthritis), in which severity and incidence increase perceptibly in the latter part of life. Even here, the contribution of biologic aging to the clinical picture is poorly understood. The absence of information undoubtedly stems from the basic complexity in handling of mineralized tissues, from the long-term commitment required of aging experiments, the inherent difficulty in biological design, finding suitable mammalian models, inadequacy in supply of aged animals, and economics. Nevertheless, selective pressure will increase to eliminate the "porosity" in our understanding of the role of biologic aging of the skeleton and its contribution to skeletal disorders. As the size of the aged population increases, an awareness of the phenomenon of aging will be necessary to the maintenance and well-being of the older segment of our population.

REFERENCES

1. **Nordin, B. E. C.,** International patterns of osteoporosis, *Clin. Orthoped.,* 45, 17, 1966
2. **Bartley, M. H., Jr. and Arnold, J. S.,** Sex differences in human skeletal involution, *Nature (London),* 214, 908, 1967.
3. **Morgan, D. B. and Newton-Jones, H. F.,** Bone loss and senescence, *Gerontologia,* 15, 140, 1969.
4. **Dequeker, J., Remans, J., Franssen, R., and Waes, J.,** Aging patterns of trabecular and cortical bone and their relationships, *Calc. Tiss. Res.,* 7, 23, 1971.
5. **Trotter, M., Broman, G. E., and Peterson, R. R.,** Densities of bones of White and Negro skeletons, *J. Bone Jt. Surg.,* 42A, 50, 1960.

6. **Dunnill, M. S., Anderson, J. A., and Whitehead, R.,** Quantitative histological studies on age changes in bone, *J. Pathol. Bacteriol.*, 94, 275, 1967.

7. **Bromley, R. G., Dockum, N. L., Arnold, J. S., and Jee, W. S. S.,** Quantitative histological study of human lumbar vertebrae, *J. Gerontol.*, 21, 537, 1966.

8. **Atkinson, P. J.,** Variation in trabecular structure of vertebrae with age, *Calc. Tiss. Res.*, 1, 24, 1967.

9. **Arnold, J. S.,** The quantitation of bone mineralization as an organ and tissue in osteoporosis, in *Dynamic Studies of Metabolic Bone Disease*, Pearson, O. H. and Joplin, G. F., Eds., Blackwell, Oxford, 1964, 59.

10. **Sissons, H. A., Holley, K. J., and Heighway, J.,** Normal bone structure in relation to osteomalacia, in *L'osteomalacie*, Hioco, D. J., Ed., Masson, Paris, 1967, 19.

11. **Atkinson, P. J.,** Structural aspects of ageing bone, *Gerontologia*, 15, 171, 1969.

12. **Jowsey, J.,** Quantitative microradiography: a new approach in the evaluation of metabolic bone disease, *Am. J. Med.*, 40, 485, 1966.

13. **Bartley, M. H., Arnold, J. S., Aaslam, R. K., and Jee, W. S. S.,** The relationship of bone strength and bone quality in health, disease and aging, *J. Gerontol.*, 21, 517, 1966.

14. **Frost, H. M.,** *Bone Remodelling Dynamics*, Charles C Thomas, Springfield, Ill., 1963, 5.

15. **Seldin, E. D., Villanueva, A. R., and Frost, H. M.,** Age variations in the specific surface of Howship's lacunae as an index of human bone resorption, *Anat. Rec.*, 146, 201, 1963.

16. **Jowsey, J.,** Age changes in human bone, *Clin. Orthoped.*, 17, 210, 1960.

17. **Jowsey, J.,** Microradiography of bone resorption, in *Mechanisms of Hard Tissue Destruction*, Sogannes, R. F., Ed., American Association for the Advancement of Science, Washington, D.C., 1963, 447.

18. **Silberberg, M. and Silberberg, R.,** Ageing changes in cartilage and bone, in *Structural Aspects of Ageing*, Bourne, G. H., Ed., Hafner, New York, 1961, 85.

19. **Jeffery, M. R.,** The waning joint, *Am. J. Med. Sci.*, 239, 104, 1960.

20. **Sokoloff, L,** *The Biology of Degenerative Joint Disease*, University of Chicago Press, Chicago, 1969, 24.

21. **Jebens, E. H. and Monk-Jones, M. E.,** On the viscosity and pH of synovial fluid and the pH of blood, *J. Bone Jt. Surg.*, 41B, 388, 1959.

22. **Hansen, H. J.,** Studies of the pathology of the lumbosacral disk in female cattle, *Acta Orthopaed. Scand.*, 25, 161, 1956.

23. **Fox, H.,** Chronic arthritis in wild animals, *Trans. Am. Philos. Soc.*, 31, 73, 1939.

24. **Silberberg, M., Frank, E. L., Jarrett, S. R., and Silberberg, R.,** Aging and osteoarthritis of the human sternoclavicular joint, *Am. J. Pathol.*, 35, 851, 1959.

25. **Collins, D. H.,** The pathology of articular and spinal diseases, in *Osteoarthritis*, William & Wilkins, Baltimore, 1949, 74.

26. **Silberberg, M. and Silberberg, R.,** Age changes of bones and joints in various strains of mice, *Am. J. Anat.*, 68, 69, 1941.

27. **Silberberg, R., Silberberg, M., and Fier, D.,** Life cycle of articular cartilage cells: an electron microscope study of the hip joint of the mouse, *Am. J. Anat.*, 114, 17, 1964.

28. **Davies, D. V., Barnett, C. H., Cochrane, W., and Palfrey, A. J.,** Electron microscopy of articular cartilage in the young adult rabbit, *Ann. Rheum. Dis.*, 21, 11, 1962.

29. **Meachim, G.,** The histology and ultrastructure of cartilage, in Cartilage Degeneration and Repair, Bassett, C. A. L., Ed., National Academy of Sciences, National Research Council, Washington, D.C., 1967, 3.

30. **Hjertquist, S.-O. and Lamperg, R.,** Identification and concentration of the glycosaminoglycans of human articular cartilage in relation to age and osteoarthritis, *Calc. Tiss. Res.*, 10, 223, 1972.

31. **Anderson, C. E., Ludowieg, J., Harper, H. A., and Engleman, E. P.,** The composition of the organic component of human articular cartilage, *J. Bone Jt. Surg.*, 46A, 1176, 1964.

32. **Silberberg, R. and Silberberg, M.,** Pathogenesis of osteoarthrosis, *Pathol. Microbiol.*, 27, 447, 1964.

33. **Schubert, M. and Hamerman, D.,** Aging and osteoarthritis, in *A Primer on Connective Tissue Biochemistry*, Lea & Febiger, Philadelphia, 1968, chap. 8.

34. **Sokoloff, L., Critteden, L. B., Yamamoto, R. S., and Jay, G. E., Jr.,** The genetics of degenerative joint disease in mice, *Arthritis Rheum.*, 5, 531, 1962.

35. **Tonna, E. A.,** Fracture callus formation in young and old mice observed with polarized light microscopy, *Anat. Rec.*, 150, 349, 1964.

36. **Tonna, E. A.,** The cellular complement of the skeletal system studied autoradiographically with tritiated thymidine (H[3]TDR) during growth and aging, *J. Biophys. Biochem. Cytol.*, 9, 813, 1961.

37. **Tonna, E. A. and Cronkite, E. P.,** An autoradiographic study of periosteal cell proliferation with tritiated thymidine, *Lab. Invest.*, 11, 455, 1962.

38. **Tonna, E. A. and Cronkite, E. P.,** The aging cellular phase of the skeletal system studied with ^{3}H-thymidine (H[3]TDR), in *Medical and Clinical Aspects of Aging*, Blumenthal, H. T., Ed., Columbia University Press, New York, 1962, 192.

39. **Tonna, E. A. and Cronkite, E. P.,** The periosteum: autoradiographic studies on cellular proliferation and transformation utilizing tritiated thymidine, *Clin. Orthoped.*, 30, 218, 1963.

40. **Tonna, E. A. and Pillsbury, N.,** Mitochondrial changes associated with aging of periosteal osteoblasts, *Anat. Rec.,* 134, 739, 1959.

41. **Tonna, E. A.,** Histologic and histochemical studies on the periosteum of male and female rats at different ages, *J. Gerontol.,* 13, 14, 1958.

42. **Tonna, E. A. and Severson, A. R.,** Changes in localization and distribution of adenosine triphosphatase activity in skeletal tissues of the mouse concomitant with aging, *J. Gerontol.,* 26, 186, 1971.

43. **Tonna, E. A. and Cronkite, E. P.,** Cellular response to fracture studied with tritiated thymidine, *J. Bone Jt. Surg.,* 43A, 352, 1961.

44. **Tonna, E. A. and Cronkite, E. P.,** Changes in the skeletal cell proliferative response to trauma concomitant with aging, *J. Bone Jt. Surg.,* 44A, 1557, 1962.

45. **Tonna, E. A.,** Skeletal cell aging and its effects on the osteogenetic potential, *Clin. Orthoped.,* 40, 57, 1965.

46. **Tonna, E. A.,** Response of the cellular phase of the skeleton to trauma, *Periodontics,* 4, 105, 1966.

47. **Tonna, E. A.,** The source of osteoblasts in healing fractures in animals of different ages, in *The Healing of Osseous Tissue,* Robinson, R. A., Ed., Natl. Acad. Sciences Conf., Warrington, Virginia, 1967, 93.

48. **Tonna, E. A.,** Post-traumatic variations in phosphatase and respiratory enzyme activities of the periosteum of aging rats, *J. Gerontol.,* 14, 159, 1959.

49. **Tonna, E. A. and Cronkite, E. P.,** Utilization of tritiated histidine (H^3HIS) by skeletal cells of adult mice, *J. Gerontol.,* 17, 353, 1962.

50. **Tonna, E. A., Cronkite, E. P., and Pavelec, M.,** An autoradiographic study of the localization and distribution of tritiated histidine in bone, *J. Histochem. Cytochem.,* 10, 601, 1962.

51. **Tonna, E. A., Cronkite, E. P., and Pavelec, M.,** A serial autoradiographic analysis of H^3-glycine utilization and distribution in the femora of growing mice, *J. Histochem. Cytochem.,* 11, 720, 1963.

52. **Tonna, E. A.,** An autoradiographic evaluation of the aging cellular phase of mouse skeleton using tritiated glycine, *J. Gerontol.,* 19, 198, 1964.

53. **Tonna, E. A.,** Protein synthesis and cells of the skeletal system, in *Use of Radioautography in Investigation of Protein Synthesis Symposium for Cell Biology,* Vol. 4, Leblond, C. P. and Warren, K., Eds., Academic Press, New York, 1965, 215.

54. **Tonna, E. A.,** A study of osteocyte formation and distribution in aging mice complemented with H^3-proline autoradiography, *J. Gerontol.,* 21, 124, 1966.

55. **Tonna, E. A.,** An autoradiographic comparison of the utilization of amino acids by skeletal cells concomitant with aging, in *Biology and Clinical Medicine, Proc. 7th Int. Congr. Gerontol.,* Verlag der Wiener Med-izinischen Akademie, International Association of Gerontology, 1, 225, 1966.

56. **Tonna, E. A.,** An autoradiographic study of H^3-proline utilization by aging mouse skeletal tissues. II. Cartilage cell compartments, *Exp. Gerontol.,* 6, 405, 1971.

57. **Tonna, E. A.,** An autoradiographic study of H^3-proline utilization by aging mouse skeletal tissues. III. Estimation and comparison of the turnover of different cell compartments, *Gerontologia,* 17, 273, 1971.

58. **Tonna, E. A. and Pavelec, M.,** An autoradiographic study of H^3-proline utilization by aging mouse skeletal tissues. I. Bone cell compartments, *J. Gerontol.,* 26, 310, 1971.

59. **Burchard, J., Fontaine, R., and Mandel, P.,** Métabolisme des acides ribonucléiques de l'os de lapin et de rat *in vivo, C. R. Soc. Biol. (Paris),* 153, 334, 1959.

60. **Singh, I. J. and Tonna, E. A.,** Autoradiographic evaluation of H^3-uridine utilization by aging mouse cartilage, *Gerontologist,* 12, 41, 1972.

61. **Singh, I. J. and Tonna, E. A.,** An autoradiographic study of the utilization of tritiated uridine by osteoblasts of young mice, *Anat. Rec.,* 175, 243, 1973.

62. **Tonna, E. A. and Singh, I. J.,** The uptake and utilization of ^{3}H-uridine by young mouse cartilage cells studied autoradiographically, *Lab. Invest.,* 28, 300, 1973.

63. **Singh, I. J. and Tonna, E. A.,** An autoradiographic evaluation of the utilization of H^3-uridine by osteoblasts of aging mice, *J. Gerontol.,* 29, 1, 1974.

64. **Pavelec, M., Tonna, E. A., and Fand, I.,** The localization and distribution of protein-bound sulfhydryl and disulfide groups in skeletal tissues of mice during growth and aging, *J. Gerontol.,* 22, 185, 1967.

65. **Meyer, K.,** The mucopolysaccharides of bone, in *Bone Structure and Metabolism,* Wolstenholme, G. E. and O'Connor, C. M., Eds., Little, Brown, Boston, 1956, 65.

66. **Vajlens, L.,** Glycosaminoglycans of human bone tissues. I. Pattern of compact bone in relation to age, *Calc. Tiss. Res.,* 7, 175, 1971.

67. **Casuccio, C., Bertolini, N., and Falzi, M.,** Dell'osteoporosi senile, *La Clin. Orthoped.,* 14, 1, 1962.

68. **Dziewiatkowski, D. D.,** Effect of age on some aspects of sulfate metabolism in the rat, *J. Exp. Med.,* 99, 283, 1954.

69. **Tonna, E. A. and Cronkite, E. P.,** Histochemical and autoradiographic studies on the effects of aging on the mucopolysaccharides of the periosteum, *J. Biophys. Biochem. Cytol.,* 6, 171, 1959.

70. **Tonna, E. A. and Cronkite, E. P.,** Autoradiographic studies of changes in S^{35}-sulfate uptake by the femoral diaphyses during aging, *J. Gerontol.,* 15, 377, 1960.

71. **Mirkinson, L. and Tonna, E. A.,** ^{3}H-galactose utilization by aging bone in mice studied autoradiographically, in *Proc. 33rd Annu. Meet. Gerontol. Soc.,* San Diego, November 21 to 25, 1980.

72. **Gross, J.,** Aging of connective tissues; the extracellular components, in *Structural Aspects of Aging,* Bourne, G. H., Ed., Hafner, New York, 1961, chap. 11.

73. **Sinex, F. M.,** The role of collagen in aging, in *Treatise on Collagen,* Gould, B. S., Ed., Academic Press, New York, 1968, 410.

74. **Hall, D. A.,** *The Aging of Connective Tissue,* Academic Press, New York, 1976.

75. **Cannon, D. J. and Davison, P. F.,** Aging, and crosslinking in mammalian collagen, *Exp. Aging Res.,* 3, 87, 1977.

76. **Neuberger, A., Perrone, J. C., and Slack, H. G. B.,** The relative metabolic inertia of tendon collagen in rat, *Biochem. J.,* 49, 199, 1951.

77. **Ohuchi, K. and Tsurufuji, S.,** Degradation and turnover of collagen in the mouse skin and the effects on whole body x-irradiation, *Biochim. Biophys. Acta,* 208, 475, 1970.

78. **Bailey, A. J., Robins, S. P., and Balian, G.,** Biological significance of the intermolecular cross-links of collagen, *Nature (London),* 251, 105, 1974.

79. **Jackson, D. S.,** Temporal changes in collagen-aging or essential maturation, in *Aging, Advances in Biology of Skin,* Vol. 6, Montagna, W., Ed., Pergamon Press, New York, 1965, 219.

80. **Franzblau, C., Sinex, F. M., and Faris, B.,** Chemistry of maturation of elastin, in *Perspectives in Experimental Gerontology,* Shock, N. W., Ed., Charles C Thomas, Springfield, Ill., 1966, 98.

81. **King, A. L.,** Some studies in tissue elasticity, in *Tissue Elasticity,* Remington, J. W., Ed., Waverly Press, Baltimore, 1957, 123.

82. **Kohn, R. R.,** *Principles of Mammalian Aging,* Found. Dev. Biol. Series, Prentice-Hall, Englewood Cliffs, N.J., 1971, 39.

83. **LaBella, F. S. and Lindsay, W. G.,** The structure of human aortic elastin as influenced by age, *J. Gerontol.,* 18, 111, 1963.

84. **LaBella, F. S., Vivian, S., and Thornhill, D. P.,** Amino acid composition of human elastin as influenced by age, *J. Gerontol.,* 21, 550, 1966.

85. **Murakami, H. and Emry, M. A.,** The role of elastic fibers in the periosteum in fracture healing in guinea pigs. I. Histological studies of the elastic fibers in the periosteum and the osteogenic cells and cells that form elastic fibers, *Can. J. Surg.,* 10, 359, 1974.

86. **Fahrenbach, W. H., Sandberg, L. G., and Cleary, E. G.,** Ultrastructural studies on early embryogenesis, *Anat. Rec.,* 155, 563, 1966.

87. **Greenlee, T. K., Jr. and Ross, R.,** The development of the rat flexor digital tendon, a fine structure study, *J. Ultrastruct. Res.,* 18, 354, 1967.

88. **Tonna, E. A.,** Electron microscopy of aging skeletal cells. III. The periosteum, *Lab. Invest.,* 32, 609, 1974.

89. **Yu, S. Y.,** Elastic tissue and arterial calcification, in *Cowdry's Arteriosclerosis,* 2nd. ed., Blumenthal, H. T., Ed., Charles C Thomas, Springfield, Ill., 1967, 170.

90. **Hall, D. A., Slater, R. S., and Tesal, I. S.,** The use of elastolytic enzymes as probes in the study of ageing in aortic tissue, in *Connective Tissue and Ageing,* Vol. 1, Int. Congr. Series, No. 264, Vogel, H. G., Ed., Excerpta Medica, Amsterdam, 1972, 47.

91. **Sobel, H.,** Aging of ground substance in connective tissue, in *Advances in Gerontological Research,* Vol. 2, Strehler, B. L., Ed., Academic Press, New York, 1967, 205.

92. **Balazs, E. A.,** Intracellular matrix of connective tissue, in *Handbook of the Biology of Aging,* Finch, C. E. and Hayflick, L., Eds., Van Nostrand Reinhold, New York, 1977, 222.

93. **Gardell, S., Fransson, L.-A., and Heinegard, D.,** Chemical structure of extracellular glycosaminoglycans (mucopolysaccharides) and proteoglycans, in *Aging of Connective and Skeletal Tissue,* Engel, A. and Larsson, T., Eds., Nordiska Bokhandelns Förlag, Stockholm, 1969, 49.

94. **Movat, H. Z. and Fernando, N. V. P.,** The fine structure of connective tissue. I. The fibroblast, *Exp. Mol. Pathol.,* 1, 509, 1962.

95. **Porter, K. R.,** Cell fine structure and biosynthesis of intracellular macromolecules, *Biophys. J.,* 4, 167, 1964.

96. **Ross, R. and Benditt, E. P.,** Wound healing and collagen formation. I. Sequential changes in components of guinea pig skin wounds observed in the electron microscope, *J. Biophys. Biochem. Cytol.,* 11, 677, 1961.

97. **Tonna, E. A.,** Accumulation of lipofuscin (age pigment) in aging skeletal connective tissues revealed by electron microscopy, *J. Gerontol.,* 30, 3, 1975.

98. **Tonna, E. A.,** Electron microscopic evidence of alternating osteocytic-osteoclastic and osteoplastic activity in the perilacunar walls of aging mice, *Conn. Tiss. Res.,* 1, 221, 1972.

99. **Tonna, E. A.,** An electron microscopic study of osteocyte release during osteoclasis in mice of different age, *Clin. Orthoped.,* 87, 311, 1972.

100. **Tonna, E. A.,** An electron microscopic study of skeletal cell aging. II. The osteocyte, *Exp. Gerontol.,* 8, 9, 1973.

101. **Tonna, E. A. and Lampen, N.,** Electron microscopy of aging skeletal cells. I. Centrioles and solitary cilia, *J. Gerontol.,* 27, 316, 1972.

102. **Jande, S. S. and Bélanger, L. F.,** Electron microscopy of osteocytes and the pericellular matrix in rat trabecular bone, *Calc. Tiss. Res.,* 6, 280, 1971.

103. **Jande, S. S. and Bélanger, L. F.,** The life cycle of the osteocyte, *Clin. Orthoped.,* 94, 281, 1973.

104. **Severson, A. R., Tonna, E. A., and Pavelec, M.,** Histochemical demonstration of adenosine triphosphatase activity in the femurs of young mice, *Anat. Rec.,* 161, 57, 1968.

105. **Schajowicz, F. and Cabrini, R. L.,** Histochemical localization of acid phosphatase in bone tissue, *Science,* 127, 1147, 1958.

106. **Burstone, M. S.,** Histochemical demonstration of acid phophatase activity in osteoclasts, *J. Histochem. Cytochem.,* 7, 39, 1959.

107. **Scott, B. L. and Pease, D. C.,** Electron microscopy of the epiphyseal apparatus, *Anat. Rec.,* 126, 465, 1956.

108. **Heller, M., McLean, F. C., and Bloom, W.,** Cellular transformations in mammalian bones induced by parathyroid extract, *Am. J. Anat.,* 87, 315, 1950.

109. **Levenson, D. and Tonna, E. A.,** unpublished data, 1980.

110. **Horton, J. E., Oppenheim, J. J., Mergenhagen, S. E., and Raisz, L. G.,** Macrophage-lymphocytic synergy in the production of osteoclast activating factor, *J. Immunol.,* 113, 1278, 1974.

111. **Zichner, L.,** The effect of calcitonin on bone cells in young rats. An electron microscopic study, in *Calcified Tissue Structural, Functional and Metabolic Aspects,* Menczel, J. and Harell, S., Eds., Academic Press, New York, 1971, 27.

112. **Tonna, E. A.,** Osteoclasts and the aging skeleton: a cytological, cytochemical and autoradiographic study, *Anat. Rec.,* 137, 251, 1960.

113. **Tonna, E. A.,** Periosteal osteoclasts, skeletal development and aging, *Nature (London),* 185, 405, 1960.

114. **Tonna, E. A.,** H^3-histidine and H^3-thymidine autoradiographic studies of the possibility of osteoclast aging, *Lab. Invest.,* 15, 435, 1966.

115. **Scott, B.,** The occurrence of specific cytoplasmic granules in the osteoclast, *J. Ultrastruct. Res.,* 19, 417, 1967.

116. **Tonna, E. A. and Cronkite, E. P.,** Use of tritiated thymidine for the study of the origin of the osteoclast, *Nature (London),* 190, 459, 1961.

117. **Tonna, E. A. and Cronkite, E. P.,** Skeletal cell labeling following continuous infusion with tritiated thymidine, *Lab. Invest.,* 19, 510, 1968.

118. **Young, R. W.,** Specialization of bone cells, in *Bone Biodynamics,* Frost, H. M., Ed., Little, Brown, Boston, 1964, 117.

119. **Fishman, D. A. and Hay, E. D.,** Origin of osteoclasts from mononuclear leucocytes in regenerating newt limbs, *Anat. Rec.,* 143, 329, 1962.

120. **Walker, D. G.,** Osteopetrosis cured by temporary parabiosis, *Science,* 180, 875, 1973.

121. **Büring, K.,** On the origin of the cells in heterotopic bone formation, *Clin. Orthoped.,* 110, 293, 1975.

122. **Kahn, A. J. and Simmons, D. J.,** Investigation of cell lineage in bone using a chimaera of chick and quail embryonic tissue, *Nature (London),* 258, 325, 1975.

123. **Göthlin, G. and Ericsson, J. L. E.,** The osteoclast. Review of ultrastructure, origin, and structure-function relationship, *Clin. Orthoped.,* 120, 201, 1976.

124. **Thyberg, J.,** Electron microscopic studies on the uptake of exogenous marker particles by different cell types in the guinea pig metaphysis, *Cell Tiss. Res.,* 156, 301, 1975.

125. **Heyden, G., Kindblom, L.-G., and Nielson, J. M.,** Disappearing bone disease, *J. Bone Jt. Surg.,* 59A, 57, 1977.

126. **Mundy, G. R., Altman, A. J., Gondek, M. D., and Bandelin, J. G.,** Direct resorption of bone by human monocytes, *Science,* 196, 1109, 1977.

127. **Kahn, A. J., Steward, C. C., Teitelbaum, S. L.,** Contact-mediated bone resorption by human monocytes *in vitro, Science,* 199, 988, 1978.

128. **Prockop, D. J., Pettengill, O., and Holtzer, H.,** Incorporation of sulfate and the synthesis of collagen by cultures of embryonic chondrocytes, *Biochim. Biophys. Acta,* 83, 189, 1964.

129. **Smith, P. H.,** Autoradiographic evidence for the concurrent synthesis of collagen and chondroitin sulfates by chick sternal chondrocytes, *Conn. Tiss. Res.,* 1, 181, 1972.

130. **Bornstein, P.,** The biosynthesis of collagen, *Ann. Rev. Biochem.,* 43, 567, 1974.

131. **Nimni, M. E.,** Metabolic pathways and control mechanisms involved in the biosynthesis and turnover of collagen in normal and pathological connective tissue, *J. Oral Pathol.,* 2, 175, 1973.

132. **Bhatnagar, R. S. and Prockop, D. J.,** Dissociation of the synthesis of sulfated mucopolysaccharides and the synthesis of collagen in embryonic cartilage, *Biochim. Biophys. Acta.,* 130, 382, 1966.

133. **Hall, B. K.,** Correlations between the concentrations of acid mucopolysaccharide and collagen in the tibia of the embryonic chick, *Can. J. Zool.,* 51, 771, 1973.

134. **Walton, A. G.,** Molecular aspects of calcified tissue, *J. Biomed. Mater. Res.,* 8, 409, 1974.

135. **Sokoloff, L., Malemud, C. J., and Srivastava, V. M. L., and Morgan, W. D.,** In vitro culture of articular chondrocytes, *Fed. Proc. Fed. Am. Soc. Exp. Biol.,* 32, 1499, 1973.

136. **Stockwell, R. A.,** The cell density of human articular and costal cartilage, *J. Anat.,* 101, 753, 1967.

137. **Platt, D. and Dorn, M.,** Die Bedeutung der Glycosaminoglycano-Hydrolasen in der Pathogenese des degenerativen Rheumatismus, *Z. Rheumaförsch.,* 27, 291, 1968.

138. **Beneke, G., Endres, O., Becker, H., and Kulka, R.,** Uber Wachstum und Degeneration des Trachealknorpels, *Virchows Arch. Pathol. Anat.,* 341, 365, 1966.

139. **Beneke, G.,** Cell density in cartilage and DNA content of the cartilage cells dependent on age, in *Connective Tissue and Ageing,* Int. Congr. Series, No. 264, Vogel, H. G., Ed., Excerpta Medica, Amsterdam, 1973, 91.

140. **Tonna, E. A. and Cronkite, E. P.,** A study of the persistence of the H^3-thymidine label in the femora of rats, *Lab. Invest.,* 13, 161, 1964.

141. **Mankin, H. J.,** Localization of tritiated thymidine in articular cartilage. II. Repair in immature cartilage, *J. Bone Jt. Surg.,* 44A, 688, 1962.

142. **Mankin, H. J.,** Localization of tritiated thymidine in articular cartilage of rabbits. III. Mature articular cartilage, *J. Bone Jt. Surg.,* 45A, 529, 1963.

143. **Mankin, H. J.,** Mitosis in articular cartilage of immature rabbits. A histologic, stathmokinetic (colchicine) and autoradiographic study, *Clin. Orthoped.,* 34, 170, 1964.

144. **Telhag, H. and Havdrup, T.,** Nucleic acids in articular cartilage from rabbits of different ages, *Acta Orthoped. Scand.,* 46, 185, 1975.

145. **Barnett, G. H., Davies, D. V., and MacConaill, M. A.,** *Synovial Joints. Their Structure and Mechanics,* A. H. Thomas, Springfield, Ill., 1961.

146. **Telhag, H.,** Nucleic acids in human normal and osteoarthritic articular cartilage, *Acta Orthoped. Scand.,* 47, 585, 1976.

147. **Havdrup, T., Hulth, A., and Telhag, H.,** Scattered mitosis in mature joint cartilage in rabbits after local trauma "a chalone effect?", *Clin. Orthoped. Relat. Res.,* 113, 246, 1975.

148. **Crelin, E. S. and Southwick, W. D.,** Changes induced by sustained pressure in the knee joint articular cartilage of adult rabbits, *Anat. Rec.,* 149, 113, 1964.

149. **Manning, W. K. and Bonner, W. M., Jr.,** Isolation and culture of chondrocytes from human adult articular cartilage, *Arthritis Rheum.,* 10, 235, 1967.

150. **Green, W. T., Jr.,** Behavior of articular chondrocytes in cell culture, *Clin. Orthoped. Relat. Res.,* 75, 248, 1971.

151. **Green, W. T., Jr. and Ferguson, R. J.,** Histochemical and electron microscopic comparison of tissue produced by rabbit articular chondrocytes in vivo and in vitro, *Arthrit. Rheum.,* 18, 273, 1975.

152. **Stidworthy, G., Masters, Y. F., and Shetlar, M. R.,** The effect of aging on mucopolysaccharide composition of human costal cartilage as measured by hexosamine and uronic acid content, *J. Gerontol.,* 13, 10, 1958.

153. **Kaplan, D. and Meyer, K.,** Ageing of human cartilage, *Nature, (London),* 183, 1267, 1959.

154. **Mathews, M. B.,** Glycosaminoglycans in development and aging of vertebrate cartilage, in *Connective Tissue and Ageing,* Int. Congr. Series, No. 264, Vogel, H. G., Ed., Excerpta Medica, Amsterdam, 1973, 151.

155. **Greiling, H. and Baumann, G.,** Age dependent changes of nonsulfated disaccharide groups in the proteoglycans of knee joint cartilage, in *Connective Tissue and Ageing,* Int. Congr. Series, No. 264, Vogel, H. G., Ed., Excerpta Medica, Amsterdam, 1973, 160.

156. **Shetlar, M. R. and Masters, Y. F.,** Effect of age on polysaccharide composition of cartilage, *Proc. Soc. Exp. Biol. Med.,* 90, 31, 1955.

157. **Rosenberg, L., Johnson, B., and Schubert, M.,** Protein polysaccharides from human articular and costal cartilage, *J. Clin. Invest.,* 44, 1647, 1965.

158. **Loewi, G.,** Changes in the ground substance of ageing cartilage, *J. Pathol. Bacteriol.,* 65, 381, 1953.

159. **Anderson, C. E., Ludowieg, J., Harper, H. A., and Engleman, E. P.,** The composition of the organic component of human articular cartilage, *J. Bone Jt. Surg.,* 46A, 1176, 1964.

160. **Greer, R. B., Skinner, S., Zarins, A., and Mankin, H.,** Distribution of acidic glycosaminoglycans in rabbit growth plate cartilage, *Calc. Tiss. Res.,* 9, 194, 1972.

161. **Lindner, J.,** Contributions to the intracellular localization of the synthesis of glycosaminoglycans during the ageing of cartilage, in *Connective Tissue and Ageing,* Int. Congr. Series, No. 264, Vogel, H. G., Ed., Excerpta Medica, Amsterdam, 1973, 119.

162. **Silberberg, R., Stamp, W., Lesker, P. A., and Hasler, M.,** Aging changes in ultrastructure and enzymatic activity of articular cartilage of guinea pigs, *J. Gerontol.,* 25, 184, 1970.

163. **Silberberg, R., Hasler, M., and Lesker, P. A.,** Aging of the shoulder joint of guinea pigs. Electron microscopic and quantitative histochemical aspects, *J. Gerontol.,* 28, 18, 1973.

164. **Silberberg, R. and Lesker, P. A.,** Fine structure and enzyme activity in articular cartilage of aging male guinea pigs, in *Connective Tissue and Ageing,* Int. Congr. Series, No. 264, Vogel, H. G., Ed., Excerpta Medica, Amsterdam, 1973, 98.

165. **Balazs, E. A.,** Some aspects of aging and radiation sensitivity of the intercellular matrix with special regard to hyaluronic acid in synovial fluid and vitreous, in *Aging of Connective and Skeletal Tissue,* Thule Int. Symp., Engel, A. and Larsson, T., Eds., Nordiska Bokhandelns Förlag, Stockholm, 1969, 107.

166. **Balazs, E. A., Bloom, G. D., and Swan, D. A.,** Fine structure and glycosaminoglycan content of the surface layer of articular cartilage, *Fed. Proc.,* 25, 1813, 1966.

167. **Balazs, E. A. and Gibbs, D. A.,** The rheological properties and biological function of hyaluronic acid, in *The Chemistry and Molecular Biology of the Intercellular Matrix,* Balazs, E. A., Ed., Academic Press, New York, 1970, 1241.

168. **Stockwell, R. A. and Scott, J. E.,** Distribution of acid glycosaminoglycans in human articular cartilage, *Nature (London),* 215, 1376, 1967.

169. **Hjertquist, S.-O. and Engfeldt, B.,** Chemical studies on the glycosaminoglycans of human articular cartilage in aging and in osteoarthritis, *Acta Pathol. Microbiol. Scand.,* Suppl., 187, 40, 1967.

170. **Barnett, C. H., Cochrane, W., and Palfrey, A. J.,** Age changes in articular cartilage of rabbits, *Ann. Rheum. Dis.,* 22, 389, 1963.

171. **Collins, D. H., Ghadially, F. N., and Meachim, G.,** Intracellular lipids of cartilage, *Ann. Rheum. Dis.,* 24, 123, 1965.

172. **Dahmen, G.,** Polarizing and electronmicroscopic investigations of maturing and old human articular cartilage, in *Connective Tissue and Ageing,* Int. Congr. Series, No. 264, Vogel, H. G., Ed., Excerpta Medica, Amsterdam, 1973, 109.

173. **Viidik, A.,** The aging of collagen as reflected in its physical properties, in *Aging of Connective and Skeletal Tissue,* Thule Int. Symp., Engel, A. and Larsson, T., Eds., Nordiska Bokhandelns Förlag, Stockholm, 1969, 125.

AGING LIVER CELLS

Dick L. Knook

INTRODUCTION

The intact liver and the tissue, cells, or organelles obtained from it are widely used in clinical, pharmacological, biochemical, or biological studies. It is, therefore, surprising how little information is available on alterations in the functioning of liver or liver cells with aging. There is also a wide gap in our knowledge of the relation between morphological and biochemical aging changes at the cellular level and physiological functions at the organ level.

The objective of this chapter is to review what is known of the age-related changes in the liver, with emphasis on alterations at the level of the main cell types, especially the parenchymal cells. Where possible, an attempt will be made to relate structural, metabolic, or functional cellular changes with those reported to occur with age in liver functions, using isolated parenchymal cells as a model system for studies on cellular aging. In addition, the few data available on aging changes in the sinusoidal liver cells, including Kupffer cells and endothelial liver cells, are discussed in view of the recently increased attention with respect to their role in liver function and diseases.

CELLULAR COMPOSITION OF THE LIVER AND CHANGES WITH AGING

Liver Cell Types and Their Functions

In common scientific parlance, liver cells are synonymous with hepatocytes or liver parenchymal cells. These are indeed at first sight the most prominent of the liver cells, as they account for about two-thirds of the total cell number and occupy 80 to 85% of the total liver volume. Functions of the parenchymal cells include carbohydrate, lipid, and protein metabolism, synthesis of plasma proteins, detoxification of oxygenous compounds, and bile secretion. However, several different cell types are present in the mammalian liver. The largest nonparenchymal cell population in the liver consists of the sinusoidal cells. Due to their much smaller size, they have literally been overlooked or neglected in biochemical experiments, but they constitute roughly one-third of all liver cells. Although they comprise only about 6% of the liver volume in male Sprague-Dawley rats, weighing 214 ± 6 g, they contain 38% of the total volume of liver lysosomes, 55% of all lipid droplets, and 58% of all pinocytotic vesicles.[1] Four types of sinusoidal cells have been distinguished: Kupffer, endothelial, fat-storing, and pit cells.

The functional characteristics of the sinusoidal cells have been recently reviewed.[2,3] Kupffer cells are the main tissue macrophages. They play an important role in the clearance of abnormal and foreign substances and particulate material, including senescent erythrocytes, immune complexes, lysosomes, microorganisms, and endotoxin.[2,4] Kupffer and endothelial cells also have a specific function in the removal of cholesterol esters and glycoproteins from the circulation.[4-7] In addition, endothelial cells have a filtration effect. Only particles as small as 0.1 µm, such as chylomicrons, can pass the fenestrae of the endothelial cells and thus reach the space of Disse and the parenchymal cells.

Fat-storing cells are located under the sinusoidal lining in the space of Disse and are characterized by the lipid droplets in their cytoplasm. They are considered to be involved in lipid metabolism, vitamin A storage, and intralobular fibrogenesis.[3] The function of pit cells is as yet unknown.[2]

Heterogeneity of Liver Cells

Several factors can influence the morphology and functional capacity of parenchymal cells, and result in striking differences even within parenchymal cells of the same liver lobule. Some of these factors, which may also influence other liver cell types, include the following:

1. Species, strain, and sex differences;
2. Incidence of pathological lesions, related with the above;
3. Bacteriological background; e.g., germ-free rats have lower liver weights than those raised under conventional conditions;[8]
4. Nutritional status. Histological and morphometric studies performed on human liver material obtained at autopsy from native Japanese and Japanese in Hawaii, as well as from Caucasians in the U.S. and in Costa Rica, revealed that age-related changes in number and size of parenchymal nuclei and cells are influenced by environmental, most probably nutritional, conditions.[9] There are several studies on the effect of nutrition on cytochemical, biochemical, and functional properties of parenchymal cells in rodent livers.[10-12]

Heterogeneity of parenchymal cells in individual livers can result from the following:

1. Differences between cells in the various lobes of the liver;
2. Sublobular location. The peripheral or periportal area of the liver lobule has an environment of a relatively high supply of oxygen and food components as well as toxic compounds, whereas the centrolobular or perivenous cells are more likely to suffer from deprivation of oxygen and nutrients. A clear metabolic zonation has been described.[13] The cells also exhibit differences in size and fine structure, depending on their sublobular location;
3. The presence of mono- and binucleated cells;
4. The presence of cells of various ploidy classes.

Because the aging process influences all of these factors in different ways, it is clear that the overall picture for the aging changes in parenchymal cells is rather complex.

The first indications that Kupffer cells were also heterogeneous were obtained in a study on such cells isolated from the rat liver in which two suspensions of Kupffer cells differed morphologically and in lysosomal enzyme content.[14] Heterogeneity in Kupffer cells was also observed in another study on the distributions of glucose-6-phosphate dehydrogenase over rat liver sinusoidal cells.[15] We recently demonstrated a functional heterogeneity for Kupffer cells which is related to their position in the liver acinus. Kupffer cells with high endocytic activity, large and heterogeneous lysosomes, and high lysosomal enzyme activities are found in the periportal zone.[16] Kupffer cells in the midzonal and perivenous areas are smaller and show lower lysosomal enzyme activities.

Shift to Polyploid Parenchymal Cells with Aging

The effect of aging has been relatively extensively investigated for factors which are easily measurable such as changes in ploidy. Most mammals have a mononuclear diploid parenchymal cell population at birth. During development and aging, changes occur in nuclear and cellular volume that result in an increasing polyploidy. The rate of the process of polyploidization is species-specific, and, at least in the rat, also strain-specific. At least six cytogenetic types of cells are present during postnatal development in the rat: mononuclear diploid, tetraploid, and octaploid cells, as well as their binuclear equivalents. The main changes in the ploidy state of the parenchymal cells take place within the first year of the

Table 1
DI-AND POLYPLOID PARENCHYMAL CELL PROPORTIONS IN AGING RAT LIVERS

Strain	Sex	Age (mo)	Monon. Dipl.	Bin. Dipl.	Monon. Tetra.	Bin. Tetra.	Monon. Octa.	Bin. Octa.	Ref.
Long-Evans	F	0.5	86	5	4				90
		1	55	32	7	1			
		2	33	26	35	2			
		3	21	16	52	7	1		
		12	6	1	66	11	14		
Wistar	M	0.75	79		21		0		91
		2	26		73		2		
		4	20		78		2		
		6	13		86		2		
		16	12		87		1		
		24	10		73		17		
RU	F	3	72		25		3		92
		6	54		42		4		
		12	30		61		9		
		24	29		63		8		
		27	27		66		7		
WAG/Rij	F	0.5	97	2	1				78
		1	36	48	9	4	3		
		3	9	18	53	16	4	2	
		30	11	3	46	19	18	4	
BN/BiRij	F	1	50	43	6				93
		3	4	40	50	6			
		12	1	10	73	16			
		24	4	2	63	28	2	1	
		33	1	4	63	24	6	2	

Note: All values are expressed as percentages of total number of parenchymal cells; Monon., mononuclear; Bin., binuclear; Dipl., diploid; Tetra., tetraploid; Octa., octaploid.

life of the rat (Table 1). Therefore, the shift to polyploidy cannot be considered as a real aging phenomenon.

In man, the main shift to polyploidy occurs before puberty,[17] but is far less pronounced than in rats. In the aged human liver, about 70% of the parenchymal cells are mononuclear diploid.[18-20]

In mouse liver the development of polyploid nuclei takes place early after birth. A relatively large number of octaploid cells is already present at two weeks of age and 16-, 32-, and even 64-ploid nuclei appear afterwards.[21-23]

Changes in Number of Parenchymal and Kupffer Cells

In most rat strains, the relative liver weight does not increase with age.[24] Therefore, the age-related increase in the degree of ploidy of the parenchymal cells (which means an increase in larger cells) leads to a reduction in the number of parenchymal cells per gram of liver tissue. As with the shift in ploidy, this reduction in relative number is most evident early in the life of the rat, in some strains within the first three months of life (see Table 1). Direct data on the numbers of parenchymal cells present at various ages are scarce. In morphometric studies, decreases in the number of nuclei have been reported especially during the first eight months (see Reference 24 for a review), which might be explained by a decrease in the number of binuclear cells.[24]

The absolute weight of the human liver remains constant between 20 and 65 years but shows a pronounced decline later in life.[25,26] The relative liver weight shows a tendency to a slight decrease between 20 and 70 years.[26] After 60 years, a significant decline in cell number, which is partly related with nutritional conditions, has been reported.[25]

The number of Kupffer cells in relation to the number of parenchymal cells shows an increase in conventional and gnotobiotic rats within the first 3 months of the life of the rats.[27] The number of Kupffer cells in 11-month-old rats is not changed, while the number in conventional rats shows a slight decline.[27] The observed early increase in the relative number of Kupffer cells is a result of a decline in the absolute number of parenchymal cells due to polyploidization rather than to an increase in the absolute number of Kupffer cells.

Qualitative and Quantitative Aging Changes at the Subcellular Level

Routine electron microscopy of parenchymal cells does not reveal important changes in the cellular fine structure with aging.[28] The only clear age-related change is the increase in number and size of lipofuscin-containing dense bodies.[28]

Little information is available on morphological changes occurring in the sinusoidal cells during aging of the liver.[29] One study describes the occurrence of numerous swollen sinusoidal cells with rounded nuclei and a more abundant acidophilic P.A.S.-positive cytoplasm in the liver of 24- to 30-month-old rats.[30]

As a consequence of the shift to higher ploidy levels, parenchymal cells increase in cellular volume, nuclear volume, and DNA content with aging (see Reference 31 for literature on rat liver). These are general increases which vary with species, strain, and sex.[11,32] The increase can be quite impressive, however, as illustrated by the comparison between 3-month-old and 35-month-old female WAG/Rij rats: the volume of the parenchymal cells in the livers of the old rats increases between 25% and 60%, depending on the zone of the lobule.[31] Similar increases are found for parenchymal cells after isolation (Table 2). The nuclear volume of the parenchymal cells increases between 40% and 80%.[31]

There are distinct morphological[33] and metabolic[13] differences between parenchymal cells located in the various zones of the liver lobule. It may be reasonably assumed that perivenous or centrolobular parenchymal cells situated at the venous end of the sinusoidal blood flow age under different environmental conditions than do periportal or peripheral cells at the arterial end. Morphometric studies at the subcellular level can provide insight into quantitative aging changes in the morphology of the parenchymal cells in the periportal, midzonal, and central areas of the liver lobule. Several investigators have published morphometric data on age-related changes in the fine structure of rat liver parenchymal cells,[31,32,34] and although there are some conflicting data, there are some general tendencies. For example, quantitative age-related changes in ultrastructure occur regardless of the sublobular location of the parenchymal cells, but the degree of a change is largely influenced by the location.

There is also a marked increase in the volume of the lysosomes with age, which is most pronounced in the centrolobular hepatocytes. A portion of this increase is due to accumulation of lipofuscin within these organelles. There is no direct relation with the observed changes in the specific lysosomal enzyme activities in aging parenchymal cells.

An age-related decrease, although in some cases very slight, has been observed for the total mitochondrial volume. The change in mitochondrial content of the rat parenchymal cells may not be a major critical factor in the liver physiology of old age, since biochemical studies reveal that the respiratory capacity of rat parenchymal cells remains constant during their lifespan.[35]

There are contradictory results with respect to the age-dependent changes in the amount of hepatic smooth endoplasmic reticulum (SER). Both a loss of SER,[32] which might be correlated with an age-related reduction in the drug-metabolizing capacity of the liver, and an increase in the amount of SER in old rats have been reported.[31,34] No correlation has

Table 2
GENERAL CHARACTERISTICS OF PURIFIED LIVER CELLS ISOLATED FROM YOUNG AND OLD FEMALE BN/BiRij RATS

Cell suspension	Age (mo)	Yield[a]	Viability (%)	Purity (%)	No.[b]
Parenchymal cells	3	40 ± 12	91 ± 3	98 ± 1	3
	33 — 34	22 ± 2	90 ± 1	98 ± 1	3
Endothelial cells	3	25 ± 3	93 ± 2	96 ± 1	6
	33 — 34	25 ± 2	94 ± 2	85 ± 8	3
Kupffer cells	3	5 ± 1	91 ± 1	88 ± 1	6
	33 — 34	5 ± 1	94 ± 1	88 ± 1	3

Note: Values represent the mean ± S. E.; Reference: 88.

[a] Yield (10^6 cells per gram of liver).
[b] Number of cell preparations.

been observed between possible quantitative changes in the volume of the rough endoplasmic reticulum and the changes in total protein and albumin synthesis that occur with aging in isolated parenchymal cells.

Age-Associated Pathology of the Liver

To interpret the changes observed for the aging liver, knowledge of age-associated pathological lesions that could influence the aging characteristics under study is essential. Data on the incidence of major lesions — foci or areas of cellular changes and neoplastic nodules — are available for BN/BiRij rats, WAG/Rij rats, and their F_1 hybrid.[36] For Fisher 344 rats, bile duct hyperplasia has been described as the most frequent lesion.[37]

ISOLATED PARENCHYMAL AND SINUSOIDAL CELLS AS MODELS FOR CELLULAR AGING STUDIES

Introduction

The presence of various cell types with their respective aging characteristics within the liver and the existence of procedures for obtaining several cell types in isolation make liver cells suitable models for studies on the mechanisms of cellular aging. It is possible to obtain separately of long-lived parenchymal cells and relatively short-lived sinusoidal cells, thus facilitating studies on the comparison of aging changes in cells having different life-spans. Because the various sinusoidal cells types, i.e., Kupffer cells, endothelial cells and fat-storing cells, can be purified by special centrifugation techniques, cellular aging changes can also be compared in short-lived cells with very diverse functions.

Parenchymal Cell Model System for Studies on Aging of Postmitotic Cells

Postmitotic cells are highly specialized, fixed cells which become differentiated early in life. They cannot be replaced, as they lack the capacity to divide. Accumulating evidence suggests that these cells exhibit more pronounced symptoms of cellular senescence during their long lifetime than continually dividing intermitotic cells such as fibroblasts.[38]

Parenchymal cells can be classified as a "reverting" postmitotic cell; i.e., they retain their potential to divide, but are triggered to do so only in response to a stimulus such as extensive cell loss due to partial hepatectomy. Normally cell division is very rare, with a mitotic incidence of about 0.01% or less,[39,40] or about 1 in every 10,000 parenchymal cells is in mitosis at any given time. This implies that nearly all parenchymal cells will age in the liver and live as long as the organism. Thus, parenchymal cells isolated from donors of various ages will be of the same age as the donor, and in this way provide a suitable system for studies of aging in long-lived postmitotic cells.

Aging Changes in RES Functions Studied with Isolated Kupffer and Endothelial Cells

Kupffer cells are the main tissue macrophages of the organism and as such play an important role in the functions of the reticuloendothelial system (RES). These functions include the maintenance of low levels of possible toxic compounds (such as endotoxins) and of circulating tumor cells and autologous tissue debris; the synthesis and excretion of enzymes and effectors, such as pyrogens, prostaglandins and interferon; and a participation in the immune response (see Reference 41 for a review). We have recently demonstrated that not only Kupffer cells but also endothelial liver cells contribute to several RES clearance functions.[4]

Descriptions of age-dependent changes in RES clearance functions in humans are scarce in the literature. The few studies do show a progressive decline in clearance capacity (see Reference 29 for a review). Also, the small amount of data obtained with aging mice and rats show an age-related decline in clearance capacity, but only one study included actually senescent animals.[29] The isolation and, if necessary, cultivation of Kupffer and endothelial cells obtained from donors of various ages may then be useful in studies of a cellular basis for the age-related decline in RES functions and in studies of aging characteristics of the mechanism of endocytosis and breakdown of foreign components.

Use of Isolated Kupffer Cells for Studies on Cellular Aging

The origin of Kupffer cells is far from clear. They may be derived from stem cells in the bone marrow, but mitosis of Kupffer cells has also been reported to occur in normal liver and under several experimental conditions such as after partial hepatectomy and during RES blockage and RES stimulation.[42,43] The average lifespan of Kupffer cells in the rat under normal conditions may range between 25 and 100 days.[44,45] In addition, the mitotic index has been estimated to be 0.4% for mouse Kupffer cells and 0.5% for rat Kupffer cells (see Reference 44 for review). Those data suggest that Kupffer cells are relatively short-lived as compared with parenchymal cells. Thus, isolated Kupffer cells can be used to study aging phenomena in short-lived cells and for comparison with age changes in the long-lived parenchymal cells from the same organ.

Advantages of the Model System

In addition to the possible uses already mentioned for isolated parenchymal, Kupffer, and endothelial cells for studies on cellular aging and on the cellular basis of aging changes in liver functions, some general advantages are listed below.

1. Large amounts of pure and viable parenchymal, Kupffer, and endothelial cells can be isolated. The functional capacities of the parenchymal cells resemble those of these cells in vivo. A quantitative comparison of the functions of Kupffer and endothelial cells in vivo and in vitro has not yet been made;
2. There are age-related changes in both ultrastructure and functional capacities of the parenchymal cells which may underlie aging changes in liver function; isolated cells make possible correlated studies on the cellular basis of aging changes in liver function;
3. Isolated parenchymal cells can be further separated according to the degree of ploidy and thus can be compared according to different ploidy classes;
4. Primary monolayers of parenchymal cells can be prepared for studying in vitro the effect of factors which may influence the cellular aging process in vivo. These factors include hormones, toxins, and chemical carcinogens;

5.   Isolated parenchymal and sinusoidal cells from young and old rat livers can be co-cultivated to study the effect of aging on cell contacts, cell recognition, polarity, and mutual cellular influence.

More detailed information on the advantages of the use of parenchymal cells for aging studies can be found in a recent review.[46]

ISOLATION, PURIFICATION, AND CULTURE OF PARENCHYMAL AND SINUSOIDAL CELLS

Isolation and Purification of Parenchymal Cells from Donors of Various Ages

Several procedures are now available for the isolation of parenchymal cells from livers of a variety of species. In 1974, Van Bezooijen et al. described a procedure for isolating parenchymal cells from rats of various age groups.[47] They demonstrated that the cells from young rats were not inferior to cells from old rats with respect to morphological appearance, respiratory capacity, and permeability of the cell membrane. Isolated parenchymal cells from rat liver are now used by several groups studying cellular aging processes.[46-54] The methods for cell isolation used in those studies are generally based on a preperfusion of the liver with a calcium-free medium followed by a perfusion and incubation with a collagenase-containing buffer in an oxygen and temperature-controlled environment.[55] After collagenase digestion, a cell suspension is obtained that consists mainly of parenchymal cells, which is centrifuged using various techniques to remove nonparenchymal contaminants. In most cases a differential centrifugation step at low g values (e.g., 5 min at 50 g) is sufficient to remove nonparenchymal cells and cell debris. The yield and other general characteristics of cells obtained from young and old rats are listed in Table 2.

Parenchymal cells can now be separated into subpopulations of diploid and tetraploid cells using a relatively new centrifugation technique known as centrifugal elutriation.[56]

Culture of Parenchymal Cells Isolated from Young, Adult, and Old Rats

Parenchymal cells isolated from young or adult rats have been used by many investigators to prepare primary monolayer cultures.[57] This experimental system also appears promising for the culture of cells from 2-year-old rats.[53] It is very suitable for the ultrastructural analysis of cell contacts and for studies in the field of pharmacology, toxicology, endocrinology, and metabolism. The main shortcoming of the culture system, i.e., the loss of cell viability and differentiation within a relatively short culture period, may be due mainly to the interdependence on other cell types or on extracellular matrix. These problems may be solved by culturing the parenchymal cells on connective tissue fibers isolated from normal rat liver[58] or by coculturing parenchymal and sinusoidal liver cells.[59,60]

Isolation, Purification and Culture of Kupffer and Endothelial Cells

Several methods have been described for the isolation of sinusoidal cells from rat, mouse, hamster, rabbit, and human liver,[2,61-65] that involve two main approaches, one is the preparation of a total liver cell suspension by collagenase treatment followed by the separation of sinusoidal cells by differential centrifugation;[5,59,62] the other is by the selective destruction of parenchymal cells by incubation with trypsin or pronase.[2,61,62] In both cases, the cell suspension obtained consists largely of Kupffer and endothelial cells and can be used for further purification of both cell types.

In a recent review,[2] it was concluded that only the purification method based on the use of centrifugal elutriation results in 90% to 95% pure populations of both Kupffer and endothelial cells. This method has also been employed for the purification of Kupffer and endothelial cells isolated from young and old rats (Table 2).

Liver sinusoidal cells obtained by enzymatic digestion of the liver are suitable for several biochemical and analytical studies,[5] but cannot be used for experiments on endocytosis in vitro due to the probable loss of membrane receptors.[6] However, Kupffer cells in primary culture are capable of endocytosing several substances,[66-68] suggesting a reconstitution of specific membrane receptors in culture.

Recently, a new isolation and purification procedure for rat liver endothelial cells has enabled us to achieve large-scale survival of those cells in maintenance culture.[69] The cultured cells show several ultrastructural characteristics typical of liver endothelial cells in vivo, including the fenestrated cytoplasmic projections which cover the liver sinusoids in vivo.[69] The reformation of specific membrane receptors is indicated by the reappearance of the capacity to take up horseradish peroxidase by adsorptive endocytosis, a characteristic that is generally lost during the isolation of endothelial cells.[69]

Coculture of Parenchymal and Sinusoidal Cells

The coculture of parenchymal and sinusoidal cells is an experimental model that closely resembles the in vivo situation. In these symbiotic cultures, cell contacts similar to those existing in vivo are reestablished between endothelial, Kupffer, and parenchymal cells.[59,60]

This system appears to be suitable for the maintenance of specific cell characteristics and for prolonging the survival time of parenchymal cells in culture by twice that of control cultures containing only parenchymal cells.[60] This increased survival time may be due to the clearance of toxins and other negative factors from the medium by the Kupffer and endothelial cells, the release of trophic factors by the sinusoidal cells, or to intercellular interaction that is beneficial to both cell classes. Cocultures of parenchymal cells and sinusoidal cells from 24-month-old rats also reveal a high degree of purity and a similar prolonged survival time for the parenchymal cells as observed for these cells in symbiotic cultures with sinusoidal cells obtained from young or adult rats.[53] This may indicate that the endocytic activity of the sinusoidal cells and/or the beneficial intercellular interaction between parenchymal and sinusoidal cells are not negatively affected by aging.

THE EFFECT OF AGING ON FUNCTIONS OF ISOLATED LIVER CELLS

Introduction

Any age-related change in the functions of liver cells in vivo can be the result of both intracellular and extracellular effects. The hepatic blood flow can be considered one of the most important extracellular causes for a decline in several liver functions. The blood flow in the human liver decreases with age.[24,70,71] There is only one study on hemodynamic changes in aging rats in which a sharp decline of at least 50% in heavy (300 to 400 g), and thus probably old, white male rats as compared with young and light (120 to 150 g) animals was observed.[70] Unfortunately, no exact ages were given. The influences of blood flow and other extracellular factors such as hormonal regulation can be excluded in an in vitro system using isolated liver cells.

General Characteristics of Isolated Parenchymal, Endothelial, and Kupffer Cells Obtained from Rats of Various Ages

Functional and biochemical characteristics of isolated liver cells are most often expressed on a per cell basis and per mg of cellular protein. However, in aging studies, both parameters should be considered as variable. They can give rise to individual aging patterns due to changes in size or protein composition of the cells. A shift in the ploidy state of parenchymal cells with age has already been described. General information on cell size and DNA and protein content of parenchymal, endothelial, and Kupffer cells isolated from young and old female BN/BiRij rats is presented in Table 3.

Table 3
VOLUME AND PROTEIN AND DNA CONTENTS OF LIVER CELLS
ISOLATED FROM FEMALE BN/BiRij RATS OF DIFFERENT AGES

Cell type	Age (mo)	Volume (μm^3)	Protein ($\mu g/10^6$ cells)	DNA ($\mu g/10^6$ cells)
Parenchymal cells	3	5930 ± 460	1503 ± 84	40.7 ± 0.8
	12	8340 ± 550[a]	2193 ± 161[a]	44.5 ± 3.9
	33	14790 ± 2040[a]	2880 ± 230[a]	71.1 ± 1.0[a]
Endothelial cells	3	179 ± 1	49 ± 2	2.9 ± 0.9
	33	240 ± 9[a]	52 ± 2	5.8 ± 0.4[a]
Kupffer cells	3	472 ± 18	118 ± 7	2.3 ± 0.4
	33	842 ± 20[a]	177 ± 3[a]	6.0 ± 0.4[a]

Note: The values represent the mean ± S.E. of 3 or 4 different cell preparations. Protein and DNA contents of the isolated cells were determined by standard procedures.[86] Cell diameters were measured from light microscopic photographs of purified cell suspensions with the Zeiss TGZ-3 particle size analyzer. Cell volumes were calculated by considering that all isolated cells had a spherical form. For the determination of cell volume at least 200 cells were measured per age group.

[a]   Value differs significantly ($p < 0.05$) from 3-month value.

The average size of isolated endothelial cells increases slightly with age, but there is no significant correlation with an increase in the protein content of the cells. The large variation in DNA content cannot be explained at present. In contrast to endothelial cells, Kupffer cell characteristics change greatly with age. Between 3 and 33 months of age, there is a doubling of the cell volume and in the amount of DNA per 10^6 cells. The protein content increases by 50%. The amount of protein and DNA in isolated Kupffer cells can be influenced by the presence of extracellular material which is endocytosed by the cells during the isolation procedure. A possible increase in extracellular material accumulated in Kupffer cells isolated from old rats could be due to a decreased capacity to break down these substances. However, the lysosomal proteolytic enzyme activities[72] and the DNAse activity (unpublished results) are higher in Kupffer cells from old rats as compared to those from young animals. For this reason it is unlikely that an accumulation of extracellular material accounts for the higher protein and DNA contents of cells from old rats.

The changes in characteristics of parenchymal, endothelial, and Kupffer cells should be taken into account when enzymatic or functional activities in cells isolated from old rats are interpreted and compared with activities of cells from young animals.

Age-Related Changes in Functions of Isolated Liver Cells
Albumin, Total Protein and RNA Synthesis in Isolated Parenchymal Cells

Albumin synthesis is a liver-specific function, and the capacity to synthesize this protein can be considered as a reflection of the reserve capacity of the rat liver.[73] Possible age-related changes in this synthesis were investigated with parenchymal cells isolated from female WAG/Rij rats of various ages.[74,75] Cells from 3-month-old rats synthesize an amount of albumin comparable to that synthesized in vivo.[74] Between 3 and 12 as well as 12 and 24 months, a significant decrease takes place, and cells from 24-month-old rats synthesize only 40% of the amount of albumin produced by cells from 3-month-old rats. By contrast, a sharp increase has been found after 24 months, and cells from 36-month-old animals synthesize nearly twice the amount as cells from 3-month-old rats.[75]

The capacity of isolated parenchymal cells to synthesize total protein also shows an age-related pattern which is to some extent comparable with that observed for albumin synthesis.

A decrease in protein synthesis takes place between 3 and 12 months in parenchymal cells from female WAG/Rij rats.[28] A similar decline was found by Richardson's group for female Sprague-Dawley rats[50] and male Fisher F344 rats.[76] No decline has been observed for parenchymal cells from male WAG/Rij rats,[77] female BN/BiRij rats,[77] or male Sprague-Dawley rats.[52]

Between 12 and 18 to 24 months of age no significant changes in the protein synthesis have been observed for isolated parenchymal cells from any of the strains and sexes mentioned above, with the exception of male Sprague-Dawley rats, which show a 50% decline in this age period.[52] After 24 months of age, a sharp and continuous increase in the protein synthesizing capacity occurs, which is most probably independent of sex and strain.[77] The increase in albumin and total protein synthesis in late age is not due to a compensation by the liver for increased protein excretion due to kidney insufficiency.[75] Other possible explanations have been discussed elsewhere.[77]

Cultured parenchymal cells can also be used to study the effect of aging on RNA synthesis, because in contrast to other in vitro systems, the nuclear-cytoplasmic interactions in these cells are the same in vivo and in vitro. The RNA synthesis in parenchymal cells isolated from 18- and 30-month-old male Fischer rats is 70% and 63%, respectively, of that observed for 12-month-old rats.[51]

BSP Metabolism by Parenchymal Cells Isolated from Rats of Various Ages

The clinical bromsulfophthalein (BSP) retention test is used to determine the capacity of the liver to transfer a number of organic anions (including steroids and many drugs) into the parenchymal cells and finally into the bile. Several authors have reported an age-related decrease when this test was used in groups of aged people (see Reference 78 for review). The same situation is found for aging rats.[75]

With parenchymal cells isolated from female WAG/Rij rats, a sharp decrease in the amount of BSP stored by the cells was observed between 3 and 12 months of age; this was followed by a less pronounced decline up to 36 months.[75] The age-related pattern of the storage capacity of the isolated cells completely follows that observed with the BSP retention test in aging female WAG/Rij rats.[75] This strongly suggests that the decline in the capacity of the liver to remove BSP from the blood is at least partly due to an age-related decline in the BSP storage capacity of the parenchymal cells.

Aging and Drug Metabolism by Isolated Parenchymal Cells

There are indications that the liver generally shows a decreased capacity to metabolize drugs with advancing age.[79] However, the drug metabolizing capacity in vivo is dependent not only on pharmacological variables but also on physiological variables, including neurological, endocrinological, and circulatory factors. All extrahepatic influences can be excluded when isolated parenchymal cells are used for studies on drug metabolism. The isolated cells have the advantage over the microsomal preparations used for studies on drug metabolism in that they contain the complete biological system necessary for the metabolism of drugs, including substrate and cofactor supply, which is absent in microsomal preparations.

Of course, the results obtained with isolated parenchymal cells should also be carefully interpreted in view of such variables as sex, genetic background and age-associated liver pathology.

It is well known that the rates of metabolism of compounds by isolated parenchymal cells resemble those in vivo,[80,81] but, surprisingly, only one study has been performed on the effect of aging on drug metabolism using isolated cells.[82] In this study, the capacity of parenchymal cells isolated from male BN/BiRij rats to metabolize the cardiac glycoside digitoxin was determined and preliminary results revealed no qualitative changes in the pattern of digitoxin metabolites.[82] The plasma disappearance of labeled digitoxin is prolonged in aged 12-, 24-, and 30-month-old female BN/BiRij rats, suggesting a decrease in hepatic clearance with age.[83]

The Role of Lysosomes in Cellular Aging

In nearly all cells and especially in long-lived parenchymal cells, most intracellular constituents are continuously renewed. Lysosomes play the key role in the controlled breakdown and turnover of cellular components. This degradation process is called autophagy. Lysosomes are also involved in heterophagy, the digestion of extracellular material taken up by endocytosis. The key function of lysosomes in maintaining cellular homeostasis and in hydrolysis of endocytosed material made it of interest to evaluate the role of lysosomes in cellular aging. For this purpose, our group used various classes of isolated liver cells.

In morphometric studies on parenchymal cells of young and old rats, clear alterations in the lysosomal system are observed. The cytoplasmic volume occupied by lysosomal structures increases with age in all areas of the liver lobule, and the lysosomes become progressively loaded with indigestible material.

In studies on lysosomal enzyme activities in suspensions of various liver cell classes, it was found that the patterns of aging changes differed considerably for parenchymal, Kupffer, and endothelial cells.[72,84-86] Most enzyme activities show a tendency to decline with age in parenchymal cells, with the exception of cathepsin D, which increases. Although we did not observe a real deficiency in any lysosomal enzyme, the decline in activities combined with the increased lysosomal volume may lead to a kind of "dilution" of the amount of enzymes available in each lysosome. This and the observed increase in undigestible material (lipofuscin) with aging are indications of an impairment in the autophagic degradation process in old parenchymal cells.

The activities of lysosomal enzymes in isolated Kupffer and endothelial cells show both increases (for proteolytic enzymes in Kupffer cells[72]) and decreases with aging. These variations in changes in enzyme activities do not allow a general conclusion as to the effect of age on lysosomal functions of Kupffer and endothelial cells.

Age-related changes in lysosomal enzyme activities in Kupffer and endothelial cells are related to alterations in the endocytosis and hydrolysis of certain substrates from the circulation, as shown in studies with, e.g., cultured Kupffer cells.

A complicating factor is the presence of multiple forms of acid phosphatase and cathepsin D not only in parenchymal, Kupffer, and endothelial cells, but even within individual cells, e.g., the parenchymal cells.[87] A strong increase in heterogeneity in multiple forms of acid phosphatase, which may be the result of posttranslational modifications, was observed in aging parenchymal cells.[88]

Changes in Functions of Kupffer Cells with Aging

Although there are reports on an age-related decrease in the function of the reticuloendothelial system (RES) as determined from the clearance rate of colloidal substances from the blood, little is known about the effect of aging upon the functions of the sinusoidal cells in the intact liver. One study suggests a functional decline of the Kupffer cells in the livers of 24-month-old male Sprague-Dawley rats in response to colloidal carbon.[30]

The cellular basis of the age-related changes in the functional capacities of the RES has been studied by our group using isolated Kupffer cells maintained in culture. Two major functions of these cells, i.e., endocytosis and catabolism, can be determined by using ^{125}I-labeled monomeric formaldehyde-treated albumin or heat-aggregated colloidal albumin (^{125}I-C.A.), a clinically used RES test substance.[67,89]

The capacities of Kupffer cells from 3- and 36-month-old female BN/BiRij rats to endocytose, degrade, and excrete ^{125}I-C.A. have been determined in our Institute. Preliminary results suggest that there are no significant age-related differences in the degradation and excretion of ^{125}I-C.A. On the other hand, the maximum capacity of Kupffer cells from 30- to 36-month-old rats to endocytose ^{125}I-C.A. appears significantly lower than that of Kupffer cells from 3-month-old rats.[29] This is an indication that Kupffer cells undergo functional impairment with increasing age in rats, which corresponds with the observed age-related decline in the clearance function of the RES.[29]

CONCLUDING REMARKS

This review on aging and liver cells has been largely restricted to aging changes in isolated liver cells. This is not to disregard the experimental results obtained with intact liver, or liver homogenates, slices, or isolated organelles. These results have undoubtedly provided important information on the aging of the liver and its cellular constituents. Reasons for the occasional discrepancies in the reported results have been discussed elsewhere.[24]

The intention of this paper was to stress the importance of the use of isolated liver cells for several types of aging studies. Compared with cultures of cell lines, which are often used for the study of cellular aging, the various types of isolated liver cells constitute an experimental system with a well defined "age", e.g., parenchymal cells isolated from old rats are old cells and thus findings on isolated cells are potentially comparable with in vivo results. This system is of great value not only for the understanding of the cellular basis of liver aging, but also for studies on cellular aging processes.

REFERENCES

1. **Blouin, A., Bolender, R. P., and Weibel, E. R.,** Distribution of organelles and membranes between hepatocytes and nonhepatocytes in the rat liver parenchyma, *J. Cell Biol.,* 72, 441, 1977.
2. **Wisse, E. and Knook, D. L.,** The investigation of sinusoidal cells: a new approach to the study of liver function, in *Progress in Liver Diseases,* Vol. VI, Popper, H. and Schaffner, P., Eds., Grune & Stratton, New York, 1979, 153.
3. **Knook, D. L. and De Leeuw, A. M.,** Cell types of the normal liver and their possible role in fibrogenesis, in *Connective Tissue of the Normal and Fibrotic Liver,* Pott, G. and Gerlach, U., Eds., G. Thieme Verlag, Stuttgart, 1982, 67.
4. **Praaning-van Dalen, D. P., Brouwer, A., and Knook, D. L.,** Clearance capacity of rat liver Kupffer, endothelial and parenchymal cells, *Gastroenterology,* 81, 1036, 1981.
5. **Van Berkel, Th. J. C.,** The role of nonparenchymal cells in liver metabolism, *TIBS,* September 1979, 202.
6. **Steer, C. J., Kusiak, J. W., Brady, R. O., and Jones, E. A.,** Selective hepatic uptake of human β-hexosaminidase A by a specific glycoprotein recognition system on sinusoidal cells, *Proc. Natl. Acad. Sci. U.S.A.,* 76, 2774, 1979.
7. **Hubbard, A., Wilson, G., Ashwell, G., and Stukenbrok, M.,** An electron microscope autoradiographic study of the carbohydrate recognition systems in rat liver. 1. Distribution of ^{125}I-ligands among the liver cell types, *J. Cell Biol.,* 83, 47, 1979.
8. **Knook, D. L., Barkway, C., and Sleyster, E. C.,** Lysosomal enzyme content of Kupffer and endothelial liver cells isolated from germfree and clean conventional rats, *Infect. Immun.,* 33, 620, 1981.
9. **Tauchi, H. and Sato, T.,** Effect of environmental conditions upon age changes in the human liver, *Mech. Age. Dev.,* 4, 71, 1975.
10. **Schimke, R. R.,** The importance of both synthesis and degradation in the control of arginase levels in rat liver, *J. Biol. Chem.,* 239, 3808, 1964.
11. **Ross, M. H.,** Aging, nutrition and hepatic enzyme activity patterns in the rat, *J. Nutr.,* 97, 563, 1969.
12. **Barrows, C. H., Jr. and Kokkonen, G. C.,** Diet and life extension in animal model systems, *Age,* 1, 130, 1978.
13. **Jungermann, K. and Sasse, D.,** Heterogeneity of liver parenchymal cells, *Trends in Biochem. Sci.,* 3, 198, 1978.

14. **Sleyster, E. Ch., Westerhuis, F. G., and Knook, D. L.,** The purification of nonparenchymal liver cell classes by centrifugal elutriation, in *Kupffer Cells and Other Liver Sinusoidal Cells,* Wisse, E. and Knook, D. L., Eds., Elsevier/North-Holland Biomedical Press, Amsterdam, 1977, 289.

15. **Knook, D. L., Sleyster, E. Ch., and Teutsch, H. F.,** High activity of glucose-6-phosphate dehydrogenase in Kupffer cells isolated from rat liver, *Histochemistry,* 69, 211, 1980.

16. **Sleyster, E. Ch. and Knook, D. L.,** Relation between localization and function of rat liver Kupffer cells, Lab. Invest., 47, 484, 1982.

17. **Swartz, F. J.,** The development in the human liver of multiple desoxyribose nucleic acid (DNA) classes and their relationship to the age of the individual, *Chromosoma,* 8, 53, 1956.

18. **Altmann, H. W., Loeschke, K., and Schenck, K.,** Über das Karyogramm der menschlichen Leber unter normalen und pathologischen Bedingungen, *Virchows Arch. Path. Anat.,* 311, 85, 1966.

19. **Meek, E. S. and Harbinson, J. F. A.,** Nuclear area and deoxyribonucleic acid content in human liver cell nuclei, *J. Anat.,* 101, 487, 1967.

20. **Denkhaus, W.,** Kerngrösse, DNS-Gehalt und Ploidie-Klassen menschlicher Leberzellen in Abhängigkeit vom Lebensalter, *Z. Geront.,* 3, 88, 1970.

21. **Inamdar, N. B.,** Development of polyploidy in mouse liver, *J. Morphol.,* 103, 65, 1958.

22. **Epstein, C. J.,** Nuclear ploidy in mammalian parenchymal liver cells, *Nature,* 214, 1050, 1967.

23. **Tongiani, R. and Puccinelli, E.,** Existence of hepatic cell classes and their relationship with the nuclear classes in the liver of the white mouse. A microinterferometric study, *Histochemie,* 21, 33, 1970.

24. **Knook, D. L.,** Organ ageing in relation to cellular ageing, in *Lectures on Gerontology,* Vol. 1, Viidik, A. A., Ed., Academic Press, London, 1982, 213.

25. **Tauchi, H. and Sato, T.,** Hepatic cells of the aged, in *Liver and Ageing — 1978,* Kitani, K., Ed., Elsevier/North-Holland Biomedical Press, Amsterdam, 1978, 3.

26. **Lindner, J., Grasedyck, K., Bittmann, S., Mangold, I., Schütte, B., and Ueberberg, H.,** Some morphological and biochemical results on liver ageing, esp. regarding connective tissue, in *Liver and Aging,* Platt, D., Ed., Schattauer Verlag, Stuttgart-New York, 1977, 23.

27. **Podoprigora, G. I. and Zaitsev, T. I.,** Morphological and functional status of elements of the reticuloendothelial system in gnotobiotic animals, *Fol. Microbiol.,* 24, 55, 1979.

28. **Van Bezooijen, C. F. A., Grell, T., and Knook, D. L.,** The effect of age on protein synthesis by isolated liver parenchymal cells, *Mech. Age. Dev.,* 6, 293, 1977.

29. **Brouwer, A. and Knook, D. L.,** The reticuloendothelial system (RES) and aging, *Mech. Age. Dev.,* 21, 205, 1983.

30. **Patek, P. R., De Mignard, V. A., and Bernick, S.,** Age changes in structure and responses of reticuloendothelial cells of the rat liver, *J. Reticuloendothelial Soc.,* 4, 211, 1967.

31. **Meihuizen, S. P. and Blansjaar, N.,** Stereological analysis of liver parenchymal cells from young and old rats, *Mech. Age. Dev.,* 13, 111, 1980.

32. **Schmucker, D. L., Mooney, J. S., and Jones, A. L.,** Stereological analysis of hepatic fine structure in the Fischer 344 rat. Influence of sublobular location and animal age, *J. Cell. Biol.,* 78, 319, 1978.

33. **Loud, A. V.,** A quantitative stereological description of the ultrastructure of normal rat liver parenchymal cells, *J. Cell Biol.,* 42, 68, 1969.

34. **Pieri, C., Zs.-Nagy, I., Mazzufferi, G., and Giuli, C.,** The aging of rat liver as revealed by electron microscopic morphometry. I. Basic parameters, *Exp. Gerontol.,* 10, 291, 1975.

35. **Brouwer, A., Van Bezooijen, C. F. A., and Knook, D. L.,** Respiratory activities of hepatocytes isolated from rats of various ages. A brief note, *Mech. Age. Dev.,* 6, 265, 1977.

36. **Hollander, C. F. and Burek, J. D.,** Strain and age-associated pathology of the rat liver, in *Liver and Aging — 1978,* Kitani, K., Ed., Elsevier/North-Holland Biomedical Press, Amsterdam, 1978, 39.

37. **Coleman, G. L., Barthold, S. W., Osbaldiston, G. W., Foster, S. J., and Jonas, A. M.,** Pathological changes during aging in barrier-reared Fisher 344 male rats, *J. Geront.,* 32, 258, 1977.

38. **Evans, C. H.,** On the aging of organisms and their cells, *Med. Hypotheses,* 5, 53, 1979.

39. **Nadal, C. and Zajdela, F.,** Polyploidie somatique dans le foie de rat, *Exp. Cell. Res.,* 42, 99, 1966.

40. **Greengard, O., Federman, M., and Knox, W. E.,** Cytomorphometry of developing rat liver and its application to enzymic differentiation, *J. Cell. Biol.,* 52, 261, 1972.

41. **Altura, B. M.,** Reticuloendothelial cells and host defence, *Adv. Microcirc.,* 9, 252, 1980.

42. **Wisse, E.,** Ultrastructure and function of Kupffer cells and other sinusoidal cells in the liver, in *Kupffer Cells and Other Liver Sinusoidal Cells,* Wisse, E. and Knook, D. L., Eds., Elsevier/North-Holland Biomedical Press, Amsterdam, 1977, 33.

43. **Bouwens, L. and Wisse, E.,** On the origin of the Kupffer cell, *Kupffer Cell Bull.,* 3, 23, 1980.

44. **Volkman, A.,** The unsteady state of the Kupffer cells, in *Kupffer Cells and Other Liver Sinusoidal Cells,* Wisse, E. and Knook, D. L., Eds., Elsevier/North-Holland Biomedical Press, Amsterdam, 1977, 459.

45. **Heine, W-D.,** Proliferation kinetics of Kupffer cells, in *The Reticuloendothelial System and the Pathogenesis of Liver Disease,* Liehr, H. and Grün, M., Eds., Elsevier/North-Holland Biomedical Press, Amsterdam, 1980, 27.

46. **Knook, D. L.,** The isolated hepatocyte: a cellular model for aging studies, *Proc. Soc. Exp. Biol. Med.,* 165, 170, 1980.

47. **Van Bezooijen, C. F. A., Van Noord, M. J., and Knook, D. L.,** The viability of parenchymal liver cells isolated from young and old rats, *Mech. Age. Dev.,* 3, 107, 1974.

48. **Britton, G. W., Britton, V. J., Gold, G., and Adelman, R. C.,** The capability for hormone-stimulated enzyme adaption in liver cells isolated from aging rats, *Exp. Geront.,* 11, 1, 1976.

49. **Ungemach, F. R. and Hegner, D.,** Age-dependent transport of thymidine in rat liver hepatocytes, in *Liver and Aging,* Platt, D., Ed., Schattauer Verlag, Stuttgart-New York, 1977, 163.

50. **Ricca, G. A., Liu, D. S. H., Coniglio, J. J., and Richardson, A.,** Rates of protein synthesis by hepatocytes isolated from rats of various ages, *J. Cell Physiol.,* 97, 137, 1978.

51. **Kreamer, W., Zorich, N., Liu, D. S. H., and Richardson, A.,** Effect of age on RNA synthesis by rat hepatocytes, *Exp. Geront.,* 14, 27, 1979.

52. **Viskup, R. W., Baker, M., Holbrook, J. P., and Pennial, R.,** Age-associated changes in activities of rat hepatocytes. I. Protein synthesis, *Exp. Aging Res.,* 5, 487, 1979.

53. **Mosselman, R., Wanson, J-C., Brouwer, A., and Knook, D. L.,** Isolation and culture of sinusoidal cells and hepatocytes, extracted from adult and 24 months aged rats. Behaviour of endothelial cells, Kupffer cells and fat-storing cells in single and co-cultures, in *Electron Microscopy 1980,* Vol. 2, Brederoo, P. and de Priester, W., Eds., Seventh European Congress on Electron Microscopy Foundation, Leiden, 1980, 122.

54. **Wilson, P. D., Watson, R., and Knook, D. L.,** Effects of age on rat liver enzymes. A study using isolated hepatocytes, endothelial and Kupffer cells, *Gerontology,* 28, 32, 1982.

55. **Berry, M. N. and Friend, D. S.,** High-yield preparation of isolated rat liver parenchymal cells, *J. Cell Biol.,* 43, 506, 1969.

56. **Bernaert, D., Wanson, J-C., Mosselman, R., de Paermentier, F., and Drochmans, P.,** Separation of adult rat hepatocytes into distinct subpopulations by centrifugal elutriation. Morphological, morphometrical and biochemical characterization of cell fractions, *Biol. Cellulaire,* 34, 159, 1979.

57. **Wanson, J-C., Bernaert, D., and May, C.,** Morphology and functional properties of isolated and cultured hepatocytes, in *Progress in Liver Diseases,* Vol. VI, Popper, H. and Schaffner, P., Eds., Grune & Stratton, New York, 1979, 1.

58. **Rojkind, M., Gatmaitan, Z., Mackensen, S., Giambrone, M-A., Ponce, P., and Reid, L. M.,** Connective tissue biomatrix: its isolation and utilization for long-term cultures of normal rat hepatocytes, *J. Cell Biol.,* 87, 255, 1980.

59. **Wanson, J-C., Drochmans, P., Mosselmans, R., and Knook, D. L.,** Symbiotic culture of adult hepatocytes and sinus-lining cells, in *Kupffer Cells and Other Liver Sinusoidal Cells,* Wisse, E. and Knook, D. L., Eds., Elsevier/North-Holland Biomedical Press, Amsterdam, 1977, 144.

60. **Wanson, J-C., Mosselman, R., Brouwer, A., and Knook, D. L.,** Interaction of adult rat hepatocytes and sinoisoidal cells in cocultures, *Biol. Cellulaire,* 36, 7, 1979.

61. **Knook, D. L. and Sleyster, E. Ch.,** Separation of Kupffer and endothelial cells of the rat liver by centrifugal elutriation, *Exp. Cell Res.,* 99, 444, 1976.

62. **Knook, D. L. and Sleyster, E. C.,** Preparation and characterization of Kupffer cells from rat and mouse liver, in *Kupffer Cells and Other Liver Sinusoidal Cells,* Wisse, E. and Knook, D. L., Eds., Elsevier/North-Holland Biomedical Press, Amsterdam, 1977, 273.

63. **Knook, D. L. and Brouwer, A.,** The biochemistry of Kupffer cells, in *The Reticuloendothelial System and the Pathogenesis of Liver Disease,* Liehr, H. and Grün, M., Eds., Elsevier/North-Holland Biomedical Press, Amsterdam, 1980, 17.

64. **Kirn, A., Steffan, A. M., Bingen, A., Gendrault, J. L., and Cinqualbre, J.,** Uptake of viruses by Kupffer cells isolated from human liver, *Lancet,* 585, 1980.

65. **Steffan, A.-M., Lecerf, F., Keller, F., Cinquaere, J., and Kirn, A.,** Isolement et culture de cellules endothéliales de foies humain et murin, *C. R. Acad. Sc. Paris,* 292, 809, 1981.

66. **Munthe-Kaas, A. C., Kaplan, G., and Seljelid, R.,** On the mechanism of internalization of opsonized particles by rat Kupffer cells *in vitro, Exp. Cell. Res.,* 103, 201, 1976.

67. **Brouwer, A. and Knook, D. L.,** Quantitative determination of endocytosis and intracellular digestion by rat liver Kupffer cells *in vitro,* in *Kupffer Cells and Other Liver Sinusoidal Cells,* Wisse, E. and Knook, D. L., Eds., Elsevier/North-Holland Biomedical Press, Amsterdam, 1977, 343.

68. **Brouwer, A., Wanson, J-C., Mosselmans, R., Knook, D. L., and Drochmans, P.,** Morphology and lysosomal enzyme activity of primary cultures of rat liver sinusoidal cells, *Biol. Cellulaire,* 37, 35, 1980.

69. **De Leeuw, A. M., Barelds, R. J., de Zanger, R., and Knook, D. L.,** Primary cultures of endothelial cells of the rat liver. A model for ultrastructural and functional studies, *Cell Tissue Res.,* 223, 201, 1982.

70. **Wiener, E. and Rabinovici, N.,** Liver haemodynamics and age, *Proc. Soc. Exp. Biol.,* 108, 752, 1961.

71. **Skaunic, V., Hulek, P., and Martínková, J.,** Changes in kinetics of exogenous dyes in the aging process, in *Liver and Aging — 1978,* Kitani, K., Ed., Elsevier/North-Holland Biomedical Press, Amsterdam, 1978, 115.

72. **Knook, D. L. and Sleyster, E. Ch.,** Lysosomes in Kupffer cells isolated from young and old rats, in *Liver and Aging — 1978,* Kitani, K., Ed., Elsevier/North-Holland Biomedical Press, Amsterdam, 1978, 241.

73. **Chen, J. C., Ove, P., and Lansing, A. I.,** *In vitro* synthesis of microsomal protein and albumin in young and old rats, *Biochim. Biophys. Acta,* 312, 598, 1973.

74. **Van Bezooijen, C. F. A., Grell, T., and Knook, D. L.,** Albumin synthesis by liver parenchymal cells isolated from young, adult and old rats, *Biochem. Biophys. Res. Commun.,* 71, 513, 1976.

75. **Van Bezooijen, C. F. A. and Knook, D. L.,** Aging changes in bromsulfophthalein uptake, albumin and total protein synthesis in isolated hepatocytes, in *Liver and Aging,* Platt, D., Ed., Schattauer Verlag, Stuttgart-New York, 1977, 227.

76. **Coniglio, J. J., Liu, D. S. H., and Richardson, A.,** A comparison of protein synthesis by liver parenchymal cells isolated from Fisher F344 rats of various ages, *Mech. Age. Dev.,* 11, 77, 1979.

77. **Van Bezooijen, C. F. A., Sakkee, A. N., and Knook, D. L.,** Sex and strain dependency of age-related changes in protein synthesis of isolated rat hepatocytes, *Mech. Age. Dev.,* 17, 11, 1981.

78. **Van Bezooijen, C. F. A.,** Cellular Basis of Liver Aging Studied with Isolated Hepatocytes, Ph.D. thesis, University of Utrecht, 1978.

79. **Van Bezooijen, C. F. A.,** Influence of age-related changes in rodent liver morphology and physiology on drug metabolism — A review, *Mech. Age Dev.,* 25, 1, 1984.

80. **Thurman, R. G. and Kauffman, F. C.,** Factors regulating drug metabolism in intact hepatocytes, *Pharmacol. Rev.,* 31, 229, 1980.

81. **Sirica, A. E. and Pitot, N. C.,** Drug metabolism and effects of carcinogens in cultured hepatic cells, *Pharmacol. Rev.,* 31, 205, 1980.

82. **Van Bezooijen, C. F. A., Sakkee, A. N., Kitani, K., and Knook, D. L.,** Changes in digitoxin biotransformation by the liver with age, in *Abstr. XII Int. Congress of Geront.,* Vol. 1, 1981, 59.

83. **Kitani, K., Sato, Y., and Van Bezooijen, C. F. A.,** The effect of age on the biliary excretion of digitoxin and its metabolites in female BN/Bi rats, *Arch. Geront. Geriatr.,* 1, 43, 1982.

84. **Knook, D. L. and Sleyster, E. Ch.,** Lysosomal enzyme activities in parenchymal and nonparenchymal liver cells isolated from young, adult and old rats, *Mech. Age. Dev.,* 5, 389, 1976.

85. **Knook, D. L.,** The role of lysosomes in protein degradation in different types of rat liver cells, *Acta Biol. Med. Germ.,* 36, 1747, 1977.

86. **Knook, D. L. and Sleyster, E. Ch.,** Isolated parenchymal, Kupffer and endothelial rat liver cells characterized by their lysosomal enzyme content, *Biochem. Biophys. Res. Commun.,* 96, 250, 1980.

87. **Sleyster, E. Ch. and Knook, D. L.,** Multiple forms of acid phosphatase in rat liver parenchymal, endothelial and Kupffer cells, *Arch. Biochem. Biophys.,* 190, 756, 1978.

88. **Sleyster, E. Ch. and Knook, D. L.,** Aging and multiple forms of acid phosphatase in isolated rat liver cells, *Mech. Age. Dev.,* 14, 443, 1980.

89. **Brouwer, A., Praaning-van Dalen, D. P., and Knook, D. L.,** Endocytosis of denatured albumin by rat Kupffer cells *in vitro,* in *The Reticuloendothelial System and the Pathogenesis of Liver Disease,* Liehr, H. and Grün, M., Eds., Elsevier/North-Holland Biomedical Press, Amsterdam, 1980, 107.

90. **Alfert, M. and Geschwind, I. I.,** The development of polysomaty in rat liver, *Exp. Cell Res.,* 15, 230, 1958.

91. **Post, J., Klein, A., and Hoffman, J.,** Responses of the liver to injury, *Arch. Pathol.,* 70, 314, 1960.

92. **De Leeuw-Israel, F. R.,** Aging Changes in The Rat Liver, Ph.D. thesis, University of Leiden, 1971.

93. **Knook, D. L.,** unpublished data, 1981.

THE EFFECT OF AGING ON THE KIDNEY

Leah M. Lowenstein and Gretchen H. Bean

Aging of the kidney results in decreased renal function and the decreased ability of the organism to adapt to physical stress. Underlying these effects are age-related changes in structure, cellular function, response to hormones, and the reaction to renal disease. Information on aging in the kidney is derived largely from the rat which is a convenient experimental model, as structure, function, and humoral responsiveness are altered in both the rat and human kidney during aging.

"Old" and "elderly" in this chapter generally refer to individuals over 65 years of age and to rats over 22 months of age.

STRUCTURE

In humans, renal mass is decreased during aging: approximately 25% loss in weight of the kidney occurs between age 40 to 80. Renal mass is lost primarily in the cortex, with a decrease in the total number of glomeruli, 30% decrease in the cell number of glomerular tufts, 25% loss in the number of epithelial cells of convoluted tubules, and a concomitant increase in the size of the cell.[1-3]

Vascular channels in the glomeruli are reduced, with progressive collapse of the tuft and subsequent hyalinization.[4] Shunts develop around glomeruli, especially in the juxtamedullary region.[4] The number of abnormal glomeruli increases with age, from 5% at age 40 to about 60% at age 90.[5]

The medulla, with aging, remains normal or increases in mass as it undergoes interstitial fibrosis.[6]

Many of the structural changes in aging have been associated with changes in the renal vasculature. Renal arteriosclerotic changes in aging include gradual vascular degeneration, with deposition of collagen in the muscular arteries, reduplication of elastic tissue, intimal thickening, increased tortuosity of the interlobular arteries, and arteriolosclerosis of the efferent arterioles.[7-12] These changes are usually not associated with altered luminal diameters and probably occur independently of hypertension. However, on angiography, progressive changes in the arterial pattern in old normotensive patients are similar to those found in hypertensive patients.[13,14]

During aging in rats, from 12 to 24 months, the kidney decreases 30% in weight in Wistar rats[15] but remains the same or increases in Sprague-Dawley rats.[16,17] The total number of nephrons decreases in some strains: kidneys of Wistar rats 18 months old contain 70% the number of nephrons of 12-month-old rats.[15] Serum creatinine concentrations in a variety of strains indicate that only 80% of the nephrons remain by 24 months.[18] As the number of glomeruli diminishes, the average diameter of the remaining glomeruli increases slightly.

The most striking structural change in the aging rat kidney is the development of focal glomerulosclerosis, with epithelial and mesangial cell proliferation, thickening of the basement membrane, and fusion of the epithelial cell foot processes. The focal glomerulosclerosis varies from a mild increase in mesangial matrix to global sclerosis. Eosinophilic material is deposited in Bowman's space. These lesions have been found to some degree in all of the strains of rats studied.[17,19-25] In 27-month-old Sprague-Dawley rats, the lesion is associated with the deposition of IgM and fibrin in the mesangium.[21] The deposits have the characteristics of noncomplement-fixing immune complexes; there are no circulating antibodies to renal or nuclear antigens. The lesions are present in "pathogen-free" Sprague-Dawley rats 24-months-old and in some germ-free rats, but absent in a similar set of germ-free rats.[23]

The incidence of focal glomerulosclerosis can be decreased in the rat by restriction of protein or carbohydrate in their diets.[19,26] For example, on a normal diet the incidence of glomerulosclerosis is 20% in 13-month-old Sprague-Dawley rats; on a reduced diet, the incidence is 20% at 31 months.[19]

The tubules in old rats usually appear normal. Any tubular damage present is usually correlated with the degree of glomerulosclerosis[21] and can vary from collapsed tubules to grossly dilated spaces filled with colloid casts. The renal blood vessels of old rats, unlike those of man, are remarkably free of arteriosclerosis,[21,24-26] which is consistent with the known resistance of the rat to athero- and arteriosclerosis.

FUNCTION

In humans, renal function decreases with increasing age as measured by a decline in the glomerular filtration rate.[27,33] In the absence of renal or vascular diseases, by the age of 80, only 70% of the original glomerular filtration remains.[34] One major difficulty with evaluating the effects of aging on renal function is that subjects are often compared in a cross-sectional study, i.e., the comparison of values between a current group of young and old, instead of longitudinal studies on the same individuals. The few longitudinal studies indicate that the decline in glomerular filtration rate proceeds more quickly after the age of 50 years than from 20 to 50 years. The regression line of inulin clearance with age is as follows: C_{IN} (mℓ/min) = 153.2 (0.96) (age in years).[32,34] The reduction in creatinine clearance with age is often accompanied by a reduction of total amount of creatinine excretion.[34]

Associated with the decline in the glomerular filtration rate is a similar rate of decline in tubular function,[35] which includes an impaired response to acid loading,[36] and an impaired maximal tubular reabsorption of glucose[34,37] as follows: Tm_G (mg/min) = 432.8 − (2.6) (age in years).[34]

Renal hemodynamics are altered with aging, due in part to the structural vascular changes described above.[36,38] Renal plasma flow is decreased, and the decreased renal vascular bed especially diminishes perfusion of the cortex.[14] The changes in renal blood flow parallels that of the glomerular filtration rate, so that the filtration fraction stays normal or rises slightly.[32,36]

Diminished ability to maintain water and salt homeostasis occurs in the kidney with aging. A reduction in renal concentrating ability occurs[27,39] even after optimal doses of vasopressin.[40] A diminished conservation of urinary sodium also is present.[41]

In rats, the decrease in renal function with age is not as pronounced as in humans. Elevated serum creatinine concentrations or low creatinine and inulin clearances indicate decreases in renal function of 5 to 50% in old rats of the Long-Evans[18] and Sprague-Dawley strains.[21] Wistar rats, on the other hand, have normal or increased renal function at 20 to 24 months.[42,43] Although the daily urine volume and urinary sodium excretion are unaffected by aging,[42] both the urinary sodium and potassium concentrations are 60% of normal.[43] Maximal urine osmolality is lower in old rats than in adult animals and in transplanted kidneys from 26- and 31-month-old rats compared to 3-month-old rats.[44] The urinary concentrating defect is due in part to a decrease in water permeability along the collecting duct.[45] Impaired natriuresis also occurs after volume expansion, which is unrelated to changes in the glomerular filtration rate.[46]

All strains of rats have proteinuria normally. As rats age, the protein excretion generally increases; the degree of proteinuria is associated with the degree of glomerulosclerosis (Table 1).[17,21,24]

Although membrane transport of electrolytes appears little affected by aging, the ability to take up *p*-amino hippurate and α-amino isobutyrate by renal slices of old Wistar rats is depressed by 30 to 50%, indicating a possible decrease in the membrane transport of organic anions with aging.[47,48]

Table 1

CHANGES IN SERUM AND URINE PROTEINS AND SERUM CHEMISTRIES WITH AGE

Age (mo.)	Urine		Serum					
	Proteinuria (mg/dℓ)	Proteinuria %	Urea nitrogen (mg/dℓ)	Creatinine (mg/dℓ)	Total protein (g/dℓ)	Albumin g/dℓ	Globulin g/dℓ	Cholesterol (mg/dℓ)
3	7 ± 4	0	18.4 ± 2.2	0.7 ± 0.1	7.0 ± 0.1	3.04 ± 0.15	3.94 ± 0.41	68 ± 8
6	16 ± 5	5	19.4 ± 3.6	0.7 ± 0.1	7.2 ± 0.2	3.03 ± 0.04	4.19 ± 0.39	74 ± 9
12	50 ± 14	50	17.0 ± 1.2	0.6 ± 0.1	7.1 ± 0.2	2.49 ± 0.29	4.61 ± 0.91	131 ± 24
24	140 ± 50	90	30.4 ± 6.9	1.1 ± 0.3	7.1 ± 0.3	2.20 ± 0.33	4.90 ± 0.87	189 ± 43

Note: Mean and standard error of values from 20 rats in each group (rats are male Sprague-Dawley).

Data from Couser, W. G. and Stilmant, M. M., *Lab. Invest.*, 33, 491, 1975.

METABOLISM

Changes in enzymatic activities during aging have been measured in experimental animals as indirect indicators of errors both in gene function and protein synthesis. The rat kidney in general has not been a good organ in which to evaluate these errors, as only a few of the many enzymes measured show more than a mild (30%) decrease or increase in activity. The Wistar strain has generally been used for studies of enzymatic activities. Decreases in the activities of enzymes catalyzing mitochondrial respiration, microsomal function, sodium transport, glucose metabolism, and phosphorylation are found in rats older than 20 months, compared to younger rats (Table 2).[49-54] Cathepsin and heme oxygenase activity are increased.[52] Enzymes that do not change in activity during aging include arginase,[55] lactic and malic dehydrogenases,[56] pyrophosphatase,[53] and catalase (Fisher strain).[51]

Metabolic pathways that have been studied appear to be regulated normally in the kidneys of old rats. Fatty acid oxidation,[57] protein synthesis,[53] and mitochondrial protein turnover[58] all appear unchanged from young to old rats. The turnover of catalase is slightly depressed in 24- to 30-month-old Fisher rats.[51]

Two difficulties arise in assessing changes in enzymatic activity during aging. First, activities of the same enzyme may be affected by aging differently in different strains, and few workers have used more than one strain in the measurement of enzymatic activity.[50] Second, tubular damage rather than aging per se may result in decreased enzymatic activity. This factor is difficult to assess, especially as preparations for biochemical analysis have not generally been evaluated for the degree of histologic renal damage.

HORMONAL ADAPTATION

The kidney plays a major role in hormonal balance because it synthesizes and degrades an important group of hormones. Several hormonal responses are blunted during aging. The renin-angiotensin-aldosterone system is altered: in elderly humans, plasma renin activity and plasma concentration of aldosterone are decreased.[59,60] Diminished responsiveness of plasma renin activity to salt loading or deprivation is manifested as an impaired adaptation to sodium intake.[41] However, responses of aldosterone and cortisol plasma levels to corticotropin administration are unaffected by aging.[59] Plasma arginine-vasopressin levels are increased following a hypertonic load of 3% sodium chloride to a greater extent in 60-year-olds than in 30-year-old subjects, which indicates that the osmoreceptor in the older group of patients may be more sensitive.[61]

Although renin production in old rats has not been directly measured, the granularity of juxtaglomerular cells has been used as an indirect measure of renin synthesis. The granularity decreases to one third of the adult value in Wistar rats over 18 months old.[62,63] The low granularity can be restored following mineralocorticoid deficiency, indicating that the old kidney still has normal responsiveness to aldosterone.[63]

The renal inactivation and hydrolysis of parathyroid hormone is decreased by 50% in 24-month-old Wistar rats, compared to 6-month-old rats.[64] However, there is no difference in the disappearance rate of parathyroid hormone from plasma or its uptake by the kidney in the old and young rats.[65]

Hormones in general hasten the tissue alterations that accompany aging.[66] Hypophysectomy retards the onset of proteinuria and glomerulosclerosis in old rats.[66,67] However, hypophysectomized rats eat less than control animals; and food restriction itself retards the development of proteinuria and glomerulosclerosis.[24,67,68] Pitressin deficiency has been held responsible for the minimal decrease in renal sodium excretion in 24-month-old Wistar rats.[69]

Table 2
ENZYMES THAT CHANGE WITH AGE IN RAT KIDNEYS

	Strain	Sex	% Increase (+) or decrease (−) of old (vs. young)	Level of significance (p)	Age	Ref.
Acid phosphatase: µg of p/mg of N/10 min (kidney homogenate)	Wistar	M and F	− 14	< 0.001	6—7 vs. 23—25	49
Alkaline phosphatase: µmoles p-nitrophenol released/hr/mg protein (plasma membranes)	Sprague-Dawley	M	− 33	< 0.02	3—6 vs. 24	50
Catalase: kU/g tissue (homogenate)	Fisher	M	− 71	< 0.05	3 vs. 24	51
Catalase: units/mg protein (tissue homogenates)	Sprague-Dawley	M	− 44	< 0.01	12 vs. 22—24	52
Cytochrome oxidase: k/min/mg protein (in mitochondria)	Sprague-Dawley	M	− 39	< 0.01	12 vs. 22—24	52
Cathepsin: µg trypsin released/g wet wt/hr (kidney homogenate)	Sprague-Dawley	F	+ 28	Significant	12 vs. 24	53
Heme oxygenase: µmoles/hr/mg protein (microsomal fraction)	Sprague-Dawley	M	+ 85	< 0.007	12 vs. 22—24	52
Maltase: µmoles glucose released/hr/mg protein (plasma membranes)	Sprague-Dawley	M	− 37	< 0.02	3—6 vs. 24	50
Oxidative phosphorylation: oxygen — µ atoms/150 mg wet wt/20 min (cortex homogenates)	McCollum	M	− 15	< 0.01	12—14 vs. 24—27	54
	McCollum	F	− 10	< 0.01	12—14 vs.24—27	54
Oxidative phosphorylation: phosphorus — µmoles/150 mg wet wt/20 min (cortex homogenates)	McCollum	M	− 15	< 0.01	12—14 vs. 24—27	54
	McCollum	F	− 11	< 0.05	12—14 vs. 24—27	54
Phosphodiesterase: µmoles p-nitrophenol released/hr/mg protein (plasma membranes)	Sprague-Dawley	M	−31	< 0.05	3—6 vs. 24	50
Sodium potassium ATPase: µM of phosphorus liberated/mg protein/hr (inner slices)	Wistar	M and F	− 24	< 0.01	12—14 vs. 24—28	48
Succinate dehydrogenase: µmoles succinate oxidized/min/mg protein (in mitochondria)	Sprague-Dawley	M	− 69	< 0.002	12 vs.22—24	52
Succinate dehydrogenase: as succinoxidase system: $mm^3 O_2$ utilized/100 mg wet wt/hr (whole homogenate)	McCollum	M and F	− 11	< 0.011	12—14 vs. 22—24	54

RENAL DISEASE

Over one third of the elderly population develops renal-urinary disease. In addition, the old patient is affected more severely than the young for major categories of renal disease, e.g., acute and chronic renal failure[70,71] and acute glomerulonephritis.[72,73] The increased severity may arise from a decreased capacity for renal regeneration. Although age has been regarded as a limiting factor of the development of renal enlargement as a compensation for disease or ablation of renal mass, compensatory hypertrophy does occur in patients over the age of 50, although to a lesser degree than those in young patients.[74]

In the aged rat, the capacity for renal compensatory growth is diminished. Optimal renal function entails the regulation of cellular hyperplasia and hypertrophy. The occurrence of mitoses in the kidneys of young rats is sufficiently rare (1 mitotic cell per high power microscopic field) so that a possible decrease in the normal incidence of mitosis in the old rat is difficult to use as an index of proliferation. One study reported only 25% the number of mitoses in the proximal tubules of 40-month-old albino rats, compared to 4-month-old rats,[75] whereas another study reported the same number of mitoses in 21-month-old Sprague-Dawley rats as in young rats.[76] The DNA content of kidneys is the same in 24- and 12-month-old Sprague-Dawley rats.[53] The 5-methylcytosine content of renal DNA is also the same in old and young rats;[77] thus, the degree of renal cell proliferation may be unaffected by age. Interestingly, the percentage of renal tubular cells that take up ^{3}H-thymidine in vivo is five times greater in 24-month-old albino rats than 12-month-old rats.[78] The increased uptake is associated with tubular dilatation. The compensatory response to loss of one kidney is less in old rats than in adult rats: in the compensating kidney, the compensatory increase in weight is 25 to 50% less in old Sprague-Dawley rats,[16,76] but both old and adult rats have the same (seven- to tenfold) increase in mitoses in renal tubular cells.[76] High protein diets also increased mitoses eightfold in both young and old rats, and a combination of contralateral nephrectomy plus a high protein diet provoked a greater mitotic response in old than in adult rats.[24]

The compensatory response includes an increase in the epithelial and mesangial cells of the glomerulus, thus qualitatively enhancing the degree of glomerulosclerosis.[79] However, the functional regenerative capacity of the kidney of old rats may be impaired, according to studies on the renal function of donor kidneys of old Brown Norway rats.[80] The kidneys of 26-month-old donors are less efficient than donor kidneys of younger rats in restoring blood urea nitrogen concentrations of recipients to normal.

The predominant renal lesion in old rats is glomerulosclerosis, which is associated with immune complex deposits of IgG, localized in a pattern similar to that of human mesangiopathic glomerulonephritis.[81]

REFERENCES

1. **Roessle, R. and Roulet, F., Nieren,** in *Mass und Zahl in der Pathologie,* Berlin, 1932, 63.
2. **Tauchi, H., Tsuboi, K., and Okutoni, J.,** Age changes in the human kidney of the different races, *Gerontologia,* 17, 87, 1971.
3. **Dunnill, M. S. and Halley, W.,** Some observations on the quantitative anatomy of the kidney, *J. Pathol.,* 110, 113, 1973.
4. **Takazakura, E., Sawabu, K., Handa, A., Takada, A., Shinoda, A., and Takeuchi, J.,** Intrarenal vascular changes with age and disease, *Kidney Int.,* 2, 224, 1972.
5. **Sworn, M. J. and Fox, M.,** Donor kidney selection for transplantation, *Br. J. Urol.,* 44, 377, 1972.

6. **Keresztury, S. and Megyeri, L.,** Histology of renal pyramids with special regard to changes due to aging, *Acta Morphol. Acad. Sci. Hung.,* 11, 205, 1962.

7. **Oliver, J. R.,** Urinary system, in *Problems of Ageing: Biological and Medical Aspects,* Cowdry, E. V., Ed., Williams & Wilkins, Baltimore, 1939, 257.

8. **Tauchi, H., Tsuboi, K., and Sato, K.** Histology and experimental pathology of senile atrophy of the kidney, *Nagoya Med. J.,* 4, 71, 1958.

9. **Yamaguchi, T., Omae, T., and Katsuki, S.,** Quantitative determination of renal vascular changes related to age and hypertension, *Jpn. Heart J.,* 10, 248, 1969.

10. **Darmady, E. M., Offer, J., and Woodhouse, M. A.,** The parameters of the aging kidney, *J. Pathol.,* 109, 195, 1973.

11. **Ljungqvist, A. and Lagergren, C.,** Normal intrarenal arterial pattern in adult and aging human kidney. A microangiographical and histological study, *J. Anat. (London).* 96, 285, 1962.

12. **Williams, R. H. and Harrison, T. R.,** A study of the renal arteries in relation to age and to hypertension, *Am. Heart J.,* 14, 645, 1937.

13. **Davidson, A. J., Talner, L. B., and Downs, W. M., III,** A study of the angiographic appearance of the kidney in an aging normotensive population, *Radiology,* 92, 975, 1969.

14. **Hollenberg, N. K., Adams, D. F., Solomon, H. S., Rashid, A., Abrams, H. L., and Merrill, J. P.,** Senescence and the renal vasculature in normal man, *Circ. Res.,* 34, 309, 1974.

15. **Arataki, M.,** On the postnatal growth of the kidney, *Am. J. Anat.,* 36, 399, 1926.

16. **Barrows, C. H., Jr., Roeder, L. M., and Olewine, D. A.,** Effect of age on renal compensatory hypertrophy following unilateral nephrectomy in the rat, *J. Gerontol.,* 17, 148, 1962.

17. **Berg, B. N.,** spontaneous nephrosis, with proteinuria, hyperglobulinemia, and hypercholesterolemia in the rat, *Proc. Soc. Exp. Biol.,* 119, 417, 1965.

18. **Kozma, C. K., Weisbroth, S. H., Stratman, S. L., and Conejeros, M.,** Normal biological values for Long-Evans rats, *Lab Anim. Care,* 19, 746, 1969.

19. **Berg, B. N. and Simms, H. S.,** Nutrition, onset of disease, and longevity in the rat, *Can. Med. Assoc. J.,* 93, 911, 1965.

20. **Guttman, P. H. and Kohn, H. I.,** Progressive intercapillary glomerulosclerosis in the mouse, rat, and Chinese hamster, associated with aging and X-ray exposure, *Am. J. Pathol.,* 37, 293, 1960.

21. **Couser, W. G., Stilmant, M. M., and Lowenstein, L. M.,** Mesangial lesions and the pathogenesis of focal glomerular sclerosis, *Clin. Res.,* 23, 358a, 1975.

22. **Moore, R. A. and Hellman, L. M.,** The effect of unilateral nephrectomy on the senile atrophy of the kidney in the white rat, *J. Exp. Med.,* 51, 51, 1930.

23. **Pollard, M. and Kajima, M.,** Lesions in aged germ-free Wistar rats, *Am. J. Pathol.,* 61, 25, 1970.

24. **Saxton, J. A., Jr. and Kimball, G. C.,** Relation of nephrosis and other diseases of albino rats to age and to modifications of diet, *Arch. Pathol.,* 32, 951, 1941.

25. **Wilens, S. L. and Sproul, E. E.,** Spontaneous cardiovascular disease in the rat, *J. Pathol.,* 14, 201, 1938.

26. **Bras, G.,** Age-associated kidney lesions in the rat, *J. Infect. Dis.,* 120, 131, 1969.

27. **Lewis, W. H., Jr. and Alving, A. S.,** Changes with age in the renal function in adult men, *Am. J. Physiol.,* 123, 500, 1938.

28. **Goldring, W., Chasis, H., Rangers, H. A., and Smith, H. W.,** Relations of effective renal blood flow and glomerular filtration to tubular excretory mass in normal man, *J. Clin. Invest.,* 19, 739, 1940.

29. **Rowe, J. W., Andres, R., Tobin, J. D., Norris, A. H., and Shock, N. W.,** The effect of age on creatinine clearance in man: a cross-sectional and longitudinal study, *J. Gerontol.,* 31, 155, 1976.

30. **Pelz, K. S.,** Kidney function studies in old men and women, *Geriatrics,* 20, 145, 1965.

31. **Slack, T. K. and Wilson, D. M.,** Normal renal function: C_{in} C_{pah} in healthy donors before and after nephrectomy, *Mayo Clin. Proc.,* 51, 5, 1976.

32. **Davies, D. F. and Shock, N. W.,** Age changes in glomerular filtration rate, effective renal plasma flow, and tubular excretory capacity in adult males, *J. Clin. Invest.,* 29, 496, 1950.

33. **Friedman, S. A., Raizner, A. E., Rosen, H., Solomon, N. A., and Sy, W.,** Functional defects in the aging kidney, *Ann. Int. Med.,* 76, 41, 1972.

34. **Goldman, R.,** Aging of the excretory system: kidney and bladder, in *Handbook of the Biology of Aging,* Goldman, R., Ed., 1977, 409.

35. **Shock, N. W. and Yiengst, M. J.,** Age changes in the acid-base equilibrium of the blood of males, *J. Gerontol.,* 5, 1, 1950.

36. **Miller, J. H., McDonald, R. K., and Shock, N. W.,** The renal extraction of *p*-aminohippurate in the aged individual. *J. Gerontol.,* 6, 213, 1951.

37. **Miller, J. H., McDonald, R. K., and Shock, N. W.,** Age changes in the maximal rate of renal tubular reabsorption of glucose, *J. Gerontol.,* 7, 196, 1952.

38. **Wesson, L. G., Jr.,** Renal hemodynamics in physiological states, in *Physiology of the Human Kidney,* Grune & Stratton, New York, 1959, 96.
39. **Rowe, J. W., Shock, N. W., and DeFronzo, R. A.,** The influence of age on the renal response to water deprivation in man, *Nephron,* 17, 270, 1976.
40. **Miller, J. H. and Shock, N. W.,** Age differences in the renal tubular response to antidiuretic hormone, *J. Gerontol.,* 8, 446, 1953.
41. **Epstein, M. and Hollenberg, N. K.,** Age as a determination of renal sodium conservation in normal man, *J. Lab. Clin. Med.,* 87, 411, 1976.
42. **Friedman, S. M. and Friedman, C. L.,** Salt and water balance in aging rats, *Gerontologia,* 1, 107, 1957.
43. **Gregory, J. G. and Barrows, C. H., Jr.,** The effect of age on renal functions of female rats, *J. Gerontol.,* 24, 321, 1969.
44. **Hollander, C. F.,** Functional and cellular aspects of organ aging, *Exp. Gerontol.,* 5, 313, 1970.
45. **Bengele, H. H., Mathias, R. S., Perkins, J. H., and Alexander, E. A.,** Urinary concentrating defect in the aged rat, *Am. J. Physiol.,* 240, F147, 1981.
46. **Bengele, H. H., Mathias, R. S., and Alexander, E. A.,** Impaired natriuresis after volume expansion in the aged rat, *Renal Physiol.,* 4, 22, 1981.
47. **Adams, J. R. and Barrows, C. H., Jr.,** Effect of age on PAH accumulation by kidney slices of female rats, *J. Gerontol.,* 18, 37, 1963.
48. **Beauchene, R. E., Fanestil, D. D., and Barrows, C. H., Jr.,** The effect of age on active transport and sodium-potassium-activated ATPase activity in renal tissue of rats, *J. Gerontol.,* 20, 306, 1965.
49. **Franklin, T. J.,** The influence of age on the activities of some acid hydrolase in the rat liver and kidney, *Biochem. J.* 82, 118, 1962.
50. **O'Bryan, D., and Lowenstein, L. M.,** Effect of aging on renal membrane-bound enzyme activities, *Biochim. Biophys. Acta,* 339, 1, 1974.
51. **Haining, J. L. and Legan J. S.,** Catalase turnover in rat liver and kidney as a function of age, *Exp. Gerontol.,* 8, 85, 1973.
52. **Paterniti, J. R., Jr., Ching, I., Lin, P., and Beattie, D. S.,** Regulation of heme metabolism during senescence: activity of several heme-containing enzymes and heme oxygenase in the liver and kidney of aging rats. *Mech. Aging Dev.,* 12, 81, 1980.
53. **Barrows, C. H., Jr. and Roeder, L. M.,** Effect of age on protein synthesis in rats, *J. Gerontol.,* 16, 321, 1961.
54. **Barrows, C. H., Jr., Falzone, J. A., and Shock, N. W.,** Age differences in the succinoxidase activity of homogenates and mitochondria from the livers and kidneys of rats, *J. Gerontol.,* 15, 130, 1960.
55. **Shukla, S. P. and Kanungo, M. S.,** Effect of age on the activity of arginase of the liver and kidney cortex of rat, *Exp. Gerontol.,* 4, 57, 1969.
56. **Schmukler, M. and Barros, C. H., Jr.,** Age differences in lactic and malic dehydrogenase in the rat, *J. Gerontol.,* 21, 109, 1966.
57. **Weinbach, E. C. and Garbus, J.,** Coenzyme A content and fatty acid oxidation in liver and kidney mitochondria from aged rats, *Gerontologia,* 3, 253, 1959.
58. **Menzies, R. A. and Gold, P. H.,** The turnover of mitochondria in a variety of tissues of young adult and aged rats, *J. Biol. Chem.,* 246, 2425, 1971.
59. **Weidmann, P., DeMyttenaere-Bursztein, S., Maxwell, N. H., and De Lima, J.,** Effect of aging on plasma renin and aldosterone in normal man, *Kidney Int.,* 8, 325, 1975.
60. **Crane, M. G. and Harris, J. J.,** Effect of aging on renin activity and aldosterone excretion, *J. Lab. Clin. Med.,* 87, 947, 1976.
61. **Helderman, J. H., Vestal, R. E., Rowe, J. W., Tobin, J. D., Andres, R., and Robertson, G. L.,** The response of arginine vasopressin to intravenous ethanol and hypertonic saline in man: the impact of aging, *J. Gerontol.,* 33, 39, 1978.
62. **Cain, H. and Kraus, B.,** Der juxtaglomerulare apparat der rattenniere im verlauf von wachstum, reifung und alterung, *Virchows Arch. B: Zellpathol,* 9, 164, 1971.
63. **Dunihue, F. W.,** Reversal of the reduced juxtaglomerular cell granularity (GCI) in old rats, *J. Gerontol.,* 26, 299, 1971.
64. **Fujita, T., Masahiro, O., Orimo, H., and Yoshikawa, M.,** Age and parathyroid hormone inactivation by kidney tissue, *J. Gerontol.,* 26, 20, 1971.
65. **Fujita, T., Okano, K., Orimo, H., Ohat, M., and Yoshikawa, M.,** Age and fate of parathyroid hormone, *J. Gerontol.,* 27, 25, 1972.
66. **Everitt, A. V.,** The hypothalamic-pituitary control of aging and age-related pathology, *Exp. Gerontol.,* 8, 265, 1973.
67. **Everitt, A. V. and Cavanagh, L. M.,** The aging process in the hypophysectomized rat, *Gerontologia,* 11, 198, 1965.

68. **Johnson, J. E., Jr. and Cutler, R. G.,** Effects of hypophysectomy on age-related changes in the rat kidney glomerulus: observations by scanning and transmission electron microscopy, *Mech. Ageing Dev.,* 13, 63, 1980.

69. **Friedman, S. M. and Friedman, C. L.,** Prolonged treatment with posterior pituitary powder in aged rats, *Exp. Gerontol.,* 1, 37, 1964.

70. **Kumar, R., Hill, C. M., and McGeown, M. G.,** Acute renal failure in the elderly, *Lancet,* 1, 90, 1973.

71. **Perlman, L. V., Kennedy, B. W., and Hayner, N. S.,** Primary and secondary renal failure in a total community (Tecumseh, Michigan): preponderance in the elderly and possible antecedent factors, *J. Am. Geriatr. Soc.,* 22, 25, 1974.

72. **Boswell, D. C. and Eknoyan, G.,** Acute glomerulonephritis in the aged, *Geriatrics,* 23, 73, 1968.

73. **Arieff, A. I., Anderson, R. J., Massry, S. G.,** Acute glomerulonephritis in the elderly, *Geriatrics,* Cedars-Sinai Symposium, Part I, September, 1971, 74.

74. **Ekelund, L. and Gothlin, J.,** Compensatory renal enlargement in older patients, *Am. J. Roentgenol.,* 127, 713, 1976.

75. **McCreight, C. E. and Sulkin, N. M.,** Cellular proliferation in the kidneys of young and senile rats following unilateral nephrectomy, *J. Gerontol.,* 14, 440, 1959.

76. **Konishi, F.,** Renal hyperplasia in young and old rats fed a high protein diet following unilateral nephrectomy, *J. Gerontol.,* 17, 151, 1962.

77. **Vanyushin, B. F., Nemirovsky, L. E., Klimenko, V. V., Vasiliev, V. K., and Belozersky, A. N.,** The 5-methylcytosine in DNA of rats, *Gerontologia,* 19, 138, 1973.

78. **Sworn, M. J. and Fox, M.,** Renal age changes in the rat compared with renal senescence, *Invest. Urol.,* 12, 140, 1974.

79. **Striker, G. E., Nagle, R. B., Kohnen, P. W., and Smuckler, E. A.,** Response to unilateral nephrectomy in old rats, *Arch. Pathol.,* 87, 439, 1969.

80. **Van Bezooijen, K. F. A., deLeeuw-Israel, F. R., and Hollander, C. F.,** Long-term functional aspects of syngeneic, orthotopic rat kidney grafts of different ages, *J. Gerontol.,* 29, 11, 1974.

81. **Couser, W. G. and Stilmant, M. M.,** Mesangial lesions and focal glomerular sclerosis in the aging rat, *Lab. Invest.,* 33, 491, 1975.

INTESTINAL TISSUE AND AGE

Elizabeth Hamilton

STRUCTURE AND SIZE OF THE GUT

Structure

The mammalian gut can be divided into four regions: esophagus, stomach, small intestine, and large intestine. These all have the same overall structure, consisting of:

1. Mucosa. This lines the inner, luminal surface of the gut. The mucosal surface is covered with epithelium, below which lies the *lamina propria* — a zone of connective tissue containing fibroblasts, histiocytes, and areas of lymphocytes. A layer of longitudinal smooth muscle — the *muscularis mucosae* — lies beneath the *lamina propria*.
2. Submucosa. This zone has a collagenous matrix with blood vessels, lymphatics, and nerve fibers running through it.
3. *Muscularis externa*. The major gut muscle layers run in a circular (inner layer) and a longitudinal (outer layer) direction. Both layers consist of smooth muscle.
4. Fibrosa. This is a loose layer of connective tissue which binds the gut to its surroundings.

Size of the Gut

The size and weight of the intestine has been measured in mice and rats. The intestine grows as the animal develops. After rodents reach maturity, the gut continues to grow in some species and strains but not in others.[1-7] The small intestine becomes shorter and lighter as wild house mice age,[1] but continues to grow in length in laboratory mice after maturity.[2] Several workers have measured the length of the small intestine in laboratory rats,[3-7] and their results are summarized in Table 1. As the rat matures the small intestine grows, reaching a maximum length of about 130 cm by 9 months. Subsequently, although the rat continues to gain weight, there is no further growth of the intestine. The quotient gut length (cm) per 100 g body weight therefore falls in old animals. This is shown in Figure 1, which also demonstrates that the rapid increase in body weight in the maturing animal far outstrips the rate of intestinal growth. The variations in body weight and small intestinal length between different rat strains (Table 1) is no longer evident in Figure 1, showing that there is a constant relation between rat weight and small intestine length at different ages.

The large intestine of laboratory rats grows steadily, from 15 cm in length at 3 months of age, to 18 cm at 24 months.[6] This growth, however, does not keep pace with the increase in body weight, and the length of colon per 100 g weight falls slowly.

For the rest of this chapter the gut will be considered as two zones: the mucosa and the tissues below it.

INTESTINAL MUCOSA

Structure

Esophagus

This organ is lined with stratified squamous epithelium which does not keratinize. The epithelium is replaced by upward migration of cells from the basal layer. In the esophagus of the mouse cell division is limited to the basal layer and the differentiating cells above this layer rapidly lose their nuclei.[8] The total number of nuclei per unit length of basement membrane changes little in the esophagus of mice 3 to 19 months old.[8] The stratified squamous epithelium at the base of the esophagus is sometimes replaced by columnar

Table 1
CHANGES IN SMALL INTESTINE LENGTH WITH
AGE IN MALE RATS

Age (mo)	Body weight (g)	Small intestine length (cm)	Ref.
0.5	61	79	3
1	101	92	4
1.5	170	91	5
2	254	100	4
4	259	111	7
6	550	117	6
9	616	129	3
12	703	124	5
16	636	134	3
20	707	122	4
24	670	119	6
37	700	124	4
42	763	125	4

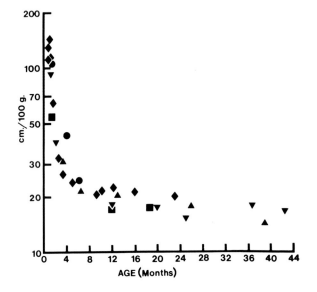

FIGURE 1.   Change in length of small intestine per 100 g body weight with age in male rats. Data from Meshkinpour et al.[3] (♦); Hollander and Morgan[4,5] (▼, ■); Ecknauer et al.[6] (▲); and Pénzes and Skála[7] (●).

epithelium in elderly human beings.[9] This may be a result of repeated epithelial denudation by gastric acid reflux.

Stomach

The lumen of this organ is lined with columnar epithelium. On the flat surface, most cells secrete mucus, but single endocrine cells, the gastrin and somatostatin cells, are also found. In the antrum of the stomach of $2^{1}/_{2}$- to 6-month-old rats, there were 3.3 to 5 × 10^5 gastrin cells and 1.3 to 2 × 10^5 somatostatin cells.[10] This gave a ratio of 2.5:1 for gastrin:somatostatin cells. Between 6 and 27 months of age, the number of gastrin cells decreased, and in the older rats the ratio was 1.5:1.

The surface epithelium of the stomach is depressed into pits. Clusters of gastric glands, containing four types of secretary cells — chief, parietal, mucus neck, and enteroendocrine cells — open into the pits. Both the surface columnar epithelium and the secretory cells are renewed by division of cells near the neck of the glands. This renewal tends to become more erratic with age in man, leading to an increased incidence of both atrophic gastritis and gastric polyps.[9] Also in aging humans, the chief and parietal cells of the gastric glands tend to be replaced by goblet cells.[9]

Small Intestine

This part of the gut is divided into three regions. The stomach opens into the duodenum, which is a short segment, followed by the jejunum and the ileum. The muscle wall is thicker in the duodenum than in the jejunum and ileum. These latter two segments are characterized by the presence of Peyer's patches, aggregations of lymphatic nodules. The columnar epithelium of the small intestine is formed into leaf-like villi, which protrude into the lumen and have an absorptive function. The villi are covered by cells formed in the crypts of Lieberkuhn, which lie below the surface of the epithelium. Five or more crypts feed cells onto each villus, the cells migrate to the tip and are sloughed off into the gut lumen. The height of the villi decreases down the small intestine. In 6-month-old rats the mean height of villi in the duodenum, jejunum, and ileum is 0.44, 0.37, and 0.22 mm, respectively.[11] The villi shrink with age and are 0.25, 0.26, and 0.10 mm tall in the duodenum, jejunum, and ileum of 24-month-old rats. In 30-month-old rats the villi are irregular in shape and atrophied.[12] In 24-month-old mice villi apparently fuse, so that their mean length is greater than at 6 months (102 vs. 86 cells, respectively).[2]

The primary function of villi is to increase the absorptive surface area of the small intestine. Calculations in which the gut is considered to be a simple tube, of known length and volume, suggest the surface area increases from 108 cm^2 at 1 month to 170 cm^2 in 16-month-old rats.[3] If the villi are also taken into account in the calculations, the small intestinal surface area is found to be very much larger, as shown in Table 2.

The total surface area per cm decreases down the gut as the villi shorten, but no firm conclusions can be drawn about changes in surface area with age in rats (Table 2). The surface area of human upper jejunum is significantly less in 60 to 73-year-old subjects than in those 16 to 30 years old.[13]

The most prominent indication of aging in the small intestinal mucosa is the deposition of amyloid in the *lamina propria* which surrounds the crypt and supports the villi. The amyloid appears as an increase in collagen and sclerotic connective tissue, while the number of fibroblast cells per unit area is decreased. Such changes have been seen in the small intestine of 22 to 27-month-old mice[14,15] and in people aged 60 to 73 years.

Large Intestine

This includes the cecum, ascending, transverse, and descending colons and the rectum. The colon and rectum are characterized by having columnar epithelium, with a flat surface and long thin crypts below it. The epithelium is renewed by cell division occurring in the lower 2/3 of the crypts. In male rats, the crypts per mm^2 of colon epithelium increase steadily with age, from 117 at 3 months to 138 at 24 months.[6] In mice, the number of crypts around a circular section of the descending colon, cut at right angles to the lumen, remains fairly constant from 4 to 24 months of age.[16] However, in 26- and 30-month-old mice there are fewer crypts per section (104 and 96, respectively) compared with the average value of 110 in younger animals. The colon epithelium appears atrophic in old mice.[15,16] However, there is no significant decrease in the number of cells per crypt (see Figure 2). Amyloid builds up in the *lamina propria* of the colon of aging mice.[15]

The mucosal structure of the esophagus and stomach changes little with age. However, in the small and large intestine there is evidence of mucosal atrophy in aging animals and man. The villi may change in number and dimensions, and the number of colon crypts may also alter.

Table 2
SURFACE AREA OF SMALL INTESTINE OF MALE RATS

Surface area in different zones (cm²/cm length)

Age (mo)	Duodenum	Jejunum	Jejuno-ileal junction	Ileum	Terminal ileum	Surface of entire small intestine (cm²)	Ref.
1	4.5		4.1	2.9		257	7
3	7.2		6.3	4.8		—	7
3		8.1			5.2	752	6
4	8.2		6.3	4.4		701	7
6	4.0		4.5	4.3		365	7
6		9.6			5.5	886	6
12		8.8			5.5	887	6
24		9.9			6.1	956	6

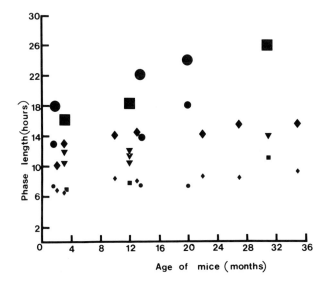

FIGURE 2.   Changes in cell kinetic parameters of mouse small intestinal crypts with age. Small symbols = length of DNA synthesis, T_s; medium symbols = length of cell cycle, T_c; large symbols = turnover time, T_t. Data from Lesher et al.[19] (▼); Lesher et al.[20] (■); Thrasher and Greulich[21] (●); and Lesher and Sacher[23] (♦).

Cell Kinetics
Unperturbed State

Many measurements have been made of the various parameters of cell division in the aging mammalian intestine. Most authors have used tritiated thymidine (³H-TdR) to label DNA synthesizing cells and visualized these labeled cells by autoradiography. Methods for cell kinetic measurement and their relative merits are discussed by Aherne et al.[17] In the gut, the parameters most often measured are (1) Labeling Index (LI) which is the percent of proliferating cells labeled by ³H-TdR and represents the cells in DNA synthesis at the time the isotope was injected; (2) mean length of DNA synthesis (T_s) which can be measured from a percentage of labeled mitosis (PLM) curve, as can (3) mean cell cycle time (T_c) which is the average time for cells to complete a division cycle, and (4) population turnover time (T_t) or the time taken to renew all the cells in the population, both those that divide (the growth fraction, GF) and those that do not. T_t may be measured directly, as the time it takes for all the cells in a population to become labeled when ³H-TdR is injected repeatedly (continuous labeling). Alternatively, T_t can be calculated from the equation:

Table 3
CELL KINETIC PARAMETERS FOR MOUSE ESOPHAGUS

Age (mo)	Labeling index LI (%)	DNA synthesis time T_s (hr)	Turnover time T_t (hr)	Ref.
1-2	8.3	8.1	95	18
3	—	—	108	8
13-14	7.0	7.8	113	18
19	—	—	126	8
19-21	6.0	7.6	120	18

$$\frac{\text{Labeled cells}}{\text{Total cells}} = \frac{\text{Length of S phase}}{\text{Turnover time}} \quad \text{or} \quad LI = T_s/T_t$$

$$T_t \text{ and } T_c \text{ are related: } \frac{T_c}{T_t} = GF$$

Esophagus

The cell division kinetics of the stratified squamous epithelium lining the esophagus have been studied in mice of various ages by Thrasher[18] and Cameron.[8] Their results are summarized in Table 3, where all the data refer to the basal layer of the epithelium. The LI is significantly lower at 19 to 21 months than at 1 to 2 months. Since T_s remains fairly constant with age, Thrasher found that T_t (calculated from T_s/LI) increased with age. Cameron[8] came to the same conclusion when he measured T_t by the continuous labeling method. Associated with this increase in T_t was a slowing of the rate of upward migration from the basal layer.[8] In 3-month-old mice, labeled cells were visible above the basal layer after 1 day of continuous labeling. In 19-month-old mice, however, it took 2 days before labeled cells were seen suprabasally.

The rate of cell turnover in mouse esophageal epithelium, therefore, decreases with age. The fall in basal layer LI may be caused by a decrease in the growth fraction or by an increase in the length of the G_1 and G_2 phases of the cycle. Either mechanism would give rise to an increased T_t.

Small Intestine.

Several workers have studied mouse duodenal crypt cell proliferation in relation to age. Some data are also available for the jejunum and ileum.[19] Those authors who measured cell cycle parameters by the PLM technique,[19-23] found that the peaks and troughs of the curve became less distinct with age. This is caused by desynchronization of the labeled population as it progresses through the cycle and suggests a greater variation in the length of the phases of the cycle in old animals.

The mean values for T_t (from T_s/LI), T_c, and T_s are plotted in Figure 2. The mean T_t increased considerably with age, from 16 to 18 hr in young mice to 26 hr in 31-month-old animals. There was a much less marked increase in T_c with age. This suggests that the change in T_t resulted from a fall in the growth fraction in the crypts. T_s also changed little with age (Figure 2), but this and a small increase in the length of G_1 phase[23] gave rise to the increased T_c.

The LI of duodenal crypts did not change with age according to Lesher et al.[20] They found values of 30, 34, and 33% for LI in the duodenal crypts of 3-, 12-, and 31-month-old mice, respectively. Thrasher and Greulich,[21] however, showed an age-related decline in LI in the duodenum, with values of 41, 35, and 30% in mice 1 to 2, 13 to 14, and 19 to 21 months old, respectively. These authors, in a later paper,[24] measured the LI at different positions up the side of the crypt. Their data are summarized in Table 4, which shows that

Table 4
MEAN % LABELED CELLS AT VARIOUS POSITIONS IN
MOUSE DUODENAL CRYPTS[a]

	Cell position								
Age (mo)	2	4	6	8	10	12	14	16	18
1 — 2	2.1	12.0	28.1	45.5	46.8	52.0	36.0	27.7	11.6
13 — 14	0.9	3.7	18.3	39.7	50.7	46.9	39.7	28.6	13.8
19 — 21	0.0	1.7	14.5	38.8	48.3	43.8	40.0	20.8	11.3

[a] Cells were scored down the sides of 160 crypts for each age group. Crypts were "normalized" to 18 cells deep and percent LI calculated for each position.

Data from Thrasher and Greulich.[24]

Table 5
RATE OF MOVEMENT OF LABELED
CELLS THROUGH MOUSE
DUODENAL EPITHELIUM[a]

Age (mo)	Crypt transit (hr)	Villus transit (hr)
3	5.5	35.5
12	6.5	41.5
31	>10	>43

[a] Crypt transit = time for labeled cells to reach 50% of villus bases; Total transit = time for labeled cells to reach 50% of villus tips; Villus transit = total minus crypt transit.

the LI decrease at nearly every cell position with age. The low LI at the base of the crypt, in the zone where Paneth cells occur, is characteristic of all small intestinal crypts.

The rate at which labeled cells move to the zone of exfoliation was measured by Lesher et al.[25] in mouse duodenum. Table 5 gives a summary of their results. It can be seen that the rate at which cells moved up both crypts and villi decreased with advancing age.

In the small intestine, therefore, the rate at which cells are formed by division and the rate at which they move to the exfoliation zone decrease with age. In the crypts the cell cycle slows little with age, but a decreasing proportion of cells are in active division as the animal grows older.

Large Intestine

In this organ, changes in cell kinetics in relation to age have only been studied in the descending colon of mice. Thrasher[26] found an age-related decline in the LI of colon crypts. The LI fell from 22% at 1 to 2 months to 20% and 16% at 13 to 14 and 19 to 21 months, respectively. However, as with the duodenum, Thrasher's results for LI[21,26] are contradicted by those of other workers. Hamilton and Franks[16] found no age-related changes in LI in mice 4 to 30 months old. These results and those for the depth of the colon crypts are plotted in Figure 3. The LI dropped between 4 and 18 months, the period over which Thrasher reported that the LI fell. However, in animals 20 to 30 months old the LI was much higher than at 18 months (Figure 3). The low LI values at 12 and 18 months were found in short colon crypts (Figure 3, top), while the crypt depth increased at 20 and 24 months. Figure 4 demonstrates the large diurnal variation in both LI and crypt depth that is found in the

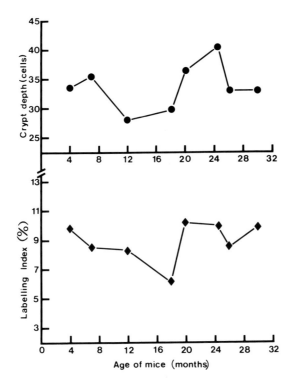

FIGURE 3.   Crypt depth (upper panel) and LI (lower panel) in the descending colon of C57BL*a¹* mice of different ages. Data from Hamilton and Franks.[16]

colon of 4-month-old mice.[27] The greatest LI occurred at the time that the crypts were longest. A comparison of Figures 3 and 4 shows that the variation in LI and crypt depth in mice of different ages is no greater than the diurnal variation found in young mice. Hamilton and Franks[16] concluded that the timing of the colonic diurnal variation may become more variable as mice age. Whether significant age-related changes are seen in the LI or not may depend on the time of day at which the mice are killed.

Thrasher[26] measured the length of the cell cycle in mouse descending colon using PLM curves. Again, as in the small intestine, the curves from older animals were more spread out than those from young ones, demonstrating a greater variation in cycle times with age. Table 6 shows Thrasher's[26] results and those of Cameron[8] who measured T_t in the colon from continuous labeling data. Both authors reported an increase in T_t with age. T_s, however, was unchanged, and T_c increased only at 19 to 21 months.

In the large intestine of mice, cell production probably slows with increasing age. In very old mice, the number of crypts is reduced (see Section II A 4), and the reduction in cell division between 2 and 19 months may be caused by a decrease in the crypt growth fraction, as LI falls but T_c changes little. Beyond 20 months the changes are unclear - LI rises but turnover data are not available.

Kinetics in Response to Injury

Several workers have studied the effects of radiation on the gut of different aged mice. Mice die within 7 days after a dose of 1,000 or more rads x-rays, from acute gastrointestinal failure. The radiation kills almost all proliferating cells and virtually halts cell production. Exfoliation of cells continues at a reduced rate, and the epithelium finally ceases to cover the mucosa. The mice die of dehydration and infection by luminal bacteria. If not all the crypt cells are sterilized, the survivors begin to divide rapidly a day after irradiation, and in 4 to 5 days form hyperplastic crypts from which the epithelium is repopulated.

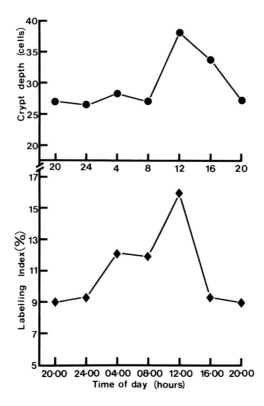

FIGURE 4.   Diurnal variation in crypt depth (upper panel) and LI (lower panel) in the descending colon of 4-month-old C57BL*a'* mice. The mice lived in a 12 hr light-dark cycle, with lights on at 08.00 hr. Data from Hamilton.[27]

Table 6
CELL KINETIC PARAMETERS IN DESCENDING COLON OF MICE

Age (mo)	DNA synthesis time T_s (hr)	Cell cycle time T_c (hr)	Turnover time T_t (hr)	Ref.
1 — 2	8.0	19	35	26
3	—	—	40	8
13 — 14	8.0	19	39	26
19	—	—	54	8
19 — 21	7.7	21	49	26

After 1100 rad whole-body x-irradiation and a transplant of unirradiated bone marrow (to support hemopoiesis) the number of mice that survive for 7 days decreases with age.[28] All 4-month-old mice survive compared with 80%, 60%, 50%, and 0% of 10-, 15-, 19-, and 26-month-old mice, respectively. Mice older than 24 months were more likely to die within a week after irradiation of a 2-cm length of descending colon than were 6-month-old mice.[16]

This decrease in radiation tolerance with age is not due to a defect in the proliferative capacity of surviving crypt cells. The colon of 24-month-old mice was able to repopulate at least four times after doses of 1250 rad x-rays, given at 6-week intervals. This dose reduced the number of proliferating cells per circumference from over 26,000 in unirradiated colon[27] to about 100.

Table 7
EFFECTS OF 12 RAD/DAY γ-IRRADIATION ON LENGTH OF CELL CYCLE IN MOUSE DUODENAL CRYPTS[a]

Age (mo)	Unirradiated T_c (hr)	1 day exposure T_c (hr)	3 day exposure T_c (hr)	10 day exposure T_c (hr)
3	13.0	12.3	11.3	12.3
13	14.0	13.3	10.8	12.7
27	14.8	14.3	12.0	12.1

[a] T_c measured from PLM curves.

Data from Lesher and Sacher.[22]

Table 8
RECOVERY OF GASTROINTESTINAL INJURY IN 24 HR AFTER A DOSE OF 400 RAD X-RAYS[a]

Age of mice (mo)	Rads "repaired"	% of 400 rad dose recovered
3	274	68.4
10	198	49.5
15	152	38.0
19	72	18.0

[a] Recovery measured as increase in $LD_{50(7)}$.

Data from Yuhas et al.[28]

Once stimulated, the crypt cells of old mice cycle as rapidly as those of young animals.[22] Table 7 shows the effect of continuous γ-irradiation to a dose of 12 rad per day, on T_c in duodenal crypt cells. In unirradiated mice, T_c was longer at 27 months than at 3 months of age. After 10 days γ-irradiation, T_c had decreased and was the same length in mice 3, 13, and 27 months old. However, as Table 4 shows, the response was slower in the older animals. In 3-month-old mice, the reduction in T_c occurred in 1 day, but it took 3 days in 27-month-old animals.[22]

Yuhas et al.[28] measured repair of radiation damage over a 24-hour period in mouse gut. The x-ray dose that killed 50% of mice in 7 days ($LD_{50(7)}$) was measured 24 hr after a single dose of 400 rad. The size of the $LD_{50(7)}$ indicated the amount of damage repaired. As Table 8 shows, the amount of recovery decreased as the mice grew older. This effect may be due to a slower response to injury in the gut of older mice.

Colon repopulation was followed in 4- and 24-month-old mice after a single dose of 1250 rad x-rays given locally to the descending colon.[29] The changes in LI in the regenerating crypts are shown in Table 9. In young mice the LI was $2^1/_2$ times the control level 5 days after 1 dose, but by 8 days it had decreased markedly. In the old mice, the 5-day LI was higher ($3^1/_2$ times control), and it was still $2^1/_2$ times the control level at 8 days. The LI only returned to normal levels 14 days after irradiation in old mice. The effect of these changes in LI on crypt size are shown in Figure 5. In young mice, the post-irradiation hyperplasia was lost by 8 days after irradiation. In old mice, the crypts became hyperplastic later and only returned to normal by 14 days.[29] Thus, the response to acute, high-dose

Table 9
MEAN LI IN MOUSE DESCENDING
COLON EPITHELIUM AFTER
DOSES OF 1250 RAD X-RAYS[a]

| Days after x-Rays | 4 month-old | | 24-month-old |
	1 dose	3 doses	1 dose
5	23.8	20.3	35.3
8	11.8	12.5	25.2
11	14.8	8.0	16.9
14	13.0	—	9.5
21	14.7	8.2	9.8

[a]   Unirradiated mice, LI 4 months 9.8; 24 months 10.2. Repeated doses given at 6-week intervals.

Data from Hamilton.[29]

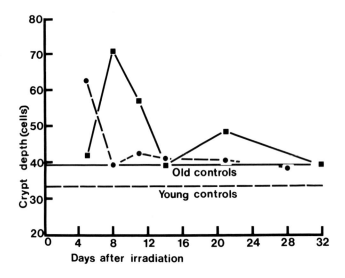

FIGURE 5.   Changes in descending colon crypt size in 4-month (young) and 24-month (old) C57BL*a*[t] mice with time after a local dose of 1250 rad x-rays. (- - Young mice; — old mice.) Data from Hamilton.[29]

irradiation was much slower in old mice, and this is probably the cause of the increase in radiation mortality with age. [28] Repeated irradiation of young mouse colon, although increasing the number of divisions undergone by the crypt cells, did not slow the repopulation response, as Table 9 shows. Young mice given 3 doses of 1250 rad X-rays at 6-week intervals responded as rapidly to the radiation as those given 1 dose.[29] The delayed response in old mice is probably not, therefore, due to an increase in "proliferative age" of the crypts.

The responses to injury — a reduction in T_c and an increase in LI — occur more slowly in the gut of old than of young mice. This probably makes old mice more sensitive to radiation-induced gastrointestinal death. Once stimulated, the gut cells in old mice cycle as rapidly as those in young mice and appear to have no "proliferative limit." The slowing of the response to injury in old mice might be related to a decrease in the crypt growth fraction.

Table 10
PRODUCTION OF TUMORS IN THE GUT OF RATS BY *N*-METHYL-*N*¹-NITRO-*N*-NITROSOGUANIDINE

Age (mo)	Site	Percent of animals with lesions			
		Adenocarcinoma	Sarcoma	Adenomatous hyperplasia	Other benign lesions
1¹/₂	Stomach	71	0	74	32
	Small intestine	10	13	0	24
	Total	*81*	*13*	*74*	*56*
5	Stomach	21	2	71	2
	Small intestine	7	5	12	7
	Total	*28*	*7*	*83*	*9*
10	Stomach	18	0	33	15
	Small intestine	0	15	3	3
	Total	*18*	*15*	*36*	*18*

Data from Kimura et al.[32]

Tumorigenesis

The incidence of spontaneous intestinal tumors was greater in 30-month-old mice (10 in 226 animals) than in those 24 months old (6 in 279 mice).[30] The incidence of benign abnormalities in the intestine also increased with age. Out of 40 6-month-old mice 11 had duodenal plaques and intestinal polyps, while 32 out of 43 24-month-old and 33 out of 43 30-month-old animals showed such defects.[30]

The susceptibility of mouse descending colon to tumor induction by weekly injections of 1-2-dimethylhydrazine was no greater at 24 months than at 6 months of age.[31] Table 10 shows the incidence of tumors and other lesions in the gut of rats of 3 ages given *N*-methyl-*N*¹-nitro-*N*-nitrosoguanidine in drinking water for 30 weeks.[32] The total tumor incidence decreased markedly in the older rats. The number of unspecified benign lesions also decreased with age, while there were fewer rats in the 10-month age group with adenomatous hyperplasia. These results suggest that the increasing incidence of spontaneous tumors with age is probably caused by old animals being in contact with carcinogens for longer than young animals. The intestinal cells of old animals do not appear to be more prone to neoplastic transformation than those of young ones.[31]

Physiology

Chemical Composition

The composition of the whole small intestine of female rats 2¹/₂, 8, and 27¹/₂ months old has been analyzed.[33] The percentage of water per unit wet weight and the amount of nitrogen per 10 mg fat-free dry weight did not change with age. The percentage of carbon, however, was slightly less in old rats. The amounts of 18 protein-bound amino acids in the gut wall were also measured.[33] Most remained constant with age; however, the levels of arginine, tyrosine, and threonine were lower in old animals, while the amount of alanine increased with age.

Enzyme Systems

The amount and activity of certain enzymes in the gut do not change with age. Fumarase, an enzyme of the respiratory cycle, was found in equal quantities in the gut of 1- and 24-month-old- mice,[34] and the activity of ornithine decarboxylase, an enzyme involved in

Table 11
CHANGES IN MEAN ACTIVITY OF ENZYMES IN
HUMAN DUODENAL MUCOSA WITH AGE

Enzyme	Activity[a] 18 — 29 yr	Percent decrease in activity		
		30 — 59 yr	60 — 74 yr	75 — 89 yr
1 γ-amylase	14.7	13	34	58
2 Maltase	52.1	6	20	32
3 Invertase	27.8	5	15	16
4 Lactase	12.2	1	8	33

Note: Values for subjects 18 to 29 years old = 100%.

[a]   1, activity in μg hydrolyzed starch/mg mucosa/min; 2 to 4, activity in μg hexoses formed/mg mucosa/min.

Data from Valenkevich et al.[37]

polyamine synthesis, was no different in young and old rats.[35] However, the quantities of other enzymes do change with age. The level of phosphomonoesterase in the luminal membrane of the small intestinal epithelium decreased between 11 and 34 months of age in mice,[36] while maltase and sucrase activities were 50% higher in 24-month-old than in 6-month-old mice.[2] Valenkevich et al.[37] studied the activity of several enzymes in biopsies of human small intestinal mucosa. Their results, summarized in Table 11, show that different gut enzyme systems begin to lose their activity at different ages. γ-Amylase activity decreased progressively from the age of 30, while lactase activity only fell in subjects over the age of 60.

Several workers have studied the activity of acid and alkaline phosphatases in the gut in relation to age. The amounts of both these enzymes varied diurnally in the small intestine of rats, but the peak levels were higher in 30-month-old than in 4-month-old animals.[12] Moog[2] found that the alkaline phosphatase concentration of whole mouse gut increased with age, but the amount per mg of protein remained constant. A decrease in the alkaline phosphatase content of mouse small intestinal mucosa with age has also been reported.[15]

Absorption from the Gut Lumen

The rate at which various amino acids are absorbed from the lumen changes with age. L-Arginine was removed most rapidly from the small intestinal lumen in young rats.[38] There was, however, no evidence that the absorption mechanism was impaired in aging: it merely became less active, presumably as the requirement for arginine lessened. The affinity of the mucosa for L-phenylalanine and L-proline fell with age, but the rate at which these amino acids were released from their carrier molecules varied.[39] Table 12 shows the amount of various amino acids found in the small intestinal mucosa of rats after a 20- or 60-min absorption period. The uptake of L-tryptophan and L-valine decreased with age. L-proline uptake increased with age, while the greatest absorption of L-phenylalanine over a 20-min period occurred in 12-month-old rats. Ning et al.[40] concluded that the decrease in L-valine uptake followed the increase in rat weight with age.

The absorption of substances other than amino acids also varies with age. As Table 13 shows, the amount of vitamin A and cholesterol absorbed by the rat small intestine rose steadily with age to 42 months.[4,5] The transport of calcium across the small intestinal mucosa of rats, however, decreased with age.[41,42] The active transport of calcium from the lumen decreased between 1 and 3 months of age in rats.[41] This decrease was correlated with the

Table 12
ABSORPTION OF VARIOUS AMINO ACIDS INTO RAT SMALL INTESTINAL MUCOSA

Rat age (mo)	L-Tryptophan (2 mM)[a]	L-Phenylalanine (5 mM)[a]	L-Proline (10 mM)[a]	L-Valine (3 mM)[b]
2	—	—	—	4.7
4$\frac{1}{2}$	—	—	—	3.2
6	7.8	17.0	24.1	—
10	—	—	—	1.9
12	7.3	20.4	25.5	—
27	7.0	16.3	36.0	—

[a]   μmol/g wet wt/20 min. Data from Pénzes and Boross.[39]
[b]   μmol/500 mg wet wt/60 min. Data from Ning et al.[40]

Table 13
ABSORPTION OF VITAMIN A AND CHOLESTEROL INTO RAT SMALL INTESTINE

Rat age (mo)	Percent perfused substance absorbed/hr	
	Vitamin A	Cholesterol
1	—	14
1.5	25	—
2	—	17
12	28	18
19	29	—
20	—	25
25	33	—
37	—	37
39	37	—
42	—	38

Note: Absorption from flow of 1 mℓ/min through lumen, containing 350 nM Vitamin A or 50 μM cholesterol, in buffer.

Data from Hollander and Morgan.[4,5]

loss of calcium-binding protein from the mucosa. Calcium was, however, still absorbed from the lumen in 24-month-old rats, although in much smaller amounts than in 2-month-old animals.[42] The transport system became saturated at a lower concentration in 24-month-old rats, suggesting that there were fewer calcium carrier molecules in the mucosa.

6-Deoxy-D-glucose was absorbed by the small intestinal mucosa of 1- and 24-month-old mice at the same rate.[43] However, less glucose was absorbed in the jejunum and ileum of 24-month-old rats than in 6-month-old animals.[11] Duodenal glucose absorption was the same at 6, 12, and 24 months of age. The villi of rat small intestine were significantly shorter at 24 than at 6 months of age.[11] In the jejunum and ileum the decrease in glucose absorption was correlated with that in height, suggesting that the absorption per unit surface area was unchanged with age.

Absorption from the intestine is affected by the thickness of the unstirred water layer (UWL) held in place by the epithelial microvilli. The UWL was probably larger in the jejunum of 3-week- and 2-month-old rabbits than in 12-month-old animals, so uptake of long-chain fatty acids was greatest in the jejunum of the 12-month-old rabbits.[44]

Table 14
TRANSPORT RATES THROUGH SMALL AND LARGE INTESTINES OF RATS OF VARIOUS AGES

Age (mo)	Body wt (g)	Small intestine			Large intestine	
		Length (cm)	Transport rate (cm/hr)		Length (cm)	Transport rate (cm/hr)
			1st quarter	4th quarter		
2	102	84	139	14	11	0.8
4	201	106	156	19	12	0.7
20	455	119	165	16	17	0.9

Data from Varga.[47]

In humans, malabsorption from the intestine is more common among the elderly. Less sugar is absorbed, the absorption of fats is slower, and passive transport from the lumen decreases with age.[9] Subjects over the age of 80 absorbed less xylose from the small intestine than did younger people.[45] Absorption was also slower, so that the peak in blood xylose was lower and occurred later in people over the age of 80 than in younger subjects.

The available data suggest that certain enzyme systems remain stable (e.g., respiratory enzymes) and that some substances are absorbed at a constant rate (e.g., 6-deoxy-D-glucose) throughout life. The activity of other enzymes changes with age (e.g., γ-amylase, sucrase) and some chemicals are not taken up at a constant rate (e.g., calcium). These varying results are probably due to the changing requirements of the body for nutrients throughout life.

INTESTINAL SUBMUCOSA AND *MUSCULARIS*

Esophagus

There are reports that the esophageal muscles of elderly patients often appear incoordinated.[9] However, in a comparison of normal subjects over the age of 80 with younger people, no increase in spontaneous contractions or in the absence of peristalsis after swallows was found in the elderly subjects.[46] The speed, time of onset, and duration of each contraction was also unchanged with age. The amplitude of the peristaltic waves, however, was much smaller in people over the age of 80. These results were interpreted as showing an intact nervous pathway, but a weakening of the esophageal muscles with age.

Intestine

The rate at which [85]Sr-labeled microspheres are propelled down the gut was measured in rats.[47] The results are summarized in Table 14. The rate of transport through the small intestine increased as the organ grew. The time for microspheres to pass through the entire small intestine, therefore, remained fairly constant. In the colon, however, the rate of transport remained fairly constant, while the length of the organ increased. The transit time through the rat colon therefore increased with age.

In humans the *muscularis externa* of the colon thickens in old age.[9] The contractions of the circular muscle become exaggerated, and at the same time the colon does not elongate properly. This gives rise to local areas of high pressure in the lumen, possibly the cause of colonic diverticula in aged patients.[9] The colonic blood supply also becomes abnormal in old age. The colic arteries are more tortuous in people over the age of 60 than in younger subjects.[48] This tortuosity alters the rate of blood flow, often causing colonic ischemia. Increased tortuosity in the arterioles and venules of the submucosa leads to blockages and bleeding into the colonic lumen.[49] This, like ischemia, is also a disorder most common among elderly people.

The musculature of the intestinal tract appears to degenerate slightly in elderly people. The nervous pathways controlling the muscles, however, are not affected. The blood vessels feeding the colon become tortuous and obstructed in aging patients, and in rats the time taken for matter to pass through the colon increases with age.

REFERENCES

1. **Barnett, S. A., Munro, K. M. H., and Stoddart, R. C.,** Growth and pathology of aged house mice, *Exp. Geront.* 9, 275, 1974.
2. **Moog, F.,** The small intestine in old mice: growth, alkaline phosphatase and disaccharidase activities, and deposition of amyloid, *Exp. Geront.,* 12, 223, 1977.
3. **Meshkinpour, H., Smith, M., and Hollander, D.,** Influence of aging on the surface area of the small intestine in the rat, *Exp. Geront.,* 16, 399, 1981.
4. **Hollander, D., and Morgan, D.,** Increase in cholesterol intestinal absorption with aging in the rat, *Exp. Geront.,* 14, 201, 1979.
5. **Hollander, D., and Morgan, D.,** Aging: its influence on vitamin A intestinal absorption *in vivo* by the rat, *Exp. Geront.,* 14, 301, 1979.
6. **Ecknauer, R., Vadakel, T., and Wepler, R.,** Intestinal morphology and cell production rate in aging rats, *J. Gerontol.,* 37, 151, 1982.
7. **Pénzes, L., and Skála, I.,** Changes in the mucosal surface area of the small gut of rats of different ages, *J. Anat.,* 124, 217, 1977.
8. **Cameron, I. L.,** Cell proliferation and renewal in aging mice, *J. Gerontol.,* 27, 162, 1972.
9. **Geokas, M. C. and Haverback, B. J.,** The aging gastrointestinal tract, *Am. J. Surg.,* 117, 881, 1969.
10. **Lehy, T. Grès, L., and Ferreira de Castro, E.** Quantitation of gastrin and somatostatin cell populations in the antral mucosa of the rat, *Cell Tissue Res.* 198, 325, 1979.
11. **Jakab, L. and Pénzes, L.,** Relationship between glucose absorption and villus height in ageing, *Experentia.,* 37, 740, 1981.
12. **Höhn, P., Gabbert, H., and Wagner, R.,** Differentiation and aging of the rat intestinal mucosa. II. Morphological, enzyme histochemic and disc electrophoretic aspects of the aging of the small intestinal mucosa, *Mech. Age Dev.,* 7, 217, 1978.
13. **Warren, P. M., Pepperman, M. A., and Montgomery, R. D.,** Age changes in small intestinal mucosa, *Lancet,* 2, 849, 1978.
14. **Andrew, W. and Andrew, N. V.,** An age involution in the small intestine of the mouse, *J. Gerontol.,* 12, 136, 1957.
15. **Suntzeff, V. and Angeletti, P.,** Histological and histochemical changes in intestines of mice with aging, *J. Gerontol.,* 16, 225, 1961.
16. **Hamilton, E. and Franks, L. M.,** Cell proliferation and aging in mouse colon, II. Late effects of repeated X-irradiation in young and old mice, *Eur. J. Cancer,* 16, 663, 1980.
17. **Aherne, W. A., Camplejohn, R. S., and Wright, N. A.,** *An introduction to cell population kinetics,* Edward Arnold, London, 1977.
18. **Thrasher, J. D.,** Age and the cell cycle of the mouse oesophageal epithelium, *Exp. Geront.,* 6, 19, 1971.
19. **Lesher, S., Fry, R. J. M., and Kohn, H. I.,** Aging and the generation cycle of intestinal epithelial cells in the mouse, *Gerontologia,* 5, 176, 1961.
20. **Lesher, S. Fry, R. J. M., and Kohn, H. I.,** Age and the generation time of the mouse duodenal epithelial cell, *Exp. Cell Res.,* 24, 334, 1961.
21. **Thrasher, J. D. and Greulich, R. C.,** The duodenal progenitor population, I. Age-related increase in the duration of the cryptal progenitor cycle. *J. Exp. Zool.,* 159, 39, 1965.
22. **Lesher, S. and Sacher, G. A.,** Changes in cell proliferation produced by 12 roentgens of ^{60}Co gamma irradiation per day in the intestinal crypt cells of 100, 400 and 825-day old BCF$_1$ mice, *Radiat. Res.,* 30, 654, 1967.
23. **Lesher, S. and Sacher, G. A.,** Effects of age on cell proliferation in mouse duodenal crypts, *Exp. Geront.,* 3, 211, 1968.
24. **Thrasher, J. D. and Greulich, R. C.,** The duodenal progenitor population, II. Age-related changes in size and distribution, *J. Exp. Zool.,* 159, 385, 1965.
25. **Lesher, S. Fry, R. J. M., and Kohn, H. I.,** Influence of age on transit time of cells of mouse intestinal epithelium, *Lab. Invest.,* 10, 291, 1961.

26. **Thrasher, J. D.,** Age and the cell cycle of the mouse colonic epithelium, *Anat. Rec.,* 157, 621, 1967.
27. **Hamilton, E.,** Diurnal variation in proliferative compartments and their relation to cryptogenic cells in the mouse colon, *Cell Tissue Kinet.,* 12, 91, 1979.
28. **Yuhas, J. M., Huang, D., and Storer, J. B.,** Residual radiation injury: hematopoietic and gastrointestinal involvement in relation to age, *Radiat. Res.,* 38, 501, 1969.
29. **Hamilton, E.,** Cell proliferation and aging in mouse colon. I. Repopulation after repeated X-ray injury in young and old mice, *Cell Tissue Kinet.,* 11, 423, 1978.
30. **Rowlatt, C., Chesterman, F. C., and Sheriff, M. U.,** Lifespan, age changes and tumor incidence in an aging C57BL mouse colony, *Lab Ans.,* 10, 419, 1976.
31. **Defries, E. A.,** The effects of age and carcinogen treatment on adult mouse colon epithelium. PhD Thesis, University of London, 1976.
32. **Kimura, M., Fukuda, T., and Sato, K.,** Effect of aging on the development of gastric cancer in rats induced by N-methyl-N^1-nitro-N-nitrosoguanidine, *Jap. J. Gann.,* 70, 521, 1979.
33. **Pénzes, L.,** Data of the chemical composition of the aging intestine, *Digestion,* 3, 174, 1970.
34. **Zorzoli, A.,** Fumarase activity in mouse tissues during development and aging, *J. Gerontol.,* 23, 506, 1968.
35. **Ball, W. J. and Balis, M. E.,** Changes in ornithine decarboxylase activity in rat intestines during aging, *Cancer Res.,* 36, 3312, 1976.
36. **Sayeed, M. M. and Blumenthal, H. T.,** Age difference in the intestinal phosphomonesterase activity of mice, *Proc. Soc. Exp. Biol. Med.,* 129, 1, 1968.
37. **Valenkevich, L. N., Morozov, K. A., and Ugolev, A. M.,** Relations between cavitary and membrane digestion during aging, *Human Physiol.,* 4, 64, 1978.
38. **Pénzes, L.,** Intestinal transfer of l-arginine in relation to age, *Exp. Geront.,* 5, 193, 1978.
39. **Pénzes, L. and Boross, M.,** Intestinal absorption of some heterocyclic and aromatic amino acids from the aging gut, *Exp. Geront.,* 9, 253, 1974.
40. **Ning, M., Reiser, S., and Christiansen, P. A.,** Variations in intestinal trasport of l-valine in relation to age, *Proc. Soc. Exp. Biol. Med.,* 129, 799, 1968.
41. **Ambrecht, J. J., Zenser, T. V., Bruns, M. E. H., and Davis, B. B.,** Effect of age on intestinal calcium absorption and adaption to dietary calcium. *Am. J. Physiol.,* 236, E769, 1979.
42. **Winter, D., Dobre, V., and Oeriu, S.,** Calcium transport across intestinal wall, as related to age, *Exp. Geront.,* 8, 17, 1973.
43. **Calingaert, A. and Zorzoli, A.,** The influence of age on 6-deoxy-D-glucose accumulation by mouse intestine, *J. Gerontol.,* 20, 211, 1965.
44. **Thomson, A. B. R.,** Effect of age on uptake of homologous series of saturated fatty acids into rabbit jejunum, *Am. J. Physiol.,* 239, G363, 1980.
45. **Guth, P. H.,** Physiologic alterations in small bowel functions with age, *Am. J. Digest. Dis.,* 13, 565, 1968.
46. **Hollis, J. B. and Castell, D. O.,** Esophageal function in elderly men, *Ann. Intern. Med.,* 80, 371, 1974.
47. **Varga, F.,** Transit time changes with age in the gastrointestinal tract of the rat, *Digestion,* 14, 319, 1976.
48. **Binns, J. C. and Isaacson, P.,** Age-related changes in the colonic blood supply: their relevance to ischaemic colitis, *Gut,* 19, 384, 1978.
49. **Boley, S. J., Sammartano, R., Adams, A., DiBiase, A., Kleinhaus, S., and Sprayregen, S.,** On the nature and etiology of vascular ectasias of the colon, *Gastroenterol,* 72, 650, 1977.

AGING OF THE REPRODUCTIVE SYSTEM

Richard F. Walker

GENERAL CONSIDERATIONS

Altered states of reproductive activity are often accompanied by significant structural and functional somatic changes in most organisms. In fact, animals from diverse phylogenetic groups often show a relationship between lifespan and reproduction.[1] For example, lifespan can be prolonged if reproductive development is retarded;[2] while physiologic changes leading to degeneration and death seem to follow the onset of reproductive activity in certain animals.[3-6] These studies suggest that aging may begin at or about puberty, and that senescence may accelerate afterwards.[7,8] Therefore, the study of reproductive system aging may well provide a foundation for gerontological studies in the whole organism.

Progress in molecular biology has fostered the tendency to search for "control mechanisms" at cellular and subcellular levels. However, the exclusive use of molecular theories to investigate reproductive aging, a systems phenomenon, without reference to other levels of biological organization may be unwise.[9,10] Finch,[11] suggested that cells age in response to intrinsic events occurring independent of other cells or tissues, as well as to extrinsic factors in their environment. Cells of the reproductive system may age in response to both factors; however true inherent defects have been difficult to define. Possibly, age-related decline in reproductive function follows certain primary or intrinsic processes in "key" cells of the neuroendocrine axis,[12-14] while the majority of senile changes are extrinsic, resulting from loss of homeostasis within the reproductive system. Therefore, the purpose of this chapter is to describe chronological neuroendocrine changes and to rank their influence on reproductive system aging.

FUNCTIONAL DECLINE OF THE REPRODUCTIVE SYSTEM

Typically, aging organisms show progressive degeneration of physiological and behavioral reproductive functions. Though gametogenesis continues until death in most males, sperm number, ejaculate volume, and testosterone production diminish significantly and accompany waning libido as correlates of advancing age.[15-17] In contrast, gamete production ceases after a period of gradually decreasing physiologic activity characterized by atypical or irregular cycles in old female animals.[18] Unlike humans, female rodents experience one or more postreproductive syndromes, identified by unique endocrine parameters.[19] In contrast to males, aging females may experience enhanced libido and show increased sex behavior[20-22] which results apparently from hormonal changes associated with increased androgen production during ovarian senescence.[23] Degenerative changes in the structure of gonadal and adnexal tissue occur in aging individuals of both sexes. General characteristics of sexual senescence are presented in Table 1; however, greater detail is available in a recently published monograph on clinical and experimental aspects of reproductive system aging.[24]

Gonads

Structural atrophy of primary and secondary sex organs may be extrinsic age changes since many are readily reversed by treatment with certain steroids.[11] Since serum profiles of gonadal hormones are modified with advancing age, it is possible that biochemical alterations in steroid synthesizing pathways and/or changes in gonadal sensitivity to trophic hormone stimulation might occur during senescence. Hypothetically, these changes in gon-

Table 1
SYMPTOMATOLOGY OF REPRODUCTIVE SYSTEM AGING

Symptom	Species	Ref.
Female		
Frequency of irregular or anovulatory reproductive cycles increases	Human	25—31
	Monkey	32
	Rat	18,33
	Mouse	34,35
Reduced fecundity; increased difficulty in maintaining pregnancy	Human	36
	Mouse	37,38
	Hamster	39
	Rat	3,40,41
General atrophy with increasing frequency of cancer of reproductive tissues	Human	42-45
	Rat	46,47
Possible increase in sexual receptivity	Human	22
	Rat	20,21
Male		
Endocrine alterations and functional impairment of sex organs	Human	48,49
	Rat	47
	Hamster, Mice (no gross disturbances)	37,50,51
Involution of reproductive structures accompanied by increased pathology	Human	17,52,
	Rat	53,54
Increasing prevalence of erectile impotence	Human	16
	Rat	55,63

Note: Aging of the reproductive system in both sexes is generally characterized by diminished functional capacity, structural degeneration, and altered sex behavior.

adal metabolism could contribute to loss of reproductive function as animals grow older. Indeed, changes in steroidogenic competence, presumably due to enzyme alterations, and in gonadotropin:gonad interaction has been reported. These findings are summarized in Table 2.

Though reports of biochemical and morphological changes in aging gonads have been confirmed, they do not seem to be intrinsic age characteristics or causative of reproductive decline in animals undergoing senescence since similar changes occur after experimental alteration of pituitary function in young animals. For example, loss of testicular 3β-hydroxysteroid dehydrogenase activity, and diminished Leydig cell mass, probably account in part for the deficiency of testosterone secretion in aging rats, as in man.[17,77,81,82] However, similar enzymatic, ultrastructural, and morphologic changes can be induced in rats by depressing LH secretion with antibodies[83] or grafts of prolactin-producing tumors.[84,85] Furthermore, we[77] found that responsivity of individual Leydig cells to LH stimulation remains fairly normal in free-cell suspensions prepared from aged testes, even though total cell mass is reduced. In contrast, Tsitouras et al.[79] recently reported a diminished response of Leydig cells, in vitro to hCG. These apparent discrepancies may be due to the fact that different rat strains (Fischer vs. Wistar) and gonadotropin preparations (LH vs. hCG) were used in the two studies. The latter authors, who also found restored testosterone secretion after hCG treatment in vivo,[73] suggested that the effect might be due to recruitment of additional Leydig cells after chronic stimulation, rather than reversal of an intrinsic cellular deficit. Reduced secretion of testoterone, which occurs as a correlate of aging, is probably not due to reduced

Table 2
ENDOCRINE CHANGES IN THE AGING GONAD

Change	Species	Ref.
Female		
Decreased estrogen production by the ovary	Human	42,43,56,57
Altered follicular and plasma steroids	Rat	58,59
Increased androgen and "nonclassical phenol-steroid" production by ovarian stroma	Human	23,60,61
	Rat	62
Increased "peripheral conversion" of androgen to estrogen	Human	63,64
Decreased and altered patterns of progesterone secretion	Human	42
	Mouse	65
	Rat	66
Depressed Δ^5-3β-hydroxysteroid dehydrogenase activity	Rat	67,68
	Hamster	69
	Mouse	70
Male		
Decreased testosterone synthesis and secretion	Human	15,17,71
	Rat	47,55,72,73
Increased estrogen/testosterone ratio in plasma	Human	15,74
Increased "peripheral conversion" of androgen to estrogen	Human	75
Reduced Leydig cell mass	Human	76
	Rat	77
Reduced responsiveness to gonadotropins	Human	74,78
	Rat	47,73
Reduced gonadotropin receptors	Rat	79,80

Note: Gonadal steroid production and secretion generally fall with advancing age. These changes may result from altered activity of biosynthetic enzymes and/or loss of steroid-producing cells. Steroidogenic decrements are accompanied by reduced sensitivity to gonadotropins and depressed numbers of gonadotropin receptors.

testicular gonadotropin receptors in old rats since fewer than 20% of the Leydig cell receptors need be saturated to produce a maximum hormone response.[86] Furthermore, the age deficit in testosterone secretion may be more apparent than real, resulting from dilution factors in older animals.[87,88] Since steroidogenic enzyme activity and normal testosterone secretion are restored in aged rats by stimulation with exogenous gonadotropin, biochemical, physiological, and morphological changes in the aging testes are probably extrinsic.

Morphological and biochemical defects in the aging female gonad may also be due to inadequate gonadotropin stimulation, since age-type changes can be produced experimentally in pituitary-altered young animals, and the old ovary shows some potential for restored function under favorable endocrine conditions. Deficiency cells, characteristic of inadequate LH stimulation, appear in the ovaries of 13-month-old rats[18] suggesting that ovarian aging may occur in response to inadequate LH stimulation.[88,89] Furthermore, nonfunctional ovaries taken from old rats, become cyclic when transplanted into young hosts, while reciprocal transplants cause young ovaries to fail in old hosts.[18,90] Recent findings also suggest that the ovary may adversely affect its own environment, since estrogen injections accelerate age changes and hasten the onset of constant estrus in relatively young rats.[91] Conversely, chronic estrogen deficiency resulting from long-term ovariectomy, renders old female mice capable of causing ovarian grafts to cycle at an age when intact animals are acyclic.[92] Ovarian

Table 3
EVIDENCE FOR EXTRINSIC GONADAL AGING

Observation	Ref.
Cycle renewed when "old" ovary transplanted to young rat	90
Exogenous gnRH stimulates renewed cycles and ovulation in old female rats	40
Depletion of gonadotropins in young rats	
Depresses steroidogenesis	93
Causes degenerative ultrastructural changes and loss of Leydig cells	83,94
Reduces sensitivity of gonads to subsequent stimulation by gonadotropins	94
Spermatogenesis is androgen dependent	95
Prolactin reduces testosterone secretion and causes regression of Leydig cells	85,96
Elevated prolactin levels lead to infertility	
Women	97
Rats	88,98,99
Biochemical, age-type changes in testes are accelerated by vasectomy	100

Note: Many of the structural and functional changes associated with gonadal aging can be simulated in young individuals by hypostimulating the gonad. This occurs following treatments which alter pituitary function so as to depress gonadotropin secretion, enhance prolactin secretion, or qualitatively change the molecular structure of gonadotropins.

steroidogenic deficits may also result from a deficient endocrine environment, since progesterone synthesis is enhanced in ovaries from old rats after prolonged hCG treatment.[62] These findings suggest that age-related loss of ovarian cycles results from extraovarian factors rather than intrinsic change associated with senescence. Table 3 summarizes the results of studies suggesting that extrinsic age changes occur in the gonads of aging animals of both sexes.

HIGHER CENTERS

Pituitary Hormones

Since gonadal dysfunction in aging animals seems to result from inadequate or improper endocrine signals, primary age lesions in the reproductive system may occur at higher centers. As indicated above, gonadal changes may occur in response to chronic understimulation, resulting from various changes in hypophyseal function including (1) altered patterns of gonadotropin and prolactin secretion, (2) degeneration of feedback mechanisms, and/or (3) qualitative changes in the molecular species of gonadotropin. For example, pituitaries from old rats of both sexes show diminished capacity to serete gonadotropins, though they can secrete more prolactin than young animals.[33,47] These changes could in turn trigger the functional and anatomical deficits seen in aging gonads, since antibody depletion of gonadotropin levels in young male rats leads not only to anticipated drops in testosterone levels and spermatogenesis, but also to reduced Leydig cell mass and to degenerative intracellular changes in Sertoli cells.[61,83,94] Hence, reduced LH secretion initially produces reversible decrements in the hormonal milieu, which if maintained, cause irreversible secondary damage to germinal elements. These manifestations of gonadotropin deficiency which mimic spontaneous age changes, may also be compounded by enhanced prolactin secretion. Though prolactin plays a positive role in steroidogenesis in young animals,[101] experimentally induced hypersecretion of the lactogen causes depressed testosterone synthesis as well as profound Leydig cell involution.[85] Prolactin also suppresses plasma LH,[102] which in turn may contribute to the basic problem of insufficient gonadal stimulation with advancing age. The validity of using these models for pituitary hormone dependent, gonadal senescence is demonstrated by the fact that prolactin-secreting adenomas are commonly found in old animals and humans.[54,103,104]

Table 4
PITUITARY CHANGES WHICH MIGHT ACCOUNT FOR
GONADAL DYSFUNCTION DURING AGING

Pituitary alteration	Ref.
Reduced content or altered secretion patterns of gonadotropins and prolactin	58,88,107,112—116
Reduced sensitivity of pituitary to gnRH	89,106,115,117
Pleomorphic changes in gonadotropin molecules which may affect their biological activity	108,118
Altered feedback relationships between pituitary and other elements of the reproductive axis	
Human	26,119
Rat	106,113,114,120—122

Secretory patterns of pituitary hormones change in aged animals in response to altered feedback relationships and/or diminished sensitivity to GnRH stimulation,[47,105] effects which can be simulated by altered endocrine environments.[106,107] Altered secretory patterns of pituitary hormones are also accompanied by molecular modification of gonadotropin structure. Conn et al.[108] recently showed that a larger form of the LH molecule, apparently containing increased sialic acid moieties, is present in pituitary and serum of aged rats. These qualitative age changes in LH structure also occur in primates and may result from altered gonadotropin secretion, since they are simulated by castration of young animals and are reversible by testosterone or estrogen replacement.[108-110] Furthermore, such molecular modification of gonadotropin to a more biologically inactive form might explain why the ovaries of postmenopausal women are seemingly refractory to high levels of circulating LH.[111] Pituitary changes that could account for gonadal dysfunction during aging are presented in Table 4.

Hypothalamic Neuropeptides

Since the pituitary does not function autonomously, gonadotopic age changes may derive from extrinsic influence of gonadal or hypothalamic factors. Since structural modification and biological inactivation of LH seem to be controlled by the endocrine environment, it is unlikely that biochemical changes in gonadotropin synthesizing systems represent primary lesions of reproductive aging. Similarly, age alterations in pituitary hormone content and secretion as well as changes in sensitivity to neuropeptides may result from extrahypophyseal (extrinsic) influence. For example, even though the pituitary content of gonadotropin (↓) and prolactin (↑) of aged, constant-estrus rats is different from that of young rats showing regular estrous cycles,[123] pituitary hormone content is the same in aged and young rats showing comparable neuroendocrine states.[18]

Furthermore, LH secretion in repetitive pseudopregnancy (RPP) is greater than in constant estrus (CE), again suggesting that endocrine environment, not age, affects gonadotropin secretion. Although gonadotropin secretory capacity of old and young pituitary glands is the same if the old pituitaries are given adequate stimulation,[47] pituitary sensitivity to LH-releasing hormone stimulation may decrease with age. However these changes also vary under different endocrine conditions and the sensitivity can be modified hormonally in young experimental rats.[124] Therefore, it is doubtful that intrinsic pituitary changes account for reproductive system aging. Instead pituitary changes associated with age probably occur in response to hypothalamic alterations in content and/or storage of regulatory peptides such as gnRH and PIF. Indeed, LH-releasing and prolactin-inhibiting capacity of hypothalamic extracts from aged rats are depressed when compared with activity of comparable preparations

Table 5

FACTORS FAVORING EXTRINSIC AGING OF PITUITARY FUNCTION

Factor	Ref.
Pituitaries from old, acyclic rats support ovarian cycles in young rats	90
Hypophyseal secretory capacity varies according to reproductive state, not age	31,106,126,127
Prolactin suppresses LH secretion	102
Estrogen decreases the biologic potency of LH	128
Castration alters molecular nature of LH	108—110
Mode of gnRH presentation alters hypophyseal gonadotropin secretion and gnRH sensitivity	124,129
Reduced gnRH and PIF activity in hypothalamus of aged animal	47,123
Physiochemical properties of LHRH neurons and subneural distribution of LHRH changes with aging	125

Note: Hypothalamic alterations involving neuropeptides may lead to age-type changes in pituitary function. Changing feedback relationships between the gonad and pituitary may also contribute to hypophyseal senescence.

from young animals.[117] On the other hand, Barnea et al.[125] recently reported that the hypothalamic content of LHRH is higher in old than in young rats. These differences may result from the fact that in the latter study LHRH content was determined by RIA which does not consider bioactivity. As with LH, age-dependent pleiomorphic changes might alter the biological action of LHRH without depressing the concentration of immunoreactive release factor. In addition, there is a distinct change in the subneuronal distribution of LHRH which might account for functional changes associated with aging in the pituitary-ovarian axis[125] (Table 5).

Hypothalamic Monoamines

Although abnormal changes at any level of the reproductive axis could cause total failure of the system, it seems that none below the level of the CNS are primary defects. This can be demonstrated in aged female rats, where reproductive cycles are reinstated following treatments which restore "monoamine-balance" within the hypothalamus. Accordingly, neuroleptic compounds which enhance catecholamine and depress serotonin metabolism are most effective in reversing or delaying the onset of reproductive senescence. These treatments have a sound physiological basis since (1) monoamines regulate normal reproductive activity, (2) age-type reproductive syndromes such as RPP or CE can be produced in young rats by blockade of catecholamine or serotonin metabolism, respectively, and because (3) spontaneous modification of monoamine metabolism seems to occur in aged rats (Table 6).

Experimental studies suggest that an inhibitory dominance of serotonin coupled with diminishing facilitatory influence of catecholamines promotes reproductive system aging; however the findings may be misleading. For the most part, they do not consider the dynamics of monoaminergic control of the reproductive axis, which may be an important oversight considering that periodic serotoninergic signals coming as a component of the circadian pattern of hypothalamic serotonin metabolism are required for phasic LH secretion in the female rat.[148-150] Metabolic alterations in aging animals, leading to enhanced serotonin synthesis throughout the day, disrupts cyclic reproductive function by obliterating the circadian periodicity of serotonin rather than imposing excessive inhibitory signals on the neuroendocrine axis. This can be demonstrated experimentally in young rats using anti-ovulatory drugs having opposite effects on serotonin synthesis, but common effects on the daily serotonin rhythm. Both 5-hydroxytryptophan (5HTP), which increases serotonin synthesis and *p*-chlorophenylalanine (*p*CPA), which decreases serotonin synthesis abolish the hypo-

Table 6
POSSIBLE INVOLVEMENT OF HYPOTHALAMIC
MONOAMINES IN REPRODUCTIVE SYSTEM AGING

Observation	Ref.
Monoamine metabolism becomes altered in old animals; possibly responsible for age changes in pituitary secretion	11,37,130—135
CE or RPP occurs after differential alteration of hypothalamic monoamines by pharmacological or surgical procedures	133,136—138
Monoamine neuroleptics reinstate ovulation and estrous cycles in aged female rats	40,46,115,137,139—142
Reduced feedback capacity of hypothalamus; diminished numbers of steroid receptors in brain	105,120,143—145

Note: Since monoamines normally regulate adenohypophyseal secretion of gonadotropins and prolactin in young animals,[146,147] changes in the metabolic dynamics of these putative neurotransmitters may lead to age dysfunction in the reproductive neuroendocrine axis.

thalamic serotonin circadian rhythm and also block the LH surge.[14,149] In fact, age-type CE can be produced in young rats by local blockade of serotonin synthesis (and thereby its circadian rhythm) in the rostral hypothalamus.[133] Notably, estrogen, which may accelerate reproductive aging in the rat[91,92] also abolishes the serotonin rhythm in the hypothalamus,[151] while those treatments which reinstate reproductive function in old animals such as progesterone[152] and reduced exposure to light,[153] change the rate of serotonin metabolism and thereby may provide intermittent, rather than constant serotoninergic signals.[154] Altered patterns of serotonin metabolism in aging animals may also contribute to hypophyseal changes favoring prolactin secretion[155] and may thereby indirectly alter gonadotropin secretion.

SEROTONIN AND CNS PACEMAKERS: POSSIBLE ROLE IN REPRODUCTIVE SYSTEM AGING

The significance of serotoninergic involvement in reproductive aging comes from the fact that the suprachiasmatic nucleus (SCN), the major control nucleus for circadian organization in rodents[156] is rich in serotonin.[157] Since selective destruction of the SCN[136] or blockade of serotonin synthesis in the region of SCN[133] abolishes normal circadian reproductive rhythms, serotonin might be an essential pacemaker substance for temporal organization within the reproductive system. Furthermore, temporal patterns of serotonin circadian rhythms change during development and establish certain phase relationships with other functions showing circadian rhythms, such as phasic secretion (preovulatory surge) of luteinizing hormone.[158] For example, a temporally related pattern of serotonin metabolism in the hypothalamus and the LH surge is seen in the adult female rat, where the acrophase of the monoamine rhythm occurs in the afternoon, just prior to the onset of the "critical period" for activation of the pituitary-ovarian axis. When the serotonin rhythm is altered by acute changes in light exposure, the LH surge profiles become abnormal.[14] Interdependence of the serotonin and LH rhythms is also suggested by the work of Meyer and Quay,[159] who found that serotonin uptake in the SCN region of the hypothalamus in vitro occurs prior to the time of LH secretion. Furthermore, uptake is enhanced up to three times in hypothalamic tissue taken from rats in proestrus as compared with other stages of the estrous cycle. Since monoamine re-uptake occurs during periods of synaptic activity, these observations suggest an involvement of serotonergic neurons in generation of the LH circadian rhythm. Studies of phase changes in serotonin circadian rhythms in rats up to 66 days of age show that the changes occur postnatally as a function of development.[158] Continued change in timing of the serotonin rhythm with advancing age might lead to diminished performance in the reproductive axis if phasic secretion of LH is dependent upon serotonergic timing signals.

Experimental modification of serotonin activity by surgical or pharmacological methods has provided more convincing data demonstrating an involvement of serotonin in generation of the LH circadian rhythm (LH surge). For example, depletion of brain serotonin content with drugs such as *p*CPA or 5,7-DHT, or by surgical ablation of the mid-brain raphe nuclei, abolishes phasic secretion of LH.[160-162] Furthermore, the preovulatory LH surge can be blocked or prematurely terminated by properly timed administration of methysergide or cyproheptadine, serotonin receptor antagonists. When administered prior to the onset of the critical period, both drugs prevent phasic secretion of LH and block ovulation. Daily administration of the drugs sustains ovulatory blockade and vaginal smears remain cornified.[163] Very recently, Marko and Fluckiger[164] found that serotonin receptor antagonists given before, but not after the critical period block the LH surge. These data support and extend the contention that profiles of phasic secretion of LH are determined by a period of serotonergic neurotransmission. Conversely, altered dynamics of serotonin metabolism would lead to changes in LH secretion and could represent one component of a mechanism leading to loss of cyclic reproductive function in aging female rats. We[133] recently reported that the effect of altered serotonin metabolism on phasic secretion of LH can be localized to the rostral hypothalamus, in the area of the suprachiasmatic nuclei. Properly timed, local implants of *p*CPA crystals into this region in young cycling rats blocked the LH surge and induced CE, which was endocrinologically similar to that occuring spontaneously in old animals. Recently Coen and MacKinnon[165] showed that the facilitatory effect of serotonin on phasic secretion of LH requires the neural circuitry of the SCN. These data provide strong evidence that the serotonergic terminals which are concentrated within SCN are active in maintaining functional integrity of the female reproductive system by controlling LH phasic secretion. Therefore, changes within this region of the hypothalamus may be responsible for loss of temporal organization within the aging reproductive system. Circumstantial evidence in support of this hypothesis derives from the fact that age-like changes in LH secretion patterns occur when female rats are exposed to constant light, a treatment which produces reversible, but functional lesions of the SCN. Rats exposed to constant light for long periods develop CE,[166] after a period of estrous cycle irregularity in which the acrophase and amplitude of the LH surge are delayed and depressed, respectively.[167] Similar changes occur spontaneously in middle-age rats prior to the onset of constant estrus[46] (Table 7).

SEROTONIN, REPRODUCTIVE AGING, AND CANCER

Tumors of the reproductive organs are uncommon before puberty; however, their incidence increases thereafter; peaking at the end of the reproductive lifespan.[44] The mammary glands of noncyclic aged rats are commonly hypertrophied and secreting and often show a high incidence of tumors.[18,46,54] These changes in mammary structure and function result from pituitary prolactin hypersecretion.[18] While the high frequency of "chromophobe adenomas" in senescent rats has been known for many years, a recent study suggests that they also occur quite commonly in old humans[103] and might account for age-associated gonadal atrophy and the increased risk for developing reproductive neoplasms. Experimental studies have shown that treatments which disrupt patterns of serotonin metabolism or obliterate internal temporal order enhance tumorogenesis.[187] Conversely, those treatments which restore "serotonin-catecholamine balance" in rats, delay the onset of pituitary and mammary tumors[46] or cause them to regress.[188,189] It is possible, therefore, that functional changes possibly involving serotonin metabolism in central biorhythm pacemakers lead not only to degen-

Table 7
DEGENERATION OF TEMPORAL ORDER IN THE AGING REPRODUCTIVE SYSTEM

Evidence for pacemaker insufficiency	Ref.
Unusual gonad:pituitary hormone associations occur in women during the menopausal transition	26
Transient, episodic neuroendocrine events are associated with menopausal flush episodes	168
Timing of ovulation may be delayed in aging women	111,169
Disintegration of time relationships between hormone secretion and light dark cycle involving LH, progesterone, and prolactin	66,170,—174
Exposure to constant light, which abolishes circadian rhythm of serotonin metabolism, causes age-type changes in LH secretion, and CE in rats	166,167,175
Serontoninergic signals may be pacemakers for temporal organization in the reproductive system	14,148,150,154,159,161,163,164,176
Serotonin circadian rhythm may provide intermittent serotonin pacemaker signal	150,151,177,178
Pacemaker neural circuits may reside in SCN	136,156,162,165,177,179
Neural signals for gonadotropin release may be absent in old rats	135,180,181
Blockade of serotonin synthesis in region of SCN produces CE	133
Neonatal androgen treatment or lesions of SCN produce CE and stimulate abnormal prolactin secretion from pituitary	98,99,136,182,183
Androgen sterility is accompanied by depressed serotonin metabolism in the brain	184
Estrogen injections accelerate reproductive system aging, cause CE and neuronal degeneration in hypothalamus and loss of diurnal changes in brain serotonin levels	91,92,154,185,186
Hypothalamus serotonin circadian rhythm absent in aged CE rats	154
"Short days" or progesterone treatments, which restore estrous cycles in old female rats, also stimulate diurnal changes in hypothalamus serotonin	153,154

Note: Functional changes in the suprachiasmatic nucleus, which is rich in serotonin terminals, lead to altered time-activity relationships among different components of the reproductive axis. This loss of temporal organization can lead to functional degeneration of the reproductive system as occurs during senescence.

eration of the aging reproductive system, but also promote pathogenesis. Table 8 summarizes data in support of this hypothesis.

CONCLUSIONS

It is difficult to define intrinsic changes which might account for senescence in the reproductive system, and perhaps they do not exist. Rather than finding obvious defects in cells at specific levels, reproductive system aging seems to initiate from more subtle causes involving integration of function between brain and gonad. Perhaps, reproductive senescence occurs in two stages whereby (1) a primary loss of temporal organization is followed by (2) secondary defects and pathology developing in the resulting suboptimal endocrine environment. This argument is strengthened by the fact that aging animals show spontaneous disturbances in their biologic time structure, characterized by aberrant or attenuated circadian

Table 8
ROLE OF SEROTONIN IN PITUITARY AND MAMMARY TUMORIGENESIS

Observation	Ref.
Morphological changes in pituitary favor increased percentages of chromophobes and prolactin-secreting adenomas	
Human	103,190
Rat	18,46,54,191,192
Pituitary tumors are associated with decreased fertility, gonadal atrophy, and mammary tumors	
Human	193
Rat	194
Tumor prolactin secretion causes changes in reproductive tissue and mammary glands	195
Pituitary tumors transplanted to young rats do not proliferate until animals grow old, suggesting that the "internal state" of the aged rat is optimal for pituitary tumor growth	104,196
Drugs which enhance serotonin activity promote tumor growth	197—199
Serotonin promotes prolactin secretion	155,200—202
Animals with pituitary tumors show higher hypothalamic serotonin than age-matched controls without tumors	199
Treatments which depress serotonin synthesis delay onset of pituitary and mammary tumor development	46,188,189,203—207

Note: Pituitary age changes favoring the secretion of prolactin may result from altered temporal dynamics in serotonin metabolism. Continued hyperactivity of prolactin cells in response to continuous rather than intermittent serotonin signals leads to neoplastic transformation. Mammary tumors develop secondarily to hyperprolactinemia.

rhythms. Furthermore, experimental perturbations of temporal order cause age-type changes, including increased pathology (specifically neoplastic disease) and reduced lifespan in young animals. These experiments suggest that mechanisms for degenerative change need not be an inherent genetic property of somatic cells to account for functional decline in each physiological system during senescence. Rather, a general loss of temporal organization, perhaps mediated by changes in a population of central pacemaker cells, could represent a primary mechanism for organismal aging. This is most clearly seen in experimental rodent systems which develop age-type reproductive syndromes after exposure to conditions which perturb circadian rhythms. For example, if female rats are kept in constant light, reproductive cycles initially become irregular, then fail with the establishment of constant estrus. An interesting similarity in LH secretion occurs in light-exposed and aging rats during the period of irregular cycles, which suggests a common mechanism. In both cases, the LH surge is delayed in onset, and its amplitude is attenuated prior to being lost. Enhanced serotonin turnover, resulting in a loss of the hypothalamic serotonin circadian rhythm, occurs in both groups during the time that temporal patterns of phasic LH secretion are changing[154] The altered dynamics of neurotransmitter metabolism may actually promote the initial changes indicative of early reproductive senescence, since the serotonin rhythm, which provides intermittent signals, is required for cyclic reproductive function.[149,163] Furthermore, those treatments which restore cycling in aging female rats, such as reduced light exposure,[153] progesterone,[208] and L-DOPA,[141] are mediated in part by serotonin. Therefore, it is proposed that reproductive senescence results from intrinsic changes in central pacemaker cells, resulting in loss of temporal organization. The serotonin-rich suprachiasmatic nucleus is the major circadian pacemaker of the rat brain, and decremental neurochemical changes in this region of the hypothalamus could lead to physiological decline typical of aging. This may represent a generalized aging phenomenon, since other rhythms also show comparable phase shifts and attenuations in old and SCN-lesioned rats.[209,210]

Central to this hypothesis is the fact that multicellular animals must undergo a continuous transformation from embryo to adult during ontogeny. It is intuitively obvious that this transformation, involving continual physiologic and biochemical reorganization occurring in all tissues and organs of the body, must be regulated to guarantee normal development and survival of the animal to adulthood. In light of this consideration, certain assumptions can be made concerning the process of change and its involvement in age-related decline:(1) certain cells within the CNS, presumably in the suprachiasmatic nuclei, serve to integrate change in somatic tissue, and thereby act as pacemakers of the dynamic reorganizational continuum known as development and aging, (2) genetically regulated changes of molecular processes within the pacemaker cells are required for proper integrative control from birth to puberty and into young adulthood, and (3) postmaturational changes in the pacemaker cells, presumably residual from the program of development, leads to their reduced integrative capacity, ultimately reflected in poor organismal homeostasis and declining physiological performance.

Contemporary studies of the reproductive system have taken the first step in differentiating between the molecular and systemic contributions to gonadal senescence. Perhaps future work will deal with problems of postmaturational change in pacemaker cells in an attempt to understand and possibly modify the effect of advancing age on their integrative capacity.

REFERENCES

1. **Comfort, A.**, *The Biology of Senescence,* 3rd ed., Elsevier, New York, 1979.
2. **Bidder, G. P.**, Senescence, *Br. Med. J.,* 115, 5831, 1932.
3. **Arvay, A.**, Reproduction and aging, in *Hypothalamus Pituitary and Aging,* Everitt, A. V. and Burgess, J. A., Eds., Charles C Thomas, Springfield, Ill., 1976, 362.
4. **Bilewicz, S.**, Influence of mating on the longevity of *Drosophila melanogaster, Folia Biol.,* 1, 175, 1953.
5. **Matthes, E.**, Der Einfluss der Fortpflanzung auf die Lebensdauer lines Schmetterling, *Fumea crassiorella, Z. Vergl. Physiol.,* 33, 1, 1951.
6. **Wodinsky, J.**, Hormonal inhibition of feeding and death in octopus: control by optic gland secretion, *Science,* 198, 948, 1977.
7. **Kanungo, M. S.**, Theories of aging, in *Biochemistry of Aging,* Kanungo, M. S., Ed., Academic Press, New York, 1980, 242.
8. **Williams, G. C.**, Pleiotropy, natural selection and the evaluation of senescence, *Evolution,* 11, 398, 1957.
9. **Sacher, G. A.**, Molecular versus systemic theories on the genesis of aging, *Exp. Gerontol.,* 3, 265, 1968.
10. **Wright, B. E. and Davison, P. F.**, Mechanisms of development and aging, *Mech. Aging Dev.,* 12, 213, 1980.
11. **Finch, C. E.**, The regulation of physiological changes during mammalian aging, *Q. Rev. Biol.,* 51, 49, 1976.
12. **Finch, C. E.**, Neuroendocrine mechanisms and aging, *Fed. Proc.,* 38, 178, 1979.
13. **Timiras, P. S.**, Biological perspectives on aging, *Am. Sci.,* 66, 605, 1978.
14. **Walker, R. F. and Timiras, P. S.**, Pacemaker insufficiency and the onset of aging, in *Cellular Pacemakers,* Carpenter, D., Ed., Wiley, Interscience, New York, 396, 1982.
15. **Harman, S. M.**, Clinical aspects of aging of the male reproductive system, in *The Aging Reproductive System,* Schneider, E. L., Ed., Raven Press, New York, 1978, 29.
16. **Kinsey, A. C., Pomeroy, W. B., and Martin, C. E.**, *Sexual Behavior in the Human Male,* W. B. Saunders, Philadelphia, 1948.
17. **Talbert, G. B.**, Aging of the reproductive system in *The Biology of Aging,* Finch, C. E. and Hayflick, L., Eds., Van Nostrand Reinhold, New York, 1977, 318.
18. **Aschheim, P.**, Aging in the hypothalamic-hypophyseal ovarian axis in the rat, in *Hypothalamus Pituitary and Aging,* Everitt, A. V. and Burgess, J. A., Eds., Charles C Thomas, Springfield, Ill., 1976, 376.
19. **Meites, J., Steger, R. W., and Huang, H. H. H.**, Relation of neuroendocrine system to the reproductive decline in aging rats and human subjects, *Fed. Proc.,* 39, 3168, 1980.
20. **Cooper, R. L.**, Sexual receptivity in aged female rats, *Horm. Behav.,* 9, 321, 1977.

21. **Cooper, R. L. and Linnoila, M.,** Sexual behavior in aged non-cycling female rats, *Physiol. Behav.,* 18, 573, 1977.
22. **Kinsey, A. C., Pomeroy, W. B., Martin, C. E., and Gebgard, P. H.,** *Sexual Behavior in the Human Female,* W. B. Saunders, Philadelphia, 1953.
23. **Mattingly, R. F. and Huang, W. Y.,** Steroidogenesis of the menopausal and post-menopausal ovary, *Am. J. Obstet. Gynecol.,* 103, 679, 1969.
24. **Schneider, E. L., Ed.,** *The Aging Reproductive System,* Vol. 4, Raven Press, New York, 1978.
25. **Adamopoulos, D. A., Loraine, J. A., and Dove, G. A.,** Endocrinological studies in women approaching menopause, *J. Obstet. Gynaecol. Br. Commonw.,* 78, 62, 1971.
26. **Metcalf, M. G., Donald, R. A., and Livesey, J. H.,** Pituitary-ovarian function in normal women during the menopausal transition, *J. Endocrinol.,* 87, 191, 1980.
27. **Novak, E. R.,** Ovulation after fifty, *Obstet. Gynecol.,* 36, 903, 1970.
28. **Rakoff, A. E. and Nowroozi, K.,** The female climacteric, in *Geriatric Endocrinology,* Greenblatt, R. B., Ed., Raven Press, New York, 1978, 165.
29. **Sherman, B. M. and Korenman, S. G.,** Hormonal characteristics of the human menstrual cycle throughout reproductive life, *J. Clin. Invest.,* 55, 669, 1975.
30. **Timiras, P. S.,** *Developmental Physiology and Aging,* Macmillan, New York, 1972.
31. **Treolar, A. E. Boynton, R. E., Behn, B. G., and Brown, B. W.,** Variation in the human menstrual cycle throughout life, *Int. J. Fertil.,* 12, 77, 1967.
32. **Van Wagenen, G.,** Menopause in a subhuman primate (Macaca mulatta) *Anat. Rec.,* 166, 392, 1976.
33. **Ingram, D. K.,** The vaginal smear of senile laboratory rats, *J. Endocrinol.,* 19, 182, 1959.
34. **Nelson, J. F., Felicio, L. S., Randall, P. K., Sims, C., and Finch, C. E.,** A longitudinal study of estrous cyclicity in aging C57BL/6J mice. I. Cycle frequency, length and vaginal cytology, *Biol. Reprod.,* 27, 327, 1982.
35. **Thung, P. H., Boot, L. M., and Mühlbock, O.,** Senile changes in the oestrous cycle and in ovarian structure of some inbred strains of mice, *Acta Endocrinol.,* 23, 8, 1956.
36. **Woolf, C. M.** Stillbirths and parental age, *Obstet. Gynecol.,* 26, 1, 1965.
37. **Finch, C. E.,** Reproductive senescence in rodents: factors in the decline of fertility and loss of regular estrous cycles, in *The Aging Reproductive System,* Schneider, E. L., Ed., Raven Press, New York, 1978, 193.
38. **Talbert, G. B. and Krohn, P. L.,** Effect of maternal age on viability of ova and uterine support of pregnancy in mice, *J. Reprod. Fertil.,* 11, 399, 1966.
39. **Blaha, G. C.,** Effect of age of the donor and recipient on the development of transferred golden hamster ova, *Anat. Rec.,* 150, 413, 1964.
40. **Meites, J., Huang, H. H., and Simpkins, J. W.,** Recent studies on neuroendocrine control of reproductive senescence in rats, in *The Aging Reproductive System,* Schneider, E. L., Ed., Raven Press, New York, 1978, 213.
41. **Sopelak, V. M. and Butcher, R. L.,** Decreased amount of ovarian tissue and maternal age affect embryonic development in old rats, *Biol. Reprod.,* 27, 449, 1982.
42. **Asch, R. H. and Greenblatt, R. B.** The aging ovary: morphologic and endocrine correlations, in *Geriatric Endocrinology,* Greenblatt, R. G., Ed., Raven Press, New York, 1978, 141.
43. **Everitt, A. V.,** The female climacteric in *Hypothalamus, Pituitary and Aging,* Everitt, A. V., and Burgess, J. A., Eds., Charles C Thomas, Springfield, Ill., 1976, 419.
44. **Freedman, A.,** Cancer, aging and the pituitary, in *Hypothalamus, Pituitary and Aging,* Everitt, A. V., and Burgess, J. A., Eds., Charles C Thomas, Springfield, Ill., 1976, 431.
45. **Schiff, J. and Wilson, E.,** Clinical aspects of aging of the female reproductive system, in *The Aging Reproductive System,* Schneider, E. L., Ed., Raven Press, New York, 1978, 29.
46. **Cooper, R. L. and Walker, R. F.,** Potential therapeutic consequences of age-dependent changes in brain physiology, in *Experimental and Clinical Aspects of the CNS Aging Process,* Meier-Ruge, T. W., Ed., Karger, Basel, 1979, 1.
47. **Riegle, G. D. and Miller, A. E.,** Aging effects on the hypothalamic-hypophyseal-gonadal control system in the rat, in *The Aging Reproductive System,* Schneider, E. L., Ed., Raven Press, New York, 1978, 159.
48. **Albeaux-Fernet, M., Bohler, C. C. S., and Karpas, A.,** Testicular function in the aging male, in *Geriatric Endocrinology,* Greenblatt, R. B., Ed., Raven Press, New York, 1978, 201.
49. **Mainwaring, W. I. P. and Brandes, D.,** Functional and structural changes in accessory sex organs during aging, in, *Male Accessory Sex Organs,* Brandes, D., Ed., Academic Press, New York, 1974, 469.
50. **Swanson, L. J., Desjardins, C., Turek, F. W.,** Aging of the reproductive system in the male hamster: behavioral and endocrine patterns, *Biol. Reprod.,* 26, 791, 1982.
51. **Craigen, W. and Bronson, F. H.,** Deterioration of the capacity for sexual arousal in aged male mice, *Biol. Reprod.,* 26, 869, 1982.
52. **Brandes, D. and Garcia-Bunuel, R.,** Aging of the male sex accessory organs, in *The Aging Reproductive System,* Schneider, E. L., Ed., Raven Press, New York, 1978, 127.

53. **Cockrell, B. Y. and Garner, F. M.,** Interstitial cell tumor, *Comp. Pathol. Bull.,* 8, 2, 1976.
54. **Coleman, G. L., Barthold, S. W., Osbaldiston, G. W., Foster, S. J., and Jonas, A. M.,** Pathological changes during aging in barrier-reared Fischer 344 male rats, *J. Gerontol.,* 32, 258, 1977.
55. **Gray, G. D.,** Age-related changes in penile erections and circulating testosterone in middle-aged male rats, *Adv. Exp. Med. Biol.,* 113, 149, 1978.
56. **Grodin, J. M., Siiteri, P. K., and MacDonald, P. C.,** Source of estrogen production in postmenopausal women, *J. Clin. Endo. Metab.* 36, 207, 1973.
57. **Judd, H. L., Judd, G. E., Lucus, W. E. and Yen, S. S. C.,** Endocrine function of the postmenopausal ovary: Concentration of androgen and estrogen in ovarian and peripheral vein blood, *J. Clin. Endocrinol., Metab.,* 39, 1020, 1974.
58. **Huang, H. H., Steger, R. W., Bruni, J. F., and Meites, J.,** Patterns of sex steroid and gonadotropin secretion in aging female rats, *Endocrinology,* 103, 1855, 1978.
59. **Page, R. D. and Butcher, R. L.,** Follicular and plasma patterns of steroids in young and old rats during normal and prolonged estrous cycles, *Biol. Reprod.,* 27, 383, 1982.
60. **Dilman, V. M.,** The hypothalamic control of aging and age-associated pathology, The elevation mechanism of aging, in *Hypothalamus, Pituitary and Aging,* Everitt, A. V. and Burgess, J. A., Eds., Charles C Thomas, Springfield, Ill., 1976, 634.
61. **Schenker, J. G., Polishuk, W. Z., and Eckstein, B.,** Pathways in the biosynthesis of androgens in the post menopausal ovary in vitro, *Acta. Endocrinol.,* 66, 325, 1971.
62. **Chan, S. W. C. and Leathem, J. H.,** Aging and ovarian steriodogenesis in the rat, *J. Gerontol.,* 32, 395, 1977.
63. **Judd, H. L., Lucas, W. E., and Yen, S. S. C.,** Serum 17B-estradiol and estrone levels in postmenopausal women with and without endometrial cancer, *J. Clin. Endocrinol. Metab.,* 43, 272, 1976.
64. **Siiteri, P. K. and MacDonald, P. C.,** Role of extraglandular estrogen in human endocrinology, in *Handbook of Physiology,* Vol. 2(Part 1), Greep, R. O. and Astwood, E. B., Eds., Williams & Wilkins, Baltimore, 1973, 615.
65. **Flurkey, K., Gee, D. M., Sinha, Y. N., Wisner, J. R., and Finch, C. E.,** Age effects on luteinizing hormone, progesterone and prolactin in proestrous and acyclic C57BL/6J mice, *Biol. Reprod.,* 26, 835, 1982.
66. **Miller, A. E. and Riegle, G. D.,** Temporal changes in serum progesterone in aging female rat, *Endocrinology,* 106, 1579, 1980.
67. **Leathem, J. H. and Murone, E. K.,** Ovarian Δ^5-3B-hydroxysteroid dehydrogenase in aging rats, *Fertil. Steril.,* 26, 996, 1975.
68. **Leathem, J. H. and Shapero, B. H.,** Aging and ovarian Δ^5-3B-hydroxysteroid dehydrogenase in rats, *Proc. Soc. Exp. Biol. Med.,* 148, 793, 1975.
69. **Blaha, G. C. and Leavitt, W. W.,** Ovarian steroid dehydrogenase histochemistry and circulating progesterone in aged golden hamster during the estrous cycle and pregnancy, *Biol. Reprod.,* 11, 156, 1974.
70. **Albrecht, E. D., Koos, R. D., and Wehrenberg, W. B.,** Ovarian Δ^5-3B-hydroxysteroid dehydrogenase and cholesterol in the aged mouse during pregnancy, *Biol. Reprod.,* 13, 158, 1975.
71. **Vermuelen, A. R., Rubens, R., and Verdonek, L.,** Testosterone secretion and metabolism in male senescence, *J. Clin. Endocrinol. Metab.,* 34, 730, 1972.
72. **Chan, S. W. C., Leathem, J. H., and Esashi, T.,** Testicular metabolism and serum testosterone in aging male rats, *Endocrinology;* 101, 128, 1977.
73. **Harman, S. M., Danner, R. L., and Roth, G. S.,** Testosterone secretion in the rat in response to chorionic gonadotropin: alterations with age, *Endocrinology,* 102, 540, 1978.
74. **Vermuelen, A., Leydig-cell function in old age,** in *Hypothalamus, Pituitary and Aging,* Everitt, A. V. and Burgess, J. A., Eds., Charles C Thomas, Springfield, Ill., 1976, 458.
75. **Hemsell, D. L., Groden, J. M., Brenner, P. F., Siiteri, P. K., and MacDonald, P. C.,** Plasma precursors of estrogen II, correlation of extent of conversion of plasma androsteredrone to estrone with age, *J. Clin. Endocrinol. Metab.,* 38, 476, 1974.
76. **Tillenger, K. G.,** Testicular morphology, *Acta. Endocrinol.,* 30(Suppl.), 28, 1957.
77. **Bethea, C. L. and Walker, R. F.,** Age-related changes in reproductive hormones and in Leydig cell responsivity in the male Fischer 344 rat, *J. Gerontol.,* 34, 21, 1979.
78. **Rubens, R., Dhont, M., and Vermeulen, A.,** Further studies on Leydig cell function in old age, *J. Clin. Endocrinol. Metab.,* 39, 40, 1974.
79. **Tsitouras, P. D., Kowatch, M. A., and Harman, S. M.,** Age-related alterations of isolated rat leydig cell function: gonadotropin receptors, adenosine 3,5-monophosphate response, and testosterone secretion, *Endocrinology,* 105, 1400, 1979.
80. **Vassileva-Popova, J.,** Developmental changes in gonadal binding of gonadotropins, *Proc. 5th Asia Oceania Congr. Endocrinol.,* 242, 1974.

81. **Leathem, J. H. and Albrecht, E. D.,** Effect of age on testis Δ^5-3B-hydroxysteroid dehydrogenase in the rat, *Proc. Soc. Exp. Biol. Med.,* 145, 1212, 1974.

82. **Peng, M. T., Pi, W. P., and Peng, Y. M.,** The hypothalamic-pituitary-testicular function of the old rat, *J. Formosan Med. Assoc.,* 72, 195, 1973.

83. **Dym, M. and Madhwaraj, H. G.,** Response of adult rat serotoli cells and Leydig cells to depletion of luteinizing hormone and testosterone, *Biol. Reprod.,* 17, 676, 1977.

84. **Fang, V. S., Refetoff, S., and Rosenfield, R. L.,** Hypogonadism induced by a transplantable prolactin producing tumor in male rats: hormonal and morphological studies, *Endocrinology,* 95, 991, 1974.

85. **Perotti, M. E. and Fang, V. S.,** Ultrastructural study of the testicular interstitial cells and the prostate involution in rats bearing a transplantable prolactin and growth hormone-producing tumor, *J. Ultrastruct. Res.,* 52, 202, 1975.

86. **Catt, K. J. and Dufau, M. L.,** Gonadal receptors for luteinizing hormone and chorionic gonadotropin, *Methods Enzymol.,* 37, 167, 1975.

87. **Kaler, L. W. and Neaves, W. B.** The androgen status of aging male rats. *Endocrinology,* 108, 712, 1981.

88. **Shaar, C. J., Euker, J. S., Riegle, G. D., and Meites, J.,** Effects of castration and gonadal steroids on serum luteinizing hormone and prolactin in old and young rats, *J. Endocrinol.,* 66, 45, 1975.

89. **Watkins, B. E., Meites, J., and Riegle, G. D.,** Age-related changes in pituitary responsiveness to LHRH in the female rat, *Endocrinology,* 543, 1975.

90. **Peng, M. T. and Huang, H. O.,** Aging of hypothalamic-pituitary-ovarian function in the rat, *Fertil. Steril.,* 23, 535, 1972.

91. **Brawer, J. R., Naftolin, F., Martin, J., and Sonnenschein, C.,** Effects of a single injection of estradiol valerate on the hypothalamic arcuate nucleus and on reproductive function in the female rat, *Endocrinology,* 103, 501, 1978.

92. **Nelson, J. F., Felicio, L. S., and Finch, C. E.,** Ovarian hormones and the etiology of reproductive aging in mice, in *Aging — Its Chemistry,* Dietz, A. A., Ed., American Association of Clinical Chemists, Washington, D.C., 1980, 64.

93. **Greep, R. O. and Fevold, H. L.,** The spermatogenic and secretory functions of the gonads of hypophysectomized rats treated with pituitary FSH and LH, *Endrocrinology,* 21, 611, 1937.

94. **Gondos, B., Rao, A., and Ramachandran, J.,** Effects of antiserum to luteinizing hormonal on the structure and function of rat Leydig cells, *J. Endocrinol.,* 87, 265, 1980.

95. **Nelson, W. O. and Merckel, C. E.,** Maintenance of spermatogenesis in hypophysectomized mice with androgenic substances, *Proc. Soc. Exp. Biol. Med.,* 38, 737, 1938.

96. **Sharpe, R. M., McNeilly, A. S., Davidson, D. W., and Swanston, I. A.,** Leydig cell function in hyperprolactinaemic adult rats, *J. Endocrinol.,* 87, 28, 1980.

97. **delPozo, E., Varga, L., Wyss, H., Tolis, G., Friesen, H., Wenner, R., Vetter, L., and Uettwiler, L.,** Clinical and hormonal response to bromocriptin (CB 154) in the galactorrhea syndromes, *J. Clin. Endocrinol. Metab.,* 39, 18, 1974.

98. **Mallampati, R. S. and Johnson, D. C.,** Serum and pituitary prolactin, LH, and FSH in androgenized and normal male rats treated with various doses of estradiol benzoate, *Neuroendocrinology,* 11, 46, 1973.

99. **Ratner, A. and Peake, G. T.,** Maintenance of hyperprolactinemia by gonadal steroids in androgen sterilized and spontaneously constant estrous rats, *Proc. Soc. Exp. Biol. Med.,* 146, 680, 1974.

100. **Collins, P. M., Bell, J. B. G., and Tsang, W. N.,** The effect of vasectomy on steroid metabolism by seminiferous tubules and interstitial tissue of the rat testes: a comparison with the effects of aging, *J. Endocrinol.,* 55, 18, 1972.

101. **Hafiez, A. A., Bartke, A., and Lloyd, C. W.,** The role of prolactin in the regulation of testes function: the synergistic effects of prolactin and luteinizing hormone on the incorporation of [1-^{14}C] acetate into testosterone and cholesterol by testes from hypophysectomized rats *in vitro, J. Endocrinol.,* 53, 223, 1972.

102. **Winters, S. J. and Loriaux, D. L.,** Suppression of plasma luteinizing hormone by prolactin in the male rat, *Endocrinology,* 102, 864, 1978.

103. **Kovacs, K., Ryan, N., Horvath, E., Suiger, W., and Ezrin, C.,** Pituitary adenomas in old age, *J. Gerontol.,* 35, 16, 1980.

104. **Saxton, J. A., Jr. and Graham, J. B.,** Chromophobe adenoma-like lesions of the rat hypophysis, *Cancer Res.,* 4, 168, 1944.

105. **Huang, H. H., Steger, R. W., Sonntag, W. E., and Meites, J.,** Positive feedback by ovarian hormones on prolactin and LH in old versus young female rats; *Neurobiol. Aging,* 1, 141, 1981.

106. **Lu, J. K., Damassa, D. A., Gilman, D. P., Judd, H. L., and Sawyer, C. H.,** Differential patterns of gonadotropin responses to ovarian steroids and to LH-releasing hormone between constant-estrous and pseudopregnant states in aging rats, *Biol. Reprod.,* 23, 345, 1980.

107. **Lu, K. H., Hopper, B. R., Vargo, T. M., and Yen, S. S. C.,** Chronological changes in sex steroid, gonadotropin and prolactin secretion in aging female rats displaying different reproductive states, *Biol. Reprod.,* 21, 193, 1979.

108. **Conn, P. M., Cooper, R., McNamara, C., Rogers, D. C., and Shoenhardt, L.,** Qualitative change in gonadotropin during normal aging in the male rat, *Endrocrinology*, 106, 1549, 1980.

109. **Peckham, W. D. and Knobil, E.,** Qualitative changes in the pituitary gonadotropins of the male rhesus monkey following castration, *Endocrinology*, 98, 1061, 1976.

110. **Peckham, W. D. and Knobil, E.,** The effects of ovariectomy, estrogen replacement and neuraminidase treatment on the properties of the adenohypophysis glycoprotein hormones of the rhesus monkey, *Endocrinology*, 98, 1054, 1976.

111. **Jones, E. C.,** The aging ovary and its influence on reproductive capacity, *J. Reprod. Fertil.*, 12 (Suppl.), 17, 1970.

112. **Gosden, R. G. and Bancroft, L.,** Pituitary function in reproductively senescent female rats, *Exp. Gerontol.*, 157, 1976.

113. **Gray, G. D., Smith, E. R., and Davidson, J. M.,** Gonadotropin regulation in middle-aged male rats, *Endocrinology*, 107, 2021, 1980.

114. **Gray, G. D., Tennent, B., Smith, E. R., and Davidson, J. M.,** Luteinizing hormone regulation and sexual behavior in middle-aged female rats, *Endocrinology*, 107, 187, 1980.

115. **Riegle, G. D. and Meites, J.,** Effects of aging on LH and prolactin after LHRH, L-DOPA, methyl-DOPA and stress in male rat, *Proc. Soc. Exp. Biol. Med.*, 151, 507, 1976.

116. **Parkening, T. A., Calcote, R. D., and Collins, T. J.,** Plasma and pituitary concentrations of LH, FSH, and prolactin in reproductively senescent Chinese hamsters during various stages of the estrous cycle, *Biol. Reprod.*, 25, 825, 1981.

117. **Riegle, G. D., Meites, J., Miller, A. E., and Wood, S. M.,** Effect of aging on hypothalamic LH-releasing and prolactin inhibiting activities and pituitary responsiveness to LHRH in the male laboratory rat, *J. Gerontol.*, 32, 13, 1977.

118. **TerHaar, M. B. and Wilson, C. A.,** Evidence for pleiomorphism of luteinizing hormone in peripubertal female rats, *J. Endocrinol.*, 79, 133, 1978.

119. **Albert, E. A.,** in *Hormones and the Aging Process*, Eagle, E. T. and Pincus, G., Eds., Academic Press, New York, 1956, 49.

120. **Lu, K. H., Huang, H. H., Chen, H. T., Krucz, M., Mioduszewski, R., and Meites, J.,** Positive feedback by estrogen and progesterone on LH release in old and young rats, *Proc. Soc. Exp. Biol. Med.*, 154, 82, 1977.

121. **McPherson, J. C., Costoff, A., and Mahesh, V.,** Effects of aging on the hypothalamic hypophyseal-gonadal axis in female rats, *Fertil. Steril.*, 28, 1365, 1977.

122. **Peng, M. T. and Peng, Y. M.,** Changes in the reuptake of tritiated estradiol in the hypothalamus and adenohypophysis of old female rats, *Fertil. Steril.*, 24, 534, 1973.

123. **Clemens, J. A. and Meites, J.,** Neuroendocrine status of old constant-estrous rats., *Neuroendocrinology*, 7, 249, 1971.

124. **Badger, T. M., Rosenblum, P. M., Clement, R. E., and Loughlin, J. S.,** Effects of chronic luteinizing hormone-releasing hormone administration on gonadotropin dynamics of adult male rats, *Proc. Soc. Exp. Biol. Med.*, 165, 253, 1980.

125. **Barnea, A., Cho, G., and Porter, J. C.,** Effect of aging on the subneural distribution of luteinizing hormone-releasing hormone in the hypothalamus, *Endocrinology*, 106, 1980, 1980.

126. **Wise, P. M. and Ratner, A.,** LHRH-induced LH and FSH responses in the aged female rat, *J. Gerontol.*, 35, 506, 1980.

127. **Lu, J. K. H., Gilman, D. P., Meldrum, D. R., Judd, H. L., and Sawyer, C. H.,** Relationship between circulating estrogens and the control mechanisms by which ovarian steroids stimulate luteinizing hormone secretion in aged and young female rats, *Endocrinology*, 108, 836, 1980.

128. **Lucky, A. W. Rebar, R. W., Rosenfield, R. L., Roche-Bender, N., and Helke, J.,** Reduction of the potency of LH by estrogen, *N. Engl. J. Med.*, 300, 1034, 1979.

129. **Belchetz, P., Plant, T. M., Nakai, Y., Keogh, E. J., and Knobil, E.,** Hypophyseal responses to continuous and intermittent delivery of hypothalamic gonadotropin-releasing hormone, *Science*, 202, 631, 1978.

130. **Gudelsky, G. A., Nansel, D. D., and Porter, J. C.,** Dopaminergic control of prolactin secretion in the aging male rat, *Br. Res.*, 204, 446, 1981.

131. **Jonec, V. and Finch, C. E.,** Aging and dopamine uptake by subcellular fractions in the C57BL/6J male mouse brain, *Br. Res.*, 91, 197, 1975.

132. **Simpkins, J. W., Mueller, G. P., Huang, H. H., and Meites, J.,** Evidence for depressed catecholamine and enhanced serotonin metabolism in aging male rats: possible relation to gonadotropin secretion, *Endocrinology*, 100, 1672, 1977.

133. **Walker, R. F., Cooper, R. L., and Timiras, P. S.,** Constant estrus: role of rostral hypothalamic monoamines in development of reproductive dysfunction in aging rats, *Endocrinology*, 107, 249, 1980.

134. **Wilkes, M. M., Lu, K. H., Hopper, B. R., and Yen, S. S. C.,** Altered neuroendocrine status of middle-aged rats prior to the onset of senescent anovulation, *Neuroendocrinology*, 29, 255, 1979.

150. **Wilson, C. A., Andrews, M., Hadley, J. C., Lemon, M., and Yeo, T.,** The role of hypothalamic serotonin (5HT) before ovulation in immature rats treated with pregnant mare serum (PMS), *Psychoneuroendocrinology*, 2, 267, 1977.

151. **Yates, C. A. and Herbert, J.,** The effects of different photoperiods on circadian 5HT rhythms in regional brain areas and their modulation by pinealectomy, melatonin and oestradiol, *Br. Res.*, 176, 311, 1979.

152. **Everett, J. W.,** The restoration of ovulatory cycles and corpus luteum formation in persistent-estrous rats by progesterone; *Endocrinology*, 27, 681, 1940.

153. **Everett, J. W.,** Certain functional inter-relationships between spontaneous persistent estrus, "light estrus" and short day anestrus in the albino rat, *Anat. Rec.*, 82, 409, 1942.

154. **Walker, R. F.,** Reproductive senescence and the dynamics of hypothalmic serotonin metabolism in the female rat in *Brain Neurotransmitters and Receptors in Aging and Age-Related Disorders*, Samorajski, T., Enna, S. J., and Beer, B., Eds., Raven Press, New York, 1981, 95.

155. **Clemens, J. A., Roush, M. E., and Fuller, R. W.,** Evidence that serotonin neurons stimulate secretion of prolactin releasing factor, *Life Sci.*, 22, 2209, 1978.

156. **Mosko, S. S. and Moore, R. Y.,** Neonatal suprachiasmatic nucleus lesions: effects on development of circadian rhythms in the rat, *Br. Res.*, 164, 17, 1979.

157. **Saavedra, J. M., Palkovits, M., Brownstein, M. J., and Axelrod, J.,** Serotonin distribution in the nuclei of the rat hypothalamus and preoptic region, *Br. Res.*, 77, 157, 1974.

158. **Asano, Y.,** The maturation of the circadian rhythm of brain norepinephrine and serotonin in the rat, *Life Sci.*, 10, 833, 1971.

159. **Meyer, D. C. and Quay, W. B.,** Hypothalamic and suprachiasmatic uptake of serotonin *in vitro*: twenty-four-hour changes in male and proestrous female rat, *Endocrinology*, 98, 1160, 1976.

160. **Héry, M., LaPlante, E., and Kordon, C.,** Participation of serotonin in the phasic release of luteinizing hormone. II. Effects of lesions of serotonin containing pathways in the central nervous system, *Endocrinology*, 102, 1019, 1978.

161. **Héry, M., LaPlante, E., Pattou, E. and Kordon, C.,** Participation of serotonin in the phasic release of LH. I. Evidence from pharmacological experiments, *Endocrinology*, 99, 496, 1976.

162. **Meyer, D. C.,** Hypothalamic and raphe serotonergic systems in ovulation control, *Endocrinology*, 103, 1067, 1978.

163. **Walker, R. F.,** Serotonin neuroleptics change patterns of preovulatory secretion of luteinizing hormone in rats, *Life Sci.*, 27, 1063, 1980.

164. **Marko, M. and Fluckiger, E.,** Role of serotonin in the regulation of ovulation, *Neuroendocrinology*, 30, 228, 1980.

165. **Coen, C. W. and MacKinnon, P. C. B.,** Lesions of the suprachiasmatic nuclei and the serotonin-dependent phasic release of luteinizing hormone in the rat: effects of drinking rhythmicity and on the consequences of preoptic area stimulation, *J. Endocrinol.*, 84, 231, 1980.

166. **Browman, L. G.,** Light in its relation to activity and estrous rhythms in the albino rat., *J. Exp. Zool.*, 75, 375, 1937.

167. **McCormack, C. E. and Sridaran, R.,** Timing of ovulation in rats during exposure to continuous light: evidence for a circadian rhythm of luteinizing hormone secretion, *J. Endocrinol.*, 76, 135, 1978.

168. **Casper, R. F., Yen, S. S. C., and Wilkes, M. M.,** Menopausal flushes: a neuroendocrine link with pulsatile luteinizing hormone secretion, *Science*, 205, 823, 1979.

169. **Talbert, G. B.,** Effect of maternal age on reproductive capacity, *Am. J. Obstet. Gynecol.*, 102, 451, 1968.

170. **Cooper, R. L., Conn, P. M., and Walker, R. F.,** Characterization of the LH surge in middle-aged female rats, *Biol. Reprod.*, 23, 611, 1980.

171. **Damassa, D. A., Gilman, D. P., Lu, K. H., Judd, H. C., and Sawyer, C. H.,** The 24-hour pattern of prolactin secretion in aging female rats, *Biol. Reprod.*, 22, 571, 1980.

172. **Van der Schoot, P.,** Changing pro-oestrous surges of luteinizing hormone in aging 5-day cyclic rats, *J. Endocrinol.*, 69, 287, 1976.

173. **Wise, P. M.,** Alterations in proestrous LH, FSH and prolactin surges in middle aged rats, *Proc. Soc. Exp. Biol. Med.*, 169, 348, 1982.

174. **Fayein, N. A. and Aschheim P.,** Age-related temporal changes of levels of circulating progesterone in repeatedly pseudopregnant rats, *Biol. Reprod.*, 23, 616, 1980.

175. **Snyder, S. H., Zweig, M., Axelrod, J., and Fischer, J. E.,** Control of the circadian rhythm in serotonin content of the rat pineal gland, *Proc. Natl. Acad. Sci. U.S.A.*, 53, 301, 1965.

176. **Meyer, D. C.,** Serotonergic and noradrenergic activity during the estrous cycle in limbic and hypothalamic nuclear regions, *Endocrine Soc. Abstr.* 62nd Ann. Mtg., 100, 1980.

177. **Héry, M., Faudon, M., Dusticier, G., and Héry, F.,** Daily variations in serotonin metabolism in the suprachiasmatic nucleus of the rat: influence of oestradiol impregnation, *J. Endocrinol.*, 94, 157, 1982.

178. **Walker, R. F.,** Quantitative and temporal aspects of serotonin's facilitatory action on phasic secretion of LH in female rats, *Neuroendocrinology*, in press.

179. **Mosko, S. S., Erickson, G. F., and Moore, R. Y.,** Dampened circadian rhythms in reproductively senescent female rats, *Behav. Neural Biol.,* 28, 1, 1980.

180. **Steger, R. W. and Peluso, J. J.,** Hypothalamic-pituitary function in the old irregularly cycling rat, *Exp. Aging Res.,* 5, 303, 1979.

181. **Walker, R. F.,** Reinstatement of LH surges by serotonin neuroleptics in aging, constant estrous rats, *Neurobiol. Aging,* In press.

182. **Bethea, C. L. and Neill, J. D.,** Lesions of the suprachiasmatic nuclei abolish the cervically stimulated prolactin surges in the rat, *Endocrinology,* 107, 1, 1980.

183. **Bishop, W., Fawcett, C. P., Krulick, L., and McCann, S. M.,** Acute and chronic effects of hypothalamic lesions on the release of FSH, LH and prolactin in intact and castrated rats., *Endocrinology,* 91, 643, 1972.

184. **Ladosky, W. and Gaziri, L. C. J.,** Brain serotonin and sexual differentiation of the nervous system, *Neuroendocrinology,* 6, 168, 1970.

185. **Munaro, N. I.,** The effect of ovarian steroids on hypothalamic 5-hydroxytryptamine neuronal activity, *Neuroendocrinology,* 26, 270, 1978.

186. **Schipper, H., Brawer, J. R., Nelson, J. F., Felicio, L. S., and Finch, C. E.,** Role of the gonads in the histologic aging of the hypothalamic arcuate nucleus, *Biol. Reprod.,* 25, 413, 1981.

187. **Harker, J. E.,** Experimental production of midget tumors in *periplaneta americana L., Exp. Zool.,* 35, 251, 1958.

188. **Quadri, S. K. and Meites, J.,** Regression of spontaneous mammary tumors in rats by ergot drugs, *Proc. Soc. Exp. Biol. Med.,* 138, 999, 1972.

189. **Quadri, S. K., Kledzik, G. S., and Meites, J.,** Effects of 1-DOPA and methyl dopa on growth of mammary cancers in rats, *Proc. Soc. Exp. Biol. Med.,* 142, 22, 1973.

190. **Rasmussen, A. T.,** The proportion of the various subdivisions of the normal adult human hypophysis cerebri and the relative number of the different types of cells in pars distalis, *Proc. Assoc. Res. Nerv. Ment. Dis.,* 17, 118, 1936.

191. **Ito, A., May, P., Kaunitz, H., Kortwright, K., Clarke, S., Furth, J., and Meites, J.,** Incidence and character of the spontaneous pituitary tumors in stain CR and W/FU male rats, *J. Natl. Cancer Inst.,* 49, 701, 1972.

192. **Wolfe, J. M. Bryan, W. R., and Wright, A. W.,** Histologic observations on the anterior pituitaries of old rats with particular reference to the spontaneous appearance of pituitary adenomata, *Am. J. Cancer,* 34, 352, 1938.

193. **Boyar, R. M. and Hellman, L.,** Syndrome of benign nodular adrenal hyperplasia associated with feminization and hyperprolactinemia, *Ann. Int. Med.,* 80, 389, 1974.

194. **Wolfe, J. M., Burack, E., and Wright, A. W.,** The estrous cycle and associated phenomena in a strain of rats characterized by a high incidence of mammary tumors together with observations on the effects of advancing age on these phenomena, *Am. J. Cancer,* 38, 383, 1940.

195. **Welsch, C. W., Jenkin, T. W., and Meites, J.,** Increased incidence of mammary tumors in the female rat grafted with multiple pituitaries, *Cancer Res.,* 30, 1024, 1970.

196. **Saxton, J. A., Jr.,** The relation of age to the occurrence of adenoma-like lesions in the rat hypophysis and to their growth after transplantation, *Cancer Res.,* 1, 277, 1941.

197. **Lacassagne, A. and Duplan, J. F.,** Le mechanisme de la cancerisation de la mamelle chez la souris, considere d'apres les resultats d'experiences au moyen de la reserpine, *C. R. Acad. Sci.,* 249, 810, 1959.

198. **Welsch, C. W. and Meites, J.,** Effects of reserpine on development of carcinogen-induced mammary tumors in rats, in *24th Int. Congr. Physiol. Sci.,* Vol. 6, NIH, Washington, D.C., 1968, 466.

199. **Walker, R. F. and Cooper, R. L.,** Synergistic effects of estrogen and serotonin-receptor agonists upon the development of pituitary tumors in aging rats, *Proc. Soc., Exp. Biol. Med.* in press, 1983.

200. **Garthwaite, T. L. and Hagen, T. C.,** Evidence that serotonin stimulates a prolactin-releasing factor in the rat, *Neuroendocrinology,* 29, 215, 1979.

201. **Lu, K. H. and Meites, J.,** Effects of serotonin precursors and melatonin on serum prolactin release in rats, *Endocrinology,* 93, 152, 1973.

202. **Pavasuthipaisit, K., Norman, R. L., and Spies, H. G.,** Evidence that serotonin is involved in prolactin release by electrical stimulation of the medial basal hypothalamus in the rhesus monkey, *Neuroendocrinology,* 31, 256, 1980.

203. **Dilman, V. M. and Anisimov, V. N.,** Effect of treatment with phenformin, diphenylhydantoin or L-DOPA in life span and tumor incidence in C3H/SN mice, *Gerontolgy,* 26, 241, 1980.

204. **Quadri, S. K., Clark, J. L., and Meites, J.,** Effect of LSD, pargyline and haloperidol on mammary tumor growth in rats, *Proc. Soc. Exp. Biol. Med.,* 142, 22, 1973.

205. **Quadri, S. K., Kledzck, G. S., and Meites, J.,** Enhanced regression of DMBA-induced mammary cancers in rats by combination of ergocormine with ovariectomy or high doses of estrogen, *Cancer Res.,* 34, 499, 1974.

206. **Quadri, S. K., Lu, K. H., and Meites, J.,** Ergot-induced inhibition of pituitary tumor growth in rats, *Science,* 176, 417, 1972.
207. **Segal, P. E. and Timiras, P. S.,** Patho-physiologic findings after chronic tryptophan deficiency in rats: a model for delayed growth and aging, *Mech. Ageing Dev.,* 5, 1672, 1976.
208. **Everett, J. W.,** Reinstatement of estrous cycles in middle-aged spontaneously persistent estrous rats: importance of circulating prolactin and the resulting facilitative action of progesterone, *Endocrinology,* 106, 1691, 1980.
209. **Halberg, J., Halberg, E., Regal, P., and Halberg, F.,** Changes with age characterize circadian rhythm in telemetered core temperature of stroke-prone rats, *J. Gerontol.,* 36, 28, 1981.
210. **Mosko, S. S. and Moore, R. Y.,** Aging of circadian rhythms in female rats, *Soc. Neurosci. Abstr.,* 5, 29, 1979.

MAMMARY CELLS

Charles W. Daniel, Gary B. Silberstein, and Phyllis Strickland

INTRODUCTION

In interpreting in vitro studies on cellular aging, a central concern is to what extent aging in culture reflects the gradual accumulation of cellular changes resulting from cultureal conditions such as media composition, unusual substrates, and the lack of normal cell associations. For this reason, it is important to consider in vivo alternatives to tissue culture, and a number of animal model systems are available for such studies.[1] The most desirable experimental protocol, in terms of providing a meaningful in vivo corollary to continuous cell culture, is the serial transfer of cells or tissues between animals in such a manner that optimal conditions are provided for survival, proliferation, and function. This necessarily places certain limitations and requirements on the species and the types of cells that can be used for such studies. Some of the factors to be considered are

1. Histocompatibility matching between donor and host
2. Ability to distinguish between graft and host tissue
3. Suitable nature of transplant site (as monitored by establishment of transplant and functional competence of grafted tissue)
4. Availability of sufficient free space for growth and the lack of competitive restrictions on cell division
5. Methods of quantifying growth and of determining deviations from normal.

Mouse mammary epithelium is unusually well suited for the study of serially aged tissue because of growth characteristics of the gland and the transplantation techniques which are available.

PROPAGATION OF MAMMARY CELLS IN VIVO

Transplantation

In connection with studies on mammary tumors and preneoplastic tissues, De Ome et al.[2] devised a technique for the transplantation of mammary epithelial tissues between syngeneic mice. The procedure is based upon the observation that in prepubertal female mice of approximately 3 weeks of age, the rudimentary gland consists only of the nipple and a network of short ductal elements. The fatty stromal tissue, termed the fat pad, is well developed, however, and by surgically removing the epithelial component the remaining portion of the fat becomes available as a site for implantation of mammary epithelial transplants.

Female mice 3 weeks of age are anesthetized, short incisions are made in the ventral skin, and the inguinal (No. 4) mammary fat pads are exposed (Figure 1A). The cautery is used to destroy the nipple area (the origin of mammary ductal growth) and afferent and efferent blood vessels (Figure 1B). The mammary parenchyma is removed, and the resulting gland-free fat pads are available for immediate transplantation, or may be used at any time thereafter. It has been demonstrated by Soemarwoto and Bern[3] that this surgery does not interfere with circulation to the remaining portion of the fat pad. The procedure has no long-term effect on the hosts.

Transplants usually consist of 0.5-mm pieces of mammary gland that are removed either from a primary source or from a previously transplanted outgrowth. In some cases the donor mouse is given the intraperitoneal injection of 0.05% trypan blue 4 hr before removal of

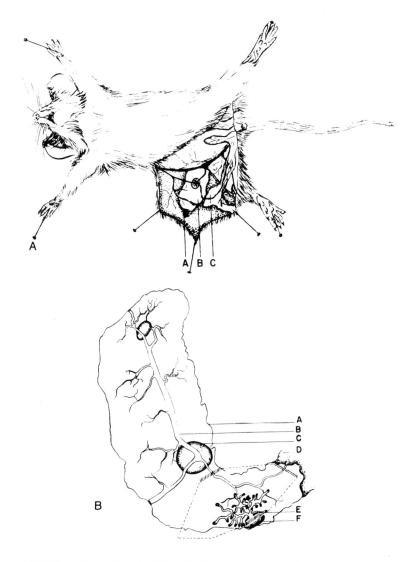

FIGURE 1.   (A) A 3-week-old female C3H mouse prepared for the removal of the mammary gland elements from the right No. 4 fat pad. (A) nipple area; (B) right No. 4 fat pad; (C) right No. 5 fat pad. (B) Drawing of a right No. 4 fat pad from a 3-week-old female mouse. The blood vessels, fat pad, and nipple area were cauterized along the slant lines. The fat pad and the surrounding connective tissue bounded by the dashed line were removed with fine scissors. (A) Boundary line of No. 4 fat pad; (B) large vein; (C) inguinal lymph node; (D) portion of No. 5 fat pad; (E) branching ducts of the No. 4 mammary gland; (F) nipple area. (From De Ome, F. B., Faulkin, L. J., Bern, H. A., and Blair, P. B., *Cancer Res.,* 19, 515, 1959. With permission.)

the transplants,[4] a procedure which makes the mammary gland more easily visible within the semiopaque fat. Transplants are placed in a small incision in the center of the hosts' No. 4 gland free fat pad.

In primary transplants DNA synthesis begins within hours of transplantation[5] and end buds (EB) are visible within 3 to 4 days of transplantation. Primary transplants placed in young or adult hosts characteristically grow to fill the available fat pad within 8 to 12 weeks. These outgrowths are normal in appearance and are indistinguishable from samples of the hosts' own glands. The ability of these outgrowths to complete the cycle of mammary

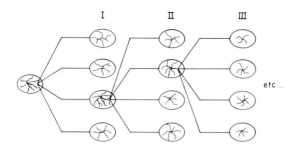

FIGURE 2. Serial transplantation of mouse mammary tissues. Primary implants are removed from a single donor and transplanted into 10 to 14 gland-free fat pads (generation I). Subsequent transplants are taken from the most vigorously growing outgrowth of the preceding generation. (From Daniel, C. W., Aidells, B. D., Medina, D., and Faulkin, L. J., Jr., *Proc. Fed. Am. Soc. Exp. Biol.*, 34, 64, 1975. With permission.)

development is normal. In addition to the formation of typical ducts in virgin hosts, the outgrowths display normal lobuloalveolar development in response to pregnancy. During lactation these lobules become secretory and, in the absence of a nipple connection, are engorged with milk.

Serial Propagation

In preparation for subsequent transplantations all host mice from the preceding transplant generation are anesthetized, and the previously transplanted fat pads are exposed and examined.[6] An outgrowth is selected for continued propagation — in most experiments this is the one showing the most vigorous growth (Figure 2). After all hosts for the new generation have received transplants, all No. 4 glands and a portion of the hosts' thoracic gland are removed, fixed in Tellyesniczky fixative,[7] extracted in acetone, stained with hematoxylin, dehydrated, and examined in methyl salicylate. The resulting preparation is optically clear, and even very subtle details of mammary growth and morphology are clearly visible. The intervals between transplant generations may vary between different experiments but is kept constant for any one serial line.

Growth Measurements

The serial propagation of the mammary epithelium at 2- or 3-month intervals allows opportunity for young, rapidly growing mammary grafts to fill the available fat pad, and standardization of the time interval between transplant generations provides a useful parameter for evaluation of growth potential. The stained No. 4 fat pads from each generation are examined under a dissecting microscope, and the amount of available fat pad filled by the outgrowth (to the nearest 10%) is reported. The mean value for percent fat pad filled of all glands in a single generation characterizes the growth potential. Only fat pads containing no mammary epithelium are excluded, since this is assumed to indicate transplant failure. This method of evaluating growth potential is comparable to the use of population density used to characterize cell culture lines in in vitro studies. A statistically significant decline in percent fat pad filled with successive generations indicates a decline in growth rate.

Since EBs represent the growth points of mammary ducts, the rate of growth is a function of both the number of EBs per gland and the growth rate of individual EBs. When evaluating growth rate over relatively short periods of time (one or two weeks) the number of EBs has proven to be an accurate measure of growth rate, that correlates well with other parameters of growth such as total number of EB cells and total number of ^{3}H-thymidine labeled EB cells per gland.

DNA Synthesis

The mammary gland grows by a process of ductal elongation in which mitotic activity is mainly confined to the growing tips, or end buds.[8] This developmental pattern is clearly seen in ^{14}C-thymidine autoradiographs made from whole-mount preparations, in which structural details may be related to DNA synthetic activity.[9] More detailed autoradiographic analyses may be carried out on sectioned gland from mice previously injected with 1 to 4 μCi ^{3}H-thymidine/g body weight. Cell-cycle analysis has also been described using squashes of primary end buds double-labeled with ^{3}H and ^{14}C thymidine.[10] However, the labeling index varies greatly between end buds in the same gland, thus making cell cycle analysis difficult by requiring large sample sizes for [3]H-[14]Ci double-labeling studies. This variation is probably the result of local influences involved in the spacing of ductal elements.[11] End buds also exhibit diurnal fluctuations in DNA synthesis and mitotic index.[12] Such fluctuations preclude the use of the wave of labeled mitoses method which assumes a uniform progression of cells through the cell cycle.[13] No published information is available on DNA synthetic activity or cell cycle characteristics of mammary cells aged during serial transplantation.

Morphology

Whole mount preparations which result after fixing, staining, and clearing the glands are optically clear and make apparent even subtle variations in ductal morphology. This makes it possible to evaluate the influence of factors within the fat pad affecting growth such as spacing of ducts, lymph node avoidance, relative size of end buds and their orientation. It also makes possible the precise orientation of individual glandular structures when preparing sectioned material for light microscope examination. A whole-mount technique has recently been developed for preparing similar samples for transmission electron microscopy,[14] an important advantage when dealing with a structure as complicated as the mammary duct system.

Elvax 40P Implants

In conjunction with a study of factors affecting normal mammary ductal morphology, Silberstein and Daniel[15] used a plastic polymer, Elvax 40P®, to slowly release bioactive molecules to localized areas of the fat pad.(Elvax 40P: ethylene-vinyl acetate copolymer; 40% vinyl acetate by weight; Dupont Chemical Co.) The implants resulted in a very local, non-systemic response, that allows evaluation of primary tissue response, to a wide variety of molecules. In the mammary gland Elvax has been used to deliver hormones, growth factors, enzymes, cyclic nucleotides, prostaglandins,and inhibitors to these agents to very specific areas of the gland. Using intact 5-week or Generation I females with actively growing EBs., factors affecting normal growth and morphology can be studied. Ovariectomized virgin mice, in which the end buds have regressed and the ducts are spindly and mitotically quiescent, are used to investigate the biomolecules involved in initiation of ductal growth.[16]

Elvax implants have been used to study serially aged mammary gland.[17] When this implant technique was used to release cholera toxin (which elevates the intracellular levels of cAMP) locally to mitotically senescent mammary outgrowths, new EBs and increased DNA synthesis resulted. This temporary reversal of the senescent phenotype suggests mammary aging results from specific changes in cell regulation rather than a generalized cellular deterioration.

Elvax 40P is washed for 1 week in several changes of 95% ethanol with continuous stirring, after which it is dissolved in dichloro methane to give a 10% solution (w/v). Dry chemical is dispersed in the Elvax solution and the mixture quick-frozen in an acetone-dry-ice bath for 10 min. The frozen pellet is transferred to a vial and kept for 2 days at −20°C, after which it is dried for 2 more days at room temperature under mild vacuum. The dried pellet is weighed and cut into small pieces of a weight that will deliver a specific amount of chemical to the implant site. A small, regular-shaped pellet of Elvax is implanted adjacent

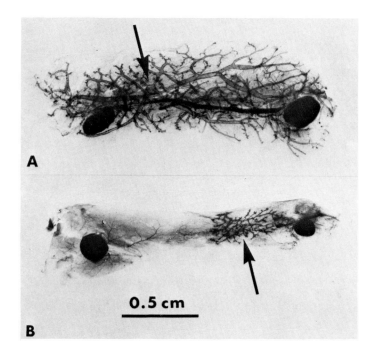

FIGURE 3.   Representative outgrowths of serially transplanted mammary tissues. Darkly staining oval structures at ends of fat pads are lymph nodes. Center of outgrowths is indicated by arrows. (A) Young outgrowth from virgin host, transplant generation I. (B) Old outgrowth from virgin host, transplant generation VI (age 24 months), which has filled only a fraction of the fat pad. (Glands stained with hematoxylin; all photomicrographs, magnification ×3.) (From Daniel, C. W., Young, L. J. T., Medina, D., and De Ome, F. B., *Exp. Gerontol.*, 6, 95, 1971. With permission.)

to the area of interest. Elvax blanks, or Elvax containing an inactive material such as bovine serum albumin can be used as a control in the contralateral glands. Elvax produces no inflammatory response. Pellets are inserted into small pockets in the fat pad made with fine-tipped jeweler's forceps, then gently pressing the opening shut. Implants are left in place for three to ten days. The glands are then excised, fixed, and stained for wholemount examination as earlier described. Thirty minutes prior to removing glands mice can receive an i.p. injection of 100 μCi ³H-thymidine (79 Ci/mmol), enabling sectioned material to be examined autoradiographically for patterns of DNA synthesis.

GROWTH SPAN

The serial transplantation of mammary tissue was first reported by Hoshino and Gardner,[18] who obtained a maximum life span of 3 years and 9.5 months (transplant generation 7); it was concluded that mammary parenchyma can survive indefinitely if conditions are favorable. It is important to note that standard transplant intervals were not used, and one host was 812 days old at necropsy.

Different results were reported by Daniel et al.[1,6] In a large-scale series of experiments, mammary tissue was passaged in young mice using constant intervals of 2 or 3 months. When growth rate, rather than simple survival, was measured by estimating percent fat pad filled, serial transplantation invariably led to a decline in proliferative capacity. After a number of serial transfers, the tissue became progressively unable to fill the available fat pad, and eventually the lines were lost due to difficulty in identification and in transplantation of the minute, slowly proliferating outgrowths (Figure 3).

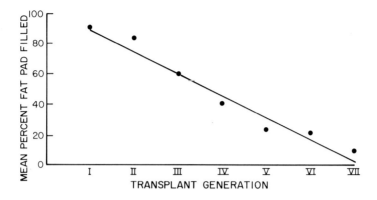

FIGURE 4.   Decline in growth rate of mammary tissue during serial propagation. This plot summarized results of eight transplant lines, and each point represents 80 to 100 transplants. The slope indicates a 15% loss of growth potential at each passage (line fitted by method of least squares). (From Daniel, C. W., Aidells, B. D., Medina, D., and Faulkin, L. J., Jr., *Fed. Am. Soc. Exp. Biol.*, 34, 64, 1975. With permission.)

Table 1
SERIAL TRANSPLANTATION OF MOUSE MAMMARY GLAND[a]

Expt. No.	Strain	Age at Transplantation (wk)	Reproductive state of Host[a]	Trypan blue used	Tissue	Total time carried (mo.)	Total generations carried
1	BALB/c	12—20	LP	No	Lobule	6	4
2	C57	3	LP	Yes	Lobule	11	4
3	C57	3	V	Yes	Duct	11	4
4	BALB/c	3	V	Yes	Duct	24	7
5	C3H	3	LP	No	Lobule	12	6

[a]   v = virgin; LP = lactating or pregnant.

After Daniel, C. W., De Ome, K. B., Young, J. T., Blair, P. B., and Faulkin, L. J., Proc. Natl. Acad. Sci. U.S.A., 61, 52, 1968. With permission.

Data from typical experiments is summarized in Figure 4. The decline in growth occurred at the rate of approximately 15% in each passage and was linear. This reduced proliferation inevitably occurred in response to serial passage even though conditions for growth were optimal, and careful selection was exercised for the most vigorously proliferating outgrowths. Growth span is independent of mouse strain, is not related to the presence of Mammary Tumor Virus, and is not affected by the use of Trypan Blue (Table 1). It is important to note that these results were obtained with tissue maintained under optimal conditions. Non-physiological experiences, such as X-irradiation[19] and even short-term primary cell culture[20] can drastically reduce growth rate.

FACTORS INFLUENCING GROWTH SPAN

Chronological Time and Cell Division

Is the limited span of somatic cells attributable to a finite and specified number of potential cell doublings, or can the passage of time account for cell senescence independently of mitotic activity? Two experimental designs employing mammary cells have been used in

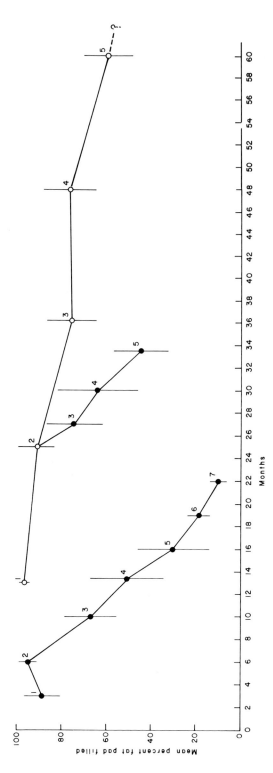

FIGURE 5. Serial transplants of mouse mammary gland at long and short transplant intervals. Two transplant lines were initiated from a single donor at time 0 and transplanted at intervals of 12 months (○ — ○) or 3 months (● — ●). At 24 months, a second short-interval line was split from the 12-month series. Vertical lines represent 95% confidence intervals. (From Daniel, C. W. and Young, L. J. T., *Exp. Cell. Res.*, 65, 27, 1971. With permission.)

attempts to answer these question.[21] The first experiment was based upon observation that 2 to 3 months is required for primary transplants to fill the fat pad. Two transplant lines were initiated from a single donor. One line was transplanted at intervals of approximately 3 months, and the tissue was thereby maintained in a state of nearly continuous growth. The other line was transplanted at yearly intervals, with growth taking place mainly at the beginning of each generation. During most of the 12 month interval the fat pad was filled, and although the tissue was metabolically active, proliferation was inhibited; this situation is comparable to confluency in monolayer cultures. Mammary tissue transplanted at 3-month intervals showed a progressive decline in growth until the line was lost at 22 months (Figure 5), while tissue transplanted at yearly intervals was growing vigorously at 24 months. At this time the line was split, one group continuing at yearly intervals and another at 3-month intervals. The short-interval transplants displayed a decline in growth that approximately paralleled that seen in the short-interval line lost at 22 months.[21,22].

Although these data strongly suggest that decline in growth rate occurs most rapidly under conditions that allow unrestricted proliferation, the result could also be interpreted as a response to the trauma of transplantation, which was more frequent in the short-interval line. A second experiment was based on the observation that ductal growths arising from mammary transplants may be considered mosaics, with tissue age varying with geographical location within the fat pad. Gland at the center of the growth is formed soon after transplantation and has experienced relatively few divisions, whereas gland toward the periphery of the outgrowth was formed later, and has a history of more mitotic events. A transplant line was initiated from a single donor. At the first passage the line was split, one group being transplanted from the center of the donor outgrowth and the other from the periphery. This pattern was continued in subsequent passages, and the transplant interval was the same for both sublines. The peripheral lines showed a substantial decrease in growth by the third transplant generation, while the line transplanted from the center continued to grow vigorously (Figure 6). These data indicate that it is the number of cell divisions experienced by mammary epithelium, rather than the passage of time, which determines its growth span.

Donor and Host Age

The mammary transplant technique is well suited to studies of the influence of donor age upon graft longevity. Young et al.[23] initiated transplant lines from 3-week-old and 26-month-old donors, which were subsequently carried in young hosts. The data (Figure 7) indicated no differences between the growth span of mammary cells taken from old and young donors. When the reciprocal experiment was performed, neither young nor old transplants grew vigorously in old hosts (Table 2). This result is expected in view of the fact that elongation of ducts in virgin females requires stimulation by mammogenic hormones released during regular estrogen cycling,[24] and slow growth of gland carried in old hosts may be due to irregular or infrequent estrous; host mice had irregular cycles ranging from 6 to 10 days or were in constant diestrous.

Hormonal Influences

The mammary gland is an endocrine target organ whose cycles of growth, differentiation, and secretion are controlled by the complex interaction of many hormones.[24] It would not be unexpected that different endocrine environments might influence the long-term growth and ultimate lifespan of serially transplanted mammary epithelium. Daniel[25] used pituitary isografts[26] to provide mammogenic stimulation during the entire time-course of serial passage. Outgrowths from late passages showed a decline in ductal elongation and, in addition, alveoli were infrequent or lacking altogether. Thus in the site of alveolar-mammogenic stimulation, these slow-growing tissues consisted of a simple network of ducts that resembled outgrowths from unstimulated transplants.

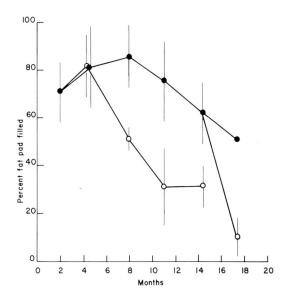

FIGURE 6.   At time 0 mice were transplanted with gland from a single 12-week donor. At the first transplant generation (3 months) transplants were removed either from the center (● — ●) or from peripheral areas (○ — ○) of the outgrowth. In subsequent generations transplants continued to be taken either from the center or periphery. At 14 months, the center-propagated line was split into a second peripheral-generated group. (From Daniel, C. W., Young, L. J. T., Medina, D., and De Ome, K. B., *Exp. Gerontol.*, 6, 95, 1971. With permission.)

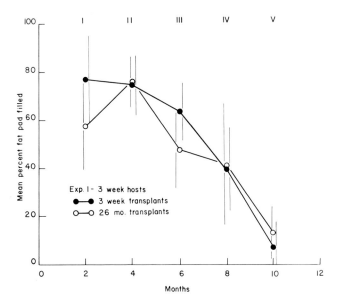

FIGURE 7.   Growth of mammary transplants from young and old donors serially passaged in young hosts; vertical lines indicate 95% confidence intervals. (From Young, L. J. T., Medina, D., De Ome, K. B., and Daniel, C. W., *Exp. Gerontol.*, 6, 49, 1971. With permission.)

In another type of experiment, mammary tissue was serially passaged in virgin hosts without supplemental pituitary stimulation until the growth rate was markedly reduced, and hosts were then placed with males and bred during the first transplant generation. The

Table 2
GROWTH OF MAMMARY TRANSPLANTS OF
VARIOUS AGES IN OLD HOST

Expt.	Host age	Age of init. transplant	Mean % fat pad filled (95% C.I.) transplant generation	
			I	II
A	25—26 months	3 weeks	37(±10)	NA
	25—26 months	25—26 months	28(±11)	NA
B	25 months	3 weeks	31(±16)	35(±48)
	25 months	29(±21)	31(±21)	31(±39)
Control				
	3 weeks	3 weeks	77(±19)	NA
	3 weeks	26 months	58(±19)	NA

After Young, L. J. T., Medina, D., De Ome, F. B., and Daniel, C. W., *Exp. Gerontol.*, 6, 49, 1971. With permission.

Table 3
EFFECTS OF HORMONAL STIMULATION ON SERIALLY PASSAGED
MAMMARY GLAND

Type of stimulation	Mammogenic effect	Response to serial passage
Virgin, continuous	Ductal proliferation	Decline in ductal proliferation
Pituitaries, isografts, continuous	Ductal and alveolar proliferation	Decline in both ductal and alveolar proliferation
Virgin, continuous, followed by one breeding in late passage	Ductal elongation; alveolar proliferation during pregnancy	Decline in ductal proliferation; alveolar proliferation normal during pregnancy

After Daniel, C. W., Young, L. J. T., Medina, D., and De Ome, K. B., *Exp. Gerontol.*, 6, 95, 1971. With permission.

outgrowths from breeding mice, although small, all displayed lobuloalveolar development and secretion comparable to that observed in the hosts' own glands. These results are summarized in Table 3. It was concluded that aging of mammary epithelium during serial passage was related to declining ability to respond to any endocrine stimulation that was constant and which continued over a long period of time. When cells had exhausted their ability to grow, in response to one stimulus, a new set of hormonal clues stimulated another round of proliferation, although along a different developmental pathway.

Neoplastic and Preneoplastic Transformations

It is well established that the genesis of mammary cancer in certain strains of mice is at least a two-step process, with benign hyperplastic alveolar nodules (HAN) as an obligatory intermediate in the transformation of normal into malignant cells.[2] HANS have altered hormonal requirements for maintenance of their alveolar cell,[24,27] such that they are maintained in nonpregnant females, under circumstances in which normal alveoli are regressed, and in which normal mammary tissue exits as a simple network of ducts.

HAN tissue may be serially transplanted, and in a long-term study the growth potential of C3H nodules was investigated.[6] The HAN lines were passaged for longer than 8 years and 30 transplant generations, and both lines were proliferating vigorously when the ex-

Table 4
SERIAL TRANSPLANTATION OF MOUSE MAMMARY NODULE OUTGROWTH LINES

Nodule line	Strain	Virus	Origin	Age of host when HAN noticed (months)	Total time carried (years)	Total generations carried
D1	BALB/c	MTV-	Retired breeder	18	10	29
D2	BALB/c	MTV-	Pituitary isografts, 11 months	18	8.5	26
D2a	BALB/c	MTV-	Pituitary isografts, 11 months	18	5.5	13
D7	BALB/c	MTV-	Pituitary isografts, 14 months	16	5.0	11
D8	BALB/c	MTV-	Pituitary isografts, 14 months	16	3.5	8
F1	C3Hf	MTV-	Retired breeder	18	4.0	13
F2	C3Hf	MTV-	Retired breeder	18	3.5	12

After Daniel, C. W., Aidells, B. D., Medina, D., and Faulkin, L. J., Jr., *Proc. Fed. Am. Soc. Exp. Biol.*, 34, 64, 1975. With permission.

periment was terminated. It was concluded that, like malignant neoplasms, HAN cells have an indefinite life span in vivo. The unlimited division potential of HANs is not influenced by the etiological agent inducing the original preneoplastic lesion[25] (Table 4). However, other dysplastic variants which arise as primary lesions or as variants within a preneoplastic population may exhibit a limited division potential. These variants include keratinized nodules, cystic alveolar nodules, and end-bud ductal hyperplasias.[29]

These studies indicated that an unlimited, or greatly extended, growth span may be associated with transformation from a normal to an intermediate preneoplastic cell type. The release from cellular aging is therefore an early event on the pathway to neoplasia and is not directly associated with malignancy. Indeed, the tumorigenic potential of many of the preneoplastic mammary tissues examined is extremely low, and carcinomas appear only very occasionally during years of serial passage.[30]

MAMMARY CELLS IN VITRO

Although many immortal cell lines derived from mammary tumors have been developed, the growth of normal, diploid mammary epithelium in culture has been disappointing and has not proved useful in aging studies, though new developments in this field appear promising.[31] Mouse mammary cells grown as monolayers on glass or plastic are capable of only limited replication, and are difficult to subcultivate. An interesting feature of such cultures, however, is their ability to form mammary outgrowths when implanted into suitable hosts.[20] It has recently been demonstrated that normal mammary cells may be cultivated in serum free collagen gels, where they proliferate as tubular structures which may be further subcultivated.[32] This method may eventually prove suitable for determining the in vitro replicative lifespan of mammary cells.

SUMMARY

Mouse mammary cells may be serially propagated in vivo until their potential for replication has been lost. In this chapter data is presented which characterizes this decline in proliferative capacity, and factors which influence it are reviewed.

REFERENCES

1. **Daniel, C. W.,** Cell longevity *in vivo,* in *Handbook on the Biology of Aging,* Finch, C. E. and Hayflick, L., Eds., Van Nostrand Reinhold Company, New York, 1977, chap 6.
2. **De Ome, K. B., Faulkin, L. J., Bern, H. A., and Blair, P. B.,** Development of mammary tumors from hyperplastic alveolar nodules transplanted into gland-free mammary fat pads of female C3H mice, *Cancer Res.,* 19, 515, 1959.
3. **Soemarwoto, I. N. and Bern, H. A.,** The effect of hormones on the vascular pattern of the mouse mammary gland, *Am. J. Anat.,* 103, 403, 1958.
4. **Hoshino, K.,** Morphogenesis and growth potentiality of mammary glands in mice. II. Quantitative transplantation of mammary glands of normal male mice, *J. Natl. Cancer Inst.,* 30, 585, 1963.
5. **Williams, M. F. and Hoshino, K.,** Early histogenesis of transplanted mouse mammary glands. I. Within 21 days following isografting, *Z. Anat. Entwickl.-Gesch.,* 132, 305, 1970.
6. **Daniel, C. W., De Ome, K. B., Young, J. T., Blair, P. B., and Faulkin, L. J.,** The *in vitro* life span of normal and preneoplastic mouse mammary glands: A serial transplantation study, *Proc. Natl. Acad. Sci. USA,* 61, 52, 1968.
7. **Lillie, R. D.,** *Histopathologic Technical and Practical Histochemistry,* 3rd ed., Blakiston, New York, 1965.
8. **Bresciani, F.,** Topography of DNA synthesis in the mammary gland of C3H mouse and its control of ovarian hormones: An autoradiographic study, *Cell Tissue Kinetics,* 1, 51, 1968.
9. **Daniel, C. W.,** Regulation of cell division in aging mouse mammary epithelium, in *Explorations of Aging,* Cristofalo, V. J. and Roberts, J., Eds., Plenum Press, New York, 1975.
10. **Bresciani, F.,** Effect of ovarian hormones on duration of DNA synthesis in cells of the C3H mouse mammary gland, *Exp. Cell Res.,* 38, 13, 1965.
11. **Faulkin, L. J., Jr., and De Ome, K. B.,** Regulation of growth and spacing of gland elements in the mammary at pad of the C3H mouse, *J. Natl. Cancer Inst.,* 24, 953, 1960.
12. **Berger, J. J. and Daniel, C. W.,** Diurnal rhythms in developing ducts of the mouse mammary gland, *J. Exp. Zool.,* 224, 115, 1982.
13. **Clausen, O. P. F., Thorud, E., and Aarnaes, E.,** Evidence of rapid and slow progression of cells through G_2 phase in mouse epidermis: A comparison between phase durations measured by different methods, *Cell Tissue Kinet.,* 14, 227, 1981.
14. **Strum, J. M.,** A mammary gland whole mount technique that preserves cell fine structure for electron microscopy, *J. Histochem. and Cytochem.,* 27, 1271, 1979.
15. **Silberstein, G. B. and Daniel, C. W.,** Elvax 40P implants: sustained, local release of bioactive molecules influencing mammary ductal development, *Dev. Biol.,* 93, 272, 1982.
16. **Silberstein, G. B., Strickland, P., Trumpbour, V., Coleman, S., and Daniel, C. W.,** *In vivo,* cAMP stimulates growth and morphogenesis of mouse mammary ducts, *Proc. Natl. Acad. Sci. USA,* in press.
17. **Daniel C. W., Silberstein, G. B., and Strickland, P.,** Reinitiation of growth and normal morphogenesis in senescent mouse mammary gland, *Science,* in press.
18. **Hoshino, K. and Gardner, W. U.,** Transplantability and life span of mammary gland during serial transplantation in mice, *Nature,* 213, 193, 1967.
19. **Faulkin, L. J., Jr.,** Mammary changes resulting from X-ray and chemical carcinogens, Ph.D. Thesis, University of California, Berkeley, 1964.
20. **Daniel, C. W. and De Ome, K. B.,** Growth of mouse mammary gland *in vivo* after monolayer culture, *Science,* 149, 634, 1965.
21. **Daniel, C. W. and Young, L. J. T.,** Life span of mouse mammary epithelium during serial propagation *in vivo:* influence of cell division on an aging process, *Exp. Cell Res.,* 65, 27, 1971.
22. **Daniel, C. W.,** Finite growth span of mouse mammary gland serially propagated *in vivo, Experientia,* 29, 1422, 1973.
23. **Young, L. J. T., Medina, D., De Ome, K. B., and Daniel, C. W.,** The influence of host and tissue age on life span and growth rate of serially transplanted mouse mammary gland, *Exp. Geront.,* 6, 49, 1971.
24. **Nandi, S.,** Endocrine control of mammary gland development and function in the C3H/He Crg1 mouse, *J. Natl. Cancer Inst.,* 21, 1039, 1958.
25. **Daniel, C. W., Young, L. J. T., Medina, D., and De Ome, K. B.,** The influence of mammogenic hormones on serially transplanted mouse mammary gland, *Exp. Gerontol.,* 6, 95, 1971.
26. **Loeb, J. and Kirtz, M. M.,** The effects of transplants of anterior lobes of the hypophysis on the growth of the mammary gland and on the development of mammary gland carcinoma in various strains of mice, *Am. J. Cancer,* 36, 56, 1939.
27. **Nandi, S.,** Effect of hormones on maintenance of hyperplastic alveolar nodules in mammary glands of various strains of mice, *J. Natl. Cancer Inst.,* 27, 187, 1961.

28. **Daniel, C. W., Aidells, B. D., Medina, D., and Faulkin, L. J., Jr.,** Unlimited division potential of precancerous mouse mammary cells after spontaneous or carcinogen-induced transformation, *Proc. FASEB,* 34, 64, 1975.

29. **Medina, D.,** Preneoplasia in breast cancer, in *Breast Cancer 2. Advances in Research and Treatment,* McGuire, W., Ed., Plenum Medical Book Co., New York, 47, 1978.

30. **Medina, D.,** Mammary tumorigenesis in chemical carcinogen-treated mice. VI. Tumor-producing capabilities of mammary dysplasias in BALB/c Crg1 mice, *J. Natl. Cancer Inst.,* 57, 1185, 1976.

31. **Danielson, K. G., Oborn, C. J., Durbam, E. M., Butel, J. S., and Medina, D.,** An epithelial mouse mammary cell line exhibiting normal morphogenesis *in vivo* and functional differentiation *in vitro, Proc. Nat. Acad. Sci. USA,* in press.

32. **Yang, J., Larson, L., Flynn, D., Elias, J., and Nandi, S.,** Serum-free primary culture of human normal mammary epithelial cells in collagen gel matrix, *Cell Biol. Intl. Rep.,* 6(10), 969, 1982.

CULTURED ENDOTHELIAL CELLS AS AN IN VITRO MODEL SYSTEM

Elliot M. Levine and Stephen N. Mueller

INTRODUCTION AND SOURCES

In vivo the vascular endothelium is comprised of a monolayer of epithelial cells forming the luminal surfaces of the circulatory and lymphatic systems.[71] It plays a crucial role in the normal physiologic functioning of higher organisms, and has been implicated in the etiology of certain age-related disorders in man, such as atherosclerosis and tumor metastasis.[75] As a model cultured cell system, endothelial cells are considered particularly advantageous for several reasons.

First, endothelial cells from many species can be obtained from various anatomic sites. Second, in contrast to other commonly studied cultured cell types such as lung and embryo fibroblast-like cells, endothelial cells have a clearly defined histologic origin. In addition, sufficient numbers of endothelial cells can be obtained for the study of primary cultures that are nearly homogeneous. Finally, in analogy to the human diploid fibroblast system that has been studied for many years, serially cultivable endothelial cells can be obtained that retain stringent growth control, normal karyotypes, and finite life-spans. The endothelial cell system has the further advantage of expressing certain differentiated endothelial specific functions under culture conditions.

These advantages of the endothelial cell system have led in recent years to an impressive number of successful isolations of endothelial cells from many different anatomic sites in various animal species (Table 1), although the cells isolated have not been serially cultivated in all cases. Not only have cells been isolated from major venous and arterial vessels, but also, most recently, from the capillary beds of various tissues and organs. Because cultured capillary endothelial cells have only just recently begun to be studied and require more complicated isolation procedures, this review will be limited to a discussion of cells cultured from "large vessel" endothelium.

The isolation of relatively pure populations of endothelial cells for primary cultures is possible, in part, because of the histologic architecture of major blood vessels. The endothelium forms a cellular monolayer lining the vessel lumen, and rests on an acellular matrix containing collagen and other components that separate it from the underlying layer, the media, which contains smooth muscle cells and perhaps some fibroblast-like cells. This separation of the endothelial layer from the medial smooth muscle layer permits detachment of endothelial cells by various means without major contamination by the other cell types.

ISOLATION METHODS

Several different methodologies have evolved over the years to release the endothelium and to isolate endothelial cells. The most commonly used are listed in Table 2. Based on the collagenous composition of the matrix, enzymatic treatments utilizing collagenase, or in some cases, trypsin, EDTA, or thrombin have been used to detach the endothelium. The vessel can be incubated with enzyme solution under either static or perfusion conditions. The endothelium also can be detached by purely mechanical means, such as scalpel blades, steel wool or cotton pledgets, and the "hautchen" method. Of course, any combination of enzymatic and mechanical methods also may be used. Finally, the vessel may be subjected to these treatments *in situ,* dissected free of the animal but intact, everted over a steel rod, or slit longitudinally and incubated "open-faced".

Table 1
**SUCCESSFUL ISOLATION (AND PROPAGATION) OF
ENDOTHELIAL CELLS FROM VARIOUS SPECIES AND
SITES**

Species (fetal and adult)	Ref.
Bovine	3,10,14—21,62,63
Chicken	45
Dog	27,28,64
Guinea pig	26,39,40
Human	1—9,11—13
Mouse	3,44
Pig	26—39
Rabbit	2,22—26,65
Rat	3,41—43,68,69
Sheep	69
Site — "Large Vessels"	
Aortic (thoracic and abdominal)	14—18,21,29,39,40,47—49,62,63
Carotid	68
Heart (endocardial)	26,42
Iliac	68
Jugular	28,64,68
Mesenteric	68
Ovarian	48
Portal	40,49
Pulmonary	19,26,33,37,41,48,60
Saphenous	16
Umbilical	4—8,11,12,49
Vena cava	28,64,68
Site — Capillaries	
Adrenal	3
Cerebral	21,43,44,51,52,61
Dermis	1,2,46,54,58
Fat pads	59
Foreskin	1,2,9,54—57
Kidney	3
Lung	3,26
Mycocardial	3,53
Spleen	3

Table 2
METHODS FOR HARVESTING ENDOTHELIAL CELLS

1. Enzymatic or chemical agents (static incubation or perfusion[26])
    a. trypsin[4,11,12,29,42,45,48]
    b. collagenase[5,8,14,18,29,29,32,48]
    c. EDTA[10]
    d. thrombin[66]
    e. balanced salts solution[67]
    f. combination of the above[28,29,30,40,65]
2. "Mechanical scraping"
    a. scalpel,[11,29] director[17]
    b. steel wool pledget,[27] cotton swab[17]
    c. "hautchen" preparations[68-71]
3. Combination of 1 and 2[29,11]

Note: In most cases the vessel is dissected free and filled with the appropriate solutions, but it may be everted over a steel rod,[16,24,28,39,40,64] treated "open-faced",[10,17,27,29,68-71] or perfused *in situ*.[26]

The collagenase method employed in our laboratory,[62,63] to isolate endothelial cells from fetal bovine aortas is based on procedures previously described.[5,8,21] Under sterile conditions, the dissected vessel with its intervertebral branches clamped off is attached to a combination of cannulas and stopcocks that is used to facilitate filling and emptying the vessel with the appropriate solutions. A sterile collagenase solution is introduced using a syringe through the stopcock. When the vessel is filled, the stopcocks are closed and the entire set-up is incubated for 10 minutes at 37°. The collagenase solution is removed gently, and the vessel is flushed with 10 mℓ of growth medium, rapidly delivered from a syringe. This treatment almost completely detaches the endothelium, and yields 2 to 4 $\times$ 10^5 cells from a 4-cm segment of vessel. The endothelium is detached in sheets or clumps of cells that adhere to the culture flask surface and begin to migrate within 6 to 10 hr after isolation. At 24 to 36 hours after isolation, most, if not all, of the cells will have migrated out from the clumps. After about one week in culture, the cells from a 4-cm aortic segment will have migrated and grown to cover 25 to 50 cm^2 of flask surface. The culture morphology in primary cultures is the classic cobblestone array of nonolayered cells that is reminiscent of the intact endothelium in vivo. There is a secondary growth pattern observed at times in some cultures (including clones) in which some endothelial cells grow under the intact monolayer; this has been termed "sprouting."[21,47,62,73] In most instances, however, endothelial cultures exhibit a particularly stringent form of growth regulation, in which refeeding confluent cultures with fresh medium and serum does not result in an increase in cell density.[7]

The purpose of all these isolation procedures is to obtain highly enriched primary cultures of endothelial cells that are as free as possible of other cell types. The most common contaminant is the medial smooth muscle cell; disruption of the endothelium by trauma during isolation is thought to contribute to the release of smooth muscle cells.[2,4,8,14,18,71,72] Although much recent work involves the study of cell lines and strains derived from primary cultures, it should be emphasized that studying primary cultures is also important. Although in primary cultures, some cellular changes may have already taken place in cells that have adapted to in vitro conditions, primaries are the closest benchmark available for comparison to the endothelium in vivo. Three different routes can be followed toward the goal of obtaining homogeneous serially passaged cells (Figure 1). Primary cultures can be cloned (A), cultured under selective conditions (B), or passaged without imposing any selective pressures (C). Most workers (see References 71 and 72 for reviews) have found that the latter approach (culture under nonselective conditions) ultimately leads to the overgrowth of the endothelial cells by contaminating vascular cell types, and, in most instances, this contaminant is the smooth muscle cell. As mentioned previously, confluent endothelial cells are refractory to a fresh serum stimulus,[7] but fresh serum is mitogenic for smooth muscle cells.[74] Therefore, repeated exposure of endothelial cultures containing smooth muscle cells to fresh medium and serum during subculture ultimately results in smooth muscle cell overgrowth. This problem of overgrowth has made the issue of endothelial cell identification a central one in attempts at developing pure endothelial cell strains and lines.

IDENTIFICATION CRITERIA

Unlike many other cultured cell systems in which in vitro senescence has been studied, endothelial cells can be distinguished by certain identifying characteristics (Table 3) from other cell types most likely to contaminate endothelial isolates (i.e., smooth muscle cells and fibroblast-like cells). Some of these differences are more useful than others in establishing endothelial identity. For instance, although there are marked differences in culture morphology at confluence, these differences are not so apparent in sparse cultures or when mixed cultures of two cell types are observed. This also is true in the case of serum stimulation where mixed cultures or the sprouting growth pattern can confuse the issue. The electron

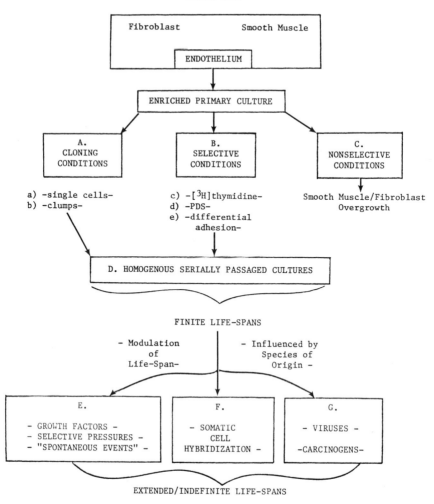

FIGURE 1.   Homogeneous serially passaged cells.

microscope can be used to detect Weibel-Palade bodies, a cytoplasmic organelle found only in endothelial cells.[71,72] In this case, the problems are small sample size and the fact that this organelle is not always present in all endothelial cells. Angiotensin-converting enzyme is present at very high levels of activity in endothelial, but not smooth muscle or fibroblast-like cells; however, even among endothelial preparations, enzyme activity varies from culture to culture, and it is difficult to use enzyme activity alone as an identifying characteristic. Factor VIII antigen, a component of the Factor VIII coagulation complex, is found in endothelial cells but not in smooth muscle or fibroblast-like cells. The presence of Factor VIII antigen is the best available identification criterion for endothelial cells in culture. Factor VIII antigen is usually detected by indirect immunofluoresence (Figure 2).

SERIALLY CULTIVATED CULTURES

With the use of Factor VIII antigen as an identification criterion for endothelial cells, various procedures are available for achieving homogeneous serially passaged cultures. Three procedures that select for survival of endothelial cells and the inhibition or elimination of

Table 3
SOME IDENTIFYING CHARACTERISTICS OF VASCULAR CELLS[a]

Cell type	Morphology in dense cultures	Serum stimulated growth	Weibel-Palade bodies	Factor VIII antigen	Angiotensin-converting enzyme activity nmol/(hr)/(10^6 cells)[b]
Fibroblastic	Fusiform cells in whorls	Yes	Absent	Absent	Low levels (<20)
Smooth muscle	Dendritic cells in hillocks and nodules	Yes	Absent	Absent	Low levels (<20)
Endothelial	Polygonal cells in cobblestone array	No	Sometimes present	Present	High levels (200—600)[c]

[a]   Adapted from References 8, 48, 60, 72, 75.
[b]   Unpublished data using ^{3}[H] hippurylglycylglycine as a substrate.
[c]   Primary cultures and most clones.

smooth muscle cells are listed in Figure 1 under the category of selective conditions (B). First, there is the tritiated-thymidine selection procedure (C) developed by Schwartz,[21] in which the growth response of smooth muscle cells to serum stimulation is coupled with incubation with high doses of tritiated-thymidine to kill the stimulated growing smooth muscle cells. The endothelial cells are refractory to serum stimulation and thus survive this treatment. A second method (D) that results in a selective growth advantage for endothelial cells involves the use of plasma-derived serum (PDS) instead of the usual growth supplement of whole blood serum (WBS). WBS contains platelet-derived growth factor (PDGF), which is essential for the multiplication of smooth muscle cells; PDS does not contain significant amounts of PDGF, but it does support the growth of endothelial cells.[7,76,77] Therefore, passaging cells in medium containing PDS instead of WBS results in endothelial cultures free of smooth muscle contamination. A third procedure (E) exploits the fact that endothelial cells usually adhere to the culture flask surface much more rapidly than smooth muscle cells or fibroblasts.[3,19,40,48] Therefore, if at subculture only 30 to 90 minutes are allowed for cells to adhere before the medium is replaced, there is a significant selective pressure in favor of endothelial cells. All these methods can be used on primary and secondary cultures or repeatedly throughout the culture life-span. By way of caution, it is important to realize that these procedures, by definition, exert selective pressures that may also eliminate certain endothelial cell populations from the culture.

Another approach to obtaining homogeneous cultures is to resort to cloning (Figure 1A). One can either isolate distinctive clumps of endothelial cells or single cells by standard ring cloning or dilution cloning techniques. Because endothelial clumps may contain adherent smooth muscle cells, one probably should combine this clump method with culturing under selective conditions. The isolation of a single cell that can be demonstrated as endothelial is the most foolproof procedure available for generating serially cultivable homogeneous endothelial cultures. There are, however, some caveats. First, selective pressures are significant, and the single cells that survive cloning may not be representative of the in vivo population. Second, because the cell strain starts from a single cell, the culture does not contain one million cells (or cover the surface of a 25-cm² culture flask) until 20 population doublings have elapsed. This precludes many types of measurements on cells at early population doubling levels, and again, may mean that the cloned culture is significantly removed from the in vivo situation. Nevertheless, many laboratories have found clonal strains to be extremely useful.[3,17,22,40,62,63]

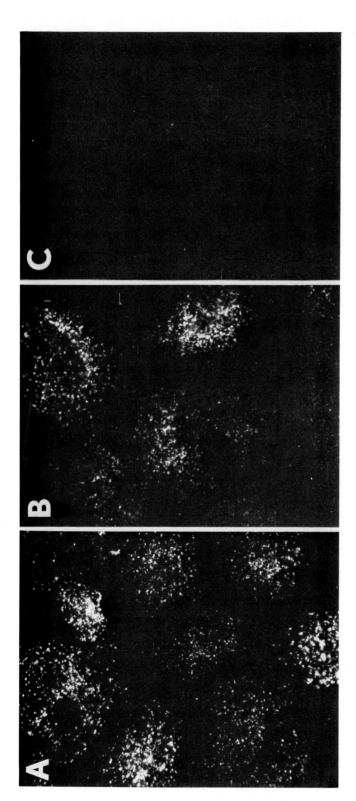

FIGURE 2.   Detection of Factor VIII antigen by indirect immunofluorescence in cloned endothelial cells. Cloned endothelial cultures at various CPDLs were grown to confluency, extracted with acetone, and incubated with rabbit anti-bovine Factor VIII antigen. 1:1000 dilution (A and B) or normal rabbit serum, 1:500 dilution (C) followed by goat anti-rabbit IgG-FITC (1:80) (see Ref. 63 for details). Panel A, Phase II culture; Panel B, Phase III culture; Panel C, Phase II culture. Magnification × 856. Note that Phase III cells had larger cell-attachment areas than did Phase II cells (see Figures 4B and C). (From Rosen, E. M., Mueller, S. N., Noveral, J. P., and Levine, E. M., *J. Cell. Physiol.*, in press. With permission.)

IN VITRO SENESCENCE

In the bovine system, most, but not all, of the cell strains and lines that have been isolated exhibit finite life-spans and in vitro cellular senescence, as has been described extensively for the human fibroblast system. Figure 3 depicts the proliferative life history of a fetal bovine aortic endothelial cell clone isolated in our laboratory.[62,63,75] At early passages after cloned cultures begin to be subcultured (i.e., Phase II; CPDL 30-60), cultures routinely proliferate rapidly from 10^4 to more than 10^5 cells in one week. After continued subculturing, the proliferation rate slows in Phase III, and the harvest cell density decreases dramatically, until at CPDL 82 (for this culture), there is no increase in cell number after a week. In fact, for as many weeks thereafter as these cultures have been kept, rapidly proliferating cells have not been detected in Phase III cultures. This phenomenon is quite reproducible in bovine aortic endothelial cells, and has been observed for almost 100 different cultures in our laboratory. An additional interesting point is that at least one differentiated function, the expression of Factor VIII antigen, occurs throughout Phase II and into Phase III (see Figure 2).

In addition to differences in proliferation, endothelial cells in Phase II and Phase III differ in other aspects. Figure 4 shows a comparison of cellular attachment sizes in Phase II and Phase III cells as revealed when cultures are prepared with silver nitrate stain. Exact measurements of individual cells from photographs such as these indicate that cellular attachment size in Phase III cells is more than three times that in Phase II cells. Other characteristics that change as the culture progresses from Phase II to Phase III are 2- to 4-fold increases in cellular protein content, rapid decreases in the growth fraction of the population measured by tritiated-thymidine labelling, and increases in the percentage of cells with chromosomal translocations.[62]

VI. MODULATION OF LIFE-SPAN

The finite life-span of normal cells in culture can be modulated by many perturbations. In Figure 1 some of these are grouped into three arbitrary categories (E, F, and G). The first, E, encompasses growth factors, unspecified selective pressures, and spontaneous events. In the case of growth factors, it has been reported that FGF, EGF, and thrombin greatly extend the life-span of human and bovine endothelial cells.[6,17] In addition, selective pressures and/or spontaneous heritable changes may be responsible for the tendency of some cell types to transform "spontaneously" into continuous lines with indefinite life-spans. This tendency is apparently dependent, to a great extent, on the animal species from which the cells are derived.[78-81] For instance, using the fibroblast system as an example, rodent cell cultures spontaneously transform into continuous cell lines almost without exception. At the other extreme, normal avian and human cells do not transform into continuous lines at all. Based on our experience, the bovine system seems to be one in which the frequency of spontaneous transformation is fairly low,[62,63] although several continuous bovine cell lines do exist.[83,84]

Somatic cell hybridization (Category F) is another useful tool for producing continuous cell lines, but this has only been explored to a limited extent with endothelial cells.[85,86] Under category G, the extension of life-span of human endothelial cells by SV40 has been reported.[87,88] In our laboratory, under the carcinogen heading, we have been studying the ability of benzo(a)pyrene treatment to produce continuous lines of endothelial cells.[82]

In summary, this overview has dealt with methods for isolating and propagating endothelial cells as clonal strains and cell lines with finite and extended life-spans. Some of the pitfalls in various approaches and the necessity for careful characterization of the cell lines being studied have been pointed out, and should be taken into consideration by the investigator planning to employ this model cell system.

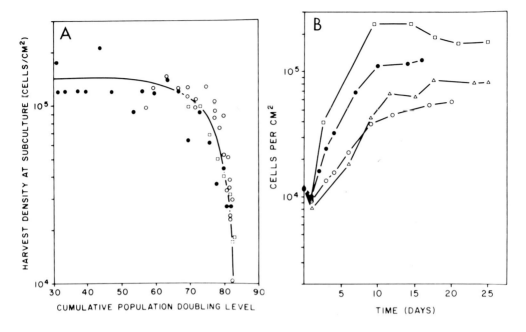

FIGURE 3.   (A) Cell density at subculture vs. CPDL (cumulative population doubling level) for endothelial cell clone BFA-1c. Data were obtained from (◯) individual A (inoculation density, 2.5×10^4 cell/cm^2), (●) individual B (1.0×10^4 cell/cm^2), and (▢) individual C (1.0×10^4 cell/cm^2). The number of population doublings (PDs) that occurred in each subculture was calculated by using the formula: PD = log$_2$[cell density at subculture/(cell density at inoculation × attachment efficiency)], where attachment efficiency was the number of cells attached to flask 24 hr after inoculation expressed as a fraction of the total number of cells inoculated (it varied from 0.70 to 0.95). The CPDL at any time, therefore, was the sum of all previously determined PDs. Weekly subculture cell density data could not be obtained until CPDL 30, because approximately 30 PDs occurred during the cloning procedure. Subcultivation was continued until inoculation cell density did not double after 2 weeks with weekly refeeding. The BFA-1c cells were frozen and stored in liquid nitrogen for various periods of time; when thawed, these cells proliferated similarly to cells that had not been frozen. All cultures remained free of detectable mycoplasma infection, as demonstrated by agar plate cultivation, fluorescent staining for cytoplasmic DNA, and immunofluorescent staining for *M. hyorhinis* antigen. (B) Growth curves as a function of the CPDL for endothelial cell clone BFA-1c. Changes in proliferative rates and plateau densities associated with cellular senescence were monitored by comparing growth curves of cell density with time for cells taken from cultures at different CPDLs. Stock 75-cm^2 cultures at the desired CPDL were trypsinized and the cells were inoculated into numbered 25-cm^2 flasks at densities of 1.0×10^4 to 1.2×10^4 cell/cm^2 (8 mℓ of medium per flask). Flasks were incubated under standard conditions and given fresh culture medium every 6 or 7 days. At various times after inoculation, duplicate cultures were selected according to a random number table and counted with a Coulter counter to determine density. Replicate cell density determinations agreed within 10%. The CPDLs studied were 55 (▢), 59 (●), 70 (△), and 79 (◯). (From Mueller, S. N., Rosen, E. M., and Levine, E. M., *Science*, 207, 889, 1980. With permission.)

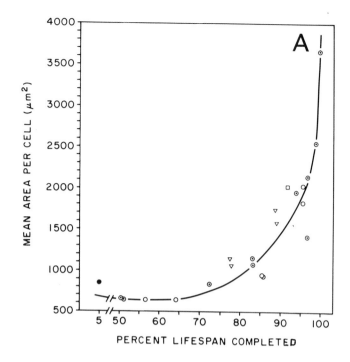

FIGURE 4. Mean cell-attachment area as a function of percent life-span completed for uncloned endothelial cultures. (A), Mean cell area of silver nitrate stained endothelial cultures was determined by ocular micrometry as described in Ref. 63. Each point represents a single determination. Data are presented for endothelial clones (⊙), BFA-1c; (□), BFA-34a; (○), BFA-37j; (▽), BFA-34m, as well as an uncloned endothelial culture at early passage (●). (B), Photomicrographs of clone BFA-1c stained with silver nitrate at CPDL 41, 51% life-span completed. (C), at CPDL 80, 100% life-span completed. Magnification × 470. (From Rosen, E. M., Mueller, S. N., Noveral, J. P., and Levine, E. M., *J. Cell. Physiol.*, in press. With permission.)

PERCENT LIFESPAN COMPLETED

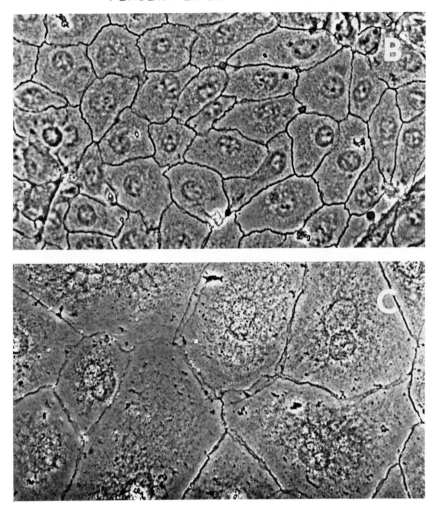

FIGURE 4B, 4C

REFERENCES

1. **Davison, P., Charlton, M., Bensch, K., and Karesek, M.,** Isolation, growth, and morphology of human skin endothelial cell *in vitro,* Abstract, Meeting of the Society for Investigative Dermatogogy, Inc., 1978.
2. **Davison, P. and Karasek, M.,** Regulation of skin endothelial cell cell growth *in vivo* and *in vitro, Clin. Res.,* 27, A136, 1979.
3. **Folkman, J., Haudenschild, C. C., and Zetter, B. R.,** Long-term culture of capillary endothelial cells, *Proc. Natl. Acad. Sci.,* 76, 5217, 1979.
4. **Fryer, D. G., Birnbaum, G., and Luttell, C. N.,** Human endothelium in cell culture, *J. Atheroscler. Res.,* 6, 151, 1966.
5. **Gimbrone, M. A., Cotran, R. S., and Folkman, J.,** Human vascular endothelial cells in culture, *J. Clin. Invest.,* 51, 46a, 1972.
6. **Gospodarowicz, D., Brown, K. D., Birdwell, C. R., and Zetter, B. R.,** Control of proliferation of human vascular endothelial cells, *J. Cell Biology,* 77, 774, 1978.

7. **Haudenschild, Chr. C., Zahniser, D., Folkman, J., and Klagsbrun, M.,** Human vascular endothelial cells in culture, *Exptl. Cell. Res.*, 98, 175, 1976.

8. **Jaffe, E. A., Nachman, R. L., Becker, C. G., Minick, C. R.** Culture of human endothelial cells derived from umbilical veins, *J. Clin. Invest.*, 52, 2745, 1973.

9. **Sherer, G. K., Fitzharris, T. P., LeRoy, E. C.,** Morphologic and functional characterization of cultured microvascular endothelial cells obtained from human preputial skin, *In Vitro*, 15, 201, 1979.

10. **Wechezak, A. R. and Mansfield, P. B.,** Isolation and growth characteristics of celllines from bovine venous endothelium, *In Vitro*, 9, 39, 1973.

11. **Lewis, L. J., Hoak, J. C., Maca, R. D., and Fry G. L.,** Replication of human endothelial cells in culture, *Science*, 181, 454, 1973.

12. **Maruyama, Y.,** The human endothelial cell in tissue culture. *Zeitschrift fur Zellforschung*, 60, 69, 1963.

13. **Davison, P. M., Bensch, K., and Karasek, M.,** Isolation and growth of endothelial cells from the microvessels of the newborn human foreskin in cell culture. *J. Invest. Derm.*, 75, 316, 1980.

14. **Booyse, F. M., Sedlak, B. J., Rafelson, M. E.,** Culture of arterial endothelial cells, *Thrombos. Diathes. haemorrh. (Stuttg.)*, 34, 825, 1975.

15. **Duthu, G. S. and Smith, J. R.,** Effects of culture media and FGF on bovine endothelial cell growth. *In Vitro*, 15, 222, 1979.

16. **Eskin, S. G., Sybers, H. D., Trevino, L., Lie, J. D., Chimoskey, J. E.,** Comparison of tissue-cultured bovine endothelial cells from aorta and saphenous vein, *In Vitro*, 14, 216, 1978.

17. **Gospodarowicz, D., Moran, J., Braun, D., and Birdwell, C.,** Clonal growth of bovine vascular endothelial cells: fibroblast growth factor as a survival agent, *Proc. Natl. Acad. Sci.*, 73, 4120, 1976.

18. **Macarak, E. J., Howard, B. V., and Kefalides, N. A.,** Properties of calf endothelial cells in culture, *Lab. Invest.*, 36, 62, 1977.

19. **Ryan, U. S., Clements, E., Habliston, D., and Ryan, J. W.,** Isolation and culture of pulmonary artery endothelial cells, *Tissue and Cell*, 10, 535, 1978.

20. **Fenselau, A. and Mello, R. J.,** Growth stimulation of cultured endothelial cells by tumor cell homogenates, *Cancer Res.*, 36, 3269, 1976.

21. **Schwartz, S. M.,** Selection and characterization of bovine aortic endothelial cells, *In Vitro*, 14, 966, 1978.

22. **Buonassisi, V. and Venter, J. C.,** Hormone and neurotransmitter receptors in an established vascular endothelial cell line, *Proc. Natl. Acad. Sci.*, 73, 1612, 1976.

23. **Colburn, P. and Buonassisi, V.,** Estrogen-binding sites in endothelial cell cultures, *Science*, 201, 817, 1978.

24. **Carnes, W. H., Abraham, P. A., and Buonassisi, V.,** Biosynthesis of elastin by an endothelial cell culture, *Biochem. and Biophys. Res. Comm.*, 90, 1393, 1979.

25. **D'Amore, P. and Shepro, D.,** Stimulation of growth and calcium influx in cultured, bovine, aortic endothelial cells by platelets and vasoactive substances, *J. Cell. Physiol.*, 92, 177, 1977.

26. **Habliston, D. L., Whitaker, C., Hart, M. A., Ryan, U. S., and Ryan, J. W.,** Isolation and culture of endothelial cells from the lungs of small animals, *Am. Rev. Respir. Dis.*, 119, 853, 1979.

27. **Herring, M., Gardner, A., and Glover, J.,** Seeding endothelium onto canine arterial prostheses, *Arch. Surg.*, 114, 679, 1979.

28. **Graham, L. M., Vinter, D. W., Ford, J. W., Kahn, R. H., Burkel, W. E., and Stanley, J. C.,** Endothelial cell seeding of prosthetic vascular grafts, *Arch. Surg.*, 115, 129, 1980.

29. **Slater, D. N. and Sloan, J. M.,** The porcine endothelial cell in tissue culture, *Atherosclerosis*, 21, 259, 1975.

30. **Atherton, A.,** Growth stimulation of endothelial cells by simultaneous culture with sarcoma 180 cells in diffusion chambers, *Cancer Res.*, 37, 3619, 1977.

31. **Barnes, M. J., Morton, L. F., and Levene, C. I.,** Synthesis of interstitial collagens by pig aortic endothelial cells in culture, *Biochem. Biophys. Res. Commun.*, 84, 646, 1978.

32. **Booyse, F. M., Quarfoot, A. J., Bell S., Fass, D. N., Lewis, J. C., Mann, K. G., and Bowie, E. J. W.,** Cultured aortic endothelial cells from pigs with von Willebrand disease: *In vitro* model for studying the molecular defects(s) of the disease, *Proc. Natl, Acad. Sci.*, 74, 5702, 1977.

33. **Dieterle, Y., Ody, C., Ehrensberger, A., Stalder, H., and Junod, A. F.,** Metabolism and uptake of adenosine triphosphate and adenosine by porcine aortic and pulmonary endothelial cells and fibroblasts in culture, *Circ. Res.*, 42, 869, 1978.

34. **De Bono, D.,** Effects of cytotoxic sera on endothelium *in vitro*, *Nature* 252, 83, 1974.

35. **Hayes, L. W., Goguen, C. A., Stevens, A. L., Magargal, W. W., and Slakey, L. L.,** Enzyme activities in endothelial cells and smooth muscle cells from swine aorta, *Proc. Natl. Acad. Sci.*, 76, 2532, 1979.

36. **Levene, C. I. and Heslop, J.,** The synthesis of collagen by cultured pig aortic endothelium and its possible role in the pathogenesis of the fibrous atherosclerotic plaque, *J. Mol. Med.*, 2, 145, 1977.

37. **Ody, C., Dieterle, Y., Wand, I., Stalder, H., and Junod, A. F.,** PGA_1 $PGF_{2\alpha}$ metabolism by pig pulmonary endothelium, smooth muscle, and fibroblasts, *J. Appl. Physiol.*, 46, 211, 1979.

38. **Pearson, J. D., Carleton, J. S., Hutchings, A., and Gordon, J. L.,** Uptake and metabolism of adenosine by pig aortic endothelial and smooth-muscle cells in culture, *Biochem. J.,* 170, 265, 1978.

39. **Blose, S. H. and Chacko, S.,** Vascular endothelium: *In vitro* differences between an artery and vein, *J. Cell. Biol.,* 63, 30a, 1974.

40. **Blose, S. H. and Chacko, S.,** *In vitro* behavior of guinea pig arterial and venous endothelial cells, *Dev. Growth Diff.,* 17, 153, 1975.

41. **Darnule, A. T., Parshley, M. S., Darnule, T. V., Likhite, V., Mandl, I., and Turino, G. M.,** Antiserum to surface antigens as a marker for cultured rat lung endothelial cells, *Immunol. Commun.,* 7, 323, 1978.

42. **Acosta, D. and Peili, C.,** Injury to primary cultures of rat heart endothelial cells by hypoxia and glucose deprivation, *In Vitro,* 15, 929, 1979.

43. **Phillips, P., Kumar, P., Kumar, S., and Waghe, M.,** Isolation and characterization of endothelial cells from rat and cow brain white matter, *J. Anat.,* 129, 261, 1979.

44. **De Bault, L. E. and Cancilla, P. A.,** γ-Glutamyl transpeptidase in isolated brain endothelial cells: induction by glial cells *in vitro, Science,* 207, 653, 1980.

45. **Murata, K., Quilligan, J. J., and Morrison, L. M.,** Growth of chick aorta endothelial cells: incorporation of tritiateduridine and thymidine, *Experientia,* 21, 637, 1965.

46. **Fleischmajer, R. and Perlish, J. S.,** [³H]thymidine labeling of dermal endothelial cells in scleroderma, *J. Invest. Dermatol.,* 69, 379, 1977.

47. **Cotta-Pereira, G., Sage, H., Bornstein, P., Ross, R., and Schwartz, S.,** Studies of morphologically atypical (''sprouting'') cultures of bovine sortic endothelial cells. Growth characteristics and connective tissue protein synthesis, *J. Cell. Physiol.,* 102, 183, 1980.

48. **Johnson, A. R.,** Human pulmonary endothelial cells in culture, *Clin. Invest.,* 65, 841, 1980.

49. **Haudenschild, C. C., Cotran, R. S., Gimbrone, M. A., Folkman, J.,** Fine structure of vascular endothelium in culture, *J. Ultrastruct. Res.,* 50, 22, 1975.

50. **Panula, P., Joo, F., and Lechardt, L.,** Evidence for the presence of viable endothelial cells in cultures derived from dissociated rat brain, *Experientia,* 34, 95, 1978.

51. **Spatz, M., Bembry, J., Dodson, R. F., Hervonen, H., and Murray, M. R.,** Endothelial cell cultures derived from isolated cerebral microvessels, *Brain Res.,* 191, 577, 1980.

52. **Williams, S. K., Gillis, J. F., Matthews, M. A., Wagner, R. C., and Bitensky, M. W.,** Isolation and characterization of brain endothelial cell morphology and enzyme activity, *J. Neurochem.,* 35, 374, 1980.

53. **Simionescu, M. and Simionescu, N.,** Isolation and characterization of endothelial cells from the heart microvasculature, *Microvas. Res.,* 16, 426, 1978.

54. **Davison, P. and Karasek, M.,** Serial cultivation of human dermal endothelium: role of serum and fibronectin, *J. Invest. Derm.,* 74, 256, 1980.

55. **Sherer, G. K., Fitzharris, T. P., Faulk, W. P., and Le Roy, E. C.,** Cultivation of microvascular endothelial cells from human preputial skin, *In Vitro,* 16, 675, 1980.

56. **Davison, P. M., Bensch, K., and Karasek, M.,** Isolation and growth of endothelial cells from the microvessels of the newborn human foreskin in cell culture, *J. Invest. Derm.,* 75, 316, 1980.

57. **Davison, P., Charlton, M., Bensch, K., and Karasek, M.,** Isolation, growth and morphology of human skin endothelial cells *in vitro, Clin. Res.,* 26, 569a, 1978.

58. **Davison, P. M., Bensch, K., and Karasek, M.,** Growth and morphology of rabbit marginal vessel endothelium in cell culture, *J. Cell. Biol.,* 85, 187, 1980.

59. **Shepro, D., Robinson, A., and Hechtman, H. B.,** Serotonin clearance by capillaries isolated from epididymal fat, *Bibliotheca Anatomica,* 18, 108, 1979.

60. **Ryan, J. W. and Ryan, U. S.,** Pulmonary endothelial cells, *Fed. Proc.,* 36, 2683, 1971.

61. **De Bault, L. E., Kahn, L. E., Frommes, S. P., and Cancilla, P. A.,** Cerebral microvessels and derived cells in tissue cultue: Isolation and preliminary characterization, *In Vitro,* 15, 473, 1980.

62. **Mueller, S. N., Rosen, E. M., and Levine, E. M.,** Cellular senescence in a cloned strain of bovine fetal aortic endothelial cells, *Science,* 207, 889, 1980.

63. **Rosen, E. M., Mueller, S. N., Noveral, J. P., and Levine, E. M.,** Proliferative characteristics of clonal endothelial cell strains *J. Cel. Physiol.,* In press.

64. **Ford, J. W., Burkel, W. E., and Kahn, R. H.,** Isolation of adult caninevenous endothelium for tissue culture, *In Vitro,* 17, 44, 1981.

65. **Buonassisi, V.,** Sulfated mucopolysaccharide synthesis and secretion in endothelial cell cultures, *Exp. Cell. Res.,* 76, 363, 1973.

66. **McDonald, R. I., Shepro, D., Rosenthal, M., and Booyse, F. M.,** Properties of cultured endothelial cells, *Ser. Hemtol.* VI (4), 469, 1973.

67. **Shepro, D., Batbouta, J. C., Robblee, L. S., Carson, M. P., and Belamarich, F. A.,** Serotonin transport by cultured bovine aortic endothelium, *Circ. Res.,* 36, 799, 1975.

68. **De Bono, D.,** *En face* organ culture of vascular endothelium, *Br. J. Exp. Pathol.,* 56, 8, 1975.

69. **Pugatch, E. M. J. and Saunders, A. M.,** A new technique for making hautchen preparations of unfixed aortic endothelium, *J. Atheroscler. Res.,* 8, 735, 1968.

70. **Smith, U. and Ryan, J. W.,** Electron microscopy of endothelial cells collected on cellulose acetate paper, *Tissue & Cell,* 5, 333, 1973.
71. **Thorgeirsson, G. and Robertson, A. L.,** The vascular endothelium — pathobiologic significance, *Am. J. Pathol.,* 93, 301, 1978.
72. **Gimbrone, M. A., Jr.,** Culture of vascular endothelium, Grune and Stratton, New York, 1976, 1. in *Progress in Hemostasis and Thrombosis,* Vol. 3, Spaet, T. H., Ed.,
73. **Gospodarowicz, D., Mescher, A. L., and Birdwell, C. R.,** The control of cellular proliferation by the fibroblast and epidermal growth factors, *Natl. Cancer Inst. Monogr.,* 48, 109, 1978.
74. **Ross, R., Nist, C. Kariya, B., Rivest, M. J., Raines, E., and Callis, J.,** Physiological quiescence in plasma-derived serum: Influence of platelet-derived growth factor on cell growth in culture. *J. Cell. Physiol.,* 97, 497, 1978.
75. **Levine, E. M. and Mueller, S. N.,** Cultured vascular endothelial cells as a model system for the study of cellular senescence. *Intern. Rev. Cytol.,* Suppl., 10, 67, 1979.
76. **Davies, P and Ross, R.,** Mediation of pinocytosis in cultured arterial smooth muscle and endothelial cells by platelet-derived growth factor, *J. Cell. Biol.,* 79, 663, 1978.
77. **Dickinson, E. S. and Slakey, L. L.,** Plasma-derived serum as a selective agent to obtain endothelial cell cultures from swine aorta, *In Vitro,* 16, 227, 1980.
78. **Martin, G. M.,** Cellular aging — clonal senescence, A Review (Part I), *Am. J. Pathol.,* 89, 484, 1977.
79. **Maciera-Coelho, A.,** Introduction, *Gerontology,* 22, 3, 1976.
80. **Cristofalo, V. J.,** Animal cell cultures as a model system for the study of aging. *Adv. Gerontol. Res.,* 4, 45, 1972.
81. **Hayflick, L.,** The cellular basis for biological aging, in, *Handbook of the Biology of Aging,* Finch, C. E. and Hayflick, L., Eds, Van Nostrand Reinhold, New York, 1977, 159.
82. **Grinspan, J. B., Mueller, S. N., and Levine, E. M.,** Transformation *in vitro* of bovine aortic endothelial cells by benzo(a)pyrene, *Cancer Res.,* submitted.
83. **Madin, S. H. and Darby, N. B.,** Established kidney cell lines of normal adult bovine and ovine origin, *Proc. Soc. Exp. Biol. Med.,* 98, 574, 1958.
84. **McClurkin, A. W., Pirtle, E. C., Caria, M. F., and Smith, R. L.,** Comparson of low- and high-passage bovine turbinate cells for assay of bovine viral diarrhea virus, *Arch. Gesamte Virusforschung,* 45, 285, 1974.
85. **Kefalides, N. A.,** Persistence of basement membrane collagen phenotype in hybrids of human vascular enthelium and rodent fibroblasts, *Fed. Proc.,* 38, 816, 1979.
86. **Kefalides, N. A.,** Expression of the gene for human Factor VIII-antigen in hybrids of human endothelium and rodent fibroblasts, *Fed. Proc.,* 40, 327, 1981.
87. **Gimbrone, M. and Fareed, G.,** Transformation of cultured human vascular endothelium by SV40 DNA, *Cell,* 9, 685, 1976.
88. **Reznikoff, C. A. and DeMars, R.,** *In vitro* chemical mutagenesis and viral transformation of a human endothelial cell strain, *Cancer Res.* 41, 1114, 1981.

AGING OF THE ERYTHROCYTE

David Danon

Investigations of cellular aging lead to a better understanding of the alterations that diminish the capacity of cells to fulfill their function as a part of tissues, organs, and organisms. They enable the study of direct or indirect cell-to-cell interactions, which are necessary for the maintenance of the harmony required for the continuation of life and the preservation of its quality. The choice of a model for such a study may be directed either by the relative importance of the cell to be investigated or by its relative simplicity in terms of cytology, biochemistry, biophysics, and physiology, and therefore the higher probability that it offers to elucidate some basic problems in the process of cellular aging. From both viewpoints, the mammalian erythrocyte seems a very suitable model for the study of aging at the cellular level.

After the initial differentiation of a stem cell in the hemopoietic centers to the erythroid line, every cell division is followed by changes that lead to the finally differentiated cell, the mature erythrocyte. After three or four divisions,[2] the nucleus is expelled.[3,4] The reticulocyte which enters the circulation contains some polyribosomes and monosomes, m-RNA, t-RNA, and essential enzymes for protein synthesis, as well as some mitochondria and other organelles.[5,6] During the process of maturation, the reticulocyte loses its intracellular organelles, its ribosomes, and its RNA. Its capacity for protein synthesis gradually decreases until finally no protein synthesis can be demonstrated. The mature erythrocyte contains no mitochondria and consumes only minute amounts of oxygen through the pentose shunt.[7] The existence of mitochondria in reticulocytes was considered to be an explanation for the high oxygen consumption and the high content of ATP in these cells. A functional tricarboxylic acid cycle has, in fact, been demonstrated in reticulocytes,[8] but as they mature, the activity of this metabolic pathway declines.[9] During this final stage of maturation of the reticulocyte, the mitochondria and other intracellular organelles are expelled or disintegrated cellularly.[10] Remodelling of the membrane takes place by endocytosis and exocytosis.[11] During this last stage of maturation, the surface charge density of the reticulocyte is slightly increased.[12-14]

The mature mammalian erythrocyte is a bidiscoidal cell composed of a cell membrane that contains hemoglobin and a variety of enzymes. These enzymes catalyze the reactions required for its physiological function and the maintenance of its structure. The cell is no longer capable of synthesizing enzymes or any other proteins *de novo*. The mature erythrocyte in the circulation has all its structural and functional components but no capacity for renewal of these components by resynthesis.

During the life span of the erythrocyte in the circulation, various changes in structure, composition, biochemical, and biophysical properties, as well as some immunological characteristics of its membrane surface occur. Several reviews dealing with the biological,[15,16] biochemical,[17-19] and biophysical[20,21] aspects of red cell aging have been published recently.

Two primary conditions are necessary for the study of the in vivo aging of erythrocytes. One is the determination of the normal life span of the erythrocytes in the circulation of the animal to be studied. The second condition is the availability of an appropriate method for separating the cells into age groups.

Table 1 contains the life spans of erythrocytes in various animals and the methods for their determination. Additional information regarding the life spans of erythrocytes in various animals and the methods used for their determination is presented in Table 2. Table 3 contains the various methods for separating erythrocytes into age groups, and is based on reviews published in 1968[22] and in 1974,[23] as well as investigations published since then. Table 4 contains the information on the biochemical aspects, Table 5 the biophysical aspects, and Table 6 the immunological aspects of the aging erythrocytes. Recently, an increased number of publications has dealt with characteristics of erythrocytes in aging humans and animals rather than aging of the cells in the circulation. These data are compiled in Table 7.

Table 1
ERYTHROCYTE LIFE-SPAN DATA

Animal	Method	Analytic method	Mean	Range	Remarks	Ref.
Mouse	[51]Cr	$t^{1}/_{2}$	20	± 2.5	—	24
	[51]Cr	$t^{1}/_{2}$	—	15—20	—	25
	[51]Cr	Ext. pt.[a]	—	50—55	—	25
	[14C]glycine	—	41	40—43	—	26
	[32P]DFP	—	42	40—51	—	27
			40.7	± 1.9	—	28
	[32P]DFP	—	46.9	—	—	29
Rat	[51]Cr	$t^{1}/_{2}$	19	—	—	25
	[51]Cr	$t^{1}/_{2}$	18	± 2.5	—	24
	[51]Cr	$t^{1}/_{2}$	17.1	± 2.33	—	30
	[51]Cr	$t^{1}/_{2}$	20.7	± 2.5	—	31
	[51]Cr	$t^{1}/_{2}$	17	—	—	32
	[51]Cr	$t^{1}/_{2}$	18	—	—	33
	[51]Cr	Ext. pt.	approx. 60	—	—	25
	[51]Cr	Ext. pt.	65	—	—	32
	Ashby diff. agglut.	—	65	50—75	—	32
	[59]Fe	—	60	—	[59]Fe cohort labeling	34
	[59]Fe	—	59	—	[59]Fe cohort labeling	31
	[59]Fe	—	54	—	[59]Fe cohort labeling	35
	[14C]glycine	—	55	—	—	36
	[32P]DFP	—	65	56—90	—	27
	[32P]DFP	—	60	± 3.2	—	28
	[32P]DFP	—	56	52—60	—	33
	[14C]cyanate	—	66.2	± 7.6	Random hemolysis 0.67%/day	38
Hamster	[51]Cr	$t^{2}/_{2}$	14.5	—	—	39, 40
	[51]Cr	$t^{1}/_{2}$	12—20	—	—	41
	[51]Cr	Ext. pt.	60—70	—	—	41
	[51]Cr	Ext. pt.	78.5	—	—	42

Table 1 (continued)
ERYTHROCYTE LIFE-SPAN DATA

Animal	Method	Analytic method	Mean	Range	Remarks	Ref.
Squirrel	51Cr	Ext. pt.	60	—	—	43
Guinea pig	51Cr	t1/2	20	—	—	44
	51Cr	t1/2	16	—	Autologous, isologous	45
	51Cr	t1/2	12	—	Homologous	45
	51Cr	Ext. pt.	65	—	—	44
	51Cr	Ext. pt.	80—90	—	—	45
	51Cr	Ext. pt.	80	—	—	27
	[32P]DFP	Ext. pt.	79	—	—	27
	DFP	Ext. pt.	54—64	—	—	46
	[14C]glycine		57—70	—	—	46
Rabbit	51Cr	t1/2	12	8—17	—	47
	51Cr	t1/2	13.5	—	—	48
	51Cr	t1/2	19.0	+5.9	—	49
	51Cr	t1/2	21.3	18.8—24.2	—	50
	51Cr	Ext. Pt.	68	±5	—	51
	51Cr	Ext. pt.	—	55—60	—	47
	51Cr	Ext. pt.	65	60—68	—	48
	59Fe		65	50—80	59Fe cohort labeling	51
	59Fe		60	—	59Fe cohort labeling	52
	59Fe		64	Random destruct. 1% day	59Fe cohort labeling	53
	59Fe		57	±0.2	59Fe cohort labeling	54
	[14C]glycine		50	±0.2		54
	[32P]DFP		50.4	41.6—59.7	5 animals	50
Cat	[14C]glycine		—	66—78		55
Dog	[14C]glycine		108	97—133		56
	51Cr	t1/2	24.3	±2.8		56
	51Cr	t1/2	18.7	±1.0		57
	51Cr	t1/2	—	22—35	Intraven. administr.	58
	51Cr	t1/2	—	24—32	Intraperitoneal administr.	58
	51Cr	Ext. pt.	—	96—110	Intraperitoneal administr.	58

Animal	Label	Method	Value	Ext. pt. / Range	Remarks	Ref.
	^{51}Cr	Ext. pt.	—	102—110	Intraperitoneal administr.	58
	^{51}Cr	Ext. pt.	104	±10.4	Autologous cell intraperitoneal	59
	—	Ext. pt.	98	8.9	Autologous cell, intraperitoneal	59
	—	Ext. pt.	96.5	11.6	Isologous cell	59
	—	Ext. pt.	98.8	10.5	Puppy	59
	—	Ext. pt.	108.6	—	Intramedullary administr.	59
Mule deer	[^{14}C]cyanate	$t\frac{1}{2}$	27.5	±1.1		60
	[^{14}C]glycine	—	95	—	1 animal	61
White-tailed deer	[^{32}P]DFP	—	—	65—201		62
	[^{32}P]DFP	—	—	100—155		62
Llama	[^{14}C]glycine	—	225	—	3% cells destroyed randomly	63
Sheep						
Auodad	[^{14}C]glycine	—	65,170	—	2 populations of cells (20 and 80% resp.)	61
Domestic	[^{14}C]glycine	—	80—135	—	6 animals	64
Domestic	[^{14}C]glycine	—	64,94	—	2 animals	65
Domestic	[^{14}C]glycine	—	131,150,157	—	2 animals	66
Karakul	[^{14}C]glycine	—	118,130	—	2 animals	65
Bighorn	[^{14}C]glycine	—	147	—	1 animal	65
	[^{32}P]DFP	—	100	—		67
Antelope	^{51}Cr	$t\frac{1}{2}$	13.7	±3.7	5 animals	68
	^{59}Fe	—	111.7	±20.6	^{59}Fe cohort labeling	68
	Glycine	—	80	—	1 animal	69
Goat						
Himalayan	[^{14}C]glycine	—	160, 165	—	2 animals	69
Domestic	[^{14}C]glycine	—	125	—	1 animal	69
Monkey	^{51}Cr	Ext. pt.	91, 98, 101	—	5 animals	70
	^{50}Cr	$t\frac{1}{2}$	17	—		71
	^{51}Cr	$t\frac{1}{2}$	20.7	17—28	4 animals	72
Pig	[^{14}C]glycine	$t\frac{1}{2}$	86	—	Life span	73
	^{51}Cr	$t\frac{1}{2}$	17	—	Homologous cells	74
	^{51}Cr	$t\frac{1}{2}$	13.8	±5.7	Homologous cells	75
	—	—	28.0	±4.0	Autologous cells	75

Table 1 (continued)
ERYTHROCYTE LIFE-SPAN DATA

Animal	Method	Analytic method	Mean	Range	Remarks	Ref.
Man	[32P]DFP	—	127	114—136	8 subjects	76
	[32P]DFP	—	122	112—133	6 subjects	77
	[32P]DFP	—	132	118—154	5 subjects	78
	[32P]DFP	—	124	118—127	8 subjects	79
	[32P]DFP	—	117.5	±18.1	14 subjects	80
	[32P]DFP	—	111	±20.3	13 subjects	81
Males	[14C] or [3H]bilirubin	—	110	±12	14 subjects	82
Females	[14C] or [3H]bilirubin	—	94	±10	8 subjects	82
Cattle	[32P]DFP	$t^{1/2}$	107	—	—	83
	[51]Cr	$t^{1/2}$	11.7	10.5—13.0	—	84
Horse	[14C]glycine	—	140, 150	—	2 animals	61
Birds						
Duck	[14C]glycine	Ext. pt.	39	—	—	85
	51Cr	—	42	—	—	86
	—	—	42	±2	—	49
Chicken	[14C]glycine	—	20	—	Random destruction	85
	[15N]glycine	—	28	—	1 animal	87
	32Pb	—	28	—	2 animals	88
	32Pb	$t^{1/2}$	32	—	1 animal	89
Pigeon	51Cr	$t^{1/2}$	10.8	±2.9	—	49
	51Cr	Ext. pt.	44.0	±3	—	49
Turkey	51Cr	$t^{1/2}$	12.5	9—16	—	90
Alligator	[H3]DFP	—	184, 260. 437	—	3 animals	91
Frog	[32P]DFP	—	200	—	—	92
Toad	Glycine	—	1000—4000	—	—	93
Turtle	Glycine	—	600—800	—	—	93
Armadillo	51Cr	Ext. pt.	70—75	—	—	94

a   Ext. pt., extinction point.

b   Measured by incorporation of [32P]NaH$_2$PO$_4$ into cell nucleic acid.

From Berlin, N. I. and Berk, P. D., *The Red Blood Cell*, Vol. II, 2nd ed., Surgenor, D. M., Ed., Academic Press, N.Y., 1975, 957. With permission.

Table 2
ERYTHROCYTE LIFE-SPAN DATA (ADDENDUM TO TABLE 1)

Animal	Method	Analytic method	Mean	Range	Remarks	Ref.
			116	± 28	Review	95
	^{51}Cr		111.2	SD 20.3	1% elution/day	96
Man	^{59}Fe		105	SD 28	+ Computer	97
					Math. corrections for existing methods	98
	[^{15}N]Glycine	Ext. pt.	119	118—120	Shortened by thermal injury	99
	[^{14}C]Cyanate		115	± 12	No elution	100
	^{51}Cr				Math. correction of elution	101
	^{50}Cr & ^{51}Cr	$t^{1}/_{2}$	35.5 & 33.8		One normal control	102
	^{59}Fe			26—124	Study of 19 patients	103
Rabbit	^{125}I	$t^{1}/_{2}$	18.9	SD 1.5	Enzymatically labeled	104
Rabbit	^{51}Cr	$t^{1}/_{2}$	13.8	SD 2.5		
Rabbit	^{51}Cr	$t^{1}/_{2}$	50.4	up to 70d	In vitro labeling	105
	[^{32}P]DFP	$t^{1}/_{2}$	50.4	± 2 SD	In vivo labeling	105
Hamsters	^{51}Cr	Ext. pt.	55.0	± 3.5	12 animals	106
Rat	^{51}Cr				I.V. or I.P. injection of isotope	107
Mongolian gerbil	^{51}Cr & [^{32}P]DFP					108
Opossum	[^{14}C]glycine		77		5 animals	109
Sheep	^{59}Fe	$t^{1}/_{2}$	111.7	8.4 SE	^{51}Cr not appropriate	110
Horse	[^{32}P]DFP		125		First sample on 10^{3} day	111
Camel	^{51}Cr	Ext. pt.		90	One animal winter	112
Camel	^{51}Cr	Ext. pt.		120	Summer $\pm$ water (ad lib.)	112
Camel	^{51}Cr	Ext. pt.		150	Summer dehydration	112
Chicken	^{51}Cr	Ext. pt.	29	27.5—31.5		113
Frog	[^{3}H]Thymidine	$t^{1}/_{2}$	98.7	89—116	Autoradiography	114

[a] Ext. pt., extinction point.

Table 3
METHODS FOR SEPARATION OF ERYTHROCYTES INTO AGE GROUPS

Species	Anticoagulant	Suspending medium	Conditions of centrifugation			Enrichment of fraction		Ref.
			G (1000)	Time (min)	Temp (°C)	"Young" in lightest cells	"Old" in heaviest cells	
A. Centrifugation methods with data on efficiency of separation								
Human	Heparin	Plasma	2	40	—	2.5 in 25% (^{59}Fe)	1.7 in 25% (^{59}Fe)	115
Human	Heparin	Plasma	51	120	4	2.3 in 33% (^{59}Fe)	1.6 in 33% (^{59}Fe)	116
Human	ACD	Plasma	100	120	4	2.3 in 33% (^{59}Fe)	1.6 in 33% (^{59}Fe)	117
Human	Heparin	Plasma	50	20	4	4.1 in 14% (^{59}Fe)	1.6 in 14% (^{59}Fe)	118
Human	Heparin	Plasma	12	15—240	20	5.9 in 6% (^{59}Fe)	—	119
Human	Heparin	Phthalate esters	12	15	RT	14 in 2% (^{59}Fe)	—	120
Rabbit	Heparin	Plasma	15	40	RT	3.2 in 20% (^{59}Fe)	1.6 in 20% (^{59}Fe)	121
Rabbit	Heparin	Saline	1.6	30	—	2.0 in 12.5%	1.5 12.5%	122
Rabbit	Heparin	Phthalate esters	12	14	RT	—	8.0 in 10% (^{59}Fe)	123
Rabbit	Heparin	30% BSA 0.4% NaCl	12	14	RT	—	3.6 in 9% (^{59}Fe)	124
Rabbit	Heparin	BSA gradient	33	60	4	8.9 (^{59}Fe)	1.5 (^{59}Fe)	125
Rabbit	Defibrination	BSA gradient	130	60	4	5.2 in 7% (^{59}Fe); 5.3 in 9% (^{14}C)	1.5 in 13% (^{59}Fe) 5.5 in 2% (^{14}C)	125
Rat	Heparin	BSA gradient	25 RPM	30	RT	8.7 in 10% (^{14}C glycine)	2.3 in 10% (^{14}C glycine)	126
Dog	EDTA	30% BSA	5.9	40	4	6.3 in 10% (^{59}Fe)	1.3 in 10% (^{59}Fe)	127
B. Centrifugation and other methods with incomplete data								
Human	EDTA	30% BSA	10	5	4	High in 6% (^{59}Fe)	High in 6% (^{59}Fe)	128
Human	Alsever	BSA gradient	20	60	—	high (^{59}Fe)	High (^{59}Fe)	129
Human	Heparin	Saline	Threshold centrif.			—	—	130
Human	Defibrination		15	60	30	—	—	131

RPM angle head centrif.

Species	Anticoagulant	Method				Result		Ref.
Human	EDTA or heparin	Ficol-trisol gradient	35	60	4	—	—	132
Human		Counterflow Centrifuge — RPM		60		—	—	133
Human		Isopycnic dextran	40	60	25	—	—	134
Human		Hypaque gradient	—			—	—	135
Various species	ACD	Counter current distribution				—	—	136
Various species		Temperature gradient				—	—	137
Human	ACD	Density gradient electrophoresis				—	—	138
Rat								
Rat	Citrate	Plasma	—	10	—	Concentrated (⁵⁹Fe)	—	139
Rat	Heparin	Ficol gradient — RPM	15	60	—	⁵⁹Fe	⁵⁹Fe	140
Rabbit	Arabino-galactane	Isotonic	—	—	—	Mean survival 28d (^{51}Cr)	Mean survival 14d (^{51}Cr)	141
		Dextran + albumin gradient	90	60	—	—	—	142

Table 4
BIOCHEMICAL ASPECTS OF ERYTHROCYTE AGING

E.C. No.	Enzymes	Change	Remarks	Ref.
1.1.1.8	Glyceraldehyde-3-phosphate dehydrogenase	→ ↓		143 144
1.1.1.27	Lactate dehydrogenase	↓		144—147
1.1.1.41	Isocitrate dehydrogenase	↓		151, 152
1.1.1.43	G-P-Gluconate dehydrogenase	↓		145, 146
1.1.1.43	G-P-Gluconate dehydrogenase	↓	Lower activity in fractions containing young cells	148—151, 152
1.1.1.49	Glucose-G-P-dehydrogenase	↓		144—146, 148—154
1.1.5.11	Superoxide dismutase	↓		155
1.2.1.12	Triose-P-dehydrogenase	↓		147
1.13.1.1	Catalase	↓		156, 157
2.2.1.1	Transketolase	↓		149
2.2.1.2	Transaldolase	↓		149
2.4.2.1	Purine nucleoside phosphorylase	↓		143
2.4.2.1	Purine nucleoside phosphorylase	→	Lower activity in fractions containing young cells	152
2.6.1.1	Glutamic oxaloacetic transaminase	↓		151, 158
2.6.1.2	Glutamic pyruvic transaminase	↓		159
2.7.1.1	Hexokinase	↓	Post-translational change in predominance of enzyme type	145, 160—168
2.7.1.23	NAD-kinase	↓		161, 166, 169
2.7.1.23	Protein kinase	↓		150
2.7.1.40	Pyruvate kinase	↓		145, 147, 170, 171
2.7.2.3	Phosphoglycerate kinase	↓		143, 147, 172
2.7.4.3	Adenylate kinase	↓		143
2.7.4.3	Adenylate kinase	→		173
2.7.1.	Phosphofructo kinase	↓		174
2.7.4.8	Guanylate kinase	↓	Biphasic pattern of decay	175
2.7.6.1	P-Ribosyl pyrophosphate synthetase	↓	Enzyme changes from high to low M.W.	165
2.7.4.	UMP-Kinase	↓		176, 177
3.1.1.7	Acetylcholine esterase	→	Inactivation independent of erythrocyte age	178
3.1.1.7	Acetylcholine esterase	↓		143, 154, 158
3.1.1.8	Cholinesterase	↓		156, 157, 179
3.1.3.2	Acid phosphatase	↓	Biphasic form no secondary isozymes	180, 181
3.1.4.1	Phosphodiesterase	↓		182
3.2.2.5	NADase	↓		174
3.6.1.1	Pyrophosphatase	↓		183
3.6.1.3	$Na^+K^+ATPase$	→		143, 145
3.6.1.4	Mg^{++} ATPase	↓		143, 184
4.1.2.7	Aldolase	↓		147, 185, 186
4.1.2.7	Aldolase	↑		187
4.4.1.5	Glyoxylase	↓		156, 157
5.3.1.1	Triose-P-isomerase	↓		147
5.3.1.9	Hexose-P-isomerase	↓		146
	Methemoglobin reductase	↓		147, 149, 164, 189
	Methemoglobin diaphorase	↓		147,149,164
	Nucleoside phosphorylase	↓		146, 148
	Nucleotides and coenzymes (AMP, ADP, ATP, NADP, GDP, GTP, IMP, UDGP)	↓		145, 147, 153, 175, 190—197

Table 4 (continued)
BIOCHEMICAL ASPECTS OF ERYTHROCYTE AGING

Membrane constituents	Change	Remarks	Ref.
Proteins	↓		198—200
Proteins	↑		197, 201
Proteins	→	Conformational change	202—205
Glycoproteins		Change	206
Lipids	↓		191, 200, 207, 208
Linoleic acid	↑		209
Total carbohydrates	↓		154, 210
Sialic acid	↓		200, 210—220
Galactose	↓		154
Galactose amine	↓		154
Glycophorin	↓	Proportional decrease of glycophorin and sialic acid	220
Total membrane	↓	Membrane fragments probably lost in aging process	200
Total membrane	↓	Release of diacylglycerol-enriched vesicles	221
Total membrane	↓	Evaluation by toluidine blue	222

Cell content	Change	Remarks	Ref.
Hemoglobin	↓		153, 198
Hemoglobin	↑		198, 223
Hemoglobin	→	Denaturation, micro Heinz-bodies	224
Glutathione	→		191, 225
Glutathione	↓		226, 227
Calcium	↑		195, 227
Magnesium	→		202
Magnesium	↓		228
Na	↑		191
K	↓		191

Table 5
BIOPHYSICAL ASPECTS OF ERYTHROCYTE AGING

Feature	Change	Remarks	References
Size	↓		229—233
Membrane surface area	↓		234—236
Density	↑		232, 237—244
Sedimentation in hyper-tonic media	↑		245
Cell biconcavity	↓	Tendency to become spherical	229, 232, 233, 246 —
Reversible deformability	↓		233, 247—256
Elasticity of membrane	↓	Attributed to increased protein cross-linkage	255
Osmotic resistance	↓		237—239, 243, 257, 258
Osmotic resistance	→	Rat	259
Acid resistance	↑	Human	231
Acid resistance	→	Rat	260
Granularity of membrane surface	↓	Electron microscopy of shadow cast preparation of membranes	261, 262
Stromalytic forms	↓	Electron microscopy of osmot-ically lysed, negatively stained cell membranes	263
Membrane fluidity	↓	Spin label motion	233
Electrophoretic mobility	↓		264—267
Electrophoretic mobility	→		268, 269
Negative surface charge density	↓	Evaluated with cationized ferritin and or colloidal iron	270—272
Agglutination by polylysine	↑	Probably less repulsion by negative charge	243, 265, 272
Agglutination by blood group antibodies	↑	More antigenic sites and less repulsion	273, 274
Agglutination by immunoglobulins	↑	Goat anti-human antibodies, revealing of antigenic sites	275
Agglutination by soybean agglutinin	↓		276
Labeling density with soybean agglutinin	↓		276
Aggregation by polyvinyl pyrrolidone	↓	Probably higher rigidity of cells	277
Electrophoresis of hemoglobin	↑		278—282
Rate of oxygen binding	↓		283, 284
Rate of oxygen dissociation	↓		285
Chrome binding	↓		286—288

Table 6
IMMUNOLOGICAL ASPECTS OF RED CELL AGING

Feature	Changes in old cells	Remarks	Ref.
Surface antigen density	↑	Evaluated by (a) ferritin-conjugated antibodies, (b) hybrid antibodies	289
Receptors to immunoglobulin	↑	Agglutination with goat anti-human immunoglobulins	290
Agglutinability by blood group antibodies	↑	Adsorbed immunoglobulin, probably anti-T	291, 292—294
Immune lysis	↑		295, 296
Phagocytosis by macrophages	↑	Human blood monocytes derived macrophages in vitro	297, 298
		Syngeneic mouse peritoneal macrophages in vitro	299, 300

Table 7
ALTERATIONS IN ERYTHROCYTES OF AGING ANIMALS

Animal	Enzyme and Component	Change	Remarks	Ref.
A. Biochemical aspects:				
Human (H)	2,3 Diphosphoglycerate	↓	Related to lower conc. of plasma inorganic phosphate	301
H	2,3 Diphosphoglycerate	↓		302
	Na⁺K⁺ATPase	↓		303
H	Na⁺K⁺ATPase	↓		304
H	Superoxide dismutase	↓	Units/mg hemoglobin	305
H	Superoxide dismutase	↓		306
H	Superoxide dismutase	→		307
H	Superoxide dismutase	→	No change with sex	308
H	Cholesterol	→		304
Rat	Cholesterol	↑		309, 310
H	Phospholipids	↓		304
H	Fatty acids	→		304
Mouse	Glutathione	↓	Very old (37 m)	311
Mouse	Glutathione reductase	↓	Very old (37 m)	311
H	Sialic acid	↓		304
B. Biophysical aspects				
H	Mean cell volume	↑		312, 313
H	Mean hemoglobin concentration	↓		313, 314
Mouse	Density	↑		311
H	Osmotic fragility	↑		313, 315
H	Membrane fluidity	↓		316
H	Membrane fluidity	↓		304
H	Rate of lysis			309, 313, 316
H	Iron uptake			317
Pig	Amino acid transport			318
H	Cell count, hematocrit		Sharp decline after 70 in men, little or no change in women	314
H	Cell count, hematocrit		Lower serum iron	319

REFERENCES

1. **Harris, J. W. and Kellermeyer, R. W.,** *The Red Cell: Production, Metabolism, Destruction, Normal and Abnormal,* 2nd ed., Harvard University Press, Cambridge, Mass., 1970, 281.
2. **Rifkind, R. A., Bank, A., Marks, P. A., and Nossell, H. L.,** *Fundamentals of Hematology,* Year Book Medical Publishers, Chicago, Ill., 1976, 4.
3. **Bessis, M. and Bricka, M.,** Aspect dynamique des cellules du sang. Son etude par la microcinematographie en contraste de phase, *Rev. Hematol.,* 7, 407, 1952.
4. **Skutelsky, E. and Danon, D.,** An electron microscopic study of nuclear elimination from the late erythroblast, *J. Cell Biol.,* 33, 625, 1967.
5. **Lowenstein, L. M.,** The mammalian reticulocyte, *Int. Rev. Cytol.,* 8, 135, 1959.
6. **Marks, P. A., Rifkind, R. A. and Danon, D.,** Polyribosomes and protein synthesis during reticulocyte maturation *in vitro, Proc. Natl. Acad. Sci. USA,* 50, 336, 1963.
7. **Schweiger, H. G.,** Pathways of metabolism in nucleate and anucleate erythrocytes, *Internatl. Rev. Cytol.,* 13, 135, 1962.
8. **Gasko, O. and Danon, D.,** The metabolism of maturing reticulocytes. I. The existence of a functional tricarboxylic acid cycle, *Brit. J. Haematol.,* 23, 525, 1972.
9. **Gasko, O. and Danon, D.,** The metabolism of maturing reticulocytes. II. Decline in activity of the tricarboxylic acid cycle associated with reticulocyte maturation, *Brit. J. Haematol.,* 23, 535, 1972.
10. **Gasko, O. and Danon, D.,** Deterioration and disappearance of mitochondria during reticulocyte maturation, *Exp. Cell Res.,* 75, 159, 1972.
11. **Gasko, O. and Danon, D.,** Endocytosis and exocytosis in membrane remodelling during reticulocyte maturation, *Brit. J. Haematol.,* 28, 463, 1974.
12. **Stephens, J. G.,** Surface and fragility differences between mature and immature red cells, *J. Physiol.,* 99, 30, 1940.
13. **Walter, H., Wince, R. and Selby, F. W.,** Counter-current distribution of red blood cells of different ages and from different species, *Biochim. Biophys. Acta,* 109, 293, 1965.
14. **Skutelsky, E. and Danon, D.,** Electron microscopical analysis of surface charge labelling density at various stages of the erythroid line, *J. Membr. Biol.,* 2, 173, 1970.
15. **Landaw, S. A.,** Biological Aspects of Senescence in Red Cells, in *Exp. Aging Res.,* Merril, E. F., Eleftherious, B. E., and Elias, P. K., Eds., EAR, Bar Harbor, Maine, 1976, 303.
16. **Berlin, N. I. and Berk, P. D.,** The Biological Life of the Red Cell, in *The Red Blood Cell,* Vol. II, 2nd ed., Surgenor, D. M., Ed., Academic Press, N.Y., 1975, 957.
17. **Bunn, H. F.,** Erythrocyte destruction and hemoglobin catabolism, *Seminars in Hematology,* 9, 3, 1972.
18. **Fornaini, G.,** Biochemical modification during the life span of the erythrocyte, *Ital. J. Biochem.,* 16, 257, 1967.
19. **Brewer, G. J.,** General red cell metabolism, in *The Red Blood Cell,* Vol. I, 2nd ed., Surgenor, D. M., Ed., Academic Press, N.Y., 1974, 387.
20. **Danon, D.,** Biophysical aspects of red cell ageing, *Bibl. Haemat.,* 29, 179, 1966.
21. **Danon, D.,** Biophysical aspects of red cell ageing, in *Physiology and Pathology of Human Aging,* Rockstein, M., Ed., Academic Press, N.Y., 1975, 95.
22. **Van Gastel, C.,** Methods for studying the *in vivo* aging of red cells, Blood Information Service (Library of Congress Catalog Card No. 68-9821), 1968.
23. **Hjelm, M.,** Methological aspects of current procedures to separate erythrocytes into age groups, in *Cellular and Molecular Biology of Erythrocytes,* Yoshikawa, H. and Rapoport, S. M., Eds., Urban and Schwarzenberg, Muenchen-Berlin-Wien, 1974, 427.
24. **Smith, L. H. and Toha, J.,** Survival of mouse-grown rat erythrocytes, *Proc. Soc. Exp. Biol. Med.,* 98, 125, 1958.
25. **Goodman, J. W. and Smith, L. H.,** Erythrocyte life span in normal mice and in radiation bone marrow chimeras, *Amer. J. Physiol.,* 200, 764, 1961.
26. **Ehrenstein, G. V.,** The life span of the erythrocytes of normal and of tumor-bearing mice as determined by glycine-2-^{14}C. *Acta Physiol. Scand.,* 44, 80, 1958.
27. **Edmondson, P. W. and Wyburn, J. R.,** The erythrocyte life-span, red cell mass and plasma volume of normal guinea-pigs as determined by the use of ^{51}Cr, ^{32}P labelled di-isopropyl fluorophosphonate and ^{131}I-labelled human serum albumin, *Brit. J. Exp. Pathol.,* 44, 72, 1963.
28. **Van Putten, L. M.,** The life span of red cells in the rat and mouse as determined by labeling with DFP^{32} *in vivo, Blood,* 13, 789, 1958.
29. **Abbrecht, P. H. and Littell, J. K.,** Erythrocyte life-span in mice acclimatized to different degrees of hypoxia, *J. Appl. Physiol.,* 31, 443, 1972.
30. **Brown, D. V., Boehne, E. M., and Norlind, L. M.,** Anemia with positive direct Coombs' test induced by trypan blue, *Blood,* 18, 543, 1961.
31. **Belcher, E. H. and Harriss, E. B.,** Studies of red cell life span in the rat, *J. Physiol.,* 146, 217, 1959.

32. **Smith, L. H., Odell, T. T., and Caldwell, B.,** Life span of rat erythrocytes as determined by ^{51}Cr and differential agglutination methods, *Proc. Soc. Exp. Biol. Med.,* 100, 29, 1959.

33. **Thompson, J. S., Gurney, C. W., Hanel, A., Ford, E., and Hofstra, D.,** Survival of transfused blood in rats, *Amer. J. Physiol.,* 200, 327, 1961.

34. **Jones, N. C. H.,** The use of ^{51}Cr and ^{59}Fe as red cell labels to determine the fate of normal erythrocytes in the rabbit, *Clin. Sci.,* 20, 315, 1961.

35. **Owen, C. A. and Orvis, A. L.,** Elution of chromium from rat erythrocytes, *Amer. J. Physiol.,* 210, 573, 1966.

36. **Forssberg, A. and Tribukait, B.,** Bestimmung der Lebenszeit der nach akuter Blutung gebildeten Rattenerythrozyten (C[14] Hamin), *Acta Physiol. Scand.,* 54, 152, 1962.

37. **McKee, L. C., Wasson, M., and Heyssel, R. M.,** Experimental iron deficiency in the rat. The use of ^{51}Cr, DF[32]P and ^{59}Fe to detect haemolysis of iron-deficient cells, *Brit. J. Haematol.,* 14, 87, 1968.

38. **Landaw, S. A.,** The use of ^{14}C-cyanate for red blood cell survival studies, *Proc. Soc. Exp. Biol. Med.,* 142, 712, 1973.

39. **Rigby, P. G., Emerson, C. P., Betts, A., and Friedell, G. H.,** Comparison of *in vitro* and *in vivo* methods of erythrocyte tagging with Cr[51], *J. Lab. Clin. Med.,* 58, 854, 1961.

40. **Rigby, P. G., Betts, A., Friedell, G. H., and Emerson, C. P.,** Kinetics and loci of destruction of erythrocytes in tumor-bearing hamsters, *J. Lab. Clin. Med.,* 59, 638, 1962.

41. **Rigby, P. G., Emerson, C. P., and Friedell, G. H.,** RBC survival in hamsters using intraperitoneal Na$_2$Cr[51]O$_4$, *Proc. Soc. Exp. Biol. Med.,* 106, 313, 1961.

42. **Brock, M. A.,** Production and life span of erythrocytes during hibernation in the golden hamster, *Amer. P. Physiol.,* 198, 1181, 1960.

43. **Marvin, H. N.,** Some metabolic and nutritional factors affecting the survival of erythrocytes, *Amer. J. Clin. Nutr.,* 12, 88, 1963.

44. **Grönroos, P.,** The life-span of guinea pig red cells estimated by the use of Na$_2$[51]CrO$_4$, *Austral. J. Sci.,* 23, 195, 1960.

45. **Smith, L. H. and McKinley, T. W.,** Erythrocyte survival in guinea pigs, *Proc. Soc. Exp. Biol. Med.,* 111, 768, 1962.

46. **Yamanaka, W., Winchell, H. S., and Ostwald, R.,** Erythrokinetics in dietary hypercholesteremia of guinea pigs, *Amer. J. Physiol.,* 213, 1278, 1967.

47. **Gardner, E., Wright, C. S., and Williams, B. Z.,** The survival of virus-treated erythrocytes in normal and splenectomized rabbits, *J. Lab. Clin. Med.,* 58, 743, 1961.

48. **Sutherland, D. A., Minton, P., and Lanz, H.,** The life span of the rabbit erythrocyte, *Acta Haematol.,* 21, 36, 1959.

49. **Marvin, H. N. and Lucy, D. D.,** The survival of radiochromium-tagged erythrocytes in pigeons, ducks and rabbits, *Acta Haematol.,* 18, 239, 1957.

50. **Smith, G. N. and Mollison, P. L.,** Normal red cell survival in the rabbit, *Scand. J. Haematol.,* 11, 188, 1973.

51. **Jones, N. C. H. and Cheney, B.,** The use of Cr[51] and Fe[59] as red cell labels to determine the fate of normal erythrocytes in the rat, *Clin. Sci.,* 20, 323, 1961.

52. **Karle, H.,** Significance of red cell age to red cell destruction during experimental pyrexia, *Brit. J. Haematol.,* 15, 221, 1968.

53. **Sorbie, J. and Valberg, L. S.,** Splenic sequestration of stress erythrocytes in the rabbit, *Amer. J. Physiol.,* 218, 647, 1970.

54. **Gower, D. B. and Davison, W. M.,** The mechanism of immune haemolysis. I. The relationship of the rate of destruction of red cells to their age, following the administration to rabbits of immune haemolysin, *Brit. J. Haematol.,* 9, 132, 1963.

55. **Koneko, J. J., Green, R. A., and Mia, A. S.,** Erythrocyte survival in the cat as determined by glycine-2-C[14], *Proc. Soc. Exp. Biol. Med.,* 123, 783, 1966.

56. **Weissman, S. M., Waldmann, T. A., and Berlin, N. I.,** Quantitative measurement of erythropoiesis in the dog, *Amer. J. Physiol.,* 198, 183, 1960.

57. **Wellington, J. S. and Gardner, R. E.,** Survival of homologous erythrocytes, *Arch. Surg.,* 84, 491, 1962.

58. **Rochlin, D. B., Rawnsley, H., Duhring, J. H., and Blakemore, W. S.,** The selective removal of transfused compatible erythrocytes from the circulation of normal dogs, *Surg. Gynecol. and Obstet.,* 112, 675, 1961.

59. **Clark, C. H. and Woodley, C. H.,** The effects of certain diseases and transfusion methods on the life span of red blood cells, *Amer. J. Vet. Res.,* 20, 1069, 1959.

60. **Graziano, J. H., deFuria, F. G., and Cerami, A.,** The use of ^{14}C-cyanate as a method for determining erythrocyte survival, *Proc. Soc. Exp. Biol. Med.,* 144, 326, 1973.

61. **Cornelius, C. E., Kaneko, J. J., Benson, D. C., and Wheat, J. D.,** Erythrocyte survival studies in the horse, using glycine-2-C[14], *Amer. J. Vet. Res.,* 21, 1123, 1960.

62. **Noyes, W. D., Kitchem, H., and Taylor, W. J.,** Red cell life span of white-tailed deer odocoileus virginiamus, *Comp. Biochem. Physiol.,* 19, 471, 1966.
63. **Kaneko, J. and Cornelius, C. E.,** Erythrocyte survival studies in the Himalayan Tahr and domestic goats, *Amer. J. Vet. Res.,* 23, 913, 1962.
64. **Carter, M. W., Matrone, G., and Mendenhall, W.,** Estimation of the life span of red blood cells, *J. Gen. Physiol.,* 47, 851, 1964.
65. **Kaneko, J. J., Cornelius, C. E., and Heuschele, W. P.,** Erythrocyte survival studies in domestic and bighorn sheep, using glycine-2-C^{14}, *Amer. J. Vet. Res.,* 22, 683, 1961.
66. **Judd, J. T. and Matrone, G.,** Sheep erythrocyte life span in estimation of hemoglobin turnover in iron metabolism studies, *J. Nutr.,* 77, 264, 1963.
67. **Eadi, G. S., Smith, W. W., and Brown, I. W.,** The use of DFP^{32} as a red cell tag with and without simultaneous tagging with Cr^{51} in certain animals in the presence or absence of random destruction, *J. Gen. Physiol.,* 43, 825, 1960.
68. **Hildebrandt, P. K., Giles, R. C., Berman, A., and McCaffrey, R. P.,** Sheep erythrocyte survival studies using ^{51}Cr and ^{59}Fe, *Fed. Proc.,* 33, 356 (No. 853), 1974.
69. **Cornelius, C. E., Kaneko, J. J., and Benson, D. C.,** Erythrocyte survival studies in the mule deer, Aoudad sheep, and springbok antelope, using glycine-2-C^{14}, *Amer. J. Vet. Res.,* 20, 917, 1959.
70. **Marvin, H. N., Dinning, J. S., and Day, P. L.,** Erythrocyte survival in vitamin E deficient monkeys, *Poc. Soc. Exp. Biol. Med.,* 105, 473, 1960.
71. **Glomski, C. A., Pillay, S. K. K., and Hagle, R. E.,** Survival of ^{50}Cr-labelled erythrocytes as studied by instrumental activation analysis, *J. Nucl. Med.,* 12, 31, 1971.
72. **Huser, H. J., Rieber, E. E., and Berman, A. R.,** Experimental evidence of excess hemolysis in the course of chronic iron deficiency anemia, *J. Lab. Clin. Med.,* 69, 405, 1967.
73. **Bush, J. A., Berlin, N. I., Jensen, W. N., Brill, A. B., Cartwright, G. E., and Wintrobe, M. M.,** Erythrocyte life span in growing swine as determined by glycine-2-C^{14}, *J. Exp. Med.,* 101, 451, 1955.
74. **Bush, J. A., Jensen, W. N., Athens, J. W., Ashenbrucker, H., Cartwright, G. E., and Wintrobe, M. M.,** Studies on copper metabolism. XIX. The kinetics of iron metabolism and erythrocyte life-span in copper deficient swine, *J. Exp. Med.,* 103, 701, 1956.
75. **Talbort, R. B. and Swenson, M. J.,** Survival of Cr^{51} labeled erythrocytes in swine, *Proc. Soc. Exp. Biol. Med.,* 112, 573, 1963.
76. **Eernisse, J. G. and Van Rood, J. J.,** Erythrocyte survival time determinations with the aid of DF^{32}P, *Brit. J. Haematol.,* 3, 382, 1961.
77. **Garby, L. and Hjelm, M.,** Short-term assessment of red cell survival with an improved Cr^{51}-technique, *Scand. J. Clin. Invest.,* 14, 581, 1962.
78. **Brewer, G. J., Tarlov, A. R., and Kellermeyer, R. W.,** The hemolytic effect of primaquine. XII. Shortened erythrocyte life span in primaquine-sensitive male negroes in the absence of drug administration, *J. Lab. Clin. Med.,* 58, 217, 1961.
79. **Bove, J. R. and Ebaugh, F. G.,** The use of di-isopropylfluorophosphate^{32} for the determination of *in vivo* red cell survival and plasma cholinesterase turnover rates, *J. Lab. Clin. Med.,* 51, 916, 1958.
80. **Diez-Ewald, M. and Layrisse, M.,** Mechanisms of hemolysis in iron deficiency anemia. Further studies, *Blood,* 32, 884, 1968.
81. **Bentley, S. A., Glass, H. I., Lewis, S. M., and Szur, L.,** Elution correction in ^{51}Cr red cell survival, *Brit. J. Haematol.,* 26, 179, 1974.
82. **Berk, P. D., Bloomer, J. R., Howe, R. B., Blaschke, T. F., and Berlin, N. I.,** Bilirubin production as a measure of red cell life span, *J. Lab. Clin. Med.,* 79, 364, 1972.
83. **Mizuno, N. S., Perman, V., Bates, F. W., Sautter, J. H., and Schultze, M. O.,** Life-span of thrombocytes and erythrocytes in normal and thrombocytopanic calves, *Blood,* 14, 708, 1959.
84. **Baker, N. F., Osebold, J. W., and Christensen, J. F.,** Erythrocyte survival in experimental anaplasmosis, *Amer. J. Vet. Res.,* 22, 590, 1961.
85. **Brace, K. C. and Atland, P. D.,** Life span of the duck and chicken erythrocyte as determined with C^{14}, *Proc. Soc. Exp. Biol. Med.,* 92, 615, 1956.
86. **Rodnan, G. P., Ebaugh, F. G., and Fox, M. R. S.,** The lifespan of the red blood cell and the red blood cell volume in the chicken, pigeon and duck as estimated by the use of Na$_2$Cr^{51}O$_4$, *Blood,* 12, 355, 1957.
87. **Shemin, D.,** The biosynthesis of porphyrins, *Cold Spring Harbor Symp. Quant. Biol.,* 13, 185, 1948.
88. **Hevesy, G. and Ottesen, J.,** Life span of red blood corpuscles of the hen, *Nature,* 156, 534, 1945.
89. **Ottesen, J.,** Life-span of red and white blood corpuscles of the hen, *Nature,* 162, 730, 1948.
90. **Silber, R., Hedberg, S. E., Akeroyd, J. H., and Feldman, D.,** The transfusion behaviour of Avian erythrocytes: The lack of functional transplantation antigens in a nucleated cell, *Blood,* 18, 207, 1961.
91. **Cline, M. J. and Waldmann, T. A.,** Effect of temperature on red cell survival in the alligator, *Proc. Soc. Exp. Biol. Med.,* 111, 716, 1962.
92. **Cline, M. J. and Waldmann, T. A.,** Effect of temperature on erythropoiesis and red cell survival in the frog, *Amer. J. Physiol.,* 203, 401, 1962.

93. **Atland, P. D. and Brace, K. C.,** Red cell life span in the turtle and the toad, *Amer. J. Physiol.,* 203, 1188, 1962.

94. **Ebaugh, F. G. and Benson, M. A.,** Armadillo hemoglobin characteristics and red cell survival, *J. Cell Comp. Physiol.,* 64, 183, 1964.

95. **Lewis, S. M. and Szur, L. H.,** A review of red cell survival measurements, *Brit. J. Haematol.,* 25, 543, 1973.

96. **Bentley, S. S., Glass, H. I., Lewis, S. M., and Szur, L.,** Elution correction in ^{51}Cr red cell survival studies, *Brit. J. Haematol.,* 26, 179, 1974.

97. **Ricketts, C., Jacobs, A., and Cavill, I.,** Ferrokinetics and erythropoiesis in man: the measurement of effective erythropoiesis, ineffective erythropoiesis and red cell life-span using ^{59}Fe, *Brit. J. Haematol.,* 31, 65, 1975.

98. **Znojil, V. and Vacha, J.,** A theoretical model of the red cell survival followed by means of random and cohort labeling, *J. Theor. Biol.,* 66, 711, 1977.

99. **Bernat, I. and Cornides, I.,** Life span of homogeneous red cell population formed after thermal injury, *Haematologia,* 8, 153, 1974.

100. **Eschbach, J. W., Korn, D., and Finch, C. A.,** ^{14}C cyanate as a tag for red cell survival in normal and uremic man, *J. Lab. Clin. Med.,* 89, 823, 1977.

101. **Widman, J. C. and Powsner, E. R.,** Red-Cell lifetimes calculated from ^{51}Cr red-cell survival data, *Med. Phys.,* 1, 58, 1974.

102. **Glomski, C. A., Pillay, S. K. K., and MacDougall, L. G.,** Erythrocyte survival in children as studied by labeling with stable chromium-50, *Am. J. Dis. Child.,* 130, 1128, 1976.

103. **Ricketts, C., Cavill, I., and Napier, J. A. F.,** The measurement of red cell lifespan using ^{59}Fe, *Brit. J. Haematol.,* 37, 403, 1977.

104. **Weintraub, M., Gerson, K., and Silber, R.,** The use of ^{125}I as a membrane protein label for erythrocyte survival studies, *Blood,* 43, 549, 1974.

105. **Smith, G. N. and Mollison, P. L.,** Normal red cell survival in the rabbit, *Scand. J. Haematol.,* 11, 188, 1973.

106. **Meyerstein, N., Yagil, R., and Sod-Moriah, U. A.,** Red cell survival in heat-exposed hamsters, *Comp. Biochem. Physiol.,* 50A, 691, 1975.

107. **Elin, R. J.,** A simple method for determining erythrocyte survival in rats using chromium-51, *Lab. Anim. Sci.,* 23, 653, 1973.

108. **Dillon, W. G. and Glomski, C. A.,** Erythrocyte survival in the Mongolian gerbil, *J. Nucl. Med.,* 16, 682, 1975.

109. **Zinkl, J. G. and Feldman, D. B.,** The erythrocyte life-span in the American opossum *(didelphis-marsupialis-virginiana), Lab. Anim. Sci.,* 24, 500, 1974.

110. **Giles, R. C., Berman, A., Hildebrandt, P. K., and McCaffrey, R. P.,** The use of ^{51}Cr for sheep red blood cell survival studies, *Proc. Soc. Exp. Biol. Med.,* 148, 795, 1975.

111. **Manunta, G. and Cancedda, M.,** Contribution to the study of the survival time of red cells in horses using tritiated DFP, *Boll. Soc. Ital. Biol. Sper.,* 51, 771, 1975.

112. **Yagil, Y., Sod-Moriah, U. A., and Meyerstein, V.,** Dehydration and camel blood. I. Red blood cell survival in the one-humped *camelus dromedarius, Am. J. Physiol.,* 226, 298, 1974.

113. **Reddy, P. R. K., Van Krey, H. P., Gross, W. B., and Siegel, P. B.,** Erythrocyte life-span in dwarf and normal pullets from growth selected lines of chickens, *Poult. Sci.,* 54, 1301, 1975.

114. **Forman, L. J. and Just, J. J.,** The life span of red blood cells in the amphibian larvae *Rana Catesbiana, Dev. Biol.,* 50, 537, 1976.

115. **Borun, E. R., Figueroa, W. G., and Perry, S. M.,** The distribution of Fe^{59} tagged human erythrocytes in centrifuged specimens as a function of cell age, *J. Clin. Invest.,* 36, 676, 1957.

116. **Rigas, D. A. and Koler, R. D.,** Ultracentrifugal fractionation of human erythrocytes on the basis of cell age, *J. Lab. Clin. Med.,* 58, 242, 1961.

117. **Garby, L. and Hjelm, M.,** Ultracentrifugal fractionation of human erythrocytes with respect to cell age, *Blut,* 9, 294, 1963.

118. **Van Gastel, C., Vanden Berg, D., De Gier, J., and Van Deenen, L. L. M.,** Some lipid characteristics of normal red blood cells of different age, *Brit. J. Haematol.,* 11, 193, 1965.

119. **O'Connell, D. J., Caruso, C. J., and Sass, M. D.,** Separation of erythrocytes of different ages, *Clin. Chem.,* 11, 771, 1965.

120. **Brok, F., Ramot, B., Zwang, E., and Danon, D.,** Enzyme activities in human red blood cells of different age groups, *Isr. J. Med. Sci.,* 2, 291, 1966.

121. **Hoffman, J. F.,** On the relationship of certain erythrocyte characteristics to their physiological age, *J. Cell. Comp. Physiol.,* 51, 415, 1958.

122. **Prentice, T. C. and Bishop, C.,** Separation of rabbit red cells by density methods and characteristics of separated layers, *J. Cell. Comp. Physiol.,* 65, 113, 1965.

123. **Danon, D., Marikovsky, Y., Furedi, A., and Ohad, I.,** On the separation of old from young erythrocytes in hypotonic bovine albumin solutions, *Biochim. Biophys. Acta,* 104, 281, 1965.

124. **Bishop, C. and Prentice, T. C.,** Separation of rabbit red cells by density in a bovine serum albumin gradient and correlation of red cell density with cell age after in vivo labelling with Fe⁵⁹, *J. Cell. Physiol.,* 67, 197, 1966.

125. **Piomelli, S., Lurinsky, G., and Wasserman, L. R.,** The mechanism of red cell aging. I. Relationship between cell age and specific gravity evaluated by ultracentrifugation in a discontinuous density gradient, *J. Lab. Clin. Med.,* 69, 659, 1967.

126. **Winterbourn, C. C. and Batt, R. D.,** Lipid composition of human red cells of different ages, *Biochim. Biophys. Acta,* 202, 1, 1970.

127. **Coopersmit, A. and Ingram, M.,** Red cell volumes and erythropoiesis. I. Age density: volume relationship of normocytes, *Amer. J. Physiol.,* 215, 1276, 1968.

128. **Prankerd, T. A. J.,** The aging of red cells, *J. Physiol.,* 143, 325, 1958.

129. **Leif, R. C. and Vinograd, J.,** The distribution of buoyant density of human erythrocytes in bovine albumin solutions, *Proc. Natl. Acad. Sci.,* 51, 520, 1964.

130. **Rastgeldi, S.,** The threshold centrifuge, *Acta Physiol. Scand.,* 44, supp. 152, 1958.

131. **Murphy, J. R.,** Influence of temperature and method of centrifugation on the separation of erythrocytes, *J. Lab. Clin. Med.,* 82, 334, 1973.

132. **Turner, B. M., Fisher, R. A., and Harris, H.,** The age related loss of activity of 4 enzymes in the human erythrocyte, *Clin. Chim. Acta,* 50, 85, 1974.

133. **Sanderson, R. J., Palmer, N. F., and Bird, K. E.,** Separation of red cells into age groups by counterflow centrifugation, *Biophys. J.,* 15, 321A, 1975.

134. **Fitzgibbons, J. F., Koler, R. D., and Jones, R. T.,** Red cell age related changes of hemoglobin A ia-plus-b and hemoglobin A-c in normal and diabetic subjects, *J. Clin. Invest.,* 58, 820, 1976.

135. **Smith, B. D., LaCelle, P. T., and LaCelle, P. L.,** Elastic behaviour of senescent human erythrocyte membranes, *Biophys. J.,* 17, 28A, 1977.

136. **Walter, H., Wince, R., and Selby, F. W.,** Counter-current distribution of red blood cells of different ages and from different species, *Biochim. Biophys. Acta,* 109, 293, 1965.

137. **Sizova, N. A., Kamenskaya, V. V., and Fedenkov, V. I.,** The micro thermal method of separating red blood cells into age groups, *Byull. Eksp. Biol. Med.,* 84, 379, 1977.

138. **Gear, A. R. L.,** Age dependent separation of erythrocytes by preparative electrophoresis, *J. Lab. Clin. Med.,* 90, 744, 1977.

139. **Dreyfus, J. C., Schapira, G., and Kruh, J.,** Fractionation of red cells according to their age. Centrifugation of red cells labelled with radioactive iron, *Compt. Rend. Seances Soc. Biol.,* 144, 792, 1950.

140. **Rahman, Y. E., Elson, D. L., and Cerny, E. A.,** Studies on the mechanism of erythrocyte aging and destruction. I. Separation of rat erythrocytes according to age by ficoll gradient centrifugation, *Mech. Aging Dev.,* 2; 141, 1973.

141. **Piomelli, S., Seaman, C., Corash, L., Tytuna, A., and Graziano, H.,** Transfusion of red blood cells with improved survival. A new approach to the therapy of cooleys anemia, *Pediatr. Res.,* 12, 471, 1978.

142. **Schmidt, G., Gross, J., Moller, R., and Staak, R.,** Separation of red blood cells in the isopynic dextran and albumin gradient, *Acta Biol. Med. Ger.,* 34, 1621, 1976.

143. **Kadlubowski, M. and Agutter, P. S.,** Changes in the activities of some membrane association enzymes during in-vivo aging of the normal human erythrocyte, *Brit. J. Haematol.,* 37, 111, 1977.

144. **Löhr, G. W. and Waller, H. D.,** On the biochemistry of erythrocyte aging, *Folia Haematol.,* 78, 1962.

145. **Brok, F., Ramot, B., Zwang, E. and Danon, D.,** Enzyme activities in human red blood cells of different age groups, *Isr. J. Med. Sci.,* 2, 291, 1966.

146. **Marks, P. A., Johnson, A. B., and Hirshberg, E.,** Effect of age on the erythrocyte activity in erythrocytes, *Proc. Natl. Acad. Sci.,* 44, 529, 1958.

147. **Löhr, G. W., Waller, H. D., Karges, O., Schlegel, B., and Miller, A. A.,** Zur Biochemie der Alterung menschlicher Erythrocyten, *Klin. Wschr.,* 21, 1008, 1958.

148. **Marks, P. A.,** Red cell glucose-6-phosphate and 6-phosphogluconic dehydrogenase and nucleoside phosphorylase, *Science,* 127, 1338, 1958.

149. **Bonsignore, A., Fornaini, G., Fantoni, A., Leoncini, G., and Segni, P.,** Relationship between age and enzymatic activities in human erythrocytes from normal and Fava bean-sensitive subjects, *J. Clin. Invest.,* 43, 834, 1964.

150. **Fantoni, A., Leoncini, G., Calissano, P., Cartasegna, G., and Fornaini, G.,** Effect of NADP-Glycohydrolase on Glucose-6-Phosphate Dehydrogenase, *Enzymol. Biol. Clin.,* 3, 161, 1965.

151. **Bishop, C. and VanGastel, C.,** Changes in enzyme activity during reticulocyte maturation and red cell aging, *Haematologica,* 3, 29, 1969.

152. **Turner, B. M., Fisher, R. A., and Harris, H.,** The age related loss of activity of four enzymes in the human erythrocyte, *Clin. Chim. Acta,* 50, 85, 1974.

153. **Tochner, Z., Ben Bassat, J., and Hershko, C.,** Observations on the *in vivo* aging of red cells in the rats, *Scand. J. Haematol.*, 14, 377, 1975.

154. **Choy, Y. M., Wang, S. L., and Lee, C. Y.,** Changes in surface carbohydrates of Erythrocytes during *in vivo* aging, *Biochem. Biophys. Res. Comm.*, 91, 410, 1979.

155. **Gershon, D.,** personal communication.

156. **Sabine, J. C.,** Choline esterase of blood cells and plasma in blood dyscrasias, with special reference to pernicious anemia, *J. Clin. Invest.*, 19, 833, 1940.

157. **Allison, A. C. and Burn, G. P.,** Enzyme activity as a function of age in the human erythrocyte, *Brit. J. Haematol.*, 1, 291, 1955.

158. **Voyda, B., Knox, R. L., Lochner, J., and Seaman, G. V. F.,** Acetylcholineesterase and glutamic oxal transaminase as markers of red cell age, *Biophys. J.*, 21, 121A, 1978.

159. **Hubert, C. and Poon, M. C.,** Glutamic pyruvic transaminase activity related to red blood cell age, *Canad. J. Biochem.*, 53, 731, 1975.

160. **Grignani, F. and Löhr, G. W.,** Ueber die Hexokinase in menschlichen Blutzellen, *Klin. Wschr.*, 38, 796, 1960.

161. **Brewer, G. and Powell, R. D.,** Hexokinase activity as a function of age of the human erythrocyte, *Nature*, 199, 7604, 1963.

162. **Rogers, P. A., Fisher, R. A., and Harris, H.,** An examination of the age related patterns of decay of the hexokinases of human red cells, *Clin. Chim. Acta*, 65, 291, 1975.

163. **Pfeffer, S. R. and Swislocki, N. I.,** Age related decline in the activities of erythrocyte membrane adenylate cyclase and protein kinase, *Arch. Biochem. Biophys.*, 177, 117, 1976.

164. **Rijksen, G., Schoop, I., and Staal, G. E. J.,** Properties of human erythrocyte hexokinase related to cell age, *Clin. Chim. Acta*, 80, 193, 1977.

165. **Yip, L. C., Roome, S., and Balis, M. E.,** *In vitro* and *in vivo* age related modification of human erythrocyte phosphoribosyl pyrophosphate synthetase EC-2.7.6.1., *Biochemistry*, 17, 3286, 1978.

166. **Magnani, M., Stocchi, V., Bossu, M., Dacha, M., and Fornaini, G.,** Decay pattern of rabbit erythrocyte hexokinase in cell aging, *Mech. Ageing Dev.*, 11, 209, 1979.

167. **Fornaini, G., Magnani, M., Dacha, M., Bossu, M., and Stocchi, V.,** Relationship between glucose phosphorylating activities and erythrocyte age, *Mech. Ageing Dev.*, 8, 249, 1978.

168. **Fornaini, G., Dacha, M., Magnani, M., and Stocchi, V.,** Glucose phosphorylating activities and erythrocyte age, *Bull. Mol. Biol. Med.*, 4, 37, 1979.

169. **Kaplan, N. O.,** in *Metabolic Pathways*, 2nd ed., D. M. Greenberg, Ed., Academic Press, N.Y., Vol. 2, 627, 1961.

170. **Tanaka, K. R., Valentine, W. N., and Miwa, S.,** Pyruvate kinase (PK) deficiency hereditary nonsperocytic hemolytic anemia, *Blood*, 19, 267, 1962.

171. **Powell, R. D. and De Gowin, R. L.,** Relationship between activity of pyruvate kinase and age of the normal human erythrocyte, *Nature*, 205, 507, 1965.

172. **Gomperts, B. D.,** Metabolic changes in human red cells during incubation of whole blood *in vitro*, *Biochem. J.*, 102, 782, 1967.

173. **Jamil, T. P., Swallow, D. M., and Povey, S.,** A comparative study of the age related patterns of decay of some nucleoside monophosphate kinases in human red cells, *Biochem. Genet.*, 16, 1219, 1978.

174. **Bertolini, A. M., Massoni, N., and Guardamagni, C.,** Transaminase activity of the erythrocyte stroma in globular senescence, *Giorn. Gerontol.*, 9, 537, 1961.

175. **Murphy, I. R.,** Influence of temperature and method of centrifugation on the separation of erythrocytes, *J. Lab. Clin. Med.*, 82, 334, 1973.

176. **Teng, Y. S., Chen, S. H., and Giblett, E. R.,** Red cell UMP kinase effects of red cell aging on the activity of 2 UMP kinase gene products, *Am. J. Hum. Gen.*, 28, 138, 1976.

177. **Turner, B., Fisher, R. A., and Harris, H.,** in *Isozymes*, I, Academic Press, San Francisco, 1975.

178. **Herz, F.,** Acetylcholinesterase EC-3.1.1.7 inactivation in young and old human red blood cells, *Blut*, 31, 17, 1975.

179. **Sabine, J. C.,** The cholinesterase of erythrocytes in anemia, Blood 6, 151, 1951.

180. **Rogers, P. A., Fisher, R. A., and Putt, W.,** An examination of the age related patterns of decay of acid phosphatase in human red cells from individuals of different phenotypes, *Biochem. Gen.*, 16, 727, 1978.

181. **Valentine, W. N. and Tanaka, K. R.,** The glyoxalase content of human erythrocytes and leukocytes, *Acta Haemat.*, 26, 303, 1961.

182. **Bylund, D. B., Tellez-Inon, M. I., and Hollenberg, M. D.,** Age related parallel decline in β-adregenic receptors adenylate cyclase and phosphodiesterase activity in rat erythrocyte membranes, *Life Sci.*, 21, 403, 1977.

183. **Malkin, A. and Denstedt, O. F.,** The metabolism of the erythrocyte. X. The inorganic phyrophosphatase of the erythrocyte, *Canad. J. Biochem.*, 34, 121, 1956.

184. **Luthra, M. G. and Kim, H. D.,** Influence of calcium and cytoplasmic activator protein on various states of membrane calcium ion plus magnesium ion ATPase of total and density separated human red cells, *Fed. Proc.,* 38, 745, 1979.

185. **Schapira, F.,** Activite aldolasique et age des hematies, Rev. Franc. d'et., *Clin. et Biol.,* 4, 151, 1959.

186. **Mennecier, F., Weber, A., Tudury, C., and Dreyfus, J. C.,** Modifications of aldolase during *in vivo* aging of rabbit red cells, *Biochimie* (Paris), 61, 79, 1979.

187. **Bertolini, A. M., Massoni, N., Civardi, F., and Tenconi, L.,** The aldolase activity of the erythrocyte stroma globular senescence, *Giorn. Gerontol.,* 9, 551, 1961.

188. **Prankerd, T. A. J.,** Chemical and physical changes in the aging red cells, *Folia Haematol.,* 78, 382, 1962.

189. **Jalavistoe, E.,** Bleeding anemia and methemoglobin reduction in dog erythrocytes, *Acta Physiol. Scand.,* 46, 252, 1959.

190. **Hoffman, E. C. G. and Rapoport, S.,** Das Verhalten von Triphospho-pyridinnucleotid flavinademin-dinucleotid, coenzym A und Thiaminpyrophosphat in Haemolysaten von Kaninchen-Reticulozyten und Erythrocyten, *Hoppe-Seyler's Zeischr. Physiol. Chem.,* 304, 157, 1956.

191. **Prankerd, T. A. J.,** The aging of red cells, J. Physiol. London, 143, 325, 1958.

192. **Mandel, P.,** Aspect biochemique due phenomene de vieillissement et des maladies qui s'y rattachent, in *Expose's Annueles de Biochimie Medicale,* Masson et Co. Paris, 26, 1965, 299.

193. **Prentice, T. C., and Bishop, C.,** Separation of rabbit cells by density methods and characteristics of separated layers, *J. Cell. Comp. Physiol.,* 65, 113, 1965.

194. **Piomelli, S., Lurinsky, G. and Wasserman, L. R.,** I. Relationship between cell age and specific gravity evaluated by ultracentrifugation in a discontinuous density gradient, *J. Lab. Clin. Invest.,* 69, 660, 1966.

195. **La Celle, P. L., Kirkpatrick, F. H., Udkow, M. P., and Arkin, B.,** Membrane fragmentation and Ca^{++} membrane interaction; potential mechanisms of shape change in the senescent red cell, *Nuov. Rev. Fr. Hematol.,* 12, 789, 1972.

196. **Ramot, B., Brok, F., and Ben-Bassat, I.,** Alterations in the metabolism of human erythrocytes with aging, Plenary Session Papers, XII Congr. Int. Soc. Hematol., N.Y., 1968, 169.

197. **Shiga, T., Maeda, N., Suda, T., Kan, K., and Sekiya, M.,** The decreased membrane fluidity of *in vivo* aged, human erythrocytes. A spin label study, *Biochim. Biophys. Acta,* 553, 84, 1979.

198. **Ganzoni, A. M., Oakes, R., and Hillman, R. S.,** Red cell aging *in vivo, J. Clin. Invest.,* 50, 1373, 1971.

199. **Morrison, M., Michael, A. W., Phillips, D. R., and Choi, S-I.,** *Nature,* 248, 763, 1974.

200. **Cohen, N. S., Ekholm, J. E., Luthra, M. G., and Hanahan, D. J.,** Biochemical characterisation of density-separated human erythrocytes, *Biochim. Biophys. Acta,* 419, 229, 1976.

201. **Kadlubowski, M. and Harris, J. R.,** The appearance of a protein in the human erythrocyte membrane during aging, *FEBS Letters,* 47, 252, 1974.

202. **Smith, B. D., La Celle, P. T., and La Celle, P. L.,** Elastic behavior of senescent human erythrocyte membranes, *Biophys. J.,* 17, 28A, 1977.

203. **Moore, G. L., Cooper, D. A., Antonoff, R. S., and Robinson, S. L.,** Changes in human erythrocyte membrane proteins during storage, *Vox Sang.,* 20, 239, 1970.

204. **Bienzle, U. and Pjura, W. J.,** Alteration of membrane protein during erythrocyte aging, *Clin. Chim. Acta,* 76, 183, 1977.

205. **Lutz, H. U., Liv, S. C., and Palek, J.,** Release of spectrin-free vesicles from human erythrocytes during ATP depletion, *J. Cell Biol.,* 73, 548, 1977.

206. **Balduini, C., Brovelli, A., Balduini, C. L., and Ascari, E.,** Structural modifications in membrane glycoproteins during the erythrocyte life span, *Ric. Clin. Lab.,* 9, 13, 1979.

207. **Shohet, S. B. and Holby, J. E.,** in *Red Cell Shape,* Bessis, M., Weed, R. I., and LeBlond, P. F., Eds., Springer Verlag, N.Y., 1973, 41.

208. **Van Deenen, L. L. M. and De Gier, J.,** in, *The Red Blood Cell,* 2nd ed., Surgenor, D. M., Ed., Academic Press, N.Y., 1974, 198.

209. **Phillips, G. B., Dodge, J. T., and Howe, C.,** The effect of aging of human red cells *in vivo* on their fatty acid composition, *Lipids,* 4, 544, 1969.

210. **Greenwalt, T. J. and Steane, E. A.,** Quantitative haemagglutination. IV. Effect of neuraminidase treatment on agglutination by blood group antibodies, *Brit. J. Haematol.,* 25, 207, 1973.

211. **Balduini, C., Balduini, C. L., and Ascari, E.,** Membrane glycopeptides from old and young human erythrocytes, *Biochem. J.,* 140, 557, 1974.

212. **Durocher, J. R., Payne, C., and Conrad, M. E.,** Role of sialic acid in erythrocyte survival, *Blood,* 45, 11, 1975.

213. **Gattegno, L., Bladier, D., and Cornillot, P.,** Ageing *in vivo* and neuraminidase treatment of rabbit erythrocytes: Influence on half-life as assessed by ^{51}Cr labelling, *Hoppe-Seyler's Z. Physiol. Chem.,* 356, 391, 1975.

214. **Baxter, A. and Beely, J. G.,** Changes in surface carbohydrate of human erythrocyte aged *in vivo, Biochem. Soc. Trans.,* 3, 134, 1975.

215. **Bocci, V.,** The role of sialic acid in determining the life span of circulating cells and glycoproteins, *Experientia,* 32, 135, 1976.

216. **Gattegno, L., Bladier, D., Garnier, M., and Cornillot, P.,** Changes in carbohydrate content of surface membranes of human erythrocytes during ageing, *Carbohydr. Res.,* 52, 197, 1976.

217. **Luner, S. J., Szklarek, D., Knox, R. J., Seaman, G. V. F., Josefowicz, J. Y., and Ware, B. R.,** Red cell charge is not a function of cell age, *Nature,* 269, 719, 1977.

218. **Seaman, G. V. F., Knox, R. J., Nordt, F. J., and Regan, D. H.,** Red cell aging, Part 1: Surface charge density and sialic-acid content of density fractionated human erythrocytes, *Blood,* 50, 1001, 1977.

219. **Gattegno, L., Fabia, F., Bladier, D., and Cornillot, P.,** Physiological aging of red blood cells and changes in membrane carbohydrates, *Biomedicine,* 30, 194, 1979.

220. **Lutz, H. V. and Fehr, J.,** Total sialic acid contents of glycophorins during senescence of human red blood cells, *J. Biol. Chem.,* 254, 11177, 1979.

221. **Allan, D., Billah, M. M., Finean, J. B., and Michell, R. H.,** Release of diacylglycerol-enriched vesicles from erythrocytes with increased intracellular (Ca^{2+}), *Nature,* 261, 58, 1976.

222. **Greenwalt, T. J. and Lau, F. O.,** Evaluation of toluidine blue for measuring erythrocyte membrane loss during *in vivo* ageing, *Brit. J. Haematol.,* 39, 545, 1978.

223. **Fitzgibbons, J. F., Koler, R. D., and Jones, R. T.,** Red cell age related changes of hemoglobin A-IA-Plus-B and hemoglobin A-IC in normal and diabetic subjects, *J. Clin. Invest.,* 58, 820, 1976.

224. **Lessin, L. S.,** Denaturation of Hb., in *Red Cell Shape,* Bessis, M., Weed, R. I., Leblond, P. F., Eds., Springer Verlag, N.Y., 1973, 151.

225. **Harris, J. W. and Kellermeyer, R. W.,** in *The Red Cell,* Harvard University Press, Cambridge, Mass., 1970.

226. **Rigas, D. A. and Koler, R. D.,** Erythrocyte enzymes and reduced glutathione (GSH) in hemoglobin H disease: relation to cell age and denaturation of hemoglobin H, *J. Lab. Clin. Med.,* 58, 471, 1961.

227. **Abraham, E. C., Fuller, J., Taylor, F. J., and Lang, C. A.,** Influence of mouse age and erythrocyte age on glutathione metabolism, *Biochem. J.,* 174, 819, 1978.

228. **Allan, D. and Michell, R. H.,** Elevation of intracellular calcium ion concentration provokes production of 1,2-diacylglycerol and phosphatiolate in human erythrocytes, *Biochem. Soc. Transact.,* 3, 751, 1975.

229. **Nakao, M., Nakao, T., and Yamazoe, S.,** Adenosine triphosphate and maintenance of shape of the human red cell, *Nature,* 187, 945, 1960.

230. **Brecher, G., Jakobek, E. F., Schneiderman, M., Willisam, G. Z., and Schmidt, P. J.,** Size distribution of erythrocytes, *Ann. N.Y. Acad. Sci.,* 99, 242, 1962.

231. **Danon, D.,** *Biophysical aspects of red cell aging,* Xth Int. Soc. Haematol., Sydney, 1966, 1.

232. **Rachman, J. E., Elson, D. L., and Cerny, E. A.,** Studies on the mechanism of erythrocyte aging and destruction. I. Separation of rat erythrocytes according to age by ficoll gradient centrifugation, *Mech. Ageing Dev.,* 2, 141, 1973.

233. **Shiga, T., Maeda, N., Suda, T., Kon, K., and Sekiya, M.,** The decrease membrane fluidity of *in vivo* aged human erythrocytes, *Biochim. Biophys. Acta,* 553, 84, 1979.

234. **Canham, P. B.,** Difference in geometry of young and old human erythrocytes explained by a filtering mechanism, *Circulation Res.,* 25, 39, 1969.

235. **Cohen, N. S., Ekholm, J. E., Luthra, M. G., and Hanahan, D.,** Biochemical characterization of density separated human erythrocytes, *Biochim.,* 419, 229, 1976.

236. **Greenwalt, T. J. and Lau, F. O.,** Evaluation of toluidine blue for measuring erythrocyte membrane loss during *in vivo* aging, *Brit. J. Haematol.,* 39, 545, 1978.

237. **Chalfin, D.,** Differences between young and mature rabbit erythrocytes, *J. Cell. Comp. Physiol.,* 47, 215, 1956.

238. **Simon, E. R. and Topper, Y. J.,** Fractionation of human erythrocytes on the basis of their age, *Nature,* 180, 1211, 1957.

239. **Marks, P. A. and Johnson, A. B.,** Relationship between the age of human erythrocytes and their osmotic resistance: a basis for separating young and old erythrocytes, *J. Clin. Invest.,* 37, 1542, 1958.

240. **Prankerd, T. A. J.,** The aging of red cells, *J. Physiol.* (London), 143, 325, 1958.

241. **Danon, D. and Marikovsky, Y.,** Determination of density distribution of red blood cells, *J. Lab. Clin. Med.,* 64, 668, 1964.

242. **Bishop, C. and Prentice, T. C.,** Separation of rabbit red cells by density in a bovine serum albumin gradient and correlation of red cell density with cell age after *in vivo* labelling with Fe[59], *J. Cell Physiol.,* 67, 197, 1966.

243. **Knyszynski, A. and Danon, D.,** Membrane characteristics of old and Rauscher leukemia virus infected mouse red blood cells, *Exp. Cell Res.,* 100, 303, 1976.

244. **Leif, R. C.,** in *Automated cell identification and cell sorting,* Wied, G. L. and Bahr, G. F., Eds., Academic Press, N.Y., 1970, 21.

245. **Knox, R. J., Zukoski, C. F., and Seaman, G. V. F.,** Divergent responses of young and old red cells to hypertonic media, *Biophys. J.,* 21, 121A, 1978.

246. **Jacob, H. S. and Jandl, J. H.,** Effect of membrane sulfhydryl deficiency on the shape and survival of red cells, *Clin. Res.,* 9, 162, 1961.

247. **Jandl, J. H., Simmons, R. L., and Castle, W. B.,** Red cell filtration and the pathogenesis of certain hemolytic anemias, *Blood,* 18, 133, 1961.

248. **Bergentz, S. E. and Danon, D.,** Alterations in red blood cells of traumatized rabbits. I. Decreased filterability, *Acta Chir. Scand.,* 130, 165, 1965.

249. **Danon, D.,** Reversible deformability and mechanical fragility as function of red cell age, *First International Conf. Hemorheology,* Pergamon Press, Elmsford, N.Y., 497, 1967.

250. **Weed, R. L., LaCelle, P. L., and Merril, E. W.,** Metabolic dependence of red cell deformability, *J. Clin. Invest.,* 48, 795, 1969.

251. **Weed, R. L. and LaCelle, P. L.,** in *Red Cell Membrane Structure and Function,* Jamieson, G. A. and Greenwalt, T. J., Eds., Lippincott, Philadelphia, 1969, 318.

252. **Burton, A. C.,** in *Permeability and Function of Biological Membranes,* Bolis, L., Katchalsky, A., Keynes, R. D., Loewenstein, W. R., and Pethica, B. A., Eds., North Holland, Amsterdam, London, 1970, 1.

253. **LaCelle, P. L., Kirkpatrick, F. H., Udkow, M. P., and Arkin, B.,** Membrane fragmentation and Ca^{++} membrane interaction: potential mechanisms of shape change in the senescent red cell, *Nuov. Rev. Franc. Haematol.,* 12, 789, 1972.

254. **Levander, O. A., Morris, V. C., and Ferretti, R. J.,** Effect of aging on the filterability of red blood cells from vitamin E deficient lead poisoned rats, *Fed Proc.,* 36, 1168, 1977.

255. **Smith, B. D., LaCelle, P. T., and LaCelle, P. L.,** Elastic behaviour of senescent human erythrocyte membranes, *Biophys. J.,* 17, 28A, 1977.

256. **Williams, H. G., Escoffery, C. T., and Gorst, D. W.,** The fragility of normal and abnormal erythrocytes in controlled hydrodynamic shear field, *Brit. J. Haematol.,* 37, 379, 1977.

257. **Jaffé, E. R., Lowy, B. A., Vanderhoff, G. A., Aisen, P., and London, I. M.,** Effects of nucleosides on the resistance of normal human erythrocytes to osmotic lysis, *J. Clin. Invest.,* 36, 1498, 1957.

258. **Danon, D.,** Osmotic hemolysis by a gradual decrease in the ionic strength of the surrounding medium, *J. Cell. Comp. Physiol.,* 57, 111, 1961.

259. **Tochner, Z., Benbassat, J., and Hershko, C.,** Observations on the *in vivo* aging of red cells in the rats, *Scand. J. Haematol.,* 14, 377, 1975.

260. **Sadovnikova, I. P. and Obukhova, L. K.,** Use of the method of acid erythrograms for analyzing erythrocyte populations in experimental leukosis, *Izv. Akad Nauk SSSR Ser. Biol.,* 5, 744, 1974.

261. **Danon, D., and Marikovsky, Y.,** Morphologie des membranes des erythrocytes jeunes et ages, *Comp. Rend. Soc. Biol.,* 155, 12, 1961.

262. **Danon, D. and Perk, K.,** The age population distribution of erythrocytes in domestic animals. An electron microscopic study, *J. Cell. Comp. Physiol.,* 59, 117, 1962.

263. **Marikovsky, Y. and Danon, D.,** Structural differences between old and young negatively stained red cell membranes, *J. Ultrastr. Res.,* 20, 83, 1967.

264. **Danon, D. and Marikovsky, Y.,** Difference de charge electrique de surface entre erythrocytes jeunes et ages, *Comp. R. Acad. Sci.,* (Paris) 253, 1271, 1961.

265. **Marikovsky, Y., Danon, D., and Katchalsky, A.,** Agglutination by polylysine of young and old red blood cells, *Biochim. Biophys. Acta,* 124, 154, 1966.

266. **Yaari, A.,** Mobility of human red blood cells of different age groups in an electric field, *Blood,* 33, 159, 1969.

267. **Gear, A. R. L.,** Age dependent separation of erythrocytes by preparative electrophoresis, *J. Lab. Clin. Med.,* 90, 744, 1977.

268. **Luner, S. J., Szklarek, D., Knox, R. J., Seaman, G. V. F., Josefowicz, J. Y., and Ware, B. R.,** Red cell charge is not a function of cell age, *Nature (London),* 269, 719, 1977.

269. **Seaman, G. V. F., Knox, R. J., Nordt, F. J., and Reagan, D. H.,** Red cell aging, Part 1: Surface charge density and sialic acid content of density fractionated human erythrocytes, *Blood,* 50, 1001, 1977.

270. **Marikovsky, Y. and Danon, D.,** Electron microscope analysis of young and old red blood cells stained with colloidal iron for surface charge evaluation, *J. Cell Biol.,* 43, 1, 1969.

271. **Danon, D., Goldstein, L., Marikovsky, Y., and Skutelsky, E.,** Use of cationized ferritin as a label of negative charges on cell surfaces, *J. Ultrastr. Res.,* 38, 500, 1972.

272. **Marikovsky, Y. and Danon, D.,** Changes in surface negative charge and agglutination kinetics in red blood cells, *Biorheology,* 11, 349, 1974.

273. **Cohen, I.,** The Production and Use of Heteroantibodies Reacting with Proteins of the Anti-Gamma Globulin Type, Ph.D. thesis, The Hewbrew University, Jerusalen, 1965.

274. **Marikovsky, Y. and Danon, D.,** Agglutination of young and old human red cells by blood group antibodies, *Vox Sng.,* 20, 174, 1971.

275. **Tannert, C., Schmidt, G., Klatt, D., and Rapoport, S. M.,** Mechanisms of senescence of red blood cells, *Acta Biol. Med. Ger.,* 36, 831, 1977.

276. **Marikovsky, Y., Lotan, R., Lis, H., Sharon, N., and Danon, D.,** Agglutination and labelling density of soybean agglutinin on young and old human red blood cells, *Exp. Cell Res.,* 99, 453, 1976.

277. **Sewchand, L. S., Dixon, S. J., and Canham, P. B.,** Effects of pH and cell age on polyvinyl pyrrolidone induced doublet formation of human red cells, *Nouv. Rev. Fr. Hematol. Blood Cells,* 19, 727, 1977.

278. **Kunkel, H. G. and Bearn, A. G.,** Minor hemoglobin components of normal human blood, *Fed. Proc.,* 16, 760, 1957.

279. **Meyering, C. A., Israels, A. L. M., Sebens, T., and Huisman, T. H.,** Studies on the heterogeneity of hemoglobin. II. Quantitative aspects, *J. Clin. Chim. Acta,* 5, 208, 1960.

280. **Rosa, J., Schapra, G., and Dreyfus, J. C.,** Vieillissement moleculaire de l'hemoglobine de lapin pendant la vie du globule rouge, *Bull. Soc. Chim. Biol.,* 43, 555, 1961.

281. **Walter, H.,** Macromolecular ageing *in vivo, Nature,* 198, 189, 1963.

282. **Fitzgibbons, J. F., Koler, R. D., and Jones, R. T.,** Red cell age related changes of hemoglobin A ia-plus-b and hemoglobin A-c in normal and diabetic subjects, *J. Clin. Invest.,* 58, 820, 1976.

283. **Haidas, S., Labie, D., and Kaplan, J. C.,** Increased oxygen affinity with cell age, *Blood,* 38, 463, 1971.

284. **Edwards, M. J. and Staub, N. C.,** Kinetics of O_2 uptake by erythrocytes as a function of cell age, *J. Appl. Physiol.,* 21, 173, 1976.

285. **Edwards, M. J., Koler, R. D., Rigas, D. A., and Pitcairn, D. M.,** Increased oxygen affinity in HB with cell age, *J. Clin. Invest.,* 40, 630, 1961.

286. **Heisterkamp, D., and Ebauch, F. G.,** Site of attachment of the chromate ion to the haemoglobin molecule, *Nature,* 193, 1253, 1962.

287. **Walter, H.,** Chemical reactivity of a macromolecule as a function of its age, *Biochim. Biophys. Acta,* 69, 410, 1963.

288. **Danon, D., Marikovsky, Y., and Gasko, O.,** [51]Cr uptake as a function of red cell age, *J. Lab. Clin. Med.,* 67, 70, 1977.

289. **Skutelsky, E., Marikovsky, Y., and Danon, D.,** Immuno-ferritin analysis of membrane antigen density: A. Young and old human red blood cells. B. Developing erythroid cells and extruded erythroid nuclei, *Eur. J. Immunol.,* 4, 512, 1974.

290. **Tannert, C., Schmidt, G., Klatt, D., and Rapoport, S. M.,** Mechanism of senescence of red blood cells, *Acta Biol. Med. Ger.,* 36, 831, 1977.

291. **Alderman, E. M., Fundenberg, H. H., and Lovins, R. E.,** The isolation and characterization of the immunoglobulin receptor on aged human red cell membranes, *Fed. Proc.,* 38, 806, 1979.

292. **Stratton, F., Renton, P. H., and Rawlinson, V. T.,** Serological difference between old and young cells, *Lancet,* 25, 1388, 1960.

293. **Cohen, I.,** The production and use of heteroantibodies reacting with protein of the anti-gamma globulin type, Ph.D. Thesis, The Hebrew University, Jerusalem, 1965.

294. **Marikovsky, Y. and Danon, D.,** Agglutination of young and old human red cells by blood group antibodies, *Vox Sang.,* 20, 174, 1971.

295. **London, I. M.,** Metabolism of the erythrocyte, in *The Harvey Lectures,* Academic Press, New York, London, 1960, 151.

296. **Danon, D.,** Biophysical aspects of red cell aging, Xth Intl. Soc. Haematol., Sydney, p. 1, 1966.

297. **Kay, M. M. B.,** Mechanism of removal of senescent cells by human macrophages *in situ, Proc. Nat. Acad. Sci.,* 72, 352, 1975.

298. **Kay, M. M. B.,** Role of physiologic autoantibody in the removal of senescent human red cells, *J. Supramol. Struct.,* 9, 555, 1978.

299. **Knyszynski, A., Leibovich, S. J., and Danon, D.,** Phagocytosis of ''old'' red blood cells by macrophages from syngeneic mice *in vitro, Exp. Hematol.,* 5, 480, 1977.

300. **Knyszynski, A., Leibovich, S. J., Skutelsky, E., and Danon, D.,** Macrophages from unstimulated mineral-oil and thioglycollate stimulated mice: A comparative study of surface anionic sites and phagocytosis of ''old'' red blood cells, *J. Reticuloendothel. Soc.,* 24, 205, 1978.

301. **Purcell, Y. and Brozovic, B.,** Red cell 2,3-diphosphoglycerate concentration in man decreases with age, *Nature,* 251, 511, 1974.

302. **Kalofoutis, A., Paterakis, S., Koutselinis, A., and Spanos, V.,** Relationship between erythrocyted 2,3 di-phosphoglycerate and age in a normal population, *Clin. Chem.,* 22, 1918, 1976.

303. **Naylor, G. J., Dick, D. A. T., Worrall, E. P., Dick, E. G., Dick, P., and Boardman, L.,** Changes in erythrocyte sodium pump with age, *Gerontology,* 23, 256, 1977.

304. **Hegner, D., Platt, D., Heckers, H., Schloeder, J., and Breuninger, V.,** Age-dependent physiochemical and biochemical studies of human red cell membranes, *Mech. Ageing Dev.,* 10, 117, 1979.

305. **Ueda, K.,** The levels of human erythrocyte superoxide-dis-mutase activity classified by age and *in vitro* inhibition by some oxide-dis-mutase activity classified by age, *Okayama Igakkai Zasshi*, 90, 451, 1978.

306. **Ueda, K., and Ogata, M.,** Levels of erythrocyte super oxide-dis-mutase activity in Japanese people, *Acta Med. Okayama,* 32, 393, 1978.

307. **Joenje, H., Frants, R. R., Arwert, F., and Eriksson, A. W.,** Specific activity of human erythrocyte superoxide dismutase as a function of donor age, *Mech. Ageing Dev.*, 8, 265, 1978.

308. **Stevens, C., Goldblatt, M. J., and Freedman, J. C.,** Lack of erythrocyte superoxide dismutase change during human senescence, *Mech. Ageing Dev.*, 4, 415, 1975.

309. **Araki, K. and Rifkind, J. M.,** Erythrocyte membrane cholesterol: An explanation of the aging effect on the rate of hemolysis, *Life Sci.*, in press.

310. **Malhotra, S. and Kritchevsky, D.,** Cholesterol exchange between the red blood cells and plasma of young and old rats, *Mech. Ageing Dev.*, 4, 137, 1975.

311. **Abraham, E. C., Fuller, J., Taylor, F., and Lang, C. A.,** Influence of mouse age and erythrocyte age on glutathione metabolism, *Biochem. J.*, 174, 819, 1978.

312. **Stenhose, N. and Woodliff, H.,** Red cell indices and age, *Med. J. Austral.*, 2, 614, 1974.

313. **Koji, A. and Rifkind, J. M.,** Age dependent changes in osmotic hemolysis of human erythrocytes, *J. Gerontol.*, in press.

314. **Kelly, A. and Munan, L.,** Haematologic profile of natural populations: red cell parameters, *Brit. J. Haematol.*, 35, 153, 1977.

315. **Detraglia, M., Cook, F. B., Stasiw, D. M., and Cerny, L. C.,** Erythrocyte fragility in aging, *Biochim. Biophys. Acta,* 345, 213, 1974.

316. **Araki, K. and Rifkind, J. M.,** A relationship between membrane lipid fluidity and the rates of osmotic hemolysis of human erythrocytes, *Biophys. J.*, 25, 189a, 1979.

317. **Marx, J. J. M.,** Normal iron absorption and decreased red cell iron uptake in the aged, *Blood,* 53, 204, 1979.

318. **Grey, J. E. and Lee, P.,** Age dependence of transport of aminoacids in pig red blood cells, *Fed. Proc.,* 36, 563, 1977.

319. **Dvilansky, A., Bar-Am, J., Nathan, I., Kaplan, H., and Galinsky, D.,** Hematologic values in healthy older people in the Negev area, *Isr. J. Med. Sci.*, 15, 821, 1979.

CELLS OF THE IMMUNE RESPONSE

J. E. Nagel, R. H. Yanagihara, and W. H. Adler

INTRODUCTION

In mammalian species the progenitors of cells that are ultimately responsible for an immune response are first found in the yolk sac. During fetal life, hemopoiesis shifts to the liver, then, with further growth, to the bone marrow where regeneration of lymphoid and auxiliary cell precursors continues throughout life. These cells are part of the lymphoreticular system, a network composed of formed, encapsulated lymphoid organs, unencapsulated collections of lymphoid cells, and circulating cells with diverse functions.[1]

One method which is useful in examining the lymphoreticular system is to divide it into central and peripheral elements. The central, encapsulated organs consist of the spleen and thymus. Both have a well-defined architecture encompassing collections of lymphoid cells. However, the spleen and thymus are functionally quite different. Antibody synthesis, for instance, can take place in the spleen, but is rarely associated with the thymus. Another difference is the ability of thymic cortical tissue to atrophy or markedly involute under stress or with age. This reactive pattern of involution is not found in the spleen, although areas of white pulp can diminish during therapy with corticosteroid drugs. Peripheral lymphoreticular tissue can be divided into both encapsulated and nonencapsulated lymphoid tissue and circulating lymphocytes. Encapsulated lymph nodes are found throughout the body and are the sites of regional responses to infection and immunization. Their structure is well defined, with paracortical areas containing a high proportion of T lymphocytes, and cortical areas containing germinal centers, the sites of antibody synthesis. Nonencapsulated lymphoid tissues are found associated with the gut wall and lung tissue. Their organization parallels that of lymph nodes, but the absence of a capsule allows more diffuse blending of the lymphoid elements into the tissues of the organ. The peripheral blood contains lymphoid cells that are fed into the vascular bed from the bone marrow, spleen, thymus, and lymphatic channels. These circulating lymphocytes are not a homogeneous population as may be deduced from their multiple origins. Peripheral blood lymphocytes can migrate to specific organs, lymphoid tissues, or to sites of infection or inflammation. They are also the source of cells seen in body fluids (urine, CSF, breast milk), gut contents, and respiratory secretions. In man, peripheral blood lymphocytes are the most studied of the lymphoreticular tissues and provide the basis for the majority of knowledge of human cellular immune function. It is important to realize that the immune function of these circulating lymphocytes may not mirror the function of either the central lymphoid organs or the lymph node tissues.

Among the earliest detectable cells of the lymphoreticular system are the colony forming units in the spleen (CFU-S), identifiable as macroscopic splenic nodules in lethally irradiated mice 8 to 10 days following the injection of syngeneic bone marrow. Studies tracing radiation-induced chromosomal markers have established that a single CFU-S can give rise to precursors of myeloid-monocytoid, erythroid, and megakaryocytoid lines. Lymphoid precursors have not been isolated from CFU-S. There is, however, indirect support for the proposal that a common lymphoid precursor exists, which can give rise to more than one functional lymphocyte subclass, and that a single totipotent stem cell may be the precursor to all hematopoietic cells.[2]

Early work established the functional diversity of the immune response by experimental extirpation of one or another central lymphoid organ. In the chicken, for instance, neonatal thymectomy results in the loss of the ability to initiate both antibody- and cell-mediated immune responses, whereas neonatal removal of the bursa of Fabricius results only in defective antibody synthesis.[3] Research in experimental animals and observations of human immunodeficiency syndromes have reinforced the concept that effective immunity often represents an equilibrium between separate but complementary lymphoid functions.

By convention, two major lymphocyte subclasses have been defined. One cell type, which depends on the thymus for full expression of its phenotypic and functional repertoire, has been termed the "thymus-derived" or T lymphocyte. T cell-associated immune functions include allograft rejection, delayed hypersensitivity responses, some types of cellular cytotoxicity, and the production, elaboration, and recognition of various soluble mediator substances. The other major lymphocyte subclass, which is derived from precursor cells in the bone marrow or avian bursa, has been termed the B lymphocyte. The major role of B lymphocytes is to synthesize antibody.[4] In addition to functional and phenotypic differences among lymphocytes, other characteristics expressing their heterogeneity have been described. T and B lymphocytes, for example, differ in their relative sensitivity to the effects of corticosteroids, cytotoxic drugs, or X-irradiation. Detailed study of functional lymphocyte subpopulations has been facilitated by the recognition and identification of specific cell surface alloantigens. While many of these cell membrane markers have recently proven to be less specific or unique than was once thought, a constellation of several markers can often identify or phenotype a cell with a high degree of accuracy.

T LYMPHOCYTES

Thymus-derived or T lymphocytes comprise the majority (60 to 80%) of the circulating pool of lymphoid cells in humans and account for 20 to 30% of murine splenic lymphocytes. While T cells are morphologically homogeneous, they may be separated by numerous physical and immunological techniques from other lymphoid cells and into distinct subpopulations. A simple classification based on function recognizes helper/inducer, suppressor, and cytotoxic activities among T lymphocytes.

As lymphocyte subpopulations differentiate, discrete sets of genes, which control the expression of cell surface antigens, are expressed. Using congenic and noncongenic mouse strains, it has been possible to prepare many alloantisera to these cell membrane macromolecules. Some of the antisera, such as those of the Ly series, identify lymphocyte-specific antigenic systems, while other antisera react widely with both lymphocytes and nonlymphoid tissue. Among the more widely recognized of the nonlymphocyte-specific surface markers are the H-2, Qa, G_{ix}-gp70 ($Gross_{ix}$-gp70), TL (thymic leukemia), and Thy-1 (theta) antigens. Each of these markers was originally defined by alloantisera and each, like the Ly antigens, are the product of structural and occasionally regulatory genes on specific chromosomes. As noted, most antibodies to murine cell surface antigens were originally produced by immunization of congenic mice; however this source of antisera has now been largely superceded by monoclonal antibodies produced by cell hybridization. Monoclonal antibodies to the Thy-1, H-2, Qa, Lyt, and Lyb antigens, as well as new markers such as LFA-1 (lymphocyte function associated antigen-1), have now been produced.[5] The use of monoclonal antibodies has generally resolved the common problems inherent to oligospecific alloantisera raised in congenic mice: low and unpredictable titers; contamination by autoantibodies, soluble immune complexes, or antiviral antibodies; or antibodies with multiple specificities. Two inherent properties of monoclonal antibodies, (1) their ability to be produced against weak or difficult to isolate immunogenic determinants and (2) the limitation of their antibody specificity to a single defined idiotype, allow examination of specific cell populations with a much greater degree of precision than was possible using cytotoxic depletion with even the best conventional antisera. Many of the results from recent studies utilizing monoclonal antibodies are inconsistent with prior dogma regarding the distribution of specific cell surface antigens on functional cell populations. For example, the Lyt-1, Lyt-

2, and Lyt-3 antigens were once considered to be T lymphocyte specific. However, severalgroups have demonstrated that a population of large, Thy-1$^+$, IgM$^+$ splenic B cells, and several murine B cell lymphomas express the Lyt-1 phenotype.[6-8] While discrete stages of T cell differentiation are still thought to correlate with the relative or absolute acquisition or loss of specific antigenic markers, it is now clear that this is not the case among functional cell populations. For example, prothymocytes in adult bone marrow have H-2 antigens, but are otherwise phenotypically G_{ix}-gp70$^-$TL$^-$Thy-1$^-$Lyt-1$^-$23$^-$. Under appropriate inductive circumstances in the thymus, prothymocytes develop additional surface antigenic markers as the next step of differentiation. All peripheral blood T cells have the Lyt-1 antigen on their surface. The density of this antigen is proportional to the maturity of the cell, with approximately four times as much Lyt-1 antigen present on the surface of peripheral blood lymphocytes as on thymocytes. When thymocytes leave the thymus as immunocompetent T cells, they continue to express Thy-1 antigen but at a lower level. Therefore it has been suggested that the amounts of Lyt-1 and Thy-1 antigen expressed by a cell are inversely related.[9] The G_{ix}-gp70 and TL antigens are no longer detectable on thymocytes, while the amount of H-2 antigen has increased. Discrete functional subpopulations that are Lyt-1$^+$23$^-$, Lyt-1$^-$23$^+$, or Lyt-1$^+$23$^+$ may be identified.[10] It is presently unclear whether differentiated T cells with various Lyt phenotypes derive from a single Lyt-1$^+$23$^+$ precursor or whether precursor lineages exist that are Lyt-1$^+$23$^+$ and Lyt-1$^+$23$^-$.[11] As previously mentioned, immunocompetent T cells were thought to retain both their Lyt phenotype and function over time; however, recent experiments indicate this is not the case. Lyt-1$^+$2$^-$ T helper cells can differentiate into Lyt-1$^-$2$^+$ T helper cells during an immune response.[12] Likewise, Lyt-1$^+$23$^+$ cytotoxic precursor cells differentiate, following in vitro activation, into Lyt-1$^-$23$^+$ cytotoxic effector cells.[13] These findings suggest that the Lyt phenotypes of functionally differentiated T cells are not unalterable and may change during cell activation.

Other murine cell membrane alloantigenic systems (e.g., Qa), as well as additional Lyt and Lyb determinants continue to be described. While work is ongoing to characterize the cellular and functional distribution of the surface membrane antigens, their actual purpose, function, and cellular source remain uncertain. Detailed analysis of several membrane antigens, including H-2K, H-2D, and Qa-2, indicate that a striking structural similarity exists among these molecules, suggesting that they have all evolved from a common primordial gene.[14] Restriction enzyme probes and peptide sequencing experiments also demonstrate that Thy-1 glycoprotein is very likely a homologue of the variable-region immunoglobulin domains.[15] While present evidence strongly suggests that surface membrane antigens directly regulate cell and antigen recognition, cell activation, cell secretion, cell-to-cell interaction, or function as receptors for soluble immunoregulatory signals, information at the molecular level about these interactions is currently lacking.

Until recently, the variety and number of cell membrane antigens available to identify, study, and separate populations of human T lymphocytes was much less diverse and sophisticated than what was available in the murine system. Before the development of monoclonal antibodies, progress in developing high quality, specific antihuman antisera, which could be used to define functional lymphocyte subpopulations, was very difficult. Because of the lack of widely available antisera, separation of human mixed cell populations was generally done using physical techniques based on rosetting with sheep or murine erythrocytes, or erythrocytes coated with IgM or IgG. While these latter techniques yield cell subpopulations with helper and suppressor/cytotoxic functions, they are time consuming and yield cells that subsequent analysis has shown have little correlation with T cell subsets defined with monoclonal antisera.[16] Using hybridoma technology, a large number of laboratories have now successfully generated monoclonal antibodies to cell membrane antigens of both human normal and tumor cells. Because of the demonstration that the representation of helper and suppressor cells, or their ratio, is altered in many diseases, there is great interest among clinical researchers in monoclonal antibodies that identify antigens on functional cell subpopulations. However due to a lack of understanding of how these markers relate to the etiology or subsequent pathology of a particular disease, there is disagreement as to their present clinical usefulness and significance.

One function of helper T cells is to interact with B cells to trigger the production of antibody by these cells to most antigens. T helper cell activity may be mediated by intact cells or their soluble products, and may be antigen specific or nonspecific. T helper lymphocytes synthesize and secrete numerous peptides which cause differentiation and activation of other T lymphocytes, monocytes, B lymphocytes, or hematopoietic precursor cells.[17] In the mouse, antigens may be classified as thymus dependent or thymus independent on the basis of their requirement for antigen-sensitized T helper cells in mounting an immune response.[18] Historically, murine T cells with helper activity were characterized by their expression of the antigen Lyt-1. The presence of Lyt-1 seems to be preprogramed and is not antigen specific.[19] Recently, however, flow cytometry studies employing a monoclonal antibody to the 67 kd Lyt-1 antigen have demonstrated that the expression of Lyt-1 is not restricted to Lyt-2$^-$ helper cells, but is clearly present on Lyt-2$^+$ cytotoxic/suppressor cells.[9,20] This finding has also been confirmed by functional analysis of the allospecific T cell line C.C3.11.75, which is homogeneous for Lyt and Thy-1 expression and expresses only Lyt-1, by demonstrating that proliferation, allogeneic help, and allogeneic cytotoxicity are all carried out by Lyt-1$^+$ cells. Since the cytotoxic activities of these Lyt-1 cells do not depend upon the presence of Lyt-2 molecules, it has been suggested that Lyt-1 may be a marker for Class II (I region) MHC antigens, while Lyt-2 is a marker for Class I (H-2) antigens.[21] To further complicate matters, carrier-primed T helper cells have been found to be capable of changing their Lyt phenotype from Lyt-1$^+$2$^-$ to Lyt-1$^-$2$^+$ following secondary antigen challenge.[12] Since it has been generally held that helper T cells do not, on the basis of their resistance to X-irradiation and metabolic inhibitors, undergo cell division, the mechanism by which this phenotypic change occurs is presently unclear.

In humans, it was initially demonstrated that cells bearing the F_c receptor for IgM (T_μ cells) displayed T helper cell activity. While helper cell activity of T_μ cells is less conclusive than parallel observations of Lyt-1 cellular function in the mouse, this is probably attributable to the finding that the T_μ subset contains both helper/inducer (OKT4$^+$) and cytotoxic/suppressor (OKT5$^+$) populations and is similar to unfractionated T cells.[16] Recently several monoclonal antibodies (OKT-4, Leu-3a, Leu-3b, T4A) have been produced that identify 55-64 kd antigen(s) on human helper/inducer T lymphocytes. Functional studies with these and other antibodies demonstrate, as in the murine system, a heterogeneity within the OKT4$^+$ T helper cell population, with some helper cells apparently being capable, under the influence of appropriate inductive signals, of exerting suppressive or cytotoxic activity.[22-24] The suppressor activity within the T helper cell population has been theorized to function as a feedback control mechanism to down regulate other helper/inducer T cells and thereby limit an immune response.[25] The T helper cells capable of exerting suppressor activity may be distinguished from "pure" helper T cells by their radiosensitivity, the absence of IL-2 receptors on their surface, and their reactivity with the monoclonal antibody OKT-17.[23,24] The cytotoxic effector cells bearing the "helper" OKT4$^+$ or Leu-3$^+$ phenotype have been shown to recognize class II (HLA-DR) antigens which allows their functional distinction from T8$^+$ or Leu-2$^+$ cytotoxic cells, which are directed against class I (HLA-A,B) antigens.[26] Clearly, additional research is needed to define the antigen phenotypes and precise functional characteristics of the helper cell subpopulations in both mice and humans.

Suppressor T lymphocytes regulate immune responses by initiating inhibitory signals. Phenotypically, murine suppressor cells have been reported to have a variety of Lyt antigens expressed on their surface. Lyt antisera have defined the suppressor subpopulation as cells which alternatively bear the Lyt-1$^-$23$^+$, Lyt-1$^+$23$^-$, or Lyt-1$^+$23$^+$ phenotype. While these findings have raised considerable doubt regarding the association of a single Lyt phenotype with a given functional T cell population, further investigation has demonstrated that T suppressor cells are not functionally homogeneous, but are composed of phenotypically different interacting cell subpopulations. It is not known whether suppressor activity is carried out by several distinct, phenotypically diverse sets of suppressor T lymphocytes functioning in concert, or by a single cell type undergoing successive changes in phenotype. Like helper/inducer function, suppressor function also may be mediated by soluble factors. For instance, antigen-specific cloned T cells with the phenotype Thy-1$^+$, Lyt-1$^+$23$^+$, in addition to carrying antigen receptors on their surface, produce a group of 70 kd antigen-binding peptides which directly inhibit the activity of Lyt-1$^+$ helper cell.[27] In the normal immune response, the "activation" by antigen of Lyt-1$^+$ helper cells also results in the induction of Lyt-1$^+$23$^+$, Qa-1$^+$ cells which feedback inhibitory signals to the Lyt-1$^+$ helper cells. Aged or autoimmune prone mouse strains appear to have abnormalities in this Lyt-1$^+$23$^+$ suppressor cell population.[28] Suppressor cell inducers can be activated by the plant mitogen concanavalin A. These inducer cells actively synthesize DNA, are Lyt-1$^+$ and Ia$^+$, and carry the surface antigens for the I-J region of the mouse histocompatibility complex. Following mitogen activation, these inducer cells in turn activate a population of suppressor cells that are I-J$^+$ and Lyt-1$^-$23$^+$. While Con A-generated suppressor cells lack specificity, they are capable of inhibiting both T and B cell responses in vitro and have been extensively used as a laboratory model to study suppressor activity.[29]

In humans, the functional counterpart to murine Lyt-1$^-$23$^+$ suppressor cells was initially characterized as the population of T cells with Fc receptors for IgG (T$_\gamma$). Since then the production of monoclonal antibodies to human T cell surface antigens has led to the recognition that cells with helper/inducer or cytotoxic/suppressor function have distinct, mutually exclusive surface phenotypes. Using a number of monoclonal antibodies (OKT-8, Leu-2, T8A, OKT5, UCHT-4) researchers have found that human suppressor/cytotoxic cells are characterized by the presence of a group of 30- to 34-kd glycoproteins on their surface. While both anti-HLA-A,B cytotoxic activity, and suppressor function are mediated by OKT-8$^+$ or Leu-2$^+$ cells, differences in the ability of various monoclonal antibodies to inhibit cell-mediated lympholysis, while leaving suppressor function intact, strongly indicate heterogeneity within the human suppressor/cytotoxic cell population.[30] In preliminary experiments, the anti-Leu-8 monoclonal antibody, which reacts with only a portion of Leu-2$^+$ suppressor cells, has demonstrated that at least two phenotypically distinct cell types of suppressor lineage interact to produce suppression of an immune response.[31]

The functional assay for cytotoxic T lymphocytes (CTL) is the quantitation of specific killing of syngeneic and allogeneic cells in vitro. T cell cytotoxic activity is dependent on specific antigenic sensitization and is mediated by intact cells in the absence of antibody or complement. CTL are derived from precursor cells (CTL-P) found in the spleen and lymph nodes, as well as in the peripheral blood. While CTL-P were known to express the Thy-1 antigen, it has only recently become clear that CTL-P are small Lyt-1$^+$2$^+$ lymphocytes.[32] Following triggering by antigen in the presence of a soluble growth factor(s) that is not interleukin-2 (IL-2), and without apparent DNA synthesis, CTL-P are metabolically activated and develop cytolytic activity. This initial antigen recognition and triggering may be blocked by anti-Lyt-2 monoclonal antibodies.[33] Cellular activation is followed in turn by several rounds of proliferation which are regulated by IL-2.[34-36] The mature CTL appear morphologically as large lymphoblasts with some of the same surface antigens (Thy-1$^+$, Lyt-1$^+$2$^+$) as CTL-P. With time, CTL lose their cytolytic activity; however some remain as "memory" CTL which may be quickly reactivated by exposure to the appropriate antigen. Until recently, all CTL were characterized by the Lyt-1$^-$23$^+$ phenotype; however Lyt-1$^+$23$^-$ cells are now recognized as distinguishing class II major histocompatibility complex antigens and exerting allogeneic cytotoxicity.[21] Skin graft rejection, for both H-2 and non-H-2 loci, is entirely dependent on Lyt-1$^+$ cells.[37] While CTL express a variety of surface antigens, including Thy-1, H-2, Ala, T145, and Lyt-5,6, it appears obligatory that cells with the Lyt-1$^+$, Lyt-123$^+$, and Lyt-1$^-$23$^+$ phenotype must all collaborate for the induction of both H-2 and allogeneic cytotoxicity. Monoclonal antibodies to the alloantigens Qat-4 and Qat-5 seem to be capable of distinguishing between H-2-restricted and alloreactive CTL effector cells, respectively.[38] In the murine system a homologue for the human T3 surface receptor complex has not yet been identified. Some, but not all, antibodies to Thy-1 antigen are capable of inducing IL-2 release from T cells and enhancing IL-2 induced proliferation of cloned T cells; however, the surface density of Thy-1 is many times greater than T3. Therefore, it has been suggested that Thy-1 and T3 may share epitopes and that homology exists between T3, Thy-1 and the T cell receptor Ti. Several models to explain the interactions between antigen and murine cell surface receptors have been proposed.[39]

The events involved in human CTL activity are equally complex. As previously noted, alloreactive CTL may be T4$^+$ or T8$^+$ depending on whether they are directed, respectively, against DR or HLA-A,B MHC antigens on the target cells. Present evidence indicates that in addition to the T4 or T8 surface glycoproteins, the surface molecule T3 (a marker for mature T lymphocytes) and Ti (the human T cell antigen receptor) are necessary for the development of CTL activity. The T3-Ti receptor complex is triggered by antigen and the appropriate MHC gene product, and results in modulation of the T3-Ti complex, decreased numbers of antigen receptors, and induction of IL-2 receptor expression. This, in turn, leads to endogenous IL-2 production. In the absence of continued antigen stimulation, surface T3-Ti receptor is re-expressed and the number of IL-2 receptors decreases, down regulating clonal proliferation.[40,193] The recognition that enhanced allograft survival is achieved by matching a donor and recipient for both HLA-A,B and DR antigens has led to the experimental clinical use of monoclonal antibodies as a treatment for renal allograft rejection.[42] Similarly, monoclonal antibodies have been used in vitro to remove undesired cell populations from allogeneic or autochthonous bone marrow used in the treatment of immunodeficiency syndromes or neoplasia.[43,44]

The precise lineage of both murine and human natural killer (NK) cells is controversial. Despite the fact that human, rat, and murine NK cells may be morphologically distinguished as large granular lymphocytes (LGL), their phenotypic characterization with monoclonal antibodies indicates that they share antigens with both T lymphocytes and monocytes. For example, at least a portion of murine NK cells bear the T cell surface alloantigens Thy-1, Lyt-1, Qat-3,4,5, H11, 7.2, Lyt-5,6, T200, or Lyt-10,11. Likewise, at least some human NK cells are reactive with a number of monoclonal antibodies (9.6 3A1, OKT-10) which are thought to detect only T cell-associated antigens. While the OKT-10 monoclonal antibody, which identifies mainly thymocytes, reacts with virtually all human LGL, it is unreactive with most peripheral blood lymphocytes. In contrast, the OKT-3, OKT-11, and 9.3 monoclonal antibodies, which identify mature T lymphocytes, are nonreactive with LGL. A portion of LGL is OKT-8$^+$, but none appear to be OKT-4$^+$, or OKT-6$^+$.[45-47] As can readily be appreciated, the phenotype of human LGL is clearly different from most T lymphocytes, yet NK cells still express many T cell-associated surface antigens. Evidence linking NK activity with cells of the monocyte lineage is slightly less convincing. While LGL and monocytes do appear morphologically similar, treatment of NK cells with anti-monocyte antibodies has yielded variable results. Human LGL are known to react with several monoclonal antibodies (OKM-1, Mac-1, 63D3, 4F2) that also react with a high proportion of monocytes and granulocytes, but that are unreactive with most T lymphocytes.[45-47] It is clear that the receptors on NK cells differ from those on CTL. While CTL recognize MHC antigens, NK cells have no MHC restrictions. NK cells demonstrate highly efficient killing of target cells such as K562 which do not express MHC determinants. NK cytotoxic cells also have surface antigens that are distinct from T and B lymphocytes and monocytes. In the mouse, two nonallelic alloantigens designated NK-1 and NK-2 are considered to be specific for NK cells.[48-49] Several monoclonal antibodies (HNK-1, anti-N901, VEP 13, NK8, anti-Leu 11a,b, anti-Leu 15) have been produced which identify human LGL.[50,53]

B LYMPHOCYTES

The expression of intrinsic surface and/or cytoplasmic immunoglobulin is considered to be a trait restricted to B lymphocytes. Such cells comprise 10 to 25% of murine splenic or human peripheral blood lymphocytes. In addition to surface (S) or cytoplasmic (C) immunoglobulin (Ig), mature B lymphocytes frequently have other cell surface markers. While these markers are in many cases not specific or universal, they remain traditional indices for the identification of B lymphocytes, and include complement and Fc receptors, and Ia, Ia-like, and DR antigens encoded by genes of the murine or human histocompatibility complex. Other lymphoid cells, particularly monocytes, may also variably express one or several of these nonimmunoglobulin surface markers.

At approximately the 12th day of embryonic development, cells with cytoplasmic μ heavy chains may be detected at multiple sites in the murine fetal liver. By birth and in adult life, the production of B lymphocytes is generally confined to the bone marrow. Cytoplasmic Ig^+, surface Ig^- (C_μ^+, S_μ^-) pre-B cells are presently the earliest identifiable precursors of B lymphocytes, although the monoclonal antibody DNL 1.9 does react with a portion of C_μ^-, S_μ^- cells which may be earlier B cell precursors.[54] The C_μ^+, S_μ^- precursor cells occur in two sizes: large and small. The small cells, which are nonproliferating and are the most numerous of the μ^+ cells, outnumber the large C_μ^+, S_μ^- cells by approximately 2:1. Some of the small cells are the progeny of the large C_μ^+, Su_μ^- cells; however others appear to be the direct offspring of C_μ^-, S_μ^- precursor cells. The small C_μ^+, S_μ^- cells ultimately give rise to small primary C_μ^+, S_μ^+ B cells.[55] The subsequent events that lead to further differentiation of these small doses of C_μ^+, S_μ^+ mature B lymphocytes are less well understood. The first step of this process is cell activation, which is generally considered to be the effect on the cell that causes it to go from the G_0 to the G_1 phase of the cell cycle, or in this case from a small quiescent B lymphocyte to a large, metabolically active B lymphoblast. Many different agents such as anti-immunoglobulins, lipopolysaccharides, lectins, calcium ionophores, anti-β_2 microglobulin, and carbohydrate polymers such as dextran or Ficoll may all induce B lymphocytes to proliferate.[18] Because of the diverse nature of these stimuli, it would seem doubtful that they would all function by generating the same activating signals. This idea is supported by the findings that some agents are only capable of effecting specific B cell subpopulations, that some activators can only stimulate B cells partially through the cell cycle, and that no activator can stimulate all B cells.

One model that has proven extremely useful to study the activation requirements and heterogeneity of adult B lymphocytes is the CBA/N xid mouse, which has a mutant X chromosomal gene that leads to distinct defects in B lymphocyte differentiation. Expression of this genetic defect results in adult xid mice that lack B lymphocytes with the Lyb-3, Lyb-5, and Lyb-7 antigens that are found in normal mice. These adult mice therefore have a B cell surface antigen phenotype similar to that of normal neonatal mice. Despite the absence of several surface antigens, the immunologic abnormalities of xid mice are principally accounted for by their virtual lack of B cells with the Lyb-5 phenotype. While spleen cells from normal neonatal mice also lack the Lyb-5 marker, they rapidly acquire it during the first several weeks of life.[56,57] The acquisition of Ly-5 (and Lyb-7) antigen parallels responsiveness to certain types of thymus-independent (TI) antigens, and hence the usefulness of xid mice in investigating B lymphocyte proliferation. While recent experimental data indicate that many (perhaps all) TI antigens may not be truly T cell independent, they still serve as useful experimental probes. In general TI-1 or Type 1 antigens (those capable of activating CBA/N xid and neonatal B cells, i.e., Lyb-5$^-$) are bacterial products such as lipopolysaccharides, whereas TI-2 or Type 2 antigens, which activate only mature B cells (i.e., Lyb-5$^+$), are poorly metabolized, high-molecular-weight polymers such as dextran, Ficoll, or levan. Unlike most Type II antigens, Type I TI antigens are all strong polyclonal B cell activators. Differences between Lyb-5$^+$ and Lyb-5$^-$ cells are primarily expressed by the different activation requirements of the two cell types. Lyt-5$^+$ cells respond to a combination of antigen and soluble helper factors derived from T helper cells or macrophages and are therefore not restricted by the major histocompatibility complex (MHC). Activation of Lyb-5$^-$ cells requires the presence of MHC-restricted, antigen-specific T helper cells.[58-61] Presently, it is unclear whether antigen-processing cells such as macrophages interact directly with B lymphocytes or whether they interact with T helper cells which in turn produce soluble factors that lead to B cell activation. It is thought, however, that the T helper cell population that interacts with both the Lyb-5$^+$ and Lyb-5$^-$ cell populations may be the same, since high antigen concentration leads to MHC-unrestricted B cell activation controlled by soluble factors, while low antigen concentration produces MHC-restricted T helper-B cell interactions leading to B cell activation.[62]

Using the model of anti-immunoglobulin induced cell activation, it has been known for some time that antibodies directed against either surface IgM or IgD, but not their monovalent Fab fragments, are sufficient to initiate B cell proliferation.[63-67] Initially it was thought that anti-IgM-induced B cell proliferation proceeded independently of T helper or accessory cells; however utilizing more rigorous experimental protocols, this idea has subsequently been disproven. While immunoglobulin cross-linking is a common requirement for the activation of Lyb-5$^+$ B lymphocytes, this signal is not sufficient to drive the cells to secrete immunoglobulin. Cross-linking has recently been found to be quantitative, since low levels of antibody are only capable of producing cell enlargement ($G_0 \rightarrow G_1$), while larger amounts produce progressively more proliferation and may stimulate late G_1 cells to enter the S phase of the cell cycle.[68] This information, coupled with the knowledge that tolerance will result, particularly with TI antigens, unless signals other than Ig cross-linking are received by the B cells, led to investigations that have demonstrated that at least three separate stimuli are required to cause small Lyt-5$^+$ lymphocytes to enter S phase. Different stimuli appear necessary to induce cells from the $G_0 \rightarrow G_1$ phase and from the $G_1 \rightarrow S$ phase of the cell cycle. In addition to Ig cross-linking, a second signal in the form of a T cell-derived factor (BCGF?), followed by a third signal in the form of a macrophage-derived factor (probably IL-1) are required.[69] It is presently not clear whether the activating signal and BCGF work sequentially or whether they both need to be continuously present. After these stimuli drive the cell into the late G_1 phase, IL-1 is required for the cells to progress to the S phase. Although there is a strong suspicion that they exist, no specific receptors for either BCGF or IL-1 have yet been demonstrated on B lymphocytes. It should be perhaps pointed out that the substance BCGF has not, to date, been biochemically characterized and may in fact represent several biologically active substances that effect the proliferation of B lymphocytes.[63] This question is being intensively investigated in several laboratories.

The nature and source of the signals, particularly at the DNA and cell mitosis level that initially begin the transition of small B lymphocytes into large activated B lymphocytes, are known to involve the binding of a ligand (antigen or polyclonal activator) to receptors on the cell membrane and involve Ca^{++} transport.[70,71] Better information is available, however, about the immunomodulating signals that control the maturation and ultimately the immunoglobulin secretion of the plasma cell. This is largely due to modern cell hybridization technology that has permitted the construction of specific T cell and macrophage lines that secrete mixtures of antigen-nonspecific (an in a few cases antigen-specific) lympho- and monokines. Laboratory created cell lines are now capable of providing large quantities of homogeneous material that may be utilized to structurally, biochemically, and genetically investigate mediator substances and their effect on immunologic function.[72] Additionally, specific T cell lines, which depend on an exogenous source of growth factor, have been established in several laboratories. These lines greatly facilitate the quantitation of certain mediator substances (IL-2 and IL-3). While the precise roles of many factors are not completely understood, it is generally agreed that once Lyb-5$^+$ B cells are proliferating, the factors required for their maturation into immunoglobulin secreting cells are the same.[63] There are presently under investigation several discrete and independent T cell replacing factors that convert cycling B lymphocytes into antibody-secreting plasma cells. These factors, B15-TRF, EL-TRF, BCDF$_\mu$, BCDF$_\gamma$ and BCGF appear critical in regulating clonal expansion and differentiation of B cells.[63,73,74,192] The DBA/2Ha mouse has proven useful to investigate the functional properties of TRF and the genetic basis of the activation of B cells. This mouse strain has an X-linked recessive gene defect which controls the acceptor site for TRF.[75] Not only are T cell factors responsible for transforming activated B cells into antibody-secreting cells, they can also influence the isotype, idiotype, and affinity of the antibody expressed. This ability to influence the spectrum of antibody synthesized by B cell clones (their repertoire) is thought to be related to both the nature of the activating antigen (i.e., TI or TD) and to the V$_H$ region genes expressed by either the Lyb-5$^+$ or Lyb-5$^-$ cell populations. The fundamental question is, however, at what stage of development does a B cell clone become irreversibly and unalterably committed to an ultimate antibody specificity? If this occurs relatively late in B cell maturation, a heterogeneity would be expected in the antibody repertoire expressed by a single B cell clone. Because it is not generally possible to know the genotype of a B cell clone, it is presently impossible to differentiate between a genetically programed repertoire and one that is the consequence of extrinsic regulatory influences.[76]

Most of the cell activation requirements discussed above are applicable to only the Lyb-5$^+$ B cell subpopulation. As previously noted the activation of the Lyb-5$^-$ cell population depends on the interaction of T helper cells and B cells in an antigen-specific, MHC-restricted manner. The T helper cells must recognize both the antigen and a class II MHC molecule on the B lymphocyte. Whether this interaction occurs via direct cell to cell contact or through the liberation of antigen-specific factors by the T helper cells, and whether there is direct antigen binding to B cell membrane immunoglobulin receptors is uncertain. During the past few years several laboratories have produced monoclonal antibodies that identify antigens on individual clones of murine and human T cells.[194-198] These antigens identify the long-sought after T cell receptor (Ti) for antigen and the MHC. Continued work has shown the Ti to be a 90 kD disulfide linked heterodimer which expresses clonotypic idiotypes. It is composed of 2 glycoprotein chains, an acidic α chain of 49-51 kD, and a basic β chain of approximately 43 kD.[199,200] Molecular cloning and amino acid sequencing have identified both the Tiβ and Tiα genes and demonstrated that while there is a general organizational similarity (20 to 40% homology) between Ti and immunoglobulin variable and constant region genes, they are clearly distinct.[201-208] Complementary DNA probes for the human Tiβ sequence have shown the Tiβ gene to reside on chromosome 7, making it distinct from the MHC and immunoglobulin loci.[209] In the mouse the Ti gene has been mapped in the same region, but not closely linked with the kappa locus on chromosome 6.[210] The human and murine genes are highly homologous (82%) in their constant regions.[211] Despite considerable information soluble factors that influence Lyb-5$^+$ cells, there is much less known about the antigen-specific, MHC-restricted factors that regulate the proliferation and differentiation of Lyb-5$^-$ cells. Many laboratories have now succeeded in making stable T cell hybridoma lines which can provide antigen-specific helper or suppressor factors.[63,72] Whether some of the nonspecific factors such as BCGF or TRF participate in the activation and development of Lyb-5$^-$ cells is at this time unknown.

Considering all information, the evidence seems to indicate that there are several populations of B lymphocytes with different activation requirements and distinct, but perhaps overlapping, surface antigen phenotypes. In spite of the recognition of Lyb-5$^+$ and Lyb-5$^-$ cell populations, it is unclear whether they represent only different maturational stages of a single lineage of B cell differentiation that is extrinsically controlled by regulatory signals from T helper cells, monocytes, or other accessory cells, or whether there are inherent B lymphocyte subpopulations.[63] Once cells can be induced to enter the G_1 phase of the cell cycle, the stimuli for further maturation and ultimately immunoglobulin secretion may be relatively uniform for all B lymphocytes.

As previously described, cell surface antigens have assumed a major role in the identification and characterization of T lymphocyte subpopulations. T cells irreversibly differentiate into populations with discrete functional capabilities such as help or suppression, and which have distinct surface antigen phenotypes. While numerous murine B cell specific monoclonal antibodies (Lyb-2, Ly-m20, PC.2, DNL 1.9, RA3-2C2, RA3-3A1, 14.8) have been described, they have not as yet contributed to experimental results which suggest that B lymphocytes, like T lymphocytes, differentiate into subpopulations with distinct functions.[54,78-85]

Unlike the situation with human T lymphocytes, the characterization of human B lymphocytes has progressed more slowly. This is principally because, until recently, there have really been no widely available, specific anti-human-B cell monoclonal antibodies that are recognized as exclusively reacting with large populations of normal B lymphocytes. Additionally, many antibodies that are available identify antigens that have proven difficult to correlate with traditional markers of B cells such as Ia or Fc receptors, or surface immunoglobulin. Four monoclonal antibodies, B1, HC11, anti-Leu 12, and FMC-1, are generally considered to only react with peripheral blood B lymphocytes, but they are not markers for any functionally defined B lymphocyte subpopulations.[86-88] Other anti-B cell antibodies (AB-89, Y29, 55, B2, BA-1, J5, and others) react with subpopulations of B cells such as those found in tissue but not in peripheral blood, or with B cell leukemias.[85,89] Some monoclonal antibodies also cross-react with non-B cell lineage cells such as monocytes, granulocytes, or platelets. Unfortunately, an understanding of the significance of the distribution patterns of human B cells with varying antigen phenotypes is at present totally lacking. Over the next few years, it is hoped that additional anti-human B cell monoclonal antibodies will be produced that will facilitate analysis of human B cell differentiation and clarify some of the above questions.

Unlike murine B lymphocytes, human B cells have proven more difficult to activate and induce to immunoglobulin secretion with specific antigens. Therefore most functional studies of human B cells have involved polyclonal B cell activators such as pokeweed mitogen, wheat germ agglutinin, Streptolysin O, Staphylococcal phage lysate, Nocardia water soluble mitogen or Epstein-Barr virus. Using these activators, human B cells also have been shown to be either T cell dependent or independent (Nocardia, EBV); however, to date, no phenotypic marker analogous to Lyb-5 has been found on human B cells.[90] Recently several laboratories have developed experimental models using T suppressor cell depleted cell populations that can be activated by specific antigens such as tetanus toxoid or pneumococcal polysaccharide, and that may prove useful in future studies of human B lymphocyte function.[91,92]

MONOCYTES-MACROPHAGES

An increased understanding of the mechanisms of immune function has enhanced the appreciation of the monocyte-macrophage as a critical regulatory cell in the immune response. Monocytes are known to play an integral part in the growth and differentiation of lymphoid cells. They are intimately involved in T lymphocyte proliferation to lectins and antigen, T-B cell interactions, the differentiation of B lymphocytes, monocyte-mediated cellular cytotoxicity, and produce a vast number of immunomodulating substances.

Monocytes, along with other phagocytic cells, arise from precursor in the CFU-S and comprise 2 to 7% of the circulating human leukocytes. Both murine and human monocytes are classically characterized by their relatively distinct morphologies, their phagocytic abilities, and their propensity to adhere to glass or plastic surfaces.[93,94] Specific histochemical staining of lysosomal enzymes such as acid phosphatase or β glucuronidase also permit identification of monocytes.[95] In addition, macrophages may also variably express one or several surface membrane markers that are shared with B lymphocytes and include complement Fc receptors and Ia antigens. Until recently there were no specific monoclonal antibodies available which reacted only with monocytes or macrophages, and therefore the separation of these cell types from other lymphoid cells has generally relied on physical techniques such as passage through nylon wool, adherence to plastic Petri dishes or tissue culture flasks, or special centrifugation techniques.[96,97] A number of laboratories have now produced monoclonal antibodies that react with monocytes or macrophages.[98-104] While few, if any, of these antibodies are totally monocyte specific, they should undoubtedly facilitate the study of mononuclear phagocytes.

Until the development of monoclonal antibodies, little was known about the intermediary stages of monocyte development besides their origin from a self-renewing bone marrow precursor cell and their sequential acquisition of lysosomes, increased Fc and C3 receptors, and various functional capabilities such as phagocytosis or pinocytosis.[105] Although our understanding of the development of the monocyte lineage from precursor cell to tissue macrophage remains incomplete, it is clear that there exists both a functional and phenotypic heterogeneity within the monocyte-macrophage cell population.[105-108] Besides differences in the presence or absence of surface receptors and Ia antigens, discrete subpopulations of monocytes have been identified with monoclonal antibodies.[99,101] It is presently unknown whether the differences in surface antigen reflect varying functional capabilities or whether they represent varying stages of differentiation.

Another powerful tool for the investigation of mononuclear phagocytes is the continuous macrophage or macrophage/granulocyte cell line. Although these cell lines are often quite different from normal cells, they remain useful. This is particularly true in the case of macrophage cell lines which produce sizeable quantities of immunomodulating substances like IL-1 or colony-stimulating factor (CSF), and thereby serve as a source of these materials for other immunological or biochemical studies. Continuous macrophage cell lines are also useful to study antibody-dependent cellular cytotoxicity (ADCC) and cytotoxicity induced by tumoricidal factors secreted by macrophages.[109]

Recent interest in macrophage function has focused in several broad areas: (1) the secretion of biologically active products or monokines, (2) the effects of monocytes on the proliferative responses of T and B lymphocytes, and (3) the selective tumoricidal capacities of activated macrophages. Macrophages are now known to synthesize and secrete a large number of regulatory substances which influence the growth and development of both lymphocytes and hematopoietic cells. Additionally, other products which have important biological properties such as interferon, complement proteins, prostaglandins, superoxide, hydrogen peroxide, and fibronectin are secreted by macrophages.[105,109,110]

It has been known for perhaps a decade that adherent cells release a factor(s) that promotes the proliferation of lymphocytes. Despite this knowledge, it was not until relatively recently that lymphocyte researchers began to seriously consider the effects of contaminating monocytes or macrophages on studies of lymphocyte function. Several points have emerged. Considerable effect on lymphocyte function can be produced by a very small number of monocytes, and the effects can be either positive or negative, depending on the macrophage concentration.[105] The promotive effects appear to be more nonspecific than the suppressive-type effects.

Perhaps the most widely studied effects of macrophages are those associated with the induction of T lymphocyte proliferation by antigen and the control of this response by the immune-response gene. Considerable evidence supports the idea that macrophages function in the capacity of antigen-presenting cells. Early investigations clearly demonstrated that the response of T cells did not involve a simple antigen-T cell receptor interaction, but involved antigen and a specific Ia product interacting with the T cells.[111,112] Despite a tremendous amount of research, the precise molecular events that occur between soluble protein antigens, macrophages, and T lymphocytes is not certain. Initially T cells bind to the macrophages bearing the antigen molecules, and physically form small clusters of cells. Rather than the T cells directly recognizing the antigen, they recognize or recruit macrophages bearing Ia antigens of the appropriate histocompatibility haplotype. This is followed by the secretion by the macrophage of the lymphocyte activating factor IL-1. The activated T cells then begin to proliferate and secrete other lymphokines such as IL-2.[113] Other cell types such as the Langerhans cells of the skin, the dendritic cells of the spleen, B cells, and endothelial cells are also capable of antigen presentation.[114]

Infection of mice with *Mycobacterium bovis, Listeria monocytogenes,* or similar organisms causes hyperplasia of the reticuloendothelial system, and elaboration of peritoneal macrophages that display an enhanced in vitro ability to kill microorganisms. These "activated" macrophages demonstrate, under appropriate culture conditions, enhanced nonspecific killing of tumor cells. The cytotoxicity does not require phagocytosis, or sensitization, and is selective for spontaneously transformed target cells. A number of toxic molecules have been proposed to mediate the tumoricidal activity of macrophages.[115,116]

EFFECTS OF AGING

The classical view of the apparent economy of the immune response has been superceded by a picture of increasing complexity. The inclusion of a broad variable such as aging tends to quickly compound the analysis of immune function. While it has been generally shown that immune function declines with age, there are undoubtedly many possibly interrelated mechanisms underlying this decline. The impaired function may be secondary to a relative or absolute increase or decrease in a particular type of cell that is necessary for optimal function. Alternatively, age-related deficits in immune function may be due to a decrease in the functional ability of cells still present in normal numbers, but which are no longer able to perform adequately. A combination of altered cell populations and a decreased (or increased) functional ability of particular cells offers yet another broad mechanism of immunosenescence. At present it is not possible to assign any cell population changes, any cellular function deficits, or a combination of the two as the reason(s) for the immune deficits associated with aging. A few basic observations that have emerged are summarized in Table 1.

The most obvious and well-studied effect of age on the cells and tissues of the immune system is seen in the thymus. In mammalian species the cortical thymus undergoes a marked involution with age. In man, this decrease begins at approximately the time of puberty and insidiously continues until about 60 years of age, a point at which the gland is generally fibrotic. In addition to serving as a site of lymphocyte differentiation, the thymus also functions as an endocrine organ with several thymic hormones being recognized as potent inducers of T cell differentiation.[117-119] Direct measurement of thymic serum factors has shown that they diminish with age and that the age-associated changes in thymic endocrine function precede the impairment of peripheral blood T lymphocyte function.[120,121] While lymphocyte differentiation is not completely understood, it is known that both the microen-

Table 1
SUMMARY OF AGE-ASSOCIATED CHANGES IN CELLS AND FUNCTION OF THE IMMUNE SYSTEM

Tissues and cells	Change	Ref.
Lymphoid tissue	Thymic involution begins at puberty	165, 166
	Spleen and lymph node mass decreases	167
	Lymphoid follicles in marrow, other organs may increase	168
Stem cells	No quantitative change	169
	Possible defective differentiation probably represents genetic variability	170
T lymphocytes	Possible mild quantitative decrease (results variable)	158, 171, 172, 173
	In vivo decrease in functional capacities	
	Impaired allograft rejection	174
	Impaired ability to induce GvH response	175
	Impaired ability to mount DTH response	176, 177
	Impaired resistance to tumor transplant	178
	In vitro decrease in functional capacity	
	Impaired proliferative response to allogeneic cells (MLR)	124, 152
	Impaired proliferative response to lectins	158, 175, 179, 139, 180
	Impaired cell-mediated cytotoxicity	181, 182
	Impaired proliferative response to syngeneic non-T cells (SMLC)	183
B lymphocytes	Probably no quantitative change (results variable)	158, 171, 124, 173, 151
	In vivo functional capacities reflect aberrancies	
	No change in serum immunoglobulin levels	139, 158, 159, 160
	Increased incidence of monoclonal immunoglobulins and autoantibodies	184, 144, 167
	Impaired primary antibody response to T-dependent antigens	185
	Impaired primary antibody response to T-dependent antigens	141
	In vitro functional capacity reflect abberancies	
	No change in proliferative response to polyclonal B cell activators	186
	Impaired primary antibody response to T-independent antigens	187, 157
	Possibly no change in polyclonal induction of immunoglobulin (results variable)	158, 151
Non-T, Non-B cells	In mice (spleen cells), impaired NK activity at birth and after 12 weeks of age; no change in peripheral blood	188
	In man (peripheral blood) no change	189, 190, 191, 212, 213
Monocyte-macrophage	In mouse, probable quantitative increase	187
	In man, no quantitative or functional change (in phagocytic ability, enzyme content, or induction of antibody response); production of prostaglandin is enhanced	162, 214

vironment of the thymus and the humoral factors produced by the thymus are important. Until recently, the influence of the thymus on lymphocyte differentiation was thought to be confined to T cells; however, there is now evidence indicating that the thymus may also play a role in the differentiation of B lymphocytes.[122-124] Despite the involution of the thymus gland and its effects on T cell differentiation, there does not seem to be any marked diminution in humans in the relative or absolute numbers of peripheral blood T cell.[125-128] Likewise, in mice the number of splenic T lymphocytes does not change with age.[129] This contrasts with the results obtained when T cell populations are examined with murine antihuman monoclonal antibodies. There is a decrease found in the percent representation of cells reactive with any of several antibodies which identify mature T cells (OKT3, Leu-4, Lyt-3, T101), and in the representation of suppressor/cytotoxic cells reactive with the OKT8 or Leu-2a antibodies.[130,131] While many T lymphocyte-associated functions such as allograft rejection, graft-vs.-host and delayed hypersensitivity reactions, T cell-mediated cytotoxicity, mixed lymphocyte reactivity, and cell activation by plant lectins have also been shown to undergo age-related alterations, it is only in the case of cell activation by mitogens that fairly definitive information is available regarding the mechanisms responsible for the changes. The number of T cells capable of activation by the mitogen are decreased and those that are activated proliferate poorly. The T cells produce a decreased amount of the lymphokine IL-2 and also lose the ability to respond to its effects. T cells from aged individuals also display altered responses to immunomodulators such as prostaglandins.[132-135]

In humans there occurs with age an increase in the frequency of homogeneous serum immunoglobulins, autoantibodies, and circulating immune complexes.[136-141] These auto-antibodies are not associated with an increased incidence of clinical autoimmune disease in the elderly.[142] Similarly in aged mice, there is an increase in monoclonal immunoglobulins and monoclonal serum proteins which may be increased by thymectomy.[143,144] One unanswered question is how the presence of autoantibodies is related to the mechanisms of self-tolerance. Autoantibodies have been interpreted by some to represent a breakdown in the induction and maintenance of self-tolerance which is known to be impaired by age.[145-147] Whether the alterations leading to autoantibody production are the result of defects in helper T cells, suppressor T cells, or in B cell regulation by anti-idiotype antibodies is unclear.[141-148] In vitro analysis of murine B cell function demonstrates that with advanced age the population of B lymphocytes responding to an antigen becomes less diverse and the antibody synthesized becomes less avid. Repeated immunization does not enhance this response. These changes may reflect alterations in the regulatory cell populations that direct and influence the differentiation and maturation of B lymphocytes or changes in the B cells themselves.[149-151] Most studies in the mouse indicate that B cell numbers do not decrease with age, but the responses to both TI and TD antigens decrease markedly.[152-158] Irrespective of these alterations in B lymphocyte responses to antigen, there is no significant alteration in serum immunoglobulin concentration.[139,158-160]

As indicated in the previous section, the monocyte-macrophage plays a critical role in the immune response. Most studies to date indicate that the functional capacities of the monocyte are preserved and in some cases increased with aging. This includes phagocytic ability, enzyme content, ability to induce antibody synthesis, and soluble factor production.[161-164]

In order to more precisely trace the immunological changes that accompany human aging, investigators at several centers have begun longitudinal studies on large groups of men and women of different ages. In the Baltimore Longitudinal Study of Aging conducted by the National Institute on Aging, NIH, immunologic data has accrued on over 1200 subjects for the past 6 years. Although many of the subjects have been tested on more than one occasion, the brief follow-up period necessitates that most analyses be made on cross-sectional rather than true longitudinal data. The results generally confirm prior observations by others on the trend with age towards a decrease in T cell number and function. B cell and monocyte numbers and function do not appear to change markedly with age.

The study of the mechanisms of immune function and their modulation by aging is an important area of continued research. As an increasing proportion of the population becomes elderly, a better understanding of the basis of age-related illness and physiologic aging are especially important. While it remains to be proved that altered immune function is the cause of the diminished physiological competence observed with aging, certainly various components of the immune system decrease in function with age. Ultimately an enhanced understanding of the mechanisms underlying these changes may permit the reconstitution, augmentation, or substitution for age-related immune deficits.

REFERENCES

1. **Owen, J. T. T.,** The origin and development of lymphocyte populations, in *Ontogeny of Acquired Immunity,* Porter, R. and Knight, J., Eds., Elsevier, Amsterdam, 1972, 35.
2. **Quesenberry, P. and Levitt, L.,** Hematopoietic stem cells, *N. Engl. J. Med.,* 301, 755, 819, 868, 1979.
3. **Cooper, M. D., Peterson, R., South, M. A., and Good, R. A.,** The function of the thymus system and the bursa system in the chicken, *J. Exp. Med.,* 123, 75, 1966.
4. **Katz, D. H. and Benacerraf, B.,** The regulatory influence of activated T cells on B cell responses to antigen, *Adv. Immunol.,* 15, 1, 1972.
5. **Kurzinger, K. and Springer, T. A.,** Purification and structural characterization of LFA-1, a lymphocyte function-associated antigen, and Mac-1, a related macrophage differentiation antigen associated with the type three complement receptor, *J. Biol. Chem.,* 257, 12412, 1982.
6. **Manohar, V., Brown, E., Leiserson, W. M., and Chused, T. M.,** Expression of Lyt-1 by a subset of B lymphocytes, *J. Immunol.,* 129, 532, 1982.
7. **Hayakawa, K., Hardy, R. R., Parks, D. R., and Herzenberg, L. A.,** The 'Ly-1B' cell population in normal, immunodefective, and autoimmune mice, *J. Exp. Med.,* 157, 202, 1983.
8. **Lanier, L. L., Warner, N. L., Ledbetter, J. A., and Herzenberg, L. A.,** Expression of Lyt-1 antigen on certain B cell lymphomas, *J. Exp. Med.,* 153, 998, 1981.
9. **Ledbetter, J. A., Rouse, R. V., Micklem, H. S., and Herzenberg, L. A.,** T cell subsets defined by expression of Lyt-1,2,3 and Thy-1 antigens. Two parameter immunofluorescence and cytotoxicity analysis with monoclonal antibodies modifies current views, *J. Exp. Med.,* 152, 280, 1980.
10. **McKenzie, I. F. C. and Potter, T.,** Murine lymphocyte surface antigens, *Adv. Immunol.,* 27, 179, 1979.
11. **Mathieson, B. J., Sharrow, S. O., Rosenberg, Y., and Hammerling, U.,** Lyt 1$^+$23$^-$ cells appear in the thymus before Lyt 123$^+$ cells, *Nature (London),* 289, 179, 1981.
12. **Thomas, D. B. and Calderon, R. A.,** T-helper cells change their Lyt-1,2 phenotype during an immune response, *Eur. J. Immunol.,* 12, 16, 1982.
13. **Simon, M. M., Edwards, A. J., Hammerling, U., McKenzie, I. F. C., Eichmann, K., and Simpson, E.,** Generation of effector cells from T cell subsets. III. Synergy between Lyt-1 and Lyt-123/23 lymphocytes in the generation of H-2 restricted and alloreactive cytotoxic T cells, *Eur. J. Immunol.,* 11, 246, 1981.
14. **Soloski, M. J., Uhr, J. W., Flaherty, L., and Vitetta, E. S.,** Qa-2, H-2K, and H2D alloantigens evolved from a common ancestral gene, *J. Exp. Med.,* 153, 1080, 1981.
15. **Williams, A. F. and Gagnon, J.,** Neuronal cell Thy-1 glycoprotein: homology with immunoglobulin, *Science,* 216, 696, 1982.
16. **Reinherz, E. L., Moretta, L., Roper, M., Breard, J. M., Mingari, M. C., Cooper, M. D., and Schlossman, S. F.,** Human T lymphocyte subpopulations defined by Fc receptors and monoclonal antibodies, *J. Exp. Med.,* 151, 969, 1980.
17. **Fresno, M., Simonian, H. D., Nabel, G., and Cantor, H.,** Proteins synthesized by inducer T cells: evidence for a mitogenic peptide shared by inducer molecules that stimulate different cell types, *Cell,* 30, 707, 1982.
18. **Mosier, D. E. and Subbarao, B.,** Thymus-independent antigens: complexity of B lymphocyte activation revealed, *Immunol. Today,* 3, 217, 1982.
19. **Cantor, H. and Boyse, E. A.,** Functional subclasses of T lymphocytes bearing different Ly antigens. I. The generation of functionally distinct T-cell subclasses as a differentiative process independent of antigen, *J. Exp. Med.,* 141, 1376, 1975.
20. **Ledbetter, J. A. and Herzenberg, L. A.,** Xenogeneic monoclonal antibodies to mouse lymphoid differentiation antigens, *Immunol. Rev.,* 47, 63, 1979.
21. **Swain, S. L., Dennert, G., Wormsley, S., and Dutton, R. W.,** The Lyt phenotype of a long-term allospecific T cell line. Both helper and killer activities to IA are mediated by Lyt-1 cells, *Eur. J. Immunol.,* 11, 175, 1981.

22. **Thomas, Y., Rogozinski, L., Irigoyen, O. H., Friedman, S. M., Kung, P. C., Goldstein, G., and Chess, L.,** Functional analysis of human T cell subsets defined by monoclonal antibodies. IV. Induction of suppressor cells within the OKT4$^+$ population, *J. Exp. Med.*, 154, 459, 1981.

23. **Thomas, Y., Rogozinski, L., Irigoyen, O. H., Shen, H. H., Talle, M. A., Goldstein, G., and Chess, L.,** Functional analysis of human T cell subsets defined by monoclonal antibodies. V. Suppressor cells within the activated OKT4$^+$ population belong to a distinct subset, *J. Immunol.*, 128, 1386, 1982.

24. **Uchiyama, T., Nelson, D. L., Fleisher, T. A., and Waldman, T. A.,** A monoclonal antibody (anti-Tac) reactive with activated and functionally mature human T cells. II. Expression of Tac antigen on activated cytotoxic killer T cells, suppressor cells, and on one of two types of helper cells, *J. Immunol.*, 126, 1398, 1981.

25. **Eardley, D. D., Hugenberger, J., McVay-Boudreau, L., Shen, F. W., Gershon, R. K., and Cantor, H.,** Immunoregulatory circuits among T-cell sets. I. T-helper cells induce other T-cell sets to exert feedback inhibition, *J. Exp. Med.*, 147, 1106, 1978.

26. **Meuer, S. C., Schlossman, S. F., and Reinherz, E. L.,** Clonal analysis of human cytotoxic T lymphocytes: T4+ and T8+ effector cells recognize products of different major histocompatibility complex regions, *Proc. Natl. Acad. Sci. U.S.A.*, 79, 4395, 1982.

27. **Fresno, M., Nabel, G., McVay-Boudreau, L., Furthmayer, H., and Cantor, H.,** Antigen-specific T lymphocyte clones. I. Characterization of a T lymphocyte clone expressing antigen-specific suppressive activity, *J. Exp. Med.*, 153, 1246, 1981.

28. **Cantor, H., McVay-Boudreau, L., Hugenberger, J., Naidorf, K., Shen, F. W., and Gershon, R. K.,** Immunoregulatory circuits among T-cell sets. II. Physiologic role of feedback inhibition in vivo: absence in NZB mice, *J. Exp. Med.*, 147, 1116, 1978.

29. **Dwyer, J. M. and Johnson, C.,** The use of concanavalin A to study immunoregulation of human T cells, *Clin. Exp. Immunol.*, 46, 237, 1981.

30. **Reinherz, E. L., Hussey, R. E., Fitzgerald, K., Snow, P., Terhorst, C., and Schlossman, S. F.,** Antibody directed at a surface structure inhibits cytolytic but not suppressor function of human T lymphocytes, *Nature (London)*, 294, 168, 1981.

31. **Gatenby, P. A., Kansas, G. S., Xian, C. Y., Evans, R. L., and Engleman, E. G.,** Dissection of immunoregulatory subpopulations of T lymphocytes with the helper and suppressor sublineages in man, *J. Immunol.*, 129, 1977, 1982.

32. **Cerottini, J-C. and MacDonald, H. R.,** Limiting dilution analysis of alloantigen reactive T lymphocytes. V. Lyt phenotype of cytolytic T lymphocyte precursors reactive against normal and mutant H-2 antigens, *J. Immunol.*, 126, 490, 1981.

33. **Gullberg, M. and Larsson, E.-L.,** Selective inhibition of antigen-induced 'step one' in cytotoxic T lymphocytes by anti-Lyt-2 antibodies, *Eur. J. Immunol.*, 12, 1006, 1982.

34. **MacDonald, H. R. and Lees, R. K.,** Dissociation of differentiation and proliferation in the primary induction of cytolytic T lymphocytes by alloantigens, *J. Immunol.*, 124, 1308, 1980.

35. **MacDonald, H. R.,** Differentiation of cytolytic T lymphocytes, *Immunol. Today*, 3, 183, 1982.

36. **Conzelmann, A., Corthesy, P., Cianfriglia, M., Silva, A., and Nabholz, M.,** Hybrids between rat lymphoma and mouse T cells with inducible cytolytic activity, *Nature (London)*, 298, 170, 1982.

37. **Loveland, B. E., Hogarth, P. M., Ceredig, Rh., and McKenzie, I. F. C.,** Cells mediating graft rejection in the mouse. I. Lyt-1 cells mediate skin graft rejection, *J. Exp. Med.*, 153, 1044, 1981.

38. **Zahn, G., Hammerling, G. J., Eichmann, K., and Simon, M. M.,** Expression of Qat-4 and Qat-5 alloantigens on cytotoxic precursor and effector cells: different surface phenotypes of alloreactive and H-2 restricted cytotoxic T cells, *Eur. J. Immunol.*, 12, 43, 1982.

39. **Lancki, D. W., Ma, D. I., Havran, W. L., and Fitch, F. W.,** Cell surface structures involved in T cell activation. *Immunol. Rev.*, 81, 65, 1984.

40. **Reinherz, E. L., Acuto, O., Fabbi, M., Bensussan, A., Milanese, C., Royer, H. D., Meuer, S. C., and Schlossman, S. F.,** Clonotypic surface structure on human T lymphocytes: functional and biochemical analysis of the antigen receptor complex. *Immunol Rev.*, 81, 95, 1984.

41. **Reinherz, E. L., Meuer, S. C., and Schlossman, S. F.,** The delineation of antigen receptors on human T lymphocytes, *Immunol. Today*, 4, 5, 1983.

42. **Cosimi, A. B., Colvin, R. B., Burton, R. C., Rubin, R. H., Goldstein, G., Kung, P. C., Hansen, W. P., Delmonico, F. L., and Russell, P. S.,** Use of monoclonal antibodies to T-cell subsets for immunologic monitoring and treatment in recipients of renal allografts, *N. Engl. J. Med.*, 305, 308, 1981.

43. **Reinherz, E. L., Geha, R., Rappeport, J. M., Wilson, M., Penta, A. C., Hussey, R. E., Fitzgerald, K. A., Daley, J. F., Levine, H., Rosen, F. S., and Schlossman, S. F.,** Reconstitution after transplantation with T lymphocyte-depleted HLA haplotype-mismatched bone marrow for severe combined immunodeficiency, *Proc. Natl. Acad. Sci. U.S.A.*, 79, 6047, 1982.

44. **Ritz, J. and Schlossman, S. F.,** Utilization of monoclonal antibodies in the treatment of leukemia and lymphoma, *Blood*, 59, 1, 1982.

45. **Golightly, M. C., Haynes, B. F., Brandt, C. P., and Koren, H. S.**, Phenotypic and functional characterization of natural killer cells by monoclonal antibodies, in *NK Cells and Other Natural Effector Cells,* Herberman, R. B., Ed., Academic Press, New York, 1982, 79.

46. **Perussia, B., Fanning, V., and Trinchieri, G.**, Phenotypic characterization of human natural killer and antibody-dependent killer cells as a homogeneous and discrete cell subset, in *NK Cells and Other Natural Effector Cells,* Herberman, R. B., Ed., Academic Press, New York, 1982, 39.

47. **Herberman, R. B.**, Natural killer (NK) cells and their possible roles in resistance against disease, *Clin. Immunol. Rev.,* 1, 1, 1981.

48. **Glimcher, L., Shen, F. W., and Cantor, H.**, Identification of a cell surface antigen selectively expressed on the natural killer cell, *J. Exp. Med.,* 145, 1, 1977.

49. **Pollack, S. B. and Emmons, S. L,.** Anti-NK 2.1: an activity of NZB anti-BALB/c serum, in *NK Cells and Other Natural Effector Cells,* Herberman, R. B., Ed., Academic Press, New York, 1982, 113.

50. **Abo, T. and Balch, C. M.**, A differentiation antigen on human NK and K cells identified by a monoclonal antibody (HNK-1), *J. Immunol.,* 127, 1024, 1981.

51. **Griffin, J. D., Hercend, T., Beveridge, R., and Schlossman, S. F.**, Characterization of an antigen expressed by human natural killer cells, *J. Immunol.,* 130, 2947, 1983.

52. **Nieminen, P., Paasivuo, R., and Saksela, E.**, Effect of a monoclonal anti-large granular lymphocyte antibody on human NK activity, *J. Immunol.,* 128, 1097, 1982.

53. **Rumpold, H., Kraft, D., Obexer, G., Bock, G., and Gebhart, W.**, A monoclonal antibody against a surface antigen shared by human large granular lymphocytes and granulocytes, *J. Immunol.,* 129, 1458, 1982.

54. **Dessner, D. S. and Loken, M. R.**, DNL 1.9: a monoclonal antibody which specifically detects all murine B lineage cells, *Eur. J. Immunol.,* 11, 282, 1981.

55. **Landreth, K. S., Rosse, C., and Clagett, J.**, Myelogenous production and maturation of B lymphocytes in the mouse, *J. Immunol.,* 127, 2027, 1981.

56. **Scher, I.**, The CBN/N mouse strain: an experimental model illustrating the influence of the X-chromosome on immunity, *Adv. Immunol.,* 33, 1, 1982.

57. **Scher, I.**, CBA/N immune deficient mice; evidence for the failure of a B cell subpopulation to be expressed, *Immunol. Rev.,* 64, 117, 1982.

58. **Mond, J. J., Scher, I., Cossman, J., Kessler, S., Mongini, P. K. A., Hansen, C., Finkelman, F. D., and Paul, W. E.**, Role of the thymus in directing developing of a subset of B lymphocytes, *J. Exp. Med.,* 155, 924, 1982.

59. **Asano, Y., Singer, A., and Hodes, R. J.**, Role of the major histocompatibility complex in T cell activation of B cell populations. MHC restricted and unrestricted B cell responses are mediated by distinct B cell subpopulations, *J. Exp. Med.,* 154, 1100, 1981.

60. **Singer, A., Morrissey, P. J., Hathcock, K. S., Ahmed, A., Scher, I., and Hodes, R. J.**, Role of the major histocompatibility complex in T cell activation of B cell subpopulations. Lyb-5$^+$ and Lyb-5$^-$ B cell subpopulations differ in their requirement for major histocompatibility complex-restricted T cell recognition, *J. Exp. Med.,* 154, 501, 1981.

61. **Asano, Y. and Hodes, R. J.**, T cell regulation of B cell activation. T cells independently regulate the responses mediated by distinct B cell subpopulations, *J. Exp. Med.,* 155, 1267, 1982.

62. **Asano, Y., Shigeta, M., Fathman, C. G., Singer, A., and Hodes, R. J.**, Lyb5$^+$ and Lyb$^-$ B cells differ in their requirements for restricted activation by cloned helper T cells, *Fed. Proc.,* 41, 721, 1982.

63. **Howard, M. and Paul, W. E.**, Regulation of B-cell growth and differentiation by soluble factors, *Ann. Rev. Immunol.,* 1, 307, 1983.

64. **Sieckmann, D. G., Asofsky, R., Mosier, D. E., Zitron, I. M., and Paul, W. E.**, Activation of mouse lymphocytes by anti-immunoglobulin. I. Parameters of the prolfierative response, *J. Exp. Med.,* 147, 814, 1978.

65. **Zitron, I. M. and Clevinger, B. L.**, Regulation of murine B cells through surface immunoglobulin. I. Monoclonal anti-δ antibody induces allotype specific proliferation, *J. Exp. Med.,* 152, 1135, 1980.

66. **Parker, D. C.**, Stimulation of mouse lymphocytes by insoluble anti-mouse immunoglobulins, *Nature (London),* 258, 361, 1975.

67. **Lamers, M. C., Heckford, S. E., and Dickler, H. B.**, Monoclonal anti-Fc IgG receptor antibodies trigger B lymphocyte function, *Nature (London),* 298, 178, 1982.

68. **DeFranco, A. L., Kung, J. T., and Paul, W. E.**, Regulation of growth and proliferation in B cell subpopulations, *Immunol. Rev.,* 64, 161, 1982.

69. **Howard, M., Mizel, S. B., Lachman, L., Ansel, J., Johnson, B., and Paul, W. E.**, Role of interleukin 1 in anti-immunoglobulin-induced B cell proliferation, *J. Exp. Med.,* 157, 1529, 1983.

70. **Greaves, M. F. and Bauminger, S.**, Activation of T and B lymphocytes by insoluble phytomitogens, *Nature (London),* 235, 67, 1972.

71. **Feldmann, M., Greaves, M. F., Parker, D. C., and Rittenberg, M.,** Direct triggering of B lymphocytes by insolubilized antigen, *Eur. J. Immunol.,* 4, 591, 1974.

72. **Altman, A. and Katz, D. H.,** The biology of monoclonal lymphokines secreted by T cell lines and hybridomas, *Adv. Immunol.,* 33, 73, 1982.

73. **Nakanishi, K., Howard, M., Muraguchi, A., Farrar, J., Takatsu, K., and Paul, W. E.,** Soluble factors involved in B cell differentiation: identification of two distinct T cell replacing factors, *J. Immunol.,* 130, 2219, 1983.

74. **Pure, E., Isakson, P. C., Takatsu, K., Hamaoka, T., Swain, S. L., Dutton, R. W., Dennert, G., Uhr, J. W., and Vitetta, E. S.,** Induction of B cell differentiation by T cell factors. I. Stimulation of IgM secretion by products of a T cell hybridoma and a T cell line, *J. Immunol.,* 127, 1953, 1981.

75. **Takatsu, K. and Hamaoka, T.,** DBA/2Ha mice as a model of an X-linked immunodeficiency which is defective in the expression of TRF-acceptor site(s) on B lymphocytes, *Immunol. Rev.,* 64, 25, 1982.

76. **Klinman, N. R., Wylie, D. E., and Teale, J. M.,** B-cell development, *Immunol. Today,* 2, 212, 1981.

77. **Jensenius, J. C. and Williams, A. F.,** The T lymphocyte antigen receptor-paradigm lost, *Nature (London),* 300, 583, 1982.

78. **Hammerling, U., Chin, A. F., and Abbott, J.,** Ontogeny of murine B lymphocytes: sequence of B-cell differentiation from surface-immunoglobulin-negative precursors to plasma cells, *Proc. Natl. Acad. Sci. U.S.A.,* 73, 2008, 1976.

79. **Yakura, H., Shen, F-W., Kaemmer, M., and Boyse, E. A.,** Lyb-2 system of mouse B cells. Evidence for a role in the generation of antibody-forming cells, *J. Exp. Med.,* 153, 129, 1981.

80. **Kimura, S., Tada, N., Nakayama, E., Liu, Y., and Hammerling, U.,** A new mouse cell-surface antigen (Ly-m20) controlled by a gene linked to the Mls locus and defined by monoclonal antibodies, *Immunogenetics,* 14, 3, 1981.

81. **Tada, N., Kimura, S., Hoffmann, M., and Hammerling, U.,** A new surface antigen (PC.2) expressed exclusively on plasma cells, *Immunogenetics,* 11, 351, 1980.

82. **Coffman, R. L. and Weissman, I. L.,** A monoclonal antibody that recognizes B cells and B cell precursors in mice, *J. Exp. Med.,* 153, 269, 1981.

83. **Coffman, R. L. and Weissman, I. L.,** B220: a B cell-specific member of the T200 glycoprotein family, *Nature (London),* 289, 681, 1981.

84. **Kincade, P. W., Lee, G., Watanabe, T., Sun, L., and Scheid, M. P.,** Antigens displayed on murine B lymphocyte precursors, *J. Immunol.,* 127, 2262, 1981.

85. **McKenzie, I. F. C. and Zola, H.,** Monoclonal antibodies to B cells, *Immunol. Today,* 4, 10, 1983.

86. **Stashenko, P., Nadler, L. M., Hardy, R., and Schlossman, S. F.,** Characterization of a human B lymphocyte-specific antigen, *J. Immunol.,* 125, 1678, 1980.

87. **Brooks, D. A., Beckman, I., Bradley, J., McNamara, P. J., Thomas, M. E., and Zola, H.,** Human lymphocyte markers defined by antibodies derived from somatic cell hybrids. I. A hybridoma secreting antibody against a marker specific for human B lymphocytes, *Clin. Exp. Immunol.,* 39, 477, 1980.

88. **Brooks, D. A., Beckman, I. G. R., Bradley, J., McNamara, P. J., Thomas, M. E., and Zola, H.,** Human lymphocyte markers defined by antibodies derived from somatic cell hybrids. IV. A monoclonal antibody reacting specifically with a subpopulation of human B lymphocytes, *J. Immunol.,* 126, 1373, 1981.

89. **Stashenko, P., Nadler, L. M., Hardy, R., and Schlossman, S. F.,** Expression of cell surface markers after human B lymphocyte activation, *Proc. Natl. Acad. Sci. U.S.A.,* 78, 3848, 1981.

90. **Waldmann, T. A. and Broder, S.,** Polyclonal B-cell activation in the study of the regulation of immunoglobulin synthesis in the human system, *Adv. Immunol.,* 32, 1, 1982.

91. **Cavagnaro, J. and Osband, M.,** In vitro primary immunization of human peripheral blood mononuclear cells and its role in the development of human-derived monoclonal antibodies, *Biotechniques,* 1, 30, 1983.

92. **Volkman, D. J., Allyn, S. P., and Fauci, A. S.,** Antigen-induced *in vitro* antibody production in humans: tetanus toxoid-specific antibody synthesis, *J. Immunol.,* 129, 107, 1982.

93. **van Furth, R., Raeburn, J. A., and vanZwet, T. L.,** Characterization of human mononuclear phagocytes, *Blood,* 54, 485, 1979.

94. **van Furth, R., Diesselhoff-DenDulk, M. M. C., Raeburn, J. A., vanZwet, T. L., Croften, R., and Blusse van Oud Albas, A.,** Characteristics, origin and kinetics of human and murine mononuclear phagocytes, in *Mononuclear Phagocytes. Functional Aspects,* van Furth, R., Ed., Martinus Nijhoff, The Hague, 1979.

95. **Yam, J. T., Li, C. Y., and Crosby, W. H.,** Cytochemical identification of monocytes and granulocytes, *Am. J. Clin. Pathol.,* 55, 283, 1971.

96. **Julius, M. H., Simpson, E., and Herzenberg, L. A.,** A rapid method for the isolation of functional thymus-derived murine lymphocytes, *Eur. J. Immunol.,* 3, 645, 1973.

97. **Yasaka, T., Mantich, N. M., Boxer, L. A., and Baehner, R. L.,** Functions of human monocyte and lymphocyte subsets obtained by countercurrent centrifugal elutriation: differing functional capacities of human monocyte subsets, *J. Immunol.,* 127, 1515, 1981.

98. **Dimitriu-Bona, A., Burmester, G. R., Waters, S. J., and Winchester, R. J.,** Human mononuclear phagocyte differentiation antigens. I. Patterns of antigenic expression on the surface of human monocytes and macrophages defined by monoclonal antibodies, *J. Immunol.,* 130, 145, 1983.

99. **Shen, H. H., Talle, M. A., Goldstein, G., and Chess, L.,** Functional subsets of human monocytes defined by monoclonal antibodies: a distinct subset of monocytes contains the cells capable of inducing the autologous mixed culture, *J. Immunol.,* 130, 698, 1983.

100. **Breard, J., Reinherz, E. L., Kung, P. C., Goldstein, G., and Schlossman, S. F.,** A monoclonal antibody reactive with human peripheral blood monocytes, *J. Immunol.,* 124, 1143, 1980.

101. **Eisenbarth, G. S., Haynes, B. F., Schroer, J. A., and Fauci, A. S.,** Production of monoclonal antibodies reacting with peripheral blood mononuclear cell surface differentiation antigens, *J. Immunol.,* 124, 1237, 1980.

102. **Nunez, G., Ugolini, V., Capra, J. D., and Stastny, P.,** Monoclonal antibodies against human monocytes. II. Recognition of two distinct cell surface molecules, *Scand. J. Immunol.,* 16, 515, 1982.

103. **Haynes, B. F., Hemler, M. E., Mann, D. L., Eisenbarth, G. S., Shelhamer, J., Mostowski, H. S., Thomas, C. A., Strominger, J. L., and Fauci, A. S.,** Characterization of a monoclonal antibody (4F2) that binds to human monocytes and to a subset of activated lymphocytes, *J. Immunol.,* 126, 1409, 1981.

104. **Todd, R. F., III, Nadler, L. M., and Schlossman, S. F.,** Antigens on monocytes identified by monoclonal antibodies, *J. Immunol.,* 126, 1435, 1981.

105. **Unanue, E. R.,** The regulatory role of macrophages in antigenic stimulation. II. Symbiotic relationship between lymphocytes and macrophages, *Adv. Immunol.,* 31, 1, 1981.

106. **Sorg, C. and Neumann, C.,** A developmental concept for the heterogeneity of macrophages in response to lymphokines and other signals, in *Lymphokines,* Vol. 3, Pick, E. and Landy, M., Eds., Academic Press, New York, 1981, 85.

107. **Walker, W. S.,** Functional heterogeneity of macrophages: subclasses of peritoneal macrophages with different antigen-binding activities and immune complex receptors, *Immunology,* 26, 1025, 1974.

108. **Arenson, E. B., Jr., Epstein, E. B., and Seeger, R. C.,** Volumetric and functional heterogeneity of human monocytes, *J. Clin. Invest.,* 65, 613, 1980.

109. **Ralph, P.,** Continuous macrophage cell lines — their use in the study of induced constitutive macrophage properties and cytotoxicity, in *Lymphokines,* Vol. 4., Pick, E. and Landy, M., Eds., Academic Press, New York, 1981, 175.

110. **Bentley, C., Zimmer, B., and Hadding, U.,** The macrophage as a source of complement components, in *Lymphokines,* Vol. 4. Pick, E. and Landy, M., Eds., Academic Press, New York, 1981, 197.

111. **Waldron, J. A., Jr., Horn, R. G., and Rosenthal, A. S.,** Antigen-induced proliferation of guinea pig lymphocytes in vitro: obligatory role of macrophages in the recognition of antigen by immune T-lymphocytes, *J. Immunol.,* 111, 58, 1973.

112. **Rosenthal, A. S. and Shevach, E. M.,** Function of macrophages in antigen recognition by guinea pig lymphocytes. I. Requirement for histocompatible macrophages and lymphocytes, *J. Exp. Med.,* 138, 1194, 1973.

113. **Unanue, E. R.,** Symbiotic relationships between macrophages and lymphocytes, in *Macrophages and Natural Killer Cells,* Normann, S. J. and Sorkin, E., Eds., Plenum Press, New York, 1982, 49.

114. **Lipsky, P. E., and Kettman, J. T.,** Accessory cells unrelated to mononuclear phagocytes and not of bone marrow origin, *Immunol. Today,* 3, 36, 1982.

115. **Aksamit, R. R. and Kim, K. J.,** Macrophage lines produce a cytotoxin, *J. Immunol.,* 122, 1785, 1979.

116. **Russell, S. W., Gillespie, G. Y., and Pace, J. L.,** Comparison of responses to activating agents by mouse peritoneal macrophages and cells of the macrophage line RAW 264, *J. Reticuloendothel. Soc.,* 27, 607, 1980.

117. **Bach, J. F., Dardenne, M., Goldstein, A. L., Guha, A., and White, A.,** Appearance of T cell markers in bone marrow rosette-forming cells after incubation with thymosin, a thymic hormone, *Proc. Natl. Acad. Sci. U.S.A.,* 68, 2734, 1971.

118. **Basch, R. S. and Goldstein, G.,** Induction of T cell differentiation in vitro by thymin. A purified polypeptide hormone of the thymus, *Proc. Natl. Acad. Sci. U.S.A.,* 71, 1474, 1974.

119. **Lewis, V. M., Twomey, J. J., Bealmear, P., Goldstein, G., and Good, R. A.,** Age, thymic involution and circulating thymic hormone activity, *J. Clin. Endocrinol. Metab.,* 47, 145, 1978.

120. **Bach, J. F., Dardenne, M., Pleau, J. M., and Bach, M. A.,** Isolation, biochemical characterization and biological activity of circulating thymic hormone in the mouse and in the human, *Ann. N.Y. Acad. Sci.,* 248, 186, 1975.

121. **Weksler, M. E., Innes, J. B., and Goldstein, G.,** Immunological studies of aging. IV. The contribution of thymic involution to the immune deficiencies of aging mice and reversal with thymopoietin 32-36, *J. Exp. Med.,* 148, 996, 1978.

122. **Szewczuk, N. R., Sherr, D. H., and Siskind, G. W.,** Ontogeny of B lymphocyte function. VI. Ontogeny of thymus cell capacity to facilitate the functional maturation of B lymphocytes, *Eur. J. Immunol.,* 8, 370, 1978.

123. **Szewczuk, N. R., DeKruyff, R. H., Goidl, E. A., Weksler, M. E., and Siskind, G. W.,** Ontogeny of B lymphocyte function. VIII. Failure of thymus cells from aged donors to induce the functional maturation of B lymphocytes from immature donors, *Eur. J. Immunol.,* 10, 918, 1980.

124. **Weksler, M. E. and Hütteroth, T. H.,** Impaired lymphocyte functions in aged humans, *J. Clin. Invest.,* 53, 99, 1974.

125. **Siskind, G. W. and Weksler, M. E.,** The effect of aging on the immune response, *Ann. Rev. Gerontol. Geriatr.,* 3, 3, 1982.

126. **Portaro, J. K., Glick, G. I., and Zighelboim, J.,** Population immunology: age and immune cell parameters, *Clin. Immunol. Immunopathol.,* 11, 339, 1978.

127. **Cobleigh, M. A., Braun, D. P., and Harris, J. E.,** Age-dependent changes in human peripheral blood B cells and T cell subsets: correlation with mitogen responsiveness, *Clin. Immunol. Immunopathol.,* 15, 162, 1980.

128. **Gupta, S. and Good, R. A.,** Subpopulations of human T lymphocytes. X. Alterations in T, B, third population cells, and T cells with receptors for immunoglobulin M (T_μ) or $G(T_\gamma)$ in aging humans, *J. Immunol.,* 122, 1214, 1979.

129. **Stutman, O.,** Lymphocyte subpopulations in NZB mice: deficit of thymus-dependent lymphocytes, *J. Immunol.,* 109, 602, 1972.

130. **Nagel, J. E., Chrest, F. J., and Adler, W. H.,** Enumeration of T lymphocyte subsets by monoclonal antibodies in young and aged humans, *J. Immunol.,* 127, 2086, 1981.

131. **Nagel, J. E., Chrest, F. J., Pyle, R. S., and Adler, W. H.,** Monoclonal antibody analysis of T cell subsets in young and aged adults, *Immunol. Commun.,* 12, 223, 1983.

132. **Inkeles, B., Innes, J. B., Kuntz, M. M., Kadish, A. S., and Weksler, M. E.,** Immunological studies aging. III. Cytokinetic basis for the impaired response of lymphocytes from aged humans to plant lectins, *J. Exp. Med.,* 145, 1176, 1977.

133. **Hefton, J. M., Darlington, G. J., Casazza, B. A., and Weksler, M. E.,** Immunologic studies of aging. V. Impaired prolfieration of PHA responsive human lymphocytes in culture, *J. Immunol.,* 125, 1007, 1980.

134. **Gillis, S., Kozak, R., Durante, M., and Weksler, M. E.,** Immunological studies of aging. Decreased production of and response to T cell growth factor by lymphocytes from aged humans, *J. Clin. Invest.,* 67, 937, 1981.

135. **Goodwin, J. S. and Messner, R. P.,** Sensitivity of lymphocytes to prostaglandin E_2 increases in subjects over age 70, *J. Clin. Invest.,* 64, 434, 1979.

136. **Delfraissy, J. F., Galanaud, P., Wallon, C., Balavoine, J. F., and Dormont, J.,** Abolished in vitro antibody response in elderly: exclusive involvement of prostaglandin-induced T suppressor cells, *Clin. Immunol. Immunopathol.,* 24, 377, 1982.

137. **Radl, J., Sepers, J. M., Skvaril, F., Morell, A., and Hijmans, W.,** Immunoglobulin patterns in humans over 95 years of age, *Clin. Exp. Immunol.,* 22, 84, 1975.

138. **Cammarata, R. J., Rodnan, G. P., and Fennell, R. H.,** Serum anti-γ-globulin and antinuclear factors in the aged, *JAMA,* 199, 455, 1967.

139. **Hallgren, H. M., Buckley, C. E., Gilbertsen, V. A., and Yunis, E. J.,** Lymphocyte phytohemmaglutinin responsiveness, immunoglobulins, and autoantibodies in aging humans, *J. Immunol.,* 111, 1101, 1973.

140. **Riesen, W., Keller, H., Skvaril, F., Morell, A., and Barandun, S.,** Restriction of immunoglobulin heterogeneity, autoimmunity and serum protein levels in aged people, *Clin. Exp. Immunol.,* 26, 280, 1976.

141. **Blanckwater, M. J., Levert, L. A., and Hijmans, W.,** Age-related decline in the antibody response to *E. coli* lipopolysaccharide in New Zealand black mice, *Immunology,* 28, 847, 1975.

142. **Pandey, J. P., Fudenberg, H. H., Ainsworth, S. K., and Loadholt, C. B.,** Autoantibodies in healthy subjects of different age groups, *Mech. Aging Dev.,* 10, 399, 1979.

143. **Radl, J., DeGlopper, E., Van den Berg, P., and Van Zwieten, M. J.,** Idiopathic paraproteinemia. III. Increased frequency of paraproteinemia in thymectomized aging C57BL/KaLwRij and CBA/BrARij mice, *J. Immunol.,* 125, 31, 1980.

144. **Radl, J. and Hollander, C. F.,** Homogenous immunoglobulins in the sera of mice during aging, *J. Immunol.,* 112, 2271, 1974.

145. **DeKruyff, R. H., Rinnooy Kan, E. A., Weksler, M. E., and Siskind, G. W.,** Effect of aging on T-cell tolerance induction, *Cell Immunol.,* 56, 58, 1980.

146. **Dobken, J., Weksler, M. E., and Siskind, G. W.,** Effect of age on ease of B-cell tolerance induction, *Cell. Immunol.,* 55, 66, 1980.

147. **Fujiwara, M. and Cinader, B.,** Cellular aspects of tolerance. V. The in vivo cooperative role of accessory and thymus cells in responsiveness and unresponsiveness in SJL mice, *Cell. Immunol.,* 12, 194, 1974.

148. **Klinman, N. R.,** Antibody-specific immunoregulation and the immunodeficiency of aging, *J. Exp. Med.,* 154, 547, 1981.

149. **Goidl, E. A., Innes, B., and Weksler, M. E.,** Immunological studies of aging. II. Loss of IgG and high avidity plaque-forming cells and increased suppressor, *J. Exp. Med.,* 144, 1037, 1976.

150. **Doria, G., D'Agostaro, G., and Poretti, A.,** Age-dependent variations of antibody avidity, *Immunology,* 35, 601, 1978.

151. **Rosenberg, J. S., Gilman, S. C., and Feldman, J. D.,** Activation of rat B lymphocytes. II. Functional and structural changes in 'aged' rat B lymphocytes, *J. Immunol.,* 128, 656, 1982.

152. **Adler, W., Takiguchi, T., and Smith, R. T.,** Effect of age on primary alloantigen recognition by mouse spleen cells, *J. Immunol.,* 107, 1357, 1971.

153. **Callard, R. E., Basten, A., and Waters, L. K.,** Immune function in aged mice. II. B cell function, *Cell. Immunol.,* 31, 26, 1977.

154. **Haaijman, J. J. and Hijmans, W.,** Influence of age on the immunological activity and capacity of the CBA mouse, *Mech. Ageing Dev.,* 7, 375, 1978.

155. **Makinodan, T. and Peterson, W. J.,** Relative antibody forming capacity of spleen cells as a function of age, *Proc. Natl. Acad. Sci. U.S.A.,* 48, 234, 1962.

156. **Makinodan, T., Ching, F., Lever, W. E., and Brewen, B. G.,** The immune system of mice reared in clean and dirty conventional laboratory farms. II. Primary antibody forming activity of young and old mice with long life spans, *J. Gerontol.,* 26, 508, 1971.

157. **Nordin, A. A. and Buchholz, M.,** The effect of age on the *in vitro* immune response of C57BL/6 mice to a T-independent antigen, in *Immunological Aspects of Aging,* Segre, D. and Smith, L., Eds., Academic Press, New York, 1981, 91.

158. **Nagel, J. E., Chrest, F. J., and Adler, W. H.,** Human B-cell function in normal individuals of various age. I. *In vitro* enumeration of pokeweed induced peripheral blood lymphocyte immunoglobulin synthesizing cells and the comparison of the result with numbers of peripheral B and T-cells, mitogen responses, and levels of serum immunoglobulins, *Clin. Exp. Immunol.,* 44, 646, 1981.

159. **White, C. S., Adler, W. H., and McGann, V. G.,** Repeated immunization: possible adverse effects. Re-evaluation of human subjects at 25 years, *Ann. Int. Med.,* 81, 594, 1974.

160. **Buckley, C. E., Buckley, E. G., and Dorsey, F. C.,** Longitudinal changes in serum immunoglobulin levels in older humans, *Fed. Proc.,* 33, 2036, 1974.

161. **Callard, R. E.,** Immune function in aged mice. III. Role of macrophages and the effect of 2-mercaptoethanol in the response of spleen cells from old mice to phytohemagglutinin, lipopolysaccharide and allogeneic cells, *Eur. J. Immunol.,* 8, 697, 1978.

162. **Heidrick, M. L. and Makinodan, T.,** Presence of impairment of humoral immunity in nonadherent spleen cells from old mice, *J. Immunol.,* 111, 1502, 1973.

163. **Heidrick, M. L.,** Age-related changes in hydrolase activity of peritoneal macrophages, *Gerontologist,* 12, 28, 1972.

164. **Perkins, E. H.,** Phagocytic activity of aged mice, *J. Reticuloendothel. Soc.,* 9, 642, 1971.

165. **Boyd, E.,** The weight of the thymus gland in health and disease, *Am. J. Dis. Child.,* 43, 1162, 1932.

166. **Pepper, F. J.,** The effects of age, pregnancy, and lactation on the thymus gland and lymph nodes of the mouse, *J. Endocrinol.,* 22, 335, 1961.

167. **Weiss, L.,** in *The Cells and Tissues of the Immune System,* Prentice-Hall, Englewood Cliffs, N.J., 1972, 49.

168. **Walford, R.,** *The Immunologic Theory of Aging,* Munksgaard, Copenhagen, 1969.

169. **Price, G. B. and Makinodan, T.,** Immunologic deficiencies in senescence. I. Characterization of intrinsic deficiencies, *J. Immunol.,* 108, 403, 1972.

170. **Farrar, J. J., Loughman, B. E., and Adler, W. H.,** Lymphopoietic potential of bone marrow cells from aged mice: comparison of cellular constituents of bone marrow from young and aged mice, *J. Immunol.,* 112, 1244, 1974.

171. **Alexopoulos, C. and Babitis, P.,** Age-dependence of T lymphocytes, *Lancet,* 1, 426, 1976.

172. **Diaz-Jouanen, E., Williams, R. C., and Strickland, R. G.,** Age-related changes in T and B cells, *Lancet,* 1, 688, 1975.

173. **Augener, W., Cohnen, G., Reuter, A., and Brittinger, G.,** Decrease in T lymphocytes during aging, *Lancet,* 1, 1164, 1974.

174. **Gelfand, M. C. and Steinberg, A. D.,** Mechanism of allograft rejection in New Zealand mice. I. Cell synergy and its age-dependent loss, *J. Immunol.,* 110, 1652, 1973.

175. **Walters, C. S. and Claman, H. N.,** Age-related changes in cell mediated immunity in Balb/c mice, *J. Immunol.,* 115, 1438, 1975.

176. **Kishimoto, S., Shigemoto, S., and Yamamura, Y.,** Immune response in aged mice. Change of cell mediated immunity with aging, *Transplantation,* 15, 455, 1973.

177. **Giannini, D. and Sloan, R. S.,** A tuberculin survey of 1285 adults with special reference to the elderly, *Lancet,* 1, 525, 1957.

178. **Stjernswärd, J.,** Age-dependent tumor-host barrier and effect of carcinogen-induced immunodepression of rejection of isografted methylcholanthrene-induced sarcoma cells, *J. Natl. Cancer Inst.,* 37, 505, 1966.

179. **Hori, Y., Perkins, E. H., and Halsall, M. K.,** Decline in phytohemagglutinin responsiveness of spleen cells from aged mice, *Proc. Soc. Exp. Biol. Med.,* 144, 48, 1973.

180. **Pisciotta, A. V., Westring, D. W., Deprey, C., and Walsh, B.,** Mitogenic effect of phytohemagglutinin at different ages, *Nature (London),* 215, 193, 1967.

181. **Goodman, S. A. and Makinodan, T.,** Effect of age on cell-mediated immunity in long-lived mice, *Clin. Exp. Immunol.,* 19, 533, 1975.

182. **Stutman, O.,** Cell-mediated immunity and aging, *Fed. Proc.,* 33, 2028, 1974.

183. **Gutowski, J. K. and Weksler, M. E.,** Studies of the syngeneic mixed lymphocyte reaction. II. Decline in the syngeneic mixed lymphocyte reaction with age, *Immunology,* 46, 1982

184. **Makinodan, T. and Yunis, E., Eds.,** *Immunology and Aging, Comprehensive Immunology, Vol. 1,* Plenum Press, New York, 1977.

185. **Nordin, A. A. and Makinodan, T.,** Humoral immunity in aging, *Fed. Proc.,* 33, 2033, 1974.

186. **Nariuchi, H. and Adler, W. H.,** Dissociation between proliferation and antibody formation by old mouse spleen cells in response to LPS stimulation, *Cell. Immunol.,* 45, 295, 1979.

187. **Nordin, A. A. and Adler, W. H.,** Effect of aging on *in vitro* cellular interactions, in *Developmental Immunobiology,* Siskind, G. W., Litwin, S. D., and Weksler, M. E., Eds., Grune & Stratton, New York, 1979, 215.

188. **Lanza, E. and Djeu, J. Y.,** Age-independent natural killer cell activity in murine peripheral blood, in *NK Cells and Other Natural Effector Cells,* Herberman, R. B., Ed., Academic Press, New York, 1982, 335.

189. **Pross, H. F. and Baines, M. G.,** Studies of human natural killer cells. I. *In vivo* parameters effecting normal cytotoxic function, *Int. J. Cancer,* 29, 383, 1982.

190. **Nagel, J. E., Collins, G. D., and Adler, W. H.,** Spontaneous or natural cytotoxicity of K562 erythroleukemic cells in normal patients, *Cancer Res.,* 41, 2284, 1981.

191. **Penschow, J. and MacKay, I. R.,** NK and K cell activity of human blood: differences according to sex, age, and disease, *Ann. Rheum. Dis.,* 39, 82, 1980.

192. **Pure, E., Isakson, P. C., Kappler, J. W., Marrack, P., Krammer, P. H., and Vitetta, E. S.,** T cell derived B cell growth and differentiation factors. Dichotomy between the responsiveness of B cells from adult and neonatal mice, *J. Exp. Med.,* 157, 600, 1983.

193. **Weiss, A., Imboden, J., Wiskocil, R., and Stobo, J.,** The role of T3 in the activation of human T cells, *J. Clin. Immunol.,* 4, 165, 1984.

194. **Allison, J. P., McIntyre, B. W., and Bloch, D.,** Tumor specific antigen of murine T-lymphoma defined with monoclonal antibody, *J. Immunol.,* 129, 2293, 1982.

195. **Meuer, S. C., Fitzgerald, K. A., Hussey, R.E., Hodgdon, J. C., Schlossman, S. F., and Reinherz, E. L.,** Clonotypic structures involved in antigen-specific human T cell function, *J. Exp. Med.,* 157, 705, 1983.

196. **Kappler, J., Kubo, R., Haskins, K., White J., and Marrack, P.,** The mouse T-cell receptor: comparison of MHC restricted receptors on two T cell hybridomas, *Cell,* 34, 724, 1983.

197. **Samelson, L. E., Germain, R. N., and Schwartz, R. H.,** Monoclonal antibodies against the antigen receptor on a cloned T-cell hybrid, *Proc. Natl. Acad. Sci. USA.,* 80, 6972, 1983.

198. **Haskins, K., Kubo, R., White, J., Pigeon, M., Kappler, J., and Marrack, P.,** The major histocompatibility complex-restricted antigen receptor on T cells. I. Isolation with a monoclonal antibody, *J. Exp. Med.,* 157, 1149, 1983.

199. **Meuer, S. C., Aceto, O., Hussey, R. E., Hodgdon, J. C., Fitzgerald, K. A., Schlossman, S. F., and Reinherz, E. L.,** Evidence for the T3-associated 90K heterodimer as the T-cell antigen receptor, *Nature (London),* 303, 808, 1983.

200. **Meuer, S. C., Cooper, D. A., Hodgdon, J. C., Hussey, R. E., Fitzgerald, K. A., Schlossman, S. F., and Reinherz, E. L.,** Identification of the receptor for antigen and major histocompatibility complex on human inducer T lymphocytes, *Science,* 222, 1239, 1983.

201. **Yanagi, Y., Yoshikai, Y., Leggett, K., Clark, S. P., Aleksander, I., and Mak, T. W.,** A human T cell-specific cDNA clone encodes a protein having extensive homology to immunoglobulin chains, *Nature (London),* 308, 145, 1984.

202. **Hedrick, S. M., Cohen, D. I., Nielsen, E. A., and Davis, M. M.,** Isolation of cDNA clones encoding T cell-specific membrane associated proteins, *Nature (London),* 308, 149, 1984.

203. **Hedrick, S. M., Nielsen, E. A., Kavaler, J., Cohen, D. I., and Davis, M. M.,** Sequence relationships between putative T-cell receptor polypeptides and immunoglobulins, *Nature (London),* 308, 153, 1984.

204. **Acuto, O., Hussey, R. E., Fitzgerald, K. A., Protentis, J. P., Meuer, S. C., Schlossman, S. F., and Reinherz, E. L.,** The human T cell receptor: appearance in ontogeny and biochemical relationship of α and β subsets on IL-2 dependent clones and T cell tumors, *Cell,* 34, 717, 1983.

205. **Jones, N., Leiden, J., Dialynas, D., Fraser, J., Clabby, M., Kishimoto, T., Strominger, J. L., Andrews, D., Lane, W., and Woody, J.,** Partial primary structure of the alpha and beta chains of human tumor T-cell receptors, *Science,* 227, 311, 1985.

206. **Kavaler, J., Davis, M. M., and Chien, Y.,** Localization of a T-cell receptor diversity-region element, *Nature (London),* 310, 421, 1984.

207. **Chien, Y., Gascoigne, N. R. J., Kavaler, J., Lee, N. E., and Davis, M. M.,** Somatic recombination in a murine T-cell receptor gene, *Nature (London),* 309, 322, 1984.

208. **Acuto, O., Fabbi, M., Smart, J., Poole, C. B., Protentis J., Royer, H. D., Schlossman, S. F., and Reinherz, E. L.,** Purification and N-terminal amino acid sequence of the β subunit of a human T cell antigen receptor, *Proc. Natl. Acad. Sci. USA.* 81, 3851, 1984.

209. **Barker, P. E., Ruddle, F. H., Royer, H.D., Acuto, O., and Reinherz, E. L.,** Chromosomal location of human T-cell receptor gene Tiβ, *Science,* 226, 348, 1984.

210. **Lee, N. E., D'Eustachio, P., Pravtcheva, D., Ruddle, F. H., Hendrick, S. M., and Davis, M. M.,** Murine T cell receptor beta chain is encoded on chromosome 6, *J. Exp,. Med.,* 160, 905, 1984.

211. **Mak, T. W., and Yanagi, Y.,** Genes encoding the human T cell antigen receptor, *Immunol. Rev.,* 81, 221, 1984.

212. **Fernandes, G. and Gupta, S.,** Natural killing and antibodydependent cytotoxicity by lymphocyte subpopulations in young and aging humans, *J. Clin. Immunol.,* 1, 141, 1981.

213. **Onsrud, M.,** Age dependent changes in some human lymphocyte subpopulations. Changes in natural killer cell activity, *Acta Pathol. Microbiol. Scand.,* 89, 55, 1981.

214. **Gardner, I. D., Lim, S. T. K., and Lawton, J. W. M.,** Monocyte function in aging humans, *Mech. Ageing Dev.,* 16, 233, 1981.

BIOLOGY OF THE AGING LUNG

Ronald H. Goldstein and Jerome S. Brody

INTRODUCTION

The main function of the lung is that of gas exchange — uptake of oxygen and elimination of carbon dioxide. For this reason most basic investigation of the lung has concentrated on studies of structure and physiologic function. Until recently, exploration of the cell biology of the lung has generally been limited to studies of the connective tissue components which influence the mechanical and thereby the physiologic function of the lung in its gas-exchange role. Most of the information that exists about lung aging is therefore morphologic or physiologic and is biochemical only as it relates to lung connective tissue.

In the past decade there has been a growing interest in biochemical aspects of the lung in relation to maintenance of lung integrity and performance of the gas-exchange function and in relation to metabolic functions of the lung.[1] However, aging of the lung remains a neglected area of the study of lung cell biology.

The rapid maturation of the lung immediately before birth in preparation for the change from a liquid-containing relatively anaerobic state, to an air-containing aerobic state, and the early postnatal restructuring of the lung as it forms its gas-exchange surface have stimulated investigations of the cell biology of lung growth and maturation.[2] Considerable new information has become available relative to cell proliferation,[3] antioxidant defenses,[4] phospholipid synthesis,[5] levels of various lung enzymes,[6,7] and morphogenesis[8] in developing lung, but as yet, none of these studies have been extended to the aging lung.

In most mammalian species studied to date, all of the gas-conducting airways and their component cells are present at birth, but the gas-exchanging alveoli are poorly developed or present only in small numbers.[9] There follows a period of poorly understood lung restructuring which extends for several weeks in the mouse and rat[10,11] and for 6 to 12 years in humans,[9,12] resulting in formation of most pulmonary alveoli, the distal gas-exchange units of the lung. After this period of alveolar proliferation, lung growth occurs by a process of alveolar and airway enlargement which proceeds in association with body growth. The period of hypertrophy (rather than hyperplasia) of the pulmonary gas-exchange surface ends in early adult life or approximately 16 to 20 years in the human.[9,12] After this period, the lung is capable of compensatory and hormone-induced growth,[13-15] but its morphology, physiology, and cell biology remain stable except for changes associated with lung injury or lung aging. Physiologic, morphologic, and some cell biologic aspects of lung development or growth have been covered in several recent reviews.[2,9] Physiologic aspects of lung aging have received some attention[16,17] and are the subject of a recent review.[18] We will discuss cell biologic aspects of lung aging as they occur in the period following lung hypertrophy, touching on physiologic aspects of aging only to illustrate the effects of changes in lung cell biology. It will become apparent to the reader how little information relating to the cell biology of lung aging is available and how much work remains to be done in investigating the developing and mature lung and the aging lung.

PHYSIOLOGIC CHANGES

Virtually every mammalian species that has been studied displays a common set of physiological changes with aging. These changes have been attributed almost exclusively to the age-related alterations in lung connective tissues. With increasing age, there is loss of lung elastic recoil.[19] This decrease in lung retractive forces results in airway closure occurring at higher lung volumes[20] and produces an increase in residual volume,[16,17] diminished maximum expiratory air flow rates,[21] and an altered regional ventilation-perfusion relationship[22] that leads to widened alveolar-to-arterial oxygen gradients.[23] Most of these changes occur in a fashion linear with increasing age. Balancing this decrease in lung elastic recoil (i.e., decreasing stiffness of the lung) is an increase in chest wall recoil (i.e., increasing stiffness) in humans, and this total lung capacity remains relatively constant with age.[19] Because total lung capacity stays constant yet residual volume increases, vital capacity must fall. Diffusing capacity also decreases with age apparently as a result of decreased surface area, decreased membrane permeability, and decreased capillary blood volume.[24]

MORPHOLOGIC CHANGES

Morphologic and sterologic studies have defined several changes in the structure of the aging lung that presumably result from age-related alterations in lung connective tissue components. Initial studies of human lungs were complicated by difficulties in separating the effects of normal aging from those of pulmonary emphysema. Indeed it was initially assumed that emphysema (i.e., destruction of alveolar walls) was a natural consequence of aging.[25] Subsequent studies have established more rigid criteria, eliminating from analysis lungs with emphysema. Such human studies, plus studies of aging dogs, horses, and rodents have generated similar results.

The number of pulmonary alveoli does not change after maturity, although average size tends to increase[9,12,26,27] (Table 1). The internal surface area of the lung, i.e., the alveolar surface available for gas exchange, decreases with age. In man, it decreases an average of 30% by 80 years.[28-30] This decrease in surface area results from an increase in the volume proportion of alveolar ducts.[31-34] It appears that as the ducts increase in size (alveolar duct ectasia), adjacent alveoli flatten, resulting in a decrease in alveolar surface area. The cause of age-related duct ectasia is unclear, but several authors have noted age-related accumulation of pigment and macrophages, suggesting that chronic inflammation may alter the smooth muscle and/or connective tissue elements that form rings at the mouth of alveolar ducts.[31,33]

Early reports suggested that the pores in alveolar walls (pores of Kohn) increase in number with age in humans,[25] but studies in horses,[34] dogs,[35] and most recently a sterologic study done in mice by scanning electron microscopy,[36] have shown no such change with age. It has been suggested that there is an increase in size of these interalveolar pores although this observation has not be quantitated.

Since elastin is one of the major connective tissue elements influencing lung recoil and therefore lung morphology, investigators have sought to define age-related changes in elastin. Morphologists have described degenerative changes in the elastic fibers of aging lungs,[25,37] but careful ultrastructural analysis of these fibers and quantitative sterologic studies have not corroborated earlier suggestions that elastic fibers change with age.[38,39] In man, the length of elastic fibers per unit volume, the total number of elastic fibers, and the number of fibers less than 1 μm all change between birth and age 20 during the period of postnatal lung growth, with fiber length and number increasing and number of small fibers decreasing.[39] After age 20, no quantifiable structural changes in these fibers occur to explain the morphologic or physiologic alterations noted in the aging lung. Similar morphologic studies in aging mice have shown no change in total elastin fiber length.[40] It remains possible that scattered focal fragmentation of small numbers of elastic fibers occurs with age. Similar quantitative structural studies of other connective tissues such as collagen and proteoglycans have not been carried out, although it has been stated that no age-related ultrastructural changes occur in collagen or in the ground substance.[38] A recent study of alveolar and capillary basement membranes failed to document significant thickening of this connective tissue component with age.[41]

Table 1
AGE-RELATED CHANGES IN LUNG MORPHOMETRY

	Human (30)			Dog (33)		
	20 years	70 years	% Change	1 year	10 years	% Change
% Alveoli	48	44	−8	51	43	−14
% Alveolar ducts	36	42	+17	29	41	+41
Surface/volume				39	30	−23
ISA	73.4	59.9	−18			

Pulmonary blood vessels have been studied less extensively than alveoli. Both arteries and veins lose distensibility with age, particularly in the elastic phase of their stress/strain curves, but these physiologic changes are unaccompanied by major morphologic changes.[42] Histologic examination of muscle strips of casted pulmonary vessels suggest little or no change with age in the thickness of the vascular media.[42,43] Intimal fibrosis is found to increase with age although it appears in a minority of vessels.

CONNECTIVE TISSUE

The macromolecules of the extracellular space have long been considered to play a role in the aging process. The question of whether alterations in lung connective tissue elements are primary or secondary age-related events has been a subject of interest and controversy. As a result of their large size and location, these macromolecules may be uniquely susceptible to age-related changes since they may trap reactive molecules and may be difficult to replace once damaged.[44] Several studies have revealed persistent synthesis and turnover of both collagen and elastin in the old rat lung,[45,46] although the fate of these newly formed molecules in the old lung is uncertain. As an illustration, Kuhn and co-workers have shown that despite brisk elastin synthesis following elastase-induced injury to the lung, lung architecture remains disordered.[47]

Collagen

Collagen has been the only lung connective tissue component studied in detail. The levels of collagen have been measured in the aging human lung by several investigators (Table 2). Both an increase and no change in levels have been reported. However, the studies express collagen content as a percent of lung dry weight rather than as an absolute quantity. As other lung components may also be changing, this approach is potentially misleading.[48] For example, in experimental pulmonary fibrosis, it is more meaningful to express connective tissue content per lung than per milligram dry weight.[48,49] An additional problem in evaluating studies of lung collagen relates to varying extraction and analytical methodology. The collagen content of rat lung increases with age,[50] but this animal grows throughout life. In the mouse, the collagen content remains constant.[40] In the hamster, a rodent whose growth plateaus early in life, collagen content per lung increases 85% between 4 and 12 months of age.[51] The ratio of type III to type I collagen has been reported to be greater in skin from young than skin from old animals.[52] It is not known if a similar shift occurs in the aging lung. It has been shown that, in humans, the alveolar basement membranes (types IV and V collagen) do not change in thickness with age.[41]

Qualitative changes in collagen occur with aging in several tissues and in a variety of animal species.[53] These changes involve increased molecular stability and decreased solubility. Collagen becomes more resistant to thermal and chemical denaturation,[54,55] less sus-

Table 2
HUMAN LUNG CONNECTIVE TISSUE ALTERATIONS WITH AGING

Collagen	Effect of aging	Ref.
1	No change as percent dry weight	100
2	Increased levels, 8.7 mg% dry wt → 15 mg%	72
3	Decreased rate of collagenase digestion at 40°C (92 (meqKOH/min/mg × 10^3) → 44)	58
Elastin		
1	Increased levels	99—101
	Increased levels in pleura (8 → 15%) interlobular septa and no change in parenchyma	100
2	Decreased desmosine (1.6 → 1.0)[a] and isodesmosine (1.0 → 0.8)[a]	65
3	Amino acid composition	59, 65
	Increased levels	65
	Aspartic (3.8 → 16.5)[a]	
	Glutamic (17.2 → 26.3)[a]	
	Carbohydrate (0.23 → 0.45 g/100 g elastin)	

[a] Amino acid residues per 1000 residues.

ceptible to digestion by collagenase, and less soluble in weak acids.[56] The resistance to collagenase and decreased acid solubilization have been observed in aging human lung.[57-59] These changes have not been shown to influence collagen extractability with age in the mouse.[60] These changes have been attributed by some authors to age-related variations in intermolecular cross-links. It has been demonstrated that the number of cross-links reducible with sodium borohydride decreases with age although the nature of the cross-link that replaces them is unknown.[61] Other qualitative changes in collagen such as decreases in the relative amounts of hydroxylysine and glucosylated hydroxylysine have been described in tissues other than the lung.[62,63] These alterations may account for the increased stiffness and decreased tensile strength of tissues when they age and contribute to the changes in lung mechanical properties.

Elastin

Elastin becomes insoluble early in life and turns over very slowly.[45] Elastin content expressed per gram dry weight remains constant in human lung parenchyma but increases in the interlobar septa and pleura.[64] In the mouse, elastin content has been reported to decrease with age.[40] Increased amounts of polar amino acids such as aspartic and glutamic acids occur in elastin with age,[64] although the significance of these changes is uncertain. Methodologic problems in the analysis of elastin are even more complex than with collagen; indeed until recently elastic tissue, which is composed of an amorphous elastin and a microfibrillar component, has been defined not in absolute terms but by the method of isolation.[65-67] Since newer methods for examining elastin have become available,[66-69] the content and composition of elastin in the aging lung must be reassessed. As is the case with collagen, elastin becomes more resistant to enzymatic digestion with age.[70] The old hamster lung develops less emphysema for a given dose of intratracheally instilled pancreatic elastase than does the young adult lung.[51]

OTHER COMPONENTS

A few studies have examined changes in other lung components as a function of age. Recently Schmid et al.[71] have evaluated glycosaminoglycans (GAGs) as a function of age in the human lung. The most impressive changes occurred in the arteries and bronchioles in which a 21% increase in total GAGs was found. An increase in chondroitin-6-sulfate and a relative decrease in dermatan sulfate were shown to account for this increase in total GAGs. No significant changes occurred in GAG content or composition in lung parenchyma.

Total lipid content has been reported to decrease in the aged lung,[22] yet the percent of saturated lecithin is constant throughout adult life and old age.[73] These observations imply that lung surfactant does not change with age, although no studies have been done to determine whether alveolar surface forces change with age.

There are essentially no data regarding the level of activity of extracellular degradative enzymes such as collagenase of elastase as a function of age. One study of lysosomal acid hydrolase activity in the mouse lung reported an increase in β-glucuronidase and N-acetyl-β-D-glucosaminidase but no change in acid phosphatase.[74] An earlier study by Wilson found marked age, sex, and strain differences in the lungs of 6-, 18-, and 30-month-old mice. The pattern of change was complex; acid phosphatase was decreased in males at 18 months of age but not in females. However, β-glucuronidase was decreased at 18 months and increased at 30 months in both sexes. The explanation for these divergent results as well as the significance of these findings are uncertain.

CONNECTIVE TISSUE AND PHYSIOLOGY

Investigation of lung strips has yielded information relating connective tissue changes to physiology. Sugihara and co-workers[76] examined the length-tension properties in the human alveolar wall and found that the maximal extensibility ratio (max = predicted final length/initial length) decreases with age resulting primarily from an increase in resting, or initial tissue length. The change in initial length is a permanent deformation which the authors postulate results from elastin deformation. The changes in length-tension relationships are not consistent among different species.[77] Although seen in humans, no decrease in maximal extensibility was found in the rat, rabbit, or horse. Sugihara and associates also have attempted to induce similar aging changes in vitro by using an aldehyde to increase inter-molecular cross-links in lung preparations.[78] Maximal extensibility of lung strips decreased, and resting length increased as was seen in aging lung. However, the tissue hysteresis ratio decreased, a finding not observed in normal aging. The authors suggest that this model of increasing cross-link formation does not fit their clinical observations and that increased cross-linking and proteolytic action on connective tissue would provide a better model of lung aging. As noted above, morphologic studies of elastic fiber length and total elastic fibers have shown no change with age in the human. Studies of mouse lung have also shown no change in total fiber length with age.[40] Thus, there is at present no clear morphologic or biochemical correlate of the decreased elastic recoil and increased functional residual capacity measured in aging animals and humans.

LUNG DEFENSE MECHANISMS

The response of the aging lung to injury depends upon local as well as systemic factors. The local lung immune system involves potential for antibody production and for macrophage and lymphocyte cell-mediated immune responses, which may function either independent of or in relation to the systemic immune system.[79-81] Information regarding age-related changes in the local immune system and in the relation between systemic and local systems is minimal. One recent study of pulmonary lung lavage lymphocyte subpopulations has shown that the relative numbers of T and B lymphocytes do not change with age in NZB mice but that the percent of lavage null cells increases at the expense of T and B cells in C57BL/6 mice.[82] Clearly, more information about age effects on local immune systems and interactions with systemic systems is needed.

Table 3
AGING AND EXPERIMENTAL LUNG INJURY

Agent	Effect on aging lung	Ref.
NO$_2$	Decreased onset of cell proliferation and increased type II cell proliferation (adult and old rat)	89, 90
Silicosis	Increased silicotic nodules and increased collagen formation (adult and old rat)	96
Air pollution	Increased epithelial injury and extensive cytoplasm disruption (adult and old mice)	95
Ozone	Temporary inhibition of thymidine cell labeling in aging mice	97

Local lung and systemic interaction also occurs in relation to protection of lung connective tissue from enzymatic digestion. Studies of experimental and human emphysema have shown that leukocytic elastase is capable of destroying lung connective tissue components and inducing emphysema.[83,84] Systemic protection against the degradative effects of leukocytic enzymes is provided by circulating antiproteases, primarily α_1-antitrypsin and α_2-macroglobulin. In addition, local lung macrophages play a role in uptake and inactivation of leukocyte elastase.[85-87] How either of these protective systems is influenced by age is uncertain.

Acute lung injury induces a wave of cell proliferation that involves endothelial cells, alveolar type 2 cells, and Clara cells of the terminal bronchioles.[87] The role of endothelial cell proliferation in lung repair is unclear, but the type 2 cell proliferation provides new cells that differentiate into type 1 cells and resurface the alveolar epithelium. Clara cells serve a similar function in the terminal bronchioles. While baseline cell renewal, as measured by mitotic index, is similar in mature and old mice,[88] the proliferative response to injury appears to be slowed in older animals. Experiments with nitrogen dioxide inhalation (Table 3) have shown that the onset of type 2 cell proliferation is slower, and the extent of proliferation is greater in old compared to young rats. Furthermore, mortality and extent of injury were greater in old rats; survivors of the initial insult repaired and developed tolerance to nitrogen dioxide in a fashion similar to young rats.[89,90]

IN VIVO AGING OF LUNG FIBROBLASTS

Cell lines derived from embryonic and adult lung tissue have been used in studying age-related changes in cultured human cells. Those studies fortuitously provide data regarding the characteristics of cells derived from adult and old lungs. The original studies of in vitro aging used embryonic lung cells and subsequently adult cell lines for comparing in vivo and in vitro aging. Fetal cell lines have an average of 48 population doublings prior to cessation of mitotic activity as compared with an average of 20 obtained from adult lines derived from donors between 27 and 87 years of age.[91] Several attempts have been made to relate embryonic cells aged in vitro with cells derived from adult or old lungs. Cells originating from an adult lung exhibit increased sensitivity to low-dose radiation, as evidenced by diminished population doublings, in a fashion similar to cells that have been aged in vitro.[92] A higher level of acid phosphatase activity was found in adult diploid cultures derived from a 58-year-old donor at autopsy than in fetal lung cell cultures. Acid phosphatase levels showed a small but significant increase in senescent cells aged in vitro. Lung fibroblasts from adults had acid phosphatase, alkaline phosphatase, and lactic acid dehydrogenase activity similar to that of aged embryonic cells.[93] Since lung cells are exposed to higher concentrations of oxygen than other cells in vivo, it is of interest that both oxygen toxicity and free radical reactions are not significant in limiting lifespan of human embryonic lung diploid cells as assessed by population doublings under atmospheric conditions.[94] These studies suggest that although differences can be found in lung fibroblasts derived from adult and old individuals, the overall impression is one of similarity at least in early passage.

Though there is presently very little available information regarding the cell biology of lung aging, several generalizations can still be made. From a physiologic standpoint, lung aging is a slow process, particularly in view of the lung's vast reserve capacity. Incompletely defined alterations in the collagen, elastin, and GAG matrix occur, but at the present time it is not possible to distinguish the effects of endogenous aging from those of exogenous injury. Further, these changes do not appear to seriously hamper the lung's main function as a gas-exchange organ even at the extremes of age. Differences exist in the ability of the lung to respond to injurious agents, but again, despite some modifications, the old lung appears able to heal itself. The rate of aging of a particular organ may relate in part to the degree of specialization of its resident cell population. One possible explanation for the minor age-related changes in the lung may be the absence of a nonreplicating specialized population of cells in the lung.

REFERENCES

1. **Crystal, R. G., Ed.,** The *Biochemical Basis of Pulmonary Function,* Marcel Dekker, New York, 1976.
2. **Hodson, W. A., Ed.,** *Development of the Lung,* Marcel Dekker, New York, 1977.
3. **Kaufman, S., Burri, P. H., and Weibel, E. R.,** The postnatal growth of the rat lung. II. Autoradiography, *Anat. Rec.,* 180, 63, 1974.
4. **Yam, J., Frank, L., and Roberts, R. J.,** Age-related development of pulmonary antioxidant enzymes in the rat, *Proc. Soc. Exp. Biol. Med.,* 157, 293, 1978.
5. **Van Golde, L. M. G.,** Metabolism of phospholipids in the lung, *Am. Rev. Respir. Dis.,* 114, 977, 1976.
6. **Hamosh, M., Simon, M. R., Canter, H., and Hamosh, P.,** Lipoprotein lipase activity and blood triglyceride levels in fetal and newborn rats, *Pediatr. Res.,* 12, 1132, 1979.
7. **Powell, J. T. and Whitney, P. L.,** Postnatal development of rat lung, *Biochem. J.,* 188, 1, 1980.
8. **Burri, P. H.,** Postnatal growth of the rat lung. II. Morphology, *Anat. Rec.,* 180, 77, 1974.
9. **Thurlbeck, W. M.,** Postnatal growth and development of the lung, *Am. Rev. Respir. Dis.,* 111, 803, 1975.
10. **Burri, P. H., Dbaly, J., and Weibel, E. R.,** Postnatal growth of the rat lung. I. Morphometry, *Anat. Rec.,* 178, 711, 1974.
11. **Amy, R. W. M., Bowers, D., Burri, P. H., Haines, J., and Thurlbeck, W. M.,** Postnatal growth of the mouse lung, *J. Anat.,* 124, 131, 1977.
12. **Dunnill, M.,** Postnatal growth of the lung, *Thorax,* 17, 329, 1962.
13. **Buhain, W. J. and Brody, J. S.,** Compensatory growth of the lung following pneumonectomy, *J. Appl. Physiol.,* 35, 898, 1973.
14. **Brody, J. S., Fisher, A. B., Gocmen, A., and DuBois, A. B.,** Acromegalic pneumonomegaly: lung growth in the adult, *J. Clin. Invest.,* 49, 1051, 1970.
15. **Brody, J. S. and Buhan, W. J.,** Hormone induced growth of the adult lung, *Am. J. Physiol.,* 233, 1444, 1972.
16. **Muieson, G., Sorbini, C., and Grassi, V.,** Respiratory function in the aged, *Bull. Physiol. Pathol. Respir.,* 7, 973, 1971.
17. **Klocke, R.,** in *Handbook of the Biology of Aging,* Finch, C. B. and Hayflick, L., Eds., Van Nostrand Reinhold, New York, 1977, 432.
18. **Mauderly, J. L.,** in *Handbook of Physiology in Aging,* Masoro, E. J., Ed., CRC Press, Boca Raton, Fla., 1981, 197.
19. **Tuner, J., Mead, J., and Wohl, M. E.,** Elasticity of human lungs in relation to age, *J. Appl. Physiol.,* 25, 664, 1968.
20. **LeBlanc, P., Ruff, F., and Milic-Emili, J.,** Effects of age and body position on "airway closure" in man, *J. Appl. Physiol.,* 28, 448, 1970.
21. **Knudson, R. J., Slatin, R. C., and Lebowitz, M. D.,** The maximum expiratory flow volume curve. Normal standards, variability and effects of age, *Am. Rev. Respir. Dis.,* 113, 587, 1976.
22. **Holland, J., Milic-Emili, J., Macklem, P. T., and Bates, D. V.,** Regional distribution of pulmonary ventilation and perfusion in elderly subjects, *J. Clin. Invest.,* 47, 81, 1962.

23. **Sorbini, C. A., Grassi, V., Salinas, E., and Muiesan, G.**, Arterial oxygen tension in relation to age in health subjects, *Respiration*, 25, 3, 1968.

24. **Hamer, N. A. J.**, The effect of age on the components of the pulmonary diffusing capacity, *Clin. Sci.*, 23, 85, 1962.

25. **Pump, K. K.**, The aged lung, *Chest*, 60, 571, 1971.

26. **Richards, D. W.**, The aging lung, *Bull. N.Y. Acad. Med.*, 32, 407, 1956.

27. **Angus, G. E. and Thurlbeck, W. M.**, Number of alveoli in the human lung, *J. Appl. Physiol.*, 32, 483, 1972.

28. **Thurlbeck, W. M.**, The internal surface area of nonemphysematous lungs, *Am. Rev. Respir. Dis.*, 95, 765, 1967.

29. **Hasleton, P. S.**, The internal surface area of the adult human lung, *J. Anat.*, 112, 391, 1972.

30. **Thurlbeck, W. M. and Angus, G. E.**, Growth and aging of the normal human lung, *Chest*, 67, 35, 1975.

31. **Heppleston, A. G. and Leopold, J. G.**, Chronic pulmonary emphysema. Anatomy and pathogenesis, *Am. J. Med.*, 31, 279, 1961.

32. **Ryan, S. F., Vincent, T. N., Mitchell, R. S., Filley, G. F., and Dart, G.**, Ductectasia: an asymptomatic pulmonary change related to age, *Med. Thoracatis*, 22, 181, 1965.

33. **Robinson, N. E. and Gillespie, J. R.**, Morphologic features of the lungs of aging beagle dogs, *Am. Rev. Respir. Dis.*, 108, 1192, 1973.

34. **Hyde, D. M., Robinson, N. E., Gillespie, J. R., and Tyler, W. S.**, Morphometry of the distal airspaces in lungs of aging animals, *J. Appl. Physiol.*, 43, 86, 1977.

35. **Martin, H. B.**, The effect of aging on the alveolar pores of Kohn in the dog, *Am. Rev. Respir. Dis.*, 88, 773, 1963.

36. **Ranga, V. and Kleinerman, J.**, Interalveolar pores in mouse lungs. Regional distribution and changes with age, *Am. Rev. Respir. Dis.*, 122, 477, 1980.

37. **Wright, R. R.**, Elastic tissue of normal and emphysematous lungs, *Am. J. Pathol.*, 39, 355, 1961.

38. **Adamson, J. S.**, An electron microscopic comparison of the connective tissue from the lungs of young and elderly subjects, *Am. Rev. Respir. Dis.*, 98, 399, 1968.

39. **Niewoehner, D. E. and Kleinerman, J.**, Morphometric study of elastic fibers in normal and emphysematous human lungs, *Am. Rev. Respir. Dis.*, 115, 15, 1977.

40. **Ranga, V., Kleinerman, J., Ip, M. P. C., and Sorensen, J.**, Age-related changes in elastic fibers and elastin of the lung, *Am. Rev. Respir. Dis.*, 119, 369, 1979.

41. **Vracko, R., Thorning, D., and Huang, T. W.**, Basal lamina of alveolar epithelium and capillaries: quantitative changes with aging and diabetes mellitus, *Am. Rev. Respir. Dis.*, 120, 973, 1979.

42. **Warnock, M. L. and Kunzmann, A.**, Changes with age in muscular pulmonary arteries, *Arch. Pathol. Lab. Med.*, 101, 175, 1977.

43. **Mackay, E. H., Banks, J., Sykes, B., and Lee, G. DeJ.**, Structural basis for the changing physical properties of human pulmonary vessels with age, *Thorax*, 33, 335, 1978.

44. **Bornstein, P.**, Disorders of connective tissue function and the aging process: a synthesis and review of current concepts and findings, *Mech. Aging Dev.*, 5, 305, 1976.

45. **Pierce, J. A., Resnick, H., and Henry, P. H.**, Collagen and elastin metabolism in the lungs, skin, and bones of adult rats, *J. Lab. Clin. Med.*, 69, 485, 1967.

46. **Kao, K. Y. T., Hilker, D. M., and McGavacke, T. H.**, Connective tissue. V. Comparison of synthesis and turnover of collagen and elastin in tissues of rat at several ages, *Proc. Soc. Exp. Biol. Med.*, 106, 335, 1961.

47. **Kuhn, C., Yu, S., Chraplyvy, M., Linder, M., and Senoir, R. M.**, The induction of emphysema with elastase. II. Changes in connective tissue, *Lab. Invest.*, 34, 372, 1976.

48. **Karlinsky, J. B. and Goldstein, R. H.**, Fibrotic lung disease — a perspective, *J. Lab. Clin. Med.*, 96, 939, 1980.

49. **Goldstein, R. H., Lucey, E. C., Franzblau, C., and Snider, G. L.**, Failure of mechanical properties to parallel changes in lung connective tissue composition in bleomycin-induced pulmonary fibrosis in hamsters, *Am. Rev. Respir. Dis.*, 120, 67, 1979.

50. **Juricova, M. and Deyl, Z.**, Aging processes in collagens from different tissues of rats, *Adv. Exp. Med. Biol.*, 53, 351, 1975.

51. **Goldstein, R. H.**, Response of the aging hamster lung to elastase injury. *Am. Rev. Respir. Dis.*, 125, 295, 1982.

52. **Miller, E. J.**, Biochemical characteristics and biological significance of genetically-distinct collagens, *Mol. Cell Biochem.*, 13, 165, 1976.

53. **Prockop, D. J., Kivirikko, K. I., Tuderman, L., and Guzman, N.**, The biosynthesis of collagen and its disorders, *N. Engl. J. Med.*, 301, 77, 1979.

54. **Takacs, I. and Verzar, F.**, Macromolecular aging of collagen. I. Experiments in vivo and in vitro with different animal races, *Gerontologia*, 14, 15, 1968.

55. **Boros-Farkas, M. and Everitt, A. V.,** Comparative studies of age tests on collagen fibers, *Gerontologia,* 13, 37, 1967.
56. **Viidik, A.,** Connective tissues possible implications of the temporal changes for the aging process, *Mech. Aging Dev.,* 9, 267, 1979.
57. **Rickert, W. S. and Forbes, W. F.,** Changes in collagen with age. VI. Age and smoking related changes in human lung connective tissue, *Exp. Gerontol.,* 11, 89, 1976.
58. **Rickert, W. S. and Forbes, W. F.,** Changes in collagen with age. IV. The extraction of soluble collagen from human lung, *Exp. Gerontol.,* 7, 387, 1972.
59. **Fitzpatrick, M. and Hospelhorn, V. D.,** Studies of human pulmonary connective tissue. II. Amino acid composition of residues following collagenase digestion of lung connective tissues, *Am. Rev. Respir. Dis.,* 92, 792, 1965.
60. **Schofield, J. D.,** Connective tissue aging: differences between mouse tissues in age-related changes in collagen extractability, *Exp. Gerontol.,* 15, 113, 1980.
61. **Bailey, A. J., Robins, S. P., and Balian, G.,** Biological significance of the intermolecular crosslinks of collagen, *Nature (London),* 251, 105, 1974.
62. **Barnes, M. J., Constable, B. J., Morton, L. F., and Royce, P. M.,** Age-related variations in hydroxylation of lysine and proline in collagen, *Biochem. J.,* 139, 461, 1974.
63. **Royce, P. M. and Barnes, M. J.,** Comparative studies on collagen glycosylation in chick skin and bone, *Biochim. Biophys. Acta,* 498, 132, 1977.
64. **John, R. and Thomas, J.,** Chemical compositions of elastins isolated from aortas and pulmonary tissues of humans of different ages, *Biochem. J.,* 127, 261, 1972.
65. **Ross, R. and Bornstein, P.,** The elastic fiber. I. The separation and partial characterization of its macromolecular components, *J. Cell. Biol.,* 40, 366, 1969.
66. **Hance, A. J. and Crystal, R. G.,** The connective tissue of lung, *Am. Rev. Respir. Dis.,* 112, 657, 1975.
67. **Richmond, V.,** Lung parenchymal elastin isolated by non-degradative means, *Biochim. Biophys. Acta,* 351, 173, 1974.
68. **Keeley, F. W.,** A convenient method for the identification and estimation of soluble elastin synthesis in vitro, *Connec. Tiss. Res.,* 4, 193, 1976.
69. **Narayanan, A. S. and Page, R. C.,** Demonstration of a precursor-product relationship between soluble and cross-linked elastin, and biosynthesis of the desmosines in vitro, *J. Biol. Chem.,* 251, 1125, 1976.
70. **Fitzpatrick, M.,** Studies of human pulmonary connective tissue. III. Chemical changes in structural proteins with emphysema, *Am. Rev. Respir. Dis.,* 96, 254, 1967.
71. **Schmid, K.,** personal communication.
72. **Briscoe, A. M., Loring, W. E., and McClement, J. H.,** Changes in human lung collagen and lipids with age, *Proc. Soc. Exp. Biol. Med.,* 102, 71, 1959.
73. **Yasnoka, S., Manabe, H., Ozoki, T., and Tsubura, E.,** Effect of age on the saturated lecithin contents of human and rat lung tissues, *J. Gerontol.,* 32, 387, 1977.
74. **Traurig, H. H.,** Lysosomal acid hydrolase activities in the lungs of fetal, neonatal, adult and senile mice, *Gerontology,* 22, 419, 1976.
75. **Wilson, P. H.,** Enzyme patterns in young and old mouse livers and lung, *Gerontologia,* 18, 36, 1972.
76. **Sugihara, T., Martin, C. J., and Hildebrandt, J.,** Length-tension properties of alveolar wall in man, *J. Appl. Physiol.,* 30, 874, 1971.
77. **Martin, C. J., Chihara, S., and Chang, D. B.,** A comparative study of the mechanical properties in aging alveolar wall, *Am. Rev. Respir. Dis.,* 115, 981, 1977.
78. **Sugihara, T. and Martin, C. J.,** Simulation of lung tissue properties in age and irreversible obstructive syndromes using an aldehyde, *J. Clin. Invest.,* 56, 23, 1975.
79. **Kaltreider, H. B.,** Expression of immune mechanisms in the lung, *Am. Rev. Respir. Dis.,* 113, 347, 1976.
80. **Waldman, R. H. and Henney, C. S.,** Cell-mediated immunity and antibody responses in the respiratory tract after local and systemic immunization, *J. Exp. Med.,* 134, 482, 1971.
81. **Pennington, J. E.,** Bronchoalveolar cell response to bacterial challenge in the immunosuppressed lung, *Am. Rev. Respir. Dis.,* 116, 885, 1977.
82. **Gong, H., Clements, P. J., and Eisenberg, H.,** Pulmonary lymphocyte subpopulations. Variation in New Zealand black/white and C57BL/6 mice with age, *Am. Rev. Respir. Dis.,* 120, 821, 1979.
83. **Marco, V., Moss, B., Meranze, D. R., Weinbaum, G., and Kimbel, P.,** Induction of experimental emphysema in dogs using leukocyte homogenates, *Am. Rev. Respir. Dis.,* 104, 595, 1971.
84. **Moss, B., Ikeda, T., Meranze, D. R., Weinbaum, G., and Kimble, P.,** Induction of experimental emphysema: cellular and species specificities, *Am. Rev. Respir. Dis.,* 106, 384, 1972.
85. **Campbell, E. J., White, R. R., Senior, R. M., and Rodriguiez, R. J.,** Receptor mediated binding and internalization of leukocyte elastase by alveolar macrophages in vitro, *J. Clin. Invest.,* 64, 824, 1979.
86. **McGowan, S. E., Stone, P. J., Calore, J. D., Snider, G. L., and Franzblau, C.,** Alveolar macrophage incorporation of human leukocytic elastase. The fate of neutrophil elastase incorporated by human alveolar macrophages, *Am. Rev. Respir. Dis.,* 127, 449, 1983.

87. **Kuhn, C.,** The cells of the lung and their organelles, in *Biochemical Basis of Pulmonary Function,* Crystal, R., Ed., Marcel Dekker, New York, 1976, 3.

88. **Sinnet, J. D. and Heppleston, A. G.,** Cell renewal in the mouse lung. The influence of sex, strain and age, *Lab. Invest.,* 15, 1793, 1966.

89. **Evans, M. J., Cabral-Anderson, L. J., and Freeman, G.,** Effects of NO$_2$ on the lungs of aging rats. II. Cell proliferation, *Exp. Mol. Pathol.,* 27, 366, 1977.

90. **Cabral-Anderson, L. J., Evans, M. J., and Freeman, G.,** Effects of NO$_2$ on the lungs of aging rats, *Exp. Mol. Pathol.,* 27, 353, 1977.

91. **Hayflick, L.,** The limited in vitro lifetime of human diploid cell stains, *Exp. Cell Res.,* 37, 614, 1965.

92. **Azzarone, B., Diatloff-Zito, C., Billard, C., and Macieira-Coelho, A.,** Effect of low dose rate irradiation on the division potential of cells in vitro, *In Vitro,* 16, 634, 1980.

93. **Cristofalo, V. J., Parris, N., and Kritchevsky, D.,** Enzyme activity during the growth and aging of human cells in vitro, *J. Cell Physiol.,* 69, 263, 1967.

94. **Balin, A. K., Goodman, D. B. P., Rasmussen, H., and Cristofalo, V. J.,** Oxygen-sensivity stages of the cell cycle of human diploid cells, *J. Cell Biol.,* 78, 390, 1978.

95. **Bils, R. F.,** Ultrastructural alterations of alveolar tissue of mice. I. Due to heavy Los Angeles smog, *Arch. Environ. Health,* 12, 689, 1966.

96. **Shanker, R., Sahu, A. P., Dogra, R. K. S., and Zaidi, S. H.,** A study of the age factor in experimental silicosis in rats, *Environ. Physiol. Biochem.,* 5, 158, 1975.

97. **Evans, M. J. and Bils, R. F.,** Effects of ozone on cell renewal in pulmonary alveoli of aging mice, *Arch Environ. Health,* 22, 450, 1971.

98. **Boucek, R. J., Noble, N. L., and Marks, A.,** Age and the fibrous proteins of the human lung, *Gerontologia,* 5, 150, 1961.

99. **Johnson, R. and Andrews, F. A.,** Lung scleroproteins in age and emphysema, *Chest,* 57, 239, 1970.

100. **Pierce, J. A. and Ebert, R. V.,** Fibrous network of the lung and its change with age, *Thorax,* 20, 469, 1965.

101. **Pierce, J. A. and Hocott, J. B.,** Studies on collagen and elastin content of the human lung, *J. Clin. Invest.,* 39, 8, 1960.

LUNG-DERIVED FIBROBLAST-LIKE HUMAN CELLS IN CULTURE

James R. Smith and Olivia M. Pereira-Smith

INTRODUCTION

For the past several years, human diploid fibroblasts (especially those derived from fetal lung) have been extensively investigated in regard to in vitro cellular aging. Hayflick and Moorhead[1] first showed that diploid fibroblasts derived from human fetal lung had a finite proliferation potential in vitro. Hayflick[2] later suggested that the inability of these cells to grow indefinitely was an expression of cellular aging in vitro. Hayflick[2] reported that cells derived from embryonic tissue could achieve a greater number of doublings in vitro than those derived from adult tissue. Other investigators[3,4] have extended this observation to show that, for human skin fibroblasts, the proliferative potential of a culture was inversely related to the age of the donor and that this relationship continued throughout adult life. The rationale for using cells in culture as a model of cellular aging and the possibility of extrapolating the results to in vivo aging is strengthened by this correlation between the age of the cell culture donor and the proliferative potential of the cell cultures in vitro.

In recent years a number of cell culture systems have been developed for the study of in vitro cellular aging. Still the most exhaustively studied cell type is the fibroblast-like cell derived from human fetal lung. Studies with other cell types have confirmed the generality of the findings on human lung fibroblasts. The tables below represent a compilation of the quantitative studies and present the results in a format that should be useful to investigators in the field of in vitro cellular aging.

As some of the studies did not lend themselves to the quantitative data that could be presented in the format used, it is likely that some important data have been overlooked, and for this we must apologize. The data presented are limited to studies conducted on lung-derived fibroblast-like cells in which some cellular parameter was studied as a function of in vitro age. Studies conducted only at one point during the in vitro lifespan or at unspecified in vitro ages have not been included.

In vitro age has been expressed in many ways in the literature, the most common measure being either population doubling level (i.e., the number of population doublings achieved since initiation of the culture) or passages in vitro. The passage number given in the tables usually corresponds roughly to population doubling level. Whenever possible, the split ratio used to passage the culture and the in vitro age at which the cultures stopped dividing are given.

ACKNOWLEDGMENTS

The authors are grateful to Mrs. Cynthia Gendron for excellent secretarial assistance.

During the compilation of this material, JRS was supported by NIA Grants No. AG 00338, AG 01984, and NIA contract AG 72117. OPS was supported by NIA Grant No. AG 01984.

Table 1
DOUBLING TIME AND CELL CYCLE TIME

Aspect studied	Cell type	In vitro age: PDL[a]	Result					Ref.
Doubling time (average)[b]	WI-38	19	17.83 hr					4
		35	19.14					
		51	34.00					
	WI-38	≤25	23.3 hr					5
		≥48	Could not be done					
	WI-38	18	17 hr					6
		42	21—22 hr					
	WI-38	7—32% unlabeled nuclei	17.4 hr ± 2.8					7
		38—70% unlabeled nuclei	30.5 hr ± 10.4					8
	WI-38	30	19.0					
		50	19.4					
		60	30.7					
Cell cycle time[b]	WI-38	25	19—20 hr[c]					9
		50	18 hr[c]					
			G₁	**S**	**G₂**	**M**		
			(Time in hours)					
	WI-38	7—32% unlabeled nuclei	2.4	10.5	3.3	1.2		7
		38—70% unlabeled nuclei	14.3	11.0	3.8	1.5		
			G₁	**S**	**G₂**	**M**		
	WI-38	18	6	6	4	30 min		6
		42	8	8	4	30 min		
			G₁	**S**	**G₂**			
	WI-38	30	4.1	10.4	4.5			8
		50	4.2	10.6	4.6			
		60	14.8	10.5	5.4			

		G₁	S	G₂ and M	Ref.
IMR-90	32	50.2%	22%	27.8%	10
Cell distrib.ᵈ	50	80.8%	4.5%	14.7%	

a In vitro age: population doubling level, unless otherwise indicated.
b Time in hours, except where precent is indicated.
c Modal intermitotic time of cells obtained by mitotic shakeoff.
d Distribution of cells in cell cycle phases (flow microfluorimetry).

Table 2
MORPHOLOGICAL CHANGES

Aspect studied	Cell type	In vitro age: PDLᵃ	Result (mode) (diameter)	Ref.
Cell size				
1 unit = 34 μm dia.	WI-38	20	0.5—0.6 units	11
		21	0.4—0.6 units	
		43	0.5—1.0 units	
		50	0.5—1.0 units	
Modal volume (μm³)	WI-38	22—24	2713	12
		37—38	2472	
		48—55	3437	
	WI-38	20/44ᵇ	1378	13
		24/44	1462	
		36/44	1527	
		40/44	1773	
		43/44	1824	
	HE-125	11/25	1257	13
		19/25	1370	
		22/25	1470	
		29/31	1820	
	HE-388	10/22	1318	13
		16/22	1446	

Table 2 (continued)
MORPHOLOGICAL CHANGES

Aspect studied	Cell type	In vitro age: PDL[a]	Result	Ref.
Nuclear size	WI-38	20/22	1654	4
		19	1860	
		35	2140	
		51	3722	

Aspect studied	Cell type	In vitro age: PDL[a]	Modal size μm²	Max size μm²	Ref.
	WI-38	19	100—125	325—350	14
		35	125—150	450—475	
		45	125—150	575—600	

Aspect studied	Cell type	In vitro age: PDL[a]	Mean nucleolar area 10⁻⁷ cm²	Mean nucleolar dry mass 10⁻¹² g	% Cells with nucleoli/cell 1	2	3	4	5
Nucleolar changes	WI-38	27.3	1.892	9.783	17	40	17	20	6
		33.5	1.810	8.383	23	25	15	25	12
		39.0	3.325	23.325	51	40	3	3	3
		41.2	6.372	66.815	66	34	0	0	0

Mean nucleolar area is expressed in units of $10^{-7}\,\text{cm}^2$; mean nucleolar dry mass in units of $10^{-12}\,\text{g}$.

Aspect studied	Cell type	In vitro age: PDL[a]	Number/cell	Qualitative result	Ref.
Light, transmission and scanning electron microscopy — Mitochondria	WI-38	≤26	42.8 ± 10.6	Cristae traverse organelle.	16
		≥40	49.2 ± 15.0	Condensed mitochondria with few complete cristae.	
	IMR-90	18		Between PDL 18 and 49 decrease in no. mitochondria. More bizarre-shaped.	17
		49			

	Cell line	PDL	Description	Ref.
Endoplasmic reticulum	WI-38	≤26	Dilated	16
		>40	Becomes constricted and empty.	17
	IMR-90	18 / 49	PDL 49 decreased in amount and disarrayed.	17
Golgi	WI-38	≤26	Few vacuoles.	26
		>40	Many vacuoles.	
	IMR-90	18 / 49	PDL 49 increased in vesicles.	17
Filaments	WI-38	≤26	Mostly at edges. Increase in amount, present through cytoplasm.	16
		>40		
	WI-38	Early / Late	No change in filaments with age.	18
	WI-38	≤25 / >48	PDL 48 lacks prominent bundles of microfilaments.	5
	IMR-90	18 / 49	Filaments appear as the cells degenerate.	17
Lysosomes	WI-38	≤26 / >40	Increase in number. Autophagic vacuoles present. 6 ± 3.3 / 40 ± 3.9	16
	WI-38	Early / Late	Primary lysosomes predominate. Change to autophagic vacuoles. Increase in no. lysosomes.	19
	WI-38	Early / Late	Increase in no. and size lysosomes with culture age.	20
	IMR-90	18 / 49	Increase in no. and size lysosomes with culture age.	17
Inclusions	WI-38	Early / Late	Increase in no. inclusions with culture age.	20
	WI-38	Early / Late	Increase in no. inclusions with culture age.	18
	IMR-90	18 / 49	Increase in no. inclusions with culture age.	17

Table 2 (continued)
MORPHOLOGICAL CHANGES

Aspect studied	Cell type	In vitro age: PDL[a]	Result					Ref.
			Auto-fluorescent structures[d] Cell fluorescence % total cells					
			0	1	2	3	4	
	WI-38	31	79	18	2	2	1	21
		36	50	38	16	5	2	
		44	45	28	19	14	5	
		48	0	20	45	20	15	
		49	0	7	30	43	20	
Nucleus	WI-38	≤26	Round, few invaginations.					16
		≥40	Highly invaginated, pleiomorphic. No change in chromatin condensation.					
	IMR-90	18	Increased invagination.					17
		49	Increased chromatin condensation at PDL 49.					
Surface Blebs and microvilli	WI-38	Early Late	Increase in no. with culture age.					18

Percentage of Cells[e]

Aspect studied	Cell type	In vitro age: PDL[a]	−	+	++	+++	Ref.
Blebs and Microvilli	IMR-90	18	30	60	7	3	17
		30	16	54	27	3	
		49	3	46	39	12	
Spreading	IMR-90	Early Late	More time required for spreading and formation of gap junctions as culture ages.				10
Cell to cell contacts/ communication	IMR-90	Early Late	Lowered communication (detected by ^{3}H uridine transfer from cell to cell) as culture ages.				10
	WI-38	≤25	As cultures age, decreased ability to make con-				5

Property	Cell line		F.C.[g]	R.C.[h]	Ref.
Migration rate	WI-38	≥48	4.6 μm/hr		5
		≤25	2.6 μm/hr		
		≥48	contacts evaluated by scanning E.M.		
HL-A antigens	WI-38		HL-A2		22, 23
	WI-26		HL-A1, HL-A2, W10		
	MRC-5		HL-A2, HL-A7[f]		
Red blood cell con A binding	WI-38	32	0.48	4.56	24
	IMR-90	37	0.47	4.88	
		51	2.74	6.72	
		20	0.54	2.44	
		45	0.90	4.76	
		54	4.4	6.52	
	TIG-1	10	0.40	1.04	
		45	0.66	4.4	
		58	1.48	5.92	
		64	2.36	6.32	

Property	Cell line		No. specific [^{3}H] dexamethasone binding sites fmole/10^6 cell[i]	Ref.
Glucocorticoid receptors	WI-38	27	7000 ± 1000	25
		54	4000 ± 460	

Property	Cell line		Break and gaps (mean %)	Total polyploidy (mean %)	Ref.
Cytogenetics	WI-38	18—25	3.5	1.6	26
		26—34	3.2	2.0	
		35—46	3.2	3.1	

Table 2 (continued)
MORPHOLOGICAL CHANGES

Aspect studied	Cell type	In vitro age: PDL[a]	Result									Tetraploidy (%)	Ref.
			Chromosome counts										
			<	43	44	45	46	47	53	64	92		
	WI-38	4	2	1	3	7	183	3	—	1	—	1.0	27
		14	—	—	—	—	19	—	—	1	1	0.8	
		31	—	—	1	2	47	—	—	—	—	—	
		41	1	2	4	9	87	1	—	—	—	2.7	
		46	1	1	—	7	11	—	—	—	—	—	

a   In vitro age: population doubling level, unless otherwise indicated.
b   Passage/passage at senescence.
c   Senescence at PDL 41.2.
d   Degree of fluorescence: increasing from 0 to 4.
e   − = Smooth surface; + = few, approximately 5; + + = rather numerous; + + + = surface covered.
f   Antigens present throughout lifespan; no change in quantity.
g   F.C.; RBC absorption with the fibroblast coating method.
h   R.C.; RBC absorption with the RBC coating method.
i   The reduction in numbers occurs in binding sites in the nuclei.

Table 3
METABOLIC AND BIOSYNTHETIC CHANGES

Aspect studied	Cell type	In vitro age: PDL[a]	Result	Ref.
			Q value in $\mu\ell$/mg/hr (mean values)	
Respiration and glycolysis				
O_2 uptake without glucose	WI-38	20—29	6.0	28
		30—44	5.0	
O_2 uptake with glucose		20—29	5.0	
		30—44	4.8	

383

Lactate production

	Pop. doublings	Value	Ref.
Lactate production	20—29	3.5	
	30—44	5.0	
Lactate production with glucose present	20—29	2.5	
	30—44	3.0	

Protein and protein degradation

	Cell	Pop. doublings before senescence		Ref.
Proteins (gel electrophoresis)	WI-38	26	PDL 43 compared with PDL 26. Decrease in total number of proteins in the slow moving band. 3 new fast moving bands seen.	29
Protein degradation rates	WI-38	43		

Degradation rate by approach to equilibrium method / Degradation rate measured by intermittent perfusion

Pop. doublings before senescence	T^b	^{3}H label 25 min	^{14}C label 40 hr	Ref.
0—1	13 ± 2.1	2.7 ± 0.6	38 ± 15	30
2—10	20 ± 3.8	4.8 ± 0.4	50 ± 17	
> 25	21 ± 3.6	5.0 ± 0.4	58 ± 15	

Rate of altered protein degradation

	Cell	Pop. doublings before senescence	(Ratio of initial slopes)	Δ canavanine/Δ arginine)	Ref.
Protein degradation rates^b	WI-38	1—2	1.2	1.3	30
		> 25	1.7	2.0	

	Cell	Pop. doublings	Half time degradation of total protein in days	% Initial radioactivity present on day 4	Ref.
	MRC-5	26^c	3	50	31
		26^d	2	45	
		61	1	33	

	Cell		Ref.
Susceptibility of protein to exogenous proteases	WI-38	^{3}H and ^{14}C release after pronase treatment is the same throughout the life-span. Only at the last PDL is there an increase in susceptibility to proteolysis.	32

Table 3 (continued)
METABOLIC AND BIOSYNTHETIC CHANGES

Aspect studied	Cell type	In vitro age: PDL[a]	Homogenate whole cell	From density gradient centrifugation fractions				Ref.
				Low speed	1	2	3—6	
Proteolytic activity ^{14}C release/hr/mg protein	MRC-5	27[c]	232	0	665	560	0	33
		58	482	343	1665	720	0	

Cyclic AMP Stimulation

PGE₁ stim. cAMP/basal cAMP
Time in culture (days)

Aspect studied	Cell type	In vitro age: PDL[a]	3	5	6	8	Ref.
PGE₁ stimulation of cAMP	WI-38	18	7.1	20.9	23.9	19.9	34
		43	2.92	10.6	13.4	13.1	

Epinephrine stim. cAMP/basal cAMP

Aspect studied	Cell type	In vitro age: PDL[a]	Result	Ref.
L-epinephrine stimulation of cAMP	WI-38	25	10.8	34
		40	18.4	
		55	17.7	
	IMR-90	11	2.2	
		18	9.5	

Prostaglandins
Prostaglandin production

			Days after seeding					
			1	**2**	**3**	**4**	**5**	
Conc. PGF_2 in ng/mℓ in medium	WI-38	30^c	0.5	0.3	1.0	0.9	0.75	35
		36	0.5	0.2	0.4	0.5	0.4	
		42	1.0	1.2	0.7	1.0	1.2	
		48	1.8	2.5	2.3	1.7	2.1	

			Days after seeding					
			1	**2**	**3**	**4**	**5**	
Conc. PGE_2 in ng/mℓ in medium		30^c	0.25	0.4	0.75	0.7	0.5	
		36	0.6	0.4	0.4	0.4	0.2	
		42	0.7	0.8	0.5	0.5	0.4	
		48	1.5	1.5	1.4	1.0	1.4	

Ratio of 6 keto prostaglandin $F_1\alpha$; Prostaglandin $F_2\alpha$; thromboxane B_2; prostaglandin E_2 in cell homogenate

6 Keto prostaglandin $F_1\alpha$-production	TIG-1	12	24 : 2 : 1 : 5	36
		48	4 : 5 : 3 : 4	
		60	2 : 9 : 4 : 12	

Polysaccharide

Polysaccharide uptake	WI-38	$20—27^c$ / 41—48	No diff. in uptake/cell of insulin, dextrans of diff. mol. wt. and DEAE dextran	37

Macromolecular biosynthesis
Labeled compound added to growth medium

			Radioactivity in cells^f	
Protein hydrolysate	WI-38	22	21.5	38
		69	11.0	
Uridine		22	17.5	
		69	3.5	

Table 3 (continued)
METABOLIC AND BIOSYNTHETIC CHANGES

Aspect studied	Cell type	In vitro age: PDL[a]	Result	Ref.
Oleic acid		22	13.0	
		69	11.8	
Acetate		22	9.9	
		69	7.1	
³HTdF		22	74.0	
		69	4.6	
j				
			³H methionine/¹⁴C ethionine	
Ratio of incorporation of methionine and ethionine	MRC-5	25[c]	82.4	39
		35	109.4	
		46	40.2	
		55	2.1	
Synthesis of glycosaminoglycans Incorporation cpm/mg cell protein (incubation 40 hr)	WI-38			

		^{35}S		^{14}C glucosamine	
Doublings before senescence		Cellular	Extracellular	Cellular	Extracellular
WI-33	0	100	100	100	100
	2	500	250	450	120
	4	400	325	500	200
	6	600	475	600	300

% Of last possible subculture (incubation 40 hr)

Ref. 40

Lipids

WI-38 — Last subculture — PDL 40—45 (cpm)

11	650	425	—	300	41
14	630	—	650	300	
	53530 ± 352	175350 ± 9476	3587 ± 57	28150 ± 2774	

Passage	Total lipid[a]	Neutral lipid	Phospholipid	
Early	19 ± 1.5	6.0 ± 0.4	13 ± 0.4	41
Late	25 ± 2.6	8.0 ± 0.6	17 ± 0.6	

% Total neutral lipids

WI-38

Passage	Colesterol ester	Triglycerides	Free fatty acids	
<35	9.3 ± 2.6	11 ± 1.9	29 ± 1.9	41
>35	11 ± 1.4	14 ± 1.7	32 ± 4.1	

Passage	Diglycerides	Cholesterol	Monoglycerides
<35	11 ± 2.6	32 ± 1.7	7.1 ± 1.3
>35	7.7 ± 0.9	30 ± 2.2	10 ± 4.4

% Total neutral lipids

WI-38

Passage	Cholesterol ester	Triglycerides	Free fatty acids	
18	27	10	10	34
48	31	11	9	

Passage	Diglycerides	Cholesterol	Monoglycerides
18	7	34	12
48	6	32	11

% Total phospholipids

WI-38

Passage	PE	PI	P. serine	Lecithin	
<35	12 ± 1.7	15 ± 2.7	8.0 ± 0.9	53 ± 3.7	41
>35	6.9 ± 1.7	8.4 ± 1.7	8.9 ± 3.4	70 ± 3.8	

Table 3 (continued)
METABOLIC AND BIOSYNTHETIC CHANGES

Aspect studied	Cell type	In vitro age: PDL[a]	PE	PI	Sphingomyelin	Lecithin	Lysolecithin	Sphingo-myelin	Ref.
	WI-38	<35	29	8	9.6 ± 1.9	47	2.5 ± 0.5	16	34
		>35	28	8	9.6 ± 0.9	47	2.5 ± 0.4	17	

Aspect studied	Cell type	In vitro age: PDL[a]	μg glycogen/cell × 10^6	Range	Ref.
Glycogens	WI-38	24—28	7.69	(3.3—13)	42
		31—39	5.31	(2.5—11.9)	
		42—45	13.83	(1.6—20)	

Collagen biosynthesis
Collagen Induction

Aspect studied	Cell type	In vitro age: PDL[a]	Hydroxyproline solubilized from insoluble collagen (μg/4 × 10^6 cells)	Ref.
Oxyphenybutazone induction of collagenolytic and proteolytic enzymes	WI-38	13—35	21—45	43
		37	8	
		39—44	0	

Aspect studied	Cell type	In vitro age: PDL[a]	[3H] hydrolyproline counts/30 min/10^6 cells	Ref.
Induction by ascorbic acid	WI-38	20, 26, 33	170	43
		44	90	

Collagen synthesis

Collagen associated with cells

		mg Collagen/mg DNA (Ascorbic acid)		% Collagen intracellular (Ascorbic acid)		
		(+)	(−)	(+)	(−)	
WI-38	26	0.8	0.3	23	5	44
	41	0.9	0.5	25	13	
	41	0.7	0.4	18	29	

Collagen content of medium from 1 week at confluence

		mg Collagen/mg DNA (Ascorbic acid)		% Collagen in medium/total collagen produced (Ascorbic acid)		
		(+)	(−)	(+)	(−)	
WI-38	26	2.6	6.0	77	95	44
	41	2.8	3.4	75	87	
	41	3.2	1.0	82	71	

Ascorbate stimulation of collagen synthesis — ³H proline substrate

Activation of prolyl hydroxylase

			dpm tritium released/ 100 µg protein (Ascorbic acid)		
			(−)	(+)	
WI-38	38ⁱ				45
		Log	1524	6440	
		Stationary	2676	7092	
		Passage 55ʲ			
		Log	1788	2164	
		Stationary	2296	2616	

Table 3 (continued)
METABOLIC AND BIOSYNTHETIC CHANGES

Aspect studied	Cell type	In vitro age: PDL[a]	Result	Ref.
Activation of lysyl hydroxylase — ^{3}H lysine substrate	WI-38	38[i]		45
Ascorbate stimulation of collagen synthesis — OH-proline	IMR-90	27—29		46

Result (Ref. 45): dpm tritium released/100 µg protein (Ascorbic acid)

	(−)	(+)
Log	1768	1940
Stationary	1564	1496

Result (Ref. 46): OH-proline + proline (cpm × 100)

	Medium (%) Ascorbic acid		Cells (%) Ascorbic acid	
	(+)	(−)	(+)	(−)
Pulsed without ascorbic acid	24	4.7	10	2
Pulsed with ascorbic acid	30	32	13	8

Heat lability of enzymes

Aspect studied	Cell type	In vitro age: PDL[a]	Result	Ref.
Glucose-6-phosphate dehydrogenase	MRC-5	22[c] 48 61	Fraction that is heat labile (%)[k]: 5 10 15—25	47
	Fetal lung fibroblast[l]	34[c] 61	Whole homogenate: 10% 35% (Heat lability increases with age)	48
	WI-38	22—25 42—45	Purified enzyme: No difference in heat lability with age	49

Fraction that is heat labile

Enzyme	Cell line	In vitro age[a]	% Labile	Ref.
6-phosphogluconate dehydrogenase	MRC-5	25[c] 64	25% Labile (Not sure if any is labile)	47
N-acetyl-β-D-glucosaminidase	WI-38	22—25 42—45	NC	49
α-Dglucosidase			NC	
N-acetyl α-D galactosaminidase			NC	
Sulfite cytochrome C reductase			NC	
Glutamine synthetase				

% Initial activity time (min) at 45°C

	In vitro age	10	20	30	40	50	Ref.
WI-38	30	100	80	—	—	40	50
	60	100	100	—	—	90	

Time at 60°C

	In vitro age	10	20	30	40	
	52	80	55	10	1.5	
	45	60	25	0	—	

Variant enzyme activity — altered G6PD activity by histochemical staining

	In vitro age	Heavily stained variant Per 10⁴ cells[f]	Ref.
MRC-5	20	20	51
PDL at senescence 60	30	30	
	40	50	
	50	80	
	60	200	

a In vitro age: population doubling level, unless otherwise indicated.

b The half life = ln k/2 where k is the rate constant for protein fraction degraded per hour.

c Split ratio 1:2.

d p-Fluorphenyl alanine treated.

e PDL at senescence 50.

f Average number of grains per square. Obtained by counting grains in 150 squares (63 μm^2 each) over surface of 50 cells. Background counts were subtracted.

Table 3 (continued)
METABOLIC AND BIOSYNTHETIC CHANGES

g   Mg/100 mg dry weight.

h   Collagen in cells/total collagen produced.

i   60% of life-span completed.

j   95% of life-span completed.

k   Heat labile enzyme is different from that of young cells.

l   From Le Centre d'Etudes de Biologie Prenatale.

m   No change in heat lability with age.

Table 4
VIRAL INFECTIVITY AND REPLICATION

Aspect studied	Cell type	In vitro age: PDL[a] Cells in which the virus was grown	Result — PFU/mg viral protein	Result — Total virus (PFU)	Result — Yield of guanidine-resistant mutants	Result — Fraction of mutants in yield	Ref.
Specific infectivity of vesicular stomatitis virus	WI-38	23 61[b]	2.1×10^{11} 1.9×10^{11}				52
Mutation rates of poliovirus	WI-38	23 61[b]		1.8×10^{9} 2.6×10^{9}	8×10^{4} 7×10^{4}	4.4×10^{-5} 2.7×10^{-5}	52
Poliovirus Type 2 and Herpes Virus Type 1	WI-38	18 20 53 54	Amount of virus produced, pattern of cytopathology, and analysis of viral proteins by polyacrylamide gels: no difference with age of culture.				53
Vesicular stomatitis virus[c]	WI-38	29[c] 53 54[b]	Absorption and replication is same for passage 29 and 53. Replication may be decreased at passage 54.				54
Poliovirus[c]	WI-38	26[c] 47	Newly synthesized proteins are the same in confluent cultures at these passage levels after viral infection. Post-confluent pattern different from confluent — but no age diff. seen.				55

[a] In vitro age: population doubling level, unless otherwise indicated.
[b] Senescent.
[c] 1:2 split ratio.

Table 5
ENZYME ACTIVITIES

Enzyme studied	Cell type	In vitro age: PDL[a]	Result	Ref.
Alkaline phosphatase			nmol substrate[b] reacting/mg protein/min	
	A-11-L[c]	6—10	2.9	56
	WI-38	15—25	3.0	
		25—33	2.7	
		33—45	3.0	
	WI-1006[d]	13—15	2.2	
			mol substrate[b]/hr/mg protein	
	WI-38	26	31—34	29
		43	Not detectable	
	WI-38	22—30	0.47	57
		31—36	0.32	
		37—42	0.34	
		43—48	0.45	
		49—54	0.55	
Acid phosphatase			nmol substrate[b] utilized/mg protein/min	
	WI-38	26	31—34	29
		43	46—61[c]	
	A-11-L	6—10	38	56
	WI-38	15—25	35	
		25—33	45	
		33—45	50	
	WI-1006	13—15	50	

nmol substrateᵇ utilized/mg protein/min

		Whole cell	Nuclear	Lyso-somal	Micro-somal	Super-natant	
WI-38	<25	31.6	21.3	64.1	22.2	14.7	58
	25—34	34.1	24.5	70.0	35.7	17.3	
	> 35	55.3	37.4	126.9	59.7	22.5	

nmol substrateᵇ utilized/mg protein/min

		Nuclear	Lyso-somal	Micro-somal	Super-natant	
WI-38	17—25	21.33	64.07	22.2	14.68	59
	26—35	24.48	69.58	35.7	17.31	
	36—50	27.28	89.59	41.86	21.9	

μmol substrateᵇ utilized/hr/mg protein

WI-38	22—30	3.75 ± 0.22	57
	31—36	3.15	
	37—42	3.92 ± 0.49	
	43—48	3.01 ± 0.32	
	49—54	3.00 ± 0.04	

nmol substrateᵇ utilized/min/mg protein

		Homogenate	Peakᶠ	
MRC-5	27	34	70	33
	58	54	82	

Table 5 (continued)
ENZYME ACTIVITIES

Enzyme studied	Cell type	In vitro age: PDL[a]	Result	Ref.
	WI-38	20	**Immunochemical quantitation enzyme units/mg protein[g]**	60
		30	0.7	
		40	0.9	
		45	0.7	
		47	0.0	
		49	1.5	
			0.0	
N-acetyl β-glucosaminidase	WI-38	25	**nmol/min/mg protein**	61
		57	90	
			70	
ATPase	WI-38	26	**μg P_i/hr/mg protein**	29
		43	4.5	
			8.9	
Glucose 6-phosphate dehydrogenase	MRC-5	26	**Specific activity of enzyme in extracts (O.D. units/min/mg/mℓ)**	51
		60	0.104	
			0.166	

	Strain	Passage	Value	Ref.
Mean specific activity (nmol/min/mg/protein)				
Catalase	WI-38	20—25	113.6	
		26—33	152.1	
		33—46	163.8	62
nmol/min/mg protein × 10⁻³				
	WI-38	20	4.5	
		57	4.0	61
Oxidation of NADH (nmol substrate reacting/min/mg protein)				
Lactic dehydrogenase	A-11-L	6—10	2000	
	WI-38	15—25	2100	
		25—33	2200	
		33—45	2200	56
	WI-1006	13—15	2000	
	WI-38	20—25	2152	
		26—33	2200	
		33—46	2152	62
µmol NADH oxidized/hr/mg protein				
	WI-38	22—30	24.13 ± 3.69	
		31—36	23.73 ± 4.26	
		37—42	20.86 ± 1.73	57
		43—48	11.25	
		49—54	10.68 ± 0.80	
Immunochemical assay (relative specific activity)				
β-Glucuronidase	WI-38	18—30	0.825 ± 0.005	
		38—48	1.11 ± 0.063	60

Table 5 (continued)
ENZYME ACTIVITIES

Enzyme studied	Cell type	In vitro age: PDL[a]	Result				Ref.
			μg/hr/mg protein				
			Nuclear fraction	Lysosomal fraction	Microsomal fraction	Supernatant	
	WI-38	17—25	11.86	42.28	12.42	5.99	59
		26—35	13.92	58.84	17.71	11.67	
		36—50	18.32	70.24	25.15	14.04	

			nmol/hr/mg protein	Ref.
	WI-38	26	19—22	29
		43	37—72	

			μmol/hr/mg protein glutamylhydroxamate formed	Ref.
Glutamine synthetase	WI-38	25	0.53	50
		30	0.5	
		40	0.4	
		50	0.38	
		60	0.28	

			pmol glutamine formed/min/mg protein during growth cycle (days)				Ref.
			2	4	6	8	
	WI-38	31	550	350	600	625	50
		60, 70	400	400	300	400	

			nmol/min/mg protein		
			homogenate	4th fraction	
Succinic dehydrogenase	MRC-5	27	4.8	27.3	33
		58	4.7	29.0	
Malate dehydrogenase	WI-38		μmol NAD reduced/hr/mg protein		57
		22—30	55.67 ± 4.07		
		31—36	46.38 ± 14.63		
		37—42	56.90 ± 3.25		
		43—48	41.68		
		49—54	30.58 ± 15.98		
Isocitrate dehydrogenase	WI-38		μmol NADP reduced/hr/mg protein		57
		22—30	2.43 ± 0.71		
		37—42	1.21 ± 0.04		
		43—48	1.91		
		49—54	1.69 ± 0.35		
Glutamic pyruvic transaminase	WI-38		μmol NADH$_2$ oxidized/hr/mg protein		57
		22—30	0.21 ± 0.01		
		31—36	0.27 ± 0.07		
		37—42	0.17 ± 0.01		
Glutamic oxalacetic transaminase	WI-38		μmol NADH$_2$ oxidized/hr/mg protein		57
		22—30	5.08 ± 0.33		
		31—36	5.8 ± 0.86		
		37—42	4.12 ± 0.09		
		43—48	3.95		
		49—54	3.22 ± 0.4		

Table 5 (continued)
ENZYME ACTIVITIES

Enzyme studied	Cell type	In vitro age: PDL[a]	Result	Ref.
Glutamate dehydrogenase	WI-38	22—30 31—36 37—42 43—48 49—54	µmol $NADH_2$ oxidized/hr/mg protein 0.94 ± 0.11 1.19 ± 0.11 0.89 ± 0.55 1.28 1.03 ± 0.43	57
Cytochrome c oxidase	WI-38	28—54	No change in specific activity of the enzyme with age	63
	WI-38	25 57	nmol cyto c oxidized/min/mg protein 22 30	61
Esterase	WI-38	26 43	µmol ^3H released/min/mg/protein 74 135 — a new slow and fast moving electrophoretic band	29
Cathodal esterase isoenzyme	WI-38	18—30 30—48	Immunochemical assay (relative specific activity) 0.81 ± 0.052 1.135 ± 0.099	60
Glucosaminidase	WI-38	18—30 38—48	Immunochemical assay (relative specific activity) 8.62 ± 1.09 12.2 ± 1.17	60

Enzymes in glucose metabolism			Mean specific activity (nmol/min/mg protein)	
Hexokinase	WI-38	20—25	9.93	62
		26—33	13.99	
		33—46	6.85	
Phosphoglucose-isomerase	WI-38	20—25	735.3	
		26—33	753.2	
		33—46	385.2	
Phosphofructokinase			Mean specific activity (nmol/min/mg protein)	
	WI-38	20—25	61.8	62
		26—33	54.1	
		33—46	47.0	
Fructose 1-6 diphosphatase	WI-38	20—25	3.84	
		26—33	5.9	
		33—46	2.63	
6-Phosphogluconate dehydrogenase	WI-38	20—25	17.28	
		26—33	16.70	
		33—46	8.47	
Transaldolase	WI-38	20—25	43.38	
		26—33	45.66	
		33—46	33.41	
Transketolase	WI-38	20—25	8.03	
		26—33	4.61	
		33—46	4.80	
Ratio of catalytically active lactic dehydrogenase and enzyme protein detected immunologically	MRC-5	10[b]	0.9	39
		2	0.1	

Table 5 (continued)
ENZYME ACTIVITIES

Enzyme studied	Cell type	In vitro age: PDL[a]	Result	Ref.
Glucose-6-phosphatase	WI-38	22—30	μmol substrate/liberated hr/mg/protein: 0.49 ± 0.03	57
		37—42	0.85 ± 0.3	
		43—48	0.40	
		49—54	0.45 ± 0.1	

Ornithine decarboxylase — WI-38

ODC activity (pmol/μg DNA/hr)
(Time after medium change (hr))

In vitro age: PDL[a]	3	5	7	9	11	16
24	0	10	45	—	30	0
30	0	23	20	55	30	0
37	0	2.5	5.0	4.0	7.5	0
42	0	0	2.5	4.0	—	0

Enzyme studied	Cell type	In vitro age: PDL[a]	Result	Ref.
Superoxide dismutase	WI-38		No difference in enzyme activity with age of culture	

Enzyme studied	Cell type	In vitro age: PDL[a]	No. of experiments	Activity[i] mg protein	Ref.
	IMR-90	22—35	11	29 ± 2	66
		36—50	11	42 ± 3	

Enzyme studied	Cell type	In vitro age: PDL[a]	μg P_i released/ hr/mg protein	Ref.
5' nucleotidase	WI-38	26	3 Isoenzymes 30.8—36.2	29
		43	4 Isoenzymes[k] 45.2—67.4	

		nmol P$_i$/min/mg protein	
IMR-90	10	180	67
	30	420	
	50	950	
WI-38	20	50	61
	30	50	
	50	100	
	57—59	350	

		KM (μM) ± SE	
		Quiescent	22 hr after serum stimulation
cAMP phospho-diesterase activity			
WI-38	25	2.6 ± 0.29	1.3 ± 0.11	34
WI-38	43	2.13 ± 0.18	1.2 ± 0.09	
IMR-90	10	2.18 ± 0.25	1.5 ± 0.13	

a   In vitro age: population doubling level, unless otherwise indicated.
b   p-nitrophenyl phosphate.
c   Human fetal lung.
d   Human adult lung.
e   Increase in number of isoenzymes.
f   From density gradient centrifugation.
g   Undiluted urine fraction = 100 units.
h   Passage before death.
i   One unit of SOD specific activity = amount of enzyme giving 50% reduction in the rate of epinephrine autooxidation.
j   Senescence at PD 46 ± 4.
k   Different electrophoretic pattern.

Table 6
RNA, DNA AND PROTEIN CONTENT

Aspect studied	Cell type	In vitro age: PDL[a] Passage/passage at senescence	RNA/cell pg	Result DNA/cell pg	Ref.
RNA and DNA	HE-125	11/25	28.7	9.2	13
		19/25	31.7	9.2	
		22/25	34.6	9.2	
		29/31	35.4	9.5	
	HE-388	10/22	31.8	9.3	13
		16/22	35.2	9.2	
		20/22	38.8	8.8	
	WI-38	20/44	29.2	9.1	13
		25/44	31.8	9.2	
		36/44	31.5	8.8	
		40/44	35.3	9.9	
		43/44	39.2	8.0	

		Passage	RNA µg/10⁶ cells	DNA µg/10⁶ cells	
	WI-38	20—29	21.5 ± 1.6	9.6 ± 3.7	68
		30—39	23.2 ± 1	9.7 ± 2.8	
		40—49	45.6 ± 3.3	10.6 ± 3.0	
		50—59	64.5 ± 10	8.4 ± 2.5	

			RNA µg/10⁶ cells		
	WI-38	22—24	21.4		12
		37—38	20.6		
		48—55	47.2		

WI-38 [11]

	RNA µg/10⁶ Cells	RNA µg/mg Protein	DNA µg/10⁶ Cells	DNA µg/mg protein
19—27	20.4 ± 3.8	73.5 ± 21.5	7.83 ± 1.19	29.8 ± 7.15
28—34	23.8 ± 3.1	105 ± 13.4	7.58 ± 0.898	39.6 ± 9.42
35—54	30.8 ± 4.3	199.9 ± 44.8	8.45 ± 0.887	50.5 ± 18.2

Protein and DNA content

WI-38 [25]

	Protein µg/10⁶ cells	DNA µg/10⁶ cells
27	460 ± 32	9.3 ± 2.8
54	1080 ± 120	8.9 ± 2.2

DNA content

MRC-5 [33]

	µg DNA/mg protein	
	Homogenate	Low speed fraction
27	36	38
54	80	83

Nuclear proteins

TIG-1 [69]

	mg cellular proteins/mg DNA	mg nuclear proteins/mg DNA
20	18.0	3.0
40	20.0	4.0
60	28.0	5.0

Table 5 (continued)
ENZYME ACTIVITIES

Enzyme studied	Cell type	In vitro age: PDL[a]	Result	Ref.
RNA/DNA			RNA/DNA	
	WI-38	19—27	2.924 ± 0.54	11
		28—34	3.262 ± 0.303	
		35—54	3.67 ± 0.408	
	Human embryonic lung	20—30	2.5	70
		50—60	15.0	
	WI-38	20—24	2.26 ± 0.1	68
		35—39	2.43 ± 0.12	
		40—44	4.07 ± 0.35	
		45—49	5.31 ± 0.22	
		50—54	6.02 ± 0.43	
		55—59	8.41 ± 1.16	
	PAL II	5—9	2 ± 0.11	

Relative content of cytoplasmic RNA species

		In vitro age: PDL[a]	mRNA / rRNA	28S rRNA / 18S rRNA	4S RNA / 18S RNA	Ref.
RNA species	Human embryonic lung	20—30	1.8	2.3	1 03	70
		50—60	1.1	2.0	1 74	

[a] In vitro age: population doubling level, unless otherwise indicated.

Table 7
DNA REPAIR

Aspect studied	Cell type	In Vitro Age: PDL[a]	Result ³H Thymidine incorporation ³H cpm/μg DNA at U.V. dose ergs/mm²			Ref.
			50	100	200	
Repair replication after U.V. irradiation	WI-38	20	73	110	130	71
		41	68	111	133	
		24		35.3		
		44[b]		23.0		

Aspect studied	Cell type	In Vitro Age: PDL[a]	BrdU incorporation ³H cpm/μg DNA (Dose krad)		Ref.
X-ray irradiation			10	50	
Repair replication	WI-38	29	6.2, 5.0	15.0, 20.6	72
		50	5.6, 7.3	18.7, 19.5	

Aspect studied	Cell type	In Vitro Age: PDL[a]	Strand rejoining at 120 min post irrad.	Unirrad. control
Rate of strand rejoining		27	15.4	18.0
		55	15.6	17.7

Table 7 (continued)
DNA REPAIR

Aspect studied	Cell type	In Vitro Age: PDL[a]	Result	Ref.
γ irradiation	WI-38	23 37, 39 43, 47, 48[c]	**Excision of damaged thymine residue from γ-irradiated exogenous DNA in 1 hr** 23—33% 23—33% None	73
Co-irradiation	IMR-90	Phase 2 Phase 3	(see table below)	74

Total no. cells produced after irradiation (Dose in rads)

	0	100	200	300	400	500
Phase 2	7.5×10^7	7.5×10^7	7.5×10^7	7.5×10^7	7.5×10^7	7.5×10^7

	0	100	200	300	400	500
Phase 3	2.5×10^7	2.2×10^7	2.0×10^7	2.0×10^7	1.5×10^7	1.5×10^7

a   In vitro age: population doubling level, unless otherwise indicated.
b   Passage at senescence.
c   Four to five passages before senescence.

Table 8
CHROMATIN

Aspect studied	Cell type	In Vitro Age: PDL[a]	Result			Ref.
		% Unlabeled Nuclei	2C	4C	8C	
Frequency of total cells with varying amounts of DNA	WI-38	7	52	22	—	7
		32	34	14	—	
		38	48	32	10	
		41	57	39	2	
		56	53	23	—	
		70	40	24	3	
	WI-38	30	100	0	0	8
		50	91	7	2	
		60	73	22	5	

Ratios of DNA, histone and RNA in chromatin

Aspect studied	Cell type	Passage/passage at senescence	DNA	Histone	RNA	Ref.
Histone ratio	WI-38	19/53	1	1.12	0.39	75
		22/50	1	1.16	0.37	
		29/50	1	1.05	0.35	
		31/53	1	1.20	0.44	
		40/50	1	1.38	0.61	
		44/45	1	1.30	0.64	
		45/50	1	1.34	0.55	

	Cell type	PDL	Histone/DNA	Ref.
	WI-38	20—29	1.5	76
		30—39	1.7	
		40—49	1.8	

Table 8 (continued)
CHROMATIN

Aspect studied	Cell type	In Vitro Age: PDL[a]	Result				Ref.
			Degree of protein labeling				
			P_{6a}	P_8	P_{10}		
DNA binding proteins	WI-38	18—28	Greater	Less	Greater		77
		48—58	Less	Greater	Less		
			Circular dichroism				
Chromatin changes	WI-38	19—23	Decreased ellipticity at PDL 42—53 compared with PDL 19—23				78
		42—53					
			Binding of ethidium bromide				
		19—23	Fewer number of binding sites at PDL 42—53 compared with PDL 19—23				
		42—53					
			Gel electrophoresis 0.25 M NaCl (non-histone extractable proteins)				
		19—23	Different pattern obtained at PDL 42—53 compared with PDL 19—23				
		42—53					
			$H_1:[H_{2a} + H_{2b} + H_3]:H_4$				
Nuclear proteins ratio of histones	TIG-1	15	0.45:3:1.1				79
		56	0.26:3:1.27				
		29	0.40:3:0.81				
		60	0.23:3:0.84				

Histone acetylation

[³H] Acetate incorp. cpm/μg histone

WI-38[b]	19—25	21 ± 1.64
	45	8 ± 1.1
		76

[³H] acetate incorporated dpm/μg histone

		Resting cells (min[c])				Cells stim. by a medium change (min[c])				
		10	30	60	90	10	30	60	90	
WI-38	26—29	580	200	200	—	600	325	325	—	12
	35—37	450	200	200	—	500	220	220	—	
	48—52	350	350	300	280	250	230	220	200	

[¹⁴C] acetate incorporated in resting cells dpm/10⁶ nuclei

		Acetylation Incub. time (min)				Deacetylation Incub. time (min)				
		20	40	60	80	20	40	60	80	
Acetylase and deacetylase activity in isolated nuclei	WI-38	25—29	320	300	300	280	320	100	100	—	12
		37—40	300	300	280	250	300	150	100	—	
		44—45	350	300	270	230	350	250	220	200	

Table 8 (continued)
CHROMATIN

Aspect studied	Cell type	In Vitro Age: PDL[a]	Result [14C] acetate incorp. into stimulated cells dpm/10^6 nuclei — Acetylation Incub. time (min) 20	40	60	80	Deacetylation incub. time (min) 20	40	60	80	Ref.
		25—29	250	250	230	—	250	200	180	—	80
		37—40	210	220	200	170	210	190	150	130	
		44—45	350	350	320	300	350	240	220	200	
% DNA template available for transcription/µg chromatin	WI-38	17—18	Non-dividing cells 11.3				1 hr after med. change 26.0				80
		41—48	13.3				25.5				
		17—18	1 hr after med. change + actinomycin D 24.4								
		41—48	13.6								
Basal transcription	WI-38		As doubling level increases, basal transcription rate is 3× that of young cells								12

Aspect studied	Cell type	In Vitro Age: PDL[a]	Length of pulse 10 min	20 min	30 min	Ref.
Rate of DNA replication average track length (µm) on DNA fiber after 3H pulse	MRC-5	16	10	16	23	81
		58	7	12	19	

413

	Passage/passage at senescence	Enzyme units/mg DNA in chromatin				
		RNAase	DNAase	Protease	ATPase	
Enzymes (chromatin-associated)	WI-38					75
	19/53	1.4	0.3	0.7	0.8	
	22/50	1.5	0.3	1.0	0.7	
	31/53	2.6	0.6	1.2	1.2	
	40/50	2.4	1.2	1.2	2.1	
	45/50	2.5	1.0	3.3	2.8	
		CTPase, GTPase, UTPase	DPN pyrophosphorylase			
	19/53	1.2	1.2			
	22/50	1.0	1.6			
	31/53	1.7	2.1			
	40/50	2.2	2.7			
	45/50	2.6	2.7			
DNA polymerase (chromatin-associated)	WI-38 19/53 45/50	No change with age				75
	MRC-5 19 56	With age, activity level dropped.[d]				82

Logarithmically growing cultures (cpm/µg DNA)

	Passage/passage at senescence	Bound RNA polymerase activity	Free RNA polymerase activity	
Nucleolar changes — RNA synthesis	HE-125			83
	10/25	1.00	14.524	
	24/26	0.89	12.400	
	26/27	0.46	2.130	
	Quiescent cultures			
	10/25	1.00	8.136	
	24/26	0.93	6.474	
	26/27	0.73	1.882	

Table 8 (continued)
CHROMATIN

Aspect studied	Cell type	In Vitro Age: PDL[a]	Result			Ref.
			[³H] UTP: Mean no. silver grains over:			
			nuclei	nucleoplasm	nucleoli	
Nucleolar changes — RNA synthesis	WI-38	33	55 ± 17	22 ± 8	33 ± 12	84
		53	41 ± 17	22 ± 9	19 ± 10	

Aspect studied	Cell type	In Vitro Age: PDL[a]	Result	Ref.
			Incorporation of [³H] UMP into RNA cpm/10⁶ cells	
Chromatin template activity	WI-38	20—21	1435 ± 256 (n = 6)	85
		41—51[c]	701 ± 120 (n = 10)	

Aspect studied	Cell type	In Vitro Age: PDL[a]	Result						Ref.
			Incorp. of [³H] UMP into RNA (% control) Time after medium change (hr)						
			0.5	1.0	2.0	4.0	8.0	10.0	
	WI-38	22—26	294	183	168	273	305	322	12
		35—37	405	193	130	—	321	—	
		45	207	125	120	161	—	163	

Aspect studied	Cell type	In Vitro Age: PDL[a]	Result		Ref.
			[³H] Uridine incorp. into RNA		
			cpm × 10⁻⁶/mg RNA	cpm/cell	
	WI-38	27	7.4	0.184	86
		42	3.5	0.266	

[³H] uridine cpm/µg RNA

		non-dividing	non-div. + actinomycin D	
WI-38	17—18	3771	73	80
	41—48	3862	98	

		1 hr. after med. change	1 hr. after med. change + actinomycin D
WI-38	17—18	8605	108
	41—48	9022	139

Actinomycin D treatment followed by a med. change for 1 hr

WI-38	17—18	115
	41—48	180

[³H] Uridine

Incorp. of [³H] uridine into RNA 24 hr after med. change (cpm/µg RNA × 10⁻²)

WI-38	19—25	5.8	88
	45	2.8	

[³H] Uridine

	Passage/passage at senescence	Uptake into TCA-soluble fraction	Incorp. into TCA-insoluble fraction	
WI-38	20/44	1.0'	1.0	13
	25/44	1.0	1.0	
	36/44	1.0	1.0	
	40/44	1.2	1.1	
	43/44	1.9	1.4	

Table 8 (continued)
CHROMATIN

Aspect studied	Cell type	In Vitro Age: PDL[a]	Result		Ref.
	HE-388	10/22	1.0	1.0	13
		16/22	1.3	1.2	
		20/22	1.5	1.5	
	HE-125	11/25	1.0	1.0	13
		19/25	1.2	1.0	
		22/25	2.4	2.1	

[3H] Uridine — cpm incorp. into RNA/μg chromatin DNA (15 min incub.)

	Cell type	In Vitro Age: PDL[a]	Result	Ref.
	WI-38	19/53	555	75
		19/50	531	
		22/50	652	
		31/53	620	
		40/50	340	
		45/50	185	
	WI-38	18	Incorp. of [3H] uridine into RNA. No diff. with age.	
		42		

[3H] uridine uptake in pmol/10^5 cells/min

Cell type	In Vitro Age: PDL[a]	Quiescent	18 hr after serum stim.	Ref.
WI-38	23	14 ± 0.98	25 ± 2.2	34
	43	20 ± 1.3	38 ± 2.9	

Incorp. of [3H] UMP cpm/μg DNA

Cell type	In Vitro Age: PDL[a]	Logarithm. growing culture	Quiescent culture	Ref.
WI-38	20	1101	590	13

[³H] UMP incorporation

Cell		[³H] UMP incorp. cpm/µg DNA × 10³		Ref.
		Logarithm. growing cultures	Quiescent cultures	
HE-388	25	1028	547	13
	36	1033	565	
	40	1153	553	
	43ᵍ	720	157	
	10	1478	954	
	16	1456	948	
	20ʰ	987	526	
HE-125	9—11	2.8	0.8	83
	20—23	2.9	0.9	
	23—24	2.3	0.8	
	24—26	1.6	0.6	

Cell		Incorp. of [¹⁴C] AMP into RNA cpm/7.94 µg DNA	Ref.
WI-38	24	686 ± 97	85
	34ⁱ	235 ± 3	

Non-histone chromosomal protein synthesis

Cell		[³H]ʟ-Tryptophan cpm/µg DNA		Ref.
		Quiescent culture	1 hr after med. change	
WI-38	17—18	17800	51310	80
	41—48	18022	52110	

cpm/µg DNA 1 hr after med. change + actinomycin D

	17—18	49220
	41—48	16846

Table 8 (continued)
CHROMATIN

Aspect studied	Cell type	In Vitro Age: PDL[a]	Result	Ref.
Premature chromosome condensation	WI-38	Unlabeled nuclei 34% 76%	Frequency of PCC/total cells after fusion with HeLa 1.0% 1.3%	88
Thermal denaturation of DNA and chromatin	WI-38	20—60	No change in mean temperature of denaturation (tm) with increasing in vitro age	89
Sister chromatid exchange	IMR-90			90

SCE freq. — PRdu conc. μg/mℓ

PDL	0.25	0.5	1.0	2.5
23	7.6 ± 0.4	7.8 ± 0.3	8.8 ± 0.3	10.0 ± 0.6
24	7.6 ± 0.4		8.6 ± 0.4	
35	7.8 ± 0.2	7.9 ± 0.4	8.3 ± 0.3	8.6 ± 0.3
46	8.3 ± 0.5		8.7 ± 0.5	
51	8.6 ± 0.4		9.1 ± 0.4	
52	7.4 ± 0.3	7.9 ± 0.3	8.4 ± 0.5	8.7 ± 0.4

SCE freq. PRdu conc. μg/mℓ

PDL	5.0	10.0	25.0
23	8.9 ± 0.5	8.9 ± 0.5	16.7 ± 1.0
24	9.5 ± 0.5	10.2 ± 0.7	22.1 ± 2.6
35	9.0 ± 0.5	8.7 ± 0.3	18.1 ± 1.6
46	9.2 ± 0.6	10.8 ± 0.7	19.5 ± 1.8
51	11.3 ± 0.5	11.4 ± 0.7	18.9 ± 3.6
52	8.4 ± 0.5	8.4 ± 0.4	13.3 ± 1.3

DNA Changes

MRC-5[a]	Av. mol. wt. (× 10^6)	No. single strand breaks
		91
28	213[j]	0.00
	101	1.1
41	207[j]	0.03
	95	1.24
43	215[j]	0.00
	48	3.41
44	168[j]	0.26
	30	5.92
48	173[j]	0.23
	90	1.36
	38	4.59
54	160[j]	0.33
	62	2.42
56	224[j]	0.00
	102	1.09
	65	2.37
	24	7.83
58	214[j]	0.00
	26	7.08

a   In vitro age: population doubling level, unless otherwise indicated.
b   WI-38 grown in the presence of sodium acetate for 48 hr.
c   Time after addition of chase, in minutes.
d   As cultures aged, activity level dropped. On the basis of monitoring fidelity of polymerization, the enzyme was found to be more error-prone. Misincorporation in this case was greater only if Mn++ was used (which is known to exaggerate misincorporation in bacteria), not when Mg++ (natural) ion was used.
e   Senescence at ∼ 52.
f   Values are radioactivity taken up or incorporated per cell and normalized to unity for the youngest cultures used for each line.
g   Senescence was at passage 44.
h   Senescence was at passage 22.
i   Senescence at PDL ∼ 52.
j   Mn (number average molecular weight) of main peak; other values correspond to secondary peaks.

REFERENCES

1. **Hayflick, L. and Moorhead, P. S.,** The serial cultivation of human diploid cell strains, *Exp. Cell Res.,* 25, 585, 1961.
2. **Hayflick, L.,** The limited *in vitro* lifetime of human diploid cell strains, *Exp. Cell Res.,* 37, 611, 1965.
3. **Martin, G. M., Sprague, C. A., and Epstein, C. J.,** Replicative lifespan of cultivated human cells: effect of donor age, tissue, and genotype, *Lab. Invest.,* 23, 86, 1970.
4. **Mitsui, Y. and Schneider, E. L.,** Relationship between cell replication and volume in senescent human diploid fibroblasts, *Mech. Age. Dev.,* 5, 45, 1976.
5. **Bowman, P. D. and Daniel, C. W.,** Aging of human fibroblasts in vitro: surface features and behavior of aging WI-38 cells, *Mech. Age. Dev.,* 4, 147, 1975.
6. **Macieira-Coehlo, A., Ponten, J., and Phillipson, L.,** The division cycle and RNA synthesis in human cells at different passage in vitro, *Exp. Cell Res.,* 42, 673, 1966.
7. **Yanishevsky, R., Mendelsohn, M. L., Mayall, B. H., and Cristofalo, V. J.,** Proliferative capacity and DNA content of aging human diploid cells in culture: a cytophotometric and autoradiographic analysis, *J. Cell Physiol.,* 84, 165, 1974.
8. **Grove, G. and Cristofalo, V. J.,** Characterization of the cell cycle of cultured human diploid cells: effect of aging and hydrocortisone, *J. Cell Physiol.,* 90, 415, 1977.
9. **Kapp, L. N. and Klevecz, R. R.,** The cell cycle of low passage and high passage human diploid fibroblasts, *Exp. Cell Res.,* 101, 154, 1976.
10. **Kelley, R. O., Vogel, K. G., Crissman, H. A., Lujan, C. J., and Skipper, B. E.,** Development of the aging cell surface: Reduction of gap junction mediated metabolic cooperation with progressive subcultivation of human embryo fibroblasts (IMR 90), *Exp. Cell Res.,* 119, 127, 1979.
11. **Cristofalo, V. J. and Kritchevsky, D.,** Cell size and nucleic acid content in the diploid human cell line WI-38 during ageing, *Med. Exp.,* 19, 313, 1969.
12. **Pochron, S. F., O'Meara, A. R., and Kurtz, M. G.,** Control of transcription in aging WI-38 cells stimulated by serum to divide, *Exp. Cell Res.,* 116, 63, 1978.
13. **Hill, B. T., Whelan, R. D., and Whatley, S.,** Evidence that transcription changes in aging cultures are terminal events occurring after the expression of a reduced replicative potential, *Mech. Age. Dev.,* 8, 85, 1978.
14. **Mitsui, Y. and Schneider, E. L.,** Increased nuclear sizes in senescent human diploid fibroblast cultures, *Exp. Cell Res.,* 100, 147, 1976.
15. **Bemiller, P. M. and Lee, L. H.,** Nucleolar changes in senescing WI-38 cells, *Mech. Age. Dev.,* 8, 417, 1976.
16. **Lipetz, J. and Cristofalo, V. J.,** Ultrastructural changes accompanying the aging of human diploid cells in culture, *J. Ultrastruct. Res.,* 39, 43, 1972.
17. **Johnson, J. E., Jr.,** Fine structure of IMR 90 cells in culture as examined by scanning and transmission electron microscopy, *Mech. Age. Dev.,* 10, 405, 1979.
18. **Wolosewick, J. J. and Porter, K. K.,** Observations on the morphological heterogeneity of WI-38 cells, *Am. J. Anat.,* 149, 197, 1977.
19. **Brandes, D., Murphy, D. G., Anton, E. B., and Barnard, S.,** Ultrastructural and cytochemical changes in cultured human lung cells, *J. Ultrastruct. Res.,* 39, 465, 1972.
20. **Robbins, E., Levine, E. M., and Eagle, H.,** Morphologic changes accompanying senescence of cultured human diploid cells, *J. Exp. Med.,* 131, 1211, 1970.
21. **Deamer, D. W. and Gonzales, J.,** Autofluorescent structures in cultured WI-38 cells, *Arch. Biochem. Biophys.,* 165, 421, 1974.
22. **Brautbar, C., Payne, R., and Hayflick, L.,** Fate of HL-A antigens in aging cultured human diploid cell strains, I. *Exp. Cell Res.,* 75, 31, 1972.
23. **Brautbar, C., Pellagrino, M. A., Perrine, S., Reisfield, R. A., Payne, R., and Hayflick, L.,** Fate of HL-A antigens in aging cultured human diploid cell strains. II. Quantitative absorption studies, *Exp. Cell Res.,* 78, 367, 1973.
24. **Aizawa, S., Mitsui, Y., and Kurimoto, F.,** Cell surface changes accompanying aging in human diploid fibroblasts: Two types of age related changes revealed by concanavalin A-mediated red cell absorption, *Exp. Cell Res.,* 14, 4827, 1980.
25. **Kalimi, M. and Seifter, S.,** Glucocorticoid receptors in WI-38 fibroblasts: Characterization and changes with population doubling in culture, *Biochim. Biophys. Acta,* 583, 352, 1979.
26. **Kadanaka, Z. K., Sparkes, J. D., and MacMorine, H. G.,** A study of the cytogenetics of the human cell strain WI-38, *In Vitro,* 8, 353, 1973.
27. **Sakasela, E. and Moorhead, P. S.,** Aneuploidy in the degenerative phase of serial cultivation of human cell strains, *Genetics,* 50, 390, 1963.
28. **Cristofalo, V. J. and Kritchevsky, D.,** Growth and glycolysis in the human diploid cell strain WI-38, *J. Cell Physiol.,* 67, 125, 1966.

29. **Turk, B. and Milo, G. E.,** An in vitro study of senescent events of human embryonic lung (WI-38) cells, *Arch. Biochem. Biophys.,* 161, 46, 1974.

30. **Bradley, M. O., Hayflick, L., and Schimke, R. T.,** Protein degradation in human fibroblasts (WI-38). Effects of aging, viral transformation and amino acid analogs, *J. Biol. Chem.,* 25, 3521, 1976.

31. **Shakespeare, V. and Buchanan, J. H.,** Increased degradation rates of protein in aging human fibroblasts and in cells treated with amino acid analogs, *Exp. Cell Res.,* 100, 1, 1976.

32. **Bradley, M. O., Dice, J. F., Hayflick, L., and Schimke, R. T.,** Protein alterations in aging WI-38 cells as determined by proteolytic susceptibility, *Exp. Cell Res.,* 96, 103, 1975.

33. **Shakespeare, V. and Buchanan, J. H.,** Evidence for increased proteolytic activity in aging human fibroblasts, *Gerontology,* 25, 305, 1979.

34. **Polgar, P., Taylor, L., and Brown, L.,** Plasma membrane associated metabolic parameters and the aging of human diploid fibroblasts, *Mech. Age. Dev.,* 7, 151, 1978.

35. **Mets, T., Korteweg, M., and Verdonk, G.,** Increased prostaglandin $F_2\alpha$ and E_2 production in late passage WI-38 diploid fibroblasts, *Cell Biol. Internl. Rep.,* 3, 691, 1979.

36. **Murota, S., Mitsui, Y., and Kawamura, M.,** Effect of in vitro aging on 6-keto-prostaglandin $F_1\alpha$ producing activity in cultured human diploid lung fibroblasts, *Biochim. Biophys. Acta,* 574, 351, 1979.

37. **Press, G. D. and Pitha, J.,** Aging changes in uptake of polysaccharides by human diploid cells in culture. A brief note, *Mech. Age. Dev.,* 3, 323, 1974.

38. **Razin, S., Pfendt, E. A., Matsumura, T., and Hayflick, L.,** Comparison by autoradiography of macromolecular biosynthesis in ''young'' and ''old'' human diploid fibroblast cultures. A brief note, *Mech. Age. Dev.,* 6, 379, 1977.

39. **Lewis, C. M. and Tarrant, G. M.,** Error theory and aging in human diploid fibroblasts, *Nature,* 239, 316, 1972.

40. **Schachtschabel, D. O. and Wever, J.,** Age related decline in the synthesis of glycosaminoglycans by cultured human fibroblasts (WI-38), *Mech. Age. Dev.,* 8, 257, 1978.

41. **Kritchevsky, D. and Howard, B.,** The lipids of human diploid cell strain WI-38, *Ann. Med. Exp.,* 44, 343, 1966.

42. **Cristofalo, V. J., Howard, B. V., and Kritchevsky, D.,** The biochemistry of human cells in culture, in *Research Progress in Organic, Biological and Medicinal Chemistry,* Vol. 2, Gallo, U. and Santamaria, L., Eds., Elsevier, New York, 1970, 95.

43. **Houck, J. C., Sharma, V. K., and Hayflick, L.,** Functional failures of cultured human diploid fibroblasts after continued population doublings, *Proc. Soc. Exp. Biol. Med.,* 137, 331, 1971.

44. **Paz, M. A. and Gallop, P. M.,** Collagen synthesized and modified by aging fibroblasts in culture, *In Vitro,* 11, 302, 1975.

45. **Chen, K. H., Evans, C. A., and Gallop, P. M.,** Prolyl and lysyl hydroxylase activation and co-factor specificity in young and senescent WI-38 fibroblasts cultures, *Biochem. Biophys. Res. Commun.,* 74, 1631, 1977.

46. **Faris, B., Snider, R., Levine, A., Moscaritolo, R., Salcedo, L., and Franzblau, C.,** Effect of ascorbate on collagen synthesis by lung embryonic fibroblasts, *In Vitro,* 14, 1022, 1978.

47. **Holliday, R. and Tarrant, G. M.,** Altered enzymes in aging human fibroblasts, *Nature,* 238, 26, 1972.

48. **Kahn, A., Guillouzo, A., Leibovitch, M. P., Cottreau, D., Bourel, M., and Dreyfus, J. C.,** Heat lability of glucose-6-phosphate dehydrogenase in some senescent human cultured cells. Evidence for its postsynthetic nature, *Biochem. Biophys. Res. Commun.,* 77, 760, 1977.

49. **Houben, A. and Remacle, J.,** Lysosomal and mitochondrial heat labile enzymes in aging human fibroblasts, *Nature,* 275, 59, 1978.

50. **Viceps-Madore, D. and Cristofalo, V. J.,** Age associated changes in glutamine synthetase activity in WI-38 cells, *Mech. Age. Dev.,* 8, 43, 1978.

51. **Fulder, S. J.,** Evidence for an increase in presumed somatic mutation during the ageing of human cells in culture, *Mech. Age. Dev.,* 10, 101, 1979.

52. **Holland, J. J., Kohne, D., and Doyle, M. V.,** Analysis of viral replication in ageing human fibroblast cultures, *Nature,* 245, 316, 1973.

53. **Tomkins, G. A., Stanbridge, E. J., and Hayflick, L.,** Viral probes of aging in the human diploid cell strain WI-38, *Proc. Soc. Exp. Biol. Med.,* 146, 385, 1974.

54. **Pitha, J., Adams, R., and Pitha, P. M.,** Viral probe into the events of cellular (in vitro) aging, *J. Cell Physiol.,* 83, 211, 1974.

55. **Pitha, J., Stork, E., and Wimmer, E.,** Protein synthesis during aging human cells in culture: Direction by polio virus, *Exp. Cell Res.,* 94, 310, 1975.

56. **Cristofalo, V. J., Parris, N., and Kritchevsky, D.,** Enzyme activity during the growth and aging of human cells in vitro, *J. Cell Physiol.,* 69, 263, 1967.

57. **Wang, K. M., Rose, N. R., Bartholomew, E. A., Balzer, M., Berde, K., and Foldvary, M.,** Changes of enzyme activities in human diploid cell line WI-38 at various passages, *Exp. Cell Res.,* 61, 357, 1970.

58. **Cristofalo, V. J. and Kritchevsky, D.,** Enzyme activity during the growth and aging of human cells, *Prog. Immunobiol. Stand.,* 3, 99, 1969.

59. **Cristofalo, V. J. and Kabakjiian, J.,** Lysosomal enzymes and aging in vitro: Subcellular enzyme distribution and effect of hydrocortisone on cell lifespan, *Mech. Age. Dev.,* 4, 19, 1975.

60. **Milisauskas, V. and Rose, N. R.,** Immunochemical quantitation of enzymes in human diploid cell line WI-38, *Exp. Cell Res.,* 81, 279, 1973.

61. **Sun, A. S., Aggarwal, B. B., and Packer, L.,** Enzyme levels of normal human cells: Aging in culture, *Arch. Biochem. Biophys.,* 170, 1, 1975.

62. **Cristofalo, V. J.,** Metabolic aspects of aging in diploid human cells, in *Aging in Cell and Tissue Culture,* Holeckova, E. and Cristofalo, V. J., Eds., Plenum Press, New York, 1970, 83.

63. **Packer, L., Nolan, J. S., Katyare, S., and Smith, J. R.,** Mitochondrial biogenesis in human fibroblasts, *Bioenergetics,* 5, 85, 1973.

64. **Duffy, P. E. and Kremzner, L. T.,** Ornithine decarboxylase activity and polyamines in relation to aging of human fibroblasts, *Exp. Cell Res.,* 108, 435, 1977.

65. **Yamanaka, N. and Deamer, D.,** Superoxide dismutase activity in WI-38 cell cultures: Effects of age, trypsinization and SV_{40} transformation, *Physiol. Chem. Physics,* 6, 95, 1974.

66. **Duncan, M. R., Dell Orco, R. T., and Kirk, K. D.,** Superoxide dismutase specific activities in cultured human diploid cells of various donor ages, *J. Cell Physiol.,* 98, 437, 1979.

67. **Sun, A. S., Alvarez, L. J., Reinach, P. S., and Rubin, E.,** 5'-nucleotidase levels in normal and virus transformed cells: Implications for cellular aging in vitro, *Lab Invest.,* 41, 1, 1979.

68. **Schneider, E. L. and Shorr, S. S.,** Alteration in cellular RNAs during the in vitro lifespan of cultured human diploid fibroblasts, *Cell,* 6, 179, 1975.

69. **Sakagami, H., Mitsui, Y., Murota, S., and Yamada, M.,** Two dimensional electrophoretic analysis of nuclear acidic proteins in senescent human diploid cells, Cell Struc. Func., 4, 215, 1979.

70. **Johnson, L. F., Abelson, H. T., Penman, S., and Green, H.,** The relative amounts of the cytoplasmic RNA species in normal, transformed and senescent cultured cell lines, *J. Cell Physiol.,* 90, 465, 1976.

71. **Painter, R. B., Clarkson, J. J., and Young, B. R.,** Ultraviolet-induced repair replication in aging diploid human cells (WI-38), *Radiat. Res.,* 56, 560, 1973.

72. **Clarkson, J. M. and Painter, R. B.,** Repair of X-ray damage in aging WI-38 cells, *Mutat. Res.,* 23, 107, 1974.

73. **Mattern, M. R. and Cerutti, P. A.,** Age-dependent excision repair of damaged thymine from γ irradiated DNA by isolated nuclei from human fibroblasts, *Nature,* 254, 450, 1975.

74. **Macieiro-Coehlo, A., Diatloff, C., Billard, M., Fertil, B., Malaise, E., and Fries, D.,** Effect of low dose rate irradiation on the division potential of cells in vitro. IV. Embryonic and adult human lung fibroblast like cells, *J. Cell Physiol.,* 95, 235, 1978.

75. **Sahai Srivastava, B. I.,** Changes in enzymic activity during cultivation of human cells in vitro, *Exp. Cell Res.,* 80, 305, 1973.

76. **Ryan, J. and Cristofalo, V. J.,** Histone acetylation during aging of human cells in culture, *Biochem. Biophys. Res. Commun.,* 48, 735, 1972.

77. **Stein, G. H.,** DNA-binding proteins in young and senescent normal human fibroblasts, *Exp. Cell Res.,* 90, 237, 1975.

78. **Maizel, A., Niccolini, C., and Baserga, R.,** Structural alteration of chromatin in phase III WI-38 human diploid fibroblasts, *Exp. Cell Res.,* 96, 351, 1975.

79. **Mitsui, Y., Sakagami, H., Murota, S., and Yamada, M.,** Age related decline in histone H_1 fraction in human diploid fibroblast cultures, *Exp. Cell Res.,* 14, 4827, 1980.

80. **Stein, G. S. and Burtner, D. L.,** Gene activation in human diploid cells: Age-dependent modifications in the stability of messenger RNAs for non-histone chromosomal proteins, *Biochim. Biophys. Acta,* 390, 56, 1975.

81. **Petes, T. D., Farber, R. A., Tarrant, G. M., and Holliday, R.,** Altered rate of DNA replication in ageing human fibroblast cultures, *Nature,* 251, 434, 1974.

82. **Linn, S., Kairis, M., and Holliday, R.,** Decreased fidelity of DNA polymerase activity isolated from aging human fibroblasts, *Proc. Natl. Acad. Sci.,* USA, 73, 2818, 1976.

83. **Whatley, S. A. and Hill, B. T.,** Influence of growth state on relationship between nuclear-template activity and in vitro "aging", *Gerontology,* 26, 129, 1980.

84. **Bowman, P. D., Meek, R. L., and Daniel, C. W.,** Decreased synthesis of nucleolar RNA in aging human cells in vitro, *Exp. Cell Res.,* 101, 434, 1976.

85. **Ryan, J. and Cristofalo, V. J.,** Chromatin template activity during aging in WI-38 cells, *Exp. Cell Res.,* 90, 456, 1975.

86. **Schneider, E. L., Mitsui, Y., Tice, R., Shorr, S. S., and Braunschweiger, K.,** Alteration in cellular RNAs during the in vitro lifespan of cultured human diploid fibroblasts. II. Synthesis and processing of RNA, *Mech. Age. Dev.,* 4, 449, 1975.

Table 1
CUMULATIVE REPLICATIVE ABILITIES OF HUMAN SKIN-DERIVED FIBROBLASTS

Parameter measured	Determined by	Young subjects (20—35 years)	Old subjects (65+ years)	Ref.
Onset of senescent phase (CPD)[a]	Failure to reach confluency in 1 week after 1:4 split	35.2 ± 2.1(23)[b]	22.5 ± 1.7(21)	5
Onset of senescent phase (CPD)	Failure to reach confluency in 2 weeks after 1:4 split	41.6 ± 2.4(23)	29.6 ± 2.1(21)	5
In vitro life span (CPD)	Failure to reach confluency in 1 month	44.6 ± 2.5(23)	33.6 ± 2.1(21)	5
In vitro life span (days)	Failure to reach confluency in 1 month	273 ± 11(23)	236 ± 12(21)	5

[a] CPD, Cumulative population doublings.
[b] Numbers in parentheses indicate number of cell cultures examined; values are mean ± SEM.

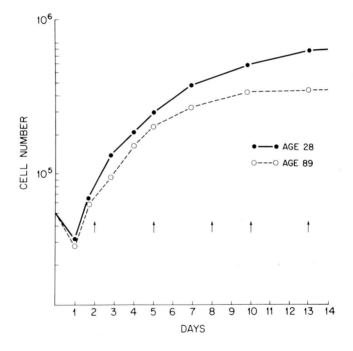

FIGURE 1.   Cell population growth curves of skin fibroblast cultures derived from a young (●) and an old (○) donor. Arrows indicate change of medium. (From Schneider, E. L. and Mitsui, Y., *Proc. Natl. Acad. Sci. U.S.A.*, 73, 3584, 1976.)

their data in the adult age range demonstrates only minimal differences. Hollenberg and Schneider,[11] focusing on the adult age range, found no changes in receptor characteristics with aging.

Measurements of EGF receptors also did not reveal any alterations as a function of the age of the cell culture donor (Table 4).[11]

Another approach to examining changes in membrane function with aging is to measure red blood cell adsorption to cultured human skin-derived fibroblasts. Ohashi et al.[12] have found a significant increase in red blood cell adsorption as a function of the age of the cell culture donor[12].

Table 2
ACUTE REPLICATIVE ABILITIES OF SKIN-DERIVED HUMAN FIBROBLASTS

Replication parameter	Determined by	Young subjects (20—35 years)	Old subjects (65+ years)	Ref.
Cell population replication rate (hr)	Cell growth curves	20.8 ± 0.8(18)[a]	24.3 ± 0.9(18)	5
% replicating cells	Autoradiography after 24 hr exposure to tritiated thymidine	87.7 ± 1.6(7)	79.6 ± 2.5(7)	5
Cell number at confluency (10^4 cells/cm^2)	Cell growth curves	7.31 ± 0.42(18)	5.06 ± 0.52(18)	5

[a] Numbers in parentheses indicate number of cell cultures examined; values are mean ± SEM.

Table 3
MACROMOLECULAR CONTENTS AND SYNTHESIS OF HUMAN SKIN-DERIVED FIBROBLASTS

Parameter measured	Determined by	Young subjects (20—35 years)	Old subjects (65+ years)	Ref.
Cellular RNA content (pg/cell)	Orcinol reaction	28.7 ± 1.9(16)[a]	29.3 ± 1.7(13)	5
Cellular protein content (pg/cell)	Lowry procedure	573 ± 33(16)	576 ± 38(13)	5
Cellular DNA content (pg/cell)	Diphenylamine reaction	11.2 ± 0.7(18)	10.6 ± 1.0(11)	—
RNA synthesis[b] (^{3}H-Uridine, cpm/cell)	Uridine incorporation into VSV RNA	0.148 ± 0.012(7)	0.135 ± 0.015(7)	9
RNA and[c] protein synthesis	VSV plaque-forming units per cell	83.7 ± 37.6(9)	223.2 ± 61.0(9)	9

[a] Numbers in parentheses indicate number of cell cultures examined; values are mean ± SEM.
[b] Cell cultures incubated for 10 hr with VSV at multiplicity of infection (MOI) of 73. Intrinsic RNA synthesis inhibited by pretreatment with actinomycin D.
[c] Cell cultures infected with VSV at low MOI for 24 hr and then medium assayed for plaque-forming units.

Table 4
MEMBRANE FUNCTIONS OF SKIN-DERIVED HUMAN FIBROBLASTS

Parameter measured	Determined by	Young subjects (20—35 years)	Old subjects (65+ years)	Ref.
Insulin binding	$C_{1,2}$ max (nM)[a]	1.6 ± 0.5(5)[b]	1.6 ± 0.5(4)	11
Insulin binding	B max (10^{-3} sites/cell)[c]	6.7 ± 3.4(5)	9.5 ± 3.4(4)	11
Insulin binding	ED_{50} (nM)[d]	0.75 ± 0.55(5)	0.74 ± 0.23(4)	11
EGF/URO binding	$C_{1,2}$ max (nM)	0.70 ± 0.11(5)	0.77 ± 0.12(4)	11
	B max (10^{-3} sites/cell)	78 ± 58(5)	106 ± 90(4)	11
	ED_{50} (nM)	0.046 ± 0.020(5)	0.030 ± 0.014(4)	11
Red blood cell adsorption				12

[a] Concentration at which insulin binding was estimated to be $^1/_2$ maximal.
[b] Numbers in parentheses indicate number of cell cultures examined; values are mean ± SEM.
[c] Maximum binding per cell.
[d] Concentration causing a half-maximal effect.

Table 5
OTHER INVESTIGATIONS OF MASS POPULATIONS OF HUMAN SKIN-DERIVED FIBROBLASTS

Parameter measured	Determined by	Young subjects (20—35 years)	Old subjects (65+ years)	Ref.
SCE background[a]	BrdU-diff. staining technique	$7.8 \pm 0.9(1)$[b]	$7.0 \pm 0.7(1)$	14
SCE by 7.5 ng/mℓ mitomycin C	BrdU-diff. staining technique	$67.9 \pm 1.6(7)$	$56.1 \pm (6)$	14
SCE induced by 1.0 g/mℓ AAAF[c]	BrdU-diff. staining technique	$25.0 \pm 1.4(1)$	$18.7 \pm 1.2(1)$	14
Cell volume	Coulter counter and channelizer	$2935 \pm 88(18)$	$3131 \pm 109(17)$	5
Prostaglandinin synthesis				16

[a] Fifteen cells from each culture were analyzed for SCEs/genome.
[b] Values are mean ± SEM; numbers in parentheses are the number of cultures examined.
[c] AAAF, *N*-acetoxy-2-acetylaminofluorene.

OTHER INVESTIGATIONS OF MASS POPULATIONS OF HUMAN SKIN-DERIVED FIBROBLASTS (SISTER CHROMATID EXCHANGE ANALYSES, CELL VOLUME, PROSTAGLANDININ SYNTHESIS)[5]

Sister chromatid exchanges (SCE) have been widely utilized as a measure of induced DNA damage.[13] Examination of background SCE indicates that no significant difference exists between young and old donor skin-derived fibroblasts.[14] However, a significant difference was observed in the frequencies of mutagen-induced SCE between young and old donor cells (Table 5). The decline in mutagen-induced SCE in old donor skin-derived fibroblasts agrees with previous findings obtained in vivo in mouse and rat bone marrow cells as a function of aging.[15]

Cell volumes of young and old skin-derived fibroblasts were measured by Coulter analyses.[5] While old donor skin-derived fibroblasts had slightly larger cell volumes, this difference was not statistically significant.

The synthesis of various prostaglandinin metabolites was measured in human skin-derived fibroblasts after stimulation of these cells with bradykinin and histamine. The results of these studies indicate a significant decline in the synthesis of prostaglandinins as a function of the age of the cell culture donor.[16]

CLONED SKIN-DERIVED FIBROBLASTS

The above studies were performed on mass cultures of human skin-derived fibroblasts. It is equally important to examine the behavior of individual cells. The ability of cloned individual human cells to form substantial colonies is inversely related to the in vitro passage level of a cell culture.[1] It was therefore interesting to examine the colony size distribution of cloned individual human derived fibroblasts from young and old human subjects at equal levels of early in vitro passage.[17] Typical colony size distributions derived from a young and old donor cell culture are seen in Figure 2. The results of colony size determinations on 17 young and old donor cell cultures are summarized in Table 6.[17] As is the case in mass cell populations, diminished cell replicative abilities were found in the old donor skin-derived cell cultures when compared to parallel cultures derived from younger subjects.

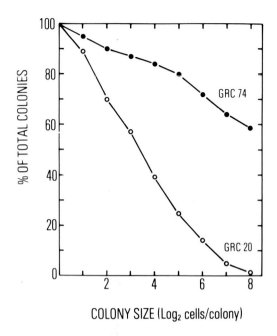

COLONY SIZE (Log₂ cells/colony)

FIGURE 2.   Percentage of colonies able to attain at least a specified size vs. colony size. Colony size is expressed as $\log_2$ of the number of cells per colony. Adult human skin cultures from a young (GRC74, age 33 years, — ●) and an old (GRC20, age 80 years, — ○) donor were cloned at the 10th PD in vitro. (From Smith, J. R., Pereira-Smith, O. M., and Schneider, E. L., *Proc. Natl. Acad. Sci. U.S.A.*, 75, 1353, 1978.)

Table 6
STUDIES OF CLONED HUMAN SKIN-DERIVED FIBROBLASTS

Parameter measured	Determined by	Young subjects (20—35 years)	Old subjects (65+ years)	Ref.
Colony-forming ability	% cells able to form colony of 16 cells 2 weeks after cloning	69.0 ± 3.3(9)[a]	48.0 ± 4.4(8)	17

[a]   Numbers in parentheses represent the number of cultures examined; values are mean ± SEM.

DISCUSSION

In these studies of human skin-derived fibroblasts, considerable overlap was observed between results obtained from young and old donor cell cultures. Physiologic studies performed on human subjects have also revealed similar degrees of overlap.[18]

Some of this variation may be related to the use of chronologic age as the index of in vivo age of the cell culture donors. Research scientists as well as clinicians have long been aware that biologic age may not be adequately represented by chronologic age. Another probable source of variability in studies of cultured human skin-derived fibroblasts is the unavoidable genetic heterogeneity of human subjects.

Despite this observed variability, the above studies in skin-derived fibroblasts clearly demonstrated statistically significant decreases in both acute and chronic replicative abilities, prostaglandinin synthesis, and mutagen-induced SCE levels with the age of the cell culture donor. Our ability to detect these differences between old and young donor cell cultures may have been related to the emphasis placed on standardization of skin biopsy procedures, explantation and subcultivation protocols, utilization of the same media, and the performance of all determinations on parallel old and young donor cell cultures. The importance of obtaining cell cultures from a nonhospitalized, normal population should also be emphasized since disorders such as diabetes have been known to alter in vitro life span as well as other in vitro parameters.[3]

It is likely that in cross-sectional studies of this nature, a relatively vigorous old population has been selected (less vigorous individuals have been removed by death before age 65). Therefore, the results may be conservative underestimates of the in vitro alterations that may occur as a function of in vivo aging. It will be of great interest if the in vitro measurements obtained on cells from human volunteer members of the Baltimore Longitudinal Study can be of predictive value in vivo. Will the old donor whose cells replicate well in tissue culture have relatively good immune function? Will the young donor whose cells proliferate poorly have impaired immune function? These are some of the questions that can be addressed with our integrated in vitro and in vivo studies of human aging.

During the past 9 years, over 800 skin-derived fibroblast cultures have been established from members of the Baltimore Longitudinal Study. Many of these cell cultures will be incorporated into the Aging Cell Bank at the Institute for Medical Research at Camden, N.J. and will be available to researchers interested in studying cellular aging.

REFERENCES

1. **Smith, J. R.,** Lung-derived fibroblast-like human cells in culture (in this volume).
2. **Stanbury, J., Wyngaarden, B., and Fredrickson, D.** Metabolic Basis of Inherited Diseases, Vol. 4, McGraw-Hill, New York, 1978.
3. **Goldstein, S., Littlefield, J. W., and Soeldner, J. S.,** Diabetes mellitus and aging: diminished plating efficiency of cultured human fibroblasts, Proc. Natl. Acad. Sci. U.S.A., 64, 155, 1969.
4. **Martin, G. M., Sprague, C. A., and Epstein, C. J.,** Replicative lifespan of cultivated human cells: effect of donor age, tissue, and genotype, Lab. Invest., 23, 86, 1970.
5. **Schneider, E. L. and Mitsui, Y.,** The relationship between in vitro cellular aging and in vivo human age, Proc. Natl. Acad. Sci. U.S.A., 73, 3584, 1976.
6. **Schneider, E. L., Braunschweiger, K., and Mitsui, Y.,** The effect of serum batch on the in vitro lifespans of cell cultures derived from old and young human donors, Exp. Cell Res., 115, 47, 1978.
7. **Goldstein, S.,** Studies of age-related diseases in cultured skin fibroblasts, J. Invest. Dermatol., 73, 19, 1979.
8. **Orgel, L. E.,** The maintenance of the accuracy on protein synthesis and its relevance to aging, Proc. Natl. Acad. Sci. U.S.A., 49, 517, 1962.
9. **Danner, D. B., Schneider, E. L., and Pitha, J.,** Macromolecular synthesis in human diploid fibroblasts: a viral probe examining the effect of in vivo aging, Exp. Cell Res., 114, 63, 1978.
10. **Rosenbloom, A. L., Goldstein, S., and Yip, C. C.,** Insulin binding to cultured human fibroblasts increases with normal and precocious aging, Science, 193, 412, 1976.
11. **Hollenberg, M. D. and Schneider, E. L.,** Receptors for insulin and epidermal growth factor-urogastrone in adult human fibroblasts do not change with donor age, Mech. Age. Dev., 11, 37, 1979.
12. **Ohashi, S., Mitsui, Y., Smith, J. et al.** personal communication.
13. **Perry, P. and Evans, H. J.,** Cytological detection of mutagen-carcinogen exposure by sister chromatid exchange, Nature (London), 258, 121, 1975.
14. **Schneider, E. L. and Gilman, B.,** Sister chromatid exchanges and aging. III. The effect of donor age on mutagen induced sister chromatid exchange, Hum. Genet., 46, 57, 1979.
15. **Kram, D., Schneider, E. L., Tice, R. R., and Gianas, P.,** Aging and sister chromatid exchange. I. The effect of aging on mitomycin-C induced sister chromatid exchange frequencies in mouse and rat bone marrow cells in vivo, Exp. Cell Res., 114, 471, 1978.

16. **Polgar, P., Taylor, L., Schneider, E. L. et al.,** personal communication.
17. **Smith, J. R., Pereira-Smith, O. M., and Schneider, E. L.,** Colony size distributions as a measure of *in vivo* and *in vitro* aging, *Proc. Natl. Acad. Sci. U.S.A.,* 75, 1353, 1978.
18. **Shock, N. W. and Morris, A.,** Aging and variability, *Ann. N.Y. Acad. Sci.,* 13, 591, 1966.

NONHUMAN FIBROBLAST-LIKE CELLS IN CULTURE

Jerry R. Williams and Kerry L. Dearfield

INTRODUCTION

Cells explanted from a variety of animals have been grown in culture. Such cells grow in culture either for finite periods, or, for some cultures, experimentally infinite periods of time. The purpose of this article is to describe patterns of growth exhibited by such cells.

The nature of most experiments forming the basis of this article is similar. Tissue is explanted from a nonhuman host; cells grow from that tissue; this cell population is passaged in culture until it fails to maintain itself or until cells with altered characteristics become dominant. The parameters measured from such experiments all relate to the temporal histories of cell populations in culture, the changes in cellular characteristics as they grow, and the proportion of cells that either eventually fail to maintain growth or that assume new biological properties. For the purpose of this article, the term *senescence* will be used to indicate a loss of growth potential under serial passage conditions. This criterion will define senescence, since many characteristics others would include in defining the term, e.g., diploid karyology and rate of macromolecular synthesis, have not been measured in many of the experiments described. Cell populations that fail to senesce coincident with the major component of the attendant or similar cell population, and cell populations that appear to be able to maintain indefinite growth will be termed "growth transformed". Such cells or clones of cells producing a malignant tumor when injected into a suitable host will be termed "malignantly transformed".

The study of senescing populations of cells centers not only on the nature of such senescence, i.e., whether it is an artifactual senescence due to culture conditions, or whether it represents a program such as terminal differentiation, but also on whether the senescing population selects or discriminates against a small population of cells that have extended growth or are progenitors of cells capable of exhibiting growth transformation or malignant transformation. The comparison of nonhuman fibroblasts is especially useful, not only in discerning general patterns in senescence and transformation, but also in approaching the question of whether senescence is an expression of the lifespan of the donor animal, i.e., do cells from short-lived animals senesce more rapidly than those from long-lived species. This article will present an abbreviated discussion of the essential questions in the phenomena of senescence and transformation. The experimental data are discussed for the mouse, rat, hamster, and other less well-studied model systems. Finally, overall similarities and differences observed in patterns from various species are discussed and conclusions drawn.

Markers of Senescence, Growth Transformation, and Malignant Transformation

The major marker that distinguishes the senescing cell from the transformed cell is growth. Senescing cells stop proliferating at some time during their serial passage, whereas transformed cells do not. For most nonhuman mammals, cultures of explanted cells contain some cells that stop growing, and other cells that might not. The proportion of these two populations determines whether cell growth measurements alone can identify a senescing population. An example of partially senescing populations is given in Figure 1, in which the data represent the general patterns observed when cells from explanted mammalian tissue are serially passaged in culture. The three patterns of growth are observed even though all three cultures are derived from explanted 17-day-old Sprague-Dawley rat embryos. The variation in these three patterns results from different conditions for tissue explant, and from passaging the cells at different cell number. These three patterns illustrate the characteristics common to

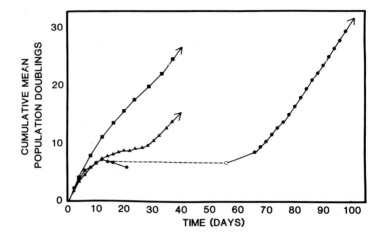

FIGURE 1.   Growth patterns of cell populations derived from rat embryos and cultured under several conditions. ■, cells grown from explanted tissue, diluted to 1×10^5 cells per 100-mm petri dish at each passage; ▲, cells grown from tissue pieces, diluted to 1×10^6 cells per 100-mm petri dish at each passage; ●, cells grown from embryo tissue pipeted to single-cell suspension, diluted to 5×10^5 cells per 100-mm petri dish; --- ○ --- ●, same cells passage 8 maintained for 6 weeks in confluency before beginning serial passage with dilution to 5×10^5 cells per 100-mm petri dish.

mammalian fibroblasts. All three patterns exhibit, over the first several passages, a rapid, approximately exponential growth, which is followed by a decline in growth. This loss of vigor is the onset of senescence in a proportion of the population and corresponds to the beginning of phase III as defined by Hayflick[1] for human cells in which senescence occurs in all cells. The major experimental differences in the three populations in Figure 1 is the time at which an apparently growth-transformed population dominates the culture. It is not clear whether the appearance of growth-transformed populations, an observation more common to cells derived from certain mammalian species, is more rapid in some cultures because the conditions select preexisting cells, or whether conditions enhance transformation of cells that would otherwise senesce.

Cell populations have been studied in which growth-transformed cells have not been observed, or where their frequency is too low to permit the detailed description of the senescing populations. The major studies of the finite lifetime of serially cultured cells have dealt with cells explanted from human tissue. These descriptions began with the pioneering work of Hayflick and Moorhead in 1961.[2] Many studies since then appear to support Hayflick and colleagues' finding that human diploid fibroblasts senesce in culture (see Reference 3 for discussion). Though human cell strains have been utilized for most studies on senescence, this article is concerned primarily with this work as a model to compare nonhuman fibroblast-like cells in culture. The term "fibroblast-like" describes cells that normally dominate most of the in vitro cultured cell populations. It does not necessarily mean that cells are descended from fibroblasts that existed in vivo, although this may be true to some degree.

Theories of Senescence

Hayflick's studies demonstrate that the history of human diploid cell strains in vitro reflects a particular potential for a finite number of cell doublings; 50 ± 10 doublings for serial passage with a 2:1 subcultivation ratio. He and co-workers defined three growth-related phases as follows: phase I, the growth from the primary explant; phase II, the apparently steady, logarithmic growth phase of the culture; and phase III, the loss of proliferative potential to an unrecoverable state. These growth kinetics provide a major marker to examine mass cultures of cells. In addition to these growth characteristics, certain cellular and molecular characteristics also have been described in senescing cell populations. Hayflick has recently reviewed many of the functional decrements that accompany phase III in human cell strains.[3] Many of these alterations also may be applicable as markers of senescence for nonhuman fibroblast-like cells in culture. Major among these markers are increased generation time, size and volume, decreased rates of DNA and RNA synthesis, and changes in karyology.

Hayflick proposed an exceptionally important hypothesis, suggesting that the progress to a finite in vitro lifespan of these cell strains represents an expression of aging in vivo. Other investigators suggest alternative mechanisms of senescence in cell cultures. Holliday and co-workers present a theory of commitment in cellular aging.[4,5] They propose that a diploid cell population contains a subpopulation of uncommitted cells which they define as potentially immortal. These uncommitted cells are hypothesized to have a given probability to divide and to become committed to a path toward senescence. Once committed, they senesce. If this commitment probability is high, then the number of uncommitted cells declines with each culture passage until such cells are lost from the population; hence, senescence is a culture technique artifact. If the cell number is sufficiently large, however, the probability of losing all uncommitted cells drops during dilution through serial cultivations, and an immortal cell population may be established. Hirsch calculates, based on Holliday's theory, that the population of cells would survive for much longer periods if a sufficient number of uncommitted cells is retained and not diluted out of the population. The probability that this might occur is greatly enhanced when the cell number exceeds 10^{10}.[6] This large population size would increase the chances for obtaining cultures surviving beyond the 50 ± 10 population doublings demonstrated by Hayflick. The large scale of such experimentation, however, would also make the substantiation experiments very difficult.

Bell et al.,[7] on the other hand, propose that fibroblasts in culture are cells forced to proliferate as if they were stem cells. Fibroblasts in vivo are part of a nondividing population that proliferate only in response to wounds or to tissue damage. Once such stimuli prompt division, fibroblasts undergo several morphological, biochemical, and physiological changes that suggest to Bell and co-workers a state of differentiation where the fibroblasts leave the cell cycle.[7] Their reasoning, similar to Smith and Whitney,[8] of a heterogeneity in the tendency for cells to leave the cycle, suggests to them that aging of fibroblast cultures may be due to such cells leaving the cycle as a process of differentiation. They contend that differentiation underlies culture senescence. Whether the phase III phenomenon seen in human fibroblast cultures is due to differentiation is yet to be determined.[9]

Differentiation is an example of a so-called deterministic model, one in which preexisting determinants are manifested as senescence. Smith and Whitney[8] present evidence suggesting that a stochastic mechanism might operate during senescence. Their results indicate that clones from human embryonic lung fibroblasts demonstrate a heterogeneity in remaining population doubling potential; even sister cells from a single mitotic event can differ in residual proliferation potential by as much as eight population doublings. Based on these data, these authors conclude that neither a precise ''counting mechanism'' nor a commitment theory for aging fit their findings. Other authors have suggested theories that contain both deterministic and stochastic processes. For instance, Schmookler et al.[10] present four possible sources for the variability in doubling potentials: unequal partition of cytoplasmic components, imperfect duplication of nuclear division counter, differential response to experimental manipulations, and distribution of generations in the parent clones. These variables are difficult to define experimentally.

A proposed mechanism that is more clearly amenable to experimental investigation is that which speculates that DNA repair is a longevity assurance mechanism, and that the decline of repair function is associated with senescence. Hart and Setlow[11] suggest a correlation between repair capacity of UV-induced damage in cell cultures and the lifespan of the donor species. Kato et al.[12] found a lack of correlation in an extensive study and Regan and co-workers[13] found a trend but no strong correlation between DNA repair and lifespan. During in vitro senescence, the amounts and rates of DNA repair as cells move through their in vitro lifespan, measured by a variety of techniques, produce conflicting results (for review see Reference 14).

TRANSFORMATION AS A VARIABLE IN SERIALLY CULTURED CELLS

Senescence may be one of two alternative fates of cells explanted from an animal, the other being transformation to indefinite growth. Although an intermediate state in which prolonged growth is maintained by exogenously supplied factors is possible, it has not yet been experimentally demonstrated. It is necessary to consider transformation as a discrete phenomenon which prevents cells from senescing. The rate at which cell cultures of a given type transform is important in the study of transformation, because transformed cells and their offspring perturb the study of senescence. For instance, rat cell cultures appear to transform at a much higher rate than human cells. This fact is important in itself, but it also makes the study of senescence more difficult in rat cells. Macieira-Coelho and co-workers[15] have presented a scalar representation of the relative propensity of fibroblast-like cells derived from various animal species to transform. Chicken fibroblast-like cells are placed at one end, as such cells invariably senesce after a number of population doublings, whereas mouse fibroblasts are placed at the other end due to their disposition invariably to transform to permanent cell lines. Other fibroblast-like cells are placed intermediate to these extremes, depending on their frequency of transformation to a cell line. Macieira-Coelho[16] suggests that these varying degrees of in vitro behavior may be due to reorganization of the cell genome at cell division. These authors propose that a "genomic reorganization", determined by stochastic events, alters properties of the cell such as repair enzyme efficiency and disposes the cell to transformation. Macieira-Coelho et al.[17] consider the role of environmental factors on cell genome reorganization important, after observing that irradiated mouse fibroblasts demonstrated an accelerated acquisition of infinite growth, whereas chicken fibroblasts, which do not "spontaneously" transform, had a shortened phase II lifespan. The division potential of phase II human fibroblasts was apparently unaffected, however, raising questions as to the general effect of external agents on senescence.

The acquisition of indefinite growth by cultures, although sometimes referred to as "spontaneous transformation", is not well understood. It usually occurs as a change in a small proportion of cells, perhaps only one, which eventually dominates the culture. Much effort and thought has been applied to the examination of growth transformation on the basis that it may be a precedent or concomitant of neoplasia. Hayflick very strongly suggests that established cell lines share various properties with cancerous cells. Most established cell lines are heteroploid or "pseudodiploid" and possess altered morphological and growth properties. Most also form tumors when placed in proper hosts.

Although Hayflick proposes that no cell population is able to proliferate indefinitely without a change in karyotype, this is impossible to verify experimentally, since small changes in karyology undetectable by current banding techniques can always be proposed to account for abnormal characteristics. Whether a change in karyotype occurs as a precedent or sequel of acquisition of infinite growth is still unclear. Terzi and Hawkins[18] describe karyotype changes that can occur soon after explantation to culture. Their cell fusion work with mouse, chick, and hamster cells suggests that polyploidy is often an important correlate for the establishment of cell lines. This genomic reorganization, as suggested by Macieira-Coelho, might allow evolution of cell types capable of infinite growth. Terzi and Hawkins[18] obtained established cell lines from mouse and hamster cultures, and these lines had other than the diploid karyotype, whereas no permanent cell line could be established from the chicken fibroblasts and maintain a diploid karyotype. The dichotomy, if such there be, between diploid cells destined to senesce and other cells that are the result of stochastic or deterministic processes, will be proven when cell markers are available that can be detected in small proportions of a culture.

Several sequential events may be necessary for neoplastic development in vivo as described by Foulds.[19,20] Barrett and Ts'o[21,22] propose that progression towards transformation in vitro is also a multistep process. They have identified several markers associated with the neoplastic process in Syrian hamster embryo cells exposed to chemical carcinogens. Among these phenotypic markers of neoplastic or preneoplastic cells were assays for morphological alterations, enhanced fibrinolytic activity, loss of anchorage-dependent growth, and tumorigenicity. Their data suggest a multistep progression in which cells placed in culture sequentially demonstrated morphological alterations and then enhanced fibrinolytic activity within 2 weeks of carcinogen treatment.[21] After 32 to 75 population doublings, anchorage-independent growth is observed by Barrett and Ts'o, and this appears to correlate well with tumorigenicity.[23] Morphological alterations and enhanced fibrinolytic activity, however, do not have an absolute correlation with tumorigenicity. Further evidence for the multistep progression of transformation is the identification of preneoplastic cells in culture. Barrett[24] has isolated a number of colonies from cultures of Syrian hamster cells that are aneuploid and that have enhanced fibrinolytic activity. These cells are not tumorigenic when injected into appropriate hosts, but if continually cultured further, do spontaneously become neoplastic. If these cells, however, are attached to plastic boats before spontaneously attaining neoplastic characteristics in vitro, they produce tumors in vivo when implanted. Boone et al.[25,26] demonstrated similar results when the aneuploid cell lines 3T3 and 10T$^1/_2$ are attached to substrates. They produce tumors in appropriate hosts when attached, yet do not produce tumors when injected in solution.

An important observation is that transformed cells generally require less serum in culture medium than do their nontransformed counterparts.[27] This suggests that the role of serum (or growth factors in the serum) is important in the transformation process. Reviews on the various classes of growth factors, growth-stimulating substances such as epidermal growth factor (EGF), fibroblast growth factor (FGF), insulin, and somatomedins can be found elsewhere.[28] Work by Cherington et al.[29] is a good representation of the research done on growth factor requirement and transformation. Their results suggest that cells with a low EGF requirement form tumors when injected into suitable hosts, but cells with a high EGF requirement for growth do not.

Dulak and Temin[30,31] have been able to isolate and characterize multiplication-stimulating activity (MSA) from the conditioned medium of cultures of Buffalo rat liver cells that grow in the absence of serum. This observation suggests that cells in vitro might be capable of producing their own growth factor(s); their failure to do so, or respond to such factors, might influence senescence or, conversely, transformation. Todaro, DeLarco and co-workers have been able to isolate a family of transforming polypeptides from the conditioned serum-free medium of murine sarcoma virus-infected 3T3 cells[32,33] and from these transformed cells themselves.[34] These isolated sarcoma and transforming growth factors apparently induce normal fibroblasts to express phenotypic characteristics of transformed cells and cause growth in soft agar. The existence of endogenously produced growth factors in transformed cells suggests a possible reason for their reduced serum requirement.

The proposed role of growth factors in maintaining growth and encouraging transformation is extremely important in considering the experimental evidence of senescence and transformation patterns in cell cultures from the various animal species. If the capacity of cells to produce or to respond to growth factors after explantation changes in culture, it would certainly be a major effect. Whether it is a dominant effect remains to be determined.

FIBROBLAST-LIKE CELLS IN CULTURE

Mouse Cells

Mouse cells explanted from embryonic and adult tissue have been the most studied of nonhuman cells. Their propensity for growth transformation has limited their usefulness in the study of senescence. An early indication that mouse fibroblast-like cells in culture have a high propensity for obtaining unlimited growth potential comes from Earle et al.,[35] who investigated methylcholanthrene-induced transformation in C3H mouse cells. Control cultures developed a slight but distinguishable change in morphology approximately 650 days after explantation from a 100-day-old C3H mouse. These "spontaneously" growth-transformed cells were able to develop tumors (sarcomas) when injected into susceptible mice.

Two reports document the development of established cell lines from cells of mouse embryos placed in culture. Rothfels et al.[36] describe the growth events in cells from explanted mouse embryos. Following multiple cultures, a decline in mitotic index from 20% in the first 4 passages to almost zero at passages 9 to 12 (50 days) was observed, indicating senescence. Some cultures senesced completely, but most evidenced increased mitotic activity from passage 13 until passage 20 (3 months), indicating growth transformation. The karyology was essentially diploid until the period of slow or no growth, but cells experienced a rapid loss of the diploid state with the subsequent rise in mitotic rate.[36] Other cell lines established by those authors from skin, lung, and kidney of inbred C3H mice were injected into isologous baby mice, and 12 of the 15 lines produced tumors,[36] indicating malignant transformation, as well as unlimited growth potential. Todaro and Green[37] examined the growth properties of fibroblasts from 17- to 90-day-old Swiss mouse embryos initiated in culture. These investigators demonstrated that mouse embryo cells undergo a decline in proliferative vigor that appears to be dependent on inoculum density. Cultures inoculated with 1×10^5 cells declined more rapidly than those inoculated with 3×10^5, and more so than those inoculated with 6×10^5 cells. By 10 to 20 generations, however, all cultures had doubling times over 70 hr. All cultures reversed this decline, and the growth rate rose subsequently to values similar to initial cultures (14 to 24 hr doubling time); this was seen in some cultures passed every 3 days as early as 30 to 45 days of culture. Once established, these cell lines acquired the ability to grow at a low inoculation density, but exhibited an altered chromosome number. The cultures remained essentially diploid through their growth decline and reversal, but later shifted to a tetraploid range of chromosomes. The method by which cell populations were carried influenced the properties of the cell line. Cells that never reached confluency from the beginning of culture establishment resulted in a cell line sensitive to contact inhibition of growth; the 3T3 cell line.[37] The other established cell lines demonstrated the ability to grow over each other and to form multilayers.

3T3 cells represent a class of rodent cells that apparently have indefinite growth characteristics but may not be malignantly transformed. It is useful to describe these cells in detail, since such cultures have many "normal" characteristics though some "abnormal" characteristics. Todaro et al.[38] examined saturation density cultures to study the mechanism of growth control in 3T3 cells. After a medium change occurs, a single division by a small population of cells in the culture occurs, apparently to replace cells that have died since the last change or trypsinization. A tenfold increase in RNA synthesis follows medium change by 30 min, protein synthesis increasing several hours later, and finally DNA synthesis is committed to follow after 12 hr. The authors suggest a serum factor that appears to release contact inhibition by the 3T3 cells via effects on the rate of RNA synthesis.[38] Steck et al.[39] suggest that 3T3 contact inhibition is regulated at high densities by a soluble inhibitory factor. Sparse cultures of 3T3 cells grown in medium conditioned by density-inhibited 3T3 cultures became growth inhibited. Fresh serum serves to dilute this inhibitory factor when cells are passaged. This evidence suggests growth inhibition might be due to inhibitory factors in the culture medium, rather than a depletion of medium components.[39] This does not rule out the possibility of cell-to-cell contact regulation of density inhibition of growth.

Aaronson and Todaro[40,41] in two reports characterize the development of 3T3-like cells from BALB/c mouse embryo cells. They state that the 3T3 cells from the outbred Swiss mouse strain would not be as useful as a cell line isolated from the inbred BALB/c strain.[40] Mouse embryo cultures followed very similar patterns as reported earlier by Todaro and Green for 3T3.[37] A decline in growth rate was progressive until 10 to 15 passages, when the decline was reversed until growth rates similar to initial cultures were seen, at about passages 20 to 25 (about 22 hr). The chromosome number after growth transformation was found to be hypotetraploid.[40] Cells passaged similarly to 3T3 cultures did not produce tumors when injected into newborn or irradiated mice, whereas cells passaged at higher densities produced tumors readily.[41] A study by Tuffery and Baker[42] reports on certain altered features of BALB/c embryo cultures during the period of growth decline before attaining unlimited growth. They found extensive parallel arrays of cytoplasmic fibrils correlating with an increase in cell size and cell spreading, similar to observations concerning human cells in phase III.

Two studies by Boone et al.[25,26] demonstrate that subcutaneously inoculated BALB/3T3 mouse embryo cells can produce tumors in mice when the cells are attached to glass beads or polycarbonate platelets. No tumors were found when BALB/cells or the substrates were inoculated alone. These authors conclude that BALB/3T3 cells are preneoplastic in what would correspond to growth transformation, but still remain anchorage dependent, preventing formation of tumors when placed in vivo.[26]

Meek et al.[43] examined the establishment of 15- to 17-day BALB/cCRL mouse embryos into culture. Again, a growth decline occurred up to the eighth in vitro passage when establishment of growth-transformed cells occurred. A concomitant decline in the percentage of cells incorporating ^{3}H-thymidine during loss of proliferative vigor as well as an increase of ^{3}H-thymidine incorporation when cells became growth transformed was noted. Cell size increased with slower growth and decreased to a stable size once established as a cell line. Interdivision times increased to 21 hr during growth decline only to obtain an interdivision time of 15 hr or less with continuous growth. The ability to obtain rapid growth potential did not correspond to malignant transformation, as cells injected into mammary fat pads of BALB/c mice did not form tumors until well after continuous growth was evident.[43] In another study with the same cell system, Meek et al.[44] noted that the ability of these cells to perform unscheduled DNA synthesis was correlated with the fraction of cells able to perform scheduled DNA synthesis.

Many other reports demonstrate a similar pattern of apparent senescence of mouse embryo cells in culture where a growth "crisis" occurs about the eighth or ninth passage with establishment of a cell line following. These include Beaupain et al.,[45] examining DNA alkali-sensitive sites during this pattern; Danot et al.,[46] assaying for altered enzymes; Weisman-Shomer et al.,[47] comparing replicative activity of isolated chromatin of early and late passages; and Paffenholz,[48] who measured DNA repair. Kakunaga[49] used the apparent preneoplastic property of 3T3 cells to select a subline as an assay for the rate of their conversion to neoplastic growth induced by chemical carcinogens.

Another cell line similar to 3T3 has been found useful for quantitative studies of transformation. This is the 10T$^1/_2$ cell line,[50,51] established by Reznikoff et al.,[52] in a manner similar to that used in establishing the BALB/3T3 line. These cells from the C3H mouse embryo undergo similar crises, and a cell line was selected which was sensitive to postconfluence inhibition of cell division. The spontaneous expression of C-type RNA viruses is not detected; tests for tumor-producing ability were negative. A somewhat stable karyotype of 81 chromosomes is noted for 10 to 60% of the cells, this number being slightly higher than the hypotetraploid number reported for Swiss/3T3 and BALB/3T3 cells. The 10T$^1/_2$ cell line is now used extensively by many laboratories for carcinogen testing.

Sacher and Hart[53] contrasted cell cultures from two mouse species, *Mus musculus* and *Peromyscus leucopus,* in an attempt to evaluate the theory that the in vitro lifespan of cultured cells appears to be dependent on that species' maximum achievable lifespan (MALS) and on DNA repair levels. *Peromyscus* has a MALS approximately 2.5 times that of *Mus.* Cell cultures from *Peromyscus* embryo tissue also evidenced approximately 2.5 times more extended growth in culture before partial senescence and the emergence of growth-transformed cells. Further, the embryonic cells from the longer-lived mouse showed significantly more DNA repair (unscheduled DNA synthesis) after 10 J/m^2 of 254-nm radiation than *Mus,* although both showed similar rates of rejoining single-strand DNA breaks.

Rat Cells

Rat fibroblast-like cells, generally obtained from embryos at 14 to 17 days of gestation, appear to produce growth-transformed cells readily. The establishment of two continuously growing cell lines from a common pool of rat embryo cells is described by Petursson et al.[54] These two lines were exposed to different media conditions during early in vitro passages and subsequently exhibited different morphologies and karyotypes. One cell line initiated in Eagle's Basal Medium with 15% fetal bovine serum (FBS) showed several morphologies, suggesting a mixed population. The karyotype at the 40th passage demonstrated a hyperdiploid state (modal number 45 chromosomes) as well as a scattered variability at higher values (60 to 90). The second cell line, initiated in 40% ^{15}N, 15% FBS, 4% NCTC 109, and 41% saline G, had a different morphology than the first, with mostly stellate or polygonal outlines. The karyotype was more stable with a large portion of the cells having the diploid or exact tetraploid number of chromosomes. The authors suggest that the genetic stability of cultured cells is influenced by growth conditions.[54] It should be noted, however, that the number of cultures examined is too small to draw general conclusions.

Jackson et al.[55] examined ALB/N rat embryo cells in culture to determine whether these cells undergo spontaneous neoplastic conversion and chromosomal alteration. Cells cultured in chemically defined medium without serum grew at very low rates for up to 4 years, appearing healthy and viable, yet did not spontaneously convert to a neoplastic state as determined by intraocular implants into adult rats. Horse serum-supplemented conditions allowed for reproducible neoplastic conversion within 98 to 188 days, whereas FBS supplemented conditions allowed conversion over 9 to 30 months. Cells grown in FBS had less than 100% neoplastic conversion, as was seen with cells grown in horse serum. When cells were carried with regular media changes, but no or little subculturing, no apparent neoplastic conversion occurred. The authors suggest that one possible requirement for neoplastic conversion is many cell divisions. Karyotypes for each cell line remained relatively stable for up to 61 to 110 days, but a considerable loss of diploidy occurred after this period. No chromosomal alteration or pattern of chromosome change was found that could distinguish between a neoplastic vs. a nonneoplastic cell population.[55]

Freeman et al.[56] describe the series of stages rat embryo cell cultures undergo in terms of spontaneous transformation and expression of endogenous C-type RNA viruses. Early cells were mostly diploid and the C-type virus undetected in 35 to 60 passages (cumulative population doublings). Subsequently, cells became heterploid over the next 50 to 60 passages, yet remained contact inhibited, failing to produce tumors in newborn isogenic rats. C-type virus could now be detected at this stage by complement fixation tests. From 90 to 120 population doublings in vitro, these cells were heteroploid, demonstrated detectable virus, and were also malignant. These authors suggest that all rat cells contain genomes for coding of endogenous C-type RNA viruses, and the expression of these viruses becomes more pronounced with time in culture.[56]

Meek et al.[57] present evidence that rat embryo cells from Fischer rats undergo in vitro cellular senescence similar to human fibroblasts before growth transformation. A concomitant decline in the fraction of cells replicating DNA, and a decline in growth prior to the tenth in vitro passage was observed. Time-lapse microcinematographic analysis revealed increasing interdivision times of cells in fifth through ninth passages, as well as a drop in the number of dividing cells. After the tenth passage, rat embryo cells evidenced conversion to indefinite growth as measured by increased proliferation and DNA synthesis. Cultures remained diploid up to 20 passages (50 population doublings), and none of these cells produced tumors when transplanted into syngeneic Fischer rats.

The work by Kontermann and Bayreuther[58,59] also supports the evidence that rat embryonic fibroblasts undergo cellular senescence similar to human cells before they show unlimited growth. Three morphologically and biochemically distinct differentiating compartments (fibroblasts, fibrocytes I and II, with separate genetic programs I, II, III, respectively) were defined in the mass cultures of rat embryo cells from strains L.BN and Lewis.[58] All three cells types were predominantly diploid. During phase I and phase II of growth, the number of fibroblasts and fibrocytes I decline. Phase III occurs after 25 cumulative population doublings with a predominance of fibrocyte II cells. Fibrocyte II cells have two possible fates: cellular degeneration into senescence or cellular neoplastic transformation. In the cell cultures that acquired permanent growth, cells that are growth transformed grow out of the fibrocyte II population with the characteristics of growth rate increase, a change to a hypodiploid state, and an association with production of endogenous type-C RNA tumor viruses.[59]

In our laboratory we have examined cell cultures initiated from 17-day-old embryos of the Sprague-Dawley rat for patterns of cell growth, senescence, and growth transformation. Figure 1 presents representative data from these experiments. The fraction of cells that senesce between 9 and 15 population doublings could be greatly influenced by methods used to initiate and to carry the cell cultures. All but an extremely small fraction of cells ($<10^{-4}$) in cultures derived from single-cell suspensions that never experience confluency undergo apparent senescence at about nine cumulative mean population doublings. A much smaller fraction of cells senesce both in cultures derived from single cell suspensions that become confluent early in their in vitro growth and in cultures derived from tissue fragments. For cultures in which the vast majority of cells senesce, cultures remained >95% diploid until senescence. Cultures obtaining unlimited growth demonstrated differing patterns of ploidy. Some cultures became hyperdiploid quickly (10 to 20 doublings) while others remained over 90% diploid for extended growth (>50 doublings). We concluded that ploidy did not correlate with growth potential.

Hamster Cells

Hamster fibroblast-like cells in culture do not possess as high a rate of transformation toward unlimited growth potential as do mouse cells. Early work with Syrian hamster embryo cells demonstrated a limited in vitro lifetime, whereas subsequent studies demonstrate that cell lines can be growth transformed if proper conditions are maintained.

Todaro et al.[60] examined cell cultures from 13-day-old Syrian hamster embryo cells. They found that growth rates declined with serial passage and ceased by the tenth generation. Todaro and Green,[61] however, were able to maintain hamster embryo cell growth for over 100 generations, although such cells did not possess markers of established lines. The addition of crystallized serum albumin to the growth medium was hypothesized to allow this extended growth span. Chromosome analysis revealed that the cells were hyperdiploid (45 to 46 chromosomes) as compared to earlier passages with a diploid complement.[61]

Two studies describe the transformation to unlimited growth potential of cells from 10- to 13-day hamster embryos. Gotlieb-Stematsky et al.[62] found after several in vitro passages distinct colonies of cells in several of their dispersed embryo cultures. These colonies were isolated and grown until enough cells were available to inoculate into hamsters. These cells proved to be tumorigenic when injected into hosts. The cells from primary cultures (early and late passages) were not malignantly transformed. The authors suggest selection of spontaneous malignantly transformed cells took place in their cultures, but do not postulate as to how this selection took place. Diamond[63] was able to establish two growth-transformed cell lines from a single pool of 13-day Syrian hamster embryos. The parent culture was passaged for up to 14 weeks (27 passages) until apparent senescence. Subcultures taken from passages 12 and 16 of the parent culture evolved into two cell lines. Both have been carried for well over 100 passages. The parent culture was not tumorigenic when inoculated into newborn hamsters, whereas one cell line produced tumors by passage 35 and the second by passage 81.[63] Both cell lines demonstrated hypodiploid chromosome numbers and all cells had abnormal karyology by their 100th passage. Diamond has reported that from 22 separate embryo cultures, no other cell lines were established,[63] indicating that hamster cells, as a class, appear to have less tendency to transform to unlimited growth than mouse cells.

Williams[64] describes a methodology to obtain reproducible patterns of senescence of Syrian golden hamster embryo cells. The predisposition of these cells to transform to indefinite in vitro growth appears to depend upon the density in which the cells are maintained during growth. Cells not allowed to reach confluency senesced regularly after 7 passages (9 to 14 mean population doublings). Apparent senescence of cultures was not seen when cells were allowed to reach confluency between passages, and these cells have been carried for over 100 mean population doublings.[93]

A Syrian hamster fibroblast cell line, BHK21, was established from cultures of kidney cells from 1-day-old Syrian hamsters by Stoker and MacPherson.[65] These cells underwent a progressive growth decline until day 65, when a sudden increase in the growth rate was noted. The mean doubling time was 12 hr and the cells appeared to grow past confluency. The karyotype of BHK21 is diploid, male, and has not changed significantly over 450 generations. A large number of cells ($> 1 \times 10^6$) is needed to induce tumor growth in 50% of inoculated hamsters. The fact that these cells induced tumors, yet appear essentially diploid (possibly "pseudodiploid"), suggests that a detectable alteration in chromosome number is not a dependable marker for classification of growth potential. The long-term growth of Chinese hamster cells in culture, demonstrated by Yerganian and Leonard,[66] without a marked alteration toward aneuploidy supports the suggestion that unlimited growth is not always associated with chromosomal alterations.

Chick Cells

Chick fibroblast-like cells cultivated in vitro appear to demonstrate a limited potential to proliferate in culture, and an extremely low capacity, if any, to transform. This resistance to transformation makes chick cells a good model by which to study senescence. Chick cells apparently senesce in a manner similar to human cells in vitro, though on a shorter time scale, and with a few differences in assayed parameters. Studies examining chick fibroblasts in culture usually use 9- to 13-day embryos from White Leghorn chickens.

Hay and Strehler[67] isolated fibroblasts from skin and muscle of chick embryos and observed several properties. Cells undergo a maximum of 25 population doublings over a period of 2 months, and this property did not differ from cells isolated either from 1- to 12-day embryos or from 18- to 20-day embryos. As the cultures are serially passaged, a decrease in the number of cells at confluency is noted, as well as a slight decrease in DNA content at late passages. Further, as the inoculum density is increased there is a decrease in the growth span. Ponten[68] also grew chick embryo cells to a maximum of 20 to 27 passages before observing senescence in phase III similar to human cell culture. The cell number did not increase, mitotic activity decreased, and considerable chromosomal heteroploidy arose, mainly subtetraploid.[68] Weismann-Shomer and Fry[69] present evidence that the lifespan of chick embryo fibroblasts is a function of calendar time and not mitotic function. Fibroblasts inoculated at densities from 2.1×10^3 to 3.1×10^4 cells/cm^2 have population doublings between 39 and 17, respectively, while all cultures lasted 57 ± 3 days. The more dense the inoculum, the earlier the cells enter phase II, and the longer their generation times in phase III. Ryan[70] rejects this interpretation, as described below.

Other investigators describe several other parameters of senescence for chick fibroblasts. Lima and Macierira-Coelho[71] examined saturation density, doubling time, time between subcultures, percentage of cells synthesizing DNA for each passage, and cell size. Phase I cultures exhibited the highest densities before subcultivation, and the densities declined rapidly through phase III. Most cell cultures could be carried for 30 to 35 population doublings. Time between subculture was 1 to 2 days until the last passages, where the time increased to 3 days. As measured by ^{3}H-thymidine uptake, almost 100% of the cells entered DNA synthetic phases throughout the cellular in vitro lifespan. The cell size increased during the first passages, then remained stable until increasing further at the later passages. Further studies by these investigators state that incorporation of radioactive precursors into RNA and protein begins to decline in early passages and that an increase in acid phosphatase is not age related.[72] Ryan[73] examined the percentage of labeled nuclei and the ability of cells to form colonies as a function of in vitro age. The incorporation of ^{3}H-thymidine into DNA was an age-associated decline in contrast to Lima and Macieira-Coelho. The discrepancy is hypothesized to occur due to differences in labeling regimes. Ryan labeled for only 2 or 24 hr, whereas Lima and Macieira-Coehlo labeled for 2 to 3 days, perhaps missing cells that proliferate more slowly.[73] Ryan also noted an age-associated decline in the number of colonies formed from a constant inoculum from successive passages; these changes occurring before changes in cell number at confluency declined in later passages. Initially Ryan supported Weismann-Shomer and Fry's contention that the lifespan of chick embryo fibroblasts in culture is calendar time dependent (51 ± 1 days), yet in subsequent work, does not continue to support this contention.[70]

Kaji and Matsuo[74,75] examined cytological aspects of aging in chick fibroblast cells. As the cells declined from rapid growth, they observed an increase of multinuclear cells (mostly binucleates) to a maximum of 20 to 25% of the culture at senescence.[74] The mean cell volume also increased with the most dramatic increase at the last passages. Most of the early cultures retained their diploid number, whereas the later cultures demonstrated a loss of diploidy up to 58% heteroploid cells.[75] The proportion of polyploid cells in later chick cultures is greater than that seen in late passage human cells.[75] In contrast to other studies, these investigators are able to carry their cultures up to 130 days in secondary culture, much more than the 50 to 60 calendar days already described.[69,73]

Further work in Ryan's laboratory reevaluates the underlying basis for the limited growth pattern of chick embryo cells in vitro. Studies by Nielsen and Ryan[70] do not support the notion that calendar time (or metabolic time) is the prime determinant of the limited in vitro lifespan of chick cells, but rather the cumulative number of population doublings. Chick embryo fibroblasts were held at confluency in minimal amounts of FBS or horse serum for periods up to 35 days without cell loss. Cultures were then passaged at weekly intervals and the proliferative history followed. All cultures underwent 28 to 31 doublings over a period of 98 to 114 days, apparently independent of time in confluency. Controls (not maintained at confluency) senesced after 29 doublings in about 91 days. This suggests cumulative population doublings as the basis for chick cell lifespan in vitro in direct opposition to the earlier work.

Ultrastructure studies on chick fibroblasts in culture are described by Brock and Hay.[76] Early passage cells contain oval nuclei with peripheral chromatin and oval mitochondria. Late passage cells (18 population doublings) show nuclear lobation, absence of peripherally located chromatin in the nucleus and ellipsoidal mitochondria, and the presence of secondary lysosomes and residual bodies.

Some of the culture medium components and the local cell concentrations have been examined for their effects on chick embryo fibroblast growth in vitro. Ryan[77] has shown that as the concentration of FBS in medium increased from 5 to 30%, there is a resultant increase in the rate of cell proliferation and cell density at stationary phase. Also, chick embryo fibroblasts grown in 5% FBS demonstrated a reduced calendar time lifespan of 35 days compared to 50 days obtained by cells carried in higher serum concentrations (10 to 30%). Ryan suggests that while culture conditions can influence the kinetics of cellular senescence, it is still controlled by intrinsic properties in the cell.[77] Rein and Rubin[78] have demonstrated that the density of cells per unit area determines the optimal growth kinetics of chick embryo fibroblasts. Cells apparently secrete a conditioning factor, they suggest, that allows for growth enhancement and is a slowly diffusing molecule. This creates concentration gradients of this factor near cells secreting it. Therefore, the denser the area with cells, the greater the enhancement of growth. This is supported by their studies in which a small number of cells grow out better when grown in medium preconditioned by a dense culture, or when there is a feeder layer present.[78] It is unclear how this may reflect on the findings by Hay and Strehler, Weismann-Shomer and Fry, and Ryan that suggest that chick embryo fibroblasts carried at high density or low FBS concentration may be caused to enter a slowly or nonproliferating state and thus reduce the growth span of the culture. Other studies show (1) chick fibroblasts evidenced reduced proliferation or quiescence when lower than physiological concentrations of calcium and/or magnesium are used;[79] (2) FBS induces more cell proliferation at low cell densities than horse serum (HS) plus hemoglobin (Hb) whereas HS + Hb is better at high cell densities to promote cell growth,[80] though Nielsen and Ryan observed that chick cell division was stimulated by adding medium containing horse serum (up to 10%);[70] (3) trypsin is able to stimulate proliferation of secondary chick embryo fibroblasts.[81]

Other Nonhuman Cells

Studies on fibroblast-like cells cultures derived from other mammals have been reported in the literature, including bovine, canine, and rabbit cells. Stenkvist[82] reports the cultivation of bovine lung cultures started from newborn calf or 2-month fetus. Cultures are maintained at passage intervals of 4 to 6 days up to 60 passages. This interval increased to 7 to 10 days until over 70 passages when the cells entered a "growth crisis", began losing proliferative vigor and eventually most were lost. Bovine cells apparently do not have a high capacity for growth transformation. Chromosome studies by Lithner and Ponten[83] showed that the chromosome number was altered as bovine cells began growth decline. The cells at passage 75 were predominantly hypodiploid, and the authors suggest that this number is reached by a fusion of centromeres of the telocentric chromosomes.

A canine cell line (Cf2Th) was cultivated from thymus tissue of a newborn pup of a mongrel bitch by Nelson-Rees et al.[84] This line was carried to at least passage 146 and the female diploid karyotype was altered toward aneuploidy (67 to 75 chromosomes) by passage 37. An increase in growth rate was noted when chromosomal alterations were first observed at passage 20. Canine cells obtained from passage 123 were malignantly transformed, as they produced tumors in inoculated ATST mice.[84] An alteration to hypodiploid karyology during culture of cells from a normal inbred beagle embryo is reported by Rhim et al.[85]

A rabbit cell line, designated Calg-ARLC, has been established and characterized by Davidenas et al.[86] The cells were obtained from lung tissue of a New Zealand adult female rabbit. The cultures were maintained until a proliferating colony appeared that was cloned to form Calg-ARLC. This growth-transformed fibroblast-like cell line has been passaged for over 2 years, has a high split ratio, a generation time of 12 hr, and a relatively stable heteroploid chromosome number (35).[86]

Stanley et al.[87] in studies of cell growth from several species, suggest that the final cell doubling number for diploid cells from a species is not correlated to that species' lifespan. Cultures from fetal or newborn sources were started and cells from man, horse, wallaroo X kangaroo hybrid, kangaroo, wallaby, and potoroo all exhibited a growth decline and failure to attain unlimited growth. Cells from monkey, cat, and rabbit continued to grow and a loss in diploidy was noted.[87] Rhim et al.[88] also found a shift in chromosome number in cultured feline embryo cells to a hypodiploid number of around 36. The stability of a diploid karyotype in cells from skin of African green monkey fetus was studied by Earley et al.,[89] who observed that the emergence of polyploidy depended on the type of medium and serum concentration during the long-term culture of these cells.

In a recent paper, Rohme[90] demonstrated a strong correlation between maximal lifespan of eight mammalian species and the total population doublings of fibroblast cell lines derived from those species at the time of "growth crisis". His conclusions were different from those of Stanley et al. described above. The endpoints examined, however, were different for some cultures. Stanley used either the exhibition of a growth crisis or loss of the diploid number of chromosomes as a measure of senescence; Rohme used only the growth crisis. Rohme also presented data that the lifespan of erythrocytes from the same eight mammalian species also showed a correlation with maximal lifespan. Since erythrocytes are postmitotic cells, he postulated a role for the cell membrane in their cellular lifespan.

CONCLUSIONS AND DISCUSSION

The most important question to be considered is whether the growth patterns of cells derived from several mammalian species demonstrate a correlation between any intrinsic cellular property and the lifespan of the donor species. Two possible correlates have been raised in our analysis: growth potential before senescence and level of DNA repair capacity.

The relationship between growth potential and lifespan can be analyzed in a general way by considering the differences in data observed between several laboratories. Figures 2 and 3 show data that contrast the maximum achievable lifespan for 17 mammalian species with the cumulative mean population doublings at senescence of cultures of fibroblast-like cells derived from explanted tissue from an individual for each species. Most of the data are from Rohme[90] and Stanley.[87] The dashed lines on these figures is that from Rohme's study, a line that he describes as demonstrating a strong power function relationship between these two variables. The dotted lines connect measurements for the same species from different laboratories. For larger values of growth for cultures from the horse, cat, rabbit, and rat, as well as for the monkey and dunnart, the level of cumulative mean population doublings at senescence is defined as the value at the time of loss of the diploid chromosome number. When these six points are omitted, the greatest variation in doublings at senescence from different laboratories is less than a factor of 3. Although this seems a large effect for a response whose entire range is less than 10, the patterns do appear to define a line that represents a minimum senescence period. To us, these data seem to indicate a clear correlation between the number of doublings of explanted fibroblast-like cells and the lifespan of the host species, provided that the measure of senescence is defined as a loss of growth vigor, that is, the occurrence of a growth crisis or partial growth crisis. Senescence defined through other endpoints, such as loss of diploid chromosome number, appears to extend the values for the time to senescence and thus decrease correlation from the line on Figure 3 that best fit Rohme's data. In unpublished data, we have compared two endpoints for senescence: loss of diploid karyology and growth crisis. In multiple experiments with rat embryo cultures the range for apparent senescence varies from 9 to over 50 mean population doublings for loss of diploidy, but varies only from 7 to 10 for the time necessary to observe the beginning of a growth crisis.

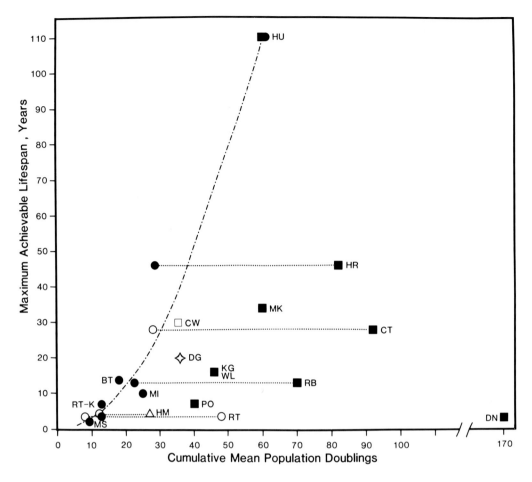

FIGURE 2.   Correlation between the in vitro lifespan of fibroblast-like cells derived from mammals and the maximum achievable lifespan of the species from which they are derived. Closed circles are values according to Rohme.[90] Closed squares are values according to Stanley.[87] The open symbols are values obtained from reports discussed in the test. HU, human; HR, horse; CW, cow; MK, monkey; CT, cat; DG, dog; BT, bat; KG, kangaroo; WL, wallaby; RT, rat; RT-K, rat-kangaroo; PO, potoroo; HM, hamster; DN, dunnart; RB, rabbit; MS, mouse.

Certain cautions should be expressed. As seen in Figure 1, in a culture for which a large proportion of cells appears to growth transform, the definition of a growth crisis can be difficult. Further, as shown by Kontermann and Beyreuther,[58] and confirmed in our work, senescing rat populations contain multiple subpopulations of cells that senesce sequentially. It is not clear whether cell populations derived from other species or tissues also contain multiple populations, but it does indicate that problems exist in defining an exact point of senescence. Nonetheless, even with these cautions, the pattern in Figure 3 seems to be significant, and we conclude that a strong correlation does exist between time to senescence of cultured fibroblast-like cells and the lifespan of the species of the host from which the cells were derived.

Another postulated correlation between lifespan and an intrinsic cellular characteristic is that of DNA repair level following UV light (254 nm) exposure.[11] As stated earlier, there is a lack of unanimity as to the strength of this correlation.[12,13] These three studies were based on the measurement of the total repair synthesis after UV. Our studies in this area have centered on more detailed measurement of DNA repair in senescing Syrian golden hamster embryo cells,[64] in senescing rat embryo cells,[91] and in V79 aneuploid Chinese

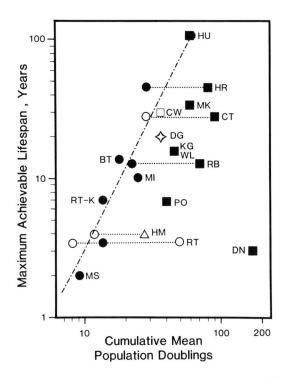

FIGURE 3.   Same as Figure 2 with data plotted on logarithmic scales.

hamster fibroblasts. The factors that influence the multiple pathways of DNA repair are not exactly defined, so it has been most difficult to define in precise terms the difference in repair between cells from short-lived and long-lived mammals. Our data would, however, imply that stages of repair that occur in the first 2 hr after acute UV exposure appear quite similar between rodent and human cells. It is only in a later phase, extending from times later than 2 hr, that cells from the short-lived rodent fall behind. It should be noted that similar levels of UV-induced cytotoxicity and mutation are seen in both rodent and human cells despite differences in these later repair levels. This may imply that the early phases of repair, similar for these two cell types, may be the most important for determining cellular fate after acute radiation for cytotoxic and mutational response. In our opinion the difference in rates of repair in later phases could correlate with lifespan, although far too few data now exist to make any such statement. Further, we remain unconvinced that DNA repair after acute doses of damage is, in general, a model for doses of DNA damage experienced under normal physiological conditions. Finally, we believe there is some evidence,[92] although not conclusive, that intrinsic cellular processes induce lesions in DNA that might be repaired by the same system that repairs DNA damage induced by UV light. If this proves to be true, a theory of aging could be postulated in which cellular aging could result from the action of programed, site-specific DNA damage induction. Such a theory would be most attractive in uniting theories of somatic mutation and theories of programmed aging.

In conclusion, our analysis of patterns of growth in nonhuman fibroblasts as it may relate to aging suggests four statements:

1.   The growth potential of fibroblast-like cell cultures established from nonmalignant tissue from a particular mammalian species seems characteristic of that species and is experimentally reproducible under careful conditions of explant and culture.

2.   There appears to be a strong correlation between the number of cell divisions in a cell population before senescence and the lifetime of the animal (mammal) from which the cell cultures are derived. Other endpoints of senescence such as loss of diploidy are not so useful in such studies.

3.   There exists a trend between the extent of DNA repair in cultured cells and the lifetime of the species from which they derive; however, lack of exact and detailed knowledge of the mechanisms of DNA repair and their biological sequelae make it difficult to make any stronger association. The possibility of intrinsically induced DNA damage exists.

4.   Cells derived from short-lived mammals may offer a facile and rapid system for studying senescence.

REFERENCES

1. **Hayflick, L.,** The limited *in vitro* lifetime of human diploid cell strains, *Exp. Cell Res.,* 37, 614, 1965.
2. **Hayflick, L. and Moorhead, P.,** The serial cultivation of human diploid cell strains, *Exp. Cell Res.,* 25, 585, 1961.
3. **Hayflick, L.,** Recent advances in the cell biology of aging, *Mech. Aging Dev.,* 14, 59, 1980.
4. **Kirkwood, T. and Holliday, R.,** Commitment to senescence: a model for the finite and infinite growth of diploid and transformed human fibroblasts in culture, *J. Theor. Biol.,* 53, 481, 1975.
5. **Holliday, R., Huschstscha, L., Tarant, G., and Kirkwood, T.,** Testing the commitment theory of cellular aging, *Science,* 198, 366, 1977.
6. **Hirsch, H.,** Commitment theory of cellular aging: possibility of an immortal diploid cell strain, *Mech. Aging Dev.,* 12, 25, 1980.
7. **Bell, E., Marck, L., Sher, S., Merrill, C., Levinstone, D., and Young, I.,** Do diploid fibroblasts in culture age?, *Int. Rev. Cytol.,* Suppl. 10, 1, 1979.
8. **Smith, J. and Whitney, R.,** Intraclonal variation in proliferative potential of human diploid fibroblasts: stochastic mechanism for cellular aging, *Science,* 207, 82, 1980.
9. **Hornsby, P., Gill, G., Bell, E., Marck, L., Merrill, C., Levinstone, D., Young, T., Eden, M., and Sher, S.,** Loss of division potential in culture: aging or differentiation?, *Science,* 208, 1482, 1980.
10. **Schmookler, R., Resi, R., Goldstein, S., and Harley, C.,** Is cellular aging a stochastic process?, *Mech. Aging Dev.,* 13, 393, 1980.
11. **Hart, R. and Setlow, R.,** Correlation between deoxyribonucleic acid excision-repair and life-span in a number of mammalian species, *Proc. Natl. Acad. Sci. U.S.A.,* 71, 2169, 1974.
12. **Kato, H., Harada, M., Tsuchiya, K., and Moriwaki, K.,** Absence of correlation between DNA repair and ultraviolet irradiated mammalian cells and lifespan of the donor species, *Jpn. J. Genet.,* 55, 99, 1980.
13. **Francis, A., Lee, W., and Regan, J.,** The relationship of DNA excision-repair of UV-induced lesions to the maximum lifespan of mammals, *Mech. Aging Dev.,* 16, 181, 1981.
14. **Williams, J. and Dearfield, K.,** DNA damage and repair in aging mammals, in *Handbook of Biochemistry in Aging,* Florini, J. R. Ed., CRC Press, Boca Raton, Fla., 1981.
15. **Macieira-Coelho, A., Diatloff, C., and Malaise, E.,** Concept of fibroblast aging *in vitro:* implications for cell biology, *Gerontology,* 23, 1977.
16. **Macieira-Coelho, A.,** Implications of the reorganization of the cell genome for aging or immortalization of dividing cells *in vitro, Gerontology,* 26, 276, 1980.
17. **Macieira-Coelho, A., Diatloff, C., and Malaise, E.,** Doubling potential of fibroblasts from different species after ionizing radiation, *Nature (London),* 261, 1976.
18. **Terzi, M. and Hawkins, T.,** Chromosomal variation and the establishment of somatic cell lines *in vitro, Nature (London),* 253, 361, 1975.
19. **Foulds, L.,** *Neoplastic Development,* Vol. 1, Academic Press, London, 1975.
20. **Foulds, L.,** *Neoplastic Development,* Vol. 2, Academic Press, London, 1975.
21. **Barrett, J. and Ts'o, P.,** Evidence for the progressive nature of neoplastic transformation *in vitro, Proc. Natl. Acad. Sci. U.S.A.,* 75, 3761, 1978.
22. **Barrett, J.,** The progressive nature of neoplastic transformation of Syrian hamster embryo cells in culture, *Prog. Exp. Tumor Res.,* 24, 17, 1979.

23. **Kakunaga, T. and Kamahor, J.,** Properties of hamster embryonic cells transformed by 4-nitroquinoline-1-oxide *in vitro* and their correlations with the malignant properties of the cells, *Biken J.,* 11, 313, 1968.

24. **Barrett, J.,** A preneoplastic stage in the spontaneous neoplastic transformation of Syrian hamster embryo cells in culture, *Cancer Res.,* 40, 91, 1980.

25. **Boone, C.,** Malignant hemanigioendotheliomas produced by subcutaneous inoculation of BALB/3T3 cells attached to glass beads, *Science,* 188, 68, 1975.

26. **Boone, C. and Jacobs, J.,** Sarcomas routinely produced from putative nontumorigenic BALB/3T3 and C3H/10T½ cells by subcutaneous inoculation attached to plastic platelets, *J. Supramol. Struct.,* 5, 131, 1976.

27. **Temin, H.,** Studies on carcinogenesis by avian sarcoma viruses. III. The differential effect of serum and polyanions on multiplication of uninfected and converted cells, *J. Natl. Cancer Inst.,* 37, 167, 1966.

28. **Gospodarowicz, D. and Moran, J.,** Growth factors in mammalian cell cultures, *Ann. Rev. Biochem.,* 45, 531, 1976.

29. **Cherington, P., Smith, B., and Pardee, A.,** Loss of epidermal growth factor requirement and malignant transformation, *Proc. Natl. Acad. Sci. U.S.A.,* 76, 3937, 1979.

30. **Dulak, N. and Temin, H.,** A partially purified polypeptide fraction from rat liver cell conditioned medium with multiplication-stimulating activity for embryo fibroblasts, *J. Cell Physiol.,* 81, 153, 1973.

31. **Dulak, N. and Temin, H.,** Multiplication-stimulating activity for chicken embryo fibroblasts from rat liver cell conditioned medium: a family of small polypeptides, *J. Cell Physiol.,* 81, 161, 1973.

32. **DeLarco, J. and Todaro, G.,** Growth factors from murine sarcoma virus-transformed cells, *Proc. Natl. Acad. Sci. U.S.A.,* 75, 4001, 1978.

33. **Todaro, G. and DeLarco, J.,** Growth factors produced by sarcoma virus-transformed cells, *Cancer Res.,* 38, 4147, 1978.

34. **Roberts, A., Lamb, L., Newton, D., Sporn, M., DeLarco, J., and Todaro, G.,** Transforming growth factors: isolation of polypeptides from virally and chemically transformed cells by acid/ethanol extraction, *Proc. Natl. Acad. Sci. U.S.A.,* 77, 3495, 1980.

35. **Earle, W., Shilling, E., Stack, T., Staus, N., Brown, M., and Shelton, E.,** Production of malignancy *in vitro.* IV. The mouse fibroblast cultures and changes seen in the living cells, *J. Natl. Cancer Inst.,* 4, 165, 1943.

36. **Rothfels, K., Kuplewieser, E., and Parker, R.,** Effects of X-irradiated feeder layers on mitotic activity and development of aneuploidy in mouse embryo cells *in vitro, Can. Cancer Conf.,* 5, 191, 1963.

37. **Todaro, G. and Green, H.,** Quantitative studies of the growth of mouse embryo cells in culture and their development into established lines, *J. Cell Biol.,* 17, 299, 1963.

38. **Todaro, G., Lazar, G., and Green, H.,** The initiation of cell division in a contact-inhibited mammalian cell line, *J. Cell. Comp. Physiol.,* 66, 325, 1965.

39. **Steck, P., Voss, P., and Wang, J.,** Growth control in cultured 3T3 fibroblasts: assays of cell proliferation and demonstration of a growth inhibitory activity, *J. Cell. Biol.,* 83, 562, 1979.

40. **Aaronson, S. and Todaro, G.,** Development of 3T3-like lines from Balb/c mouse embryo cultures: transformation susceptibility to SV40, *J. Cell Physiol.,* 72, 141, 1968.

41. **Aaronson, S. and Todaro, G.,** Basis for the acquisition of malignant potential by mouse cells cultivated *in vitro, Science,* 162, 1024, 1968.

42. **Tuffery, A. and Baker, R.,** Alterations of mouse embryo cells during *in vitro* aging, *Exp. Cell Res.,* 76, 186, 1973.

43. **Meek, R., Bowman, R., and Daniel, C.,** Establishment of mouse embryo cells *in vitro, Exp. Cell Res.,* 107, 277, 1977.

44. **Meek, R., Rebeiro, T., and Daniel, C.,** Patterns of unscheduled DNA synthesis in mouse embryo cells associated with *in vitro* aging and with spontaneous transformation to a continuous cell line, *Exp. Cell Res.,* 129, 265, 1980.

45. **Beaupain, R., Icard, C., and Macieira-Coelho, A.,** Changes in DNA alkali-sensitive sites during senescence and establishment of fibroblasts *in vitro, Biochim. Biophys. Acta,* 606, 251, 1980.

46. **Danot, M., Gershon, H., and Gershon, D.,** The lack of altered enzyme molecules in "senescent" mouse embryo fibroblasts in culture, *Mech. Aging Dev.,* 4, 289, 1975.

47. **Weisman-Shomer, P., Kaftory, A., and Fry, M.,** Replicative activity of isolated chromatin from proliferating and quiescent early passage and aging cultured mouse cells, *J. Cell. Physiol.,* 101, 219, 1979.

48. **Paffenholz, V.,** Correlation between DNA repair of embryonic fibroblasts and different life span of 3 inbred mouse strains, *Mech. Aging Dev.,* 1, 131, 1978.

49. **Kakunaga, T.,** Requirement for cell replication in the fixation and expression of the transformed state in mouse cells treated with 4-nitroquinoline-1-oxide, *Int. J. Cancer,* 14, 736, 1974.

50. **Reznikoff, C., Bertram, J., Brankow, D., and Heidelberger, C.,** Quantitative and qualitative studies of chemical transformation of cloned C3H mouse embryo cells sensitive to postconfluence inhibition of cell division, *Cancer Res.,* 33, 3239, 1973.

51. **Terzaghi, M. and Little, J.,** Repair of potentially lethal radiation damage in mammalian cells is associated with enhancement of malignant transformation, *Nature (London),* 253, 548, 1975.

52. **Reznikoff, C., Brankow, D., and Heidelberger, C.,** Establishment and characterization of a cloned line of C3H mouse embryo cells sensitive to postconfluence inhibition of division, *Cancer Res.,* 33, 3231, 1973.

53. **Sacher, G. and Hart, R.,** Longevity, aging and comparative cellular and molecular biology of the house mouse, *Mus musculus,* and the white-footed mouse, *Peromyscus leucopus,* in *Genetic Effects on Aging,* Bergsma, D. and Harrison, D., Eds.,; *Birth Defects: Original Article Series,* 14, 71, 1978.

54. **Petursson, G., Coughlin, J., and Meylan, C.,** Long-term cultivation of diploid rat cells, *Exp. Cell Res.,* 33, 60, 1964.

55. **Jackson, J., Sanford, K., and Dunn, T.,** Neoplastic conversion and chromosomal characteristics of rat embryo cells *in vitro, J. Natl. Cancer Inst.,* 45, 11, 1970.

56. **Freeman, A., Igel, H., and Price, P.,** Carcinogenesis *in vitro.* I. *In vitro* transformation of rat embryo cells: correlations with the known tumorigenic activities of chemicals in rodents, *In Vitro,* 11, 107, 1975.

57. **Meek, R., Bowman, P., and Daniel, C.,** Establishment of rat embryonic cells *in vitro, Exp. Cell Res.,* 127, 127, 1980.

58. **Kontermann, K. and Bayreuther, K.,** Experimental evidence for a unifying concept of the molecular mechanisms of the cellular aging and the cellular neoplastic transformation, *Acta Gerontol.,* 8, 411, 1978.

59. **Kontermann, K. and Bayreuther, K.,** The cellular aging of rat fibroblasts *in vitro* is a differentiation process, *Gerontology,* 25, 261, 1979.

60. **Todaro, G., Nilausen, K., and Green, H.,** Growth properties of polyoma virus-induced hamster tumor cells, *Cancer Res.,* 23, 825, 1963.

61. **Todaro, G. and Green, H.,** Serum albumin supplemented medium for long term cultivation of mammalian fibroblast strains, *Proc. Soc. Exp. Biol.,* 116, 688, 1964.

62. **Gotlieb-Stematsky, T., Yaniv, A., and Gayeth, A.,** Spontaneous, malignant transformation of hamster embryo cells *in vitro, J. Natl. Cancer Inst.,* 36, 477, 1966.

63. **Diamond, L.,** Two spontaneously transformed cell lines derived from the same hamster embryo culture, *Int. J. Cancer,* 2, 143, 1967.

64. **Williams, J.,** Cell density dependence of senescence in embryonic Syrian hamster cells, manuscript in preparation.

65. **Stoker, M. and MacPherson, I.,** Syrian hamster fibroblast cell line BHK21 and its derivatives, *Nature (London),* 203, 1355, 1964.

66. **Yerganian, G. and Leonard, M.,** Maintenance of normal *in situ* chromosomal features in long-term tissue cultures, *Science,* 133, 1600, 1961.

67. **Hay, R. and Strehler, B.,** The limited growth span of cell strains isolated from the chick embryo, *Exp. Gerontol.,* 2, 123, 1967.

68. **Ponten, J.,** The growth capacity of normal and Rous-virus-transformed chicken fibroblasts *in vitro, Int. J. Cancer,* 6, 323, 1970.

69. **Weissman-Shomer, P. and Fry, M.,** Chick embryo fibroblasts senescence *in vitro:* pattern of cell division and lifespan as a function of cell density, *Mech. Aging Dev.,* 4, 159, 1975.

70. **Nielsen, P. and Ryan, J.,** Cumulative population doublings as the determinant of chick cell lifespan *in vitro, J. Cell Physiol.,* 107, 371, 1981.

71. **Lima, L. and Macieira-Coelho, A.,** Parameters of aging in chicken embryo fibroblasts cultivated *in vitro, Exp. Cell Res.,* 70, 279, 1972.

72. **Macieira-Coelho, A. and Lima, L.,** Aging *in vitro:* incorporation of RNA and protein precursors and acid phosphatase activity during the lifespan of chick embryo fibroblasts, *Mech. Aging Dev.,* 2, 13, 1973.

73. **Ryan, J.,** The kinetics of chick cell population aging *in vitro, J. Cell Physiol.,* 99, 67, 1979.

74. **Kaji, K. and Matsuo, M.,** Ageing of chick embryo fibroblasts *in vitro.* II. Relationship between cell proliferation and increased multinuclear cells, *Mech. Aging Dev.,* 8, 233, 1978.

75. **Kaji, K. and Matsuo, M.,** Aging of chick embryo fibroblasts *in vitro.* III. Polyploid cell accumulation, *Exp. Cell Res.,* 119, 231, 1979.

76. **Brock, M. and Hay, R.,** Comparative ultrastructure of chick fibroblasts *in vitro* at early and late stages during their growth span, *J. Ultrastruct. Res.,* 36, 291, 1971.

77. **Ryan, J.,** Effect of different fetal bovine serum concentrations on the replicative life span of cultured chick cells, *In Vitro,* 15, 895, 1979.

78. **Rein, A. and Rubin, H.,** Effects of local cell concentrations upon the growth of chick embryo cells in tissue culture, *Exp. Cell Res.,* 49, 666, 1968.

79. **Balk, S., Poliment, P., Hoon, B., LeStourgeon, D., and Mitchell, R.,** Proliferation of Rous sarcoma virus-infected, but not of normal, chicken fibroblasts in a medium of reduced calcium and magnesium, *Proc. Natl. Acad. Sci. U.S.A.,* 76, 3913, 1979.

80. **Verger, C.,** Proliferation and morphology of chick embryo cells cultured in the presence of horse serum and hemoglobin, *In Vitro,* 15, 587, 1979.

81. **Carney, D. and Cunningham, D.,** Initiation of chick cell division by trypsin action at the cell surface, *Nature (London),* 268, 602, 1977.

82. **Stenkvist, B.,** Long-term cultivation of human and bovine fibroblastic cells morphologically transformed *in vitro* by Rous sarcoma virus, *Acta Pathol. Microbiol. Scand.,* 67, 67, 1966.

83. **Lithner, F. and Ponten, J.,** Bovine fibroblasts in long-term tissue culture: chromosome studies, *Int. J. Cancer,* 1, 579, 1966.

84. **Nelson-Rees, W., Owens, R., Arnstein, P., and Kniazeff, J.,** Source, alterations, characteristics and use of a new dog cell line (Cf2Th), *In Vitro,* 12, 665, 1976.

85. **Rhim, J., Park, D., Arnstein, P., and Nelson-Rees, W.,** Neoplastic transformation of canine embryo cells *in vitro* by N-methyl-N'-nitro-N-nitrosoquanidine, *Int. J. Cancer,* 22, 441, 1978.

86. **Davidenas, J., Fritzler, M., Giblak, R., and Church, R.,** Characteristics of a new rapidly proliferating rabbit cell line, *In Vitro,* 9, 223, 1974.

87. **Stanley, J., Pye, D., and MacGregor, A.,** Comparison of doubling numbers attained by cultured animal cells with life span of species, *Nature (London),* 255, 158, 1975.

88. **Rhim, J., Nelson-Rees, W., and Essex, M.,** Transformation of feline embryo cells in culture by a chemical carcinogen, *Int. J. Cancer,* 24, 336, 1979.

89. **Earley, E., Petricciani, J., Wallace, R., and McCoy, D.,** The influence of media on the stability of the diploid karyotype of cell lines from an African green monkey, *In Vitro,* 9(Abstr.), 375, 1974.

90. **Rohme, D.,** Evidence for a relationship between longevity of mammalian species and lifespans of normal fibroblasts *in vitro* and erythrocytes *in vivo, Proc. Natl. Acad. Sci. U.S.A.,* 78, 5009, 1981.

91. **Dearfield, K. L., D'Arpa, P., and Williams, J.,** Comparative assays of DNA damage and repair associated with rat embryonic fibroblasts in culture, *Environ. Mutag.,* 4, 386, 1982.

92. **Dell'Orco, R. T. and Whittle, W. L.,** Evidence for an increased level of DNA damage in high doubling level human diploid cells in culture, *Mech. Aging Dev.,* 15, 141, 1981.

93. **Robertson, J. B.,** personal communication.

GENOME INTERACTIONS IN THE PATHOLOGY OF AGING PROTOZOA

Joan Smith-Sonneborn

INTRODUCTION

The protozoans represent a diverse evolutionary spectrum of eukaryotic cells which exhibit enormous variation in longevity. The lifespan duration, the nucleocytoplasmic influences, and senescent changes of the most thoroughly studied species are highlighted. Senescence in unicells and multicells is compared.

Perspectives

The lifespans of unicellular organisms range from days to apparent immortality.[1] The evolutionary distance separating protozoa, even within one class, the Ciliata, has been estimated to be as great as the distance between *Salmo* (bony fish) and man.[2] It is, therefore, not surprising that diversity is found both in the strategy for survival and the length of life required to maintain the myriad of species. The protozoa represent a unique position as eukaryotic unicellular forms between the prokaryotic and multicellular organisms, and as such represent a research resource available[3] for comparative studies of lifespan determination at the cellular level.

Since the time for evolution of eukaryotes was 0.6 to 2 billion years, and multicellular evolution occupied only 500 million years,[4,5] it seems unlikely that multicells would shed regulatory mechanisms so dearly won through selection during evolution.[6] It is not necessary that diverse organisms employ fundamentally similar mechanisms in identical ways; these mechanisms may be meshed in unique variations to achieve different biological objectives.[7]

Aging is a multistep, multifaceted phenomenon, with senescence occurring at all levels of organization in multicellular systems. It is certain that interactions among these levels can occur at the onset of senescence.[8] Nevertheless, the similarity of molecular structure and function shared by all eukaryotic cells would indicate that some uniformity in functional changes can be expected. A universal aging mechanism is not a prerequisite for comparisons of aging in protozoa and man. Rather, protozoans should reveal the various diverse strategies employed by cells to maintain and regulate cell lifespan and should shed light on similar mechanisms operative or inducible in cell types of higher organisms.

Although unicellularity is a dominant feature of the protozoans, some multicellularity is found in collective ameba, colonial flagellates, and some ciliates. See Grell[9] for the general biology of protozoans, and Coleman[10] for review of colonial green flagellates.

The protozoans included in this report are restricted to free-living forms, although very interesting life cycle changes are known to occur in the parasitic species in response to the age of the host.[11] Host-cell interactions in these species complicate interpretation of intrinsic life cycle events. The haploid protozoans are, in general, excluded from this review since they normally do not age,[12] but a brief consideration of colonial forms, which exhibit finite and infinite lifespan cell types, is included.

Various authors have used the term "aging" in the protozoa as an umbrella to encompass any condition that results in cell death or death of a culture. The term aging must be used more precisely if meaningful comparisons are to be attempted. "Clonal aging" refers to the decreased probability that a cell will give rise to viable progeny at the next cell division as the time (measured in cell divisions) increases since the last fertilization process. The environment in which the age deterioration is observed must not be effectively different from conditions that support indefinite multiplication (if repeated fertilizations provide new generations). "Individual aging" is expressed by the decreased function, proliferative capacity, and death of single identifiable cells, cell types, or cell lineages within the clone. Individual aging can occur at different rates for different cell types within the clone. "Cultural aging" is the termination of a cell culture resulting from prolonged stationary phase and may be accompanied by accumulation of toxic products and nutritional deprivation. Death of the culture will result regardless of the biological age of the cells, and, because of its external causation (or determination), is fundamentally different from intrinsic clonal and individual aging. Environmentally induced effects in epigenotypes (pattern of gene expression) are known to influence intracellular processes (see Alteration of Lifespan) but should not be confused with the distinctly different event that occurs when a culture "ages" by running out of food.

Since starved or malnourished cells exhibit changes similar to those seen in aged cells, there is the theoretical possibility that "aging" is due to unfavorable external conditions and that all cells would be immortal if supplied with the "proper" growth medium. See Jennings[13] for a thorough exploration of protozoan aging vs. cultural conditions. Diet-induced immortality is still an elusive dream, but it is still possible that increased longevity could result from certain external conditions.

Clonal aging is widespread among ciliates, though extensive variation in lifespan durations are documented (see Aspects of Clonal Aging). Individual aging is seen in cell types within the ciliates as well as within the immortal colonial flagellates (see Aspects of Individual Aging).

A clone is initiated at fertilization as a single cell and includes all the asexual division products. The clone is a multicellular unit with the member cells dispersed in space. This is in contrast to colonial forms in which the cell products can remain together. The members express age-dependent changes in phenotype and pass through immaturity (when cells cannot mate), senescence (when the probability that a given cell will give rise to viable progeny dramatically declines), and death.[13-15] A prerequisite for clonal aging studies is thorough investigation of the life cycle of the organism under consideration. The "Paramecium Methuselah" strain, once thought to be immortal,[16,17] was discovered to be a succession of clones undergoing autogamy or self-fertilization.[14]

In *Paramecium,* aging was established as an intrinsic process since the same medium could support infinite vigorous growth of successive lines of young fertilized cells. In the absence of fertilization, a clone in the same medium terminated with a species-specific lifespan.[14] Some interesting aspects of aging of individual cells within the clone in *Paramecium,*[18] *Tokophrya,*[19,20] and Volvacaceae[10] will also be considered as examples of intraclonal variations in lifespan duration. Precise individual aging studies should include consideration of the clonal age of the individual.

Environmentally induced changes in infinite cell lines or infinite lifespan organisms will be reviewed, since these changes may represent interference with lifespan regulatory mechanisms. Although the genotype of the organism provides the heritable cellular component, the cytoplasm and external conditions constitute the environment for stable determinations of gene expression,[21] which could influence longevity (see Alteration of Lifespan). Cell composites of young nuclei and old cytoplasm, or young cytoplasm and old nuclei can be constructed to assess the role of each in the function of the whole cell.[22-26] Constructs of old and young cells can be made and the respective role of cell parts in the cellular aging process determined (see Lifespan Determinations). The similarities and differences in aging of unicells vs. multicells will be reported (see Comparative Aspects of Intracellular Sites For Aging in the Protozoa).

General Biological Background

Clonal lifespan studies are reported here exclusively for the ciliates. These ciliates show mating types and can undergo meiosis and fertilization during the sexual phase of the life cycle. Reproduction in the asexual phase is normally by binary fission, but certain ciliates reproduce by budding. Extensive information is available on clonal life cycle events in the *P. aurelia* complex,[14,15,27-39] *P. bursaria*,[13,39-44] *P. caudatum*,[44-48] *Euplotes*,[15,49-51] *Tokophrya*,[19,52] and *Tetrahymena*.[2,53-55] Emphasis of studies in *Stylonychia* has been on molecular events and morphological changes during nuclear differentiation.[56-60]

Normally, the ciliates contain two kinds of nuclei, the micronucleus or germ line, and macronucleus or somatic line. Although both occupy a common cytoplasm, delegation of function to the different nuclei is maintained. The micronuclei are the repository for genetic information for the offspring after fertilization and are inactive transcriptionally during the asexual vegetative binary fission cycle,[61-65] though some activity is found.[66] Like the sex cells of higher organisms, the micronuclei undergo meiosis to generate gamete nuclei which function during fertilization.[67] The micronuclear gametes fuse to form a zygote nucleus, whose division products differentiate into the filial micronuclei and macronuclei for the new generation. The micronucleus is analogous to the immortal germ line of higher organisms; it is the immortal component passed to future generations representing the continuum of the species.[68] The ancestral somatic nucleus is normally discarded, but can function during development of the filial nucleus.[69-71] The RNA produced by the ancestral macronucleus represents a "maternal" cytoplasmic environment reminiscent of stored RNA in oocytes of higher organisms, but, unlike stored RNA, it is still being produced.

The macronucleus differentiates from division products of the zygote nucleus and the position of the developing nuclei within the cell determines the differentiation.[53,68] If the new macronucleus fails to form or is lost, an ancestral macronuclear fragment can regenerate to provide a macronucleus for the progeny cells.[15]

The strategy for differentiation of micronuclei and macronuclei from the zygote varies widely in the ciliates. Chromosome polytenization and chromatin diminution followed by rapid resynthesis have been observed in lower ciliates,[72] as well as in the hypotrichs.[56-60,73,74] *P. bursaria* also show certain strains which amplify segments of chromosomes while the rest undergo reabsorption or condensation.[75] Certain *Tetrahymena* and *P. tetraurelia* seem to retain most of the micronuclear genome and large DNA structure in the macronucleus, though differences are found[66,76,80] and the question of DNA size must still be critically examined.

Methodology — Clonal Lifespan Determination

Clonal lifespan is defined as the length of life measured in days or cell divisions which elapse from the origin of the clone at fertilization until clonal death. After fertilization, the products of the single cell are maintained either in daily or serial reisolation lines, or mass cultures. In either case, the cell line is subcultured and fed. The total number of successive cell divisions or days, from origin until the representatives of the clone are no longer capable of division and die, is determined as the lifespan duration. The specific procedures used vary among studies and contribute to differences reported for a given species; some procedures are designed to favor selection of the most vigorous representatives of the clone. Some form of serial reisolations has been used to determine the lifespan duration or time of onset of life cycle events in the *P. aurelia* complex,[14,26,81,82,83] *Euplotes*,[23,51] *P. caudatum*,[46,47] *P. bursaria*,[43,44] *Tokophrya lemnarum*,[19] and *Tetrahymena*.[7] Mass culture subcultivation has been used for lifespan duration estimates in *P. bursaria*,[13,84] *P. multimicronucleatum*,[15] *Stylonychia*,[56-60] and *Tetrahymena*.[85]

ASPECTS OF CLONAL AGING: CILIATES

P. aurelia Complex

P. aurelia was first divided into varieties, then syngens and more recently into species, to classify stocks (the progeny of individual cells collected from many parts of the world) into groups which could mate and yield viable progeny.[38] Although there are 14 species in the *P. aurelia* complex, only *P. primaurelia, P. biaurelia,* and *P. tetraurelia* have been utilized for clonal aging studies (Table 1). *P. tetraurelia* has the shortest lifespan and is the most thoroughly investigated. Since all the species undergo autogamy, genetic markers are required to unambiguously detect and eliminate these sublines.[14,81,82] In contrast to other paramecia (*P. caudatum, P. bursaria,* and *P. multimicronucleatum*), the *P. aurelia* complex exhibits short immaturity and longevity (Table 1). The studies of Sonneborn[14] established the intrinsic finite lifespan of clones of these cells in a constant external environment. The clones of paramecia show ordered sequential changes in cellular properties passing through immaturity, maturity, and senescence, and measure time in cell divisions.[87] The age-associated traits are summarized in Table 2.

The capacity for fertilization to provide offspring with a full vigorous lifespan declines with increased parental age.[25,26,93-98] The decline in progeny survival is due both to age-induced damage in the germ line nucleus and the hostile cytoplasmic environment.[25,26,39,89-91,97,104] In older cells, the meiotic process occurs in a differentiated aged cytoplasm, and the cumulative effect of nuclear and cytoplasmic damage is a reduced lifespan in the survivors.

P. caudatum

P. caudatum has been utilized recently for clonal aging studies[45-48] and information is available for other aspects of their life cycle.[111-113] *P. caudatum* can undergo selfing (mating within a clone) and extreme care must be taken to eliminate such clones from lifespan determinations.[47] The capacity of any strains of *P. caudatum* to undergo autogamy normally has not been established.[15] The lifespan of *P. caudatum* is intermediate between the short lifespan of species in the *P. aurelia* complex so far examined and the long-lived *P. bursaria* and *P. multimicronucleatum* (Table 1).

P. bursaria

Information on the breeding systems and life cycle events in *P. bursaria* was provided by Jennings,[117] and later by Siegel.[39,42-44] The establishment of periods of immaturity, maturity, senescence, and death as a normal expression of clonal life was made.[84] Aged clones showed high mortality and most or all exconjugants died.[118,119] There was reduced capacity to complete the fertilization process. Jennings observed the phenomena that inbreeding (matings between sibs, sister clones from the same parent clones) resulted in high mortality, but self-fertilization (matings between cells which differentiated complementary types within the same clone) showed high viability. He concluded that there is some biological relation that prevents self-fertilization from resulting in high mortality.[118,119] Genetic erosion, as the sole cause of mortality after inbreeding by forming lethal homozygotes, is difficult to reconcile with high viability of selfers. The data still defy comprehension but indicate that intraclonal and interclonal nucleocytoplasmic interactions can be different. There are biological mechanisms to protect against expression of some age damage and nucleocytoplasmic incompatibility.

The *P. bursaria* studies revealed that different clones exhibit enormous variability in their ability to undergo conjugation successfully even at similar ages.[13,119] Mortality in aged clones mimicked that of interspecies crosses, and anomalies were observed in both, prior to exchange of nuclei.[119-121] A role of aged and "foreign" cytoplasm was suggested as contributory to variability in survival after conjugation. *P. bursaria* is among the long-lived protozoa.

P. multimicronucleatum

P. multimicronucleatum has not yet been used extensively for lifespan determinations, but information is available on mating type inheritance and diurnal periodicity of mating

Table 1
LIFE CYCLE EVENTS IN *PARAMECIUM*

P. aurelia complex

	P. primaurelia	P. biaurelia	P. tetraurelia	P.caudatum	P. bursaria	P. multimicronucleatum
References	14, 15	14, 15	14, 15, 81, 82	46—48	15, 83	15, 86
Mating system	Binary	Binary	Binary	Binary	Multiple	—
Sexual process						
Conjugation	+	+	+	+	+	+
Selfing	+	+	+	+	+	+
Autogamy	+	+	+	+	0	?
Immaturity period						
Fissions (days)	0—35 (0—9)	0—27 (0—7)	0 (0)	0—58 —	(12 —1000)	60 + (80)
Maturity						
Fissions (days)	40—70 (10—20)	10—40 (3—13)	25 (6)		(1000)	90 + (3600)
Lifespan						
Fissions (days)	350 (130)	303 (160)	186—325 (40—60)	600—700 (200)	(3000)	(5400)

Note: Conjugation is the mutual exchange of gamete nuclei during mating of complementary mating types. Selfing refers to conjugation within a clone. Autogamy is the self-fertilization process which can occur within one cell; since only one haploid gamete nucleus is retained, and undergoes duplication and then fusion to form the zygote nucleus, the result is homozygosity. A binary system of mating is defined as only two complementary mating types within a species. A multiple mating system refers to a series of mating types that are capable of interbreeding with one another. The duration of lifespan was determined with genetic markers and daily reisolations for the *P. aurelia* complex. *P. caudatum* lifespan estimates were made from serial reisolations, and *P. bursaria* and *P. multimicronucleatum* lifespan estimates were made from serial cultivation of mass cultures.

+ , observed, 0, not observed.

Table 2
SENESCENT CHANGES IN THE *P. AURELIA*
COMPLEX

Changes	Ref.
Nuclear Abnormalities	
Aberrations increased	14, 25, 88—91
Variable numbers increased	91, 92
Progeny viability decreased	25, 26, 93—98
Somatic line macronucleus	
Invaginations increased	99
Fused chromatin increased	99
Nucleolar volumes decreased	100
DNA amount decreased	101, 102
Abnormalities increased	39, 99, 103
Cytoplasmic changes	
Toxicity increased	24, 26, 104, 105
Lysosomal activity increased	105
Abnormal mitochondria increased	105
Age pigments increased	105
Food vacuoles decreased	106
Functional changes	
Fission rate decreased	14, 95, 98, 107
Viability declined	14, 81, 95
DNA synthesis decreased	106
RNA synthesis decreased	101
Transcription decreased	101
Ribosomes decreased	109
UV sensitivity increased	110

type expression.[120] The morphological characteristics of micronuclei of *P. multimicronu-cleatum* are not identical with but similar to those of the *P. aurelia* complex.[15] The *P. multimicronucleatum* are reported to have a lifespan of at least 5400 days were all ami-cronucleated when examined.[15] Reduction in the number (from four to one) and eventual loss of micronuclei is a characteristic of aging in *P. multimicronucleatum* and is not associated with marked decline of fission rate.[86,123]

Euplotes

Species of *Euplotes* are widely distributed in marine and fresh water environments. The marine species include, for example, *E. vannus*,[124] *E. minuta*,[125-127] *E. crassus*,[128] *E. cris-tatus*,[129] and *E. woodruffi*,[125] which all exhibit multiple mating type systems (the existence of a number of interbreeding types).[130] The fresh water species of *E. eurystomas*[131] have a binary system (only two complementary types); *E. woodruffi*[132,133] exhibits autogamy, and *E. patella*[23,50] and *E. aediculatus*[134] show multiple mating systems.

The lifespan durations in *Euplotes* range from 235 to 1300 fissions (Table 3). Lifespan durations have also been reported for *E. crassus*[128] (1000 fissions) and for *E. minuta*[135] (500 to 700 fissions). *E. minuta* show strain-specific ability to undergo autogamy (self-fertiliz-ation), and strain differences were found to be due to presence or absence of a single gene.[127] Both *E. crassus*[128] and *E. patella* manifest age-dependent intraclonal mating or selfing. *E. minuta* and *E. patella* also exhibit age-dependent cell surface changes with age; the former species shows reduction in ciliary row numbers,[135] and the latter exhibits an increased number of bristles.[23]

Different species of *Euplotes* maintain different options for sexual processes, i.e., auto-gamy, conjugation, and old age selfing to assure that there is a mate, providing a sexual process to "rejuvenate" the clone.

Table 3
LIFE CYCLE EVENTS OF *EUPLOTES*

	E. woodruffi		*E. patella*	
	Fresh water	Marine	Diploid	Haploid and amicronucleates
References	132, 133	51, 130	23, 48	23, 48
Mating system		Multiple	Multiple	Multiple
Sexual processes				
Conjugation	0	+	+	+
Selfing	0	+	+	0
Autogamy	+	0	0	0
Immaturity period				
Fissions (days)	0—60 (30)	48—149 (24—74)	163—189 (150—350)	58—75 (60—75)
Maturity				
Fissions (days)	60—208 (30—104)	231—232 (120—140)	700 (688—780)	377—442 (365—410)
Lifespan				
Fissions (days)	208A (100)	235—442 (120—240)	1300 (1900)	600 (900)

Note: The estimates of lifespan duration in all these *Euplotes* were made using serial isolations (isolation of single cells at 2 to 3 day intervals into fresh food after fertilization). The lifespan of the fresh water *E. woodruffi* is not the maximal lifespan since autogamy (self-fertilization) occurred in all clones when they entered old age.[132] Since selfing (mating within the same clone) is found in *E. patella*, care was taken to observe the cultures for any mating pairs.[48]

+, observed, 0, not observed.

In *E. patell*[a], the lifespan of amicronucleate and haploids was half that of the diploids.[23] The amicronucleates and haploids arose from abnormal conjugants which failed to form normal diploid nuclei. It would be of interest to determine the macronuclear DNA content of the shorter lifespan variants to assess whether the functional macronucleus also showed reduction in genetic material. No difference in macronuclei DNA content has been found in haploid, diploid, or triploid micronucleated *Tetrahymena*,[136] but extreme caution must be exercised in any extrapolation from *Tetrahymena* to *Euplotes*. Aged strains of both *E. patella*[23] and *E. woodruffi*[51,133] show loss of viability in progeny after conjugation (see Lifespan Determinations for old-young crosses in *E. patella* and *E. woodruff*).

Other Hypotriches

The lifespan of *Stylonychia pustula* was found to be 316 fissions, or 133 days.[13] *Stylonychia mytilus* has exhibited age deterioration at 800 to 1000 fissions and gene loss was considered an attractive hypothesis for aging in these ciliates.[137] Aging in *Stylonychia notophora* has also been observed.[132] *Oxytricha bifaria*, a fresh water species, manifests multiple mating type systems and selfing though no autogamy has been observed.[139] The immaturity period lasted from a few months to 2 years, with maturity spanning 2 to 3 years. Some stocks of *O. bifaria* died while others showed no sign of loss of vigor.

Tokophryn

Clonal aging studies in *Tokophrya* indicate a long-lived strain which shows age-dependent deterioration.[19] (See Under Aspects of Individual Aging for individual cellular aging in *Tokophrya*.) *Tokophrya lemnarum* was suitable for clonal lifespan studies after discovery of mating types[52] and characterization of its binary mating type system. The life cycle was initiated by conjugation and followed by estimating the number of fissions (buds) that could be formed after conjugation. The estimates included the succession of adult-bud transitions throughout the life cycle (see Colgin-Bukovsan[19] for thorough explanation of life cycle). No autogamy was found though there is a suggestion that autogamy may occur in *Suctoria* unable to find a mate. Immaturity was 0 to 60 cell generations and maturity lasted over 800 fissions. Senescent characters included increased numbers of pairs that fused completely and decreased survival after conjugation.[19]

Tetrahymena — Finite Lifespan

The *Tetrahymena pyriformis* species complex illustrates enormous molecular diversification despite morphogenetic uniformity. The numerous known species have remarkably different DNA nucleotide compositions, sequences,[140-141] and isozymic patterns.[142-144]

The various biologically distinct species have been separated and renamed.[145] *Tetrahymena thermophila* (formerly *T. pyriformis*, sygnen 1) presents an example of the phenomenon of "inbreeding deterioration".[146] (See recent review of aging in *Tetrahymena*.)[55] Conjugation can result in: (1) exconjugants which die within the first fission, or within 10 fissions postconjugation, (2) nonconjugants in which a new macronucleus does not develop, or (3) accelerated onset of micronuclear anomalies. The latter group are considered "prematurely aged clones". Normal clones may not show micronuclear anomalies until after 500 to 1500 fissions.[55] The prematurely aged lines exhibit cell death at 160 to 200 fissions.[81] Not all strains, however, show the same rate of loss of viability.[55,147] When semi-amicronucleated strains (macronuclei with only a few chromosomes) (SA) mate the SA member of the pair fails to function but receives a normal haploid gamete nucleus from its partner. The haploid gamete nucleus undergoes endo-reduplication to form a normal diploid nucleus. Macronuclear differentiation fails to occur, and a heterokaryon with normal micronuclei and ancestral macronucleus is formed. The "normal" micronuclei undergo erosion characteristic of the SA strain.[85] If, however, the same normal micronucleus develops in a normal cell with a new macronucleus from the micronuclear zygote, the homokaryon does not show early micronuclear deterioration.[85] Micronuclear erosion in *Tetrahymena*, therefore, is manifested by death at conjugation and macronuclear retention (abortion of the new macronucleus, with maintenance of the ancestral macronucleus). Recent studies indicate that although micronuclear deterioration seems independent of nutritional maintenance, temperature can affect the timing of the phenomenon.[148] Evidence for an age-correlated increase in nuclear damage was not evident, though breeding deterioration was found to occur at constant specific rates for different inbred strains. A mutational basis for aging was considered consistent with the data. Certain other genotypes of *Tetrahymena* do seem to show age-dependent micronuclear deterioration.[85]

Tetrahymena — Infinite Lifespan

The absence of somatic deterioration in certain strains of *Tetrahymena*, especially those that lack micronuclei, implies that somatic immortality is possible.

Certain micronucleated *Tetrahymena* exhibit timed orderly expression of genotype but not aging.[55] Models for the structure of the *Tetrahymena* nucleus as an explanation for its genetic behavior are beyond the scope of this review (see References 55, 148—150). Of particular interest in the present review is the observation that members of the clone show locus-specific timed assortment of heterozygous genes in *Tetrahymena*. The macronucleus becomes functionally homozygous. The mechanisms for allelic assortment are not understood, but temporal segregation of alleles is observed,[150] and the possibility of "purification" of the aging macronucleus has been suggested.[151] Damaged DNA could be eliminated from the somatic nucleus by "assortment". A daughter cell receiving genetic damage would die or be selected against, while the other daughter is damage free. The relatively transcriptionally inert micronucleus would not be subjected to selective pressures and could accumulate damage.[151] A caveat to this attractive hypothesis is that it has not yet been subjected to experimental analysis, and different strains, because of their evolutionary diversity, need not age by the same mechanism.

In immortal micronucleated strains, several different genotypic combinations obtained by inbreeding result in finite lifespans.[123] In nature, the multiplicity of mating types and long immaturity periods (allowing members of a clone to become separated from one another) would favor mating with a stranger (outbreeding). Inbreeding-induced deterioration of the germ line, such as observed in the laboratory, may be rare in nature and, therefore, selection for survival under an inbreeding economy has not evolved. The micronucleus, in contrast with the macronucleus, is relatively inert and nonfunctional,[140,142,152,153] but can still cause mating inability and death when it is lost.[53,142,146,154] Oral morphogenesis may depend on events coordinated with micronuclear division.[155] The association of micronuclei and oral development may be important in life maintenance, but it is premature to causally relate the two.[151] Although amicronucleate strains are known and are immortal, these strains may have been amicronucleate from their origin.[55] Amicronucleates, if they arise from micronucleates, may uncouple dependence on survival from the micronucleus.

The ability of *Tetrahymena* to renew its oral structure during vegetative growth, in contrast to the requirement for fertilization for oral apparatus renewal in *Paramecium*, has been cited as a possible contributor to differences in their lifespan duration.[15]

Since damage to nonrenewable cell surface structures that are essential for life maintenance can surely lead to death of an organism, the *Tetrahymena* structural repair capacity clearly favors longer life.

ASPECTS OF INDIVIDUAL AGING

A. *Tokophrya* — Ciliate

In contrast to clonal aging, the fate of individual members within the clone provides

Table 4
DIFFERENCES IN YOUNG AND OLD INDIVIDUAL *TOKOPHRYA*
ADULTS[157,158]

Character	Young adults	Old adults
Size	25—40	50
Tentacles	50	15
Buds produced	12 per day	1—2 per day or none
Mitochondria	Rich in microvilli	Reduced microvilli
	Plentiful numbers	Reduced numbers
	Randomly dispersed	Concentrated at cell periphery
Age pigments	Absent	Plentiful
Endoplasmic reticulum	Rich in particle-lined and smooth vesicles	Reduced vesicle numbers
Macronucleus	5-μm diameter	15—20-μm diameter
	Spherical shape	Irregular shape
		Invaginations of membrane fragmentation
	50 Chromatin bodies	100 chromatin bodies, some with hollow centers

Note: Distinction between young and old adults was monitored by their position in a tube culture. The young organisms settled near the surface and the old were in the remainder of the culture. In contrast to clonal aging studies,[19] the age of the "young" adult (as defined by the number of successive buds and adults that were formed since the previous fertilization) was not reported.[157,158] A "young" adult was one that had recently metamorphosed from a bud.

insight into intraclonal variations and life maintenance strategies. *Tokophrya*, since it reproduces by budding and exhibits striking differences between the parent cell and progeny, offers an opportunity to monitor individuals within the clone. Both *T. infusionum*[20,156-159] and *T. lemnarum*[19] have been exploited for "individual" aging studies. The mature *Tokophrya* produces a swimming bud which will metamorphose into the stalked adult. The adult lives 10 days in *T. infusionum*[157] and 16 days in *T. lemnarum*.[19] After formation of the adult, there is a short immaturity period when buds are not produced, a mature period of bud production, and a "senescent" period when no buds can be formed. The adult *Tokophrya* is a "baby machine", and is not even provided with an organelle for excretion of solid wastes.[159] The opening to the outside releases the new bud. As noted below (see Nutrition), overfeeding will accelerate "aging", and intermittent starvation will prolong the life of the adult. The adult is adapted for a somewhat sparse food supply.[19] The senile changes which accompany the demise of the adult are listed in Table 4. The loss of the organelle for feeding surely is the proximal cause of death, but the accumulation of toxic wastes appears to precede loss of organelle function. The function of the individual adult is apparently to reproduce and is not "designed" to survive long past that delegated task. The lifespan of young vs. old adults (when the age is measured in fissions since fertilization) has not yet been studied and *Tokophrya* offers an interesting model for such clonal-intraclonal variations.[19]

Paramecium — Ciliate

Although *Tokophrya* shows unique morphologies among asexual reproducing members of the clone (buds and stalked adults), paramecia do not. The division products of an asexually dividing paramecia, however, are not equal. In the formation of organelles to produce two daughters from one cell, the anterior cell (the proter) retains the old mouth, but the posterior cell (the opisthe) receives the new mouth. The fission rate was higher for the posterior line with the new oral apparatus.[18] The hypothesis was that the old gullet could not be repaired should accidental damage occur, thereby reducing the viability of these members of the clone.

The study offers a contributing element, intraclonal variation, which could influence lifespan duration. Sublines with defective feeding organelles would then be less vigorous than sisters with superior organelles.

Volvocaceae — Colonial Flagellates

Colonial green flagellates are algae by virtue of their pigments, and protozoans because the intricate architecture of their colony is derived on a chlamydomonad building block (see References 10, 160). The only diploid cell in the volvocaceae life cycle is the zygote, which can remain dormant for years. Meiosis occurs in the germination to form colonies. Colonies can be propagated indefinitely by vegetative reproduction. In the majority of genera all cells are capable of forming daughters, but some show differentiation of a proportion of generative cells (gonidia) and those which never again undergo division (somatic cells). The number of cells in a colony and the proportion of replicative cells is species specific.[10]

In the *V. carteri* group, potential reproductive cells are set aside at a particular stage by unequal division of certain cells.[161,162] These protozoans exhibit embryogenesis, including inversion stages and separation of somatic and germ line cells, and offer an elegant protozoan model for regulation of cell proliferation capacity. Mutants are available in *Volvox,* which have arisen spontaneously or have been chemically induced.[163,164] Of particular interest are developmental mutants, which fail to segregate replicating and nonreplicating cells. The array of mutants found suggests a genetic contribution to proliferative capacity involving several regulatory genes.[164]

The colonial flagellates represent, on the one hand, the antithesis of cellular aging by producing cells that can apparently multiply endlessly with no symptoms of aging. On the other hand, these flagellates show differentiation of cells that lack infinite reproduction potential. The stability and regulation of cell proliferation potential in these cells may shed light on mechanisms operative in higher forms.

ALTERATION OF LIFESPAN

Nutrition

Maintenance Diet — Spanned Amoeba

Experimental alteration of lifespan has been observed using the free-living *Amoeba proteum* and *A. discoides.*[165-167] *Amoeba* divide by binary fission and the population appears to be immortal if fed a growth diet. Exposure to a maintenance diet for 2 to 9 weeks, however, induced a finite lifespan that ranged from 30 days to 30 weeks. Four types of results were found as examples of different metastable states. The two major types of behavior found were denoted type A and B. Type A showed a stem line growth, in which a cell division gave rise to one viable and one nonviable product until all cells were dead. Type B lines gave rise to equally viable daughters at cell division, but expressed a finite lifespan.[166] The section, Lifespan Determinations describes the use of these finite cells in the construction of cell composites with immortal and finite cell parts.

Overfeeding Diet — Tokophrya

Tokophrya feeds on living prey by sucking the contents through long slender tubes (tentacles). The sessile *Tokophrya* must come into chance contact with the prey, and then seems to paralyze the victim. Although 2 to 5 tentacles are usually attached to one prey (*Tetrahymena,*) one tentacle will suffice. When given excess food, however, a *Tokophrya* can feed simultaneously with all 50 tentacles. As overeating continues, the tentacles are lost as in aged cells.[159]

It must be stressed that *Tokophrya* does not possess a specific organelle for removal of solid wastes. The most conspicuous structure in old and overfed cells is a large vacuole around the macronucleus.[159] Vacuoles take over most of the cytoplasmic area in the aged and overfed cells. The young, heavily feeding *Tokophrya* at the age of 4 days resembles a normal organism twice that age, and can assume giant proportions (120 times normal).[159]

Intermittent Starvation Diet — Tokophrya

In contrast to the results with overfeeding, intermittent starvation prolongs life and a 6-day-old (middle-aged) *Tokophrya* shows the fine structure of a young animal.[20,157] Excessive starvation, on the other hand, leads to their demise.

Starvation Diet — Miscellaneous Protozoans

The effect of stationary phase of the culture, including crowding and food depletion, on structural changes has been investigated and is not "aging". The similarities between such starved cells in culture and clonal aging in a constant environment have been noted. Some examples of organisms which have been subjected to cultural starvation are dinoflagellates,[170] *Euglena*,[171-174] *Tetrahymena*,[175] and *Ochromonas*.[176] The dinoflagellates have been found to exhibit abnormal nuclear morphology with fused chromatin bodies. *Euglena* cytoplasm contained heavily pigmented bodies and membrane fragments. *Tetrahymena* cytoplasm was vacuolized and showed increased numbers of mitochondria and lipid droplets. *Ochromonas* cytoplasm was characterized by increased lysosome activity, lipid vacuoles, and disorganization. The morphological changes seen in exogenously starved cells mimicked the structural changes seen in aged *Paramecium*, *Euplotes*, and *Tokophrya* adults (see Tables 2 and 4). The morphological changes in the macronuclei of aged paramecia[99] have also been noted in young starved cells.[100] The aged paramecia show no food vacuoles, even in the presence of excess food.[106] Starvation may not be a cause of aging, but it appears to be a proximal cause of death. Autolytic processes, stimulated at the termination of cell life, mimic those of young cells when starved.

Radiation-Induced Lifespan Changes

X-irradiation reduced the clonal lifespan and accelerated the decline in fission rate in *Paramecium primaurelia*. The average clonal lifespan of the controls was 326 fissions while at the highest dose (138,000 R), the lifespan was reduced to 249 fissions.[177] The significance of the differences observed were not subjected to critical analysis and should be investigated further. The radiation was delivered in fractional doses. At lower doses no effect was seen on lifespan as reported by previous investigators.[178] Sensitivity to X-rays was found to be age dependent.

UV damage in *P. tetraurelia* was reversed by photoreactivation.[107,180] UV irradiation reduced lifespan of *P. tetraurelia*, but when the UV treatment was followed by photoreactivation light, the lifespan duration was restored to the normal value or exceeded the control level at critical UV doses.[81] The cumulative effect of two cycles of UV irradiation and then photoreactivation at older ages resulted in a very significant lifespan extension[81] and these results have now been repeated.[83] Lifespan extension represented a 296% increase in remaining clonal lifespan, or a 27% increase in the entire lifespan. The aged clones normally exhibit age-dependent sensitivity to UV irradiation.[119] The cells pretreated with the UV and photoreactivation exhibited radiation resistance characteristic of younger cells.[81] Recent studies indicate that UV light-induced DNA polymerase may represent a repair enzyme which could contribute to the observed increased longevity of these cells.[181] If the amount of DNA damage induced by UV irradiation exceeds the capacity of cells to repair the lesions, the excessive damage can result in reduced lifespan. If the induced damage is repaired by the alternative system, photoreactivation repair,[182] any UV-induced increased residual repair may prevent some age damage and prolong clonal lifespan.[81] A longer lasting change in repair capacity, presently not understood, is believed to contribute to the observed results.

Carcinogen-Induced Lifespan Extension

Methylcholanthrene induced lifespan extension has been reported in *P. multimicronucleatum*.[183] Six control populations died after 7.5 and 8.0 years, while populations exposed to the carcinogen for 3.5 years and then transferred to growth medium died between 7.75 to 9.0 years. Unfortunately, all the studies were conducted in mass cultures and no attempts were reported to detect selfing in the respective populations. The extension, therefore, cannot be considered established.

LIFESPAN DETERMINATION: EFFECT OF YOUNG AND OLD CELL PARTS

There are essentially three methods to construct cells with different compositions of young and old cell components. These are (1) microsurgical introduction of cell parts into recipient cells, (2) exchange of sets of cell components during the fertilization process, and (3) unidirectional transfer of parts from one to the other during mating. Reassembly of cell components for aging research has been reviewed recently[24] and pertinent aspects are included here.

Acetabularia

This uninucleated, unicellular alga has been used to assess the function of cytoplasm in the determination of structure and function of this organism.[184] An "old" cell with a maximum diameter cap became "young" when its cap was removed. When a young rhizoid was grafted to an old stalk with a full-sized cap, the production of cysts which is an age-dependent event was earlier than when the reciprocal graft was made. Nuclear transfers indicated that cytoplasm from a young cell could affect the old nucleus within less than 10 days, and old cytoplasm induced the morphology of an aged nucleus on the young nucleus. The rejuvenation was slower than the aging.[184]

Amoeba

The studies have been conducted using spanned cells from *A. proteus* and *A. discoides*.[166] (Recall that Type A cells produced one viable and one nonviable cell at division. Type B cells showed both daughter viable, but with a finite lifespan.) When a normal nucleus was placed into the cytoplasm of a Type A or B cell, all exhibited Type B behavior. The puzzling question is why A and B types differ since both contain an A type nucleus and B type cytoplasm. Nutritional changes can induce heritable changes in cell lifespan. The heritable change resides both in the nucleus and cytoplasm, and either a nucleus or cytoplasm can induce finite lifespan in normal immortal recipients.

Paramecium

The various options for fertilization can involve unique sets of cell parts. For example, it is possible to monitor, and to some degree control, the amount of cytoplasm exchanged between mating pairs. Cytogamy (self-fertilization within both of the two paired cells), without exchange of germ nuclei, can be contrasted with pairs that undergo true conjugation (cross-fertilization). Cytogamy, therefore, allows a test of the impact of cytoplasm and membrane contact in the absence of nuclear exchange. Conjugation permits the assessment of nuclear exchange with various amounts of cytoplasm (with or without cytoplasmic bridges which persist for varying lengths of time).

In autogamy, self-fertilization occurs in the absence of any cell-cell contact. In any of the fertilization processes, the zygote nucleus normally forms the new filial somatic macronucleus. The ancestral macronucleus can regenerate forming a heterokaryon with an old macronucleus and young micronucleus if the filial macronucleus fails to form. With the use of genetic and cytoplasmic markers, Sonneborn and colleagues have identified each option and established the respective role of cytoplasm and nuclei in the expression of genotypes.[15,21,53] The techniques of nucleocytoplasmic composites have been utilized in lifespan studies.

The presence of dominant lethals and chromosomal aberrations in aged cells is indicated by a large fraction of pairs in young-old crosses which showed death in both members of the pair, though some death could have been due to limited exchange of aged cytoplasm.[25,26] Dominant detrimentals were considered contributors to the reduced vigor and lifespan of progeny in both the young and old mate in true exconjugants. Recessive lethal mutations were revealed by the autogamous F_2 generation from normal F_1 (while they were young) as the proportion of nonviable clones produced from both members of the young by old pairs of exconjugants.[25,26] Lethal mutations were due in large part to chromosomal aberrations since segregation disorder has been detected.[25,82,88] Persistence of heterozygosity for one marker in the expected homozygous F_2 and loss of a member of an allelic pair has also been found.[25,82] Since age-associated cytological anomalies, such as anaphase bridges, are known, there is a chromosomal basis for the irregularities observed.

Abnormalities, traceable to the action of old cytoplasm, were found in studies of "merogones", old cells whose nuclei fail to function but have young nuclei introduced during mating.[25] The young nuclei usually become damaged. If, however, the nuclei escape damage, complete restoration of the cytoplasm occurs. The toxicity of the aged cytoplasm is further indicated by: (1) reduced survival and lifespan of the old partner in old-young conjugants and (2) reduced lifespan of the young mate after cytogamy (when only cytoplasm is transferred).[25,26]

Autogamous progeny are known to exhibit reduced lifespan as clonal age increases, indicating that regardless of the mode of fertilization, age deterioration in parent cells is "carried over" to offspring. The increase of mortality after autogamy was not linear with age. The viability of the progeny was maintained for 60 to 80 fissions, and thereafter the frequency of death increased. The nonlinearity of age-induced damage was further supported by the observation that cells which are maintained as a young series by successive autogamies, each within 6 to 7 days, showed no increased mortality or mutations over the course of 1500 nuclear divisions. The rare mutations which may occur in young clones may be below our level of detection, or can be repaired during the autogamy process. The accumulated damage in old cells far exceeds the amount of damage in young clones.[25] The capacity of any repair processes in the aged fertilization system is insufficient to correct the age deterioration.

Euplotes

In old-young crosses of *E. patella,* progeny viability declined.[23] Aged toxic cytoplasm was indicated by the failure of the mature nucleus to function only in the aged cytoplasm.[23] The mature mate was seriously damaged by contact with the aged mate although no nuclear exchange occurred. The old mate showed loss of ability to differentiate new micronuclei and loss of capacity to complete normal stomatogenesis.

In marine *E. woodruffi* crosses between young and old stocks result in high mortality for both the young and old mate.[51] There were, among 50 pairs, 6 surviving clones that exhibited reduced fission rates. These studies could not distinguish between nuclear and cytoplasmic contributions.

COMPARATIVE ASPECTS OF INTRACELLULAR SITES FOR AGING IN THE PROTOZOA

Since the cell has evolved as a composite of its parts in relation to its environment, an integrative role of parts in harmony with its ecogenetic needs seems likely. The relative functions of the micronuclei, macronuclei, cytoplasm, cell surface, and adaptive requisites for life maintenance and longevity in the different genera are considered.

Micronuclei

Micronuclear function in life maintenance varies among ciliates and species of the same genera. Decline in viability or death is correlated with loss of micronuclei in *Tetrahymena thermophila*,[55,142] *Paramecium caudatum*,[186] *P. tetraurelia*,[68] and *Stylonychia mytilus*.[57] Viable strains with no micronuclei have been observed in *P. bursaria*,[187] *P. multimicronucleatum*,[86] *Euplotes patella*,[23] and *Tetrahymena pyriformis*.[55] *T. pyriformis* may have been amicronucleated since its origin, but cells without micronuclei have arisen from micronucleated strains in *P. bursaria* and *P. multimicronucleatum*.[86,187] Loss of micronuclei seems to be correlated with loss of the gullet or gullet function.[142,154,155] The micronuclei are coupled to a life maintenance function in some strains, but not in all strains even within the same genus.

In the reproductive cycle, loss of integrity of the micronuclear genome (the germ line) will result in imperfect genetic information for the progeny. Age-induced damage to the germ line would be expected to result in: (1) increased mortality during fertilization and (2) reduced function and longevity of the resulting generation. Loss of progeny viability has been observed in all clonal aged protozoans examined for this parameter. Progeny viability decline is documented in some *T. thermophila*,[85] *P. bursaria*,[13,84] *P. tetraurelia*,[14,26] *P. caudatum*,[47] *Euplotes woodruffi*,[51] *E. patella*,[23] and *Tokophrya lemnarum*.[19] A common feature of the aged zygote is the increased frequency of the failure to differentiate the normal number of new macronuclei leading to retention of the ancestral macronuclei.[188] "Rejuvenation" of the cells by formation of the filial generation must occur prior to a given parental age or progeny will experience age damage.[95] Loss of progeny viability with increased parental age cannot be attributed solely to damaged micronuclei since aged cytoplasm can also play a role.[91]

Macronucleus

A vital macronuclear function is implied since cells without macronuclei die.[36,68,189] The macronucleus is known to regulate the phenotype of cells[36,39,64,190] and must develop from a young micronuclear zygote for production of infinite succession of long-lived vigorous progeny in *Paramecium*. *P. busaria* and *P. tetraurelia* show a decrease of DNA content[101,102] and function[99] with age in the macronuclei, whereas *Tetrahymena* does not show macronuclear DNA loss.[85] There is no doubt that the macronucleus is necessary for life maintenance and is involved in aging of some but not all species.

Cytoplasm

Aged cytoplasm causes deleterious effects on normal nuclear development and maintenance of species-specific lifespans. Decline in progeny lifespan or proliferative capacity traceable, at least in part, to cytoplasm in the *P. aurelia* complex,[25,39,104] *P. bursaria*,[118-121] *E. patella*,[23] and *Amoeba*.[24] The aged or spanned cytoplasm can affect development and proliferative capacity of a young or normal nucleus.[24]

The cytoplasm can influence the nucleus, especially during the sensitive period of nuclear differentiation, but can yield to the influence of the nucleus[21] after differentiation. The young nuclei, which survive the hostile aged cytoplasm can, after nuclear development, rejuvenate the aged cytoplasm.[25,96] Short lifespanned progeny from old parents can generate vigorous second generation offspring.[96] The evidence indicates that the cytoplasm can affect the nucleus, and the nucleus in turn can affect the cytoplasm. Determination of alternative phenotypes by the nucleus vs. the cytoplasm is interchangeable and is probably dependent on nuclear genotype, the stage of nuclear differentiation, and cytoplasmic state. Cytoplasmic incompatibility with normal nuclear function is seen not only in aged cytoplasm but in interspecies crosses.[119,120] Cytoplasmic barriers to progeny survival may be an enormous contributor to speciation; lethality occurs prior to exchange of nuclei.[120] At present, it is not possible to separate cytoplasmic effects from cell surface effects in all cases.

Cell Surface

Changes in the aged cell surface have been seen in addition to cytoplasmic and nuclear changes.[23,135] Although there are no direct studies which correlate regulation of aging in protozoa to surface changes, membrane changes do initiate intracellular events.[191-194] Indirect evidence indicates that membrane alterations may initiate intracellular abnormalities characteristic of aged cells.[195] The absence of stabilizing surface structures in mutant paramecia cells has been considered responsible for abnormal positioning of the nucleus within the cell, promoting missegregation of the nuclear material at cytokinesis.[196] Reproduction and organization of surface structures are under the influence both of the position and orientation of neighboring organelles,[197-198] and gradient fields.[199] Stable gradients are most likely in the external membrane since these cells lack internal partitions and show rapid cytoplasmic flow.[200] Thus, conveyance of morphogenetic substances would be restricted to the surface. Different regions of the ciliate surface show differential responses to mechanical stimuli,[201] and the membrane itself has been suggested as the basis for morphogenetic gradients using configurational transitions.[202] Changes in the surface during aging could be expected to stimulate intracellular changes at a regulatory level, though ciritical experiments are yet to be done with respect to aging.

At a strictly functional level, surface changes induced by accidental damage or ''age-dependent'' intracellular deterioration could lead to cell starvation.[18] Endocytic function is impaired both in aged *Paramecium*[106] and in adult *Tokophrya*.[159] The ability of *Tetrahymena* to renew its oral structure at cell division, and its immortality, contrast to the lack of regenerative capacity for oral morphogenesis during the vegetative cycle in *Paramecium*,[15] The causes of the age-induced structural changes are not known. Fertilization can, however, revert the surface structural changes to the youthful state, implying that renewal of regulatory processes, both at the level of the nucleus and at the cell surface, is possible under appropriate physiological conditions (during the fertilization process).

Ecogenetic Considerations

The enormous variation in protozoan lifespan among species may reflect special selective pressures for their survival. Marine and fresh water species can show the same variation in lifespan and diversity for fertilization capacity.[23,51] In general, species with multiple mating types and inability to undergo autogamy show significantly longer lifespans than inbreeders.[15] Long immaturity periods allow members of a clone to become widely separated from one another before they mate, and multiple mating types increase the probability of complementary mating types when strangers meet.

The outbreeder can accumulate and recombine recessive mutations by carrying the genes in the heterozygous condition for many generations whereas an inbreeder, especially after autogamy, immediately expresses the phenotype due to the resultant homozygosity. The homozygotes may die in response to the existing environmental conditions to achieve genetic variety with a lower mutation rate than an inbreeder. The capacity for maintenance of correct repair of DNA may be greater in outbreeders than inbreeders (See Nanney[55]). It should be recalled that the evolutionary distance separating the various genera and species can be enormous. Differences that emerge in the relative function of cell parts may reflect their particular ecogenetic needs. Variation in survival and life maintenance strategy in protozoa should be no more surprising than differences observed in the life strategies employed by fish and man.

Table 5
SIMILARITES IN AGE-RELATED CHANGES IN PROTOZOA
AND MULTICELLULAR ORGANISMS

	Multicell (Ref.)	Unicell (Ref.)
Parameters that decrease		
DNA content	206	101—103
DNA synthesis	206	108
RNA synthesis	206	108
Ribosomal genes, nucleolar volume	206	100—110
DNA template activity	206	101
Endocytic activity	236, 237	101
Somatic viability	206	14, 23, 92, 106, 133
Progeny viability	210—213	23, 25, 26, 90—95, 115, 133
Parameters that increase		
Abnormal mitochondria	206	105, 157
Age pigments	206	105—157
Chromosomal aberrations	206, 207	14, 25, 82, 88—91
Lysosomal activity	206	105, 157
Sensitivity to x-rays	238	179

Note: The references for the multicellular organisms are generally from extensive reviews.[206,207] The bulk of the unicellular references are from the *P. aurelia* complex since this group has been studied most extensively. Only clonal aged protozoans or aged adult individuals have been included in this Table.

AGING IN MULTICELLS VS. UNICELLS

The uniformity of the strucure of the DNA molecule and almost universality of the genetic code implies a common origin for all living forms.[12] The unicells, as eukaryotes, are the "missing link" between prokaryotes and eukaryotic multicellular organisms. In general, diploid protozoan cells age and die, though exceptions have been found. Normal human cells in tissue culture also age and die.[203-204] Many similarities have been noted between human cells in culture and aging paramecia.[205-206] Similarities in age-dependent changes in vitro between aging mammals and protozoa are found (Table 5). Since the bulk of the studies on age-correlated changes has been done on mammalian systems, other possible similarities and differences can be expected to emerge when more protozoan data become available.

DNA damage and repair have been implicated as contributory to aging in both protozoans and mammals.[8] Age-accumulated damage and loss of repair are found in some aging human cells, though experimental design and cell types influence the results obtained.[207] Loss of repair in aged paramecia seems to occur and the amount of genetic damage can influence lifespan duration.[81] The ability to repair DNA and the duration of lifespan are known correlates.[208,209] Age-dependent germinal damage is widespread in the protozoans and is seen both in older human mothers and fathers (though there is controversy with respect to the origin of damage in human mothers).[210-214] Higher rates of mortality and abnormality are found in offspring from older parents.

A determinative role for both the nucleus and cytoplasm in aging and lifespan duration is implied from the protozoan and mammalian cell studies. It been found that aged cytoplasm could damage a young nucleus but a young nucleus could not rejuvenate an old cytoplasm in human cells.[215] The determinative role of the nucleus in human cells suggested in earlier experiments[216-218] is not always observed.[24] In hybrids formed between the immortal HeLa cells and finite lifespan epithelial cells, the immortality was found to be dominant.[219] When fibroblasts contained only the nuclei from immortal and finite lifespan cells in binucleates, the limited-lifespan type was now dominant.[218] The intriguing possibility is that the cytoplasm and/or cell surface may contribute necessary elements required to confer immortality.[220] In *Paramecium*, the aged cytoplasm can damage the nucleus. If, however, some nuclei survive, they will rejuvenate the cytoplasm.[25,96] Induced damage to nuclei by cytoplasm is also seen when a foreign cytoplasm is introduced both in frogs[222] and paramecia.[118,227] Cytoplasmic effects in mammalian cell differentiations are well established,[223-227] and each kind of cell possesses its specific epigenotype.[227] Change of cytoplasm in the differentiated cell can influence the expression of genes in cybrids (the introduction of a particular type nucleus into the cytoplasm of an alternative type).[223] Cytoplasm from a nonerythroid cell into an erythroid cell can cause the extinction of the ability to induce the typical hemoglobin.[229] Not only negative control (loss of gene expression) but positive control by cytoplasm has been found.[230] Liver-specific enzyme function can be induced in fibroblasts by cytoplasm from a liver cell and the induced gene expression can last 75 to 100 generations. Likewise, in paramecia, cytoplasm can determine differences in expression of genes in homozygotes and the control may be at the level of the gene or through epigenetic interactions.[21] Injection of cytoplasm from an immature paramecia to a mature paramecia results in reversion to immaturity.[231] Like the mammalian cells, cytoplasm can cause a negative effect on gene expression.

Although these examples are not specifically cytoplasmic effects on aging, these cases raise the possibility that some differentiations that may affect lifespan may be modulated by the cyptoplasm or the cell surface not only during development of nuclei, but also after nuclear differentiation.

Environmental conditions and radiation may also induce changes in gene expression and could contribute to the observed radiation-induced lifespan extension. Radiation-induced repair is found in paramecia,[81] insect,[232] and mammals.[233] The UV treatment in prokaryotes is known to initiate a complex induction process culminating in derepression of a group of metabolically diverse but coordinated functions that could promote survival[182] or long life.

The protozoans are as remarkably different from multicellular organisms as they are from each other. It may be most profitable to concentrate on similarities since herein lie the fundamental properties of cells.

Many ciliates can undergo self-fertilization, a process apparently lost in mammals. Recent studies indicate, however, that the mouse, under appropriate conditions can be induced to undergo duplication of the oocyte nucleus and proceed to the blastocyst state.[234] This amazing result implies that these higher cells have not completely lost the potential for ''parthenogenesis''.

Another striking difference between protozoans and multicells is the apparent molecular diversity with conservative morphology in the protozoans,[55] and molecular similarity and morphological diversity in primates.[235] If the cell surface represents boundaries for gradient fields[199] for differentiation of new nuclei, changes in the surface may be lethal in unicells. If different cell types have different functions (necessary for life maintenance of all cells in the clone) the members must remain together. Morphological variation in higher forms would result from the ability to ''glue'' the eukaryote cells into functional architectural frameworks with collagen. Regulatory mechanisms to program cell death in specific regions sculpture form, creating for example, the fingers in man. It should be remembered that the fundamental unit of the multicellular organism is the eukaryote cell. Patterns of gene expression can contribute to lifespan duration of specific cell types. Different cell types may have a role in the determination of the organismal longevity in different genera.

SUMMARY OF FINDINGS AND CONCLUDING REMARKS

1. Intrinsic clonal aging has been demonstrated in the ciliated protozoa. The most extensive studies have been carried out in the *P. aurelia* complex and *P. bursaria*, with

evidence for clonal aging found also in other ciliates: *Euplotes, Tokophrya,* and *Stylonychia.*

2. Erosion of the germ line nuclei during clonal aging has been demonstrated in studies with *Paramecium, Euplotes,* and *Tetrahymena,* and is manifested by decreased survival after conjugation as clonal age increases. Hostile, aged cytoplasm can contribute to the observed results.

3. The germ line nuclei can have a role in somatic life maintenance in certain strains of *Paramecium, Tetrahymena,* and *Stylonychia,* but cells without micronuclei are known to be viable in other species of *Paramecia, Tetrahymena,* and *Euplotes.* Therefore, the presence of micronuclei can be coupled with vegatative survival, only in certain strains.

4. Aged or spanned cytoplasmic states can induce damage in cells and affect lifespan duration in *Paramecium, Euplotes,* and *Amoeba.*

5. Rejuvenation of an aged cell by young nuclei is not always observed, but has been demonstrated under restricted conditions in *Acetabularia, Paramecium,* and human cells. Further studies are required.

6. Reversion of a nucleus by cytoplasm to a more youthful or alternative state is implied in *Acetabularia, Paramecium,* and mammalian cells.

7. Immortality of certain protozoans has been observed. In ciliates, certain *Tetrahymena* exhibit apparent immortality. Immortality can be lost by inbreeding, resulting in several different genotypes.

8. Immortality in *Amoeba* can be lost by dietary changes.

9. Since protozoans can exhibit species-specific lifespans, genetic regulation of lifespan is implied. The term genetic is meant to include both the genotype of the cell and the regulation of gene expression imposed by the environment of the nucleus, i.e., the cytoplasm, the membrane, and external environment.

10. Plasticity of lifespan duration is indicated by the multiplicity of factors found to influence longevity. Lifespan is known to be reduced by diet and radiation, and can also be extended by regulation of these same factors.

11. Among protozoans longevity is, in part, selected as a response to the need for variation and adaptation to a changing environment. Organisms with long lives tend to accumulate variation by mating outside the clone and may achieve variation with lower mutation rates. Inbreeders, confined to mating with close relatives, may achieve variation with higher mutation rates and have shorter lifespans. Repair capacity should be considered a possible contributor to lifespan duration as a byproduct of the need for genome variations. There is evidence for relation of lifespan and repair in paramecia.

12. Remarkable similarities in age-associated changes are found in certain protozoans and finite lifespan mammalian cells.

The intrinsic lifespan potential of the genome is immortality in all living species of unicells and multicells in the germ line. If fertilization occurs prior to the onset of genetic damage with aging, an infinite succession of generations can be produced with species-specific lifespans. Any DNA damage which interferes with life maintenance can shorten lifespan (genetic diseases, if you will). Reduction of deleterious effects to the genome or gene structure by preventing damage, or increasing correct repair of damage which may occur, seems promising as an approach to promote longevity.

The enormous variation in protozoan lifespans reflects the diverse strategies for survival among species which have long been separated widely from each other in evolutionary time. The informational reservoir is concomitantly large, though basic similarities can be considered indicative of fundamental processes.

The failure to replace or repair any damaged cell part required for survival will limit lifespan. The immortal strains of *Tetrahymena* seem capable of repair and replacement during their somatic life cycle, whereas *Paramecium* seems capable of complete restoration of all cell parts only during fertilization and activation of the germ line.

Lifespan duration is most likely determined in multiple ways in different cells within the same organism, as well as in different organisms. The genome, however, possesses enormous plasticity for variations in lifespan duration. A relatively small number of fundamental mechanisms may be operative and subject to modulation by physical, chemical, and environmental conditions that could alter lifespan and delay aging even if no single mechanism, or set of mechanisms is known to cause aging.

The prokaryotes have served as a valuable research resource for gene regulation, and eukaryotes can be expected to provide a wealth of valuable information as well.

REFERENCES

1. **Sonneborn, T. M.,** Enormous differences in length of life of closely related ciliates and their significance, in *The Biology of Aging,* American Institute of Biological Symposium, 6, Strehler, B. L., Ed., Waverly Press, Baltimore, 1960, 289.
2. **Nanney, D. L.,** Ciliate genetics: patterns and programs of gene action, *Annu. Rev. Genet.,* 2, 121, 1968.
3. **Sonneborn, T. M.,** Methods in *Paramecium* research, in *Methods in Cell Physiology,* Vol. 4, King, R. C., Ed., Plenum Press, New York, 1970, 241.
4. **Margulin, L.,** *Origin of Eukaryotic Cells,* Yale University Press, New Haven, 1970.
5. **Simpson, G.,** *The Major Features of Evolution,* Columbia University Press, New York, 1953.
6. **Cutler, R. G.,** Nature of aging and life maintenance processes, *Interdiscip. Top. Gerontol.,* 9, 83, 1976.
7. **Nanney, D. L.,** Vegetative mutants and clonal senility in *Tetrahymena, J. Protozool.,* 6, 171, 1959.
8. **Hart, R. W., Smith-Sonneborn, J., and Brash, D. E.,** Aging in paramecia and man, in *Structural Pathology in DNA and the Biology of Ageing,* Freiburg, Germany, 1979.
9. **Grell, K. G.,** *Protozoology,* Springer-Verlag, New York, 1973.
10. **Coleman, A.,** Sexuality in colonial green flagellates, in *Biochemistry and Physiology of Protozoa,* 2nd ed. Levandowsky M., and Hutner, S. H., Eds., Academic Press, New York, 1979, 307.
11. **Cleveland, L. R., Burke, A. W., and Karlson, P.,** Modifications induced in the sexual cycles of Protozoa of *Cryptocercus* by change of host, *J. Protozool.,* 7, 240, 1960.
12. **Sonneborn, T. M.,** The origin, evolution and causes of aging, in *The Biology of Aging,* Behnke, J. A., Finch, C. E., and Moment, G. B., Eds., Plenum Press, New York, 1978, 361.
13. **Jennings, H. S.,** Genetics of the protozoa, *Bibliog., Genet.,* 5, 105, 1929.
14. **Sonneborn, T. M.,** The relation of autogamy to senescence and rejuvenescence in *Paramecium aurelia, J. Protozool.,* 7, 38, 1954.
15. **Sonneborn, T. M.,** Breeding systems, reproductive methods, and species problems in Protozoa, in *The Species Problem,* Mayr, E., Ed., American Asociation of the Advancement of Science, Washington, D.C., 1957, 155.
16. **Woodruff, L. L.,** Two thousand generations of *Paramecium, Arch. Protistenkd.,* 21, 263, 1911.
17. **Woodruff, L. L.,** *Paramecium aurelia* in pedigree culture for twenty-five years, *Trans. Am. Microsc. Soc.,* 51, 196, 1982.
18. **Siegel, R. W.,** Organellar damage as a possible basis for intraclonal variation in *Paramecium, Genetics,* 66, 305, 1970.
19. **Colgin-Bukovsan, L. A.,** Life cycles and conditions for conjugation in the *Suctorian Tokophrya lemnarum, Arch. Prostistenkd.,* 121, 223, 1979.
20. **Rudzinska, M. A.,** Differences between young and old organisms in *Tokophrya infusionum, J. Gerontol.,* 10, 469, 1955.
21. **Sonneborn, T. M. and Schneller, M. V.,** A genetic system for alternative stable characteristics in genomically identical homozygous clones, *Dev. Genet.,* 1, 21, 1979.
22. **Danielli, J. F.,** Cellular inheritance as studied by nuclear transfer in *Amoebae,* in *New Approaches in Cell Biology,* Walker, P. M. B., Ed., Academic Press, New York, 1960, 15.

23. **Katashima, R.,** Several features of aged cells in *Euplotes patella,* syngen 1, *J. Sci. Hiroshima Univ. Ser. B. Div.,* 1, 23, 59, 1971.

24. **Muggleton-Harris, A. L.,** Reassembly of cellular components for the study of aging and finite life span, *Int. Rev. Cytol.,* 9(Suppl.), 279, 1979.

25. **Sonneborn, T. M. and Schneller, M. V.,** Age-induced mutations in *Paramecium,* in *The Biology of Aging,* American Institute of Biological Science Symposium, 6, Strehler, B. L., Ed., Waverly Press, Baltimore, 1960, 286.

26. **Williams, T. J.,** Determination of Clonal Lifespan in *Paramecium tetraurelia,* Ph.D. thesis, University of Wyoming, Laramie, 1980.

27. **Beale, G. H.,** *The Genetics of Paramecium aurelia,* Cambridge University Press, London, 1954.

28. **Beale, G. H.,** The genetic control of cell surfaces, *Recent Prog. Surf. Sci.,* 2, 261, 1964.

29. **Hanson, E. D.,** Protozoan development, in *Chemical Zoology,* Vol 1, Florkin, M., Scheer, B. J., Eds., Academic Press, New York, 1967, 395.

30. **Hiwatashi, K.,** *Paramecium,* in *Fertilization,* Vol 2., Metz, C. B. and Monroy, A., Eds., Academic Press, New York, 1969, 225.

31. **Jurand, A. and Selman, G. G.,** *The Anatomy of Paramecium aurelia,* Macmillan, London, 1969.

32. **Kimball, R. F.,** The relation of repair to differential radiosensitivity in the production of mutations in *Paramecium,* in *Biochemistry and Physiology of Protozoa,* Vol. 3, Hutner, S. H., Ed., Academic Press, New York, 1964, 243.

33. **Soldo, A. T. and Godoy, G. A.,** The kinetic complexity of *Paramecium* macronuclear deoxyribonucleic acid, *J. Protozool.,* 19, 673, 1972.

34. **Preer, J. R., Jr.,** Nuclear and cytoplasmic differentiation in the protozoa, in *Developmental Cytology,* Rudnick, W., Ed., Ronald Press, New York, 1959, 3.

35. **Preer, J. R., Jr.,** Genetics of the protozoan, in *Research in Protozoology,* Vol. 3, Chen, T. T., Ed., Pergamon Press, Oxford, 1969, 129.

36. **Sonneborn, T. M.,** Recent advances in the genetics of *Paramecium* and *Euplotes, Adv. Genet.,* 1, 263, 247.

37. **Van Wagtendonk, W. J.,** *Paramecium: A Current Survey,* Elsevier, Amsterdam, 1974.

38. **Sonneborn, T. M.,** The *Paramecium aurelia* complex of fourteen species, *Trans. Am Microsc. Soc.,* 94, 155, 1975.

39. **Siegel, R. W.,** Genetics of ageing and the life cycle in Ciliates, *Symp. Soc. Exp. Biol.,* 21, 127, 1967.

40. **Jennings, H. S.,** Inheritance in protozoa, in *Protozoa in Biological Research,* Calkins, C. N. and Summers, F. M., Eds., Columbia University Press, New York, 1941, 710.

41. **Jennings, H. S.,** Senescence and death in protozoa and invertebrates, in *Problems of Ageing,* 2nd ed., Cowdry, W. V., Ed., Williams & Wilkins, Baltimore, 1942, 32.

42. **Siegel, R. W.,** Heredity factors controlling development in *Paramecium. Brookhaven Symp. Biol.,* 18, 55, 1965.

43. **Siegel, R. W. and Larison, L. L.,** The genetic control of mating types in *Paramecium bursaria, Proc. Natl. Acad. Sci. U.S.A.,* 46, 344, 1960.

44. **Siegel, R. W. and Cole, J.,** The nature of origin of mutations which block a temporal sequence for genic expression in *Paramecium, Genetics,* 55, 607, 1967.

45. **Myohara, K. and Hiwatashi, K.,** Temporal patterns in the appearance of mating type instability in *Paramecium caudatum, Jpn. J. Genet.,* 50, 133, 1975.

46. **Takagi, Y.,** Expression of the mating type trait in the clonal life history after conjugation in *Paramecium multimicronucleatum* and *Paramecium caudatum, Jpn. J. Genet.,* 45, 11, 1970.

47. **Takagi, Y. and Yoshida, M.,** Clonal life-span and clonal aging in *Paramecium caudatum, Biomed, Gerontol.,* 1, 20, 1977.

48. **Takagi, Y. and Yoshida, M.,** Clonal death coupled with the number of fissions in *Paramecium caudatum,* unpublished.

49. **Katashima, R.,** Breeding system of *Euplotes patella* in Japan, *Jpn. J. Zool.,* 13(1), 39, 1961.

50. **Kimball, R. F.,** Mating types in *Euplotes, Am. Nat.,* 73, 451, 1939.

51. **Kosaka, T.,** Age-dependent monsters or macronuclear abnormalities, the length of life, and a change in the fission rate with clonal aging in Marine *Euplotes woodruffi, J. Sci. Hiroshima Univ. Ser. B. Div.,* 1, 25, 173, 1974.

52. **Colgin-Bukovsan, L. A.,** The genetics of mating types in the suctorian *Tokophrya lemnarum, Genet. Res.,* 27, 303, 1976.

53. **Nanney, D. L.,** Nucleo-cytoplasmic interaction during conjugation in *Tetrahymena, Biol. Bull.,* 105, 133, 1953.

54. **Sonneborn, T. M.,** *Tetrahymena pyriformis,* in *Handbook of Genetics,* Vol. 2, King, R. C., Ed., Plenum Press, New York, 1975, 433.

55. **Nanney, D. L.**, Aging and long term temporal regulation in ciliated protozoa. A critical review, *Mech. Ageing Dev.*, 3, 81, 1974.

56. **Ammerman, D.**, Cytologische und genetische untersuchungen an dem ciliaten *Stylonychia mytilus* Ehrenberg, *Arch. Protistenkd.*, 108, 109, 1965.

57. **Ammerman, D.**, The micronucleus of the ciliate *Stylonychia mytilus*, its nucleic acid synthesis and function, *Exp. Cell Res.*, 61, 6, 1970.

58. **Ammermann, D., Steinbruck, G., von Berger, L., and Hennig, W.**, Development of the macronucleus in the ciliated protozoan *Stylonchia mytilus, Chromosoma*, 45, 401, 1974.

58. **Ammermann, D., Steinbruck, G., von Berger, L., Hennig, W.**, Development of the macronucleus in the ciliated protozoan *Stylonchia mytilus, Chromosoma*, 45, 401, 1974.

59. **Bostock, C. J. and Prescott, D. M.**, Evidence of gene diminution during the formation of the macronucleus in the protozoan, *Stylonchia, Proc. Natl. Acad. Sci. U.S.A.*, 69, 139, 1972.

60. **Riewe, M. and Lipps, H. J.**, Template activity of macronuclear and micronuclear chromatin of the ciliate *Stylonychia mytilus, Cell Biol. Intern. Report*, 1(6), 517, 1977.

61. **Nobili, R.**, I dimorfismo nucleare dei Ciliate: inerzia vegetativa del minronucleo, *Acca. Nazi. Lincei Rend. Class Sci. Fisiche Mat. Nat. Ser.*, 8, 32, 329, 1962.

62. **Pasternak, J.**, Differential genic activity in *Paramecium aurelia, J. Exp. Zool.*, 165, 395, 1967.

63. **Sonneborn, T. M.**, Inert nuclei: inactivity of micronuclear genes in variety 4 of *Paramecium aurelia, Genetics*, 31, 231, 1946.

64. **Sonneborn, T. M.**, Is gene K active in the micronucleus of *Paramecium aurelia?, Microb. Genet. Bull.*, 11, 25, 1954.

65. **Sonneborn, T. M.**, Genetics of cellular differentiation: stable nuclear differentiation in eucaryote unicells, *Annu. Rev. Genet.*, 11, 349, 1977.

66. **Klass, H.**, DNA Template Activity of Ethanol-Fixed Micronuclei of *Paramecium aurelia*, Ph.D. thesis, University of Wyoming, Laramie, 1974.

67. **Diller, W. F.**, Nuclear reorganization processes in *P. aurelia* with descriptions of autogamy and hemixis, *J. Morphol.*, 59, 11, 1936.

68. **Sonneborn, T. M.**, Patterns of nucleocytoplasmic integration in *Paracemium, Caryologia*, Vol. Suppl., 307, 1954.

69. **Berger, J. D.**, Nuclear differentiation and nucleic acid synthesis in well-fed exconjugants of *Paramecium aurelia, Chromosoma*, 42, 247, 1973.

70. **Berger, J. D.**, Selective autolysis of nuclei as a source of DNA precursors in *Paramecium aurelia* exconjugants, *J. Protozool.*, 21, 145, 1974.

71. **Berger, J. D.**, Gene expression and phenotypic change in *Paramecium* exconjugants, *Genet. Res.*, 27, 123, 1976.

72. **Kovaleva, V. G. and Raikov, I. B.**, Diminution and resynthesis of DNA during development and senescence of the diploid macronuclei of the ciliate *Trachelonema sulcata, Gymnostomata karyorelictida, Chromosoma*, 67(2), 177, 1978.

73. **Lauth, M. R., Spear, B. B., Heumann, J. M., and Prescott, D. M.**, DNA of ciliated protozoa: DNA sequence diminution during macronuclear development of *Oxytricha, Cell*, 7, 67, 1976.

74. **Lawn, R. M., Herrick G., Heumann, J. N., and Prescott, D. M.**, The gene-sized molecules in *Oxytricha, Cold Spring Harbor Symp. Quant. Biol.*, 42, 483, 1977.

75. **Schwartz, V. V.**, Structure and development of the macronucleus in *Paramecium bursaria, Arch. Protistenkd.*, 120, 255, 1978.

76. **Doerder, F. P. and Debault, L. E.**, Cytofluorimetric analysis of nuclear DNA during meiosis, fertilization and macronuclear development in the ciliate *Tetrahymena pyriformis*, syngen 1, *J. Cell Sci.*, 17, 471, 1975.

77. **Gorovsky, M. A.**, Macro and micronuclei of *Tetrahymena pyriformia:* a model system for studying the structure and function of eucaryotic nuclei, *J. Protozool.*, 20, 19, 1973.

78. **Yao, M. C. and Gorovsky, M. A.**, Comparison of the sequence of macro and micronuclei on DNA of *Tetrahymena pyriformis, Chromosoma*, 48, 1, 1974.

79. **Yao, M. C. and Gall, J. G.**, Alteration of the *Tetrahymena* genome during nuclear differentiation, *J. Protozool.*, 26, 10, 1979.

80. **Cummings, D. J.**, Studies on macronuclear DNA from *Paramecium aurelia, Chromosoma*, 53, 191, 1975.

81. **Smith-Sonneborn, J.**, DNA repair and longevity assurance in *Paramecium tetraurelia, Science*, 203, 1115, 1979.

82. **Nyberg, D.**, Copper tolerance and segregation distortion in aged *Paramecium, Exp. Gerontol.*, 13, 431, 1978.

83. **Smith-Sonneborn, J.**, unpublished data, 1980.

84. **Jennings, J. S.**, *Paramecium barsaria:* life history. I. Immaturity, maturity and age, *Biol. Bull.*, 86, 131, 1944.

85. **Weindruch, R. H. and Doerder, F. P.,** Age-dependent micronuclear deterioration in *Tetrahymena pyriformis,* syngen 1, *Mech. Ageing Dev.,* 4, 263, 1975.
86. **Sonneborn, T. M. and Rafalko, M.,** Aging in the *P. aurelia-multimicornucleatum* complex, *J. Protozool.,* 4, 21, 1957.
87. **Smith-Sonneborn, J. and Reed, J. C.,** Calendar life-span versus fission life-span of *Paramecium aurelia, J. Gerontol.,* 31, 2, 1976.
88. **Dippell, R. V.,** Some cytological aspects of aging in variety 4 of *Paramecium aurelia, J. Protozool.,* 2(Suppl.), 7, 1955.
89. **Fukushima, S.,** Clonal age and the proportion of defective progeny after autogamy in *Paramecium aurelia, Genetics,* 79, 377, 1975.
90. **Rodermel, S. R. and Smith-Sonneborn, J.,** Age-correlated changes in expression of micronuclear damage and repair in *Paramecium tetraurelia, Genetics,* 87, 259, 1977.
91. **Sonneborn, T. M. and Dippell, R. V.,** Cellular changes with age in *Paramecium,* in *The Biology of Aging,* American Institute of Biological Science Symposium 6, Strehler, B. L., Ed., Waverly Press, Baltimore, 1960, 285.
92. **Mitchison, N. A.,** Evidence against micronuclear mutations as the sole basis for death at fertilization in aged, and in the progeny of ultraviolet irradiated *Paramecium aurelia, Genetics,* 40, 61, 1955.
93. **Pierson, B. F.,** The relation of mortality after endomixis to the prior endomictic interval in *Paramecium aurelia, Biol. Bull.,* 74, 235, 1938.
94. **Raffell, D.,** The occurrence of gene mutations in *Paramecium aurelia, J. Exp. Zool.,* 63, 371, 1932.
95. **Smith-Sonneborn, J., Klass, M., and Cotton, D.,** Parental age and life span versus progeny life-span in *Paramecium, J. Cell Sci.,* 14, 691, 1974.
96. **Sonneborn, T. M., and Schneller, M. V.,** Are there cumulative effects of parental age transmissible through sexual reproduction in variety 4 of *Paramecium aurelia?, J. Protozool.,* 2(suppl.), 6, 1955.
97. **Sonneborn, T. M. and Schneller, M. V.,** Genetic consequences of aging in variety 4 of *Paramecium aurelia, Rec. Genet. Soc. Am.,* 24, 596, 1955.
98. **Sonneborn, T. M. and Schneller, M. V.,** Measures of the rate and amount of aging on the cellular level, in *The Biology of Aging,* American Institute of Biological Science Symposium 6, Strehler, B. L., Ed., Waverly Press, Baltimore, 1960, 290.
99. **Sundararaman, V. and Cummings, D. J.,** Morphological changes in aging cell lines of *Paramecium aurelia.* II. Macronulcear alterations, *Mech. Ageing Dev.,* 5, 325, 1976.
100. **Heifetz, S.,** Ultrastructure and Dynamics of the Macronucleus of *Paramecium tetraurelia* During Aging, Ph.D. thesis, University of Wyoming, 1979.
101. **Klass, M. and Smith-Sonneborn, J.,** Studies in DNA content, RNA synthesis, and DNA template activity in aging cells of *Paramecium aurelia, Exp. Cell Res.,* 98, 63, 1976.
102. **Schwartz, V. and Meister, H.,** Eine alterveranderung des makronucleus von *Paramecium, Z. Naturforsch. Sect. B.,* 28c, 232, 1973.
103. **Schwartz, V. and Meister, H.,** Aging in the macronucleus of *Paramecium, Z. Naturforsch.,* 28, 34, 1973.
104. **Sonneborn, T. M. and Schneller, M. V.,** Physiological basis of aging in *Paramecium* in *The Biology of Aging,* American Institute of Biological Science Symposium 6, Strehler, B. L., Ed., Waverly Press, Baltimore, 1960, 283.
105. **Sundararaman, V. and Cummings, D. J.,** Morphological changes in aging cell lines of *Paramecium aurelia.* I. Alterations in the cytoplasm, *Mech. Ageing Dev.,* 5, 139, 1976.
106. **Smith-Sonneborn, J. and Rodermel, S. R.,** Loss of endocytic capacity in aging *Paramecium:* the importance of cytoplasmic organelles, *J. Cell Biol.,* 71, 575, 1976.
107. **Sonneborn, T. M. and Schneller, M. V.,** The basis of aging in variety 4 of *Paramecium aurelia, J. Protozool.,* 2(Suppl.), 6, 1955.
108. **Smith-Sonneborn, J. and Klass, M.,** Changes in the DNA synthesis pattern of *Paramecium* with increased clonal age and interfission time, *J. Cell Biol.,* 61, 591, 1974.
109. **Smith-Sonneborn, J.,** Age correlated sensitivity in ultraviolet radiation in *Paramecium, Radiat. Res.,* 46, 64, 1971.
110. **Sundararaman, V. and Cummings, D. J.,** Morphological changes in aging cell lines of *Paramecium aurelia.* III. The effects of emetine in polysome formation, *Mech. Ageing Dev.,* 6, 393, 1977.
111. **Hiwatashi, K.,** Determination and inheritance of mating type in *Paramecium caudatum, Genetics,* 58, 373, 1968.
112. **Hiwatashi, K. and Myohara, K.,** A modifier gene involved in the expression of the dominant mating type allele in *Paramecium caudatum, Genet. Res.,* 27, 135, 1976.
113. **Hiwatashi, K.,** Genetic and epigenetic control of mating type in *Paramecium caudatum, Jpn. J. Genet.,* 44 (Suppl.), 383, 1979.
114. **Sonneborn, T. M.,** Herbert Spencer Jennings, *Biog. Mem. Natl. Acad. Sci.,* 47, 1974.
115. **Siegel, R. W.,** Nuclear differentiation and transitional cellular phenotypes in the life cycle of *Paramecium, Exp. Cell Res.,* 24, 6, 1961.

116. **Jennings, H. S.,** *Paramecium bursaria:* life history. II. Age and death of clones in relation to the results of conjugation, *J. Exp. Zool.,* 96, 17, 1944.

117. **Jennings, H. S.,** *Paramecium bursaria:* life history. III. Repeated conjugations in the same stock at different ages, with and without inbreeding, in relation to mortality at conjugation, *J. Exp. Zool.,* 96, 243, 1944.

118. **Jennings, H. S.,** *Paramecium bursaria:* life history. IV. Relation of inbreeding to mortality of exconjugant clones, *J. Exp. Zool.,* 97, 165, 1944.

119. **Jennings, H. S.,** *Paramecium bursaria.* life history. V. Some relations of external conditions, past or present, to aging and to mortality of exconjugants, with summary of conclusions in age and death, *J. Exp. Zool.,* 99, 15, 1945.

120. **Chen, T. T.,** Conjugation in *Paramecium bursaria.* II. Nuclear phenomena in lethal conjugation between varieties, *J. Morphol.,* 79, 125, 1946.

121. **Chen, T. T.,** Conjugation in *Paramecium bursaria,* IV. Nuclear behavior in conjugation between old and young clones, *J. Morphol.,* 88, 293, 1951.

122. **Barnett, A.,** A circadian rhythm of mating type reversals in *Paramecium multimicronucleatum,* sygnen 2, and its genetic control, *J. Cell. Physiol.,* 67 (2), 239, 1966.

123. **Sonneborn, T. M.,** personal communication, 1980.

124. **Heckmann, K.,** Paarungssystem und genabhängige Paarungstypdifferenzierung bei dem hypotrichen Ciliaten *Euplotes vannus* O. F. Muller, *Arch. Protistenkd.,* 106, 393, 1963.

125. **Luporini, P.,** Life cycle of autogamous strains of *Euplotes minuta, J. Protozool.,* 17 (2), 324, 8, 1970.

126. **Nobili, R.,** Mating types and mating type inheritance in *Euplotes minuta Yocom, (Ciliata, Hypotrichida), J. Protozool.,* 13, 38, 1966.

127. **Siegel, R. W. and Heckmann, K.,** Inheritance of autogamy and killer trait in *Euplotes minuta, J. Protozool.,* 13 (1), 34, 1966.

128. **Heckmann, K.,** Age-dependent intraclonal conjugation in *Euplotes crassus, J. Exp. Zool.,* 165, 269, 1967.

129. **Wichterman, R.,** Mating type, breeding system, conjugation and nuclear phenomena in marine *Ciliata cristatus* Krahe from the Gulf of Naples, *J. Protozool.,* 14, 49, 1967.

130. **Kosaka, T.,** Mating types of marine stocks of *Euplotes woodruffi* (ciliata) in Japan, *J. Sci. Hiroshima Univ. Ser. B. Div.,* 1, 24(2), 135, 1973.

131. **Katashima R.,** Mating types in *Euplotes eurvstomas, J. Protozool.,* 6, 75, 1959.

132. **Kosaka, T.,** Autogamy in fresh water *Euplotes woodruffi,* (Ciliata), *Zool. Mag.,* 79, 302, 1970.

133. **Kosaka, T.,** Effect of autogamy on clonal aging in *Euplotes woodruffi* (Ciliata), *Zool. Mag.,* 81 (8), 184, 1972.

134. **Okawa, H.,** Successive self-conjugation in exconjugant clones of *Euplotes aediculatus* (Ciliata), *Zool. Mag.,* 82, 388, 1973.

135. **Frankel, J.,** Dimensions of control of cortical patterns in *Euplotes:* the role of pre-existing structure, the clonal life cycle, and the genotype *J. Exp. Zool.,* 183, 71, 1973.

136. **Seyfert, H. M., Debault, L. E., and Preparata, R. M.,** unpublished, cited in **Nanney, D. L. and Preparata, R. M.,** Genetic evidence concerning the structure of the *Tetrahymena thermoophila* macronucleus, *J. Protozool.,* 26, 2, 1979.

137. **Lipps, H. J., Nock, A., Riewe, M., and Steinbrück, G.,** Chromatin structure in the macronucleus of the ciliate *Stylonychia mytilus, Nucl. Acids Res.,* 5 (12), 4699, 1978.

138. **Kaushal, R. L. and Saxena, D. M.,** Some morphological and cytological anomalies in the aged cultures of *Stylonychia notophora, Acta Protozool.,* 17, 69, 1978.

139. **Siegel, R. W.** Mating types in *Oxytricha* and the significance of mating type systems in ciliates, *Biol. Bull.,* 110, 352, 1956.

140. **Allen, S. L. and Gibson, J.,** Genetics of *Tetrahymena,* in *The Biology of Tetrahymena,* Elliot, A. M., Ed., Dowden, Hutchinson & Ross, Stroudsburg, Pa. 1973.

141. **Allen, S. L. and Li, C. J.,** Nucleotide sequence divergence among DNA fractions of different syngens of *Tetrahymens pyriformis, Biochem. Genet.,* 12, 213, 1974.

142. **Allen, S. L. and Weremiuk, S. L.,** Intersyngenic variations in the esterases and acid phosphatases of *Tetrahymena pyriformis, Biochem. Genet.,* 5, 119, 1971.

143. **Borden, D., Whitt, G. S., and Nanney, D. L.,** Electrophoretic characteristics of classical *Tetrahymena pyriformis* strains, *J. Protozool.,* 20, 293, 1973.

144. **Borden, D., Miller, E. T., Whitt, G. S., and Nanney, D. L.,** Electrophoretic analysis of evolutionary relationships in *Tetrahymena, Evolution,* 31, 91, 1977.

145. **Nanney, D. L. and McCoy, W.,** Characterization of the species of the *Tetrahymena pyriformis* complex, *Trans. Am. Micros. Soc.,* 95, 664, 1976.

146. **Nanney, D. L.,** Inbreeding degeneration in *Tetrahymena, Genetics,* 42, 137, 1957.

147. **Shabatura, S. K.,** Age-Associated Changes in the Nuclear Cycles of *Tetrahymena thermophila,* Masters thesis, University of Pittsburgh, 1979.

148. **Simon, E. M. and Nanney, D. L.,** Germinal aging in *Tetrahymena thermophila, Mech. Ageing Dev.,* 11, 252, 1979.

149. **Nanney, D. L. and Preparata, R. M.,** Genetic evidence concerning the structure of the *Tetrahymena thermophila* macronucleus, *J. Protozool.,* 2, 1979.

150. **Bleyman, L.,** *Temporal Patterns in the Ciliated Protozoan in Developmental Aspects of the Cell Cycle,* Cameron, I. L., Padilla, G. M., and Zimmerman, A. M., Eds., Academic Press, New York, 1971, 67.

151. **Pitts, R. A.,** Age Associated Micronuclear Defects of *Tetrahymena thermophila:* Genetic and Cytogenetic Studies, Masters thesis, University of Pittsburgh, 1979.

152. **Gorovsky, M. A. and Woodard, J.,** Studies in the nuclear structure and function in *Tetrahymena pyriformis,* I. RNA synthesis in macro and micronuclei, *J. Cell Biol.,* 42, 673, 1969.

153. **Murti, K. G. and Prescott, D. M.,** Macronuclear ribonucleic acid in *Tetrahymena pyriformis, J. Cell. Biol.,* 47, 460, 1970.

154. **Wells, C.,** Evidence for micronuclear function during vegetative growth and reproduction of the ciliate *Tetrahymena pyriformis, J. Protozool.,* 8, 284, 1961.

155. **Frankel, J. L., Jenkins, M., and Debault, L. E.,** Causal relations among cell cycle processes in *Tetrahymena pyriformis:* an analysis employing temperature-sensitive mutants, *J. Cell Biol.,* 71, 242, 1976.

156. **Rudzinska, M. A.,** Giant individuals and vigor of populations in *Tokophrya infusionum, Ann. N.Y. Acad. Sci.,* 56, 1087, 1953.

157. **Rudzinska, M. A.,** The use of protzoa for studies on ageing. I. Differences between young and old organisms of *Tokophrya infusionum* as revealed by light and electron microscopy, *J. Gerontol.,* 16, 213, 1961.

158. **Rudzinska, M. A.,** The use of protozoan for studies in ageing. II. Macronucleus in young and old organisms of *Tokophrya infusionum:* light and electron microscopy, *J. Gerontol.,* 16, 326, 1961.

159. **Rudzinska, M. A.,** The use of a protozoan for studies in aging. III. Similarities between young overfed and old normally fed *Tokophrya infusionum:* a light and electron microscope study, *Gerontolgia,* 6, 206, 1962.

160. **Weise, L.,** Genetic aspects of sexuality in *Volvocoles,* in *The Genetics of Algae,* Lewin, R. A., Ed., University of California Press, Berkeley, 1976, 174.

161. **Kochert, G.,** Differentiation of reproductive cells in *Volvox cateri, J. Protozool.,* 15, 438, 1968.

162. **Starr, R. C.,** Structure, reproduction and differentiation in *Volvox cateri, J. nagariensis* Iyengar, strains HK 9 and 10, *Arch. Protistenkd.,* 111, 204, 1969.

163. **Starr, R. C.,** *Volvox pocokiae,* a new species with dwarf males, *J. Phycol.,* 6, 234, 1970.

164. **Sessoms, A. H. and Huskey, R. J.,** Genetic control of development in *Volvox:* isolation and characterization of morphogenetic mutants, *Proc. Natl. Acad. Sci. U.S.A.,* 70(5), 1335, 1973.

165. **Danielli, J. F.,** The cell-to-cell transfer of nuclei in *Amoebae* and a comprehensive cell theory, *Annu. Rev. N.Y. Acad. Sci.,* 675, 1979.

166. **Muggleton, A. and Danielli, J. F.,** Aging of *Amoeba proteus* and *A. discoides* cells, *Nature (London),* 181, 1783, 1958.

167. **Muggleton, A. and Danielli, J. F.,** Inheritance of the ''life-spanning'' phenomenon in *Ameoba proteus, Exp. Cell Res.,* 49, 116, 1968.

168. **Rudzinska, M. A.,** The influence of amount of food in the reproduction rate and longevity of suctorian *(Tokophrya infusionum), Science,* 113, 10, 1951.

169. **Rudzinska, M. A.,** The influence of starvation and overfeeding in the fine structure of *Tokophrya infusionum* as revealed by electron microscopy, in *4th Cong. Int. Assoc. Gerontol.,* Italy, 1957, 242.

170. **Sousa, E. and Silva, E.,** Ultrastructural variations of the nucleus in dinoflagellates throughout the life cycle, *Acta Protozool.,* 31, 277, 1977.

171. **Gomez, M. P., Harris, J. B., and Walne, P. L.,** Ultrastructual cytochemistry of *Euglena gracilis* Z: from aging cultures, *J. Protozool.,* 20, 515, 1973.

172. **Gomez, M. P., Harris, J. B., and Walne, P. L.,** Studies of *Euglena gracilis* in aging cultures, II. Ultrastructure, *Br. Phycol. J.,* 9, 175, 1974.

173. **Palisano, J. R. and Walne, P. L.,** Acid phosphatase activity and ultrastructure of aged cells of *Euglena granulata, J. Phycol.,* 8, 81, 1972.

174. **Gomez, M. P., Harris, J. B., and Walne, P. L.,** Studies of *Euglena gracilis* in aging cultures. I. Light microscopy and cytochemistry. *Br. Phycol. J.,* 9, 163, 1974.

175. **Elliot, A. M. and Bak, I. J.,** The fate of mitochondria during aging in *Tetrahymena pyriformis, J. Cell Biol.,* 20, 113, 1964.

176. **Grusky, G. E. and Aaronson, S.,** Cytochemical changes in aging *Ochromonas:* evidence for an alkaline phosphatase, *J. Protozool.,* 16, (4), 686, 1969.

177. **Fukushima, S.,** Effect of X-irradiations on the clonal life-span and fission rate in *Paramecium aurelia, Exp. Cell Res.,* 84, 267, 1974.

178. **Kimball, R. F. and Gaither, N.,** Lack of an effect of a high dose of X-rays on aging in *Paramecium aurelia,* variety I, *Genetics,* 37, 977, 1954.

179. **Tixador, R., Richoilley, G., and Planel, H.,** Radiosensitivity variation of *Paramecium aurelia* as a function of the clonal age, *J. Protozool.,* 19 (Suppl.), 73, 1972.

180. **Kimball, R. F. and Gaither, N.,** The influence of light upon the action of ultraviolet on *Paramecium aurelia. J. Cell Comp. Physiol.,* 37, 211, 1951.

181. **Williams, T. J.** Induction of DNA Polymerase Activity after UV Irradiation, Ph.D. thesis, University of Wyoming, Laramie, 1980.

182. **Witken, E. M.,** Ultraviolet mutagenesis and inducible DNA repair in *Escherichia coli, Bacteriol. Rev.,* 40, 869, 1976.

183. **Spencer, R. R. and Melroy, M. P.,** Studies of survival of unicellular species. I. Variations in life expectance of a *Paramecium* under laboratory conditions, *J. Natl. Cancer Inst.,* 10, 1, 1949.

184. **Berger, S. and Schweiger, H. G.,** Cytoplasmic induction of changes in the ultrastructure of the *Acetabularia* nucleus and perinuclear cytoplasm, *J. Cell Sci.,* 17, 517, 1975.

185. **Widdus, R., Tayler, M., Powers, L., and Danielli, J. R.,** Characteristics of the "life spanning" phenomenon in *Amoeba proteus, Gerontologia,* 24, 208, 1978.

186. **Mikami, K.,** Stomatogenesis during sexual and asexual reproduction in an amicronucleated strain of *Paramecium caudatum, J. Exp. Zool.,* 208, 121, 1979.

187. **Golikova, M. N.,** Morphology and viability of amicronucleate clones of the ciliate *Paramecium bursaria, Acta Protozool.,* 17 (1), 89, 1978.

188. **Nobili, R.,** The effect of macronuclear regeneration in vitality in *Paramecium aurelia,* syngen 4, *J. Protozool.,* 7 (Suppl.), 15, 1960.

189. **Sonneborn, T. M.,** Gene-controlled, aberrant nuclear behavior in *Paramecium aurelia, Microb. Genet. Bull.,* 11, 24, 1954.

190. **Sonneborn, T. M.,** Macronuclear control of the initiation of meiosis and conjugation in *Paramecium aurelia, J. Protozool.,* 2 (Suppl.), 12, 1955.

191. **de Terra, N.,** Macronuclear DNA synthesis in *Stentor:* regulation by a cytoplasmic initiator, *Proc. Natl. Acad. Sci. U.S.A.,* 57 (3), 607, 1967.

192. **de Terra, N.,** *Cytoplasmic Control of Macronuclear Events in the Cell of Stentor,* Cambridge University Press, London, 1970.

193. **de Terra, N.,** Cortical control of cell division, *Science,* 184, 4136, 1974.

194. **Rappaport, R.,** Cytokinesis in animals, *Int. Rev. Cytol.,* 31, 169, 1971.

195. **Brachet, J. and Hubert, E.,** Studies, on nucleocytoplasmic interactions during early amphibian development, *J. Embryol. Exp. Morphol.,* 27, 121, 1972.

196. **Ruiz, F., Adoutte, A., Rossignol, M., and Beisson, J.,** Genetic analysis of morphogenetic processes in *Paramecium, Genet. Res.,* 27, 109, 1976.

197. **Beisson, J.,** Determinants nucleaires el cytoplasmiques dans la biogenese des structures chez les protozaires, *Ann. Biol.,* 11, 401, 1972.

198. **Beisson, J. and Rossignol, M.,** Movements and position of organelles in *Paramecium aurelia,* in *Nucleocytoplasmic Relationships during Cell Morphogenesis in some Unicellular Organisms,* Ruiseux-Dao, S., Ed., Elsevier, Amsterdam, 1975.

199. **Frankel, J.,** Positional information in unicellular organisms, *J. Theor. Biol.,* 47, 439, 1974.

200. **Tartar, V.,** *The Biology of Stentor,* Pergamon Press, New York, 1961, 281.

201. **Naitoh, Y. and Eckert, R.,** Ionic mechanisms controlling behavioral responses of *Paramecium* to mechanical stimulation, *Science,* 164, 963, 1969.

202. **Kaczanowska, J.,** The pattern of morphogenic control in *Chilodonella cucullulua, J. Exp. Zool.,* 187, 47, 1974.

203. **Hayflick L.,** The limited *in vitro* lifetime of human diploid cell strains, *Exp. Cell Res.,* 37, 614, 1965.

204. **Hayflick, L. and Moorhead, P. S.,** The serial cultivation of human diploid cell strains, *Exp. Cell Res.,* 25, 585, 1961.

205. **Hayflick, L.,** Cell biology of aging, *BioScience,* 25, 629, 1975.

206. **Hayflick L.,** The cellular basis for biological aging, in *Biology of Aging,* Hayflick, L. and Finch, C. E., Eds., Van Nostrand Reinhold, New York, 1977, 159.

207. **Williams, J. R. and Dearfield, K. L.,** DNA damage and repair in aging mammals, in *Handbook of Biochemistry in Aging,* Florini, J., Ed., CRC Press, Boca Raton, Fla., 1982, 25.

208. **Hart. R. W. and Setlow, R. B.,** Correlation between deoxyribonucleic acid excision-repair and life-span in a number of mammalian species, *Proc. Natl. Acad. Sci. U.S.A.,* 71, 2169, 1974.

209. **Sacher, G. A. and Hart, R. W.,** Longevity, aging, and comparative cellular and molecular biology of the house mouse, *Mus musculus,* and the white-footed mouse, *Peromyscus leucopus, Birth Defects Orig. Artic.* Ser. 14, 71, 1978.

210. **Thompson, J. S. and Thompson, M. W.,** *Genetics in Medicine,* 2nd ed., W. B. Saunders, Philadelphia, 1973.

211. **Sonneborn, T. M.,** The human early foetal death rate in relation to age of father, in *The Biology of Aging,* American Institute of Biological Science Symposium 6, Strehler, B. L., Ed., Waverly Press, Baltimore, 1960, 288.

212. **Penrose, L. S. and Smith, G. F.,** *Down's Anomaly,* Little, Brown, Boston, 1966, 218.
213. **Turpin, R. and Lejeune, J.,** *Human Afflictions and Chromosome Abnormalities,* Pergamon Press, Oxford, 1969, 57.
214. **Kram, D. and Schneider, E.,** Parental age effects: increased frequency of genetically abnormal offspring, in *The Genetics of Aging,* Schneider, E., Ed., Plenum Press, New York, 1978, 225.
215. **Muggleton-Harris, A. L. and Hayflick, L.,** Cellular aging studies by the reconstruction of replicating cells from nuclei and cytoplasms isolated from normal human diploid cells, *Exp. Cell Res.,* 321, 1976.
216. **Wright, W. E. and Hayflick, L.,** Contributions of cytoplasmic factors to *in vitro* cellular senescence. I., *Fed. Proc.,* 34, 76, 1975.
217. **Wright, W. E. and Hayflick, L.,** Nuclear control of cellular aging demonstrated by hybridization of anucleate and whole cultured normal human fibroblasts, *Exp. Cell Res.,* 96, 113, 1975.
218. **Wright, W. E. and Hayflick, L.,** Use of biochemical lesions for selection of human cells with hybrid cytoplasms, *Proc. Natl. Acad. Sci. U.S.A.,* 72, 1812, 1975.
219. **Stanbridge, E. J.,** Suppression of malignancy in human cells, *Nature (London),* 260, 17, 1976.
220. **Muggleton-Harris, A. L. and Palumbo, M.,** Nucleocytoplasmic interactions in experimental binucleates formed from normal and transformed components, *Somat. Cell Genet.,* 5(3), 397, 1979.
221. **Norwood, T. H.,** Somatic cell genetics in the analysis of in *vitro* senescence, in *The Genetics of Aging* Schneider, E., Ed., Plenum Press, New York, 1978, 337.
222. **Hennen, S.,** Chromosomal and embryological analysis of nuclear changes occurring in embryos divided from transfers between *Rana pipiens* and *Rana sylvatica, Dev. Biol.,* 6, 133, 1963.
223. **Gurdon, J. B., Laskey, R. A., and Reeves, O. R.,** The developmental capacity of nuclei transplanted from keratinized skin cells of adult frogs, *Embryol. Exp. Morphol.,* 34, 93, 1975.
224. **Illmensee, K. and Mahowald, A. P.,** Transplantation of posterior polar plasm in *Drosophila,* Induction of germ cells at the anterior pole of the egg, *Proc. Natl. Acad. Sci. U.S.A.,* 71, 1016, 1974.
225. **Whittaker, J. R.,** Segregation during ascidian embryogenesis of egg cytoplasmic information for tissue-specific enzyme development, *Proc. Natl. Acad. Sci. U.S.A.,* 70, 2096, 1973.
226. **Brothers, A. J.,** Stable nuclear activation dependent on a protein synthesized during oogenesis, *Nature (London),* 260, 112, 1976.
227. **Paul, J., Gilmour, R. S., Threlfall, G., and Kohl, D.,** Organ-specific gene masking in mammalian chromosomes, *Proc. R. Soc. London Ser. B,* 176, 227, 1978.
228. **Gopalakrishnan, T. V., Thompson, E. B., and Anderson, W. F.,** Extinction of hemoglobin inducibility in Friend erythroleukemia cells by fusion with cytoplasm of enucleated mouse neuroblastoma or fibroblast cells, *Proc. Natl. Acad. Sci. U.S.A.,* 74, 1642, 1977.
229. **Gopalakrishnan, T. V. and French Anderson, W.,** Epigenetic activation of phenylalanine hydroxylase in mouse erythroleukemic cells by the cytoplast of rat hepatoma cells, *Proc. Natl. Acad. Sci. U.S.A.,* 76, 3932, 1979.
230. **Lipsich, L. A., Kates, J. R., and Lucas, J. J.,** Expression of liver specific function by mouse fibroblast nuclei transplanted into rat hepatoma cytoplasts, *Nature (London),* 281, 74, 1979.
231. **Miwa, I., Haga, N., and Hiwatashi, K.,** Immaturity substances: material basis for immaturity in *Para-mecium, J. Cell Sci.,* 19, 369, 1975.
232. **Ducoff, H. I.,** Radiation-induced increase in lifespan of insects, in Biological and Environmental Effects of Low Level Radiation, Vol. 1., International Atomic Energy Agency, Vienna, 1976, 103.
233. **Calkins, J. and Greenlaw, R. H.,** Activated repair of skin: a damage induced radiation repair system, *Radiology,* 100, 389, 1971.
234. **Markert, C. L. and Petters, R. M.,** Homozygous mouse embryos produced by microsurgery, *J. Exp. Zool.,* 201, 295, 1977.
235. **King, M. C. and Wilson, A. C.,** Evolution at two levels in humans and chimpanzees, *Science,* 188, 107, 1975.
236. **Stiffel, G.I., Biozzi, G., and Benacerraf, B.,** Étude des modifications du pouvoir phagocytaire du système réticuloendothelial en fonction de l'âge chez le rat et le lapin, *C.R. Seances Soc. Biol. Fil.,* 150, 1075, 1956.
237. **Wagner, H.N., Migha, T., and Soloman, U.,** Effect of age on reticuloendothelial function in man, *J. Gerontol.,* 21, 57, 1966.
238. **Icard, C., Beaupain, R., Diatloff, C., and Macieira-Coelho, A.,** Effect of low dose rate irradiation on the division potential of cells *in vitro.* VI. Changes in DNA and in radiosensitivity during aging of human fibroblasts, *Mech. Ageing Dev.,* 11, 269, 1979.

AGING STUDIES IN *CAENORHABDITIS ELEGANS* AND OTHER NEMATODES

Thomas E. Johnson and Victoria J. Simpson

INTRODUCTION

At first sight the inclusion of a chapter on *Caenorhabditis elegans* in a volume on cell biology may seem unusual. However this nematode has been a superb model system for a number of cell biology studies as well as a useful model of aging. This widespread interest in *C. elegans* is engendered in large part by its genetic system and its optical clarity in Normarksi phase-contrast optics (Figure 1).

Nematodes have long been a system in wide use among experimental gerontologists, and with the introduction of *C. elegans* by Brenner in 1974,[1] this species has become the nematode of choice for most aging studies. We concentrate primarily on *C. elegans* in this review although a number of other species, including *Caenorhabditis briggsae, Turbatrix aceti*, and *Panagrellus redivivus*, have been used in aging studies also. Other reviews on aging in *C. elegans* have appeared recently,[2-5] including a more detailed review in another volume of this series.[6]

BIOLOGY

Culture Conditions

C. elegans is a free-living nematode feeding on bacteria normally found in its natural habitat of decaying vegetable material in the soil. It thus was easy to adapt this nematode to growth in a laboratory environment where it is now maintained on standard nematode growth medium (NGM) plus agar in petri plates which have been spotted with *E. coli* strain OP50 as a source of food. Growth media made from chicken egg provide a convenient and inexpensive way of preparing biochemical quantities of nematodes. Growth is also possible in media completely lacking any other living organisms (axenic), using either defined or undefined media.

Embryonic Development

C. elegans is a self-fertilizing hermaphroditic species. Adult animals contain both mature sperm and oocytes; fertilization is internal. The newly fertilized eggs begin cell division within an hour after fertilization at 20°C. Eggs are laid a few hours after fertilization after a few cell divisions so that most of the embryonic phase of development occurs outside of the mother. Cell division continues without growth until 558 somatic and two germ line cells are present in the first larval stage nematode at the time of hatch. These developmental stages have been followed in exquisite detail both through observations on living eggs using Nomarski phase-contrast optics[7] and through the isolation of temperature-sensitive (TS) mutants which do not undergo development at the nonpermissive temperature of 25°C.[8] As a result of the optical clarity of the worm, it has been possible to construct an entire cell lineage from the one-cell stage through adulthood.[7] This lineage is highly deterministic both in time and direction of cell division and in the differentiated products of the mitotic events. All adult somatic cells are postmitotic.

Larval Development

After hatch the first larval stage worm begins ingesting food, and both longitudinal and radial enlargement proceed until the worm is about 1 mm long at the time of adulthood. There are four larval stages of *C. elegans*, each more or less uniformly spaced throughout

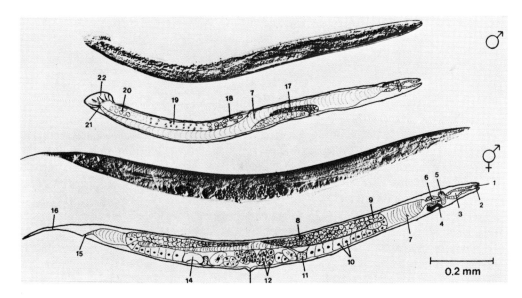

FIGURE 1.   Adults of *Caenorhabditis elegans:* male (above), hermaphrodite (below). Lateral views, anterior, right. Male: ventral, top. Hermaphrodite: ventral, bottom. Nomarski optics. (1) Buccal cavity, (2) one of six lips; (3) metacorpus of the pharynx; (4) excretory cell; (5) nerve ring; (6) terminal bulb of the pharynx; (7) intestine; (8) distal arm of the gonad, ovary; (9) loop of the gonad; (10) proximal arm of the gonad with maturing oocytes; (11) spermatheca; (12) uterus with cleaving eggs; (13) vulva; (14) mature oocyte before spermatheca; (15) anus; (16) tail; (17) testis; (18) spermatocytes; (19) vas deferens; (20) mature sperm; (21) cloaca; (22) copulatory bursa with rays and copulatory spicules. (From Schierenberg, E., Die embryonalentwicklung des nematoden *Caenorhabditis elegans* als Modell. Ph.D. Thesis, University of Gottingen, 1978. With permission.)

the first 2 days of growth after hatch. The worm becomes sexually mature after the fourth larval molt.

Anatomy

The adult nematode is essentially cylindrical with pharynx and gut running centrally the entire length of the animal (Figure 1). Food is ingested by the muscular pumping action of the two-chambered pharynx and is passed into the gut. The adult hermaphrodite is composed largely of gonadal material. There are two bilobed arms to the hermaphrodite gonad; each arm contains oocytes and sperm. Oocytes are formed distal to the vulva and pass proximally through the spermatheca (the site of sperm storage) where fertilization occurs. Muscle tissue is also found in adults, and there are 81 muscle cells lying in 4 longitudinal paired rows positioned dorsally and ventrally. Alternate contraction and release of these muscles together with the high osmotic turgor of the worm propel the worm forward by a series of dorsal-ventral waves.

Locomotion, food detection, egg lay, and other behavior are controlled by a nervous system of some 300 cells. Between the two lobes of the pharynx is a cluster of cells termed the nerve ring. Neurons run anteriorly innervating specialized sense organs. Two nerve cords run dorsally on either side of the worm to innervate the body wall muscles and the posterior parts of the worm. The animal is surrounded by a rigid cuticle which is shed at each larval molt and replaced by newly synthesized cuticle.

ADVANTAGES AND DISADVANTAGES

Life Cycle Advantages

C. elegans has a 2.5 day generation time at 20°C and a mass doubling time on the order of 8 hr. For aging studies its most outstanding characteristic is its brief lifespan which varies between 9 and 30 days depending on culture conditions and temperature (Table 1).

Table 1
LIFESPAN VARIATION AND CULTURE CONDITIONS

Species	Lifespan[a] (days)	Growth conditions	Method of establishment and maintenance of synchronous cultures	Ref.
C. elegans	58	Axenic; Sayre medium[9] plus cholesterol (0.1 gm/ℓ), bovine hemoglobin (0.5 gm/ℓ) replacing undefined extract	Filtration on stainless steel mesh	10
C. elegans	16 21	Monoxenic; E. coli OP50, on agar[1] 16°C	Eggs washed free of larvae on agar, collected and allowed to hatch for 2 hr	11
C. elegans	17.6 12.3	Axenic; Sayre medium Monoxenic; E. coli OP50 on 1% Czapekdox agar	Manual selection and transfer	12
C. elegans	21.1 29.4 15.3 16.2	Axenic; Rothstein medium[13] + 400 µM FUdR Monoxenic; E. coli OP50 on agar + 400 µM FUdR in agar	Egg wash on agar plates, manual selection, or by growth on agar or media containing 400 µM FUdR	14
C. elegans	33.5	Axenic; Sayre medium 23°C	Eggs washed free of larvae on agar, collected, and allowed to hatch; cultures maintained by manual selection	15
C. elegans	23 25	Axenic; Rothstein medium plus hemoglobin (500 µg/mℓ) +25 µM FUdR	Eggs isolated by sucrose flotation or glass wool filtration; cultures maintained by growth in 25 µM FUdR or by manual selection	16
C. elegans	11.4 11.5	Monoxenic; E. coli OP50 on agar + 400 µM FUdR	Manual selection or by growth on agar containing 400 µM FUdR	17
C. elegans	12.7	Monoxenic; E. coli OP50 on agar, 25°C, ± 5 mM FUdR	Eggs washed free of larvae on agar, allowed to hatch, and larvae collected; cultures maintained at 25°C with TS mutant or 5 mM FUdR	18
C. elegans	12.6 ± 2.0[b] 14.7 ± 1.3[b]	Monoxenic; E. coli OP50 on agar, 25°C, TS mutant + 30 µM FUdR in agar, 25°C	Eggs washed free of larvae on agar followed by glutaraldehyde treatment; maintained at 25°C with TS mutant or 30 µM FUdR	19
C. elegans	19.9 ± 0.5[b,c] 17.7 ± 0.3	Monoxenic; E. coli OP50 on agar	Manual selection	20
C. briggsae	30 30.8 26.2	Axenic; Sayre medium, 17°C 22°C 27°C	Eggs washed free of larvae on agar, collected, and hatched; cultures maintained by manual selection	21
C. briggsae	35 ± 2[b]	Axenic; 0.1% yeast extract, 4% soy peptone, 1% hemoglobin	Manual selection	22
C. briggsae	28 17	Axenic; Sayre medium 22°C 22°C, + 50 µg/mℓ FUdR	Eggs washed free of larvae on agar, collected, and allowed to hatch; cultures maintained by manual transfer or growth in 50 µg/mℓ FUdR	23
T. aceti	25	Axenic; Rothstein and Cook medium[24]	Filtration through 0.2-mm glass bead column	

Table 1 (continued)
LIFESPAN VARIATION AND CULTURE CONDITIONS

Species	Lifespan[a] (days)	Growth conditions	Method of establishment and maintenance of synchronous cultures	Ref.
T. aceti	25	Axenic; Castillo and Krusberg medium[26] 27°C	Manual selection; cultures maintained by manual transfer or	23
	16	27°C, + 50 μg/mℓ FUdR	growth in 50 μg/mℓ FUdR	
T. aceti	16	Axenic; Rothstein medium, 36°C +25 μg/mℓ cholesterol, +25 μg/mℓ β-sitosterol, +500 μg/mℓ myoglobin	Stainless steel mesh	27
	26	+100 μg/mℓ FUdR, 36°C		

[a]   50% survival at 20°C unless otherwise indicated.
[b]   Mean ± SEM.
[c]   Mean lifespans of hermaphrodites first and of males of the Brenner wild-type strain, N2.

There are a limited number of cell types in the nematode, and in *C. elegans,* all of the 959 somatic nuclei of the adult hermaphrodite are postmitotic. For cell biological studies it is possible to bring the great precision of the known developmental cell lineages to bear on the analysis of aging.

Genetic Advantages

C. elegans has a well-defined genetic system with about 300 genes mapped on the five autosomes and the X chromosome (Figure 2). This makes *C. elegans* one of the best studied genetic systems among the metazoans.

Genetics is particularly convenient because of the self-fertilizing nature of the hermaphrodite. Thus 50% of the genome is homozygous at each generation, facilitating the detection of recessive mutations. The mapping of these mutations and the construction of new stocks is expedited by the use of rare (0.1%) males. These males occur spontaneously and are obligate outcrossers.

A number of genetic studies on aging have been performed in *C. elegans*. These will be described in more detail in the next section. A final genetic advantage of *C. elegans* is the apparent lack of inbreeding depression for lifespan. Unlike other systems that have been analyzed, the F1 progeny of a cross between two different strains show lifespans that are intermediate to those displayed by the two parental strains.[20] Other species show significant extensions of lifespan in the F1. This so called "heterosis" is probably due to the inbreeding of many recessive sublethal genes which are known to be maintained in a heterozygous state within the populations of sexually reproducing organisms. Since *C. elegans'* most prevalent form of reproduction is through self-fertilization, wild populations of *C. elegans* are themselves almost completely homozygous and cannot therefore harbor these sublethal recessive genes that shorten the lifespan of inbred lines.

Finally, there are two ways of maintaining stocks in *C. elegans* that offer significant advantages for studies on the genetics of aging. First, *C. elegans* can be maintained for years frozen in liquid nitrogen and then revived for use.[1] This means that stocks remain genetically unchanged and identical reference stocks can be kept at many different locations. Second, *C. elegans* can be maintained in an alternate state called a dauer larvae. This is an extended alternative third larval stage that is resistant to starvation; dauer larvae can survive for months without food.[28] The dauer is especially important in genetic screens for long-lived mutants because progeny of putative long-lived stocks or siblings to test stocks can be maintained during testing without loss of viability or ability to reproduce.

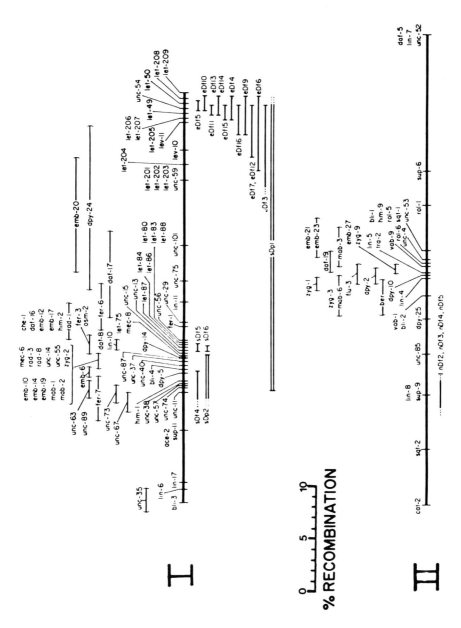

FIGURE 2. Genetic map of *Caenorhabditis elegans*. (Courtesy Riddle, D. R., Director *C. elegans* Genetics Center, Columbia, Missouri.)

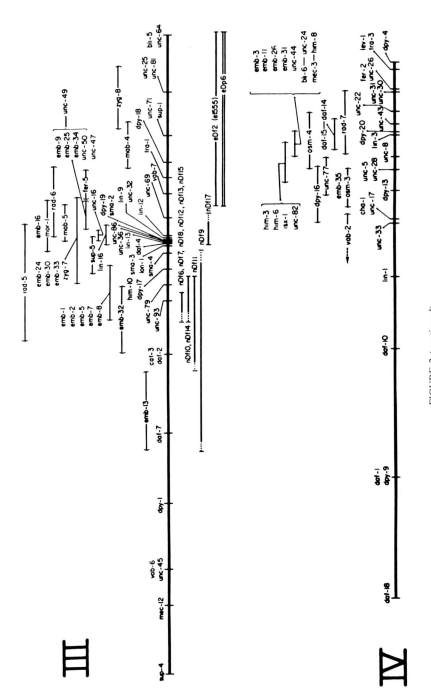

FIGURE 2 (continued)

487

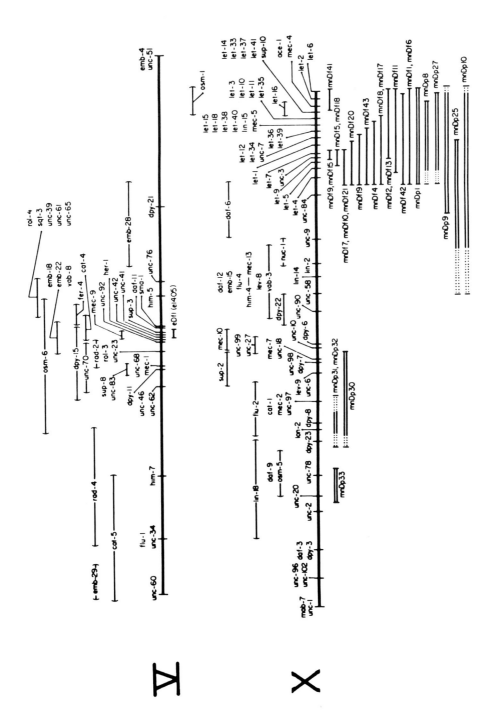

FIGURE 2 (continued)

Disadvantages

C. elegans is small and as yet no long-term methods for the isolation and maintenance of particular cell types in culture have been worked out. The rigid tough cuticle surrounding the worms makes the isolation of intact cell organelles difficult.

Finally *C. elegans* is phylogenetically far removed from humans. Thus specific mechanisms of aging in *C. elegans* may differ in part from those responsible for human aging. The presence of similar "S-shaped" survival curves in *C. elegans* as are found in mammalian populations is consistent with the notion that central aging processes may be similar.[29] However the ultimate answer to these problems can only come about through the understanding of the basic mechanisms of senescence which is much more likely to occur first through the study of *C. elegans* than through the use of humans as an experimental system.

STUDIES ON AGING USING THE NEMATODE

Length of Life

A tabulation of reported lifespans in *C. elegans*, *C. briggsae* and in *T. aceti* are given in Table 1. It is obvious that there is a wide variety of median lifespans reported. This variability in lifespan is due primarily to the variety of culture conditions, methods of obtaining a synchronous cohort for aging studies, and the techniques of maintaining this synchrony. These techniques have been reviewed in detail elsewhere and will be mentioned only briefly here.

Two methods of culture have been used. Aging populations have been maintained either axenically using one of two possible media, or monoxenically on living or heat-killed *E. coli*. In general, worms aging in axenic media live somewhat longer than those grown monoxenically perhaps due to partial starvation, a phenomenon similar to caloric restriction. Lowered bacterial concentrations have been shown to extend the lifespan of monoxenic cultures.

Methods of cohort synchrony involve the isolation of young animals either before hatch while still in the eggshell, at the time of hatch, or as larvae. It is unlikely that these techniques impose much variance in the reported median lifepans.

A second large source of lifespan variation is in the method of maintaining synchrony. Three techniques have been used: manual transfer of the aging populations to keep the adults free of the large numbers of their progeny, synchrony by filtration which separates smaller larval forms from the larger adults, and sterilization of the aging population either by the use of drugs or through the use of a TS spermatogenesis mutant. An analysis of these methods indicates that high levels of FUdR (400 μM) cause a prolongation of life but lower levels (25 μM for axenic cultures[16] and 30 μM for monoxenic cultures[19]) had little effect on lifespan. Furthermore the filtration method of preventing progeny contamination is somewhat unreliable because under some conditions significant numbers of adults die by the hatching of the eggs within the body of the mother; larvae then proceed to consume the mother forming bags of worms which are difficult to remove from synchronous populations by filtration. This latter mode of contamination is probably responsible for the much longer lifespans reported by Tilby and Moses.[10]

Effects of Temperature and Bacterial Concentration of Lifespan

Since *C. elegans* is poikilothermic (cold-blooded), there is an inverse correlation between temperature growth and the length of life (Table 2) as has been reported in other cold-blooded organisms. There is also a correlation between the bacteria concentration and the length of life (Table 3). Longest life is seen at a bacterial concentration of 10^8 bacteria per milliliter conditions which are far from optimal for egg production. Only about 25% as many eggs are produced at 10^8 bacteria per milliliter as at 10^9 bacteria per milliliter the optimal concentration for egg production.

Table 2
EFFECT OF TEMPERATURE ON LIFESPAN, DEVELOPMENT, AND FECUNDITY[a]

Temperature (°C)	Mean lifespan[b] (days)	Mean no. of fertile eggs	Egg viability (%)	Mean no. unfertile eggs	Duration of growth phase (days)	Duration reproductive phase (days)
6 ± 1	17.8 ± 1.1	0	—	0	—	—
10 ± 0.5	34.7 ± 3.0	84	58	1	10	14
14 ± 0.5	20.8 ± 0.9	206	92	45	5	7.5
16 ± 0.5	23.0 ± 1.6	250	93	43	4	7
20 ± 0.5	14.5 ± 1.0	273	95	125	3	6
24 ± 0.5	9.9 ± 0.4	269	99	135	2	4
25.5 ± 0.5	8.9 ± 0.6	103	93	51	2	4

	50% survival[c] (days)	Mean lifespan[d] (days)[b]
16	19.7	19.5 ± 0.8
20	12.6	16.7 ± 0.6
25	8.2	14.5 ± 0.8

[a]   From Table 1 of Klass[11].
[b]   Mean ± SEM.
[c]   From Table 2 of Hosono et al.[18]
[d]   Johnson, T. E., unpublished (for conditions, see Johnson and Wood[20]).

Table 3
EFFECT OF BACTERIAL CONCENTRATION OF LIFESPAN

Bacterial Concentration	Lifespan (days) (Mean ± SEM)	Number fertilized eggs
0	4 ± 1	0
10^4	5 ± 1	0
10^6	5 ± 1	0
5×10^7	15.1 ± 1.5	14
10^8	25.9 ± 2.4	63
5×10^8	19.4 ± 1.2	206
10^9	16.0 ± 1.0	273
10^{10}	15.0 ± 1.0	26

From Table III of Klass[11]

Biological Markers of Aging

A number of behavioral, biochemical, and biological markers have been studied in *C. elegans* for variation with age (Table 4). The ability to move, to ingest food, to defecate, and to produce offspring all decline with age. In the case of the decline in movements and defecation reported by Bolanowski,[17] there was no correlation with the rate of this decrease in individual animals and their individual lifespans.

Fluorescent material has been shown to accumulate over the lifespan of the nematode in an almost linear fashion. This material is presumed to be lipofuscin based on its spectral properties. A number of different enzymes show changes in activity per worm vs. age (Table 4). However, not all enzymes show such changes; choline acetyltransferase, acetylcholinesterase, and β-D-glucosidase show no significant changes after early adulthood, suggesting that there is no general pattern of change with age.

Table 4
MARKERS OF AGING IN *CAENORHABDITIS ELEGANS*

Behavioral Markers

Marker	3 days	6 days	9 days	12 days	17 days	Lifespan (days)	Ref.
Movement wave freq. (waves/min)	82	64	50	38	16	11 (median)	17
Defecation freq. (defec./min)	1.25	0.55	0.50	0.20	0.05	11 (median)	17
Pharyngeal pump rate[a] (pumps/min)	230	220	160	140	—	16 (mean)	30

Marker	2 days	4 days	6 days				Lifespan (days)	Ref.
Accum. rate[b] (per min $\times 10^2$)	11	9	6				—	31

Marker	4 days	6 days	8 days	10 days	12 days		Lifespan (days)	Ref.
Male mating behavior (fraction mating)	0.6	0.5	0.2	0.1	0.1		20[d]	32

Marker	3 days	10 days	17 days				Lifespan (days)	Ref.
Movement wave freq. (waves/min, total)[c]	98	24	15	—	—		20[d]	32
Pharyngeal pump rate (pumps/min)	193	73	65	—	—		20[d]	32

Biochemical Markers

Marker	3 days	6 days	9 days	12 days	15 days	18 days	Ref.
Fluorescent material[e] (rel. fluorescence per worm)	1.37	3.68	6.00	8.32	10.63	12.95	11

	2 days	4 days	6 days	8 days	10 days	12 days		
Acid phosphatase[f,g] (activity per animal $\times 10^4$)	4	16	16	25	29	27	8.9 (mean)	33
β-D-glucosidase[f] (activity per animal $\times 10^4$)	1	5	7	19	19	16	8.9 (mean)	33

	2 days	4 days	6 days	8 days	10 days	12 days		
β-N-acetyl-D-glucosaminidase[f] (activity per animal $\times 10^3$)	0.1	1.2	2.8	5.6	5.4	5.2	8.9 (mean)	33
α-D-mannosidase[f] (activity per animal $\times 10^3$)	0.6	4.0	6.0	8.0	6.2	1.8	8.9 (mean)	33
Protein content (μg per worm $\times 10^{-1}$)	0.4	1.2	1.7	1.6	2.3	1.7	8.9 (mean)	33

Biological Markers

	(LI)	5 days	10 days	15 days		
Single strand breaks (cpm $\times 10^4$)	<0.001	4.0	10.6	16.6	16.0[h]	34
5-Methylcytosine content (%)[i]		0.43	2.72	13.98	16.0[h]	34
Transcription capacity (cpm $\times 10^3$)	4.1	3.6	2.9	2.6	16.0[h]	34

	3 days	6 days	9 days	12 days	15 days		
Volume (nanoliters)	3.6	6.2	7.8	8.0	8.4	11 (median)	17

	5 days	10 days	15 days		
Viable fertilized eggs by males					
Virgin (%)	92	45	13	16.0[h]	34
Mated (%)	90	62	27	16.0[h]	34

Table 4 (continued)
MARKERS OF AGING IN *CAENORHABDITIS ELEGANS*

Behavioral Markers

	Age					Ref.
	1—12 hr	13—24 hr	23—36 hr	37—48 hr	44—60 hr	
Recombination freq.						
dpy-5 unc-15 (%)	2.86	1.62	1.58	1.42	1.05	—	35
dpy-5 dpy-14 (%)	2.26	1.58	1.35	1.00	1.07	—	35
dpy-11 unc-42 (%)	3.00	2.98	2.00	1.26[i]		—	35
Males/1000 total progeny	0.9	0.1	0.5	2.3	3.9		35

[a] Data are a subset of the data for the entire population and may therefore be nonrepresentative.

[b] Calculations of these accumulation rates is unclear, so that only relative values can be interpreted.

[c] Combines whole body waves in both posterior and anterior directions.

[d] Mean survival expectancy, calculated using Statistical Package for the Social Sciences, Survival subroutine.

[e] Aqueous homogenate in M9.

[f] Used the spermatogenesis mutant strain DH26.

[g] Activity of isozyme 1 shows 100-fold increase in activity between 3 and 10.

[h] Not reported, but conditions are comparable to 11.

[i] Data has been retracted after publication.

[j] 37 to 60 hr.

There is a general decrease in protein synthesis over the lifespan of the worm. These decreases in protein synthetic rate may be correlated with the increase in single strand breaks in genomic DNA and a decrease in transcriptional capacity of isolated DNA. Klass has also reported an exponential increase in the amount of methylated cytosine residues in DNA of worms as they age (Table 4); 15 day-old worms show more than 10^5 times the amount of 5-methylcytosine as is present in the DNA of young worms.

Several biological markers also show age-related changes (Table 4). The total volume as calculated by ^3H_2O efflux data show large increases up through day 9 and smaller increases thereafter. The proportion of viable progeny after mating of males show decreases over male life. Worms lose the ability to lay eggs by day 10 (20°C) and the frequency of recombination shows age-related decreases that are associated with corresponding increases in the fraction of males produced by spontaneous nondisjunction of the X chromosomes.

Genetic Studies

Klass and Hirsh[28] showed that the dauer larvae do not undergo senescence during the time spent as a dauer. Upon the resumption of normal growth these animals show completly normal lifespans. Klass[11,36] has also reported an attempt to isolate long-lived mutants. Although he isolated several mutants that showed lifespans more than 20% longer than the wild-type, these mutants could all be related to a decreased rate of food consumption and therefore were presumably due to the indirect life extension effect seen in caloric-restricted animals.

Johnson and Wood[20] approached the problem of obtaining longer-lived strains using another genetic approach, selective breeding. They crossed two different wild-type strains obtaining an F1 hybrid which was allowed to self-fertilize to produce an F2 showing individual differences for those genetic loci different in the two parental stocks. By three independent methods of analysis, those authors concluded that about 30 to 40% of the total variation in length of life seen in these populations was due to genetic variations. They also report the lack of heterosis effects.

Johnson[29] pointed out that inbreeding of the F2 from these crosses produces inbred lines which are themselves a rich source of genetic variation in length of life, showing a range of 22 days in mean lifespan from a low of 9 days to a high of almost 31 days.

ACKNOWLEDGMENTS

Supported in part by National Science Foundation PCM 8208652 to Thomas E. Johnson.

REFERENCES

1. **Brenner, S.,** The genetics of *Caenorhabditis elegans, Genetics,* 77, 71, 1974.
2. **Zuckerman, B. M., Ed.,** *Nematodes as Biological Models,* Vols. 1 and 2, Academic Press, New York, 1980.
3. **Rothstein, M.,** Effects of aging on enzymes, in *Nematodes as Biological Models,* Vol. 2, Zuckerman, B. M., Ed., Academic Press, New York, 1980, 29.
4. **Zuckerman, B. M. and Himmelhoch, S.,** Nematodes as models to study aging, in *Nematodes as Biological Models,* Vol. 2, Zuckerman, B. M., Ed., Academic Press, 1980, 3.
5. **Johnson, T. E.,** Aging in *Caenorhabditis elegans,* in *Review of Biological Research in Aging,* Rothstein, M., Ed., Alan, R. Liss, New York, in press.

6. **Johnson, T. E.,** The nematode, *Caenorhabditis elegans:* genetic and molecular aspects of aging, in *Selected Invertebrate Models for Aging Research,* Mitchell, D. H., Ed., CRC Press, Boca Raton, Fla., 1983, in press.

7. **Sulston, J. E., Schierenberg, E., White, J. G., and Thompson, J. N.,** The embryonic cell lineage of the nematode, *Caenorhabditis elegans, Dev. Biol.,* in press.

8. **Wood, W. B., Hecht, R., Carr, S., Vanderslice, R., Wolf, N., and Hirsh, D.,** Parental effects and phenotypic characterization of mutations that affect early development in *Caenorhabditis elegans, Dev. Biol.,* 74, 446, 1980.

9. **Sayre, F. W., Hansen, E. L., and Yarwood, E. A.,** Biochemical aspects of the nutrition of *Caenorhabditis briggsae, Exp. Parasitol.,* 13, 98, 1963.

10. **Tilby, M. J. and Moses, V.,** Nematode ageing. Automatic maintenance of age synchrony without inhibitors, *Exp. Gerontol.,* 10, 231, 1975.

11. **Klass, M. R.,** Aging in the nematode *Caenorhabditis elegans:* major biological and environmental factors influencing life span, *Mech. Ageing Dev.,* 6, 413, 1977.

12. **Croll, N. A. Smith, J. M., and Zuckerman, B. M.,** The aging process of the nematode *Caenorhabditis elegans* in bacterial and axenic culture, *Exp. Aging Res.,* 3, 175, 1977.

13. **Rothstein, M.,** Practical methods for the axenic culture of the free-living nematodes, *Turbatrix aceti* and *Caenorhabditis briggsae, Comp. Biochem. Physiol.,* 49B, 669, 1974.

14. **Mitchell, D. H., Stiles, J. W., Santelli, J., and Sanadi, D. R.,** Synchronous growth and aging of *Caenorhabditis elegans* in the presence of fluorodeoxyuridine, *J. Gerontol.,* 34, 28, 1979.

15. **Willett, J. D., Rahim, I., Geist, M., and Zuckerman, B. M.,** Cyclic nucleotide exudation by nematodes and the effects on nematode growth, development and longevity, *Age,* 3, 82, 1980.

16. **Gandhi, S., Santelli, J., Mitchell, D. H., Stiles, J. W., and Sanadi, D. R.,** A simple method for maintaining large, aging populations of *Caenorhabditis elegans, Mech. Ageing Dev.,* 12, 137, 1980.

17. **Bolanowski, M. A., Russell, R. L., and Jacobson, L. A.,** Quantitative measures of aging in the nematode *Caenorhabditis elegans.* I. Population and longitudinal studies of two behavioral parameters, *Mech. Ageing Dev.,* 15, 279, 1981.

18. **Hosono, R., Mitsui, Y., Sato, Y., Aizawa, S., and Miwa, J.,** Life span of the wild and mutant nematode *Caenorhabditis elegans,* Effects of sex, sterilization, and temperature, *Exp. Gerontol.,* 17, 163, 1982.

19. **Mitchell, D. H. and Santelli, J.,** Fluorodeoxyuridine as a reproductive inhibitor in *Caenorhabditis elegans:* applications to aging research, manuscript submitted.

20. **Johnson, T. E. and Wood, W. B.,** Genetic analysis of life span in *Caenorhabditis elegans, Proc. Natl. Acad. Sci. U.S.A.,* 79, 6603, 1982.

21. **Zuckerman, B. M., Himmelhoch, S., Nelson, B., Epstein, J., and Kisiel, M.,** Aging in *Caenorhabditis briggsae, Nematologica,* 17, 478, 1971.

22. **Epstein, J. and Gershon, D.,** Studies on aging in nematodes. IV. The effect of antioxidants on cellular damage and life span, *Mech. Ageing Dev.,* 1, 257, 1972.

23. **Kiesel, M., Nelson, B., and Zuckerman, B. M.,** Effects of DNA synthesis inhibitors on *Caenorhabditis briggsae* and *Turbatrix aceti, Nematologica,* 18, 373, 1972.

24. **Rothstein, M. and Cook, E.,** Nematode biochemistry, IV. Conditions for axenic culture of *Turbatrix aceti, Panagrellus redivivus, Rhabditus anomala* and *Caenorhabditis briggsae, Comp. Biochem. Physiol.,* 17, 683, 1966.

25. **Gershon, D.,** Studies on aging in nematodes, I. The nematode as a model organism for aging research, *Exp. Gerontol.,* 5, 7, 1970.

26. **Castillo, J. M. and Krusberg, L. R.,** Organic acids of *Ditylenchus triformis* and *Turbatrix aceti, J. Nematol.,* 3, 284, 1971.

27. **Hieb, W. F. and Rothstein, M.,** Aging in the free-living nematode *Turbatrix aceti.* Techniques for synchronization and aging of large-scale axenic cultures, *Exp. Gerontol.,* 10, 145, 1975.

28. **Klass, M. and Hirsh, D.,** Nonaging developmental variant of *Caenorhabditis elegans, Nature (London),* 260, 523, 1976.

29. **Johnson, T. E.,** *Caenorhabditis elegans:* a genetic model for understanding the aging process, in *Intervention in the Aging Process,* Cristofalo, V., Roberts, J., and Baker, G., Eds., Alan R. Liss, New York, 1983, in press.

30. **Hosono, R., Sato, Y., Aizawa, S. I., and Mitsui, Y.,** Age-dependent changes in mobility and separation of the nematode *Caenorhabditis elegans, Exp. Gerontol.,* 15, 285, 1980.

31. **Hosono, R.,** Age dependent changes in the behavior of *Caenorhabditis elegans* on attraction to *Escherichia coli, Exp. Gerontol.,* 13, 31, 1978.

32. **Johnson, T. E.,** unpublished data.

33. **Bolanowski, M. A., Jacobson, L. A., and Russell, R. L.,** Quantitative measures of aging in the nematode *Caenorhabditis elegans*. II. Lysosomal hydrolases as markers of senescence, *Mech. Ageing Dev.,* in press.
34. **Klass, M., Nguyen, P. N., and DeChavigny, A.,** Age-correlated changes in the DNA template in the nematode *Caenorhabditis elegans, Genetics,* 92, 409, 1979.
35. **Rose, A. M., and Baillie, D. L.,** Effect of temperature and parental age on recombination and nondisjunction in *Caenorhabolitis elegans, Genetics,* 92, 409, 1979.
36. **Klass, M. R.,** A method for the isolation of longevity mutants in the nematode *Caenorhabditis elegans* and initial results, *Mech. Ageing Dev.,* in press.

CELLULAR ASPECTS OF AGING IN INSECTS

R. S. Sohal and M. C. McArthur

INTRODUCTION

In the past few years, there has been a considerable increase in the use of insects as models for the study of the aging process. This trend is most probably attributable to the growing acceptance by the experimental biologists of the fundamental unity and similarity in the basic biochemical processes among organisms belonging to diverse phylogenetic groups. There are no major differences at the cellular or molecular levels between insects and other organisms, including the mammals. Being highly adaptable organisms, the insects have evolved to inhabit widely divergent terrestrial niches on this planet. Since adaptation to a particular habitat is often accompanied by physiological and biochemical specialization of the concerned systems, insects provide a tremendous resource for the study of biological processes. It is often possible to find an almost ideal experimental organism among insects to suit specific needs. Readers interested in further familiarization with the insect systems should consult reviews included in *The Physiology of Insecta*,[1] *Biochemistry of Insects*,[2] *Advances in Insect Physiology*,[3] and *Annual Review of Entomology*.[4]

In the context of gerontological research, insects possess certain characteristics that are potentially very useful for the elucidation of the aging processes. They have relatively short lifespans and achieve sexual maturity at a relatively early age which allows the completion and repetition of experiments at a comparatively faster pace. Large, genetically homogeneous populations can be economically maintained with ease and housed under standardized conditions in relatively small space. Somatic cells of holometabolous insects, i.e., insects undergoing complete metamorphosis such as flies, mosquitoes, bees, etc. do not divide following differentiation and are thus of the same chronological age as the organism itself. Insect tissues are avascular and are directly bathed in the body fluid called hemolymph. Nutrients and waste products are directly exchanged between the tissues and the hemolymph without the involvement of blood vessels. Oxygen is supplied to the tissues directly, by diffusion from a network of tracheolar tubes which traverse the body and open to the outside. An advantage of the open circulatory system is that entire insect organs can be isolated and maintained in vitro for a variety of studies.

The relatively small size of insects also lends itself to an easy analysis of the entire organ or organism. This minimizes sampling errors associated with individual and regional variations. An additional feature, of insects and other poikilotherms, that is highly useful for gerontological studies, is the modifiability of the lifespan. The average as well as the maximum lifespan of poikilotherms is highly responsive to metabolic rate which can be manipulated most conveniently by variation in ambient temperature and physical activity.[5]

A considerable amount of information has been gathered concerning the aging process in insects. The objective of this article is to discuss some aspects of the cellular changes occurring with age in insects. A coherent understanding of the aging process is at present lacking, and it is therefore difficult to assess the significance or the relationship of a particular time-dependent biological alteration to the aging process. Without such a perspective, aging changes described in the biological systems unfortunately remain to some extent isolated pieces of information in search of a comprehensive explanation.

DEVELOPMENT, MATURATION, AND SENESCENCE

The postembryonic life of insects consists of several successive developmental stages or instars which are punctuated by shedding of the cuticle, a process also referred to as molting.

These juvenile stages are characterized by the progressive growth of the organism and lead to the imago or adult stage with the intervening quiescent pupal stage in the holometabolous insects. During pupation the larval tissues undergo extensive histolysis and many of the adult tissues differentiate from masses of embryonic cells known as imaginal disks. Adult insects do not undergo molting but exhibit further growth and maturation immediately following emergence from the pupae, particularly in the flight muscles and the reproductive system. Flight ability improves gradually after emergence and is accompanied by an increase in the activities of several enzymes.[6] Protein content of the flight muscles gradually increases two- or threefold in the tsetse fly *Glossina*[7], and to a lesser extent in other insects. Additional growth also occurs in the gonads and accessory reproductive glands which become functional after emergence. Undoubtedly, the other organs also undergo some degree of hypertrophy during the early phase of adult life prior to sexual maturity.

Senescent changes in insects are usually studied in the adult stage and are manifested at the gross functional level in the loss of motor function and fecundity. However, the decline in the functional capacity of other organ systems has not been well studied as in the case of man.[8] One of the most notable dysfunctional alterations in insects with age is the decline in the flight potential, especially in the strong flyers like flies and mosquitoes which possess asynchronous indirect flight muscles. These muscles are physiologically interesting in that their contraction frequency ranges up to 400 times per second and is not directly dependent on neural impulses.[9,10] The major effect of aging on the muscle function is a very significant reduction in the stamina or duration of sustained flight which, as compared to the peak levels, declines by 57% in *Aedes aegypti*,[11] 65% in *Drosophila funebris*,[12] 71% in *Calliphora erythrocephala*,[13] 85% in the male housefly,[14] and 97% in female *Oncopeltus fasciatus*.[15] The speed of flight also decreases with age in several insect species.[11,16] Reduction in the flight duration and speed would obviously limit the maximum flight distance which the insects could travel and would surely contribute to the loss of adaptative and food gathering ability.

Another feature of aging concerning the flight ability is an alteration in the wing beat frequency which has been reported to increase with age in *Phormia regina*,[17] *Calliphora erythrocephala*,[13] *Musca domestica*,[18] and *Drosophila melanogaster*.[19] The increase in wing beat frequency can be due to abrasion of wings which is known to occur in aging houseflies due to copulatory activity.[20,21] Partial removal of wings, experimentally, also results in an increase in the wing beat frequency.[22] Thus, this alteration may not be a true manifestation of senescence but rather an indication of mechanical damage.

The reproductive potential of insects, like other organisms, also declines in older organisms. The number of eggs laid in each batch and the percentage viability of the eggs is lower in older as compared to younger flies.[23,24] Other functional deteriorations with age include an increase in the number of inoperative chemoreceptor sensillas.[25,26] Although losses in functional levels of different organs in aging insects have not as yet been extensively explored, there is no reason to doubt that senescence-associated loss in the adaptability of the organisms and the resultant mortality is accompanied by widespread decline in functional capacity of the different organ systems. It is generally agreed that age-associated deaths occur due to the inability of the organism to reestablish homeostasis following destabilization by environmental challenges. In order to understand the basis of physiological decline, efforts have been made to identify the loci of age-dependent deterioration at different biological levels. Functional decline at the organ level is obviously an indication of the alterations occurring at the cellular and molecular level.

CHANGES IN BODY COMPOSITION

Some insects have a tendency to store within their cells organic and inorganic waste products of metabolism. The primary sites for storage are fat body, pericardial cells, midgut,

Malphighian tubules, epidermis, and the wings. Urates accumulate progressively with age in specialized cells within the fat body of a variety of insect species.[27] Uric acid is also stored in some species as an excretory product. Several other organic waste products such as pteredines, carotinoids, biliverdin, and a variety of pigments (some of undetermined nature) also accumulate in certain tissues as end-products of metabolisms.[27] Although these organic wastes have not been specifically associated with aging and are not considered to be toxic, their rate of accumulation may act as a cellular indicator of metabolic activity.[28] Investigators should be cognizant of the potential interference of the storage-excretion products.

Inorganic wastes such as metallic cations, sulfur, and phosphorous progressively accumulate as intracellular concretions in the epithelium and midgut of several insects.[29-31] Experimental supplementation of the diet of the housefly with different concentrations of copper sulfate, iron gluconate, calcium phosphate, and zinc sulfate was found to result in a corresponding increase in the levels of minerals within the tissues. The intracellular minerals were bound to an organic matrix rich in acid mucopolysaccharides within the secondary lysosomes and lipofuscin-like granules. Progressive accumulation of minerals in the housefly appeared to be an excretory phenomenon since the older organisms contained relatively greater amount of minerals. It was postulated that this phenomenon may help in the conservation of water.[32,33]

Investigations on the changes in the relative composition of the body generally indicate that there are no significant alterations with age in the total weight, water, protein, DNA, and RNA content in the adults of *Anopheles*[34] and *Drosophila melanogaster*.[35] In contrast, studies on the adult male housefly, conducted in this laboratory, indicated that the total weight of the sucrose-fed flies decreased about 30% between 12 hr to 18 days of adult age (Table 1), whereas flies fed on a protein-supplemented diet showed a 20% decrease in total body weight during the same period (Table 2). With age, the average dry weight of the houseflies declined with age, paralleling the loss in total body weight. The concentration of base-soluble proteins, extracted from the houseflies by the procedure of Burcomb and Hollingsworth,[35] did not alter significantly with age in sugar-fed flies. However, protein-fed flies showed a gradual increase of about 25% in protein content until 10 days of age followed by a slight decline after 12 days of age, but the 18-day-old flies still exhibited a greater concentration of base-soluble proteins than the newly emerged flies. The concentration of protein-free α-amino acids in the housefly determined by the method outlined by Burcomb and Hollingsworth,[35] showed a 60% decline prior to the achievement of the average lifespan in sucrose-fed houseflies, wherease the total amino acid levels in the protein-fed flies were highly variable at different ages. Concentration of proline and hydroxyproline, which are a major source of energy for flight in some insects, was found to decline by about 55% during the first week of life in sugar-fed flies. In the protein-fed flies, a similar decrease occurred, but more gradually. Other alterations in body composition with age include a linear increase in the citrate concentration in the housefly[36] and a decline in thiamine phosphate levels.[37] In conclusion, studies on the body composition of insects have yielded at least two important pieces of information: (1) there are considerable species-specific differences and (2) that diet has a major effect on the pattern of age-associated changes in the concentration of macromolecules.

AGING CHANGES IN TISSUES

Various tissues of the insects have been examined in order to elucidate the cellular basis of the age-associated decline in physiological efficiency.

Flight Muscle

The flight muscles have been very extensively investigated perhaps due to the fact that the flight performance declines significantly with age in most species. Age-associated changes

Table 1
CHANGES IN BODY COMPOSITION OF SUCROSE-FED ADULT MALE HOUSEFLIES[a]

Age (days)	Total body wt.	Dry body wt.	Protein	Amino acids	Proline and hydroxyproline
0.5 ± 0.5	100	100	100	100	100
2—3	91.9	—	90.1	75	90
4—5	83.7	82.3	99.6	86	92
5—6	80.5	91.7	107.6	66	56
7—8	74.5	80.6	99.3	82	56
9—10	74.7	78.0	89.4	70	59
11—12	75.6	75.8	84.5	54	56
13—14	68.4	70.0	109.9	67	—
15—16	72.9	71.2	105.6	47	63
17—18	69.9	65.6	103.5	40	—

[a]   Expressed as percent of the content at 0.5 days of adult age.

Table 2
CHANGES IN BODY COMPOSITION OF PROTEIN-FED ADULT MALE HOUSEFLIES

Age (days)	Total body wt.	Dry body wt.	Protein	Amino acids	Proline and hydroxyproline
0.5 ± 0.5	100	100	100	100	100
2—3	88.3	93.8	105.6	92	—
4—5	91.3	—	112.4	114	—
5—6	82.3	87.7	117.8	68	97
7—8	85.9	—	122.3	76	—
9—10	86.2	86.6	124.8	116	—
11—12	—	—	142.6	97	—
13—14	80.0	87.2	137.2	67	75
15—16	—	—	115.6	114	78
17—18	80.3	87.6	109.4	63	74

[a]   Expressed as percent of the content at 0.5 days of adult age.

in the structure of the flight muscles have been examined in the housefly,[38-40] and in the blowflies *C. erythrocephala*[41] and *P. regina*,[42] and in *D. melanogaster*[43-44] (see reviews, References 45 and 46). Some of the age-associated changes, particularly those occurring during the first week of adult life, are maturational in nature, whereas those occurring later may be associated with senescence. One of the most prominent maturational alterations in the asynchronous flight muscles during the first week of life is an increase in the size of the mitochondria. This increase results from a dual process involving fusion of individual organelles and biosynthesis of additional mitochondrial mass, culminating in the formation of large structures often referred to as ''giant'' mitochondria. As a result of fusion, the number of mitochondria decreases during this period.[39,45] In older insects, mitochondria show focal degeneration characterized by whorl-like structures[38,42,47-49] which lack cytochrome oxidase activity.[42] Focal degeneration of the myofibrils occurs in the flight muscles of the housefly.[45] Decline in glycogen content has been correlated with declining flight ability.[12] In general, the structural changes in the flight muscles observed thus far are not of sufficient magnitude to reflect the tremendous decline of flight ability occurring with age.

Several studies have been done on the oxidative metabolism of insects in order to elucidate the mechanism of flight loss. Oxygen consumption rates of the resting adult blowfly *C. erythrocephala* was found to increase with age and was thought to be associated with a decrease in the oxidative phosphorylation ratios and respiratory control values of mitochondria from the flight muscles.[50] Aging was considered to have a deleterious effect on the tightness of the coupling between oxidation and phosphorylation. In another blowfly species, *P. regina,* no differences were found in P:O ratios, but the respiratory control indices in older flies were lower than in the younger flies. There was also a decrease in the in vitro ability of mitochondria of old flies to oxidize substrates such as pyruvate-proline and α-glycerophosphate.[51] Although the nature of the factors responsible for a decline in the respiratory efficiency of mitochondria are yet to be elucidated, these changes suggest possible alterations in the inner mitochondrial membrane. It should be pointed out, however, that the mitochondria in the asynchronous flight muscles are large, irregular structures which are extremely susceptible to damage during isolation procedures. Since the increase in mitochondrial size, due to fusion, is an age-associated phenomenon, the possibility exists that some of the reported alterations in the in vitro mitochondrial respiratory efficiency may, in part, be due to mechanical damage to the mitochondria.

Attempts have been made to correlate aging changes in the substrate levels, enzyme activities, and protein synthesis rates to the flight muscle function. The initial stimulus for these studies was provided by Williams et al.[12] who reported a correspondence between flight stamina and glycogen levels in drosophila. Inability to utilize all the available glycogen prior to flight exhaustion has been noted in older mosquitoes in contrast to young mosquitoes which expend all the available glycogen before exhaustion.[11,16]

In a series of studes, Rockstein and co-workers[6] have presented data indicating an integrated pattern of enzymatic and substrate changes associated with flight loss in the housefly. Four enzymes, namely, arginine phosphotransferase, extramitochondrial α-glycerophosphate dehydrogenase, magnesium-activated ATPase, and cytochrome *c* oxidase undergo an age-associated pattern of activity in a highly sequential manner. A similar pattern in the activity of the first two enzymes has also been reported in *Phormia* and *Drosophila.*[52] The activity of magnesium-activiated ATPase was found to be associated with the loss of wings in the housefly.[53]

The flight muscles of dipteran insects seem to have highly stable proteins with relatively little protein synthesis occurring during the postmaturational phase of adult life.[54] It has suggested that the low rate of protein turnover and the resultant lack of repair mechanisms in the mitochondria of flight muscle of *Calliphora* may be responsible for deterioration of flight performance with age.[41] It seems that a variety of deleterious alterations occur in the flight muscles of dipteran insects with age, but it is not known which of these changes is directly responsible for the functional decline.

It is possible that flight loss is due to alterations in other organs interacting with the flight muscles.

Brain

Age-associated changes in the nerve cells are of interest due to integrative role of the brain. A major concern in mammals as well as insects has been the role of the nerve cell loss with age. A 33% loss in the number of nerve cells in the antennal lobe has been reported in the honey bee;[55] however, measurement based on the distribution of nerve cells in various regions of the brain of housefly revealed no significant differences between the young and old flies.[56] Nerve cell depletion in mammals is not considered to be of sufficient magnitude to explain the age-associated decline in brain function.[57] In general, the view that decline in functional capacity of organs may be due to cell depletion lacks support. However, several structural alterations have been reported in the nerve cells of the brain of aging insects which include the loss of ribosomes, focal cytoplasmic degeneration, autophagy, and accumulation of lipofuscin granules.[56,58] A decrease in the activity of Mg^{++} ATPase and $Na^+ K^+$ ATPase in the brain and a greater sensitivity to DDT has been reported in aging honey bees.[59]

Fat Body

The fat body in insects is the chief site of intermediary metabolism, of synthesis of organic molecules of the hemolymph, and lipid, protein, and carbohydrate storage. Its functional role is thus somewhat analogous to the combined functions of liver and adipose tissue in mammals. Two types of cells, the fat cells and oenocytes, compose the abdominal fat body in the housefly.[60] The fat cells are characterized by the inclusion of lipid droplets, glycogen, and crystalline materials, presumably proteinous in nature. The chief cytological characteristic of oenocytes is the abundant presence of smooth endoplasmic reticulum. In the old houseflies, the fat cells showed an increase in the volume of fat droplets and a reduction in glycogen. In the oenocytes, the smooth endoplasmic reticulum was greatly reduced. The most prominent aging change was the accumulation of lipofuscin granules. The fat body in old insects was also infiltrated by phagocytic hemocytes.[60]

Heart

In insects, the pulsatile organ analogous to the vertebrate heart is the abdominal portion of the dorso-medially located dorsal vessel. The hearts of several insect species have been examined by electron microscopy, and there do not appear to be any fundamental differences from the mammalian heart cells. Fine structural changes have been studied in the heart of the housefly[61,62] and *D. repleta*.[62-64] The hearts in the old houseflies showed a 15 to 20% reduction in the contractile components as compared to the young flies. Focal areas of the heart showed fragmentation and dissolution of the myofilaments. Another notable change in the heart cells of old houseflies was the presence of autophagic vacuoles and lipofuscin granules. A peculiar aspect of aging in the heart of *D. repleta* was the accumulation of glycogen particles in the matrix of mitochondria. It thus seems that the insect heart undergoes aging changes in structure but, besides the accumulation of lipofuscin, the pattern of changes may differ between the species.

Physiological changes in the heart of the house cricket *Acheta domesticus* with age provide the main source of information in this area.[65-67] The heart rate of in vitro preparations was found to increase until 6 weeks of age and declined thereafter. The response of the heart to pharmacologic stimuli also varied between the young and the old crickets which was thought to be associated with alterations in the receptor sites on the cell surface.

Midgut

The main function of the midgut in insects is the synthesis of digestive enzymes and the absorption of nutrients. The midgut epithelium in the old insects often contains dense inclusions similar to lipofuscin in morphology.[30,68-71] X-ray microanalysis of the midgut of aging houseflies indicated that the dense bodies also contained high concentrations of phosphorous, sulfur, chlorine, calcium, iron and copper.[32] The minerals were initially deposited within the Golgi vesicles and lamellar bodies. The concretionary granules after inclusion in the autophagic vacuoles seemed ultimately to become a component of the residual bodies which also exhibit a positive localization of acid phosphatase activity and can thus be considered lysosomal in nature. It was postulated by Sohal et al.[32] that the dense bodies or lipofuscin granules in the midgut of the housefly sequester superfluous minerals and play a role in the excretory system of the housefly.

Malpighian tubules

The major function of the Malpighian tubules in insects is the excretion of nitrogenous and inorganic wastes and as such they play a functional role analogous to the kidney in the vertebrates. Fine structural studies on the Malpighian tubules of aging houseflies and drosophila have indicated a decrease in ribosomal distribution and an increase in the autophagic vacuoles and lipofuscin granules.[29,72]. An X-ray microanalysis study of the Malpighian tubules in the housefly indicated that the lipofuscin-like granules contained high concentrations of phosphorous, sulfur, zinc, calcium, iron, and copper.[29] These concretionary structures were thought to be involved in the sequestration of metal ions. The tubules are also known to accumulate yellowish-green pigment with age.[73]

ENZYME LEVELS AND PROTEIN SYNTHESIS

Because of the relationship between enzyme activities and the functional capacities of cells, studies have been directed towards the quantitative changes in enzyme activities occurring with age. Similarly, changes in protein synthesis rates with age are of interest due to the structural and enzymatic role of protein molecules. It is now well established that various protein molecules in the cell undergo a constant turnover albeit at different rates. Different proteins constituting a specific cell organelle also have an independent and characteristic turnover rate.[74] Although the process of protein breakdown is considered to be a random one, there is some evidence that "older" molecules may be more susceptible to degradation.[75] Obviously, a decline in protein synthesizing ability would have a deleterious effect on the functional capacity of cells, especially under stressful conditions.

Studies on the specific activity of enzymes in relation to aging in insects indicate the absence of a uniform pattern in the enzyme activity with age. Some enzymes, e.g., alkaline phosphatase in the flour beetle,[76] trehlase in the tobacco hornworm,[77] succinic-, isocitric-, and malate-dehydrogenases in the housefly,[78] NADP-cytochrome c reductase,[79] and glutamic aspartic transaminase in drosophila[80] do not exhibit any appreciable changes with age. A decrease in enzyme level in whole body homogenates has been reported in the case of several enzymes in the flight muscle of the housefly;[6] peroxidase,[18] catalase, trehlase, α-glycerophosphate dehydrogenase, esterase, and alcohol dehydrogenase in drosophila.[80] In contrast, other enzymes such as acid phosphatase in the housefly[82] and NADP-cytochrome c reductase in the mosquito[79] increased in activity with age. In general, it appears that several enzymes, especially those related to oxidative metabolism have reduced activity with age. Many other enzymes probably remain at about the same level throughout adult life.

A notable age-associated changes in insects is the decrease in the rate of protein synthesis in older organisms. A decline in protein synthesis has been reported in $D. melanogaster$,[83] $P. regina$,[84] and $C. erythrocephala$.[41] Smith et al.[54] reported relatively little protein turnover in the flight muscles of $D. subobscura$ and about 20% of the total protein in the body exhibited a turnover rate of about 10 days. It has been postulated that age-associated functional decline in insects may be due to the insufficiency of protein renewal and repair mechanisms.[41,54]

To recapitulate, the aging changes reported in insects suggest that some deteriorative changes may occur in the distribution and structure of cell organelles. There is, however, no severe degeneration or loss of organelles as a result of aging. The most ubiquitous aging change in insect cells is the accumulation of lipofuscin. At the biochemical level, there is some indication that protein synthesis rates decline with age and a decrease may also occur in the activity of some enzymes. These changes are essentially similar to those occurring in vertebrate cells which further supports the view that insects can be used as models for elucidating the basic and fundamental aspects of the aging process.

LIPOFUSCIN ACCUMULATION AND AGING

Although a large number of studies have attempted to define the loci of age-dependent deterioration, the nature of the mechanisms that effect age-associated deterioration of homeostatic mechanisms remains elusive. Such studies have revealed a variety of alterations in various systems of different organisms with age. However, on the basis of these descriptive studies, it is difficult to distinguish between age-associated changes that are trivial in nature and those that are specific to the aging process. The latter are presumed to be associated with functional deterioration of the organism and contribute to its decreased viability. In order to establish the specificity of a time-related alteration to the aging process, it is imperative to demonstrate its relationship to physiological rather than chronological age. Unfortunately the bulk of the gerontological investigations conducted thus far are unable to draw such a distinction.

Cytological studies on the tissues of aging eukaryotes examined thus far indicate that the most consistent feature of aging changes is the accumulation of lipofuscin or "age pigment." Considerable efforts have been made to study morphology, origin, composition, and possible functional significance of lipofuscin.[85] Lipofuscin granules are rounded or oblong in shape, 1 to 5 μm in diameter, increasing in size with age. Lipofuscin has a brownish color but exhibits characteristic autofluorescence in near UV light. It shows positive histochemical reactions for carbohydrates, lipids, and hydrolytic enzymes.[86]

Ultrastructural studies have shown that lipofuscin granules are bounded by a single limiting membrane with a heterogeneous internal organization consisting of a pleomorphic combination of granular particles, vacuoles, and aggregates of membranous material.[87] There is, however, considerable variation in the morphology of lipofuscin granules in different organisms. In insects, the lipofuscin granules, present in various tissues, can be divided into two broad categories.[56] The first type consists of membranous bodies which are composed of concentric or stacked arrays of dense lamellae. In the second category are structures with a dense matrix often associated with vacuolar material, presumably lipid in nature and/or membranous aggregates. Lipofuscin in the housefly is predominantly membranous in appearance.[56,60] Similar membranous structures have been characterized as ceroid in *D. melanogaster*.[71,87] Customarily, the term ceroid denotes the autofluorescent granules which accumulate in response to a variety of pathological conditions related to metabolic disturbances such as vitamin E deficiency, choline deficiency, hormone treatment, and feeding of unsaturated fatty acids.[88] All of these conditions interfere with normal lipid metabolism. Lipofuscin, on the other hand, is considered to be a product of the normal aging process. In fact, ceroid has also been characterized as partially oxidized lipofuscin.[86] The most readily identifiable differences in the properties of ceroid and lipofuscin in the mammals are that ceroid is chiefly membranous and is easily extractable in a chloroform-methanol mixture, whereas lipofuscin is more heterogeneous in organization and is highly resistant to extraction.[89,90] Despite some similarities between age pigment in insects and ceroid granules observed under pathological conditions in mammals, the term ceroid for the age pigments in insects is somewhat inappropriate since these structures occur as a product of normal aging rather than as a result of some experimental or pathological treatment. It is therefore desirable to retain the term lipofuscin for the sake of nomenclatural conformity.

Chemical analyses of isolated lipofuscin have indicate the presence of lipids, proteins, and acid hydrolysis-resistant residues.[89] Extracts of purified lipofuscin in chloroform contain a complex of lipids.[91] The fluorescent substances in the lipofuscin granules have been related to lipid peroxidation of membranous material and have been chemically characterized as conjugated Schiff bases.[92]

Investigations on the origin of lipofuscin have been made by morphological and biochemical approaches. At the morphological level, an overwhelming body of evidence suggests that lipofuscin originates by a process of autophagocytosis. Focal areas of the cytoplasm are enclosed within autophagic vacuoles which later fuse with lysosomes. The undigested residues of autophagy accumulate in the form of lipofuscin granules. The biochemical studies, on the other hand, have sought to identify the reactions which may lead to the formation of the fluorescent substances.[87,92] It has been postulated that the free radicals generated by univalent reduction of oxygen in aerobic organisms react with unsaturated lipids and lead, through a sequence of reactions, to the formation of malonaldehyde which can cross-link amino groups of proteins, nucleic acids, and phospholipids, producing fluorescent Schiff base products with the structure RN–CH–CH–NH–R.[92] Schiff base compounds have a fluorescent excitation maxima of 360 to 390 nm and an emission maxima of about 450 to 470 nm, which is similar to that exhibited by lipofuscin granules *in situ* and in chloroform extracts.[73]

It is thought that the free radicals, generated as a result of aerobic respiration, inflict damage on the various biological molecules in a manner which is similar to that caused by ionizing radiation.[92] Lipid peroxidation damage to the organelle membranes may thus provide the stimulus for cellular autophagy in older organisms. Lipofuscin accumulation is greatly accelerated under pathological conditions such as vitamin E deficiency[93] and some neuropathies,[94] and in one instance has been associated with the deficiency of peroxidase activity.[95] Factors that apparently affect the rate of lipofuscin accumulation also seem to have some relationship to the free radical-stimulated reactions. In insects, the average and maximum lifespan of the adults can be influenced by environmental temperature. It is well known that the poikilotherms live much longer at lower ambient temperatures than at high temperatures. Similarly, the level of physical activity has a direct effect on longevity of houseflies. Flies kept under low activity conditions live twice as long as those engaged in high level of physical activity.[5] Higher temperatures affect lifespan by increasing the level of physical activity.[21] The results of these studies suggest that lifespan and aging rates in insects are strongly influenced by metabolic rate. The nature of the exact mechanism by which metabolic rate affects lifespan is unknown; it is however possible that higher rates of oxygen utilization generate correspondingly higher levels of free radicals.

The rate of lipofuscin accumulation has been related to the metabolic rate in the housefly.[96-99] Lipofuscin accumulated at a slower rate in the long-lived, low-activity flies as compared to the high-activity, short-lived flies. Lipid peroxidation levels were also higher in the high-activity flies. A relationship thus seems to exist between metabolic rate, lifespan, and lipofuscin accumulation. Rate of lipofuscin accumulation apparently corresponds to the rate of physiological aging and can be considered as an indicator of aging rate.[97] Lipofuscin itself is neither thought to be toxic nor is there any compelling evidence to suggest that its existence within the cell has a deleterious effect. It is, however, suggested that lipofuscin is an indicator of the cellular damage and studies on its genesis may be helpful in elucidating the nature of the aging mechanism.

ACKNOWLEDGMENTS

Some of the original research discussed in this review has been supported by grants from the National Institutes of Health, National Institute on Aging (RO1AG00171). Financial support provided by Mrs. Louis Jones Memorial Gift is gratefully acknowledged.

ADDENDUM

Since the submission of this article in mid-1980, a considerable number of studies, concerning aging in insects, have appeared in the interim four-year period (for review, see Sohal[100]). Considerable evidence has been presented to indicate that enzymic and nonenzymic antioxidant defenses tend to decline with age, whereas free radical-induced damage increases in old insects (for review, see Sohal[101]). Levels of antioxidants, such as superoxide dismutase, catalase, glutathione, and vitamin E decrease with age in adult housefly; whereas levels of the products of free radical reactions such as H_2O_2, oxidized glutathione, and *n*-pentane evolution tend to increase with age in the housefly.[102]

The use of insects as model systems for the study of the aging process seems to have reached a stage of considerable popularity and acceptance as indicated by a forthcoming compendium.[103]

REFERENCES

1. **Rockstein, M., Ed.,** *The Physiology of Insecta,* Vols. 1 to 6, Academic Press, New York, 1973.
2. **Rockstein, M.,** *Biochemistry of Insects,* Academic Press New York, 1978.
3. **Treherne, J. E., Berridge, M. J., and Wigglesworth, W. B.,** *Advances in Insect Physiology,* Academic Press, London, 1979.
4. **Mittler, T. E., Radovsky, D. J., and Resh, V. H., Eds.,** *Annual Review of Entomology,* Annual Reviews Inc., California,
5. **Sohal, R. S.,** Metabolic rate and life span, in *Interdisciplinary Topics in Gerontology,* Vol. 9, von Hahn, H. P., Ed., S. Karger, Basel, 1976, 25.
6. **Rockstein, M. and Miquel, J.,** Aging in insects, in *The Physiology of Insecta,* Vol. 1, Rockstein, M., Ed., Academic Press, New York, 1973, 371.
7. **Bursell, E.,** Development of mitochondrial and contractile components of flight muscle in adult tsetse flies, *Glossina moristans, J. Insect Physiol.,* 19, 1079, 1973.
8. **Shock, N.,** Mortality and measurement of aging, in *The Biology of Aging,* Strehler, B. L., Ed., American Institute of Biological Science, Washington, D.C., 1960.
9. **Pringle, J. W. L.,** The contractile mechanism of insect fibrillar muscle, *Prog. Biophys. Mol. Biol.,* 17, 1, 1966.
10. **Usherwood, P. N. N.,** Insect neuromuscular mechanisms, *Am. Zool.,* 7, 553, 1967.
11. **Rowley, W. A. and Graham, C. L.,** The effect of age on the flight performance of female *Aedes aegypti* mosquitoes, *J. Insect Physiol.,* 4, 719, 1968.
12. **Williams, C. M., Barness, L. A., and Sawyer, W. H.,** Utilization of glycogen by flies during flight and some aspects of the physiological ageing of *Drosophila, Biol. Bull.,* 84, 263, 1943.
13. **Tribe, N. A.,** Some physiological studies in relation to age in the blowfly, *Calliphora erythrocephala, J. Insect Physiol.,* 12, 1577, 1966.
14. **Rockstein, M. and Bhatnagar, P. L.,** Duration and frequency of wing beat in the aging housefly, *Musca domestica* L., *Biol. Bull.,* 131, 479, 1966.
15. **Dingle, H.,** The relation between age and flight activity in the milkweed bug, *Oncopeltus, J. Exp. Biol.,* 42, 269, 1965.
16. **Nayar, J. K. and Sauerman, D. M.,** A comparative study of flight performance and fuel utilization as a function of age in females of Florida mosquitoes, *J. Insect Physiol.,* 19, 1977, 1973.
17. **Levenbook, L. and Williams, C. M.,** Mitochondria in the flight muscles of insects. III. Mitochondrial cytochrome C in relation to the aging and wing beat frequency of flies, *J. Gen Physiol.,* 39, 497, 1956.
18. **Rockstein, M. and Bhatnagar, P. L.,** Age changes in size and number of the giant mitochondria in flight muscle of common housefly, *J. Insect Physiol.,* 11, 481, 1965.
19. **Chadwick, L. E. and Williams, C. M.,** The effect of atmospheric pressure and composition of flight of *Drosophila, Biol. Bull. Woods Hole,* 97, 115, 1949.
20. **Patterson, R. S.,** On the causes of broken wings of the housefly, *J. Econ. Entomol.,* 50, 104, 1957.
21. **Ragland, S. S. and Sohal, R. S.,** Mating behavior, physical activity and aging in the housefly, *Musca domestica, Exp. Gerontol.,* 8, 135, 1973.
22. **Pringle, J. W. S.,** Locomotion: flight, in *The Physiology of Insecta,* Vol. 3, Rockstein, M., Ed., 1974, 433.
23. **Woke, P. A., Ally, M. S., and Rosenberger, C. R., Jr.,** The number of eggs developed related to the quantities of human blood ingested in *Aedes aegypti* L. (Diptera: Culicidae), *Ann. Entomol. Soc. Am.,* 49, 435, 1956.
24. **Richards, A. G. and Kolderie, M. Q.,** Variation in weight, developmental rate, and hatchability of *Oncopeltus* eggs as a function of the mother's age, *Entomol. News,* 68, 57, 1957.
25. **Reese, C. J. C.,** Age dependency of response in an insect chemoreceptor sensillium, *Nature (London),* 227, 740, 1970.
26. **Stoffalano, J. G., Jr.,** Effect of age and diapause on the mean impulse frequence and failute to generate impulses in labellar chemoreceptor sensilla of *Phormia regina, J. Gerontol.,* 28, 35, 1973.
27. **Wigglesworth, V. B.,** *The Principles of Insect Physiology,* Halsted, Press, New York, 1972.

28. **Clark, A. M. and Smith, R. E.,** Urate accumulation and adult life span in two species of *Habrobracon, Exp. Gerontol.,* 2, 217, 1967.
29. **Sohal, R. S., Peters, P. S., and Hall, T. A.,** Origin, structure, composition and age dependence of mineralized concretions in Malphigian tubules of housefly, *Musca domestica, Tissue and Cell,* 8, 447, 1976.
30. **Sohal, R. S., Peters, P. S., and Hall, T. A.,** Origin, structure, composition and age dependence of mineralized dense bodies (concretions) in the midgut epithelium of the adult housefly, *Musca domestica, Tissue Cell,* 9, 87, 1977.
31. **Martoja, A.,** Donees preliminaires sur les accululations de sels mineraux et de dechets du catabolisme dan quelques organes d'arthropodes, *C. R. Acad. Sci. Paris (D),* 273, 268, 1971.
32. **Sohal, R. S. and Lamb, R. E.,** Intracellular deposition of metals in the midgut of the adult housefly, *Musca domestica, J. Insect Physiol.,* 23, 1349, 1977.
33. **Sohal, R. S. and Lamb, R. E.,** Storage-excretion of metallic cations in the adult housefly, *Musca Domestica, J. Insect Physiol.,* 25, 119, 1979.
34. **Lang, C. A., Lau, H. Y., and Jefferson, D. J.,** Protein and nucleic acid changes during growth and aging in the mosquito, *Biochem. J.,* 95, 372, 1965.
35. **Burcombe, J. V. and Hollingsworth, M. J.,** The total nitrogen, protein, amino acid and uric acid content of ageing *Drosophila, Exp. Gerontol.,* 5, 247, 1970.
36. **Zahavi, M. and Tahori, A. S.,** Citric acid accumulation with age in houseflies and other diptera, *J. Insect Physiol.,* 11,, 811, 1965.
37. **Rockstein, M. and Hawkins, W. B.,** Thiamine in the aging housefly, *Musca domestica L., Exp. Gerontol.,* 5, 187, 1969.
38. **Sohal, R. S. and Allison, V. F.,** Age-related changes in the fine structure of the flight muscle in housefly, *Exp. Gerontol.,* 6, 167, 1971.
39. **Sohal, R. S., McCarthy, J. L., and Allison, V. F.,** The formation of "giant" mitochondria in the fibrillar flight muscles of the housefly, *Musca domestica, J. Ultrastruct. Res.,* 39,, 484, 1972.
40. **Rochstein, M., et al.,** An electron microscope investigation of age dependent changes in flight muscle of *Musca domestica L., Gerontology,* 21, 216, 1975.
41. **Tribe, M. S. and Ashhurst, D. E.,** Biochemical and structural variations in the flight muscle mitochondria of aging blowflies *Calliphora erythrocephala, J. Cell Sci.,* 10, 443, 1972.
42. **Sacktor, B. and Shimada, Y.,** Degenerative changes in mitochondria of flight muscle from aging blowfly, *J. Cell Biol.,* 52, 465, 1972.
43. **Takahashi, A., Philpott, D. E., and Miquel, J.,** Electron microscope studies of aging *D. melanogaster,* III. Flight muscle, *J. Gerontol.,* 25, 222, 1970.
44. **Sohal, R. S.,** Mitochondrial changes in flight muscles of normal and flightless *Drosophila melanogaster* with age, *J. Morphol.,* 45, 337, 1975.
45. **Sohal, R. S.,** Aging changes in insect flight muscle, *Gerontology,* 22, 317, 1976.
46. **Baker, G. T.,** Insect flight muscle: maturation and senescence, *Gerontology,* 22, 334, 1976.
47. **Davies, I.,** The effect of age and diet on the ultrastructure of Hymenoptera flight-muscle, *Exp. Gerontol.,* 9, 215, 1974.
48. **Webb, S. and Tribe, M. A.,** Are there major degenerative changes in the flight muscle of aging diptera?, *Exp. Gerontol.,* 9, 43, 1974.
49. **Johnson, B. G. and Rowley, W. A.,** Age related ultrastructural changes in the flight muscle of the mosquito, *Culex tarsalis, J. Insect Physiol.,* 18, 2375, 1972.
50. **Tribe, M. A.,** Some physiological studies in relation to age in the blowfly, *Calliphora erythrocephala* Meig, *J. Insect Phsiol.,* 12, 1577, 1966.
51. **Bulos, B., Shukla, L., and Sacktor, B.,** Bioenergetic properties of mitochondria from flight muscle of aging blowflies, *Arch. Biochem. Biophys.,* 149, 461, 1972.
52. **Baker, G. T.,** Identical age related pattern of enzyme activity in *Phormia regina* and *Drosophila melanogaster, Exp. Gerontol.,* 10, 231, 1974.
53. **Rockstein, M. and Brandt, K. S.,** Enzyme changes in flight muscle correlated with aging and flight ability in the male housefly, *Science,* 139, 1049, 1963.
54. **Maynard Smith, J., Bozcuk, A. N., and Tebbutt, S.,** Protein turnover in adult *Drosophila, J. Insect Physiol.,* 16, 601, 1970.
55. **Hodge, C. F.,** Changes in ganglion cells from birth to senile death. Observations on man and honey-bee, *J. Physiol.,* 17, 129, 1894.
56. **Sohal, R. S. and Sharma, S. P.,** Age-related changes in the fine structure and number of neurons in the brain of the housefly, *Musca domestica, Exp. Gerontol.,* 7, 243, 1972.
57. **Brody, H. and Vijayashankar, N.,** Anatomical change in the nervous system, in *Handbook of the Biology of Aging,* Finch, C., and Hayflick, L., Eds., Van Nostrand, New York, 1977.

58. **Herman, M. H.,** Insect brain as a model for the study of aging, *Acta Neuropathol.,* (Berlin), 19, 167, 1971.
59. **Ching, E. Y. and Cutkomp, L. K.,** Aging in the honey bee *Apis mellifera,* as related to brain ATPases and their DDT sensitivity, *J. Insect Physiol.,* 18, 2285, 1972.
60. **Sohal, R. S.,** Fine structural alterations with age in the fat body of the adult male housefly, *Musca domestica, Z. Zellforsch.,* 140, 169, 1973.
61. **Sohal, R. S., and Allison, V. F.,** Senescent changes in the cardiac myofiber of housefly, *Musca domestica,* An electron microscopic study, *J. Gerontol.,* 26, 490, 1971.
62. **Sohal, R. S.,** Mitochondrial changes in the heart of *Drosophila repleta, Exp. Gerontol.,* 5, 213, 1970.
63. **Burch, G. E., Sohal, R. S., and Fairbanks, L. D.,** Senescent changes in the heart of *Drosophila repleta, Nature (London),* 115, 386, 1970.
64. **Sohal, R. S.,** Aging changes in the structure and function of the insect heart, in *Aging,* Kaldor, G., and DiBattista, W. F., Eds., Raven Press, New York, 1978.
65. **McFarlane, J. E.,** Aging in an adult insect heart, *Can. J. Zool.,* 45, 1073, 1967.
66. **McFarlane, J. E. and Fong, K. T.,** Differences in the effect of drugs on young and old heart of the house cricket, *Acheta domesticus, Comp. Gen. Pharmacol.,* 3, 271, 1972.
67. **Srivastava, S. T.,** Aging of the Circulatory System of the House Cricket, *Acheta domesticus* (L), in Relation to Photoperiod, Doctoral dissertation, McGill University, Montreal, 1975.
68. **Hecker, H., Freyvogel, T. A., Briegel, H., and Steiger, R.,** The ultra structure of midgut epithelium in *Aedes aegypti* (L) (Insecta, Deptera) males, *Acta Trop.* (Basel), 28, 275, 1971.
69. **Gartner, L. P.,** Ultrastructural examination of aging and radiation induced lifespan shortening in adult *Drosophila melanogaster, Int. J. Radiat. Biol.,* 23, 23, 1973.
70. **Priester, W. De.,** Dense bodies and lytic activity in the digestive epithelium of a fly, *Z. Alternforsch.,* 28, 65, 1974.
71. **Miquel, J. et al.,** Fluorescent products and lysosomal components in aging *Drosophila melanogaster, J. Gerontol.,* 29, 622, 1974.
72. **Miquel, J.,** Aging of male *Drosophila melanogaster,* in *Advances in Gerontology Research,* Vol. 3, Strehler, B. L., Ed., Academic Press, New York, 1971, 39.
73. **Haydak, M. H.,** Increase of the protein level of the diet on the longevity of cockroaches, *Ann. Entomol. Soc. Am.,* 46, 547, 1953.
74. **Schimke, T. R.,** Principle underlining the regulation of synthesis and degradation of proteins in animal tissues, in *The Neurosciences III,* Schmitt, F. O., and Worden, F. G., Eds., MIT Press, Cambridge, Mass., 1974, 827.
75. **Goldberg, A. L.,** Regulation and importance of intracellular protein degradation, in *The Neurosciences III,* Schmitt, F. O. and Worden, F. G., Eds., MIT Press, Cambridge, Mass., 1974, 827.
76. **Raychaudhuri, A. and Butz, A.,** Aging. II. Changes in acid and alkaline phosphatase contents of aging male and female *Tribolium confusum* (Coleoptera: Tenebrionidae), *Ann. Entomol. Soc. Am.,* 58, 541, 1965.
77. **Dahlman, D. L.,** Age-dependent trehalase activity in adult tobacco hornworm tissues, *Insect Biochem.,* 2, 143, 1972.
78. **Beezeley, A. E., McCarthy, J. L., and Sohal, R. S.,** Changes in alpha-glycerophosphate succinic and isocitric dehydrogenases in the flight muscles of the housefly with age, *Exp. Gerontol.,* 9, 71, 1974.
79. **Lang, C. A.,** Cytochrome C reductase activities during development, *Exp. Cell Res.,* 17, 516, 1959.
80. **Burcombe, J. V.,** Changes in enzyme levels during aging in *D. melanogaster, Mech. Ageing Dev.,* 1, 213, 1972.
81. **Armstrong, D. et al.,** Changes of peroxidase with age in *Drosophila, Age,* 1, 8, 1978.
82. **Sohal, R. S. and McCarthy, J. L.,** Age related changes in acid phosphatase activity in male housefly. A histochemical and biochemical study, *Exp. Gerontol.,* 8, 223, 1973.
83. **Baumann, P. and Chen, P. S.,** Alterung und protein synthese bei *Drosophila melanogaster, Rev. Suisse Zool.,* 75, 1051, 1968.
84. **Levenbook, L. and Krishna, I.,** Effect of aging on amino acid turnover and rate of protein synthesis in blowfly *P. regina, J. Insect Physiol.,* 17, 9, 1971.
85. **Sohal, R. S., Ed.,** *Age Pigments,* Elsevier/North-Holland, Amsterdam, 1981.
86. **Pearse, A. G. E.,** *Histochemistry,* 3rd ed. Vol 2, Williams & Wilkins, Baltimore, 1972.
87. **Miquel, J., Oro, J., Bensch, I., and Johnson, J.,** Lipofuscin: fine-structural and biochemical studies, in *Free Radicals in Biology* Vol. 3, Pryor, W. A., Ed., Academic Press, New York, 1977, 133.
88. **Oliver, C., Essner, I., Zimring, A., and Haimes, S.,** Age-related accumulation of ceroid like pigment in mice with Chediak-Higashi syndrome, *Am. J. Pathol.,* 84, 225, 1976.
89. **Siakotos, A. M. and Koppang, N.,** Procedures for the isolation of lipopigments from brain, heart and liver, and their properties: a review, *Mech. Aging Dev.,* 2, 177, 1973.
90. **Elleder, M.,** So called neuronal ceroid-lipofuscinosis. Histochemical study with evidence of extractability of the stored material, *Acta Neuropathol.,* (Berlin), 38, 117, 1977.

91. **Strehler, B. L.,** On the histochemistry and ultrastructure of age pigment, in *Advances in Gerontology Research,* Strehler, B. L., Ed., 1964, 343.
92. **Tappel, A. L.,** Lipid peroxidation and fluorescent molecular damage to membranes in *Pathobiology of Cell Membranes,* Vol. 1, Trumps, B. F., and Arstila, A. V., Ed., Academic Press, New York, 1975, 145.
93. **Sulkin, N. M. and Srivany, P.,** The experimental production of senile pigment in the nerve cells of young rats, *J. Gerontol.,* 15, 2, 1960.
94. **Zeman, W.,** The neuronal ceroid-lipofuscinoses, Batten-Vogt syndrome: a model for human aging?, in *Advances in Gerontology Research,* Vol. 3, Strehler, B. L., Ed., Academic Press, New York, 1971, 147.
95. **Patel, V., Koppang, N., Patel, B., and Zeman, W.,** *P* -phenylenediamine-mediated peroxidase deficiency in English setters with neuronal ceroid-lipofuscinosis, *Lab. Invest.,* 30, 366, 1974.
96. **Sohal, R. S. and Donato, H.,** Effects of experimentally altered life spans on the accumulation of fluorescent age pigment in the housefly, *Musca domestica, Exp. Gerontol.,* 13, 335, 1978.
97. **Sohal, R. S. and Donato, H.,** Effect of experimental prolongation of life span on lipofuscin content and lysosomal enzyme activity in the brain of the housefly, *Musca domestica, J. Gerontol.,* 34, 489, 1979.
98. **Donato, H., Hoselton, M. A., and Sohal, R. S.,** Lipofuscin accumulation: effects of individual variation and selective mortality on population averages, *Exp. Gerontol.,* 14, 141, 1979.
99. **Sohal, R. S.,** Metabolic rate, aging and lipofuscin, in *Age Pigments,* Sohal, R. S., Ed., Elsevier/North-Holland, Amsterdam, 1981.
100. **Sohal, R. S.,** Aging in insects, in *Comprehensive Physiology, Biochemistry and Pharmacology,* Vol. 10, Gilbert, L. L., Ed., Pergamon Press, Oxford, 1985, 595.
101. **Sohal, R. S. and Allen, R. G.,** Relationship between metabolic rate, free radicals, differentiation and aging: a unified theory, in *The Molecular Biology of Aging,* Woodhead, A. D. et al., Eds., Brookhaven Symposium in Biology No. 33, Plenum Press, N.Y., 1985.
102. **Sohal, R. S., Farmer, K. J., Allen, R. G., and Cohen, N. R.,** Effect of age on oxygen consumption, superoxide dismutase, catalase, glutathione, inorganic peroxides, and chloroform-soluble antioxidants in the adult male housefly, *Musca domestica, Mech. Age. Dev.,* 25, 185, 1985.
103. **Collatz, K. G., and Sohal, R. S., Eds.,** *Comparative Biology of Insect Ageing: Strategies and Mechanisms,* Springer-Verlag, Berlin, in press.

AGING IN *DROSOPHILA*

**George T. Baker, III, Mark Jacobson,
and Gregory Mokrynski**

INTRODUCTION

Insects, particularly the higher Dipterans, provide a unique system for the study of the phenomenological events occurring in multicellular organisms with the passage of time. The Dipteran insect species, the most extensively studied for a number of morphological, molecular, and physiological changes with age, is *Drosophila*, in part, because of its earlier preeminence in genetic research. There are over 500 publications dealing with at least one aspect of aging in *Drosophila*. This is not surprising because, as adults, these insects are (1) endowed with a short lifespan, (2) economically reared and maintained under controlled laboratory conditions, (3) with the exception of gonadal tissues, contain virtually all post-mitotic cells, (4) are available in highly inbred or outbred strains as well as literally thousands of well-characterized mutant stocks and finally, (5) exhibit virtually all the manifestations of maturation, maturity, and senescence found in higher organisms.

Given the volume of material available, it is impossible to cover all the publications on changes in *Drosophila* with age. Rather, we present here some important experimental results, formulate general observations as to various factors which can affect longevity, and give an overview of those time-dependent morphological, physiological, and biochemical changes that occur in *Drosophila* with advancing age. The reader is referred to a compilation of references on aging in *Drosophila* by Soliman and Lints as a further information resource.[1]

This review is organized under somewhat arbitrary headings: the effects of species, strain, some mutations, and hybrid crosses on longevity; morphological, biochemical, and physiological alterations with age; the effects of various environmental factors; and some aspects of parental age on progeny.

In making an assessment of lifespan parameters in biogerontological research, it is imperative that scrupulous attention be paid to every possible environmental influence. Indeed, for a given species and strain there is hardly an environmental condition that has not been demonstrated to affect one or more parameters of lifespan. For example, aside from the obvious effects of temperature in this poikilothermic organism, dietary composition, larval density, adult density, total space, light-dark regime, light intensity, relative humidity, atmospheric composition, gravity, and levels of radiation have all been demonstrated to affect lifespan in *Drosophila*.

LONGEVITY OF VARIOUS SPECIES AND STRAINS OF DROSOPHILA

Table 1 lists the mean longevity or 50% survivorship of a number of *Drosophila* species and strains, particularly of *D. melanogaster*. For the most part, all the species and strains listed were regularly maintained at 25°C. Wherever possible we included the relative humidity. Other conditions such as housing, accessibility to food, number of individuals or lighting conditions, and sex ratio were not always available. Similarly, information on dietary composition was not always provided. These variables undoubtedly could be responsible for the wide divergence on longevity in various strains of laboratory cultures of *Drosophila*.[2] This is not to say, however, that under identical environmental conditions there are not distinct long-lived and short-lived strains of *Drosophila*. For example, an Oregon-R strain maintained under identical conditions as that of a Swedish-C exhibit distinct 50% survivorships as do Oregon-R compared to Canton-S.[3,4] The earlier work of Gowan,[5] who raised

Table 1
WILD-TYPE SURVIVORSHIP

Species, strain	Mean longevity days		Temp(°C)/humidity	Ref.
	Male	Female		
D. melanogaster				
Oregon	52.2	71.8	—	12
Oregon	49.8[a]	70.5[a]	—	7
	34.7	38.8	—	7
Oregon	78.0 ± 1.9	74.9 ± 1.7[a]	25/40 ± 60% RH	13
Oregon	29.5	29.9	25/65 ± 10% RH	14
Oregon-R	86.3	—	21	15
Oregon-R	38.0	42.0	25	16
Oregon-R	44.0[b]	—	25	17
Oregon-R	73.8 ± 2.1	60.3 ± 1.4[a]	25	18
Oregon-R	31	—	27	19
Oregon-R	40.3 ± 1.3	38.5 ± 1.4	25	20
Oregon-R	40.6 ± 1.0	44.1 ± 1.0	25	21
Oregon-R-44	35.5	36.4	—	13
Haceteppe	58.4 ± 1.3	56.0 ± 2.7	25/40 ± 60% RH	13
Kecioren	77.8 ± 2.3	76.4 ± 2.9	25/40 ± 60% RH	13
Line 107	26.8 ± 0.2	41.0 ± 0.5	18/25	22
	22.5 ± 0.3	35.8 ± 0.5	28/25	22
Line 107	49.1[b,c]	52.1[b,c]	—	23
A	42.2 ± 1.1	50.8 ± 1.4	—	24
B	41.2 ± 0.7	44.2 ± 1.1	—	24
Swedish-C	26.0[b]	—	25	17
Swedish-C	42.4 ± 1.5	35.3 ± 1.5	25	18
Berlin	39.5[b]	52.5	25	25
Sevelen	35.0[b]	—	25/60 ± 70% RH	26
Sevelen	29.0[b]	31.0	25/60 ± 70% RH	27
Kaduna	48.7 ± 0.7	41.3 ± 1.2	25	21
Kaduna	45.5	48.6	25	9
	47.0[a]	51.4[a]	25	9
Banded Manila	42.4	41.9	25	28
Bandless Manila	39.8	34.8	25	28
Old Falmouth	38.1 ± 0.4	40.6 ± 0.4	25	29
Ames I	49.6	50.9	—	30
Ames II(i)	53.6	51.5	—	31
Ames II(ii)	58.7	56.1	—	31
Princeton	46.7	48.4	—	19
Inbred 92	44.0	33.4	—	19
Florida 48	32.6	28.5	—	19
Swede-b-40	35.7	26.7	—	17

Male and Female Combined

Species, strain	Mean longevity days		Temp(°C)/humidity	Ref.
Oregon-R	49.3 ± 1.8		—	32
Swedish	50.8 ± 1.7		—	32
Lausanne	49.1 ± 1.7		—	32
Drosophila subobscura				
B	23.7[b]	38.0[b]	19	33
K	43.1[b]	23.6[b]	19	
B	25.8	33.3	20	
K	31.2	17.2	—	
NFS	42.4	40.7	—	
0	52.5	48.7	—	
D	47.5	50.2	—	

Table 1 (continued)
WILD-TYPE SURVIVORSHIP

Mean longevity (days)

Species, strain	Male	Female	Temp(°C)/humidity	Ref.
E	17.1	36.2	—	
F	69.2	53.8	—	
G	22.6	30.0	—	
M	51.8	35.3	—	
Kent (F₁)	53.4	58.6	—	
Galilee (F₁)	44.7	64.1	—	
Galilee (F₂)	32.2	59.2	—	
Drosophila repleta	60	a,b	60[a,b]	34
Drosophila pseudoobacura				
AR		35.6 ± 0.9	25	32
CH		33.3 ± 0.7	—	32

[a] Unmated.
[b] 50% survivorship.
[c] Graph extrapolation.

flies under identical conditions, demonstrated a range of mean survivorships for both males and females of various strains. Another example of distinct differences in longevity between strains and sexes of different strains kept under identical conditions is that reported for *D. subobscura* by Maynard-Smith.[6]

It might also be mentioned in surveying Table 1, one will observe that there is no general rule with respect to sex and longevity for any given strain or species of *Drosophila*. Indeed, of the 47 species and strains of *Drosophila* presented, males were the longer lived of the sexes 36% of the time vs. 44% for females while 20% were virtually identical. It should also be noted that unmated male or female flies have a greater lifespan than mated flies. The effect is generally much more marked in the females, and furthermore, the effect is strain specific as one can observe by comparing the results of Giess et al.[7] with Oregon to those of Hollingsworth and co-workers on Kaduna.[8,9]

It cannot be overstated that, aside from the more obvious environmental parameters, other conditions such as density can have dramatic effects on survivorship. For example, Miquel and co-workers rear their male flies in single vials and often have flies living considerably more than 100 days,[10] whereas Oregon-R maintained at other densities exhibit considerably reduced lifespans. Recent unpublished data from our own laboratories would indicate that the total space available to the fly, although the absolute space per fly is relatively constant, can influence longevity (see Table 2). This may be a reflection on the total area available to the organism and thus, the metabolic energy expended, as has been suggested by experiments on other Dipteran species, but is, at this point, still open to question.[11]

HYBRID SURVIVORSHIP

Table 3 lists a number of experiments wherein the longevity of hybrid strains were obtained. Most of these hybrids are derived from the inbred strains presented in Table 1, inviting a comparison between hybrid vs. inbred strain survivorship. It is without question that heterosis or hybrid vigor has a dramatic effect on survivorship in both sexes.[9,18,21,36-39] The heterotic effect is often more marked in one sex than the other. The extent of the

Table 2
EFFECT OF TOTAL AREA VS. SPACE/
FLY SURVIVORSHIP

| | | %50 Survivorship | |
		Males	Female
Absolute space	Initial space/fly		
80 cm³	4.0 cm³	45.0	43.0
500 cm³	5.0 cm³	39.0	35.0
7 × 10⁴ cm³	4.6 cm³	33.5	29.0

Note: All populations of *D. melanogaster* (Sevelen) were obtained from the same parental stock, reared at an approximately equal sex ratio (1:1) under identical environmental conditions (25 ± 0.5°C, 60 ± 5% R.H., 12 hr on-12 hr off light/dark cycle), and fed an identically prepared medium three times weekly.[35]

heterotic effect on the survivorship of offspring is dependent upon strain and set of the parental inbreds from which the hybrids were derived. The magnitude of the heterotic effect often approaches or exceeds a doubling of the lifespan of the hybrid as compared to inbred wild-type stocks.[40] It is noteworthy that, even when the hybrid is derived from a short-lived mutant such as vestigal and the long-lived, white-eyed mutant, the longevity of the hybrid significantly exceeds that of the wild type. Again, it should be pointed out that the conditions under which many of these hybrid and hybrid mutant crosses were maintained differed considerably, specifically regarding the number of individuals per housing unit. In any event, it is virtually impossible to make a direct comparison between any of the survivorship values for reason of differing experimental conditions.

MUTANT SURVIVORSHIP

In the earlier and extensive experiments of Gonzales a number of specific genes and gene combinations were examined as to their effects on longevity in *D. melanogaster*.[29] As a general rule, it can be stated that the presence of mutant genes in a homozygous condition results in a decrease in longevity. These initial observations have been, in general, upheld by a number of different reports on the effect of mutant genes on longevity (see Table 4). A more complete description of the mutants listed and symbols used can be obtained by referring to Lindsley and Grell.[48]

GENETICS AND PARENTAL AGE

Copulation in most *Drosophila* species normally takes approximately 20 min although this is considerably extended with advancing age.[55] A young male *Drosophila* has been shown to inseminate as many as ten virgin females in a single day.[56] A single ejaculation may contain up to 4000 sperm;[57] however, only 250 to 700 can be stored by the female.[58,59] These stored sperm are used with a high degree of efficiency by the female.[60] A female kept at a low temperature may store viable sperm for as long as 3 months.[61] The longer the mature sperm is stored in the female, the greater the mutation frequency. It is for these reasons that data on stored sperm are included in Table 5.

A number of variations associated with parental age have been demonstrated in *Drosophila*.[39,62-71] Earlier studies by Bridges have demonstrated that there were distinct variations

Table 3
HYBRID SURVIVORSHIP

Species, strain	Male	Female	Temp (°C)/ humidity	Ref.
D. Melanogaster				
Vestigial female/white male (F₁)	90.0 ± 2.5	80.3 ± 1.8	25/40—60%	13
White female/Vestigial male (F₁)	90.0 ± 1.4	85.3 ± 1.3	25/40—60	13
vg/+ w/+X vg/+ w/1/(F₁)	53.7 ± 1.6	57.9 ± 2.8	25/40—60%	13
Oregon-R female/Kaduna male (F₁)	48.7 ± 0.9	62.4 ± 1.3	25	41
Oregon-Kaduna female/Oregon-Kaduna male (F₁)	55.4 ± 1/2	59.0 ± 0.9	25	21
Kaduna female/Oregon male (F₁)	40.4 ± 0.9	58.5 ± 1.3	25	41
Oregon-R-vestigial	57.3 ± 1.8	48.9 ± 1.7	25	18
Oregon-R-nipped	74.9 ± 1.8	55.4 ± 2.0	25	18
Oregon-Kaduna Female/Kaduna-Oregon Male (F₂)	56.8 ± 1.2	50.9 ± 1.1	25	21
Kaduna-Oregon female/Kaduna-Oregon male (F₂)	58.4 ± 1.2	60.3 ± 1.3	25	21
Kaduna-Oregon female/Oregon-Kaduna male (F₂)	63.6 ± 1.4	65.8 ± 1.4	25	21
Gabarros 4 female/Abeele male (F₁)	—	53.9 ± 2.8	25/25	36
Abeele female/Gabarros 4 male (F₁)	—	58.3 ± 2.9	25/25	36
Swedish-C-vestigial	32.3 ± 1.4	33.1 ± 1.7	25	18
Swedish-C-nipped	51.3 ± 2.9	22.8 ± 2.2	25	18
White female/Berlin male (F₁)	77.0	55.0	25	25, 42
Swedish-bˣ female/Samarkand male (F₁)	73.9 ± 1.2	78.7 ± 1.8	25	38
Samarkand female/Swedish-bˣ male (F₁)	78.3 ± 1.1	74.2 ± 1.6	25	38
Purple Bandless Manila Female/Bandless male (F₁)	34.5	35.1	25	28
Eyeless Bandless Manila female/Bandless male (F₁)	30.9	30.3	25	28
Kaduna female/Oregon male (F₁)	40.5 ± 0.9	58.5 ± 1.3	25	21
Oregon female/Kaduna male (F₁)	58.6 ± 1.3	56.9 ± 1.6	25	21
	48.7 ± 1.0	62.4 ± 1.3	25	21
	59.2 ± 1.2	60.8 ± 1.4	25	21
Oregon-R-Basc/Basc	—	64.8 ± 2.1	25	43
Oregon-R-Basc	68.5 ± 2.2	62.4 ± 2.1	25	43
Inbred 92	55.4	50.0	25	43

Male and Female Combined

Species, strain	Male and Female Combined	Temp (°C)/ humidity	Ref.
Swedish female/Oregon male (F₁)	56.6 ± 1.4	—	32
Oregon female/Swedish male (F₁)	55.3 ± 1.3	—	
Lausanne female/Oregon male (F₁)	50.0 ± 1.1		
Oregon female/Lausanne male (F₁)	56.4 ± 1.0		

Species, strain	Male	Female	Temp (°C)/ humidity	Ref.
Drosophila subobscura				
B female/K male (F₁)	62.9[a]	72.8[a]	19	40
K female/B male (F₁)	70.7[a]	68.0[a]	19	40
B female/K male (F₁)	39.2 ± 1.0	—	24.8 ± 0.5	186
K female/B male (F₁)	47.6 ± 1.4	—	24.8 ± 0.5	186
K/B (F₁)	37.9	50.6	24.8	44
B/K (F₁)	28.9	53.4	24.8	44
B female/K male (F₁)	61.6	61.5	20	33
K × NFS (F₁)	27.0	35.5	25	45
K × NFS (F₁)	24.6 ± 1.1	30.6 ± 1.6	25	46
K female/NFS male (F₁)	67.3	55.9	20	33
Drosophila pseudoobscura				
Ar/CH* (F₁)	—	43.2 ± 1.0	25	47

[a] 50% survivorship.

Table 4
MUTANT SURVIVORSHIP

Species, strain	Male	Female	Temp (°C)/ humidity	Ref.
D. melanogaster				
Vestigial	38.8 ± 1.3	47.9 ± 2.4	25/40 ± 60% RH	13
Vestigial	10	11	25	16
Vestigial	38.5 ± 1.8	33.6 ± 1.5	25	18
Vestigial	11.3[a]	—	25/60 ± 70% RH	26
Vestigial	11.3[a]	13.1[a]	25/60 ± 70% RH	11
Vestigial	15.0 ± 0.3	21.0 ± 0.4	25	18
Vestigial	14.0[a,b]	19.0[a,b]	—	23
White	52.2 ± 1.7	56.8 ± 3.3	25/40 ± 60% RH	13
White	50.5	52.5	25	25, 42
Base	48.3 ± 2.5	—	25	43
Ebony	35.0	38.0	25	16
Black	41.0 ± 0.5	40.3 ± 0.3	25	29, 49
Nipped	63.6 ± 2.3	52.8 ± 2.2	25	18
Quintuple	9.4 ± 0.2	12.1 ± 0.3	25	29
Purple	27.4 ± 0.3	21.8 ± 0.2	25	29
Arc	25.2 ± 0.3	28.2 ± 0.4	25	29
Speck	46.6 ± 0.6	38.9 ± 0.6	25	29
Black-Purple	30.4 ± 0.3	24.1 ± 0.3	25	29
Black-Vestigial	16.4 ± 0.2	24.2 ± 0.4	25	29
Black-Arc	20.1 ± 0.4	23.2 ± 0.4	25	29
Black-Speck	32.4 ± 0.3	30.0 ± 0.3	25	29
Purple-Vestigial	11.7 ± 0.1	19.1 ± 0.3	25	29
Purple-Arc	36.0 ± 0.5	32.0 ± 0.4	25	29
Purple-Speck	23.7 ± 0.2	23.0 ± 0.2	25	29
Arc-Speck	38.4 ± 0.6	34.7 ± 0.7	25	29
Black-Purple-Arc	35.1 ± 0.6	30.6 ± 0.4	25	29
Black-Purple-Speck	31.2 ± 0.4	24.4 ± 0.4	25	29
Black-Speck	33.7 ± 0.5	26.8 ± 0.5	25	29
Purple-Arc-Speck	38.4 ± 0.6	40.7 ± 0.4	25	29
Vestigial-Arc-Speck	12.8 ± 0.3	25.2 ± 0.6	25	29
Purple-Vestigial-Speck	9.5 ± 0.2	12.2 ± 0.3	25	29
Black-Purple-Vestigial-Arc	14.8 ± 0.2	22.2 ± 0.3	25	29
Black-Vestigial-Speck	14.0 ± 0.3	19.6 ± 0.4	25	29
Black-Purple-Arc-Speck	22.7 ± 0.4	23.1 ± 0.4	25	29
Sterile (2) Adipose	38.4[c]	27.6[c]	25	50
	37.9[d]	59.6[d]	25	50
Sterile (2) Adipose-Kaduna	48.4[c]	50.1[c]	25	50
Purple (Bandless Manila)	25.1	24.5	25	28
Eyeless (Bandless Manila)	27.4	25.1	25	28
Hyperkinetic-1P	—	39.9 ± 8.9[c]	25/50% RH	51
Hyperkinetic-2T	—	65.9 ± 13.5[c]	25/50% RH	51
Shaker-5	—	40.5 ± 9.2[c]	25/50% RH	51
Ether A Go-Go	—	67.6 ± 12.7[c]	25/50% RH	31
α-Glycerophosphate	41.2[a]	33.8[a]	25/60 + 70% RH	26
Dehydrogenase deficient (GPDH[M5]spd[fg]/In (2LR) Cy, Roi) DNA repair deficient mutants				
mei-41[D5]	34.6[a]	31.6[a]	26/60 ± 5% RH	52, 53
mei-9[D2]	23.0[a]	28.0[a]	25/60 ± 5% RH	54
mei-9[D3]	20.5[a]	23.0[a]	25/60 ± 5% RH	54
mei-9[D4]	21.5[a]	23.0[a]	25/60 ± 5% RH	54
mus-(1) 104[D1]	36.0[a]	35.0[a]	25/60 ± 5% RH	52
mus-(3) 312	17.0[a]	21.0[a]	25/60 ± 5% RH	54

[a]　50% survivorship.

[b]　Graph extrapolation.

[c]　Standard deviation.

[d]　Unmated.

[e]　Vestigial also DNA repair deficient.

Table 5
EFFECTS OF GERMINAL CELL AGE ON GENETIC ABBERATIONS

Species, strain	Sex	Temp (°C)	Age (days)	Chromosome	No. tested	Mean % lethal	Ref.
D. melanogaster Oregon-R/Samarkand	M	25	1	X[a]	4,535	0.13	74
			2—3		6,268	0.05	
			4		4,026	0.07	
			5		4,024	0.02	
			6		4,035	0.07	
			7		4,241	0.07	
			8		4,422	0.02	
			9—10		5,734	0.10	
			11		3,492	0.17	
			12		3,267	0.09	
			13—18		2,133	0.19	
D. melanogaster	M	27	—	X[a]	3,490	0.401	79
		7	3 wks (in F)		8,312	0.168	
D. melanogaster	M	27	—	X[a]	4,869	0.33 ± 0.08	78
		18	3 wks		3,133	0.22 ± 0.08	
		7	3 wks		15,296	0.18 ± 0.03	
Oregon-K-Rc	M	25	1	X[a]	615	0.49	97
			2		1,010	0.59	
			3		1,036	0.29	
			4		939	0.43	
			5		849	0.35	
			6		1,000	0.10	
			7		929	0.22	
			8		821	0.22	
			10		744	0.13	
			13		617	0.18	

Table 5 (continued)
EFFECTS OF GERMINAL CELL AGE ON GENETIC ABBERATIONS

Species, strain	Sex	Temp (°C)	Age (days) Mean age at first mating	Chromosome	No. tested	Mean % lethal	Ref.
D. melanogaster							
Formosa	M	17	2.6	X[a]	7,193	0.21 ± 0.05	98
			45.4		6,321	0.27 ± 0.06	
			91.8		1,886	0.69 ± 0.19	
		27	2.1	X	5,965	0.23 ± 0.06	
			25.4		5,509	0.36 ± 0.08	
			43.6		570	0.88 ± 0.39	

Species, strain	Sex	Temp (°C)	Age (days)	Chromosome	No. tested	Recombination class[b] (%)(Ref 99)		
						I	II	Coincidence
D. melanogaster								
sc ct v(M)								
F₁ (F)	F	—	0—4	X	3,019	16.87	13.40	0.1902
sc ct+/+ + + *v*			5—8	X	3,652	16.04	11.95	0.4852
sc + *v*/ + *ct* +			9—12	X	3,186	12.20	12.26	0.5683
sc ct v/ + + +			13—16	X	2,354	11.04	10.85	0.4862
sc + + / + *ct v*(F)			0—4	X	3,021	23.66	16.85	0.3236
Same M and F genotype, but with			5—8	X	3,490	21.36	15.83	0.3680
Cy inversion on 2nd chromosome			9—12	X	3,318	19.69	14.24	0.4458
			13—16	X	2,963	17.89	13.40	0.2794

Species, strain	Sex	Temp (°C)	Age (at first mating)	Chromosome	No. tested	Mean % lethal	Ref.
D. melanogaster							
Forked	M	22	1	X[c,d]	2,131	0.05 ± 0.04	100
			2.5 months		4,225	0.05 ± 0.03	
			1	X[e]	446	2.91 ± 0.79	
			2.5 months		434	3.20 ± 0.84	

			II[i]			Paternal exception
Oregon K (M)	M	23		2—5	1,002	0.10
				5—8	1,017	0.10
				8—11	1,009	0.56
				11—14	1,112	0.54
				14—17	995	0.30
				17—20	1,059	0.66
				20—23	1,051	1.14
D. melanogaster						
w(M)	M	24 ± 0.5		1—14		0.06
ywf(F)				14—21		0.34
				21—28		0.52
				28—35		0.83
				35—38		1.60

a BASC test.

b Recombination class I defined as occurring between the sc-ct loci, and Class II as occurring between the ct-v loci.

c CIB test.

d Spontaneous mutation.

e X-irradiation induced mutation, 100 r.

f Total spots as percentage of total abdomens.

g Size Class Index, the number of spots per size class multiplied by the minimum numer of bristles per size class totaled and divided by the total number of abdomens bearing spots.

h T line — derived from 3 to 9-day-old parents.
   B line—derived from 6 to 8-week-old parents.
   Q line, control, taken directly from the wild.

i Production of XXY or XO progeny.

j y + Y − Y chromosome to which small part of an X chromosome carrying normal allele y has been translocated.

k Repeat of experimental data of 10°C pre-aged series.

l Cy/BIL method.

m Males kept with same female throughout; food changed every 6 days.

n Scute, vermilion, forked, carnation.

o Exhaustive mating; males placed with 3 to 6 virgins every 24 hr.

p Males kept with same female throughout; dead females replaced by virgins before 32nd day; after 32nd day, all old females were replaced with 3 virgins every few days for the life of the male.

with maternal age in the percent of recombination for seven loci of Chromosome III.[72] More recent work by Valentine has confirmed these earlier studies.[73] It is apparent that the frequencies of meiotic alterations increase with advancing maternal age, as evidenced by increases in cross-over frequency.

There is some evidence that might suggest an increase in the rate of spontaneous recessive lethals in the male germ line with advancing age.[74-76] It is clear that the first batch of sperm produced by the male following eclosion does demonstrate a higher incidence of sex-recessive lethals than subsequent batches of sperm produced within the first week. Thereafter there appears to be an increased incidence with advancing age. Purdom et al.[75] have reported a similar increase in the spontaneous mutations with age for the second chromosome (see Table 5).

However, with a test system similar to that of the Muller-5, no increases in the spontaneous mutation rate were found under exhaustive mating conditions, although considerable increases were noted after the first week under paired mating conditions.[77] Furthermore, it is apparent from a number of studies that the rate of sex-linked spontaneous mutations is temperature dependent, increasing at higher temperatures (see Table 6).[78,79]

Regarding the effect of parental age on other genetic variations, there is a substantial body of evidence that certain types of genetic aberrations do occur with increasing frequency with advancing age. For example, Lamy reported a definite trend in the percentage of paternal exceptions with age.[77] On the other hand, Kelsall reported no increases in nondisjunction after aging for 1 week.[80]

In addition to the data cited in Table 6, there are a number of quantitative traits which vary between the offspring of young and old parents. Alterations in quantitative traits observed in *Drosophila* include the number of abdominal bristles,[81,82] cell size and cell number,[83] wing size[84] developmental time,[85] asymmetry of sternopleural chaeta,[86,87] and variation in egg size.[83,88,89]

Regarding the effects of parental age on progeny longevity per se, the data are somewhat difficult to interpret, in part due to the different experimental protocols employed by various investigators (see Table 6).[90,91] While it is clear that parental age, particularly maternal age, can and does affect a number of quantitative characteristics, it remains to be firmly established whether or not successive lifespans are irreversibly affected. The reader is referred to Lints and co-workers for excellent reviews of genetic alterations as a function of parental age in *Drosophila*.[92-96]

EFFECTS OF MATERNAL AGE UPON FECUNDITY

There is a considerable body of information relating maternal age to fecundity.[35,112-132] Generally, there is a reported increase in the number of eggs laid by a female per day within the first week postemergence and thereafter a decline with advancing age, the onset being strain and species specific (see Table 7). Similarly, there is generally observed a corresponding decrease in the viability of eggs from older females in a number of studies.

It should be pointed out that conditions of mating, e.g., the ratio of males to females in absolute numbers can play a role in determining the number and viability of eggs (Table 7). Indeed, a number of environmental conditions have been shown to affect fecundity including temperature,[133-141] suitable oviposition sites,[142-146] and the preimaginal environmental conditions.[124,148-151] Similarly, the availablity of optimum food, in particular yeast,[152,153] can dramatically affect fecundity in *Drosophila*.

The decline in reproductive capacity in the female is evidently not attributable to a loss or an exhaustion of oocytes, as older females generally have a considerable number of potentially functional ovarioles.[154-156] It is concluded, therefore, that as is generally found in higher organisms, the primary factors in the decline of female reproductive capacity are due to alterations which, presumably, reside in the neuroendocrine axis.

Table 6
EFFECTS OF PARENTAL AGE ON LONGEVITY

Species, strain	Male	Female	Parental age (days)	Ref.
Drosophila melanogaster				
Oregon-R parental	40.3 ± 1.2	38.5 ± 1.4	—	20
F_1—F_9	38.4 ± 0.8	41.1 ± 0.6	0—2	
F_1—F_9	33.5 ± 0.6	34.7 ± 0.5	15—16	
F_1—F_9	32.7 ± 1.1	30.8 ± 0.8	ca. 40	
Canton-S Female/				
Oregon R-C Male	83[a]	92[a]	23	110
F_1 or F_2 hybrids	74	70	54	
	57	64	77	
Cugo (Chicago-Urbana-Gabarros-Oregon)				
P (young)	55.2 ± 3.9	48.4 ± 5.2	4	93
F_3 (young)	45.2 ± 3.9	38.8 ± 3.3	4	
F_9 (young)	33.1 ± 7.8	14.5 ± 1.6	4	
F_{13} (young)	61.6 ± 5.1	51.9 ± 5.4	4	
P (old)	55.2 ± 3.9	48.4 ± 5.2	26[b]	
F_3 (old)	27.9 ± 3.1	18.8 ± 1.8		
F_9 (old)	50.7 ± 5.7	51.7 ± 3.5		
F_{11} (old)	54.5 ± 4.0	57.0 ± 5.6		
Drosophila pseudoobscura				
	65[a,c]	50[a,c]	3—9	102
	83	70	42—56	
Drosophila subobscura				
F_1	23.95 ± 1.1		30	111
F_2	24.18 ± 1.1		30	
F_4	22.31 ± 1.1		30	
F_8	24.69 ± 0.8		30	

[a] 50% survivorship.
[b] Or where 80 to 90% of females laid 20 to 30 fertile eggs per day.
[c] Graph extrapolation.

MALE VIRILITY WITH AGE

Relatively few studies have been performed on the virility of male *Drosophila* with advancing age.[162] Those that have been reported, as seen in Table 8, clearly demonstrate a marked decline in the number of progeny produced as well as in the number of fertile maters.[55] *D. melanogaster* males become sexually mature some 12 hr following eclosion[163-166] and reach full potency in the first week.[166,167] As females of *D. melanogaster* demonstrate a high reproductive capacity of upwards of 3000 eggs laid in a lifetime,[130] a male can sire between 10,000 and 14,000 progeny in a lifetime.[167] It is also apparent that the loss of reproductive capacity in the male is not due to the exhaustion of primary spermatogonia,[168] again indicating that the decline in reproductive capacity is not due to intrinsic functional loss within the testes, but rather due to some other component of reproductive ability.[169]

EFFECTS OF TEMPERATURE ON LONGEVITY

An extensive body of literature exists with regards to the effects of temperature both during the developmental period and the adult lifespan on longevity in *Drosophila*.[150,170-182]

Table 7
EFFECTS OF MATERNAL AGE UPON FECUNDITY

Species, strain	Mating regime (no. females: no. males/container)	Female age (days)	Eggs/female/day	% Viability (eggs hatched/no. eggs/female)	Temp (°C)	Ref.
D. melanogaster Oregon-R	5:4/half pint bottle	5—6[a]	54.8 ± 3.18	97.1	25	119
		15—16	44.1 ± 1.37	88.8	25	
		25—26	26.7 ± 1.54	60.3	25	
		40—41	10.6 ± 1.15	23.6	25	
		5—6[b]	61.0 ± 1.15	90.3	25	
		15—16	46.7 ± 2.96	84.8	25	
		25—26	28.8 ± 2.58	45.5	25	
		40—41	7.5 ± 2.40	12.0	25	
Drosophila melanogaster		1	3[c]		25	153
		5	97		25	
		15	75		25	
		25	55		25	
		35	22		25	
		45	5		25	
		55	1		25	
Drosophila melanogaster Berlin	10:10/80cm³ vial	4	11.9		25	25
		40	0.6		25	
White-eyed		4	7.1		25	
		40	2.0		25	
Drosophila melanogaster Canton-S	1:1/vial	1		98.32		157
		7		95.80		
		13		89.08		
		16		80.67		
		19		72.27		
		22		57.14		

Strain	Treatment		Value (mean ± SD)	%	n	Total
D. melanogaster White-eyed	1:2/80cm³ vial	25		40.34	25 + 0.5	158
		28		23.53		
		31		10.08		
		34		4.20		
		40		2.52		
		49		1.68		
		52		0.84		
		55		0.00		
mus (1) 104[DS]		5	21.0 ± 1.5[c,d]			
		7	36.0 ± 2.0			
		9	30.0 ± 3.0			
		11	24.5 ± 1.5			
		13	18.0 ± 2.5			
		14	16.0 ± 2.25			
		5	6.0 ± 4.0			
		7	8.0 ± 4.0			
		9	13.0 ± 1.0			
		11	10.0 ± 3.0			
		13	4.0 ± 3.0			
		14	3.0 ± 1.5			
mus-41[DS]		5	5.0 ± 2.0			
		7	7.0 ± 2.0			
		9	7.5 ± 2.0			
		11	7.0 ± 2.0			
		13	4.0 ± 2.0			
		14	2.0 ± 1.0			
Drosophila pseudoobscura AR female/CH male		3	3[a]		25	47
		7	25		25	
		10	33		25	
		14	40		25	
		15	38		25	
		25	16		25	
		35	5		25	
		50	3		25	
		58	0		25	

528   CRC Handbook of Cell Biology of Aging

Table 7 (continued)
AFFECTS OF MATERNAL AGE UPON FECUNDITY

Species, strain	Mating regime (no. females: no. males/container)	Female age (days)	Eggs/female/day	% Viability (eggs hatched/no. eggs/female)	Temp (°C)	Ref.
D. melanogaster Sevelen (standard diet)	1:1/80 cm³ vial	1	11.7 ± 1.0[a]		25 ± 0.5	152
		6	52.0 ± 3.0			
		13	42.0 ± 1.5			
		23	31.0 ± 1.0			
		33	11.7 ± 3.5			
		45	3.5 ± 2.0			
		56	1.0 ± 1.0			
		60	1.0 ± 1.0			
Sevelen (yeast defeicient diet)		1	2.0 ± 1.0			
		6	21.7 ± 1.0			
		13	16.0 ± 1.0			
		23	10.0 ± 2.0			
		33	3.0 ± 1.0			
		45	1.0 ± 1.0			
		56	0.0			
D. melanogaster Vestigial/Champetiaras	4:5/cage	3	46.0[a]		25	159
		5	102.0			
		15	85.0			
		25	70.0			
		30	41.0			
		35	20.0			
		45	7.0			
		50	5.0			
D. melanogaster Sevelen	10:10/80 cm³ vial	5	23.2 ± 2.7		25	160
		15	26.2 ± 1.7			

				Larva-pupa		Ref.
		24	23.7 ± 1.7			
		35	7.4 ± 1.1			
		45	4.9 ± 1.3			
D. melanogaster 1g1 Cy	1:2/2 dℓ glass cylinder	4	75[a]	69(1—7d)	25	161
		7	135	53(8—14d)		
		15	95	17(15—21d)		
		25	70	6(22—28d)		
		40	10	2(29—35d)		
				0(36—42d)		
D. melanogaster Gabarros female/Abeele male		4	7[a]			124
		8	100			
		20	70			
		35	20			
		50	3			
D. melanogaster Vestigial-Champetieres	4:5/cage	5	85[a]		25	106
		10	80			
		20	63			
		30	39			
		40	16			
		50	5			
D. melanogaster Oregon	4:4/bottle	2	15.57 ± 1.53	49.0	25	19
		3	49.77 ± 4.73	60.2		
		4	82.73 ± 4.81	61.7		
		5	79.67 ± 5.81	58.1		
		6	69.69 ± 4.82	49.5		
		7	67.20 ± 5.68	46.7		
		8	52.83 ± 5.57	44.3		
		9	47.90 ± 5.14	32.2		
		10	37.63 ± 5.42	29.0		

Table 7 (continued)
AFFECTS OF MATERNAL AGE UPON FECUNDITY

Species, strain	Mating regime (no. females: no. males/container)	Female age (days)	Eggs/female/day	% Viability (eggs hatched/no. eggs/female)	Temp (°C)	Ref.
D. melanogaster Canton-S	5:5/vial	0—30		100.00		157;
		31		99.08		
		34		98.62		
		40		95.87		
		46		90.83		
		49		81.19		
		52		73.39		
		55		51.38		
		58		33.98		
		61		18.80		
		64		5.50		
		67		1.38		
		70		0.00		
D. melanogaster Cabarros (inbred)		4	4[a]			92
		8	50			
		20	23			
		35	3			
		50	2			
D. pseudoobscura AR/AR	5:8/ca. 340cm³	3	3[a]		25	47
		7	27			
		10	35			
		12	37			
		15	25			
		25	13			
		35	5			
		50	1			

CH/CH		25
	3	3
	7	24
	10	29
	15	24
	25	8
	35	4
	50	0

[a] Graph extrapolation.

[b] Offspring from 2- to 3-day-old parents.

[c] Offspring from 30- to 31-day-old parents.

[d] Standard deviation.

Table 8
MALE VIRILITY WITH AGE

Species, strain	Male age (days)	Virility[a] (No. offspring produced)	Ref.
D. melanogaster	0—5	2250	167
	10—15	1875	
	20—25	1250	
	25—30	500	
	40—45	250	
D. melanogaster Oregon-R	1—33	Marked decrease in no. progeny after first few days[b]	77
D. melanogaster		**Av. no. offspring**	55
Oregon-R	7	18.3	
	14	13.6	
	35	10.1	
	49	6.4	
	84	0.0	
D. melanogaster		**% Fertile maters**	55
Oregon-R	14	95	
	28	90	
	42	40	
	70	12	
	98	1	

[a] All data extrapolated from graphs.
[b] Under intensive breeding (3 to 6 virgin females every 24 hr).

Although the temperature at which *Drosophila* is reared does have an effect on longevity, the effects of temperature are much more dramatic during the adult stages of lifespan (Table 9). A lower temperature results in a longer developmental time, but generally, only a slight improvement in imaginal lifespan when adults are maintained at a constant temperature. On the other hand, maintaining the adult animal at lower temperatures significantly enhances longevity of this poikilothermic organism. The magnitude of the effect is once again strain and species specific, for example, an increase in mean longevity of some 8-fold can be observed between 15 and 30°C in *D. melanogaster*,[41] whereas an almost 17-fold increase in mean longevity has been reported for hybrids of *D. subobscura* between 18.5 and 30.5°C.[6]

EFFECT OF DENSITY ON LONGEVITY

From the earlier studies of Pearl and co-workers,[189-193] it is well known that larval or adult density will have dramatic effects on longevity. For example, in Line 107 of *D. melanogaster* these workers reported an optimum adult density of between 20 and 55 in a 30-cm³ vial (approximately) whereas, Miller and Thomas reported an optimum larval density at approximately 80 individuals in a 30 cm³ vial (approximately).[194]

From the work of Lints and co-workers a number of apparently causal effects of preimaginal population density, including longevity, have been observed.[195] There is a decrease in the mean thoracic size which is a linear function of the log of the preimaginal population density. In addition to the correlates previously mentioned, there has been reported to be positive correlation between population density (preimaginal) and adult longevity in hybrid

Condition			
10 larvae/ca.30cm³ vial	0.83	1.43	46.0 37.0
20 larvae/ca.30cm³ vial	0.83	1.43	44.0 39.0
40 larvae/ca.30cm³ vial	0.73	1.23	47.0 40.0
60 larvae/ca.30cm³ vial	0.69	1.03	45.0 41.0
80 larvae/ca.30cm³ vial	0.64	0.97	61.0 53.0
100 larvae/ca.30cm³ vial	0.56	0.75	44.0 38.0

Line 107 [197]

2 parents/ca.30cm³ vial	27.3 ± 0.6
10 parents/ca.30cm³ vial	36.2 ± 0.7
20 parents/ca.30cm³ vial	37.1 ± 0.6
35 parents/ca.30cm³ vial	39.4 ± 0.8
55 parents/ca.30cm³ vial	40.0 ± 0.5
75 parents/ca.30cm³ vial	32.3 ± 0.5
105 parents/ca.30cm³ vial	24.2 ± 0.3
200 parents/ca.30cm³ vial	12.0 ± 0.2

Male

Oregon-R-C [198]

Condition		
4 larvae/pint bottle	10.32	0.931
16 larvae/pint bottle	10.91	0.812
64 larvae/pint bottle	13.39	0.618
128 larvae/pint bottle	14.72	0.485
256 larvae/pint bottle	14.82	0.415
512 larvae/pint bottle	12.96	0.465
10 parents/pint bottle	11.84	1.033
100 parents/pint bottle	14.57	0.620
10 parents/pint bottle	11.84	1.033
100 parents/pint bottle	14.57	0.620
200 parents/pint bottle	14.85	0.485
400 parents/pint bottle	14.18	0.445
600 parents/pint bottle	13.88	0.425
800 parents/pint bottle	13.56	0.450

a   Standard deviation.
b   50% Survivorship for particular study.

Table 11
EFFECTS OF LIGHT:DARK CYCLES UPON LONGEVITY

Species, (strain)	Light:darkness regime (hr)	Longevity (days) mean + SEM		Temp (°C)	Ref.
		Male	Female		
Drosophila melanogaster Vestigial-Champetieres	Constant light (20-W fluorescent light)	49.29 ± 1.94	55.33 ± 1.46	25	201
	12:12	44.39 ± 2.49	57.44 ± 1.99		
	12:12 (with inversion of night-day phase every week)	43.95 ± 2.38	54.9 ± 1.56		
	95% Darkness (nonlighted incubator open several times)	44.29 ± 2.88	59.25 ± 1.80		
	Permanent darkness	54.72 ± 4.10	79.83 ± 3.09		
Princeton Wild Type	12:12	51.7[a]	46.7[a]	25	202
	10.5:10.5	46.8	43.5[a]		
	13.5:13.5	42.5	39.9[a]		
	Constant light	45.8	40.2[a]		
tu[9] (tumorous)	12:12	—	49.9[a]		
	10.5:10.5	—	42.9[a]		
	13.5:13.5	—	45.7[a]		
	Constant light	—	41.4[a]		
Oregon-R	Constant light	23[a,b]	—	25	199
	12:12	40	—		
	3 light: 9 dark cycle	43	—		
	9 light: 3 dark cycle	47	—		

[a] 50% Survivorship.
[b] Graph extrapolation.

reticulum, and free ribosomes, an increase in the number and distribution of virus-like particles and nuclear and cytoplasmic inclusions, a decrease in the number of mitochondria and an increase in the structural irregularities within the mitochondria, and alterations in cell number and morphology. The reader is referred to the following references: 10, 11, 15, 34, 88, 89, 109, 168, 204, 205, 233—235, 238, 252—285.

BODY WEIGHT WITH AGE

Table 12 lists the sex, temperature, relative humidity, and light cycle under which a number of different strains and species of *Drosophila* were kept wherein weight or some aspect thereof was assessed. In general, there has been observed either no significant change in wet weight or a decrease in wet weight with advancing age. However, an increased rate of water loss during desiccation was reported and attributed to a decrease in the functional capacity of the spiracles.[223] In addition, there has been a reported increase in the fat content and a corresponding decrease in the nonfat dry weight with advancing age.[286] Once again, the alterations in wet and dry body weights appear to be somewhat species and strain specific as well as dependent on the environmental conditions employed.

ALTERATIONS IN GLYCOGEN CONTENT WITH AGE

One of the more consistent findings with age in whole fly and in various tissues of *Drosophila* is a decrease in the glycogen content. This finding has been reported by a number of investigators with two exceptions, namely, an increase in glycogen content in the cephalic ganglionic center[168,270] and in the heart of *D. replenta*.[51,253,273,276] These data are presented in Table 13.

LOCOMOTOR ABILITY WITH AGE

In *Drosophila,* as has been demonstrated in most species so investigated, there is a decrease in primary locomotor ability with advancing age. This has been examined by a number of investigators by different and ingenious methodologies. In all cases, it is unquestionable that following a period of maturation of a primary locomotor ability, namely flight, usually during the first week of postemergent life, there is a rather steady decline with advancing age. A summary of these data are presented in Table 14.

OXYGEN CONSUMPTION WITH AGE

Although there is generally observed a decline in oxygen consumption with advancing age in *Drosophila* and a direct correlation between the environmental temperature and mean oxygen consumption, there is not an unequivocable correlation between longevity and mean oxygen consumption (see Tables 15 and 16). For example, the data reported by Trout and Kaplan primarily on various mutants suggests a direct correlation between metabolic rate and longevity.[51] However, the results of other studies on wild and inbred strains of *Drosophila* show no correlation between mean oxygen consumption and longevity.[303-311]

MITOCHONDRIAL ALTERATIONS WITH AGE

A considerable amount of information is available with respect to aging mitochondria in *Drosophila*. Although much of this work is morphological, a substantial body of biochemical information exists as well (see earlier reviews by Baker[251] and Sohal).[275] More recent information with respect to biochemical alterations in mitochondria with age have been re-

Table 12
BODY WEIGHT WITH AGE

Species (strain)	Sex	Temp (°C)	Relative humidity (%)	Lightcycle	Comments	Ref.
D. melanogaster Oregon-R	M	25	55	12:12	NSC in male wet weight	287
Oregon-R	F	25	55	12:12	DEC in mated female wet weight	287
	M	22	45	12:12	DEC in dry weight	168
D. melanogaster	M and F	25	75	—	NSC in wet weight and H_2O content; INC in fat content with corresponding DEC in nonfat dry weight; INC in water loss during desiccation	286, 288
D. melanogaster						
Sevelen	M	25 ± 0.5	60 ± 5	Dark[a]	DEC in wet weight	27
Sevelen	M	25 ± 0.5	60 ± 5	12:12	NSC in wet weight	289
	F	25 ± 0.5	60 ± 5	12:12	INC first week NSC thereafter	289
Vestigial	M	25	75	Dark	Slight DEC in wet and dry weights	290
	F	25	75	Dark	NSC in wet and dry weights	290
Canton S	M	—	—	—	NSC in wet weight	157
	F	—	—	—	INC in weight of virgin flies	157
Oregon R/Kaduna F_2 hybrid	M	25	—	—	NSC in wet and dry weights	185
D. subobscura						
B/K F_2 hybrid	M	20	—	—	NSC in wet and dry weights	185
D. melanogaster						
Berlin	M	26	—	—	INC in weight loss following desiccation	223
White Eye	M and F	25 ± 0.5	60 ± 5	12:12	DEC in wet weight	54
Mei-41^{DS}	M and F	25 ± 0.5	60 ± 5	12:12	DEC in wet weight	54
Oregon-R	M	25	—	12:12	DEC in wet weight	291
Gabarros Wild	—	—	—	—	DEC in wet weight	
Gabarros inbred	—	—	—	—	NSC in wet weight	292
Gabarros female/Abeele male hybrids	—	—	—	—	DEC in wet weight	

[a] Dark except for feeding.

Table 13
ALTERATIONS IN GLYCOGEN CONTENT WITH ADVANCING AGE

Species (strain)	Sex	Tissue	Result	Ref.
D. funebris (Fabricus)	F	Whole fly	DEC after initial INC	293
D. melanogaster	M and F	Whole fly, fat body, halteres	DEC after initial INC	294
D. melanogaster (Oregon R)	M	Cephalic ganglionic center	INC	168, 267
		Thoracic flight muscle	DEC	295
D. funebris	F	Whole fly	DEC	281—283
D. melanogaster (Oregon R)	M	Whole fly, flight muscle	DEC	34, 253, 273
D. replenta (Wollaston)	M and F	Intra- and extramitochondrial of heart	INC	296
D. melanogaster (Oregon R)	M and F	Whole fly (per mg DNA)	DEC	11
D. melanogaster (Oregon R)	M and F	Whole fly, flight muscle	DEC	334
D. melanogaster (Canton S)	M and F	Whole fly, flight muscle	DEC after initial INC	297
D. melanogaster (Oregon R)	M	Whole fly	DEC	

Table 14
LOCOMOTOR ABILITY

Species, strain	Sex	Temp (°C)	Results	Ref.
D. funebris (Fabricus)	F	20 ± 0.5	INC in duration of flight from ca.25 min at 1 day to 110 min at 6.5 days through 14.5 days; DEC to some 19 min at 33.5 days	293
D. melanogaster	M and F	20—22	INC in duration of flight from 133 min at 18 to 20 hr to 278 min by 1 week and a DEC to 100 min by 4 weeks	294
D. funebris			INC in wing beat frequency from 127 to 167 cps nd constant through 15 days; no subsequent data	298
D. melanogaster Oregon R	M	21 ± 0.5	Negative geotactic response[a] remained constant through the first 60 days and DEC markedly thereafter until 90 days	4
Canton S	M	21 ± 0.5	Negative geotactic response remained constant through the first 30 days and DEC thereafter until 90 days	4
Oregon R	M	21 -23	Negative geotactic response DEC in almost linear fashion from ca. 30 days until 70—80 days	15, 204, 239, 264
Sevelen	M	25 ± 0.5	DEC in negative geotactic response after first week	52
Canton S	M and F	25	Dec in positive phototactic response[b]	299
Ebony	M and F			
White	M and F			
Sevelen	M	25 ± 0.5	DEC in positive phototactic response after first week	217
White eye	M	25 ± 0.5	DEC in positive phototactic response after first week	217
mei-41[DS]	M	25 ± 0.5	DEC in positive phototactic response after first week	217
D. repleta (Wollasten)	M and F		Flies sluggish and unable to fly at 4 months (mean longevity approximately 2 months)	34, 253
D. melanogaster	M	25	DEC in geotaxis response marked until 4 weeks, slight until 6 weeks; no significant circadian rhythms	301
D. melanogaster	M	25	DEC in phototaxic response after peak at 1 week with marked dampening of circadian rhythms with age	301
D. melanogaster (Hk[1P], Hk[2T], Sh[5], Eag, Hk[1P]-Sh[5], Canton S)	F	25 ± 0.5	Maturational INC then DEC in buzzing activity (short flight and righting response)	51

[a] Negative geotaxis is a normal physiological trait of *Drosophila* species wherein the flies being tapped to the bottom of the cylinder will crawl or fly toward the top.

[b] Positive phototaxis is also a normal physiological trait of most *Drosophila* species wherein the orgaisms move toward light. (The reader is referred to Grossfield[300] for further discussion concerning geotaxis and phototaxis.)

Table 15
AVERAGE OXYGEN CONSUMPTION AND LONGEVITY

Species, strain	Mean O_2 consumption ($\mu\ell$)/mg wet wt/hr	Mean longevity (days)		Temp (°C)	Ref.
		Male	Female		
D. melanogaster					
Gabarros Wild	2.84 ± 0.10	—	35.3 ± 1.5	25	124
Gabarros Inbred	3.04 ± 0.10	—	29.8 ± 4.0	25	
Gabarros Inbred female/ Abeele Inbred male	3.69 ± 0.11	—	46.4 ± 2.5	25	
Gabarros female/Abeele female	3.69 ± 0.11[a]	—	46.4 ± 2.5[a]	25	
Abeele female/Gabarros male	3.40 ± 0.10[a]	—	43.7 ± 3.0[a]	25	
Gabarros female/Abeele male	3.66 ± 0.12[b]	—	58.8 ± 3.2[b]	25	
Abeele female/Gabarros male	3.51 ± 0.10[b]	—	59.1 ± 2.4[b]	25	
Hyperkinetic-1P	5.57	—	39.9 ± 8.9[c]		
Hyperkinetic-2T	3.73	—	65.9 ± 13.5[c]	25	51
Shaker-5	5.01	—	40.5 ± 9.2[c]	25	
Ether A Go-Go	3.44	—	67.6 ± 12.7[c]	25	
Canton S	3.39	—	76.2 ± 15.7[c]	25	
Hyperkinetic-1P-Shaker-5	4.75	—	59.5 ± 14.1[c]	25	
Oregon-R	1.58	30	—	18	15
	2.26	86	—	21	
	3.85	43	—	27	
D. subobscura					
K	3.63 ± 0.17	12.9 ± 0.5	—	25	312
B	3.19 ± 0.22	18.8 ± 0.3	—	25	
B female/K male	3.63 ± 0.24	38.7 ± 0.9	—	25	
K female/B male	3.77 ± 0.24	46.9 ± 1.4	—	25	

[a] 30 eggs/5.5 cm².
[b] 240 eggs/5.5 cm².
[c] Standard deviation.

ported.[313-315] These studies are in general agreement that there is a decrease in the overall functional efficiency of mitochondria with advancing age in *Drosophila*.

ENZYME CHANGES WITH ADVANCING AGE IN ADULT *DROSOPHILA*

With the exception of lysosomal enzymes and a few other enzymes involved primarily in intermediary degradative pathways, all other enzyme activities reported in various strains and species of *Drosophila* decrease with advancing age, though many of these enzymes demonstrate maturational postemergent increases in activity associated with various increased functional capacities. For example, the increase in locomotor ability during the first few days postemergence are paralleled by increases in specific enzyme systems involved in high energy metabolism (see Table 17).

With regards to the underlying molecular mechanism which must ultimately be responsible for the alterations in enzyme activities with age in *Drosophila*, there is only limited information available. Certainly, the reported alterations in protein synthesis (see Table 20) must play some role in these enzymatic changes with advancing age. A decrease in the ability to renew catalase activity following drug treatment (3-amino-1,2,4-triazole) with age has been well documented in *Drosophila*.[4,32,296] From the work of Baker and co-workers, it has been

Table 16
ALTERATIONS IN O_2 CONSUMPTION WITH AGE

Species, strain	Sex	Mean longevity (days)	Change in mean O_2 consumption	Temp (°C)	Ref.
D. melanogaster					
Gabarros female/Abeele male	F	46.4 ± 2.5[a]	ca. 5—70d: 19% decrease	25	124
Abeele female/Gabarros female	F	43.7 ± 3.0[a]	ca. 5—55d: 33% decrease	25	
Gabarros female/Abeele male	F	58.8 ± 3.2[b]	No significant change	25	
Abeele female/Gabarros male	F	59.1 ± 2.4[b]	ca. 3—65d: no significant change	25	
D. subobscura					
B	M	12.9 ± 0.5	ca. 5—25d: 30% decrease	25	312
K	M	18.8 ± 0.3	ca. 1—25d: no significant change	25	
B female/K male	M	38.7 ± 0.9	ca. 1—50d: 44% decrease	25	
K female/B male	M	46.9 ± 1.4	ca. 1—70d: 27% decrease	25	

[a]   30 eggs/vial.
[b]   240 eggs/vial.

demonstrated that the activities of two enzymes involved in the energizing of flight muscle, namely, ATP-L-arginine phosphotransferase (APK) and α-glycerophosphate dehydrogenase (α-GPDH) are altered independent of flight capability, mutant lifespan, and genetically reduced levels of the enzyme in the case of a-GPDH. These data suggest that many of the observed changes in enzyme activities with age in *Drosophila* may reflect intrinsic regulatory alterations. Furthermore, there is no evidence to suggest that enzyme alteration with age in *Drosophila* is related to a single underlying mechanism as APK does not exhibit age-related increased in the number of inactive enzyme molecules, whereas, the α-GPDH demonstrates an increase in thermal lability with advancing age.[26,27,289,316-319]

In addition to the enzymes listed in Table 17, a number of aminoacyl tRNA synthetase activities have also been reported to change with age (see Tables 19 and 20).

PIGMENT FORMATION IN *DROSOPHILA* WITH AGE

As in other higher organisms that have been examined for such, there is an increase in pigment accumulation with advancing age in *Drosophila* (see Table 18). We present in this table the parameters studied and methodologies employed and the results as they relate to the given strains of *D. melanogaster* so examined. A summary of the accumulation of age pigments primarily by Miquel and co-workers[269] would indicate that the primary tissues or cells wherein pigment accumulates are the central nervous system, malpighian tubules, gut, oenocytes, and the fat body with varying and lesser amounts of accumulation in other tissues. The rates of accumulation and content of ceroid-lipofuscin-type pigment has been demonstrated to be temperature and sex dependent but apparently, from the studies of Tappel and co-workers,[332] and Biscardi and Webster,[16] are diet and strain independent. Two other environmental parameters have been examined with respect to their influence on age pigment accumulation, namely, the effects of high O_2 tension wherein there was an increase in the rate of accumulation of ceroid-lipofuscin-type pigments concommittant with a decrease in mean lifespan,[205] and the effect of gamma radiation wherein only the oenocytes showed an increase in accumulation of electron dense bodies.[268]

Table 17
ENZYME CHANGES WITH ADVANCING AGE IN ADULT DROSOPHILA

Enzyme	Species (strain)	Sex	Specific activity	Ref.
β-Acetylglucosaminidase, E.C. 3.2.1.30	*D. melanogaster* Oregon R		INC	269
Acid phosphatase (lysosomal)	Oregon R	M and F (separate)	NSC	320
E.C. 3.1.3.2	Oregon R	M	DEC	269
Acid (para-nitrophenyl) phosphatase		M and F (mixed)	INC	321, 322
	Wild type	M and F (separate)	NSC	323
	White eyed	M and F (separate)	NSC	323
	Ebony	M and F (separate)	NSC	323
Acid ribonuclease (lysosomal)		M and F (mixed)	INC	321, 322
L-Alanine aminotransferase				
Paragonial gland.	*D. nigromelanica*	M	DEC	324
E.C. 2.6.1.2	*D. melanogaster*	M	DEC	324
Alcohol dehydrogenase. E.C. 1.1.1.1.	*D. melanogaster*			
	Oregon/Kaduna F_1 hybrid	M	DEC	325
	Canton S	M	DEC	4
	Oregon R	M	DEC	4
Alkaline phosphatase. E.C. 3.1.3.1	*D. melanogaster* Canton S	M and F (separate)	DEC	157
a-Amylase, E.C. 3.2.1.1	*D. immigrans*		DEC	326
ATP: L-Arginine phospho-transferase, E.C. 2.7.3.3	*D. melanogaster*			
	Sevelen	M	DEC	27, 289, 316
	Vestigial	M	DEC	27
	Oregon R	M and F (mixed)	DEC	315
F-ATPase (mitochondrial) E.C. 3.6.1.8	Oregon R	M and F (separate)	DEC	320
Catalase: E.C. 1.11.1.6	Canton S	M	DEC	4
	Oregon R	M	DEC	4
	Oregon R	M	DEC	327
	Oregon R	M	DEC	32
	Oregon R	M	DEC	328
	Oregon R	M and F	DEC	301

Table 17 (continued)
ENZYME CHANGES WITH ADVANCING AGE IN ADULT DROSOPHILA

Enzyme	Species (strain)	Sex	Specific activity	Ref.
Cathepsin-D (lyosomal), E.C. 3.4.23.5	*D. melanogaster*	M and F (mixed)	INC	321, 322
Deoxycytidine aminohydrolase, E.C. 3.5.4.14	*D. melanogaster*		INC	329
Esterase	*D. melanogaster*			
	Oregon/Kaduna F$_1$ hybrid	M	DEC	325
	Canton S	M and F (separate)	INC	157
	Wild type	M and F (mixed)	DEC	330
	Vestigial	M and F (mixed)	DEC	330
α-Glycerophosphate dehydrogenase, E.C. 1.1.1.8	*D. melanogaster*			
	Sevelen	M	DEC	26, 289, 316
	Oregon/Kaduna F$_1$ hybrid		DEC	325
Glucose-6-phosphate dehydrogenase, E.C. 1.1.1.49	*D. melanogaster*			
	Canton S	M and F (separate)	DEC	157
	Canton S	M	DEC	4
	Oregon R	M	DEC	4
Glutamic-aspartic transaminase, E.C. 2.6.1.1	*D. melanogaster*			
	Oregon/Kaduna F$_1$ hybrid	M	NSC	325
Hexose-phosphate isomerase E.C. 5.3.1.9	*D. melanogaster*			
	Canton S	M and F (separate)	INC	157
Peroxidase, E.C. 1.11.1.7	*D. melanogaster*			
	Oregon R	M and F (separate)	DEC	320
Superoxide dismutase	*D. melanogaster*			
	Oregon R	M	NSC	331
Total mitochondrial (Cu^{2+})	*D. melanogaster*		DEC	
Trehalase, E.C. 3.2.1.28	Oregon/Kaduna F$_1$ hybrid	M	DEC	325

Table 18
PIGMENT FORMATION IN *DROSOPHILA* WITH AGE

Species (strain)	Sex	Tissue	Parameters measured	Method	Results	Ref.
D. melanogaster						
Swedish	M and F	Sternite	Onset and intensity of pigment with age	Visual inspection	Onset at day 12 and onset intensity with age	334
Swedish	M and F	Sternite	Longevity at 25 and 30°C and onset of pigmentation	Visual inspection	Higher temperature shortens lifespan and hastens onset of pigmentation	335
Swedish/Samarkland hybrid	M and F	Sternite	Heterotic enhancement of longevity and onset of pigmentation	Visual inspection	Increased lifespan delays onset of pigmentation	335
Oregon-R	M	Flight muscle	Ultrastructural changes in 7- to 85-day-old individuals	E.M.[a]	Accumulation of dense osmophilic structures in mitochondria with age	281
Oregon-R	M	Flight muscle, goblet cells, oenocytes, fat body	Ultrastructural changes between 15- to 85-day-old individuals	E.M.	Dense body ("old age") pigment accumulation with age	281
Oregon-R	M	Brain	Ultrastructural changes in 6- to 100-day-old individuals	E.M.	Ceriod-lipofuscin accumulation in aging neuropil	264
Oregon-R	M	Oenocytes, fat body, digestive tract	Ultrastructural change in 2- to 100-day-old individuals	E.M.	"Dark body" accumulation with age	168
Oregon-R	M and F	Midgut, oenocytes, and other tissues	Changes in lifespan and ultrastructure upon irradiation and aging	E.M.	Accumulation of electron dense bodies in oenocytes and ceriod-like material in midgut cells with aging; irradiated oenocytes accumulate electron dense bodies	239
Oregon-R	M	Brain, oenocytes, Malpighian tubules, midgut	Ultrastructural changes and fluorescent pigment accumulation in 7- to 102-day individuals; acid phosphatase activity	E.M. & SPF[b]	Brain and viscera seem to accumulate pigment more similar to ceriod than lipofuscin; lysosomal involvement in pigment genesis	168

Table 18 (continued)
PIGMENT FORMATION IN *DROSOPHILA* WITH AGE

Species (strain)	Sex	Tissue	Parameters measured	Method	Results	Ref.
Oregon-R	M and F	Whole body	Fluorescent pigment accumulation vs. sex, environmental temperature (22.2, 26.7, 30°C) and diets with individuals between 4—100 days of age	SPF	Males contain more pigment but rates of accumulation are similar in both sexes; higher temperature increase pigment accumulation; fluorescence is not directly o⁻ dietary origin	336
Oregon-R	M	Fat body, oenocytes	Effects of high O_2 tensions on tissue ultrastructure	E.M.	Dense body accumulation in these structures	204
Oregon-R	M	Brain, Malpighian tubules, testis	Effect of high O_2 tensions on tissue ultrastructure and mean lifespan	E.M.	Ceriod-lipofuscin-like accumulations in those structures with O_2 induced shortening of mean lifespan	337
		Whole body	Effect of dietary vitamin E on fluorescent pigment accumulation	SPF	The antioxident feeding reduces fluorescence of chloroform-methanol extracts with age	336
Oregon-R Vestigial (ebony)	M and F	Whole body	Rates of accumulation of fluorescent pigments in 3 strains with different lifespans	SFF	The rates of pigment accumulation are nearly identicalin in all three strains; the rates correlate with chronological time rather than strain longevity; males have a higher content of pigment but the rates of accumulation are similar in both sexes	16

[a]   E.M., electron microscopic analysis.

[b]   SPF, spectrophotofluorimetric analysis of chloroform: methanol soluble fluorescent pigment.

DNA, RNA, AND PROTEIN CONTENT WITH AGE

Table 19 presents studies on DNA, RNA, and protein in *Drosophila* with age. With the possible exception of a loss of 50% in one of the A-T satellites of *D. melanogaster* (Sevelen) with advancing age, none of the studies indicate any significant loss of DNA with age on a per fly basis. The histological studies of Miquel and co-workers, however, suggest preferential cell loss, particularly in the central nervous system.[168] It is clear, however, that the extractability of DNA becomes more difficult with advancing age.[338] The observed loss in mitochondrial DNA is a reflection of the degradation and the absolute loss of mitochondria with advancing age as noted in various *Drosophila* species and strains. The mean temperature of thermal denaturation (69°C) of highly purified DNA from *D. melanogaster* (Sevelen) is not altered with age.[339] Similar to what has been observed in other higher organisms, however, there is an increase in the mean temperature of thermal denaturation when more than a few percent of chromatin protein is left on the DNA. This is particularly evident in postmitotic tissues.[338]

Total RNA apparently decreases somewhat until postemergent maturation of the organism and then remains relatively constant throughout the lifespan. It should be mentioned that these observation are based on only two known studies in the literature and that more detailed information is needed in this area.

Alterations in tRNA, on the other hand, have been more extensively examined from a number of different perspectives. It is clear from the literature that some tRNA species show a distinct quantitative loss with advancing age, whereas others do not appear to change very much. On a per weight basis, there has been reported a 35% decrease in total tRNA.[340,341] There has also been reported an increase in the Q-base containing isoacceptor species. It is not clear whether or not the increase in Q-base containing isoacceptors affects the overall efficiency and fidelity of protein synthesis with advancing age.[342]

It is evident from the literature that there is a decrease in the ribosomes from *Drosophila* with advancing age.[343,344] Loss of ribosomes is particularly noticeable in the central nervous system, gut, and the malpighian tubules whereas no significant change was observed in flight muscle.[15] In addition, it has been observed by Baker and co-workers that ribosomes extracted from older organisms lack a degree of structual integrity when compared to ribosomes extracted from younger organisms.[345] The ribosomal instability is apparently not due to the ribosomal proteins as no quantitative or qualitative differences were shown, but rather lies within the ribosomal RNA itself,[345] presumably due to an undermethylation of the ribose moieties with age.[346] There has also been observed a substantial decrease (40%) in the polysome to monosome ratio.[347]

Alterations in various protein fractions with advancing age have been reported by a number of authors in *Drosophila*. There is some ambiguity with precisely what fractions of proteins in *Drosophila* are being altered with advancing age. It is apparent that many of the discrepancies reported in the literature are based on the methodologies employed and specific protein fractions examined. In general, it may be stated that there is apparently no significant change in the total protein content from whole flies with advancing age and no significant change or slight decrease in cytosol protein with an increasing amount of insoluble proteins.

Studies primarily by Chen and colleagues indicate that there are substantial changes in amino acid composition with advancing age in *Drosophila*.[348] A number of amino acids exhibit a decrease with advancing age. On the other hand, there are those that increase and another class which demonstrates no significant change with advancing age.

DNA, RNA, AND PROTEIN SYNTHESIS WITH AGE

It is generally accepted that, with the exception of gonadal tissues, virtually all tissues in the adult *Drosophila* are postmitotic.[252,351] There is, however, incorporation of tritiated

Table 19
DNA, RNA, AND PROTEIN CONTENT

Species, age	Sex	Parameter	Results	Ref.
		Nucleic Acids		
D. melanogaster Vestigial-winged strain 5—25 days	—	RNA DNA — UV absorption	RNA/DNA Initial DEC to maturation, then NSC with age	290
D. melanogaster Oregon R, 8—79 days	M and F	RNA — Ornicol DNA — UV absorption Protein — Lowry	RNA, DNA content Males — NSC Females — lower DNA content in middle-aged than in younger or older flies correlated with fecundity Total protein/DNA Males and females — higher in middle-aged than in younger or older flies	296
D. melanogaster Oregon R, 2—70 days	M	Nuclear DNA Mitochondrial DNA lysates in CsCl and Cs_2SO_4	Nuclear DNA Mitochondrial DNA DEC after 30 days CsCl and $CaSO_4$ gradients NSC in buoyant densities with age	3,349
D. melanogaster	M	Nuclear DNA	Nuclear DNA content showed initial DEC to maturation then NSC on a μg per fly basis	17
Swedish C, 2—43 days	M	Mitochondrial DNA lysates in CsCl gradients	Mitochondrial DNA content DEC from approximately 5% of total DNA at 20 days to 0% at 43 days NSC in buoyant density of nuclear DNA (1.693 g/cm³) and mitochondrial DNA (1.680 g/cm³)	
D. melanogaster Sevelen, 4, 20, 40 days	M	Nuclear DNA, UV absorption	NSC in T_m, buoyant density (main band — 1.700 g/cm³) DEC in A-T satellite relative to main band RNA-DNA hybridization NSC in hybridization saturation values for rDNA cistrons	26, 338 339,340

Species	Sex	Method	Findings	Ref.
D. melanogaster Sevelen, 4, 31 days	M	DNA — protein complex. Hydroxyapatite column chromatography	DEC in whole fly, brain, flight muscle, and testis extracts. INC in T_m from whole fly (15°C), brain (11.75°C), and flight muscle (2.50°C) NSC in T_m from testis	338, 350
Protein				
D. subobscura (B/K) F$_2$ hybrids, 6—60 days	M	Protein; ³H-leucino labeling trichloracetic acid (TCA) precipitation, NaOH wash	NSC in total soluble protein from whole fly	325
D. melanogaster (Oregon, M Kaduna F) F$_2$ hybrids, 6—57 days	M	Protein; Whole fly homogenate fractionation, microbiuret, TCA precipitation, NaOH wash	NSC in total body protein content; DEC in soluble fraction (135,000 × g supernatant), microsomal fraction (135,000 × g precipitate), and mitochondrial fraction (18,000 × g precipitate), INC in insoluble fraction (5,000 × g precipitate)	185
D. melanogaster (Oregon, M/Kaduna. F) F$_2$ hybrids, 6—57 days	M	Protein; Whole fly homogenate fractionation; Microbiuret; TCA precipitation, NaOH wash	NSC in total body protein content; DEC in soluble fraction (135,000 × g supernatant), microsomal fraction (135,000 × g precipitate), and mitochondrial fraction (18,000 × g precipitate), IC in insoluble fraction (5,000 × g precipitate)	
D. melanogaster Oregon R, 1—100 days	M and F	Protein; Whole fly	Females contain more protein than males at most ages DEC in total protein content by 43% at 15,000 rpm (15 min) supernatant on a mg/mℓ supernatant basis; peaks of protein content at 1, 5, and 10 weeks	320
D. melanogaster (Oregon-R) 10—50 days	M	Protein; NaOH extraction, microbiuret	NSC in total protein content from whole fly extracts	297
D. subobscura (B,F/K,M) F$_2$ hybrids	M	Protein; TCA precipitate, NaOH wash, microbiuret	NSC in total nitrogen, protein, and ninhydrin-reacting substances with advancing age	325, 185
D. melanogaster Oregon/Kaduna F$_2$ hybrid 0—80 days	M	Nitrogen; Micro-Kjeldahl; Uric acid; Enzyme (uricase) spectrophotometric analysis; Amino acids and small peptides; Ninhydrin analysis of TCA supernatant	Initial DEC in free uric acid content of *D. melanogaster*, but not in *D. subobscura* hybrids	

Table 20
DNA, RNA, AND PROTEIN SYNTHESIS

Species (strain)	Sex	Parameter	Result	Ref.
D. melanogaster Nettlebed, 3 days and 3 weeks	F	DNA Incorporation of ³H-thymidine	DEC in incorporation rate in polyploid nurse cells of functional ovarioles	358
D. subobscura (B,F/K.M) F₂ hybrids, 5—14 days and 40 days	M	DNA Incorporation of ³H-thymidine following gamma irradiation	INC in incorporation without formation of polyploids in adult tissues	252
		Mitosis Colchicine treatment	Virtually all adult tissues except gonadal are postmitotic	352
D. subobscura B/K F₁ hybrids 20 days and 60 days	M	RNA Incorporation of ¹⁴C-uracil Protein Incorporation of ³H-leucine	Twofold INC in incorporation rate in whole fly homogenates of uracil and leucine with NSC in leucine pool	
D. melanogaster Nettlebed/Oregon hybrids 6—8 days and 28 days	F	RNA and precursors Incorporation of ³H-uracil Actinomycin D treatment Ornicol Optical density of barium-alcohol soluble and insoluble fractions of PCA whole fly extracts	DEC in half-life of messenger RNA in thorax and ovary based on incorporation rate following Act. D treatment; INC in ribosomal RNA synthesis in nurse cell nuclei; DEC (marginal statistical significance) in RNA content per egg; DEC in duration of egg maturation; INC in nucleoside and nucleotide pool sizes; INC in purine and pyrimidine bases	353, 356
D. melanogaster (Samarkand) Egg to 2 week adult su(s)² v:bw vermilion and brown eye mutant 3rd instar larvae and 2 week adult		rRNA Aminoacylation capacity for his, asp, asn, and tyr Reversed phase chromatography; Amount of isoaccepting species of tRNA RPC5 column chromatography	No changes in any preparations Loss of tRNA₁ᵀʸʳ with development	359
D. melanogaster (Samarkland) 1st instar, 3rd instar larvae and adult		tRNA Aminoacylation ability Reverse phase chromatography Isoaccepting tRNA species (tRNA synathotases from adults)	1st to 3rd instar larvae Increase in acceptance of cys, ile, leu, and met Decrease in acceptance of glu, gln, and gly 3rd larvae to adult	363

Organism	Age/Sex	Material/Method	Results	Ref.
D. melanogaster (Sevelen) 5 day (young); 22 day (mature adult); 35 day (old)	M	tRNA Aminoacylation Reverse phase chromatography Aminoacyl-tRNA synthetase activites	Increase in acceptance for glu, fly, and pro Decrease in acceptance of cys. met, and ser Increase in peak 6 of lysyl-tRNA between 1st and 3rd instar larvae Decrease in peaks 1 and 2 of glycyl-tRNA between 1st and 3rd instar larvae 10—25% DEC in aminoacylation of 35-day tRNA NSC in synthetase activity for Gly. Ile. Thr. & Val with age DEC in synthetase activity fro pro. ala. arg, ser, & leu with age Overall mean DEC of 18% in synthetase activity with age	340, 341
D. melanogaster (Sevelen) 5, 22, 30, 35 day	M	tRNA Isoacceptor patterns for Q base containing (Asp. Asn, His, and Tyr) and non-Q base (Ala, Leu, Met, and Ser) tRNA Reverse phase chromatography	DEC of 35% in total tRNA on a per weight basis NSC in isoacceptor pattern of tRNA for ala. leu, met. and regardless of age source of tRNA and aminoacylating enzymes; INC in Q base containing isoacceptor peaks	361
D. melanogaster Samarkand Oregon R v:bw sw(s)[2]		tRNA Isoacceptor patterns for Q base containing (asp, asn, his, and tyr), tRNA, reverse phase chromatography, cell-free, tRNA-dependent, mRNA-dependent system	Inc in ratio of Q/non-Q isoacceptors with a more rapid INC in Samarkand than Oregon-R strains; the mutant strain rates also different NSC in protein synthesis between Q base tRNA-enriched or deficient systems NSC in ttal tRNA for each amino acid	342
D. melanogaster Oregon-R 1 day and 35 days	M and F	RNA and Protein	NSC in total polysome bound RNA NSC of 70—80% in rate of protein chain elongation NSC in synthesis of some aminoacyl-tRNA species while others DEC as much as 6%	362
D. melanogaster (Oregon-R) 2— 104 days	M	Ribosomes Histological studies	DEC in ribosome content of neuronal, gut, and Malpighian tubule tissue. NSC in flight muscle	15, 268
D. melanogaster (Sevelen) 15 days, 30 days	4 days, M	Whole fly high salt wash extraction UV analysis Thermal denaturation Two dimensional polyacrylamide gel electrophoresis	DEC of 23% in extractable 80S monosomes based on a per gram wet weight basis: 5 × INC in split proteins dissociated from 80S ribosomes with 2.0 XCI DEC of 8°C in Tm of monosomes with accompanying DEC in ability to reassociate upon cooling	54, 343 363—369

Table 20 (continued)
DNA, RNA, AND PROTEIN SYNTHESIS

Species (strain)	Sex	Parameter	Result	Ref.
		Ion-exchange chromatography	No apparent quantitative or qualitative changes in ribosomal proteins with age; DEC of 4—6°C in Tm of ribosomal RNA with ability upon cooling; DEC in methylation of ribosomal RNA indicated by 0.2M KOH digests	
D. melanogaster (Oregon R) 1 day, 30 days, 60 days	M and F	Ribosomes; Microsomal preparation from whole fly homogenates and sucrose gradient centrifugation; UV analysis	DEC of 6% in ribosomes from 1—21 days; DEC in ratio of polyribosomes/monosomes from 1.90 at 1 day to 0.92 at 30 days to 0.80 at 60 days; NSC in 40 S; 60 S; 80 S ratios	347, 370
D. melanogaster (Oregon-R) 1—60 days	M and F	Protein; Microsomal fraction (100,000 × g precipitate) of whole fly or specific tissue homogenates; Ribosomes isolated from 150,000 × g centrifugation from suspended microsomal fraction; Rate of incorporation of ^{14}C-amino acid mixture and ^{14}C-phenylaline; Aminoacylation of ^{3}H-methionine with tRNA; Binding of ^{3}H-methionine to sucrose gradient isolated 40 S and 80 S initiation complexes; Biuret procedure; UV analysis	DEC at 70—75% in incorporation rate in first 14 days post emergence followed sloer DEC over rest of lifespan. DEC is compartmentalized as head tissue incorporation DEC by 15%, thoracic by 96% and abdominal by 33%; DEC in polyuridylate promoted, ^{14}C-labeled phenylalanine incorporation rate; DEC of 12% in binding ability of methional — tRNA to 40 S and 14—20% to 80 S; DEC of 23% during first 21 days in polysome-bound mRNA; DEC of 40% in polysome:monosome ratio	347, 352, 370, 371
D. melanogaster b pr/Oregon (*v*) F$_1$ hybrids (*b pr*) inbreds Oregon (*v*) inbreds		Protein; Treatment of late 3rd instar larvae with five amino acid analogs in (10^{-3} or 2 × $10^{-3}M$) for 4, 8, or 24 hr	DEC of 9—16% in 50% lifespan at $10^{-3}M$ with increasing exposure period; 40—70% mortality in maturational stage at 2 × $10^{-3}M$	372

Organism	M	Method/Analysis	Results	Ref.
		Canavanine 4-methyl tryptophan Ethionine -2-Thienylalanine -Fluorophenylalanine	Streptomycin treatment caused INC mortality in maturational stage, but extent varied	355
D. subobscura (B/K)F₂ hybrids larvae, 64 days	M	Protein Incorporation of ³H-leucine in 7- 14-day-old imagoes with activity measured from 20 days ³H-phenylalanine and ³H-leucine fed to larvae with activity measured for 64 days of adult life TCA precipitation following NaOH wash	DEC in total protein activity for initial 4 days post-emergence (confined to pupal fat body), then NSC in flies raised on labeled medium as larvae 80% of total protein is not replaced, including mitochondrial and structural proteins of flight muscles 20% of total protein turns over, including ribosomal proteins, with average half-life of approximately 10 days, rates of incorporation vary greatly in tissues: mid-gut (thoracic ventricule) (Malpighian tubules, fat body, accessory glands)) thoracic muscle	
D. melanogaster (Sevelen) 3 days, 20 days, 50 days	M	Protein Utilization and incorporation of ¹⁴C-labeled lysine, alpha-alanine and glycine Proteins precipitated with PCA from whole fly homegenated Biuret Paper electrophorisis	Incorporation rates per mg protein by 42.2—63.1% DEC in poolsize for lysine and glycine, INC for alanine DEC turnover rate for lysine and glycine, INC for alanine	303, 348 354, 373
D. melanogaster Nettlebed/Oregon F₁ hybrids, 2—35 days	M	Protein ³H-leucine incorporation PCA precipitation from homogenized thoraces and ovaries	Slight INC in incorporation rate of H-leucine into thoracic proteins with corresponding INC in turn-over rate, incorporation into ovarian protein follows egg laying with NSC in turnover time	356
D. subobscura (B/K) F₂ Hybrids, 6—90 days	M	Protein Fed 7- to 12-day-old adults amino acid analogs (DL-othionine and *p*-fluorophenylalanine) and cycloheximide	DEC of greater than 25% in mean survival time (MST) with 20 m*H* DL-ethionine; NSC in MST with 20 m*M* DL-ethionine and 1m*M* L-methionine; DEC of greater than 34—70% in MST with 2 m*M* or 20 m*M* DL-*p*-fluorophenylalanine DEC of greater than 43% in MST with 20 m*M* DL-*p*-fluorophenyl-alanine and 10 m*M* L-phenylalanine	374

Table 20 (continued)
DNA, RNA, AND PROTEIN SYNTHESIS

Species (strain)	Sex	Parameter	Result	Ref.
D. melanogaster (Oregon) 1—95 days		Protein	DEC of greater than 84% in MST with cycloheximide treatment	13, 357
		Survivor of 72-hr feeding period on amino acid analogs (20 mM) following emergence	INC of 16% in MST with 20 mM DL-p-fluorophenylalanine	
		L-ethionine	DEC of 8% in MST with both cycloheximide alone and cycloheximide with 20 mM DL-p-fluorophenylalanine	
		S-methyl-DL-tryptophan		
		DL-p-fluorophenylalanine	NSC in mean lifespan	
		Rate of incorporation for ¹⁴C-tyrosine and L-ethionine-ethyl-1-¹⁴C	Half-life of analog in ventriculus and oenocyte — 9.5—11 days with half-life of normal amino acid in same tissues 14.5—22 days	

Table 21
DIETARY SUPPLEMENTATION WITH VARIOUS COMPOUNDS

Animal	Compound	50% Lifespan		Maximum lifespan		Ref.
		Male	Female	Male	Female	
D. melanogaster Oregon-R	Beta-carotene (0.01 mg/mℓ medium)	8.9% INC	—	—	—	375—377
	1,4-diazobicyclo (2.2.2) octane (10^{-4} M)	7.2% INC	—	—	—	
	Gluconic acid (0.001—0.10 M, grown and maintained on throughout life)	0—14.3% INC	—	—	—	
	Lactic acid (1.0 M) (0.1—0.25 M) grown and maintained on throughout life	32.7% DEC	—	—	—	
		14—15% INC	—	—	—	

Strain	Treatment					Ref.
	Tartaric acid (0.01 M) grown and maintained on throughout life	6.0% INC	—	—	—	—
	Ascorbic acid (0.001—0.100 M)	NSC or slight DEC	2% INC-12% DEC	—	—	—
	Dehydroascorbic acid (0.001—0.01 M)	NSC	—	—	—	—
	Diiodemethane	14.5—29.3% INC	31% INC	—	—	—
	0.1—10.0 μg/mℓ medium reared and maintained on experimental medium	13.6% INC		11% INC	—	—
	1.0—10.0 μg/ℓ medium reared on experimental then maintained on control medium					
	0.1—10.0 μg/mℓ medium	7.0% DEC-3.6% INC		6—13% INC	—	—
	Iodoform (0.001—1 μg/mℓ medium)	1.9—23.8% DEC	—	—	—	—
	Iodine ($2 \times 10^{-4}\,M - 2 \times 10^{-9}\,M$)	NSC	—	—	—	—
	Hydrogen iodine ($4 \times 10^{-6}\,M - 4 \times 10^{-5}\,M$)	20.4—61.1% DEC	—	—	—	—
mei-41[DS]	Vitamin E (DL-alpha-tocopherol) 0.5% (w/w) of nutrient medium	11—29% INC	16—30% INC		—	52, 53, 112, 300
	Centrophenoxine 0.01% (w/w) of nutrient medium					
White eye	Vitamin E (DL-alpha-tocopherol) 0.5% (w/w) of nutrient medium	4—8% INC	4—8% INC		—	—
	Centrophenoxine 0.1% (w/w) of nutrient medium					
Sevelen	Vitamin E-acetate (DL-alpha-tocopherol-acetate)					160, 217
	0.5% (w/w) of dry food	10% DEC	13% DEC	NSC	NSC	
	1.0% (w/w) of dry food	26% DEC	29% DEC		—	
	Centrophenoxine					
	0.001% (w/w) of nutrient medium	NSC	NSC	NSC	NSC	
	0.01% (w/w) of nutrient medium	NSC	5% INC	NSC	NSC	
	0.1% (w/w) of nutrient medium	8% INC	14% INC	NSC	NSC	
D. melanogaster Canton S (F)/ Oregon-RC(M)	Corticosteroids (mg/100 mℓ) Hydrocortisone acetate (10—500)	25—39% INC	22% INC	15—20% INC	15—20% INC	110

Table 21 (continued)
DIETARY SUPPLEMENTATION WITH VARIOUS COMPOUNDS

Animal	Compound	50% Lifespan		Maximum lifespan		Ref.
		Male	Female	Male	Female	
	F₁ Hybrids Canton S	Other corticosteroids		Marginally effective INC		
	Salicylates and related agents					
	Acetylsalicylate (20—250)	22—40% INC	13—16% INC	5—27% INC	16—16% INC	
	Salicylamide (50—650)	17% INC	12% INC	16—18% INC	16—18% INC	
	Acetaminophen (650)	9% INC	5% INC	5% INC	38% INC	
	Indomethacin (31)		NSC			
	Sodium Salicylate (175)		NSC			
	Antihistamines					
	Diphenhydramine hydrochloride (25)					
	Doxylamide succinate (25—100)					
	Chloropheniramine maleate (4—16)	Marginally effective INC (males and females)				
	Tripelennamin hydrochloride (25)					
	Meclizine hydrochloride (6.3—25)	NSC (males and females)				
	Cyclizine hydrochloride (50)					
	Phenothiazines					
	Promethazine (15.6—62.5)		NSC or marginally effective INC			

					Ref.
Chlorpromazine (25)					
Miscellaneous compounds					
Chloroquine phosphate (250)	—	—	14—26% INC	14—26% INC	337, 378, 379
Colchicine (0.025)	—	—	11—25% INC	11—25% INC	
Procaine HCl (200)	—		NSC	18% INC	
Centrophenoxine (120)	7% INC	39% INC	15% INC	23% INC	
D. melanogaster Oregon-R					
Vitamin E (DL-alpha-tocopherol) 0.06—0.25% of nutrient medium	15% INC	—	12% INC	—	
Tocopherol-*p*-chlorophenoxyacetate, 0.1% of nutrient medium	13% INC	—	13% INC	—	
NDCA and Mg-tocopherol-*p*-chlorophenoxyacetate, 0.1% nutrient medium	20% INC	—	20% INC	—	
But.-hydroxytoluene (BHT), 0.1% of nutrient medium	DEC	—	DEC	—	
Vitamin C (ascorbic acid), 0.17% of nutrient medium		NSC			
Thiazolidine carboxylic acid (TCA). 0.27% (w/w) of nutrient medium fed from 26th day to death					
Thiazolidine carboxylic acid (TCA), 0.27% (w/w) of nutrient medium fed from 26th day to death					
NaTCA	8% INC	17% INC	4% INC		
MgTCA	16% INC				

Table 21 (continued)
DIETARY SUPPLEMENTATION WITH VARIOUS COMPOUNDS

Animal	Compound	50% Lifespan		Maximum lifespan		Ref.
		Male	Female	Male	Female	
D. melanogaster Canton S	Erythorbic acid (175 mg/ mℓ nutrient medium)		M and F (mixed) 10—15% INC with 3—4 day interval feeding until 20—34 days			380
D. melanogaster Oregon-R (30°C)		**M and F**				381, 382
	(1) Biotin (0.0067 g/ 1000 g medium)	NSC				
	(2) Sodium yeast nucleate (5.0 g/1000 g medium)	11.3% INC				
	(3) Pyridoxine (0.025 g/ 1000 g medium)	10.5% INC				
	(4) Combination of (1), (2), and (3)	20.3% INC				
	(5) Pantothenic acid	27.8% INC				
	(6) Combination of (4) and (5)	46.6% INC				
	Royal Jelly (6.67)	16.5% INC				
	Ether soluble fraction (0.33)	2.3% DEC				
	Phenol (0.33)	6.0% DEC				
	Organic Acids (0.33)	NSC				
	(6.67)	9.0% INC				
	Sterols, glycarides (0.33)	1.5% INC				
	Wax (0.33)	2.3% DEC				
	Phospholipids (0.33)	1.5% DEC				

32. **Nicolosi, R. J., Baird, M. B., Massie, H. R., and Samis, H. V.,** Senescence in *Drosophila.* II. Renewal of catalase activity in flies of different ages, *Exp. Gerontol.,* 8, 101, 1973.
33. **Maynard-Smith, J.,** The rate of ageing in *Drosophila subobscura, CIBA Found. Colloq. Aging,* 5, 269, 1959.
34. **Burch, G. E., Sohal, R. S., and Fairbanks, L. D.,** Senescent changes in the heart of *Drosophila repleta* (Wollaston), *Nature (London)* 225, 186, 1970.
35. **Baker, G. T.,** unpublished.
36. **Lints, F. A. and Lints, C. V.,** Influence of preimaginal environment on ageing in *Drosophila melanogaster* hybrids. III. Developmental speed and life span, *Exp. Gerontol.,* 6, 427, 1971.
37. **Sondhi, K. C.,** Studies in aging. VII. Integration of genotypes, homeostasis, and the expression of aging processes in *Drosophila, Exp. Gerontol.,* 2, 241, 1967.
38. **Sondhi, K. C.,** Studies in aging. IX. Brain transplantations in *Drosophila, Exp. Gerontol.,* 5, 77, 1970.
39. **Heuts, M. J.,** Nieuwe prolematick in de genetica, *Agricultura IV,* 3, 343, 1956.
40. **Clarke, J. M. and Maynard-Smith, J.,** The genetics and cytology of *Drosophila subobscura.* XI. Hybrid vigor and longevity, *J. Genet.,* 53, 172, 1970.
41. **Burcombe, J. and Hollingsworth, M. J.,** The relationship between developmental temperature and longevity in *Drosophila, Gerontologia,* 16, 172, 1970.
42. **Christian, R., Jacobson, M., and Baker, G. T.,** Older individuals and their roles in animal population dynamics, *Gerontologist,* 20, 38, 1980.
43. **Gould, A. B. and Clark, A. M.,** X-ray induced mutations causing adult life-shortening in *Drosophila melanogaster, Exp. Gerontol.,* 12, 107, 1977.
44. **Hollingsworth, M. J.,** Temperature and the rate of ageing in *Drosophila subobscura, Exp. Gerontol.,* 1, 259, 1966.
45. **Maynard-Smith, J., Clarke, J. M., and Hollingsworth, M. J.,** The expression of hybrid vigour in *Drosophila subobscura, R. Soc. London Proc.,* 144, 159, 1958.
46. **Maynard-Smith, J.,** Temperature tolerance and acclimation in *Drosophila subobscura, J. Exp. Biol.,* 34, 85, 1957.
47. **Nickerson, R. P. and Druger, M.,** Maintenance of chromosomal polymorphism in a population of *Drosophila pseudoobscura.* II. Fecundity, longevity, viability and competitive fitness, *Evolution,* 27, 125, 1973.
48. **Lindsley, D. L., and Grell, E. H.** *Genetic variations of Drosophila melanogaster,* Carnegie Institution, Washington, D.C., 1968.
49. **Pearl, R., Parker, S. L., and Gonzalez, B. M.,** Experimental studies on the duration of life. VII. The mendelian inheritance of duration of life in crosses of wild type and quintuple stocks of *Drosophila melanogaster, Am. Nat.,* 57, 163, 1923.
50. **Doane, W. W.,** Developmental physiology of the mutant female sterile (2) adipose of *Drosophila melanogaster.* I. Adult morphology, longevity, egg production, and egg lethality, *J. Exp. Zool.,* 145, 1960.
51. **Trout, W. E. and Kaplan, W. D.,** A relation between longevity, metabolic rate, and activity in Shaker mutants of *Drosophila melanogaster, Exp. Gerontol.,* 5, 83, 1970.
52. **Daly, R. N., Davis, F. A., and Baker, G. T.,** Effect of mutations at loci affecting DNA repair on longevity in *Drosophila melanogaster, Gerontologist,* 19, 60, 1979.
53. **Daly, R. N., Jacobson, M., Cunningham, E. T., Davis, F. A., and Baker, G. T.,** Effect of vitamin E on survivorship of a DNA-repair deficient mutant of *Drosophila melanogaster, in Neural Regulatory Mechanisms During Aging,* Adelman, R. C., et al., A., Eds., Alan R. Liss, New York, 1980, 209.
54. **Baker, G. T. and Daly, R. N.,** unpublished.
55. **Economos, A. C., Miquel, J., Binnard, R., and Kessler, S.,** Quantitative analysis of mating behavior in aging male *Drosophila melanogaster, Mech. Ageing Dev.,* 10, 233, 1979.
56. **Mossiage, J. C.,** Sperm utilization and brood patterns in *Drosophila melanogaster, Am. Nat.,* 89, 123, 1955.
57. **Kaufmann, B. P. and Demerec, M.,** Utilization of sperm by the female *Drosophila melanogaster, Am. Nat.,* 76, 445, 1942.
58. **Gugler, H. D., Kaplan, W. D., and Kidd, K.,** The displacement of first-mating by second-mating sperm in the storage organs of the female, *Drosophila Information Service,* 40, 65, 1965.
59. **Zimmering S., and Fowler, G.,** X-irradiation of the *Drosophila* male and its effects on the number of sperm transferred to the female, *Z. Vererbungsl.,* 98, 150, 1966.
60. **Lefevre, G., Jr. and Jonsson, U. B.,** Sperm relationships in twice mated *Drosophila melanogaster* females, *D.I.S.* 36, 85, 1962.
61. **Muller, H. J. and Settles, F.,** The nonfunctioning of the genes in spermatozoa, *Z. Indukt. Abstamm. Vereblehre.,* 43, 1927.
62. **Glass, B.,** The influence of immediate vs. delayed mating on the life span of *Drosophila, in Biology of Aging,* Strehler, B. L., Ed., 185, *Symp.(Am. Inst. Biol. Sci.),* Washington, D.C., 1960.

63. **Gowen, J.,** On chromosome balance as a factor in duration of life, *J. Gen. Physiol.,* 14, 447, 1931.
64. **Hannah, A.,** The effect of aging the maternal parent upon the sex ratio in *Drosophila melanogaster, Z. F. Indukt. (Ab) v-Ver,* 86, 574, 1955.
65. **Gowen, J.,** Metabolism as related to chromosome structure and the duration of life, *J. Gen. Physiol.,* 14, 463, 1931.
66. **Hildreth, P. E. and Ulrichs, P. C.,** A temperature effect on nondisjunction of the X chromosomes among eggs from aged *Drosophila* females, *Genetics,* 40, 191, 1969.
67. **Marinkovic, D., Tucic, N., Kekic, V., and Andielkovic, M.,** Age-associated changes in viability genetic loads of *Drosophila melanogaster, Exp. Gerontol.,* 8, 199, 1973.
68. **Maynard-Smith, J.,** The genetics of longevity in *Drosophila subobscura, Proc. 10th Int. Cong. Gent. Mont.,* 2, 182, 1958.
69. **Sandler, L. and Hiraizumi, Y.,** Meiotic drive in natural populations of *Drosophila melanogaster.* VIII. A heritable aging effect on the phenomenon of segregation-distortion, *Can. J. Genet. Cytol.,* 3, 34, 1961.
70. **Vassileva-Dryanovska, O. and Gencheva, E.,** Changes in the sex correlation depending on the age of parents in *Drosophila melanogaster, Bulg. Akad. Na Nauk. Sofia Dokl.,* 51, 1965.
71. **Wattiaux, J. M.,** Parental age effects in *Drosophila pseudoobscura, Exp. Gerontol.,* 3, 55, 1960.
72. **Bridges, C. B.,** The relation of the age of the female of crossing over in the third chromosome of *Drosophila melanogaster, J. Gen. Physiol.,* 81, 689, 1927.
73. **Valentine, J.,** Effect of maternal age on recombination in X in *Drosophila melanogaster, D.I.S.,* 48, 127, 1972.
74. **Ives, P. T.,** Patterns of spontaneous and radiation induced mutation rates during spermatogenesis in *Drosophila melanogaster, Genetics,* 48, 981, 1963.
75. **Purdom, C. E., Dyer, K. F., and Papworth, D. G.,** Spontaneous mutation in *Drosophila:* studies on the rate of mutation in mature and immature germ cells, *Mutat. Res.,* 5, 133, 1968.
76. **Kaufmann, B. P.,** Spontaneous mutation rate in *Drosophila, Am. Nat.,* 31, 777, 1942.
77. **Lamy, R.,** Observed spontaneous mutation rates to experimental techniques, *J. Genet.,* 43, 212, 1947.
78. **Byers, H. L.,** Thermal effects on spontaneous mutation rate in mature spermatozoa of *Drosophila melanogaster, Caryologia,* 6, Suppl. Part 1, 694, 1954.
79. **Byers, H. L. and Muller, H. J.,** Influence of ageing at two different temperatures on the spontaneous mutation rate in mature spermatozoa of *Drosophila melanogaster, Genetics,* 27, 570, 1952.
80. **Kelsall, P.,** Non-disjunction and maternal age in *Drosophila melanogaster, Genet. Res.,* 4, 284, 1963.
81. **Wattiaux, J. M. and Heuts, M. J.,** Cyclic variation of bristle number with parental age in *Drosophila melanogaster, Population Genet.,* 1, 168, 1963.
82. **Wattiaux, J. M. and Heuts, M. J.,** Cyclic variation of bristle number with parental age in *Drosophila melanogaster, Proc. 11th Int. Genet. Cong.,* 1, 168, 1963.
83. **Delcour, J.,** Cell size and cell number in the wing of *Drosophila melanogaster* as related to parental aging, *Exp. Gerontol.,* 3, 247, 1968.
84. **Delcour, J. and Heuats, M. F.,** Cyclic variations in wing size related to parental ageing in *Drosophila melanogaster, Exp. Gerontol.,* 3, 45, 1968.
85. **Delcour, J.,** Influence de l'age parental sur la dimension des oeufs, la duree de developpement et la taille thoracique des descendants chez *Drosophila melanogaster, J. Insect Physiol.,* 15, 1999, 1969.
86. **Beardmore, J. A., Lints, F., and Al-Baldawi, A. L. F.,** Parental age and veritability of sternopleural chaeta number in *Drosophila melanogaster, Heredity,* 34, 71, 1975.
87. **Parsons, P. A.,** Maternal age and developmental variability, *J. Exp. Biol.,* 39, 251, 1962.
88. **David, J.,** Influence de l'age de la femelle sur les dimensions des oeufs de *Drosophila melanogaster, C. R. Hebd. Seanc. Acad. Sci. Paris,* 249, 1145, 1959.
89. **David, J.,** Influence de l'age de la mere sur les dimensions des oeufs dan une souche vestigial de *Drosophila melanogaster* meig. Etude exprimentale du determinisme physiologique de ces variations, *Bull. Biol. Fr. Belg.,* 96, 505, 1962.
90. **Butz, A. and Hayden, P.,** The effect of age of male and female parents on the life cycle of *Drosophila melanogaster, Ann. Entomol. Soc. Am.,* 55, 617, 1962.
91. **Christian, R. and Baker, G. T.,** Influence of post-reproductive cohorts in the selection for increased longevity, *Gerontologist,* 19, 56, 1979.
92. **Lints, F. A. and Lints, C. V.,** Influence of preimaginal environment on fecundity and ageing in *Drosophila melanogaster* hybrids, I. Preimaginal population density, *Exp. Gerontol.,* 4, 231, 1969.
93. **Lints, F. A. and Hosts, C.,** The Lansing effect revisited. I. Lifespan, *Exp. Gerontol.,* 9, 51, 1974.
94. **Lints, F. A. and Hoste, C.,** The Lansing effect revisited. II. Cumulative and spontaneously reversible parental age effects on fecundity in *Drosophila melanogaster, Evolution,* 81, 387, 1977.
95. **Lints, F. A. and Soliman, M. H.,** Growth rate and longevity in *Drosophila melanogaster* and *Tribolium castaneum, Nature (London)* 266, 624, 1977.
96. **Lints, F. A.,** *Interdisciplinary Topics in Gerontology: Genetics and Ageing, Vol. 14,* Von Hahn, H. P., Ed., S. Karger, Basel, 1978, 1.

97. **Sheldon, B. L.,** The effect of temperature on mutation rate in *Drosophila melanogaster, Aust. J. Biol. Sci.,* 11, 36, 1958.

98. **Kunz, W. F.,** Spontaneous Mutation and the Ageing Process, Ph. D. Thesis, Columbia University, New York, 1964.

99. **Rendel, J. M.,** The effect of age on the relationship between coincidence and crossing over on *Drosophila melanogaster, Genetics,* 43, 207, 1958.

100. **Olenov, J. M.,** Relation between age and mutation process in *Drosophila melanogaster, Genetics,* 49, 598, 191.

101. **Brown, S. W. and Welshons, W.,** Maternal aging and somatic crossing over of attached X-chromosomes, *Proc. Natl. Acad. Sci. U.S.A.,* 41, 209, 1955.

102. **Wattiaux, J. M.,** Cumulative parental age effects in *Drosophila subobscura, Evolution,* 22, 406, 1968.

103. **Redfield, H.,** Regional association of crossing over in nonhomologous chromosomes in *D. melanogaster* and its variation with age, *Genetics,* 49, 319, 1964.

104. **Tokunga, C.,** The effects of temperature and aging of *Drosophila* males on the frequency of XXY and XO progeny, *Mutat. Res.,* 13, 155, 1971.

105. **Rinehart, R. R.,** Spontaneous sex-linked recessive lethal frequencies for aged spermatozoa of *Drosophila melanogaster, Mutat. Res.* 7, 417, 1969.

106. **David, J., Biemont, C., and Fouillet, P.,** Sur la forme des courbes de ponte de *Drosophila melanogaster* et leur adjustment a des modeles mathematiques, *Arch. Zool. Exp. Genet.,* 115, 263, 1974.

107. **Timofeeff-Ressovsky, N. W.,** *Zeit. F. Induct. Abstam u Vererb.,* 70, 125, 1935.

108. **Sacharov, W. W.,** The mutation process in ageing sperm of *D. melanogaster* and the problem of the specificity of the action of the factors of mutation, *D.I.S.,* 15, 37, 1941.

109. **Gartner, L. P.,** Ultrastructural studies of senescence in the visceral musculature of the fruit fly, *Anat. Rec.,* 181, 360, 1975.

110. **Hochschild, R.,** Effect of membrane stabilizing drugs on mortality in *Drosophila melanogaster, Exp. Gerontol.,* 6, 133, 1971.

111. **Comfort, A.,** Absence of a Lansing effect in *Drosophila subobscura, Nature (London),* 172, 83, 1953.

112. **Ayala, F. J.,** Dynamics of populations. I. Factors controlling population growth and population size in *Drosophila serrata, Am. Nat.,* 100, 333, 1966.

113. **Bauer, E.,** Lebensdauer. Assimilationsgrenze. Rubnerische. Konstante and Evolution, *Biol. Zbl.,* 51, 74, 1931.

114. **Chiang, H. C. and Hodson, A. C.,** An analytical study of population growth in *Drosophila melanogaster, Ecol. Meneg.,* 2, 173, 1950.

115. **Chiang, H. C. and Hodson, A. C.,** The relation of copulation to fecundity and population growth in *Drosophila melanogaster, Ecology,* 31, 255, 1950.

116. **Chinnici, J. P.,** The effect of age on crossing over in the X-chromosome of *Drosophila melanogaster, D.I.S.,* 48, 82, 1972.

117. **Crozier, W. J. and Enzmann, E. V.,** Concerning critical periods in the life of adult *Drosophila, J. Gen. Physiol.,* 20, 595, 1936.

118. **David, J. J. and Fouillet, P.,** Enregistrement continu de la ponte chez *Drosophila melanogaster* et importance des conditions experimentales pour l'etude du rhythme circadien d'oviposition, *Rev. Comp. Anima,* 7, 197, 1972.

119. **David, J., Cohet, Y., and Fouillet, P.,** The variability between individuals as a measure of senescence: a study of the number of eggs laid and the percentage of hatched eggs in the case of *Drosophila melanogaster, Exp. Gerontol.,* 17, 1975.

120. **Erk, F. C., Samis, H. V., Baird, M. B., and Massie, H. R.,** A method for the establishment and maintenance of an aging colony of *Drosophila, D.I.S.,* 47, 130, 1971.

121. **Gartner, L. P. and Sonnenblick, B. P.,** *Drosophila* husbandry and extension of life span mean and life span range, *D.I.S.* 43, 172, 1968.

122. **Gowen, J. W. and Johnson, L. E.,** On the mechanism of heterosis. I. Metabolic capacity of different races of *Drosophila melanogaster* for egg production, *Am. Nat.,* 80, 149, 1946.

123. **Hiraizumi, Y. and Crow, J. F.,** Heterozygous effects on viability fertility, rate of development, and longevity of *Drosophila* chromosomes that are lethal when homozygous, *Genetics,* 45, 1071, 1960.

124. **Lints, F. A. and Lints, C. V.,** Influence of preimaginal environment on fecundity and ageing in *Drosophila melanogaster* hybrids, *Exp. Gerontol.,* 4, 231, 1969.

125. **McMillan, I., Fitz-Earle, M., and Robson, D. S.,** Quantitive genetics of fertility. I. Life time egg production of *Drosophila melanogaster:* theoretical aspects, *Genetics,* 65, 349, 1970.

126. **Minamori, S. and Morihira, K.,** Multiple mating in females of *Drosophila melanogaster, J. Sci. Hiroshima Univ. Ser. B Div.* 1 (Zoology), 22, 1, 1969.

127. **Manning, A.,** A sperm factor affecting the receptivity of *Drosophila melanogaster* females, *Nature* London, 194, 252, 1962.

128. **Narain, P.,** Effect of age of female on the rate of egg production in *D. melanogaster, D.I.S.,* 36, 96, 1962.

129. **DiPasquale, A. and Santibanez, S. K.,** Fecundity in several lines of *Drosophila simulans* and *Drosophils melanogaster, Atti. Assoc. Genet. Ital.,* 5, 93, 1960.

130. **Shapiro, H.,** The rate of oviposition in the fruit fly, *Drosophila, Biol. Bull. Woods Hole,* 63, 456, 1932.

131. **Tantawy, A. O. and Vetukhiv, M. O.,** Effects of size on fecundity, longevity, and viability in populations of *Drosophila pseudoobscura, Am. Nat.,* 94, 395, 1960.

132. **Tsien, H. C. and Wattiaux, J. M.,** Effect of maternal age on DNA and RNA content of *Drosophila* eggs, *Nat. New Biol.,* 230, 147, 1972.

133. **Alpatov, W. W.,** Egg production in *Drosophila melanogaster* and some factors which influence it, *Exp. Zool.,* 63, 85, 1932.

134. **Biemont, C.,** Interactions between ageing and inbreeding effects on development of *Drosophila melanogaster* embryos, *Mech. Ageing Dev.,* 5, 315, 1976.

135. **Bilewicz, S.,** Experiments on the effect of reproductive functions on the length of life in *Drosophila melanogaster, Gerontologist,* 22, 389, 1953.

136. **Bonnier, G.,** Temperature and time of development of the two sexes in *Drosophila, Am. Nat.,* 63, 186, 1929.

137. **Bucheton, A.,** Non-Mendelian female sterility in *Drosophila melanogaster:* influence of ageing and thermic treatments. I. Evidence for a partly inheritable effect of these two factors, *Heredity,* 41 (3), 357, 1978.

138. **David, J. and Clavel, M. F.,** Influence de la temperature sur le nombre le pourcentage d'eclosion et la taille des oeufs fondus par *Drosophila melanogaster, Am. Soc. Entomol. Fr.,* 5, 161, 1969.

139. **Harries, F. H.,** Some temperature coefficients for insect oviposition, *Ann. Entomol. Soc. Am.,* 32, 758, 1929.

140. **Kaliss, N. and Graubard, M. A.,** The effect of temperature on oviposition in *Drosophila melanogaster, Biol. Bull. Woods Hole,* 70, 385, 1936.

141. **Solima-Simmons, A. and Levens, H.,** Effect of age and temperature on matings of *Drosophila paulistorum, D.I.S.,* 40, 47, 1965.

142. **Guyenot, E.,** Etudes biologiques sur une mouche, *Drosophila ampelophila* Low. VII. Le determinisme de la ponte, *C. R. Soc. Biol. Paris,* 74, 443, 1913.

143. **Masing, R. A.,** Egg-laying in *Drosophila melanogaster* as influenced by sugar content in the food, *C. R. (Dokl.) Acad. Sci. U.R.S.S., N.S.,* 47, 296, 1945.

144. **del Solar, E. and Palomino, H.,** Choice of oviposition in *Drosophila melanogaster, Am. Nat.,* 100, 127, 1966.

145. **David, J., Fouillet, P., and Arens, M. F.,** Influcence repulsive de la levure vivante sur l'oviposition de la *Drosophila:* importance de la salissure progressive des cages et des differences innees entre les femelles, *Rev. Comp. Anim.,* 5, 277, 1971.

146. **Sameoto, D. D. and Miller, R. S.,** Factors controlling the productivity of *Drosophila melanogaster* and *D. simulans, Ecology,* 47, 695, 1966.

147. **Cohet, Y.,** Reduction de la fecondite et du potential reproducteur de la *Drosophila* adulte consecutive au developpement larvaire a basse temperature, *C.R. Acad. Sci. Paris,* 277D, 2227, 1973.

148. **Cohet, Y. and Bouletreau-Merle, J.,** Influences epigenetique sur la reproduction d'une insecte: variations de la reactivite des femelles de *Drosophila melanogaster* a la copulation en fonction de leur temperature de developpement, *C.R. Acad. Sci. Paris,* 278D, 3235, 1974.

149. **Alpatov, W. W.,** Experimental studies on the duration of life. XIII. The influence of different feeding during the larval and imaginal stages on the duration of life of the imago of *Drosophila melanogaster, Am. Nat.,* 690, 37, 1929.

150. **Strehler, B. L.,** *Time, Cells, and Aging,* Academic Press, New York, 1977.

151. **Tantawy, A. O.,** Developmental homeostasis on populations of *Drosophila melanogaster, Evolution,* 15, 132, 1961.

152. **Brandt, M. and Baker, G. T.,** Effects of fecundity and survivorship of brewer's yeast in the diet of *Drosophila melanogaster,* (unpublished).

153. **Robertson, F. W. and Sang, J. H.,** The ecological determinants of population growth in a *Drosophila* culture. I. Fecundity of adult flies, *Proc. R. Soc. Lond. B,* 132, 258, 1944.

154. **Bouletreau-Merle, J.,** Fonctionement ovarian comare des femelles vierges et des femelles inseminees de *Drosophila melanogaster, Ann. Soc. Entomol. Fr. (N.S.)* 9, 181, 1973.

155. **Wattiaux, J. M.,** Influence de l'age sur le fonctionement ovarien chez *Drosophila melanogaster, J. Insect Physiol.,* 13, 1279, 1967.

156. **Wattiaux, J. M.,** Parental age effects in *Drosophila subobscura, Evolution,* 22, 406, 1968.

157. **Hall, J. C.,** Age-dependent enzyme changes in *Drosophila melanogaster, Exp. Gerontol.,* 4, 207, 1969.

158. **DePolo, M. E., Daly, R. W., and Baker, G. T.,** Effect of mutations at loci affecting DNA repair on fecundity, viability, and longevity in *Drosophila melanogaster, Gerontologist,* 20 (5), 92, 1980.

159. **David, J., Van Herrewege, J., and Fouillet, P.,** Quantitative under-feeding of *Drosophila:* effects on adult longevity and fecundity, *Exp. Gerontol.,* 6, 249, 1971.

160. **Cellucci, M. D. and Baker, G. T.,** Effects of Centrophenoxine on the life span of *Drosophila melanogaster, Age,* 4, 133, 1981.

161. **Hadorn, E. and Zeller, H.,** Fertilitatsstudien an *Drosophila melanogaster.* I. Untersuchungen zun altersbedingten Fertilitatsabtall., *Arch. Entw. Mech. Org.,* 142, 276, 1943.

162. **Ehrman, L.,** Sexual behavior, in *The Genetics and Biology of Drosophila,* Vol. 2b, Ashburner, M. and Wright, T. R. F., Eds., Academic Press, London, 11, 127, 1978.

163. **Chandley, A. C. and Bateman, A. J.,** Timing of spermatogenesis in *Drosophila melanogaster* using tritiated thymidine, *Nature (London),* 193, 299, 1962.

164. **Khishin, A. F. E.,** The response of immature testis of *Drosophila* to the mutagenic action of X-rays, *A. Verebangl.,* 87, 967, 1955.

165. **Stromnaes, O.,** Sexual maturity in *Drosophila, Nature (London),* 183, 409, 1959.

166. **Sturtevant, A. H.,** Culture methods for *Drosophila,* in *Culture Methods for Invertebrate Animals,* Galtsoff, P. S., Lutz, F. E., Welch, P. S., and Needham, J. G., Eds., Comstock Publishing, New York, 1937, 437.

167. **Stromaes, O. and Kvelland, I.,** Sexual activity of *Drosophila melanogaster* males, *Hereditas,* 48, 442, 1962.

168. **Miquel, J.,** Aging of male *Drosophila melanogaster:* histological, histochemical, and ultrastructural observations, *Adv. Gerontol. Res.,* 3, 39, 1971.

169. **Fowler, G. L., Eroshevich, K. E. and Zimmering, S.,** Distribution of sperm in the storage organs of the *Drosophila melanogaster* female at various levels of insemination, *Mol. Gen. Genet.,* 101, 120, 1968.

170. **Atlan, H., Miquel, J., Helme, L. C. and Dolkas, C. B.,** Thermodynamics of aging in *Drosophila melanogaster, Mech. Ageing Dev.,* 5, 371, 1976.

171. **Clark, A. M. and Rockstein, M.,** Aging in insects, in *Physiology of Insecta,* Rockstein, M., Ed., Academic Press, New York, 1964, 227.

172. **Dingley, F. and Maynard-Smith, J.,** Temperature acclimation in the absence of protein synthesis in *Drosophila subobscura, J. Insect Physiol.,* 14, 1185, 1968.

173. **Lamb, M. J.,** The temperature and lifespan in *Drosophila, Nature (London),* 220, 808, 1968.

174. **Loeb, J. and Northrup, J. H.,** On the influence of food and temperature upon the duration of life, *J. Biol. Chem.,* 32, 103, 1917.

175. **MacArthur, J. W. and Baillie, W. H. T.,** Metabolic activity and duration of life, *J. Exp. Zool.,* 53, 221, 1929.

176. **MacArthur, J. W. and Baillie, W. H. T.,** Metabolic activity and duration of life, *J. Exp. Zool.,* 53, 222, 1929.

177. **Maynard-Smith, J.,** Prolongation of life of *Drosophila subobscura* by a brief exposure of adult to high temperature, *Nature (London),* 181, 496, 1958.

178. **Parsons, P. A.,** Genotype-temperature interaction for longevity in natural populations of *Drosophila simulans, Exp. Gerontol.,* 12, 241, 1977.

179. **Shaw, Richard F. and Bercaw, B. L.,** Temperature and life span in poikilothermous animals, *Nature (London),* 196, 454, 1962.

180. **Siddiqui, W. H. and Barlow, C. A.,** Population growth of *Drosophila melanogaster* (Diptera: Drosophilidae) at constant and alternating temperatures, *Ann. Entomol. Soc. Am.,* 65, 993, 1972.

181. **Strehler, B. L.,** Studies on the comparative physiology of aging. II. On the mechanism of temperature life shortening in *Drosophila melanogaster, J. Gerontol.,* 3, 1, 1938.

182. **Sekla, B.,** Experiments on the duration of life in *Drosophila, Casopsis lik ces,* 67, 85, 1928.

183. **Lints, F. A. and Lints, C. V.,** Influence of preimaginal environment on fecundity and ageing in *Drosophila melanogaster* hybrids. II. Preimaginal temperature, *Exp. Gerontol.,* 6, 417, 1971.

184. **Parsons, P. A.,** The genotypic control of longevity in *Drosophila melanogaster* under two environmental regimes, *Aust. J. Biol. Sci.,* 19, 587, 1966.

185. **Burcombe, J. V. and Hollingsworth, M. J.,** The total nitrogen, protein, amino acid and uric acid content of aging *Drosophila, Exp. Gerontol.,* 5, 247, 1970.

186. **Strehler, B. L.,** Further studies on the thermally induced aging of *Drosophila melanogaster, J. Gerontol.,* 17, 347, 1962.

187. **Hollingsworth, M. J. and Fowler, K.,** The decline in the ability to withstand high temperatures with increase in use in *Drosophila subobscura, Exp. Gerontol.,* 1, 251, 1966.

188. **Hollingsworth, M. J.,** The effect of fluctuating environmental temperatures on the length of life of adult *Drosophila, Exp. Gerontol.,* 4, 159, 1969.

189. **Pearl, R. and Parker, S. L.,** Experimental studies on the duration of life. I. Introductory discussion of the duration of life in *Drosophila, Am. Nat.,* 641, 481, 1921.

190. **Pearl, R. and Parker, S. L.,** Experimental studies on the duration of life. IV. Data on the influence of density of population on duration of life in *Drosophila, Am. Nat.,* 56, 312, 192.

191. **Pearl, R. and Parker, S. L.,** Experimental studies on the duration of life. IX. New life tables for *Drosophila, Am. Nat.,* 58, 71, 1924.

192. **Pearl, R.,** The influence of density of population upon egg production in *Drosophila melanogaster, J. Exp. Zool.,* 63, 57, 1932.

193. **Pearl, R. and Parker, S. L.,** Experimental studies on the duration of life. II. Hereditary differences in duration of life in line-bred strains of *Drosophila, Am. Nat.,* 56, 174, 1922.

194. **Miller, R. S. and Thomas, J. L.,** The effects of larval crowding and body size on the longevity of adult *Drosophila melanogaster, Ecology,* 39, 118, 1958.

195. **Lints, F. A.,** Life span in *Drosophila, Gerontologia,* 17, 33, 1971.

196. **Lints, F. A. and Gruwez, G.,** What determines the duration of development in *Drosophila melanogaster?, Mech. Ageing Dev.,* 1, 285, 1972.

197. **Pearl, R., Miner, J. R., and Parker, S. L.,** Experimental studies on the duration of life. XI. Density of population and life duration in *Drosophila, Am. Nat.,* 31, 289, 1927.

198. **Barker, J. S. F.,** Adult population density, fecundity and productivity in *Drosophila melanogaster and Drosophila simulans, Oecologia (Berlin),* 11, 83, 1973.

199. **Erk, F. C. and Samis, H. V., Jr.,** Light regimens and longevity, *D.I.S.,* 45, 148, 1970.

200. **Rensing, L. and Hardeland, R.,** Zur Wirkung der circdianen rhythmik auf die entwicklung von *Drosophila, J. Insect Physiol.,* 13, 1547, 1967.

201. **Allemand, R.,** Importance evolutive du comportement de ponte chez les insectes: comparison du rhythme circadien d'oviposition chez les six especes de *Drosophila* du sous-groupe *melanogaster, C. R. Acad. Sci. Paris,* 279D, 2075, 1974.

202. **Pittendrigh, C. S. and Minis, D. H.,** Circadian systems: longevity as a function of circadian resonance in *Drosophila melanogaster, Proc. Natl. Acad. Sci. U.S.A.,* 69, 1537, 1972.

203. **Glaser, R. W.,** The relation of microorganisms to the development and longevity of flies, *Am. J. Trop. Med.,* 41, 85, 1924.

204. **Philpott, D. E., Bensch, K. G., and Miquel, J.,** Life span and fine structural changes in oxygen-poisoned *Drosophila melanogaster, Aerospace Med.,* 45, 283, 1975.

205. **Miquel, J., Lundgren, P., and Bensch, K.,** Effects of oxygen-nitrogen (1:1) at 760 Torr on the life span and fine structure of *Drosophila melanogaster, Mech. Ageing Dev.,* 4, 44, 1976.

206. **Fenn, W. O., Henning, M., and Philpott, M.,** Oxygen poisoning in *Drosophila, J. Gen. Physiol.,* 50, 1693, 1967.

207. **Williams, C. M. and Beecher, H. K.,** Sensitivity of *Drosophila* to poisoning by oxygen, *Am. J. Physiol.,* 140, 566, 1944.

208. **Bulsma-Meeles, E.,** Viability in *Drosophila melanogaster* in relation to age and ADH activity of eggs transferred to ethanol food, *Heredity,* 42(1), 79, 1979.

209. **Pearl, R., White, F. B., and Miner, J. R.,** Age changes in alcohol tolerance in *Drosophila melanogaster, Proc. Natl. Acad. Sci. U.S.A.,* 15, 425, 1929.

210. **Starmer, W. T., Head, W. B., and Rockwood-Sluss, E. S.,** Extension of longevity in *Drosophila mojavensis* by environmental ethanol: differences between subraces, *Proc. Natl. Acad. Sci. U.S.A.,* 74, 387, 1977.

211. **David, J. and Cohet, Y.,** Accessibility of food and life span of *Drosophila* adults, *D.I.S.,* 48, 120, 1972.

212. **Kalmus, H.,** A fractorial experiment on the mineral requirements of a *Drosophila* culture, *Am. Nat.,* 77, 376, 1942.

213. **Grieff, D.,** Longevity in *Drosophila melanogaster* and its ebony mutant in the absence of food, *Am. Nat.,* 74, 363, 1940.

214. **Pearl, R. and Parker, S. L.,** Experimental studies on the duration of life. X. The duration of life of *Drosophila melanogaster* in the complete absence of food, *Am. Nat.,* 23, 193, 1924.

215. **Baumberger, J. P.,** Studies in the longevity of insects, *Ann. Entomol. Soc. Am.,* 7, 323, 1914.

216. **Northrup, J. H.,** Duration of life on an aseptic *Drosophila* culture inbred in the dark for 230 generations, *J. Gen. Physiol.,* 9, 763, 1926.

217. **Ashleigh, R. D., Cellucci, M. D., Kocis, G., Carswell, N., Atkinson, N., DiGiacomo, R., Yonsetto, M., Kuljian, D., and Baker, G. T.,** Effects of high fat diet and vitamin E supplementation on longevity in *Drosophila melanogaster, Age,* 4, 133, 1981.

218. **Levins, R.,** Thermal acclimation and heat resistance in *Drosophila* species, *Am. Nat.,* 103, 483, 1969.

219. **Stafford, E. M.,** Lethal temperatures for *Drosophila, Proc. Natl. Acad. Sci., Ann. Res. Conf. Calif. Fig Institute,* 13, 26, 1959.

220. **Samis, H. V., Baird, M. B., and Massie, H. R.,** Deuterium oxide effect on temperature survival in populations of *Drosophila melanogaster, Science,* 183, 427, 1974.

221. **Darocha, I. B., Hirsch, G. P., and Baker, G. T.,** Life-shortening effect of dimethylsulfoxide on *Drosophila melanogaster, Gerontologist,* 18 (II), 64, 1978.

222. **McKenzie, J. A. and Parsons, P. A.,** The genetic architecture of resistance to desiccation in populations of *Drosophila melanogaster* and *D. simulans, Aust. J. Biol. Sci.,* 27, 441, 1974.

223. **Perttunen, V. and Ahonen, U.,** The effect of age on the humidity reaction of *Drosophila melanogaster* (Dipt., Drosophilidae), *Suomen Hyonteistiereellinen Aikakauskirja,* 22, 63, 1956.

224. **Perttunen, V. and Salmi, H.,** The responses of *Drosophila melanogaster* (Dipt. Drosophilidae) to the relative humidity of the air, *Suomen Hyonteistiereellinen Aikakauskirja,* 22 (1), 36, 1956.

225. **Syrjamaki, J.,** Humidity perception in *Drosophila melanogaster, Ann. Zool. Soc.,* "Vanamo", 23, 1, 1962.

226. **Northrup, J.,** The effect of prolongation of the period of growth on the total duration of life, *J. Biol. Chem.,* 32, 123, 1917.

227. **Northrup, J. H.,** The influence of the intensity of light on the rate of growth and duration of life of *Drosophila, J. Gen. Physiol.,* 9, 81, 1925.

228. **Gruwez, G., Hoste, C., Lints, C. V., and Lints, F. A.,** Oviposition rhythm in *Drosophila melanogaster* and its alteration by a change in the photo-periodicity, *Experientia,* 27, 1414, 1971.

229. **Miquel, J. and Philpott, D. E.,** Effects of weightlessness on the development and aging of *Drosophila melanogaster, Physiologist,* 21 (4), 80, 1980.

230. **Pearl, R. and Parker, S. L.,** Experimental studies on the duration of life. III. The effect of successive etherizations on the duration of life of *Drosophila, Am. Nat.,* 56, 273, 1922.

231. **Gartner, L. P.,** Radiation-induced lifespan shortening in *Drosophila Gerontologia,* 19, 295, 1973.

232. **Atlan, H., Miquel, J., and Binnard, R.,** Differences between radiation-induced life shortening and natural aging in *Drosophila melanogaster, J. Gerontol.,* 24, 1, 1969.

233. **Gartner, L. P.,** Aging and Ionizing Radiation: A Study of Lifespan and Fine Structural Alterations in *Drosophila melanogaster,* Ph. D. Thesis, Rutgers University, New Brunswick, N.J., 1970.

234. **Gartner, L. P.,** Fine structure changes in the adult *Drosophila* midgut as a function of age and ionizing radiation, *D.I.S.,* 48, 106, 1972.

235. **Gartner, L. P.,** Ultrastructural examination of aging and radiation-induced life span shortening in adult *Drosophila melanogaster, Int. J. Radiat. Biol.,* 1, 23, 1973.

236. **Lamb, M. J.,** The relationship between age at irradiation and life shortening in adult *Drosophila,* in *Radiation and Ageing,* Lindop, P. J. and Sacher, G. A., Eds., Taylor μ Francis, London, 1966, 163.

237. **Lamb, M. J. and McDonald, R. P.,** Heat tolerance changes with age in normal and irradiated *Drosophila melanogaster, Exp. Gerontol.,* 7, 207, 1973.

238. **Miquel, J., Bensch, K., and Philpott, D.,** Virus-like particles in the tissues of normal and gamma-irradiated *Drosophila melanogaster, J. Invert. Pathol.,* 19, 156, 1972.

239. **Miquel, J., Lundgren, P., and Binnard, R.,** Negative geotaxis and mating behavior in control and gamma-irradiated *Drosophila, D.I.S.,* 48, 60, 1972.

240. **Ostertag, W.,** The genetic basis of somatic damage produced by radiation in third instar larvae of *Drosophila melanogaster.* I. Death before maturity, *Z. Vererbung,* 94, 143, 1963.

241. **Harris, B. B.,** The effects of aging of X-rayed males upon mutation frequency in *Drosophila, J. Hered.,* 20, 299, 1927.

242. **Slizynski, B. M.,** Sperm utilization and radiation sensitivity in *Drosophila melanogaster, Heredity,* 24, 660, 1969.

243. **Thomas, J. J., Baxter, R. C., and Fenn, W. O.,** Interactions of oxygen at high pressure and radiation in *Drosophila, J. Gen. Physiol.,* 49, 537, 1966.

244. **Uchida, I. A.,** The effect of maternal age and radiation on the rate of non-disjunction in *Drosophila melanogaster, Can. J. Genet. Cytol.,* 4, 102, 1962.

245. **Yanders, A. I.,** The effect of age on male on X-ray induced dominant lethals in *Drosophila robusta, D.I.S.,* 26, 127, 1952.

246. **Lamb, M. J. and Maynard-Smith, J.,** Radiation and ageing in insects, *Exp. Gerontol.,* 1, 11, 1964.

247. **Lamb, M. J.,** The effects of X-irradiation on the longevity of triploid and diploid female *Drosophila melanogaster, Exp. Gerontol.,* 1, 181, 1965.

248. **Duncan, F. N.,** Some observation on the biology of the male *Drosophila melanogaster, Am. Nat.,* 64, 545, 1930.

249. **Baxter, R. C. and Blair, H. A.,** Kinetics of aging as revealed by X-ray dose: lethality relations in *Drosophila, Radiat. Res.,* 30, 48, 1967.

250. **Baker, G. T. and Merkin, S. L.,** unpublished.

251. **Baker, G. T.,** Insect flight muscle: maturation and senescence, *Gerontologia,* 22, 334, 1976.

252. **Bozcuk, A. N.,** DNA synthesis in the absence of somatic cell division associated with ageing in *Drosophila subobscura, Exp. Gerontol.,* 7, 147, 1972.

253. **Burch, G. E., Sohal, R., and Fairbanks, L. D.,** Ultrastructural changes in *Drosophila* heart with age, *Arch. Pathol.,* 89, 128, 1970.

254. **Daivd, J.,** Etude quantitative du fonctionnement ovarien chez *Drosophila melanogaster* meig, *Bull. Biol. Fasc.,* 3, 34, 1961.

255. **David, J.,** Influence de l'etat physiologique des parents sur les caracteres descendants, *Ann. Genet.,* 3 (2), 1, 1961.

256. **David, J. and Merle, J.,** Influence de la fecondation de la female de *Drosophila* sur la physiologie, *Ann. Nutr. Aliment.,* 20, 332, 1966.

257. **Deak, I.,** A model linking segmentation, compartmentalization, and regeneration in *Drosophila* development, *J. Theor. Biol.,* 84, 477, 1980.

258. **Gartner, L. P.,** Unusual structures of specific fecundity in the midgut as a function of aged *Drosophila melanogaster, J. Baltimore Coll. Dent. Surg.,* 26, 45, 1971.

259. **Gartner, L. P.,** Ultrastructural alterations as a function of age in *Drosophila melanogaster* midgut, *Anat. Rec.,* 172, 313, 1972.

260. **Gartner, L. P.,** Virus-like particles in the adult *Drosophila* midgut, *J. Invert. Pathol.,* 20, 364, 1972.

261. **Gartner, L. P. and Gartner, R. C.,** Nuclear inclusions: a study of Aging in *Drosophila, J. Gerontol.,* 31 (4), 396, 1976.

262. **Gartner, L. P.,** Aging and the visceral musculature of the adult fruit fly: an ultrastructural investigation, *Trans. Am. Micros. Soc.,* 96, 48, 1977.

263. **Herman, M. M., Johnson, M., and Miquel, J.,** Virus-like particles and related filaments in neurons and glia of *Drosophila melanogaster* brain, *J. Invert. Pathol.,* 17, 442, 1971.

264. **Herman, M. M., Miquel, J., and Johnson, M.,** Insect brain as a model for the study of aging. Age-related changes in *Drosophila melanogaster, Acta Neuropathol.,* 19, 167, 1971.

265. **King, R. C., Aggarwal, S. K., and Bodenstein, D.,** The comparative submicroscopic cytology of the corpus allatum-corpus cardiacum complex of wild type and adult female *Drosophila melanogaster, J. Exp. Zool.,* 161, 151, 1966.

266. **Krumeigel, I.,** Untersuchungen uber ale einwirkung der fortpflanzung auf altern and lebensdauer der insekten, ausgopuhrt an *Carabus* and *Drosophila, Zool. Jahe.,* 51, 111, 1029.

267. **Miquel, J., Hobbisienfken, F., and Duffy, J.,** Age differences in the glycogen content of nervous and muscle tissue of *Drosophila melanogaster, Gerontologist,* 7 (3), 17, 1965.

268. **Miquel, J., Bensch, K., Philpott, D., and Atland, H.,** Natural aging and radiation: induced life shortening in *Drosophila melanogaster, Mech. Ageing Dev.,* 1, 71, 1972.

269. **Miquel, J., Tappel, A. L., Dillard, C. J., Herman, M. M., and Bensch, K.,** Fluorescent products and lysosomal components in aging *Drosophila melanogaster, J. Gerontol.,* 29 (6), 622, 1974.

270. **Miquel, J., Oro, J., Bensch, K., and Johnson, J. E., Jr.,** Lipofuscin: fine-structural and biochemical studies, in *Free Radicals in Biology, Vol. 3,* Pryor, W., Ed., Academic Press, New York, 1977, 133.

271. **Philpott, D. E. and Miquel, J.,** Ultrastructural alterations in the cephalic ganglionic center of aged *Drosophila melanogaster, Gerontologist,* 7, 17, 1967.

272. **Philpott, D. E., Weibel, J., Atlan, H., and Miquel, J.,** Virus-like particles in the fat body, oenocytes, and central nervous tissue of *Drosophila melanogaster* imagoes, *J. Invert. Pathol.,* 14, 31, 1969.

273. **Sohal, R. S.,** Mitochondrial changes in the heart of *Drosophila repleta* (Wollaston) with age, *Exp. Gerontol.,* 5, 213, 1970.

274. **Sohal, R. S.,** Mitochondrial changes in flight muscles of normal and flightless *Drosophila melanogaster* with age, *J. Morphol.,* 145, 337, 1975.

275. **Sohal, R. S.,** Aging changes in insect flight muscle, *Gerontology,* 22, 317, 1976.

276. **Sohal, R. S.,** Aging changes in the structure and function of the insect heart, *Aging,* 6, 211, 1978.

277. **Sondhi, K. C.,** Aging and its expression in *Drosophila, Am. Zool.,* 4, 286, 1964.

278. **Sondhi, K. C.,** Hemolymph and aging in *Drosophila, Am. Zool.,* 5, 243, 1965.

279. **Sondhi, K. C.,** Studies in aging. III. The physiological effects of injecting hemolymph from outbred donors into inbred hosts in *Drosophila melanogaster, Proc. Natl. Acad. Sci. U.S.A.,* 57, 965, 1967.

280. **Sondhi, K. C.,** Studies in aging. IV. Genetic control of pigment accumulation and its bearing on the adult life span in *Drosophila melanogaster, J. Hered.,* 58, 47, 1967.

281. **Takahashi, A., Philpott, D. E., and Miquel, J.,** Electron microscope studies on aging *Drosophila melanogaster, J. Gerontol.,* 25, 210, 1970.

282. **Takahashi, A., Philpott, D. E., and Miquel, J.,** Electron microscope studies on aging *Drosophila melanogaster.* II. Intramitochondrial crystalloid in fat body cells, *J. Gerontol.,* 25, 218, 1970.

283. **Takahashi, A., Philpott, D. E., and Miquel, J.,** Electron microscope studies on aging *Drosophila melanogaster.* III. Flight muscle, *J. Gerontol.,* 25, 222, 1970.

284. **Watanabe, M. I. and Williams, C. M.,** Mitochondria in the flight muscles of insects. I. Chemical composition and enzymatic content, *J. Gen. Physiol.,* 34, 675, 1951.

285. **Watanabe, M. I. and Williams, C. M.,** Mitochondria in the flight muscles of insects. II. Effects of the medium on the site, form, and organization of isolated sarcosomes, *J. Gen. Physiol.,* 37, 71, 1953.

286. **Fairbanks, L. D. and Burch, G. E.**, Rate of water loss and fat content of adult *Drosophila melanogaster* of different ages, *J. Insect Physiol.*, 16, 1429, 1970.

287. **Perttunen, V. and Erkkila, H.**, Humidity reaction in *Drosophila melanogaster*, *Nature (London)*, 169, 78, 1952.

288. **Green, P. R. and Geer, B. W.**, Changes in the fatty acid composition of *Drosophila melanogaster* during development and ageing, *Arch. Int. Physiol. Biochim.*, 87, 485, 1979.

289. **Baker, G. T.**, Identical age-related patterns of enzyme activity changes in *Phormia regina* and *Drosophila melanogaster*, *Exp. Gerontol.*, 10, 231, 1975.

290. **Balazs, A. and Haranghy, L.**, Alteration of nucleic acid content of *Drosophila melanogaster* imagoes during ripening and ageing, *Acta. Biol. Hung.*, 15, 343, 1964.

291. **Massie, H. R. and Williams, T. R.**, Increased longevity of *Drosophila melanogaster* with lactic and gluconic acids, *Exp. Gerontol.*, 14, 109, 1979.

292. **Lints, F. A. and Lints, C. V.**, Respiration in *Drosophila*. II. Respiration in relation to age by wild, inbred and hybrid *Drosophila melanogaster* imagoes, *Exp. Gerontol.*, 3, 341, 1968.

293. **Williams, C. M., Barness, L. A., and Sawyer, W. H.**, The utilization of glycogen by flies during flight and some aspects of the physiological aging of *Drosophila*, *Biol. Bull.*, 84, 263, 1843.

294. **Wigglesworth, V. B.**, The utilization of reserve substances in *Drosophila* during flight, *J. Exp. Biol.*, 26, 150, 1949.

295. **Levenbook, L. and Williams, C. M.**, Mitochondia in the flight muscles of insects. III. Mitochondrial cytochrome-c in relation to the aging and wing beat frequency of flies, *J. Gen. Physiol.*, 39 (4), 497, 1956.

296. **Samis, H. V., Jr., Erk, F. C., and Baird, M. B.**, Senescence in *Drosophila*, I. Sex difference in nucleic acid, protein and glycogen levels as a function of age, *Exp. Gerontol.*, 6, 9, 1971.

297. **Driver, C. J. I. and Lamb, M. J.**, Metabolic changes in ageing *Drosophila melanogaster*, *Exp. Gerontol.*, 15, 167, 1980.

298. **Chadwick, and Williams, C. M.**, cited in Levenbook, L. and Williams, C. M., Mitochondria in the flight muscles of insects. III. Mitochondrial cytochrome-c in relation to aging and wing beat frequency of flies, *J. Gen. Physiol.*, 39, 497, 1956.

299. **Elens, A. and Wattiaux, J. M.**, Age and phototactic reaction in *Drosophila melanogaster*, *D.I.S.*, 46, 81, 1971.

300. **Baker, G. T., Daly, R. N., and Davis, F. A.**, Effects of centrophenoxine on survivorship in wild type and DNA-repair deficient mutants of *Drosophila melanogaster*, (unpublished).

301. **Samis, H. V., Rubenstein, B. J., Zajac, L. A., and Hargen, S. M.**, Temporal organization and aging in *Drosophila melanogaster*, *Exp. Gerontol.*, 16 (2), 109, 1981.

302. **Grossfield, J.**, Non-sexual behavior of *Drosophila*, in *The Genetics and Biology of Drosophila*, Vol. 2b, Ashburner, M. and Wright, T. R. F., Eds., Academic Press, London, 10, 44, 1978,

303. **Bauman, P. A.**, Untersuchungen zum Protein-stoffwechselbei alternden Adultmannchen, Larven des Wildtype und der Letalmutanen (ltr und lme) von *Drosophila melanogaster*, *Z. Vergl. Physiol.*, 64, 212, 1969.

304. **Lamy, R.**, Production of 2-X sperm in males, *D.I.S.*, 23, 91, 1949.

305. **Maynard-Smith, J.**, Acclimatization to high temperatures in inbred and outbred *Drosophila subobscura*, *J. Genet.*, 54, 497, 1956.

306. **Northrup, J. H.**, Carbon dioxide production and duration of life of *Drosophila* cultures, *J. Gen. Physiol.*, 9, 319, 1926.

307. **Sheherbakov, A. P.**, Metabolic rate and duration of life of *Drosophila*. I. Introductory remarks and review of the literature, *Arch. Biol. Sci.*, 38, 651, 1935.

308. **Sheherbakov, A. P.**, Metabolic rate and duration of life of *Drosophila*. IV. Effects of temperature on the metabolic rate and duration of life of *Drosophila melanogaster*, *Arch. Biol. Nauk.*, 1937.

309. **Winberg, G. C.**, Metabolic rate and duration of life of *Drosophila*. Intensity of respiration, size and duration of life, *Arch. Biol. Sci.*, 38, 657, 1936.

310. **Winberg, G. C.**, Metabolic rate and the duration of life span of *Drosophila*. V. Effect of development temperature on the metabolism and duration of life, *Bull. Exp. Biol. Med.*, 1937.

311. **Kucera, W.**, Oxygen consumption in the male and female fly, *Drosophila melanogaster*, *Physiol. Zool.*, 7, 449, 1934.

312. **Bowler, K. and Hollingsworth, M. J.**, A study of some aspects of the physiology of ageing in *Drosophila subobscura*, *Exp. Gerontol.*, 2, 1, 1966.

313. **McDaniel, R. G. and Grimwood, B. G.**, Hybrid vigor in *Drosophila:* Respiration and mitochondrial energy conservation, *Com. Chem. Physiol.*, 38B, 309, 1971.

314. **Martinez, A. O. and McDaniel, R. G.**, Mitochondrial heterosis in aging *Drosophila* hybrids, *Exp. Gerontol.*, 14, 231, 1979.

315. **Vann, A. C. and Webster, G. C.**, Age-related changes in mitochondrial function in *Drosophila melanogaster*, *Exp. Gerontol.*, 12, 1, 1977.

316. **Baker, G. T., Aguet, N. J., Pelli, D. A., Crossley, K. L., and Rossnick, J. B.,** Mechanisms of age-related enzyme activity changes in *Drosophila melanogaster, Fed. Proc.,* 37, 880, 1978.

317. **Baker, G. T. and Rossnick, J. B.,** Mechanisms of enzyme changes with age, *Mech. Ageing Dev.,* (unpublished).

318. **Baker, G. T., Aguet, N. J., and Rossnick, J. B.,** Altered thermal stability of a-glycerophosphate dehydrogenase from adult *Drosophila melanogaster* with age, *Mech. Ageing Dev.,* (unpublished).

319. **Baker, G. T., Pelli, D. A., and Crossley, K. L.,** Immunochemical study of Arginine phosphokinase from *Drosophila melanogaster* with age, *Mech. Ageing Dev.,* (unpublished).

320. **Armstrong, E., Rinehart, R., Dixon, L., and Reigh, D.,** Changes of peroxidase with age in *Drosophila, Age,* 1, 8, 1978.

321. **Webster, G. C. and Webster, S. L.,** Lysosomal enzyme activity during aging of *Drosophila melanogaster, J. Cell Biol.,* 75, 199A, 1977.

322. **Webster, G. C. and Webster, S. L.,** Lysosomal enzyme activity during aging in *Drosophila melanogaster, Exp. Gerontol.,* 13, 343, 1978.

323. **Libion-Mannaert, M. and Elens, A.,** Ageing in *Drosophila melanogaster,* ebony, white, and wild: alcohol dehydrogenase and other enzyme activity changes, *D.I.S.,* 49, 77, 1972.

324. **Chen, P. S. and Baker, G. T.,** L-alanine aminotransferase in the paragonial gland of *Drosophila, Insect Biochem.,* 6, 441, 1976.

325. **Burcombe, J. V.,** Changes in enzyme levels during ageing in *Drosophila melanogaster, Mech. Ageing Dev.,* 1, 213, 1972.

326. **Kaur, K. and Parkash, R.,** Developmental analysis of amylases in *Drosophila immigrams, Indian J. Exp. Biol.,* 18, 222, 1980.

327. **Massie, H. R. and Baird, M. B.,** Catalase levels in *Drosophila* and the lack of induction by hypolipidemic compounds: a brief note, *Mech. Ageing Dev.,* 5, 38, 1976.

328. **Samis, H. V., Baird, M. B., and Massie, H. R.,** Senescence and the regulation of catalase activity and the effect of hydrogen peroxide on nucleic acids, in *Molecuar Genetic Mechanisms in Development and Aging,* Rockstein, M. and Baker, G. T., Eds., Academic Press, New York, 1972, 113.

329. **Kang, M-S.,** A study on the CdR aminohydrolase in *Drosophila melanogaster, Kor. J. Zool.,* 20 (3), 129, 1977.

330. **Sekla, B.,** Esterolytic processes and duration of life in *Drosophila melanogaster, Br. J. Exp. Biol.,* 6, 161, 1928.

331. **Masie, H. R., Aiello, V. R., and Williams, T. R.,** Changes in superoxide dismutase and copper during development and ageing in the fruit fly *Drosophila melanogaster, Mech. Ageing Dev.,* 12, 279, 1980.

332. **Tappel, A. L.,** Free-radical lipid peroxidation damage and its inhibition by vitamin E and selenium, *Fed. Proc.,* 24, 73, 1965.

333. **Deak, I. I.,** Use of *Drosophila* mutants to investigate the effect of disuse on the maintenance of muscle, *J. Insect Physiol.,* 22, 1159, 1976.

334. **Sondhi, K. C.,** development and aging in *Drosophila, Am. Zool.,* 4, 388, 1964.

335. **Sondhi, K. C.,** Relationship between the sternal pigment and the adult life span in *Drosophila, Am. Zool.,* 5, 243, 1965.

336. **Shedahl, J. A. and Tappel, A. L.,** Fluorescent products from aging *Drosophila melanogaster;* an indicator of free radical lipid peroxidation damage, *Exp. Gerontol.,* 9, 33, 1974.

337. **Miquel, J. and Johnson, J. E., Jr.,** Effects of various antioxidants and radiation protectants on the lifespan and lipofuscin of *Drosophila* and of C57BL16J mice, *Gerontologist,* 15, 25, 1975.

338. **Darocha, I. B. and Baker, G. T.,** Age-dependent alterations in thermal stability of partially purified DNA from tissues of *Drosophila melanogaster,* in *Neural Regulatory Mechanisms During Aging,* Adelman, R. C. et al., Eds., Alan R. Liss, New York, 1980, 207.

339. **Brown, H., Mormann, J., Hennig, W., and Baker, G. T.,** Apparent loss of an A-T rich DNA component from male *Drosophila melanogaster* with age, *Gerontologist,* 18 (II), 54, 1978.

340. **Hosbach, H. A., Kubli, E., and Chen, P. S.,** Altersbedingt Veranderungen der Aminoacylierung bei *Drosophila melanogaster, Rev. Suisse Zool.,* 83 (4), 964, 1976.

341. **Hosbach, H. A. and Kubli, E.,** Transfer of RNA in aging *Drosophila.* I. Extent of aminocylation, *Mech. Ageing Dev.,* 10, 131, 1979.

342. **Owenby, R. K., Stulberg, M. P., and Jacobson, K. B.,** Alteration of the Q family of transfer RNA's in the adult *Drosophila melanogaster* as a function of age, nutrition, and genotype, *Mech. Ageing Dev.,* 11, 91, 1979.

343. **Baker, G. T., Zschunke, R., and Schmidt, T.,** Age-related loss of structural integrity in 80S ribosomes, *Gerontolgist,* 17 (II), 37, 1977.

344. **Miquel, J. and Johnson, J. E., Jr.,** Senescent changes in the ribosomes of animal cells *in vivo* and *in vitro, Mech. Ageing Dev.,* 9, 247, 1979.

345. **Schmidt, T. and Baker, G. T.,** Characterization of 80s ribosomal protein from *Drosophila melanogaster* with age, *Mol. Genet.,* 1977.

346. **Mokrynski, G. and Baker, G. T.,** Undermethylation of rRNA in *Drosophila melanogaster* with advancing age, *Gerontologist,* 20, 164, 1980.

347. **Webster, G. C., Webster, S. L., and Landis, W. A.,** The effect of age on the initiation of protein synthesis in *Drosophila melanogaster, Mech. Ageing Dev.,* 16 (1), 81, 1981.

348. **Chen, P. S. and Bauman, P.,** Protein synthesis during aging of *Drosophila melanogaster, D.I.S.,* 44, 102, 1969.

349. **Massie, H. R., Baird, M. B., and McMahon, M. M.,** Loss of mitochondrial DNA with aging in *Drosophila melanogaster, Gerontologia,* 21, 231, 1975.

350. **Darocha, I. B. and Baker, G. T.,** Qualitative alterations in DNA from *Drosophila melanogaster* with age, *Gerontologist,* 19 (5), 61, 1979.

351. **Bozcuk, A. N,** Molecular Turnovers and Ageing in *Drosophila subobscura,* O. Phil. thesis, University of Sussex, 1967.

352. **Clarke, J. M. and Maynard-Smith, J. M.,** Increase in the rate of protein synthesis with age in *Drosophila subobscura, Nature (London),* 209, 627, 1966.

353. **Wattiaux, J. M. and Tsien, H. C.,** Age effects on the variation of RNA synthesis in the nurse cells in *Drosophila, Exp. Gerontol.,* 6, 235, 1971.

354. **Bauman, P. A. and Chen, P. S.,** Alterung und Proteinsyntheses bei *Drosophila melanogaster, Rev. Suisse Zool.,* 75, 1051, 1968.

355. **Maynard-Smith, J., Bozcuk, A. N. and Tebbutt, S.,** Protein turnover in adult *Drosophila, J. Insect Physiol.,* 16, 601, 1970.

356. **Wattiaux, J. M., Libion-Mannaert, M., and Delcour, J.,** Protein turnover and protein synthesis following actinomycin-D injection as a function of age in *Drosophila melanogaster, Gerontologia,* 17, 289, 1971.

357. **Bozcuk, A. N.,** Testing the protein error hypothesis of ageing in *Drosophila, Exp. Gerontol.,* 11, 103, 1976.

358. **Wattiaux, J. M. and Lamborot, M.,** Influence of aging in the rate of incorporation of tritiated thymidine in the nurse cells of *Drosophila melanogaster, Curr. Mod. Biol.,* 1, 5, 1967.

359. **White, B. K., Tener, G. M., Holden, J., and Suzuki, D. T.,** Activity of a transfer RNA modifying enzyme during the development of *Drosophila* and its relationship to the su(s) locus, *J. Mol. Biol.,* 74, 635, 1973.

360. **White, B. K., Tener, G. M., Holden, J., and Suzuki, D. T.,** Analysis of tRNA during development of *Drosophila, Dev. Biol.,* 33, 185, 1973.

361. **Hosbach, H. A. and Kubli, E.,** Transfer RNA in aging *Drosophila.* II. Isoacceptor patterns, *Mech. Ageing Dev.,* 10, 141, 1979.

362. **Webster, G. C. and Webster, S. L.,** The effect of aging in the components and stages of translation in *Drosophila melanogaster, J. Cell Biol.,* 83, 425, 1979.

363. **Baker, G. T. and Schmidt, T.,** Changes in 80S ribosomes from *Drosophila melanogaster* with age, *Experientia,* 32, 1505, 1976.

364. **Baker, G. T. and Podgorski, E. M.,** Age-dependent alterations in ribosomal RNA from *Drosophila melanogaster, 11th Int. Cong. Gerontol.,* 41, 1978.

365. **Baker, G. T., Zschunke, R. E., and Podgorski, E. M., Jr.,** Alteration in thermal stability of ribosomes from *Drosophila melanogaster* with age, *Experientia,* 35, 1053, 1979.

366. **Podgorski, E. M., Jr. and Baker, G. T.,** Age-Dependent Alterations in the Structural Integrity of 80S Ribosomes from *Drosophila melanogaster,* Ph. D. Thesis, Drexel University, Philadelphia, 1980.

367. **Schmidt, T. and Baker, G. T.,** Analysis of ribosomal proteins from adult *melanogaster* in relation to age, *Mech. Ageing Dev.,* 11, 105, 1979.

368. **Zschunke, R. and Baker, G. T.,** Alerations in the Thermal Stability of Ribosomal Ribonucleic Acid from *Drosophila melanogaster* with Advancing Age, Ph. D. Thesis, Drexel University, Philadelphia, 1979.

369. **Zschunke, R. E., Podgorski, E. M., Kubli, E., and Baker, G. T.,** Physiochemical characterization of ribosomes from *Drosophila melanogaster* with age, *Fed. Proc.,* 38, 365, 1978.

370. **Webster, G. C., Webster, R., and Webster, S. L.,** Decreased protein synthesis by microsomes from aging *Drosophila melanogaster, Exp. Gerontol.,* 14, 343, 1979.

371. **Webster, G. C., Beachell, V. T., and Webster, S. L.,** Differential decrease in protein synthesis by microsomes from aging *Drosophila melanogaster, Exp. Gerontol.,* 15, 495, 1980.

372. **Harrison, B. J. and Holliday, R.,** Senescence and the fidelity of protein synthesis in *Drosophila, Nature (London),* 213, 990, 1967.

373. **Chen, P. S.,** Amino acid pattern and rate of protein synthesis in aging *Drosophila,* in *Molecular Genetic Mechanism in Development and Aging,* Rockstein, M. and Baker, G. T., Eds., Academic Press, New York, 1972, 199.

374. **Dingley, F. and Maynard-Smith, J.,** Absence of a life-shortening effect of amino acid analogues of adult *Drosophila, Exp. Gerontol.,* 4, 145, 1969.

375. **Massie, H. R., Baird, M. B., and Piekielniak, M. J.,** Ascorbic acid and longevity in *Drosophila, Exp. Gerontol.,* 11, 37, 1976.

376. **Massie, H. R., Baird, M. B., and Williams, T. R.,** Increased longevity of *Drosophila melanogaster* with diiodomethane, *Gerontology,* 24, 104, 1978.

377. **Massie, H. R. and Williams, T. R.,** Singlet oxygen and aging in *Drosophila, Gerontologia,* 26, 16, 1980.

378. **Mlquel, J., Binnard, R., and Howard, W. H.,** Effects of Dl-a-tocopherol on the lifespan of *Drosophila melanogaster, Gerontologist,* 13, 37, 1973.

379. **Miquel, J. and Economos, A. C.,** Favorable effects of the antioxidents sodium and magnesium thiazolidine carboxylate on the vitality and life span of *Drosophila* and mice, *Exp. Gerontol.,* 14, 279, 1979.

380. **Sondhi, K. C. and Turoczi, L. J.,** The effects of erythorbic acid treatments on melanin synthesis and on the adult life span in *Drosophila melanogaster, Proc. Natl. Acad. Sci. U.S.A.,* 56, 1743, 1966.

381. **Gardner, T. S.,** The use of *Drosophila melanogaster* as a screening agent for longevity factors. I. Patothenic acid as a longevity factor in Royal Jelly, *J. Gerontol.,* 43 (1), 1, 1948.

382. **Gardner, T. S.,** The use of *Drosophila melanogaster* as a screening agent for longevity factors. II. The effects of biotin, pyridioxine, sodium yeast nucleate and pantothenic acid on the life span of the fruit fly, *J. Gerontol.,* 43, 9, 1948.

383. **Sondhi, K. C.,** Transplantation techniques for quantitative experiments on *Drosophila, Life Sci.,* 4, 57, 1965.

384. **Sondhi, K. C.,** Studies in aging. II. The effect of injecting hemolymph from younger and older donors on the feundity and the adult life span of hosts in *Drosophila, J. Exp. Zool.,* 162, 89, 1966.

385. **Sondhi, K. C.,** The effect of hormonally induced changes during development and during adult life of *Drosophila, Genet. Today,* 1, 181, 1963.

INDEX

A

Acetabularia, lifespan determination of, 465
Acetylase activity, in lung-derived fibroblast-like cells, see also Enzyme activity, 411
N-Acetyl-β-glucosaminidase, in lung-derived fibroblast-like cells, 396
Acid phosphatase, in lung-derived fibroblast-like cells, see also Enzyme activity, 394—396
Adenosine triphosphatase, in aging bone cells, see also Enzyme activity, 200
Adrenal cortical cells
  in vitro in, 130
  mitosis vs. age in, 123
Adventitial areolar connective tissue, DNA synthetic index vs. age in, 125
Adventitial fibroblasts, DNA synthetic index vs. age in, 125
Adventitial mesothelium, DNA synthetic index vs. age in, 126
"Age pigmnet", see Lipofuscin
Aging
  beginning of, 271
  cell cycle and
    in vitro, 126, 129—132, 375
    in vivo, 138
  clonal vs. individual, 453—454
  cultural, 454
  mechanism of basis for, 132
  multicellular vs. unicellular, 469—470
  selection of subjects for study, 425, 431
Alkaline phosphatase, in lung-derived fibroblast-like cells, see also Enzyme activity, 394
Alveoli
  changes in cell numbers of, 8
  DNA synthetic index vs. age in, 125
  number of cells vs. age in, 103—104
Alzheimer's disease, 150
Amino acids, effects of aging on absorption of, 266
Amoeba
  experimental alteration of lifespan of, 463
  lifespan determination of, 465
Amygdala, number of cells or fibers vs. age in, 19—20
Amyloid deposition, in small intestinal mucosa, 257
Aneuploidy, 131
Antihistamines, effect on aging *Drosophila,* 560
Antioxidants, in aging insects, 505
Apoptosis, vs. traumatic death, 142
Arteriosclerosis, vs. age-related changes, 158, 161
Articular cartilage, age-related cell loss of, 213—215
Ascorbic acid, effect on aging *Drosophila,* 559, 561
ATPase, in lung-derived fibroblast-like cells, see also Enzyme activity, 396
ATP production, age-related reduction in, 189
Auditory nerves, number of cells or fibers vs. age in, 26, 105

Autoantibodies, increase in, 354
Autophagocytosis, of lipofuscin, 504
Autoradiographic studies
  of bone cell repair, 203—207
  of utilization by cartilage of H-amino acids, 216, 217
Axons, see also Nerve cells
  of aging nerves, 187
  irregularities of, 151, 154

B

Baltimore Longitudinal Study, 425, 426, 431
Basal transcription, in lung-derived fibroblast-like cells, 412
Basophils
  age-associated changes in numbers of, 5, 64, 65
  number vs. age of, 64—65
Beta-carotene, effect on aging *Drosophila,* 558
Biochemistry, of aging muscle, 188—190
Biotin, effect on *Drosophila,* 562
Blebs, in lung-derived fibroblast-like cells, 380
Blood
  age-associated cell loss, 3—5
  number of cells vs. age in, 59—76, 106
Blood vessels
  age-associated cell loss in, 5
  arteriosclerosis vs, age-related changes in, 161
  number of cells vs. age in, 76—78
B lymphocytes
  activation of, 348
  age-associated changes in, 353
  characterization of, 350
  maturation of, 349
  number vs. age of, 73
  precursors of, 347
Body mass
  decline in, 1
  with increasing age, 10—12
Bone
  aging of, 195—197
  cell loss of, 3
  elastic tissue of, 208
  glycosaminoglycan (mucopolysaccharide) composition of, 205
  number of cells vs. age in, 37—49, 106
  senile atrophy of, 195
Bone cells, see also Osteogenic tissue
  age related changes in
    osteogenic tissue, 199—211
    qualitative vs. quantitative response of, 202
  repair of, 203—207
Bone marrow
  age-associated cell loss in, 5
  number of cells vs. age in, 79—84, 106
Bone matrix, age-related changes in, 199—211
Bone resorption, and age, 195

Bovine aortic endothelial cell, life history of, 309,
310
Bovine cells
cultured, 444
effect of aging on, 219—220
Brain
aging in, 150
blood vessels of, 158
insect, 501—502
cell loss in, 1
number of cells or fibers vs. age in, 16—25, 105
shrinkage of, 150—151
Brain stem, number of cells or fibers vs. age in, 25,
31—32, 105
Brain weight, loss of, 150, 151
Bromsulfophthalein (BSP) retention test, 238
But.-hydroxytoluene (BHT), effect on aging *Drosophila*, 561

C

Caenorhabditis
briggsae, 481
elegans, 481—493
anatomy of, 482
biology, 481—484
culture conditions for, 481
disadvantages of studying, 488
embryonic development of, 481
genetics of, 484—487
larval development of, 481—482
life cycle, 482—484
lifespan in, 488
markers of aging in, 490—492
Calcitonin, osteoclastic resorptive activity, 211
Calcium, effects of aging on absorption of, 266—
267
cAMP phospho-diesterase, in lung-derived fibroblast-like cells, see also Enzyme activity,
403
Cancellous bone, ash content of, 195
Cancer, serotonin and, see also Malignant transformation, 278—280
Canine cells, cultured, 444
Capillary density, of skeletal muscle, see also Skeletal muscle, 185
Carcinogens, lifespan altered by, 465
Cardiac conduction, and lipofuscin accumulation
and, 142—143
Cardiac muscle, number of cells or fibers vs. age
of, 95—96, 108
β-Carotene, effect on aging *Drosophila*, 558
Carpal joints, bovine, effect on aging on, 219, 220
Cartilage, see also Chondrogenic tissues
age-related changes in, 198
costal, 213—215
general, 197—199
number of cells vs. age of, 49
Catalase, in lung-derived fibroblast-like cells, see
also Enzyme activity, 397

Cathodal esterase isoenzyme, in lung-derived fibroblast-like cells, 400
CBA/N xid mouse, 117—132
Cell cycle, 117—132
aging and, 120—124
concept of, 118, 120
definition of, 117
for fibroblast-like cells, 376—377
Cell death, 137—144
in aging, 140
classification of, 140—142
Hayflick limit, 137—138
in vitro, 140
in vivo, 138
lysosomes and, 142
mechanism of, 141
mitosis and, 139—140
physiological, 140, 141
programmed apoptosis, 140
unifiying concepts of, 143—144
Cell division, in aging mammalian intestine, see
also Mitosis, 258
Cell loss
aging and, 1
in humans, 9
Cell populations, classification of, 118—119
Cell size, for lung-derived fibroblast-like cells, 377
Cell surface
aging of protozoan, 468
antigens, 342, 350
Cellular reproductive capacity, diminished, 117
Cell volumes, of skin-derived fibroblasts, 429
Central nervous system
age-related changes in blood vessels of, 161
development and aging and, 281
Centrifugal elutriation, 235
Centrophenoxine, effect on aging *Drosophila*, 559
Cerebellum, number of cells or fibers vs. age in,
23—24, 31—32, 105
Cerebral cortex, number of cells or fibers vs. age
in, 17—18, 29—30
Cerebrum, number of cells or fibers vs. age in,
27—29, 105
Ceroid, 504
CFU-S, 341
Chick cells, cultured, 442—444
p-Chlorophenylalanine (*p*CPA), 276
Cholesterol
effects of aging on absorption of, 266—267
in lung-derived fibroblast-like cells, 387
Chondrocytes, in aging cartilage, 197
Chondrogenic tissues, age-related changes in, 211—
221
biochemical activity, 214—218
cell morphological changes, 220—221
cell proliferative activity, 213—216
intercellular matrix and synovial fluid, 217—220
Chromatin, in lung-derived fibroblast-like cells,
409—419
''Chromophobe adenomas'', 278
Ciliates

clonal aging of, 454
  Euplotes, 458, 460
  Oxytricha bifaria, 460
  P. aurelia complex, 455—456
  P. bursaria, 456, 457
  P. caudatum, 456, 457
  P. multimicronucleatum, 456—458
  Stylonychia pustula, 460
  Tetrahymena, 460—461
  Tokophrya, 460, 462
  clonal lifespan determination for, 455
  clonal lifespan studies for, 454—455
  individual aging of, 461—463
  in life-span studies, 453
  nuclei of, 455
Circadian rhythm
  LH cycles and, 280
  phase changes in, 278
  serotonin metabolism and, 277
Circulating immune complexes, increase in, 354
Citrate synthase (CS), age-related reduction in, 189
Clonal lifespan, determination of, 455
Collagen
  in aging lung, 367—368, 371
  in degenerating aging muscle, 187—188
  in lung-derived fibroblast-like cells, 388—389
Collagenase, in isolation of endothelial cells, see
    also Enzyme activity, 303
Colon
  age-related changes in mucosal cells of, 128
  muscularis externa of, 268
Compensatory hypertrophy, 185
Connective tissue
  age-related increase of, 189
  disorders, 195
  lung, 365—369
  number of cells vs. age in, 34
Connective tissue matrix, age-related changes in,
    206—208
Contraction time
  of aging muscle, 180, 181
  fiber types and, 182
Cortical bone, rates of atrophy for, 195
Corticosteroids, effect on aging *Drosophila,* 559
Costal cartilage, age-related cell loss of, 213—215
Cranial nerves, number of cells or fibers vs. age in,
    15—16, 105
Creatine phosphokinase, in aging cartilage, see also
    Enzyme activity, 216
Cutaneous neural receptors, age-related changes in,
    161
Cyclic AMP stimulation of, lung-derived fibroblast-
    like cells, 384
Cytochrome oxidase
  in aging bone cells, 200
  in lung-derived fibroblast-like cells, 400
Cytogamy, 465
Cytoplasm, protozoan, aging of, 467—468
Cytoplasmic immunoglobulin (C-Ig), increase in, 5
Cytotoxic T lymphocytes (CTL), functional assay
    for, 345

D

Deacetylase activity, in lung-derived fibroblast-like
    cells, 411
Degenerative changes, 8
Dehydroascorbic acid, effect on aging *Drosophila,*
    559
Dementia, cortical lesions of, 150
Diabetes, relevance of in vitro aging to in vivo ag-
    ing in, 129
Digestive system, see also Intestinal mucosa
  age-associated cell loss in, 6—7
  number of cells vs. age in, 99—94, 107
Digitoxin, metabolilzed by parenchymal cells,
    238—239
Digylcerides, in lung-derived fibroblast-like cells,
    387—388
Diiodemethane, effect on aging *Drosophila,* 559
Dinoflagellates, experimental starvation of, 464
DNA, aging process and, 139
DNA binding proteins, in lung-derived fibroblast-
    like cells, 410
DNA content
  in aging *Drosophila,* 549—553
  of liver cells, 237
  of lung-derived fibroblast-like cells, 404—406
DNA polymerase, in lung-derived fibroblast-like
    cells, 413
DNA repair
  and life span and, 445
  in lung-derived fibroblast-like cells, 407
DNA replication, in lung-derived fibroblast-like
    cells, 412
DNA synthesis
  in aging *Drosophila,* 549, 553—558
  in aging intestinal tissue, 258
  in articular cartilage, 213
  in cultured mouse cells, 438
  in transplantation of mammary tissue, 290, 292
DNA synthetic index, as function of age, 124, 125
Doubling time, for fibroblast-like cells, 376
Down's syndrome, relevance of in vitro aging to in
    vivo aging in, 129
Drosophila, 511—565
  aging of
    alterations in glycogen content and, 539, 541
    body weight and, 539, 540
    dietary alterations and, 553, 558—564
    DNA content in, 549—553
    DNA synthesis in, 549, 553—558
    enzyme changes in, 543—546
    fecundity and, 524, 526—531
    geminal cell age and genetic aberrations, 517—
      523
    histological changes in, 535, 539
    locomotor ability and, 539, 542
    mitochondrial alterations and, 539
    morphological changes in, 535, 539
    oxygen consumption and, 539, 543, 544
    pigment formation in, 544, 547—548

protein content in, 549—553
protein synthesis in, 549, 553—558
RNA content in, 549—553
RNA synthesis in, 549, 553—558
as aging model, 564
genetics and parental age for, 514, 517—525
hybrid survivorship of, 513—515
longevity of, 511—514
    density and, 532, 535—537
    environmental factors and, 535
    light-dark cycles and, 535, 538
    temperature and, 525, 532—535
    transplantation and, 564, 565
male virility with age, 525, 532
mutant survivorship of, 514, 516
sex-linked spontaneous mutations in, 524, 525
Drug metabolilsm, by parenchymal cells, 238—239
Duodenum, age-related changes in mucosal cells of,
    127

E

Ear epithelium, mitosis vs. age in, 121
EGF receptors, in skin-derived fibroblast cells, 427
Elastic tissue, age-related changes in, 208
Elastin, in aging lung, 366, 368, 371
Elbow, age-related changes in cartilage of, 198
Elvax 40P implants, 292—293
Endomysial fibrosis, 185
Endoplasmic reticulum, in lung-derived fibroblast-
    like cells, 379
Endosteum, general cellular complement of, 199—
    200
Endothelial cells, 303—312
    aging changes in RES functions studied with, 234
    characteristics of, 233, 236—237
    CPDL (cumulative population doubling level for),
        310
    DNA synthwtic index vs. age in, 125
    factor VIII antigen in, 308
    homogeneous serially passaged, 306
    identification criteria for, 305—307
    in vitro aging of, 130, 309—312
    isolation methods for, 303, 305
    isolation, purification and culture of, 235—236
    of liver, 229
    mean cell-attachment area of, 311—312
    methods for harvesting, 304
    as model cultured cell system, 303
    as models for studies on aging, 234
    modulation of life span of, 309
    number vs. age, 78
    successful isolation and propagation of, 304
Endurance, of elderly people, 180
Enzyme activity
    in aging chondrocytes, 216
    in aging *Drosophila*, 543—546
    in aging insects, 501, 503
    in aging kidney, 248, 249
    in aging nematode, 489, 491

of intestinal mucosa, 265—266
    in lung-derived fibroblast-like cells, 391, 394—
        403
    in senile muscle, 188—189
Enzymes, in lung-derived fibroblast-like cells, 413
Eosinophils, number vs. age of, 62—64
Epidcrmal growth factor (EGF) 129, 437
Epiphyseal plate, age-related cell loss of, 213
Epithelium, mitosis vs. age in, 122
Erythorbic acid, effect on aging *Drosophila*, 562
Erythrocytes, 317—329
    aging of, 317
        alterations in, 329
        biochemical aspects of, 326—327
        biophysical aspects of, 328
        immunological aspects of, 329
    decrease in, 3—4
    life span of, 317, 319—323
    number vs. age in, 50—54, 106
    separation into age groups, 324—325
Esophageal epithelium, DNA synthetic index vs.
    age in, 125
Esophagus
    cell kinetics of, 259
    intestinal mucosa of, 255—256
    mucosal structure of, 257
    submucosa of, 268
Esterace, in lung-derived fibroblast-like cells, see
    also Enzyme activity, 400
Estrogen production, decreased, 273
Ethionine, in lung-derived fibroblast-like cells, 386
Euglena, experimental starvation of, 464
Eukaryotes, evolution of, 453, 469
Euplotes
    clonal aging of, 458, 460
    life cycle events of, 459
    lifespan determination of, 466
Eye
    cell loss in, 2
    mitosis vs. age in, 122
    number of cells vs. age in, 33, 105

F

Fast (FG) fibers
    physical exercise and, 190
    in senile muscle, 182
Fat body, in aging insects, 502
Feline cells, cultured, 445
Femur
    cell loss in, 3
    number of cells vs. age in, 38—40, 48
Fibers
    age-related splitting of, 185, 186
    degenerating and hypertrophied, 184
    loss of, 185
Fibers, nerve, age-related loss of, 152—153
Fiber types
    age and, 182
    classification of, 181

diameters of, 182
distribution of, 183
Fibroblast
    age-related changes in, 209
    in vivo aging of lung, 370—371
    skin-derived
        acute replicative abilities of, 426—428, 430
        cell membrane functions of, 426—428
        cloned, 429, 430
        cumulative replicative abilities of, 425—427, 430
        establishment of cultures, 425
        macromolecular contents and synthesis of, 426, 428
Fibroblast growth factor (FGF), 129, 417
Fibroblast-like cells
    A-11-L, enzyme activities of, 394, 397
    in culture
        bovine, 444
        canine, 444
        cat, 445
        chick, 442—444
        hamster, 441—442
        kangaroo, 445
        monkey, 445
        mouse, 438—440
        potoroo, 445
        rabbit, 444, 445
        rat, 440—441
        wallaby, 445
    exhibiting in vitro aging, 130
    HE-125, 377
    HE-388, 377
    IMR-90, 377
    lung-derived, 375
        chromatin in, 409—419
        DNA content of, 404—406
        DNA repair in, 407—408
        enzyme activites of, 394—403
        HE-125, 413, 416, 417, 494
        HE-388, 416, 417, 494
        IMR-90, 378—381, 384, 390, 418
        metabolic and biosynthetic changes in, 382—392
        morphological changes in, 377—382
        MRC-5, 381, 383, 384, 386, 390, 391, 395, 396, 399, 401, 405, 412, 413, 419
        PAL II, 406
        premature chromosome condensation in, 418
        RNA content of, 404—406
        TIG-1, 381, 385, 405, 410
        viral infectivity and replication of, 393
        WI-38, 378—392, 404—406, 409—418
    nonhuman, theories of senescence of, 434—436
    WI-38, 376, 377
        enzyme activities of, 394—403
        viral infectivity and replication of, 393
    WI-1006, enzyme activities of, 394, 397
Fibrosis, in aging cartilage, 197
Fibula, number of cells vs. age in, 40
Filaments, in lung-derived fibroblast-like cells, 379

Flagellates, colonial, individual aging of, 463
Flight ability of aging insects, 488—501
Fractures, age-related, increase of, 195, 197
Free fatty acids, in lung-derived fibriblast-like cells, 387
Fructose I-6 diphosphatase, in lung-derived fibroblast-like cells, 401

G

Gametogenesis, cessation of, 271
Gingiva, age-associated cell loss in, 6
Glial cells, 2
    exhibiting in vitro aging, 130
    number of cells or fibers vs. age in, 26—32
Glomerular filtration rate, age-associated decline in, 246
Glomerulosclerosis
    in aging rat kidney, 245
    associated with immune complex deposits of IgG, 250
    hypophysectomy and, 248
Glucocorticoid receptors, in lung-derived fibroblast-like cells, 381
Glucocorticoids, cell death generated by, 143
Gluconic acid, effect on aging *Drosophila,* 558
Glucosaminidase, in lung-derived fibroblast-like cells, see also Enzyme activity, 400
Glucose, effects of aging on absorption of, 267
Glucose-6-phosphate dehydrogenase (G6PD), in lung-derived fibroblast-like cells, 390, 396—397, 402
β-Glucuronidase, in lung-derived fibroblast-like cells, 397—398
Glutamate dehydrogenase, in lung-derived fibroblast-like cells, 400
Glutamic oxalacetic transaminase, in lung-derived fibroblast-like cells, 399
Glutamic pyruvic transaminase, in lung-derived fibroblast-like cells, 399
Glutamine synthetase, in lung-derived fibroblast-like cells, 398
Glycerol-3-phosphate dehydrogenase (GPDH), age-related reduction in, 189
Glycogen
    in aging *Drosophila,* 539, 541
    in different fiber types, 182
    in lung-derived fibroblast-like cells, 388
Glycolysis, of lung-derived fibroblast-like cells, 382
Glycolytic enzymes, in aging cartilage, 216
Glycosaminoglycans (GAGs)
    in aging lung, 368, 371
    in bone, 205
    in lung-derived fibroblast-like cells, 386
Golgi bodies, in lung-derived fibroblast-like cells, 379
Gonadotropins
    effects of depletion of, 274
    reduced responsiveness to, 273
Gonads, age-related changes in, 271

evidence for, 274
female, 272, 273
male, 272, 273
symptomatology of, 272
Granulocytes
aging and, 4
number vs. age of, 59
Granulosa cells, exhibiting in vitro aging, 130
Granulovacuolar degeneration, 158
Growth factor requirement, and transformation, 437
Growth potential, and life span, 445
Growth transformation, markers of, 433—434
Gut, 255

H

Hamster cells, cultured, 441—442
Hayflick limit, 137—138, 434
Hayflick's phases, of in vitro aging, 129
Heart
in aging insects, 502
number of cells or fibers vs. age of, 95—96, 108
shift to anaerobic metabolism of, 189
HeLa cell line, 126, 129
Hematopoietic cells
age-associated changes in numbers of, 5
number vs. age in, 79—84
serial transplantation studies of, 124, 126
Hepatocytes, see also Liver cells, 232
exhibiting in vitro aging, 130
loss of, 6—7
number vs. age in, 93
Hexokinase, in lung-derived fibroblast-like cells, see
also Enzyme activity, 401
Hexosylhydroxylysine, 207
Hexosyllysine, 207
Hip, age-related changes in cartilage of, 198
Hippocampus, number of cells or fibers vs. age in,
21—22, 31—32
Histone ratio, for lung-derived fibroblast-like cells,
409
HL-A antigens, in lung-derived fibroblast-like cells,
381
Hock joints, bovine, effect of aging on, 219, 220
Hormones, pituitary, and gonadal dysfunction,
274—276
Housefly, lipofuscin accumulation in, see also *Dro-
sophila,* 505
H-proline autoradiographic studies, of bone cell re-
pair, 203—207
Humerus, number of cells vs. age in, 40—41
Hutchinson-Gilford syndrome, relevance of in vitro
aging to in vivo aging in, 129
Hyaline cartilage, age-related cell loss of, 213
Hyaluronic acid
forming joint surface cushion, 218
in human synovial fluid, 219
Hydrogen iodine, effect on aging *Drosophila* of,
559

Hydroxyproline, in lung-derived fibroblast-like
cells, 388
5-Hydroxytryptophan (5HTP), 276
Hyperplastic alveolar nodules (HAN), 298
Hypodiploidy, 131
Hypothalamic monoamines, and age-related changes
in reproductive system, 276—277
Hypothalamic neuropeptides, and changes in endocrine environment, 275—276
Hypothalamus, number of cells or fibers vs. age in,
22—23

I

Ileum, age-related changes in mucosal cells of,
127—128
Iliac crest, sex differences in age-related changes in,
195
Immune response, 341—355
B lymphocytes, 347—350
effects of aging on, 352—355
monocytes-macrophages, 350—352
T lymphocytes, 342—347
Immune system, local lung, 369
Immunoglobulins, increase in, 354
Inclusions, in lung-derived fibroblast-like cells,
379—380
Insect cells, transplantation of, 137
Insects, see also *Drosophila,* 497—506
advantages of studying, 497
aging of
body composition, 498—500
brain of, 501—502
development and, 497—498
enzyme activity in, 501, 503
fat body changes in, 502
flight muscle, 499
heart changes in, 502
lipofuscin accumulation in, 503—505
malpighian tubules in, 502—503
maturation in, 498
midgut changes in, 502
protein synthesis in, 503
senescence in, 498
alteration of lifespan of, 497
Interdivision time, 131
Intestinal mucosa, see also Intestine
cell kinetics of, 258—265
in response to injury, 261—264
in tumorigenesis, 265
in unperturbed state, 258—261
of esophagus, 255—256
of large intestine, 257
physiology of, 265—268
absorption from gut lumen, 266—268
chemical composition, 265
enzyme systems, 265—266
of small intestine, 257, 258
of stomach, 256—257
structure of, 255—258

Intestinal submucosa
 of esophagus, 268
 of intestine, 268—269
Intestinal tissue, see also Gut
Intestine, see also Intestinal mucosa, 255—269
 age-associated cell loss in, 6—7
 age-related changes in mucosal cells of, 127—
  128
 large
  cell kinetics of, 260—262
  mucosa of, 257
 small
  age-related changes in length of, 256
  cell kinetics of, 259—260
  mucosa of, 257, 258
  number of cells vs. age in, 92, 94, 107
  submucosa of, 268
Iodine, effect on aging *Drosophila*, 559
Iodoform, effect of aging *Drosophila*, 559
Isocitrate dehydrogenase, in lung-derived fibroblast-
  like cells, see also Enzyme activity, 399

J

Jejunum, age-related changes in mucosal cells of,
 127
Joint disease, degenerative, 195, 198, 221
Joints, age-related changes in cartilage of, 198

K

Kangaroo cells, cultured, 445
Keratinocytes, exhibiting in vitro aging, 130
Kidney cells, 245—250
 DNA dynthetic index vs. age in, 125
 effect of aging on, 8
  function, 246—247
  hormonal adaptation, 248
  metabolism, 248, 249
  structure, 245—246
 mitosis vs. age in, 121
 number of cells vs. age in, 108
Knee, age-related changes in cartilage of, 198
Kohn, pores of, 366
Kupffer cells
 age-related changes in functions of, 239
 aging changes in RES functions studies with, 234
 changes in number of, 231—232
 characteristics of, 233, 236—237
 heterogeneity of, 230
 isolation, purification and culture of, 235—236
 of liver, 229
 as models for studies on aging, 234
 in studies on cellular aging, 234

L

Labeling index (LI), 258

Lactate dehydrogenase (LDH), age-related reduction
 in, 189
Lactate production, of lung-derived fibroblast-like
 cells, 383
Lactic acid, effect on aging *Drosophila*, 558
Lactic dehydrogenase, in lung-derived fibroblast-like
 cells, see also Enzyme activity, 397
Lens epithelium, exhibiting in vitro aging, 130
Leukocytes
 aging and, 3—4
 number vs. age of, 54—59, 106
Leydig cells
 age-associated loss of, 8
 age-related decrease in, 272, 273
 number vs. age of, 99—100, 108
Lifespan
 alteration of
  carcinogen-induced, 465
  nutrition and, 463—464
  radiation-induced, 464
 correlation with cellular properties of, 445—448
 determination of
  for *Acetabularia*, 465
  Amoeba, 465
  Euplotes, 466
  Paramecium, 465—466
 reproduction and, 271
Lipid metabolism, in aging insect, 504
Lipids, in lung-derived fibroblast-like cells, 387—
 388
Lipofuscin
 accumulation of, 154—156
 in aging insects, 503—504
 in aging liver cells, 232
 in aging muscle, 187
 autophagocytosis of, 504
 cell death and, 139
 characteristics of, 142—143
Liver
 age-associated cell loss in, 6
 age-associated pathology of, 233
 blood flow in, 236
 changes in weight of, 232
 DNA synthetic index vs. age in, 125
 mitosis vs. age in, 121
 number of cells vs. age in, 92—93, 107
Liver cells, 229—240
 age-related changes in functions of, 237—240
 changes in number of 231—232
 effect of aging on, 230—231
 function of, 229
 heterogeneity of, 230
 subcellular level changes, 232—233
 types of, 229
Locomotion, in aging *Drosophila*, 539, 542
Locomotor systems, 179
Lung
 aging of, 365—371
  collagen in, 367—368
  connective tissue and physiology, 369
  defense mechanisms of, 369—370

elastin in, 368
  experimental injury and, 370
  in vivo fibroblasts, 370—371
    morphologic changes, 366—367
    physiologic changes, 365—366
  changes in cell numbers of, 8
  number of cells vs. age in, 103—104, 108
Lutenizing hormone (LH) secretion, control of, 278
Lymph nodes
  age-associated cell loss in, 6
  number of cells vs. age in, 89—90, 107
Lymphocytes, see also B lymphocytes: T
      lymphocytes
  age-associated decrease in, 4
  number vs. age of, 35, 66—76
Lymphoid tissue, age-associated changes in, 353
Lymphoreticular system, 341
Lypofuscin granules, on bone surface, 210
Lysis, 140
Lysolecithin, in lung-derived fibroblast-like cells,
      388
Lysosomes, 137—144
  in aging liver, 232
  on bone surface, 210
  cell death and, 139, 142
  in lung-derived fibroblast-like cells, 379
  role in cellular aging, 239

M

Macronucleus, protozoan, aging of, 467
Macrophages
  age-associated changes in, 353
  in immune response, 351—352
Malabsorption, from intestine, 268
Malate dehydrogenase (MDH)
  age-related reduction in, 189
  in lung-derived fibroblast-like cells, 399
Malignant transformation
  defined, 433
  markers of, 433—434
Malpighian tubules, in aging insects, 502—503
Mammary cells, 289—299
  growth span for, 293—294
    chronological time and cell division and, 294—
        296
    donor and host age and, 296
    hormonal influences on, 296—298
    neoplastic and preneoplastic transformations,
        298—299
  in vitro, 299
  in vivo propagation of
    DNA synthesis, 292
    Elvax 40P implants, 292—293
    growth measurements, 291
  morphology, 292
  serial propagation, 291
  transplantation, 289—291
Mammary gland
  hormonal control of, 296—298

serially passaged, effects of hormonal stimulation
      on, 298
  serial transplantation of mouse, 294
Mandible, number of cells vs. age in, 47
Mast cells, number vs. age of, 35—36, 76—78
Medulla, number of cells or fibers vs. age in, 24
Melanocytes, number vs. age of, 36
Membrane potential, age-related changes in, 181
Mesentery, number of cells vs. age, see also Intes-
      tine, 94, 107
Metacarpals, number of cells vs. age in, 45
Methionine, in lung-derived fibroblast-like cells,
      386
Micronuclei, protozoan, aging of, 467
Microvilli, in lung-derived fibroblast-like cells, 380
Midgut, in aging insects, 502
Migration rate, in lung-derived fibroblast-like cells,
      381
Miniature end plate potentials (m.e.p.p.), age-re-
      lated changes in, 181
Mitochondria
  of aging *Drosophila*, 539
  of lung-derived fibroblast-like cells, 378
  of periosteal osteoblasts, 201, 202
Mitosis, and cell death and, 139—140
Mitotic index, age-related decline in, 120
Molecular theories, limitation of, 271
Monkey cells, cultured, 445
Monocytes
  age-associated changes in, 353
  age-associated changes in number of, 5
  in immune response, 350—352
  number vs. age, 65—66
Monoglycerides, in lung-derived fibroblast-like
      cells, 387
Morphometric changes, in aging muclse fibers, 185
Motor nerve endings, age-related changes in, 161
Motor units, 180
  fibers of, 181—182
  hypertrophy of, 181
Mouse cells, cultured, 438—440
Mucosal neural receptors, age-related changes in,
      161
Multiplication-stimulating activity (MSA), 437
Muscle
  age-associated cell loss in, 7
  collagen in, 187—188
  number of cells or fibers vs. age in, 95—99, 108
  role of lysosome in, 140
Musculature, of intestinal tract, 269
Myelin sheaths, age-related changes in, 151, 154
Myofibrillar ATPase, 190
Myokinase, in aging cartilage, 216
Myosin ATPase, 190

N

Natural killer (NK) cells, 346
Nematodes, see also *Caenorhabditis elegans*, 481
  biological marker of aging of, 489—493

genetic studies of, 493
lifespans in, 488
  bacterial concentration and, 488, 489
  temperature and, 488, 489
Neoplastic development, in vivo, 437
Nerve cells
  aging, 149—161
    accumulation of lipofuscin in, 154—156
    blood vessels of the brain, 158, 161
    changes in, 149—150
    irregularities of axons, 151, 154
    loss of, 150—153
    myelin sheaths, 151, 154
    neuritic plaques, 157—158
    neurofibrillary tangles, 154, 156—157
    physical changes in, 158, 159
    sclerosis of, 151, 153
    shrinkage of, 151, 153
    shrinkage of brain, 150—151
Nerve conduction velocities, decline in, 181
Nervous system
  cell loss in, 1—2
  number of cells or fibers vs. age in, 13—32
  number of cells vs. age in, 105
Neuraxonal dystrophy, 151, 154
Neuritic plaques, 157, 158
Neurofibrillary tangles, 154—157
Neuroglia, age-related changes in, 158, 160
Neuronal loss, 1, 9
Neutrophils
  aging and, 4
  number vs. age of, 60—62
Nomarski phase-contrast optics, 481
Nuclear proteins, of lung-derived fibroblast-like
    cells, 405, 410
Nuclear size, of lung-derived fibroblast-like cells,
    378
Nucleolar changes, in lung-derived fibroblast-like
    cells, 378
5' Nucleotidase, in lung-derived fibroblast-like
    cells, 402
Nucleus, of lung-derived fibroblast-like cells, 380
Nutrition, lifespan altered by, 463—464

O

Ochromonas, experimental starvation of, 464
Oleic acid, in lung-derived fibroblast-like cells, 386
Oocytes, number vs. age of, 102, 108
Oral cavity, number of cells vs. age in, 90—92,
    107
Organs, weight loss of, 8
Orgel hypothesis, 426
Ornithine decarboxylase, in lung-derived fibroblast-
    like cells, see also Enzyme activity, 402
Osteoarthritis, 195, 221
  age changes and, 199
  incidence of, 198
Osteoblasts, age-related decrease in, 201
Osteoclast activating factor (OAF), 211

Osteoclasts, age-related decrease in, 212
Osteocytes
  death of, 200
  degenerative changes of, 210
  number vs. age of, 48
Osteogenic tissues, age-related changes in, 199—
    212
  biochemical activity, 203, 205—207
  cell proliferative activity, 200—204
  connective tissue matrix, 206—208
  morphological, 208—212
Osteophytes, in aging cartilage, 197—198
Osteoporosis, 221
  international incidence of, 196
  senile, 195, 197
Ova, number vs. age, 101
Ovariectomy, estrogen deficiency resulting from,
    273
Oxygen consumption, of aging Drosophila, 539,
    543, 544
Oxygen diffusion, aging of muscle and, 180
Oxyphenybutazone, in lung-derived fibroblast-like
    cells, 388
Oxytricha bifaria, clonal aging of, 460

P

Pacemaker cells, and degeneration of aging repro-
    ductive system, 278, 279, 281
Palate basal cell layer, DNA synthetic index vs. age
    in, 125
Panagrellus redivivus, 481
Paramecium
  aurelia complex
    clonal aging of, 455—456
    senescent changes in, 458
  bursaria, clonal aging of, 456, 457
  caudatum, clonal aging of, 456, 457
  multimicronucleatum, clonal aging of, 456—458
  sp.
    aging in, 454
    carcinogen-induced alteration in lifespan, 465
    individual aging of, 462—463
    life cycle events in, 457
    lifespan determination of, 465—466
    radiation-induced lifespan changes in, 464
Parathyroid, mitosis vs. age in, 123
Parathyroid hormone (PTH)
  in aging renal system, 248
  osteoclastogenesis in response to, 211
Parenchymal cells
  age-related shift to polyploid, 230—231
  aging and drug metabolism by, 238—239
  BSP metabolilsm by, 238
  changes in number of, 231—232
  characteristics of, 233, 236—237
  coculture of sinusoidal cells and, 236
  culture of, 235
  heterogeneity of, 230
  isolation and purification of, 235

of liver, 229
loss of, 9
as models for studies on aging, 233—234
Passages in vitro, 375
Pericytes, number vs. age of, 78
Periodontal disease, 195
Periosteum
age-related changes in, 201—204
general cellular complement of, 199—200
number of cells vs. age in, 48
Peripheral nerves, age-related changes in, 161
Peyer's patches, 257
Phalanx, number of cells vs. age in, 47
Phenothiazines, effect on aging *Drosophila*, 560—561
Phosphofructokinase, in lung-derived fibroblast-like cells, see also Enzyme activity, 401
6-Phosphogluconate dehydrogenase, in lung-derived fibroblast-like cells, 401
Phosphoglucose-isomerase, in lung-derived fibroblast-like cells, 401
Phosphorylase, in aging cartilage, 216
Physical exercise
aging process and, 179
aversion to, 190
Pituitary gland, age-related changes in, 274—276
Platelets, number vs. age of, 76
Poikilotherms, alteration of lifespan of, 497
Poliovirus, 393
Polymorphonuclear leukocytes, number vs. age of, 59
Polysaccharide uptake, in lung-derived fibroblast-like cells, 385
Population doubling level (PDL), 129, 375
Postmitotic cells
death of, 139
fibroblasts vs., 233
lipofuscin as marker of physiological age of, 143
muscle, 179
nerve cells, 149
Potoroo cells, cultured, 445
Preosteoblasts, characteristics of, 209
Progesterone secretion, age-related changes in, 273
Prostaglandins, of lung-derived fibroblast-like cells, 384—385
Prostaglandin synthesis, in skin derived fibroblasts, 429, 430
Protein, of lung-derived fibroblast-like cells, 405
Protein-bound disulfide (PBSS) group, 205
Protein-bound sulfhydryl (PBSH) group, 205
Protein content, in aging *Drosophila* cells, 549—553
Protein degradation, of lung-derived fibroblast-like cells, 383
Protein hydrolysate, in lung-derived fibroblast-like cells, 385
Protein synthesis
in aging *Drosophila* cells, 549, 553—558
in aging insects, 503
in aging nematodes, 493
in parenchymal cells, 238

Proteinuria
in aging kidney, 246, 247
hypophysectomy and, 248
Protozoa, see also Ciliates, 453—472
age-related changes in, 469—470
alteration of lifespan, 463—465
biological background, 454—455
clonal aging, 455—461
clonal lifespan determination, 455
ecogenetic considerations, 468
individual aging, 461—463
intracellular sites for aging in, 466—468
cell surface, 468
cytoplasm, 467—468
ecogenetic considerations, 468
macronucleus, 467
micronuclei, 467
lifespan determination, 465—466
in lifespan studies, 453
longevity of, 471
multicells vs. unicells, 469—470
Pyridoxine, effect on aging *Drosophila*, 562
Pyruvate kinase, in aging cartilage, 217

R

Rabbit cells, cultured, 444, 445
Radiation
effects on aging gut of, 261—264
lifespan altered by, 454
Radius, number of cells vs. age in, 42—43
Rat cells, cultured, 440—441
Renal disease
compensatory response in, 250
in elderly population, 250
Renal function, age-associated decrease in, 246
Renal vasculature, age-related structural changes in, 245
Reproductive system, 271—281
age-associated changes in cells of, 8
aging of, possible role of serotonin in, 277—278
functional decline of
cancer and, 278—280
gonads, 271—274
hypothalamic monoamines in, 276—277
hypothalamic neuropeptides, 275—276
pituitary hormones, 274—276
number of cells vs. age in, 108
female, 101—102
male, 99—101
Respiration, of lung-derived fibroblast-like cells, 382
Reticulocyte, maturation of, 317
Reticuloendothelial system (RES), aging changes in, 234
Rib bone
age-related changes in, 195
number of cells vs. age in, 44
"Ringbinden", 185
RNA content

in aging *Drosophila* cells, 549—553
in aging muscle, 188—189
of lung-derived fibroblast-like cells, 404—406
RNA synthesis
in aging *Drosophila* cells, 549, 553—558
in articular cartilage, 213
in cultured mouse cells, 438
in isolated parenchymal cells, 237
in lung-derived fibroblast-like cells, 413, 414
in programmed cell death, 142

S

Salicylates, effect of aging *Drosophila,* 560
Sarcolemma, age-related thickening of, 185
Satellite cells, age-related decrease of, 185
Senescence, 1
beginning of, 271
clonal, 117
defined, 433
markers of, 433—434
theories of, 434—436
Senile dementia, 150
Senile plaques, 157, 158
Serial transplantation studies, 124, 126
Serotonin
in aging of reproductive system, 277—278
cancer and, 278—280
Serotonin-catecholamine balance, 278
Sertoli cells, number vs. age of, 100
Shoulder, age-related changes in cartilage of, 198
Sinusoidal cells
coculture of parenchymal and, 236
effect of aging upon functions of, 239
of liver, 229
Sister chromatid exchanges (SCE), in skin-derived
fibroblasts, 429, 430
Skeletal muscle, see also Fibers; Muscle
aging of, 179—190
adaptability of, 190
histochemical changes in, 181—184
morphological changes, 185—190
physiological changes, 180—181
number of muscle cells or fibers vs. age, 96—98,
108
Skeletal system, aging of, 195—221
bone, 195—197
cartilage, 197—199
cells and matrix, 199—221
chondrogenic tissues, 211—221
osteogenic tissues, 199—212
Skin
cell loss in, 2
mitosis vs. age in, 121
number of cells vs. age in, 34—36, 105
serial transplantation studies of, 124, 126
Skull, number of cells vs. age in, 45—46
Smooth endoplasmic reticulum (SER), hepatic, 232
Smooth muscle, number of muscle cells or fibers
vs. age in, 98—99, 108

Smooth muscle of intima, DNA synthetic index vs.
age in, 126
Sodium yeast nucleate, effect on aging *Drosophila,*
562
Spermatozoa, number vs. age of, 100
"Spheroids", 151, 154
Sphingomyelin, in lung-derived fibroblast-like cells,
388
Spinal nerves, number of cells of fibers vs. age in,
13—14, 105
Spleen
age-associated cell loss in, 6
number of cells vs. age in, 84—89, 107
Stem cells, age-associated changes in, 353
Sternoclavicular joints, age-related changes in, 198
Stomach, mucosal structure of, 256—257
Stylonychia pustula, clonal aging of, 460
Succinic dehydrogenase, see also Enzyme activity
in aging bone cells, 200
in lung-derived fibroblast-like cells, 399
Sulfatase, in aging cartilage, 217
Superoxide dismutase, in lung-derived fibroblast-
like cells, 402
Suprachiasmatic nucleus (SCN), 277, 278
Synovial fluid, age-related changes in, 218—219
Synovial membrane, age-related changes in, 198

T

3T3 cells, cultured mouse, 438, 439
Tartaric acid, effect on aging *Drosophila,* 559
Tendon, human, age-related changes in, 208
Testosterone synthesis, decreased, 273
Tetrahymena
experimental starvation of, 464
finite lifespan of, 460—461
infinite lifespan of, 461
Thiazolidine carboxylic acid (TCA), effect on aging
Drosophila, 561
Thymidine, tritiated, in cell death, 141
Thymidine incorporation, age-related changes in
mucosal cells of mammalian intestine determined by, 127—128
Thymus
age-associated cell loss in, 6
effect of age on immune system in, 352
number of cells vs. age in, 89, 107
Thyroid, mitosis vs. age in, 122—123
Tibia
cell loss, 3
number of cells vs. age in, 40, 48
T lymphocytes, 342—344
age-associated changes in, 4, 353
helper, 344
number vs. age of, 68—72
suppressor, 344—345
Tocopherol-*p*-chlorophenoxyacetate, effect on aging
Drosophila, 561
Tokophrya
clonal aging of, 460, 462

differences in young and old adults, 462
experimental overfeeding of, 463
experimental starvation of, 464
individual aging of, 461—462
Tongue basal cell layer, DNA synthetic index vs.
age in, 125
"Torpedoes", 151, 154
Trabecular bone, rates of atrophy for, 195
Transaldolase, in lung-derived fibroblast-like cells,
401
Transformation
growth factor requirement and, 437
malignant, 433—434
in serially cultured cells, 436—437
Transketolase, in lung-derived fibroblast-like cells,
see also Enzyme activity, 401
Transplantation
effects on aging *Drosophila* of, 564—565
of insect cells, 137
of mammary epithelial tissues, 289—291
Triglycerides, in lung-derived fibroblast-like cells,
387
Trichloroacetic acid-soluble fraction, in aging muscle, 187
Triosephosphate dehydrogenase (TPDH), age-related
reduction in, 189
Titriated thymidine, periosteal cells labeled with,
202—204
Turbatrix aceti, 481
Turnover times, 118, 119

U

Ulna, number of cells vs. age in, 44
Ultrastructural changes, in aging muscle fibers, 185

Ultrastructure studies, on chick fibroblasts in cul-
ture, 443
Unicellular organisms, aging in, see also Ciliates;
Protozoa, 469—470
Uridine, in lung-derived fibroblast-like cells, 385
Uridine diphosphate glucose dehydrogenase
(UDPGDH), in aging cartilage, 217

V

Vertebral bone
age-related changes in, 195
cell loss in, 3
number of cells vs. age in, 48
volume of, 197
Vesicular stomatitis virus, 393
Villi, primary function of, see also Intestinal mucosa, 257
Vitamin A, effects of aging on absorption of, 266—
267
Vitamin C, effect on aging *Drosophila,* 559, 561
Vitamin E, effect on aging *Drosophila,* 559, 561
Volume (μm^3), for lung-derived fibroblast-like
cells, 377—378
Volvocaceae, individual aging of, 463

W

Wallaby cells, cultured, 445
"Wear and tear" theories, of aging, 179
Weibel-Palade bodies, 306
Werner's sundrome, relevance of in vitro aging to
in vivo aging in, 129